Advanced Dairy Chemistry, Volume 2

Paul L. H. McSweeney
Patrick F. Fox • James A. O'Mahony
Editors

Advanced Dairy Chemistry, Volume 2

Lipids

Fourth Edition

Springer

Editors
Paul L. H. McSweeney
School of Food and Nutritional Sciences
University College Cork
Cork, Ireland

Patrick F. Fox
School of Food and Nutritional Sciences
University College Cork
Cork, Ireland

James A. O'Mahony
School of Food and Nutritional Sciences
University College Cork
Cork, Ireland

ISBN 978-3-030-48685-3 ISBN 978-3-030-48686-0 (eBook)
https://doi.org/10.1007/978-3-030-48686-0

This Springer imprint is published by the registered company Springer Nature Switzerland AG
The registered company address is: Gewerbestrasse 11, 6330 Cham, Switzerland

Preface to the Fourth Edition

Advanced Dairy Chemistry – 2 Lipids is the second volume of the fourth edition of this well-established series on advanced topics in Dairy Chemistry and follows the recent publication of Volumes 1A: *Proteins: Basic Aspects* and 1B: *Proteins: Applied Aspects*; this series of volumes is intended to be a coordinated and authoritative treatise on Dairy Chemistry. In the years since the publication of the third edition of this volume in 2006, there have been significant developments in milk lipids and these are reflected in changes to this volume. Many of the chapters from the third edition are retained, but those on the role of milk fat in specific dairy products have been consolidated, as have chapters on analysis. Chapters on nutritional significance of dairy lipids have been considerably revised and, in some cases, new authors have given their perspective on certain topics. We hope that this volume will be of assistance to senior students of Dairy Chemistry, to academics and researchers in industry, as each chapter is referenced extensively. We wish to thank sincerely the 38 contributors to the 13 chapters of this volume, whose cooperation made our task as editors a pleasure.

University College
Cork, Ireland

P. L. H. McSweeney
P. F. Fox
J. A. O'Mahony

Preface to the Third Edition

Advanced Dairy Chemistry 2: Lipids is the second volume of the third edition of the series on advanced topics in Dairy Chemistry, which started in 1982 with the publication of Developments in Dairy Chemistry. The first volume, on milk proteins, of the third edition of Advanced Dairy Chemistry was published in 2003. This series of volumes is intended to be a coordinated and authoritative treatise on Dairy Chemistry. In the decade since the second edition of this volume was published (1995), there have been considerable advances in the study of milk lipids, which are reflected in changes to this book.

Most topics included in the second edition are retained in the current edition, which has been updated and considerably expanded from 10 to 22 chapters. For various reasons, the authors of many chapters have been changed and hence, in effect, are new chapters, at least the topic is viewed from a different perspective.

The new chapters cover the following subjects: biosynthesis and nutritional significance of conjugated linoleic acid, which has assumed major significance during the past decade; formation and biological significance of oxysterols; the milk fat globule membrane as a source of nutritionally and technologically significant products; physical, chemical and enzymatic modification of milk fat; significance of fat in dairy products: creams, cheese, ice cream, milk powders and infant formulae; analytical methods: chromatographic, spectroscopic, ultrasound and physical methods.

Like its predecessor, this book is intended for academics, researchers at universities and industry, and senior students; each chapter is referenced extensively.

We wish to thank sincerely the 37 contributors to the 22 chapters of this volume, whose cooperation made our task as editors a pleasure. The generous assistance of Ms. Anne Cahalane is gratefully acknowledged.

University College
Cork, Ireland

P. F. Fox
P. L. H. McSweeney

Preface to the Second Edition

Advanced Dairy Chemistry can be regarded as the second edition of *Developments in Dairy Chemistry*. The first volume in the series, on Milk Proteins, was published in 1992; this, the second volume, is devoted to Milk Lipids. Considerable progress has been made in several aspects of milk lipids during the past 11 years which is reflected in revised versions of seven of the eight chapters included in *Developments in Dairy Chemistry 2*, most of them by the same authors. The theme of one chapter has been changed from physical properties and modification of milk fat to the crystallization of milk fat. Two new chapters have been added, i.e. chemistry and technology aspects of low-fat spreads and the significance of fat in consumer perception of food quality, which reflect the continuing consumer awareness of a healthy diet. Low-fat spreads have become increasingly significant during the past decade and are now the major type of spread in many countries. However, reducing the fat content of foods generally results in a concomitant decrease in the organoleptic quality of the food; consumer attitudes to reduced-fat dairy products are discussed in one of the new chapters.

Like its predecessor, the book is intended for lecturers, senior students and research personnel and each chapter is extensively referenced.

I would like to thank all the authors who contributed to this book and whose cooperation made my task as editor a pleasure.

University College
Cork, Ireland

P. F. Fox

Preface to the First Edition

Many of the desirable flavour and textural attributes of dairy products are due to their lipid components; consequently, milk lipids have, traditionally, been highly valued, in fact to the exclusion of other milk components in many cases. Today, milk is a major source of dietary lipids in western diets and although consumption of milk fat in the form of butter has declined in some countries, this has been offset in many cases by increasing consumption of cheese and fermented liquid dairy products.

This text on milk lipids is the second in a series entitled Developments in Dairy Chemistry, the first being devoted to milk proteins. The series is produced as a coordinated treatise on dairy chemistry with the objective of providing an authoritative reference source for lecturers, researchers and advanced students. The biosynthesis, chemical, physical and nutritional properties of milk lipids have been reviewed in eight chapters by world experts. However, space does not permit consideration of the more product-related aspects of milk lipids which play major functional roles in several dairy products, especially cheese, dehydrated milks and butter.

Arising from the mechanism of fatty acid biosynthesis and export of fat globules from the secretory cells, the fat of ruminant milks is particularly complex, containing members of all the major lipid classes and as many as 400 distinct fatty acids. The composition and structure of the lipids of bovine milk are described in Chapter 1, with limited comparison with non-bovine milk fats. Since the fatty acid profile of milk fat, especially in monogastric animals, may be modified by diet and other environmental factors, the biosynthesis of milk lipids is reviewed in Chapter 2 with the objective of indicating means by which the fatty acid profile, and hence the functional properties of the lipids, might be modified. Lipids in foods are normally present as an emulsion, stabilized by a layer of protein adsorbed at the oil-water interface. The fat in milk and cream exists as an oil-in-water emulsion with a unique stabilizing lipoprotein membrane, referred to as the milk fat globule membrane (MFGM). The inner layers of the MFGM are formed within the secretory cell and are relatively stable; however, the outer layers, which are acquired as the fat globule is exported through the apical mem-brane of the secretory cells, are unstable. Damage to the MFGM leads to chemical and physical instability of the fat phase in milk and hence the structure of the membrane has been the subject of considerable research, the results of which are reviewed in Chapter 3.

Lipids strongly influence, for good or evil, the flavour and texture of foods, especially high-fat products such as butter. The influence of various colloidal features of milk fat on the properties of milk and cream is considered in Chapter 4, while the crystallization of milk fat and how this may be controlled, modified and measured are reviewed in Chapter 5. Unfortunately, lipids are subject to chemical and enzymatic alterations which can cause flavour defects referred to as oxidative and hydrolytic rancidity, respectively. The storage stability of high-fat foods, especially mildly flavoured foods like milk, cream and butter, is strongly influenced by these changes which have been reviewed in Chapters 6 and 7.

Dietary lipids play many diverse nutritional roles, some of which are essential. However, dietary lipids, especially saturated lipids of animal origin, have been the subject of much controversy in recent years, particularly in regard to their possible role in atherosclerosis. Various aspects of the nutritional significance of lipids are discussed in Chapter 8.

Finally, I wish to thank sincerely the 14 authors who have contributed to this text and whose co-operation has made my task as editor a pleasure.

University College
Cork, Ireland

P. F. Fox

Contents

Contributors

D. E. Bauman Department of Animal Science, Cornell University, Ithaca, NY, USA

C. Bourlieu INRAE, CIRAD, UMR1208 IATE, Montpellier, France

W. Buchheim Kiel, Germany

S. V. Crowley School of Food and Nutritional Sciences, University College Cork, Cork, Ireland

H. C. Deeth School of Agriculture and Food Sciences, University of Queensland, Brisbane, QLD, Australia

Francesca Giuffrida Nestlé Research Centre, Lausanne, Switzerland

H. Douglas Goff Department of Food Science, University of Guelph, Guelph, ON, Canada

E. Guillocheau Laboratory of Biochemistry and Human Nutrition, Agrocampus Ouest, Rennes, France

K. J. Harvatine Department of Animal Sciences, Penn State University, University Park, PA, USA

W. Hoffmann Department of Safety and Quality of Milk and Fish Products, Max Rubner-Institut, Kiel, Germany

S. A. Hogan Food Chemistry and Technology Department, Teagasc Food Research Centre, Moorepark, Fermoy, Co. Cork, Ireland

T. Huppertz Wageningen University and Research, Wageningen, The Netherlands

FrieslandCampina, Amersfoort, The Netherlands

A. L. Kelly School of Food and Nutritional Sciences, University College Cork, Cork, Ireland

K. Linehan Teagasc, Moorepark, Fermoy, Co., Cork, Ireland

A. L. Lock Department of Animal Science, Michigan State University, East Lansing, MI, USA

A. Logan CSIRO Agriculture and Food, Werribee, VIC, Australia

C. Lopez INRA, UMR1253 STLO, Rennes, France

INRA, UR1268 BIA, Nantes, France

A. K. H. MacGibbon Fonterra Research and Development Centre, Fonterra Co-operative Group Limited, Palmerston North, New Zealand

A. G. Marangoni Department of Food Science, University of Guelph, Guelph, ON, Canada

K. D. Mattice Department of Food Science, University of Guelph, Guelph, ON, Canada

O. J. McCarthy Massey University, Auckland, New Zealand

D. J. McMahon Utah State University, Logan, UT, USA

M.-C. Michalski Univ Lyon, CarMeN laboratory, INRAE, INSERM, INSA Lyon, CRNH-RA, Pierre-Bénite, France

M. Milard Univ Lyon, CarMeN laboratory, INRAE, INSERM, INSA Lyon, CRNH-RA, Pierre-Bénite, France

Cian Moloney Nestlé Development Centre Nutrition, Limerick, Ireland

T. F. O'Callaghan School of Food and Nutritional Sciences, University College Cork, Cork, Ireland

J. A. O'Mahony School of Food and Nutritional Sciences, University College Cork, Cork, Ireland

Dairy Processing Technology Centre, University College Cork, Cork, Ireland

J. O'Regan Nestlé Development Centre Nutrition, Limerick, Ireland

D. Oliveira School of Food and Nutritional Sciences, University College Cork, Cork, Ireland

Dairy Processing Technology Centre, University College Cork, Cork, Ireland

Universidade Católica Portuguesa, CBQF – Centro de Biotecnologia e Química Fina – Laboratório Associado, Escola Superior de Biotecnologia, Porto, Portugal

D. L. Palmquist Department of Animal Sciences, Ohio Agricultural Research and Development Center/The Ohio State University, Wooster, OH, USA

V. Rioux Laboratory of Biochemistry and Human Nutrition, Agrocampus Ouest, Rennes, France

R. P. Ross APC Microbiome Ireland, University College Cork, Cork, Ireland

C. Stanton Teagasc, Moorepark, Fermoy, Co., Cork, Ireland

APC Microbiome Ireland, University College Cork, Cork, Ireland

R. Conboy Stevenson Teagasc, Moorepark, Fermoy, Co., Cork, Ireland

I. Celigueta Torres Applied Science Department, Nestlé Product Technology Centre Confectionery Confectionery, Nestlé, York, UK

T. Uniacke-Lowe School of Food and Nutritional Sciences, University College Cork, Cork, Ireland

D. S. Waldron School of Food and Nutritional Sciences, University College Cork, Cork, Ireland

L. Walter CSIRO Agriculture and Food, Werribee, VIC, Australia

Faculty of Veterinary and Agricultural Sciences, The University of Melbourne, Parkville, VIC, Australia

P. Watkins CSIRO Agriculture and Food, Werribee, VIC, Australia

M. Wong Massey University, Auckland, New Zealand

A. J. Wright Department of Food Science, University of Guelph, Guelph, ON, Canada

Composition and Structure of Bovine Milk Lipids

1

A. K. H. MacGibbon

1.1 Introduction

The lipids in bovine milk are present in microscopic globules as an oil-in-water emulsion. The primary purpose of these lipids is to provide a source of energy to the new-born calf. Both the fat content of the milk and the fatty acid composition of the lipids can vary considerably as a result of changes in factors like breed of cow, diet and stage of lactation. The fat content can vary from about 3.0 to 6.0%, but typically is in the range 3.5 to 4.7%. Changes in the fatty acid composition of triacylglycerols (particularly 16:0 and 18:1) can be quite marked and can lead to changes in physical properties of the fat. These changes make comparison difficult between different samples of milk fat, and ideally comparisons should be made between cows in mid-lactation and fed on similar diets. From a practical viewpoint, milk lipids are very important as they confer distinctive nutritional, textural and organoleptic properties on dairy products, such as cream, butter, whole milk powder and cheese. The composition and structure of bovine milk fat have been reviewed extensively. There are early reviews by Christie (1978, 1995), Jensen and Clark (1988) and Jensen and Newberg (1995); more recent articles include two book chapters by Vanhoutte and Huyghebaert (2003) and Zegarska (2003) and reviews by Chen *et al.* (2017a, b) on the best use of milk. Bovine milk lipids are similar to the milk lipids of other species as they are largely composed of triacylglycerols; however, there are also minor amounts of diacylglycerols, monoacylglycerols, free (unesterified) fatty acids, phospholipids and sterols. Trace amounts of fat-soluble vitamins, β-carotene and fat-soluble flavouring compounds are also present in the bovine milk lipids (Table 1.1).

A. K. H. MacGibbon (✉)
Fonterra Research and Development Centre, Fonterra Co-operative Group Limited,
Palmerston North, New Zealand
e-mail: Alastair.MacGibbon@fonterra.com

Because the triacylglycerols account for about 98% of the total fat, they have a major and direct effect on the properties of milk fat, for example hydrophobicity, density and melting characteristics. These triacylglycerols are a complex mixture, and vary considerably in molecular weight and degree of unsaturation. After milking, fresh milk contains only small amounts of diacylglycerols and monoacylglycerols and free fatty acids. The small proportion of diacylglycerols is largely *sn*-1,2 diacylglycerols and are, therefore, probably intermediates in the biosynthesis of triacylglycerols rather than the products of lipolysis (Lok, 1979). Likewise, the profile of free fatty acids in freshly drawn milk differs somewhat from the profile of the fatty acids esterified to the triacylglycerols (e.g. there appears to be very little free butanoic acid present), also indicating that they are unlikely to be the result of lipase action (Walstra and Jenness 1974).

P. L. H. McSweeney et al. (eds.), *Advanced Dairy Chemistry, Volume 2*,
https://doi.org/10.1007/978-3-030-48686-0_1

Table 1.1. Main classes of lipids in milk[a]

Lipid class	Amount (%, w/w)
Triacylglycerols	98.3
Diacylglycerols	0.3
Monoacylglycerols	0.03
Free fatty acids	0.1
Phospholipids	0.8
Sterols	0.3
Carotenoids	Trace
Fat-soluble vitamins	Trace
Flavour compounds	Trace

[a]From Walstra and Jenness (1984).

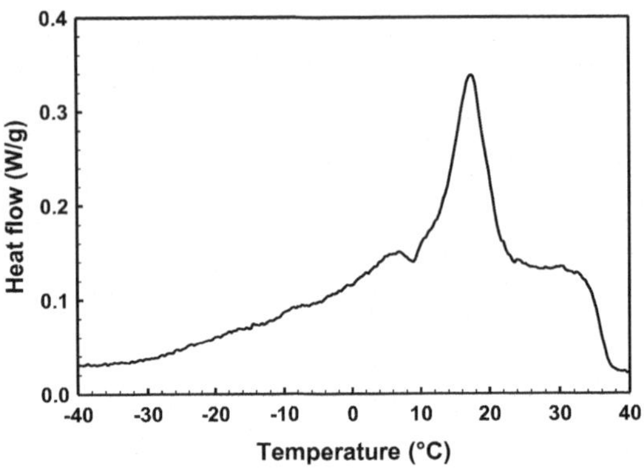

Figure 1.1. Melting profile of New Zealand milk fat, determined by differential scanning calorimetry. (From MacGibbon 1988).

Phospholipids account for only 0.8% of milk lipids. However, they play a major role in milk due to their amphiphilic properties. About 65% of phospholipids are found in the milk fat globule membrane (MFGM), whereas the rest remain in the aqueous phase. Phosphatidyl choline, phosphatidyl ethanolamine and sphingomyelin are the major phospholipids of milk, which together comprise about 80% of the total. Sterols are also a minor component, comprising about 0.3% of the fat; cholesterol, being the principal sterol, accounts for over 95% of the total sterols.

Milk fat is present in spherical droplets, which range from about 0.2 to 15.0 μm in diameter, with the bulk of the fat being in globules, 1.0–8:0 μm diameter. The MFGM, which envelopes the fat globule, consists largely of proteins and lipids. The protein of the membrane has a complex composition and over 40 polypeptides have been identified. Xanthine oxidoreductase, butyrophilin, lactadherin and adipophilin are found to be the major proteins. The lipids in the membrane are largely phospholipids and triacylglycerols. In contrast to the MFGM, the fat globule core almost exclusively consists of triacylglycerols (see Chapters 4 and 5).

The chemical properties of milk lipids can have a considerable influence on the melting characteristics of milk fat, which in turn can have a marked effect on the functional properties of a number of dairy products, such as cheese and butter (Chen *et al.* 2004; Chapter 9). Milk fat melts over a wide range, from about −35 °C to 38 °C (Figure 1.1). There is a small broad peak centred at about 7 °C, a major melting peak at about 17 °C and a plateau from 22 °C to 36 °C. It can be seen that a substantial proportion of milk fat melts between 10 °C and 20 °C. This broad melting range is directly attributable to the large number of different types of triacylglycerols present in the milk fat. Anankanbil *et al.* (2018) have described how different diets produced different fatty acids and triacylglycerols and how they affected the melting point behaviour of milk fat. How regulation of milk fat synthesis can be markedly affected by diet has been described by Harvatine *et al.* (2009), Osorio *et al.* (2016) and in Chapter 2.

1.2 Fatty Acids

Bovine milk fat is regarded as one of the most complex naturally occurring fats and oils, because of the large number of fatty acids with a variety of structures. Using a combination of chromatographic and spectroscopic techniques, researchers have identified approximately 400 fatty acids in milk fat. A listing of the various types of fatty acids has been compiled by Jensen (2002). The vast majority of these acids are present in extremely small quantities (<0.01%). However, there are about 15 fatty acids that are present at or above 1.0% of total fatty acids. The quantities of these ‘major’ fatty acids are determined relatively easily by capillary gas chromatography (GC) (IDF 2002; Chapter 13). The general procedures

Table 1.2. Composition of the major fatty acids of New Zealand milk fat

		Composition		
		Typical[a]		Range[b]
Common name		%, w/w	mole %	%, w/w
4:0	Butyric	3.0	7.8	2.6–3.6
6:0	Caproic	1.9	3.8	1.7–2.2
8:0	Caprylic	1.2	1.9	1.0–1.4
10:0	Capric	2.8	3.8	2.3–3.5
12:0	Lauric	4.5	5.1	3.1–5.5
14:0	Myristic	12.1	12.1	9.3–13.3
14:1	Myristoleic	0.9	1.2	0.5–1.3
15:0	Pentadecylic	1.1	2.7	0.9–1.4
16:0	Palmitic	30.4	27.2	26.1–34.9
16:1	Palmitoleic	1.3	1.2	1.0–1.7
18:0	Stearic	10.6	8.5	8.8–13.3
18:1 *cis*	Oleic	18.8	15.2	15.8–24.8
18:1 *trans*	Vaccenic	4.2	3.4	2.3–5.6
18:2	Linoleic	0.8	0.6	0.6–1.0
18:2 conj	CLA	1.1	0.9	0.7–1.4
18:3	α Linolenic	0.6	0.5	0.4–0.9
	Minor acids	4.6	3.9	4.0–5.1

[a]MacGibbon (unpublished 2013).
[b]Range of values for dairying season.

have been well described by Christie and Han (2010) and Amores and Virto (2019). Percentages for these fatty acids in milk fat are shown in Table 1.2. The typical values are for cows in mid-lactation, grazing on mature pasture. The range of values is for the dairying season in New Zealand, where cows graze on pasture throughout the year.

1.2.1 Origins of the Fatty Acids

The fatty acids of bovine milk fat arise from two sources: synthesis *de novo* in the mammary glands and the plasma lipids originating from the feed. The fatty acids from these two sources differ in their structure. The fatty acids that are synthesised *de novo* are short-chain and medium-chain length acids, from 4:0 to 14:0 and also some 16:0, while the C18 fatty acids and some 16:0 arise from the plasma lipids. *De novo* fatty acid synthesis accounts for approximately 45% (w/w) of the total fatty acids in milk fat, while lipids of dietary origin account for the rest (Moore and Christie 1979).

The *de novo* synthesis of fatty acids in the mammary gland utilises mainly acetate and some β-hydroxybutyrate. These precursors arise from the microbial fermentation of cellulose and related materials in the rumen. Once in the mammary gland, acetate is activated to acetyl-CoA. The mechanism of fatty acid synthesis essentially involves the carboxylation of acetyl-CoA to malonyl-CoA, which is then used in a step-wise chain elongation process. This leads to a series of short-chain and medium-chain length fatty acids, which differ by two CH_2 groups (e.g. 4:0, 6:0, 8:0, etc.) (Chapter 2). These are straight-chain, even-numbered carbon fatty acids. However, if a precursor such as propionate, valerate or isobutyrate, rather than acetate, is used, branched chain or odd-numbered carbon fatty acids are synthesised (Jenkins, 1993; Chapter 2).

Other fatty acids originate mainly from the source of diet, although these include fatty acids that can also be released from adipose tissues. Dietary lipids consist largely of glycolipids, phospholipids and triacylglycerols and the major fatty acids are linoleic acid (9c, 12c-18:2) and linolenic acid (9c, 12c, 15c-18:3). In the rumen, these lipids are hydrolysed initially to produce non-esterified fatty acids, which are then subjected to extensive biohydrogenation by micro-organisms (Jenkins 1993). The biohydrogenation sequence for linoleic acid begins with an isomerisation step, which produces conjugated linoleic acid (9c, 11t-18:2). This is followed by a reduction reaction to give vaccenic acid (11t-18:1), and then a further reduction to 18:0. Biohydrogenation of linolenic acid follows a similar pathway (Lee and Jenkins 2011; Chapter 3). The mix of fatty acids that results from biohydrogenation is esterified to triacylglycerols, which then circulate in the bloodstream within chylomicrons. These triacylglycerols are taken up by the mammary gland and cleaved to give non-esterified fatty acids. The mammary gland contains a desaturase system, which converts substantial quantities of 18:0 to oleic acid (9c-18:1).

The net result of these processes is that the fatty acids in the mammary gland, which originate from the dietary lipids, consist of substantial quantities of 16:0, 18:0 and oleic acid, small amounts of linoleic and linolenic acids and limited quantities of other monoenoic and dienoic fatty acids such as 11t-18:1 and 9c, 11t-18:2.

1.2.2 Saturated Fatty Acids

The saturated fatty acids that are present in significant quantities in milk fat are molecules with unbranched hydrocarbon chains, which vary in length from 4 to 18 carbon atoms. These fatty acids account for approximately 70–75% of the total fatty acids. The most important saturated fatty acid from a quantitative viewpoint is 16:0, which accounts for about 25–30% of the total, while two other fatty acids, 14:0 and 18:0 have values in the region 10–13% (Table 1.2). The amounts of the short-chain fatty acids, 4:0 and 6:0, are reasonably high when their proportions are expressed as molar percentages (approximately 10% and 5%, respectively — Table 1.2); appreciable amounts of medium-chain length fatty acids (C8–C12) are also present.

Short-chain and medium-chain fatty acids in milk fat have certain interesting characteristics, which may partly explain their presence. Unlike long-chain fatty acids, short-chain and medium-chain fatty acids are absorbed as non-esterified fatty acids into the portal bloodstream and are metabolised rapidly in the liver (Noble 1978). Hence, they are able to make a direct and rapid contribution to the energy metabolism of the newborn calf. Furthermore, short-chain fatty acids and, to a lesser extent, medium-chain fatty acids lower the melting point of triacylglycerols, and, thus, their presence helps keep milk fat liquid at physiological temperatures. This helps in compensating the relatively low concentration of low melting point, unsaturated fatty acids in milk fat.

1.2.3 *Cis*-Unsaturated Fatty Acids

The *cis*-monoenoic acid content of bovine milk fat is about 18–24% (Table 1.2). Oleic acid (9c-18:1) is the principal *cis*-monounsaturated fatty acid, accounting for around 15–21% of the total. There is about 0.5% of 11c-18:1, while the proportions of other *cis*-18:1 isomers are small. There are also relatively small but significant contributions from other *cis*-monounsaturated acids, namely 14:1 (about 1.0%) and 16:1 (about 1.5%).

Cis-polyenoic acids are present at low concentrations in milk fat, because of the biohydrogenation reactions that take place in the rumen. These acids are comprised almost exclusively of linoleic acid (9c, 12c-18:2), about 1.2 to 1.7% and α-linolenic acid (9c, 12c, 15c-18:3), about 0.9 to 1.2% (Table 1.2). These two fatty acids are essential fatty acids; they cannot be synthesised within the body and must be supplied by the diet. In recent times, the usage of the term 'essential' has been extended to include derivatives of these fatty acids, which are not synthesised in significant quantities (e.g. eicosapentaenoic acid, 20:5 and docosahexaenoic acid, 22:6). The proportion of α-linolenic acid appears to be affected by the cow's diet; the concentration is higher in milk from pasture-fed cows than in milk from barn-fed cows (Hebeisen *et al.* 1993; Wolff *et al.* 1995). This is also shown in a cross over trial from pasture to total mixed rations and switch back to pasture (Rego *et al.* 2016). There tends to be a decrease in linoleic acid on pasture feeding, leading to a decrease in the n-6/n-3 ratio. Indeed, the fatty acid differences between organic farming and 'conventional' farming are largely due to the supplementary feeding in the reported 'conventional' farming. Schwendel *et al.* (2017) showed that pasture feeding conventional cows removes the fatty acid differences between organic and conventionally produced milk

1.2.4 *Trans*-Unsaturated Fatty Acids

The presence of C18 *trans*-fatty acids in milk fat is the result of incomplete biohydrogenation of the unsaturated dietary lipids in the rumen. These fatty acids have attracted attention because of their adverse nutritional affects. Clinical trials have shown that *trans*-octadecenoic acids, relative to the *cis* isomer, can increase the low-density lipoprotein (LDL) cholesterol and decrease the high-density lipoprotein (HDL) cholesterol, thus, producing an unfavourable effect on the LDL: HDL ratio (Brouwer 2016).

The quantitative determination of individual isomers of *trans*-18:1 fatty acids in milk fat is not straightforward. It involves a multi-stage analytical procedure (i.e. transesterification of milk fat, argentation thin layer chromatography of the fatty acid esters to separate the *cis*-isomers and *trans*-isomers, followed by capillary GC). This method gives an almost complete separation of the 13 individual *trans*-18:1 isomers, from D4 to D16 (Precht and Molkentin 1996).

Vaccenic acid (11t-18:1) is the most important *trans* isomer with values ranging from about 30 to 60% of the total *trans*-18:1 (Table 1.3). The concentration of *trans*-18:1 varies considerably from about 2.0 to 6.0%, with mean values for milk fats from several European countries in the range 3.3–4.4% (Precht and Molkentin 2000). The higher values were for milk fat samples that were obtained from cows fed on summer pasture, whereas the lower values were associated with the feeding of concentrates and silage to cows in the winter. The feeding of fresh grass to cows appears to reduce the efficiency of the biohydrogenation reactions in the rumen, which leads to higher amounts of *trans* fatty acids. The season of calving and stage of lactation do not have any affect (Dunshea *et al.* 2008). Shingfield *et al.* (2008) has a comprehensive review of the effect of different feeding practices on *trans* fatty acid content of the milk.

Although the level of *trans*-18:1 in milk fat is significant, it is found well below the level that was present in some margarines (Ratnayake and Pelletier 1992). However, the margarine industry has changed the hard stock oil and/or changed the hydrogenation catalysts to reduce the *trans* level as a consequence of legislation. While dairy *trans*-18:1 principally has the double bond at the 11 position, partially hydrogenated vegetable oils (industrial *trans* fatty acids) have a much broader range of isomers (7–14 position). The extent to which this affects the perceived health outcomes is unclear.

Precht and Molkentin (1997) have identified and quantified a number of *trans*-octadecadienoic acids in milk fat, containing one or two *trans*-double bonds (Table 1.4). Most of these acids are present in small amounts, with only 11t, 15c-18:2 (0.33%) and 9c, 11t-18:2 (0.85%) having average concentrations above 0.30%. These *trans*-octadecadienoic acids show considerable variation; for instance, 9c, 11t 18:2 ranges from 0.25 to 1.95%. Similar to *trans*-octadecenoic acids, pasture feeding produces higher levels of *trans*-octadecadienoic acids compared to the feeding of mixed rations.

9c, 11t-18:2 is the principal isomer of the conjugated linoleic acids (CLA) in bovine milk fat, accounting for about 80–90% of the total (Parodi 1977). The term CLA refers to a mixture of positional and geometric isomers of octadecadienoic acid with conjugated double bonds (e.g. 10t, 12c-18:2, 10t, 12t-18:2). The CLA content of milk fat

Table 1.3. Composition of *trans*-octadecenoic acids of milk fat[a,b]

Trans-octadecenoic acid isomers											
Composition (%, w/w of total fatty acids)											
	Δ4	Δ5	Δ6–8	Δ9	Δ10	Δ11	Δ12	Δ13/14	Δ15	Δ16	Total
Mean values	0.05	0.05	0.17	0.24	0.17	1.75	0.21	0.48	0.28	0.34	3.74
Max values	0.13	0.12	0.30	0.31	0.26	4.00	0.31	0.73	0.47	0.51	6.34
Min values	0.02	0.02	0.03	0.16	0.00	0.52	0.11	0.25	0.10	0.16	1.91

[a]From Precht and Molkentin (1996).
[b]100 samples.

Table 1.4. Composition of *trans*-octadecadienoic acids of milk fat[a]

	Composition (%, w/w of total fatty acids)			
Trans-18:2 isomers	*n*	Mean	Max	Min
9*c*, 11*t* – 18:2	100	0.85	1.95	0.25
9*c*, 12*t* – 18:2	100	0.10	0.16	0.05
9*t*, 12*c* – 18:2	100	0.07	0.48	0.02
9*c*, 13*t* – 18:2 8*t*, 12*c* – 18:2	11	0.11	0.16	0.07
11*t*, 15*c* – 18:2	100	0.33	0.68	0.04
9*t*, 12*t* – 18:2	11	0.09	0.12	0.06
t, *t*, – NMID	11	0.19	0.38	0.10

[a]From Precht and Molkentin (1997).
[b]*NMID* Non-methylene interrupted diene.

is derived from two related sources. First, 9c, 11t-18:2 is an intermediate product of the biohydrogenation of fatty acids in the rumen. In addition, 11t-18:1 can be converted to 9c, 11t-18:2 in the mammary gland by the enzyme stearoyl-CoA desaturase (SCD), which normally catalyses the conversion of 18:0 to 9c-18:1 (see Chapter 3).

The presence of 9c, 11t-18:2 in bovine milk fat has been known for many years (Parodi 1977). However, it was only after Ha *et al.* (1987) identified 9c, 11t-18:2 as an anti-carcinogenic substance in minced beef that there was an upsurge in scientific investigation. In subsequent work, the health implications of CLA have widened to include inhibition of carcinogenesis, atherosclerosis, diabetes and weight loss induced by immune stimulation and increase in the percentage of lean body mass (Parodi 1999; Yang *et al.* 2015). These studies have shown that CLA can be regarded as unique, because it appears to provide numerous positive health effects unlike other *trans* acids. This research is discussed in Chapter 3.

An important structural difference between *cis*-unsaturated and *trans*- unsaturated fatty acids is that the *cis* configuration of the double bond puts a significant 'kink' in the hydrocarbon chain, whereas the *trans* configuration causes only a slight distortion. This difference has a major impact on the way in which triacylglycerols pack in crystal lattices when they solidify. Triacylglycerols containing *cis*-unsaturated fatty acids have a lower packing density than triacylglycerols containing either *trans*-unsaturated or saturated fatty acids, and as a result have lower melting points. Thus, the relative levels of *cis*-unsaturated and *trans*-unsaturated fatty acids can exert a considerable influence on the melting characteristics of milk fat.

1.2.5 Minor Fatty Acids

As noted earlier, there are approximately 400 minor fatty acids in bovine milk fat; about 40 are present at levels >0.01% (Table 1.5), while the remainder exist in trace amounts. Most of these fatty acids are of little practical importance and, hence, their nature and structure are of academic interest only. Among the minor saturated fatty acids are branched-chain and odd-numbered carbon fatty acids with a range of chain length from C3 to C27. Examples of odd-numbered fatty acids are 13:0 (0.19%), 17:0 (0.6%) and 19:0 (0.15%) (Table 1.5). The monomethyl branched-chain fatty acids are quite significant, accounting for about 2.5% of the total fatty acids. Examples are the C15 branched-chain fatty acids, 13-methyl tetradecanoic acid (the iso configuration) and 12-methyl tetradecanoic acid (the anteiso configuration), which together make up about 0.8% of milk fat (Table 1.5). Bainbridge *et al.* (2016) found that branch chain fatty acids were higher in Jersey cows than Holsteins and while they did not change much through lactation, they were higher at the end of lactation.

There are about 200 minor monoenoic, dienoic and polyenoic fatty acids in milk fat ranging in chain length from C10 to C24, and consisting of both positional and *cis*/*trans* isomers. A number of these fatty acids have considerable nutritional significance; for example, eicosapentaenoic acid (20:5, 0.09%) and docosahexaenoic acid (22:6, 0.01%) are present in the metabolic pathway of the n-3 fatty acids, while arachidonic acid (20:4, 0.14%) is part of the n-6 pathway and a precursor of prostaglandin synthesis.

Jensen (2002) has reported that milk fat contains about 60 hydroxy fatty acids. The C4-hydroxy and C5-hydroxy acids are of interest

Table 1.5. Composition of the minor fatty acids of milk fat[a, b]

Composition (%, w/w of total fatty acids)							
Saturated				Unsaturated			
Straight chain		Branched chain[c]		Monounsaturated		Polyunsaturated	
FA	%, w/w	FA	%, w/w	FA	%, w/w	FA	%, w/w
11:0	0.20	13:0i	0.03	10:1	0.15	20:2	0.07
13:0	0.19	14:0a	0.02	12:1	0.06	20:3	0.10
17:0	0.60	15:0i	0.40	13:1	0.03	20:4	0.14
19:0	0.15	15:0a	0.44	17:1	0.36	20:5	0.09
20:0	0.35	16:0i	0.40	19:1	0.16	22:2	0.04
21:0	0.04	17:0i	0.50	20:1	0.32	22:3	0.07
22:0	0.20	17:0a	0.52	21:1	0.04	22:4	0.03
23:0	0.12	18:0i	0.16	22:1	0.06	22:5	0.04
24:0	0.14	19:0i	0.10			22:6	0.01
25:0	0.03						
26:0	0.06						

[a]From Iverson and Sheppard (1986).
[b]Minor fatty acids present at levels ≥0.01%.
[c]i = iso, a = anteiso.

as they transform to the respective lactones with 4-carbon (γ) and 5-carbon (δ) rings, which are major contributors to the overall flavour of the milk fat. Approximately, 60 keto (oxo) acids have been isolated and identified in milk fat (Weihrauch *et al.* 1974; Brechany and Christie 1992; Yoshinaga *et al.* 2019). When milk fat is heated, β-keto acids are decarboxylated to form methyl ketones, which contribute to cooked butter flavours.

1.2.6 Variations in Fatty Acid Composition

The fatty acid composition of milk fat is not stable and is influenced by a number of factors. These include, breed of cow, stage of lactation and type and quality and quantity of feed (Moate *et al.* 2008; Petersen and Jensen 2014; Rego *et al.* 2016; Hanus *et al.* 2018). These issues are discussed in detail in Chapter 2.

In most countries, there exists a regularly recurring seasonal pattern of fatty acid variation in milk fat, which is caused largely by changes to the cow's diet. This seasonal variation can have an impact on the properties of high-fat dairy products, e.g. the hardness of butter (MacGibbon and McLennan 1987). The seasonal variation for French milk fat is presented in Table 1.6 (Wolff *et al.* 1995). It can be seen that 16:0 has a markedly lower value in spring and summer than in winter. The C6 to C14 fatty acids together show a similar trend, although the magnitude of the change is much smaller. In contrast, 18:0 and 18:1 (*cis* and *trans*) show a reverse trend with lower levels in winter. Generally, higher values of 16:0 tend to be associated with higher levels of total lipids and greater hardness of the fat. These variations are attributed to a change from winter feed of hay and concentrate to a diet of fresh grass in spring. The lipids in fresh grass contain high levels of fat (18:2 and 18:3) which, as a result of biohydrogenation and desaturation reactions in the cow, lead to increased levels of 18:0 and 18:1 in milk fat.

Similar seasonal trends in fatty acid composition have been found in other countries where the pattern of dairy husbandry practices is similar (Lindmark-Mansson 2008; Heck *et al.* 2009; O'Donnell-Megaro *et al.* 2011; Toth *et al.* 2012).

Table 1.6. Fatty acid compositions of French butters collected at different periods of the year[a]

Fatty acid	Composition (%, w/w)				
	January	March	May–June	July–August	Oct–Nov
4:0	4.0	4.4	3.8	4.2	4.3
6:0	2.5	2.7	2.4	2.5	2.6
8:0	1.5	1.6	1.4	1.5	1.5
10:0	3.5	3.5	3.1	3.1	3.3
12:0	4.0	4.0	3.6	3.5	3.8
14:0	12.1	11.9	11.0	11.2	11.3
14:1	1.1	1.0	0.9	1.1	0.9
15:0	1.2	1.2	1.2	1.2	1.1
16:0	33.3	32.5	27.1	28.3	29.3
16:1	1.5	1.6	1.4	1.5	1.7
18:0	9.0	9.0	11.0	10.5	9.6
18:1 *cis*	16.8	17.3	19.7	19.2	19.4
18:1 *trans*	2.4	2.4	4.3	3.7	3.2
18:2	1.4	1.5	1.2	1.3	1.4
18:2 conj	0.4	0.5	0.7	0.7	0.5
18:3	0.3	0.3	0.6	0.5	0.5
Minor	5.0	4.6	6.6	6.0	5.6

[a]From Wolff *et al.* (1995).

Table 1.7. General composition of triacylglycerols of milk fat

Carbon number	Triacylglycerol composition (%, w/w)			
	German milk fat[a]		New Zealand milk fat[b]	
	Average	Range[c]	Typical	Range[d]
C26	0.2	0.2–0.3	–	–
C28	0.6	0.5–0.8	0.6	0.4–0.8
C30	1.2	1.0–1.9	1.2	0.8–1.4
C32	2.6	2.1–3.2	2.5	1.8–2.9
C34	6.0	4.8–6.9	5.8	4.4–6.4
C36	10.9	9.2–12.4	11.0	9.1–11.8
C38	12.8	12.1–13.6	13.3	11.8–14.6
C40	10.1	9.5–11.2	10.7	9.7–12.1
C42	7.1	6.2–7.9	7.4	6.5–7.9
C44	6.7	5.4–7.8	6.7	5.6–7.3
C46	7.4	6.3–8.3	7.2	5.6–7.8
C48	9.1	8.0–10.7	8.6	6.9–9.9
C50	10.9	9.7–12.0	10.6	9.7–12.8
C52	9.5	7.2–12.3	9.4	7.7–12.6
C54	4.6	2.7–7.8	4.7	3.7–7.0
C56	–	–	0.4	0.4–0.6

[a]From Precht and Frede (1994).
[b]From MacGibbon (unpublished results).
[c]Range of values for different regions.
[d]Range of values over a dairying season.

1.3 Triacylglycerols

Bovine milk fat contains various triacylglycerols, which vary considerably in molecular weight and degree of unsaturation. This complexity is the direct result of the large number and wide variety of fatty acids which make up the triacylglycerols.

As noted earlier, there are some 400 fatty acids in milk fat, which means that theoretically milk fat could contain many thousands of triacylglycerols. Even if one considers only the 15 or so fatty acids that are present at concentrations above 1% (Table 1.2), and ignores the placement of these fatty acids at specific positions on the triacylglycerol molecule, there are still 680 compositionally different triacylglycerols.

The general composition of triacylglycerols can readily be determined by capillary GC. Typical triacylglycerol compositions of milk fats from Germany and New Zealand are presented in Table 1.7. The triacylglycerols show a wide molecular weight range (from acyl carbon 26 to 56), which arises from the large differences in chain length of the constituent fatty acids (from C4 to C18). The triacylglycerol composition is dominated by triacyglycerols with 36–40 acyl carbons (about 35% of the total) and 46–52 acyl carbons (about 36% of the total). The range of values for the different carbon numbers is considerable, indicating that there is significant variation in triacylglycerol composition both throughout the dairying season and between different dairying regions. Interestingly, data for the two countries are remarkably similar.

It should be noted that simple capillary GC, while convenient, just separates triacylglycerols into groups with similar molecular weight, and does not provide information on individual triacylglycerols – carbon number 38, for example, will consist of several different triacylglycerols (e.g. 4:0, 16:0, 18:0; 4:0, 16:0, 18:1; 6:0, 14:0, 18:1, etc.).

High performance liquid chromatography (HPLC) gives greater flexibility in providing separation between groups of interest. Robinson and MacGibbon (1998) used reversed-phase HPLC in which milk fat triacylglycerols were separated into 61 distinct peaks (Figure 1.4). The triacylglycerols present in each peak were identified through initial fractionation by argentation thin layer chromatography (TLC), followed by HPLC and mass spectrometry (MS). This HPLC method can be used as a single-injection routine method and appears to be sensitive enough to monitor relatively small changes in peak areas and, hence, minor changes in the amounts of small groups of triacylglycerols.

1.3.1 Structure of Triacylglycerols

Triacylglycerols are synthesised in the mammary gland by enzymatic mechanisms that exert some selectivity over the esterification of different fatty acids at each position of the *sn*-glycerol moiety (Moore and Christie 1979). A triacylglycerol molecule showing the three sn-positions is shown in Figure 1.2.

Stereo-specific analytical procedures have been developed that have enabled the determination of the positional distributions of fatty acids on the triacylglycerols. The results obtained using these procedures show that there is a highly selective stereo-specific distribution of fatty acids in the triacylglycerols of bovine milk fat (Table 1.8). For cows fed a normal diet, the fatty acids 4:0 and 6:0 are esterified almost entirely at position *sn*-3. In contrast, 12:0 and 14:0 are esterified preferentially at position *sn*-2, while 16:0 is incorporated preferentially at positions *sn*-1 and *sn*-2. 18:0 is esterified preferentially at position *sn*-1, and 18:1 shows a preference for positions *sn*-1 and *sn*-3. This overall pattern of fatty acid distribution does not change significantly either throughout lactation or between countries (Pitas *et al.* 1967; Taylor and Hawke 1975b; Parodi 1979; Christie and Clapperton 1982).

Stereo-specific analysis of milk fat fractions containing triacylglycerols of different molecular weight has shown that, for fatty acids of chain length C4–C16, the general pattern of fatty acid distribution in normal milk fat is similar to the pattern of distribution in the triacylglycerol fractions of different molecular weight. However, the pattern of distribution of 18:0 and 18:1 varies according to the molecular weight of the triacylglycerols; these fatty acids tend to be esterified preferentially at positions *sn*-1 and *sn*-3 in triacylglycerols of high molecular weight and concentrated at position *sn*-1 in triacylglycerols of medium- and low-molecular weight (Parodi 1982).

1.3.2 Composition of Triacylglycerols

As noted earlier, milk fat contains a very complex mixture of triacylglycerols. This complexity has made the identification and characterisation of

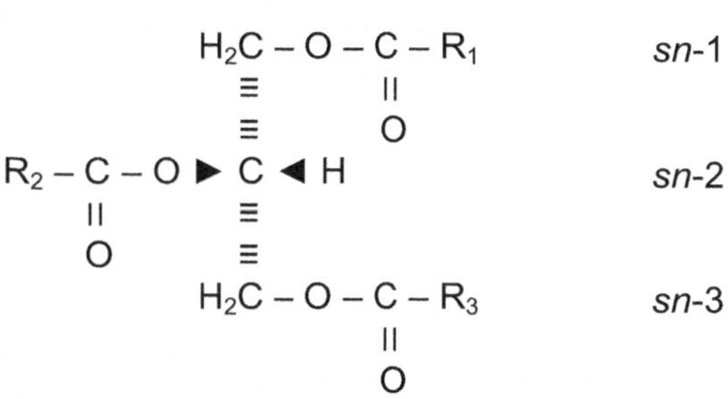

Figure 1.2. Fischer projection diagram of a triacylglycerol showing the stereospecific numbering (*sn*-) convention.

Table 1.8. Positional distribution of fatty acids in the triacylglycerols of milk fat[a]

Fatty Acid	Fatty acid composition (mole %)		
	sn-1	*sn*-2	*sn*-3
4:0	–	0.4	30.6
6:0	–	0.7	13.8
8:0	0.3	3.5	4.2
10:0	1.4	8.1	7.5
12:0	3.5	9.5	4.5
14:0	13.1	25.6	6.9
16:0	43.8	38.9	9.3
18:0	17.6	4.6	6.0
18:1	19.7	8.4	17.1

[a]Calculated from the data of Parodi (1979).

individual triacylglycerols extremely difficult. Moreover, the fact that no two batches of milk fat have exactly the same composition adds to the difficulty. As a result, the majority of the earlier studies were aimed at elucidating the general types of triacylglycerols present rather than obtaining quantitative data about individual triacylglycerols.

In a series of investigations, milk fat was fractionated into different triacylglycerol classes on the basis of molecular weight and degree of unsaturation, using a combination of chromatographic methods, namely, normal and argentation TLC and GC. This approach, in combination with stereo-specific analysis, provided detailed information on the different classes of triacylglycerols present in milk fat (Breckenridge and Kuksis 1968, 1969; Taylor and Hawke 1975a; Parodi 1980).

The high molecular weight fractions of differing degrees of unsaturation were found to consist largely of triacylglycerols containing combinations of four long-chain fatty acids, namely 14:0, 16:0, 18:0 and 18:1. The most likely placement of these fatty acids at the different positions on the triacylglycerol molecules is shown in Figure 1.3. On the other hand, the medium- and low- molecular weight fractions were comprised mainly of triacylglycerols with combinations of these four long-chain fatty acids at positions *sn*-1 and *sn*-2 and a short-chain fatty acid (either 4:0 or 6:0) esterified at position *sn*-3.

The saturated and monoene triacylglycerol classes were dominant and each comprised about 35–40% of the total milk fat, while the approximate proportions of the high-, medium- and low-molecular weight fractions were 40, 20 and 40%, respectively.

More recently, the use of more sophisticated chromatographic techniques, particularly HPLC and capillary GC, has led to the identification and quantification of individual, compositionally different triacylglycerols. In one painstaking study, Gresti *et al.* (1993) separated milk fat by reversed-phase HPLC into 47 fractions. Each fraction was then analysed for triacylglycerol and fatty acid composition by capillary GC. The data obtained were used to calculate the proportions of some 220 individual molecular species of triacylglycerols, accounting for 80% of the total triacylglycerols in the sample. The quantitatively important triacylglycerols, each present at >0.5%, are shown in Table 1.9. This list of 40 major triacylglycerols makes up about 55% of the total milk fat. An interesting aspect of the data is that some triacylglycerols are present in high proportions, for example 4:0, 16:0, 18:1 (4.2%); 4:0, 16:0, 16:0 (3.2%); 4:0, 14:0, 16:0 (3.1%); 14:0, 16:0, 18:1 (2.8%); 4:0, 16:0, 18:0 (2.5%) and 16:0, 18:1, 18:1 (2.5%). Although, this was a long and exhaustive study, it did not definitively identify the constituent triacylglycerols of milk fat as the placement of the fatty acids at the different sn- positions on the triacylglycerol

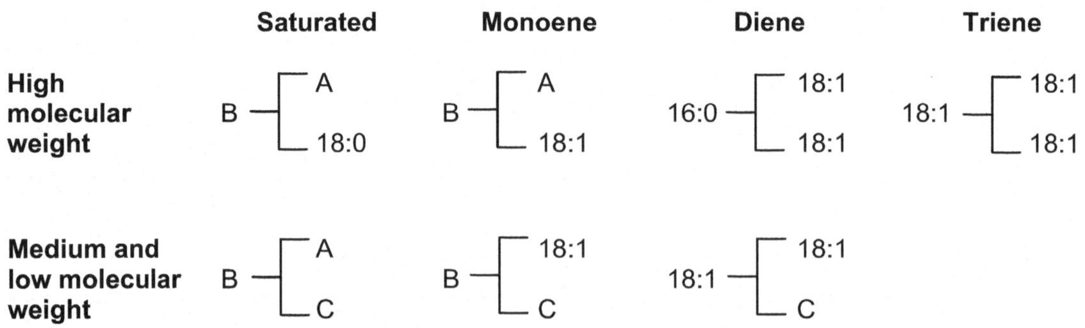

Figure 1.3. Probable composition of the major triacylglycerols of milk fat. A = 16:0 or 18:0; B = 14:0 or 16:0; C = 4:0 or 6:0.

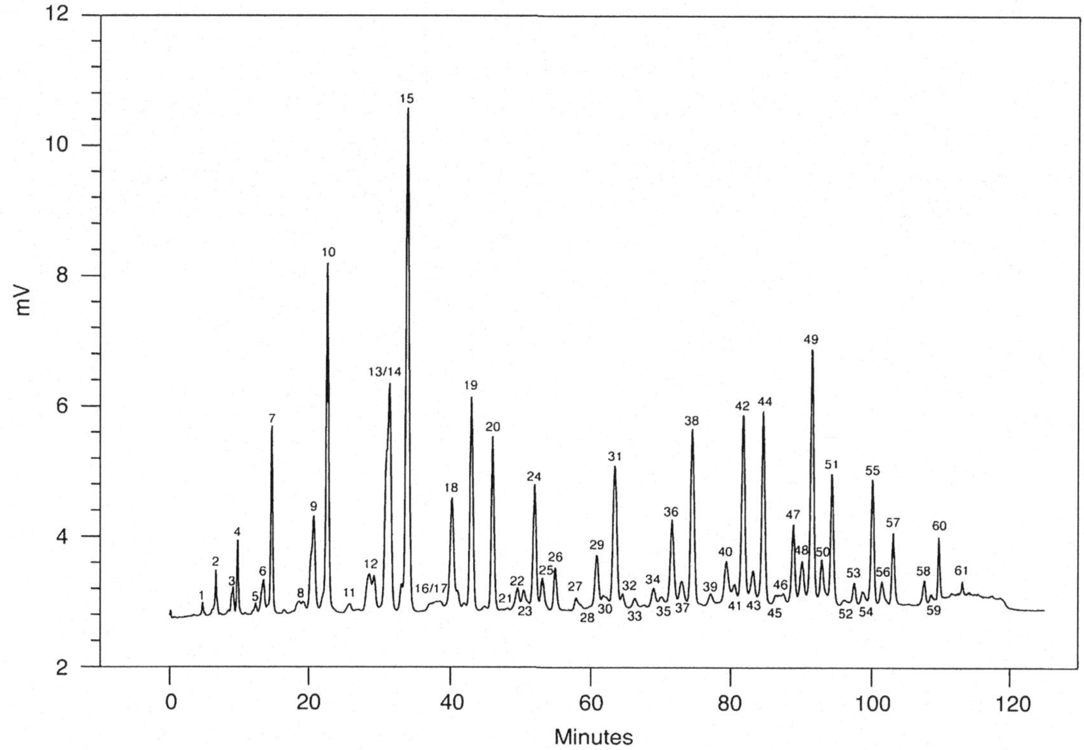

Figure 1.4. Reversed-phase HPLC trace of milk fat triacylglycerols. (From Robinson and MacGibbon 1998).

Table 1.9. Proportions of the major triacylglycerols in a sample of French milk fat[a–c]

Carbon number	Saturated triacylglycerols		Monoene triacylglycerols		Diene and triene triacylglycerols	
C30	4:0, 10:0, 16:0,	0.6%				
C32	4:0, 12:0, 16:0	0.8%				
C34	4:0, 14:0, 16:0	3.1%				
C36	4:0, 14:0, 18:0	1.3%	4:0, 14:0, 18:1	1.8%		
	4:0, 16:0, 16:0	3.2%				
	6:0, 14:0, 16:0	1.4%				
C38	4:0, 16:0, 18:0	2.5%	4:0, 16:0, 18:1	4.2%		
	6:0, 16:0, 16:0	1.5%	6:0, 14:0, 18:1	0.9%		
	6:0, 14:0, 18:0	0.6%				
C40	6:0, 16:0, 18:0	1.1%	4:0, 18:0, 18:1	1.6%	4:0, 18:1, 18:1	1.5%
	8:0, 16:0, 16:0	0.7%	6:0, 16:0, 18:1	2.0%		
	10:0, 14:0, 16:0	0.7%				
C42	10:0, 16:0, 16:0	1.0%	10:0, 14:0, 18:1	0.6%	6:0, 18:1, 18:1	0.6%
	12:0, 14:0, 16:0	0.6%				
C44	14:0, 14:0, 16:0	0.6%	10:0, 16:0, 18:1	1.6%		
C46	14:0, 16:0, 16:0	0.9%	12:0, 16:0, 18:1	1.2%	10:0, 18:1, 18:1	0.7%
			14:0, 14:0, 18:1	0.6%		
C48	14:0, 16:0, 18:0	0.7%	14:0, 16:0, 18:1	2.8%	12:0, 18:1, 18:1	0.6%
C50			14:0, 18:0, 18:1	1.4%	14:0, 18:1, 18:1	1.2%
			16:0, 16:0, 18:1	2.3%		
C52			16:0, 18:0, 18:1	2.2%	16:0, 18:1, 18:1	2.5%
					16:0, 18:1, 18:2	0.6%
C54			18:0, 18:0, 18:1	0.8%	18:0, 18:1, 18:1	1.2%
					18:1, 18:1, 18:1	1.0%

[a]From Gresti *et al.* (1993).
[b]Triacylglycerols at concentrations >0.5%.
[c]Position of fatty acid on triacylglycerol molecule not determined.

molecules was not determined. Spanos *et al.* (1995) used HPLC to isolate 58 triacylglycerol fractions and used desorption chemical ionisation MS to qualitatively describe the triacylglycerols showing similar trends to Gresti *et al.* (1993).

The mass spectroscopic analysis of milk fat has had many advances in recent years, both in the sensitivity of the instruments and the software available for analysis. Xu *et al.* (2018) recently reviewed the methods for the general area of profiling triacylglycerols in oils and fats and their applications. The method that has been found of most use for milk fat is high-resolution liquid chromatography with electrospray ionisation mass spectroscopy (ESI) or atmospheric pressure chemical ionisation (APCI). While the APCI generates abundant in-source fragments which can be used for structural determination, the complex nature of milk fat triacylglycerols and the co-elution of molecules means that the parent-product relationships can be difficult to determine. For this reason, the lower in-source fragmentation and the greater sensitivity of ESI is often used (Nagy *et al.* 2013).

One of the ultimate aims of triacylglycerol mass spectrometry is the easy elucidation of the *sn*-2 position fatty acids from the *sn*-1 and *sn*-3 without the need for the complex Grignard derivatisation and subsequent analysis of the resultant monoglyceride. It has been frequently reported that the loss of the *sn*-1(3) fatty acids is energetically more favourable than the *sn*-2, hence the *sn*-2 fatty acid is the fatty acid fragment with the lower relative abundance (Kalo *et al.* 2004). However, because of the influence of chain length, number of double bonds and experimental conditions, it is only an approximation (Byrdwell 2001). For absolute certainty, standards of the isomers are required for the appropriate correction, which is difficult to achieve. Nagy *et al.* (2013) used a number of standards and were able to obtain *sn*-2 results for palmitic acid in milk fat, pork and beef tallow which fitted with literature data.

Kalo *et al.* (2009) determined the triacylglycerols of milk fat, separated into four fractions, by HPLC and electrospray ionisation tandem mass spectroscopy. Though the initial milk fat was more saturated than the sample used by Gresti *et al.* (1993), the patterns were similar. Kalo *et al.* (2009) found the major short-chain triacylglycerols to be 16:0/16:0/4:0, 18:0/14:0/4:0, 18:1/16:0/4:0, 16:0/14:0/4:0, 18:0/12:0/4:0 and 18:0/16:0/4:0. The major medium-chain triacylglycerols were 14:0/12:0/12:0, 16:0/12:0/10:0, 16:0/14:0/8:0, 18:0/10:0/10:0, 18:0/12:0/8:0 and 14:0/14:0/10:0. The major long-chain triacylglycerols were 18:1/16:0/16:0, 18:1/14:0/18:0, 18:1/18:0/14:0, 16:0/18:0/16:1, 18:1/16:0/18:1 and 18:2/18:0/16:0.

Despite these concerns, several researchers have used MS to identify many of the constituent triacylglycerols of milk fat. These studies invariably begin with extensive fractionation of the triacylglycerols prior to mass spectral analysis, to ensure that the number of triacylglycerol species contributing to a particular fraction are as small as possible. In one of the earliest investigations, Myher *et al.* (1988) studied a milk fat fraction which was composed largely of low molecular weight triacylglycerols. After an initial separation using argentation TLC, which separated the triacylglycerols according to their degree of unsaturation, mass spectral analysis was used to identify more than 100 triacylglycerols.

Kalo *et al.* (2004) used normal-phase HPLC in combination with positive ion tandem MS to obtain quantitative information about the regioisomers of synthetic triacylglycerol mixtures and milk fat fractions containing low molecular weight triacylglycerols. In agreement with a previous study (Currie and Kallio 1993), they found that the diacylglycerol fragment ions, produced by mass spectral analysis from standard triacylglycerol mixtures, contained greater amounts of fatty acids at the *sn*-2 position than predicted. Furthermore, the ratio of fatty acids at the *sn*-2 position, relative to the fatty acids at the *sn*-1 and *sn*-3 positions, varied according to the types of fatty acids attached. From the information gained about these diacylglycerols, the regioisomers of the synthetic triacylglycerol mixtures could be identified. In a similar manner, the regioisomers of the triacylglycerols in the milk fat fractions were studied, although the fatty acids at the *sn*-1 and *sn*-3 positions could not be differentiated.

With the interest in n-3 fatty acids for health reasons, Liu *et al.* (2015) used mass spectroscopy to find the triacylglycerols of milk fat which contained long-chain n-3 fatty acids (by detections of the neutral loss of the long-chain n-3 fatty acid). They found 51 species, with docosahexanoic acid (DHA) species in lower concentration than docosapentaenoic acid (DPA) and eicosapentaenoic acid (EPA) species. The major DHA species were C22:6-C16-C4, C22:6-C16-C6; DPA species were C22:5-C16-C4, C22:5-C18:1-C4 and EPA species C20:5-C16-C4, C20:5-C16-14, C20:5-C16-C16.

Liu *et al.* (2017) studied the seasonal variation in the triacylglycerol profile of milk fat. They observed large changes in the cow-to-cow variation and in the seasonal variation illustrating the complexity of the milk fat triacylglycerols. Interestingly, the proportion of C52:2 triacylglycerol was consistent over the season. Tzompa-Sosa *et al.* (2014) studied cows with different diacylglycerol *O*-acyltransferase (DGAT) genotypes using both mass spectroscopy and Grignard positional analysis and found that the DGAT polymorphism had a significant effect at positioning the C16:0 at the *sn*-2 position. They also observed that the amount of long-chain saturated fatty acid influences the positioning of other fatty acids in the triacylglycerol. As the C16.0 concentration in the triacylglycerol increased, the percentage of C14:0, C16:0 and long-chain saturated fatty acids (C14–C18) at the *sn*-2 position (compared to *sn*-1(3)) decreased and the proportion of C18:1 at the *sn*-2 position slightly increased.

Cossignani *et al.* (2019) has provided a review of all the methods used in the direct and indirect analysis of triacylglycerols molecular species, including the methods of extraction.

With the upsurge of interest in CLA, the distribution of CLA in milk fat triacylglycerols has also become a matter of considerable interest. The distribution of CLA has been determined by a reversed-phase HPLC system, in which the eluting peaks were simultaneously detected by both evaporative light scattering detection (ELSD) and ultraviolet (UV) absorption at 233 nm (Robinson and MacGibbon 2000). The UV absorption data clearly show which peaks contain esterified CLA (the molar extinction coefficient for CLA at 233 nm is 23,360 L mol^{-1} cm^{-1}). The combined data from the two detection systems show that CLA is found in many different types of triacylglycerols, which differ in both molecular weight and degree of unsaturation. Valeille and Martin (2004) carried out a stereospecific determination of CLA in milk fat triacylglycerols and found that the CLA resided at the external positions, predominantly at the *sn*-3 position.

The elusive goal with regards to triacylglycerols is localisation of the double bond positions. This has not been carried out for milk fat; however, Hakova *et al.* (2015) did describe a method using triacylglycerol standards identified by acetonitrile adduct fragmentation and APCI collision-induced dissociation or Q-pulsed collision-induced dissociation.

1.4 Polar Lipids

The concentration of phospholipids in the milk fat ranges from 0.5 to 1.0% of the total (Patton and Jensen 1976; Table 1.1; Rombaut *et al.* 2005). Recent summaries of phospholipid content of milk and milk products have been published (Contarini and Povolo 2013; Verardo *et al.* 2017). About 60–65% of these phospholipids are associated with the intact milk fat globule membrane (MFGM). The remaining 35–40% are found in the aqueous phase associated with protein/membrane fragment material in solution, rather than still attached to the MFGM (Huang and Kuksis 1967; Patton and Keenan 1971). The MFGM that surrounds the milk fat droplets is derived from the apical plasma membrane of the secretory cells in the lactating mammary glands, and is composed of phospholipids and glycolipids, as well as proteins, glycoproteins, enzymes, triacylglycerols and minor components. Estimates of the proportion of phospholipids in the MFGM vary from 15 to 30%, depending on extraction procedure; however, values reported most commonly are at the higher end of this range. For instance, Keenan and Dylewski (1995) reported 26–31%, and Fong *et al.* (2007) found 29% of the MFGM as phospholipids (see Chapter 4). However, because of the variable extraction of ubiquitous triglycerides, cholesterol and proteins, the MFGM is probably best defined in terms of the components that are specific to the MFGM, the membrane proteins, the phospholipids and the gangliosides.

While the polar lipids constitute a very small proportion of the total milk lipids, they play an important role because of their mixed hydrophilic and hydrophobic nature. This unique characteristic of polar lipids is largely responsible for stabilising the suspension of milk fat in the aqueous environment of the milk, allowing the relatively high concentrations of milk fat and protein to coexist in the same solution (Deeth 1997). In the above process, the major structural features involved are the large non-polar (hydrophobic) fatty acid chains and the polar (hydrophilic) charged head group residue of the phospholipids. The polar lipids contain a variety of polar groups that contribute to the charged nature of the molecules. In addition to the charged headgroup, phospholipids contain a negatively charged phosphate group (Figure 1.5).

Dairy phospholipids are important structurally, because they are able to stabilise emulsions and foams, and to form micelles and membranes (Jensen and Newburg 1995). Phospholipids also have the potential to be pro-oxidants, because they contain monounsaturated and polyunsaturated fatty acids and have the ability to attract metal ions. Phosphatidyl ethanolamine binds copper strongly and is believed to be important in copper-induced oxidation in milk (O'Connor and O'Brien 1995; Deeth 1997). The polyunsaturated fatty acids and metal ions accelerate lipid oxidation, especially when heat is applied; hence, phospholipids can be degraded during the processing of milk. However, in dairy products, the situation is complex and it appears that phospholipids are able to act as either pro-oxidants or antioxidants, depending on the pH, ratio of water and phospholipid species (Chen and Nawar 1991).

As the milk is processed, the phospholipids are partitioned differently from the neutral lipids (Table 1.10). When the whole milk is separated, the phospholipids tightly bound to the MFGM go into the cream with the neutral lipids, while the phospholipids associated with the protein/membrane fragments in the aqueous phase are retained in the skim milk. Hence, the ratio of phospholipids to total fat is relatively low in cream and high in skim milk. Furthermore, during butter-making, a greater proportion of the phospholipids than the neutral lipids from the cream is retained in the buttermilk, leading to a high ratio of phospho-

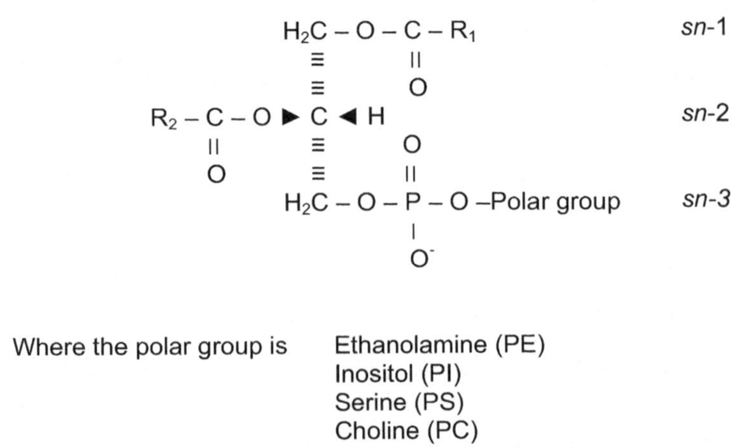

Figure 1.5. Fischer projection diagram of a glycerophospholipid showing the stereospecific numbering (*sn*-) convention.

lipid to the total fat in buttermilk (Table 1.10). Pimental *et al.* (2016) have reviewed the phospholipids in a wide range of dairy products.

1.4.1 Composition and Structure

The percentage of phospholipids in milk fat is typically within the range 0.5–1.0 % (Rombaut *et al.* 2005). Bitman and Wood (1990) found that phospholipids in milk tended to decline during lactation, but Kinsella and Houghton (1975) observed little change. While there was a change in the percentage of total phospholipids, the ratio of the major phospholipids remained relatively constant, suggesting a constant ratio of phospholipids in the MFGM.

The structures of the major polar lipids found in milk are shown in Figures 1.5 and 1.6. Glycerophospholipids [phospatidylethanolamine (PE), phosphatidylinositol (PI), phosphatidylserine (PS) and phosphatidylcholine (PC)] have fatty acids at positions *sn*-1 and *sn*-2 and a phosphate and a polar head-group on the *sn*-3 position. Of the minor phospholipids, plasmalogens have a similar structure to phosphatidylcholine and phosphatidylethanolamine but with a vinyl linkage rather than an ester linkage at the *sn*-1 position. Lysophospholipids have only one fatty acid in the glycerophospholipid.

Table 1.10. Approximate phospholipid content of different dairy products (liquid)

Product	Whole milk	Skim milk	Cream	Buttermilk
Total fat (% w/w)	4.2	0.1	42.3	0.5
Phospholipid (%, w/w)	0.04	0.02	0.19	0.16
% phospholipid in fat	1.0	15	0.4	33

Adopted from Rombaut *et al.* (2006).

The sphingophospholipid, sphingomyelin (SM) consists of a ceramide (a fatty acid linked to a long-chain sphingoloid base through an amide linkage) with a phosphorylcholine head-group (Figure 1.6). Sphingomyelin is generally included in the phospholipid group as it has similar properties (especially with phosphatidylcholine).

Phosphatidylcholine, phosphatidylethanolamine and sphingomyelin are the major polar lipids found in bovine milk and are present in similar proportions in the total phospholipids, about 22–33% (Table 1.11). Phosphatidylserine and phosphatidylinositol are present at lower levels – about 4–15% (Table 1.11). There are also significant amounts of ceramides, as glucoceramide (monohexose) and lactoceramide (dihexose). There is quite a variation in the phospholipid classes in bovine milk as shown in

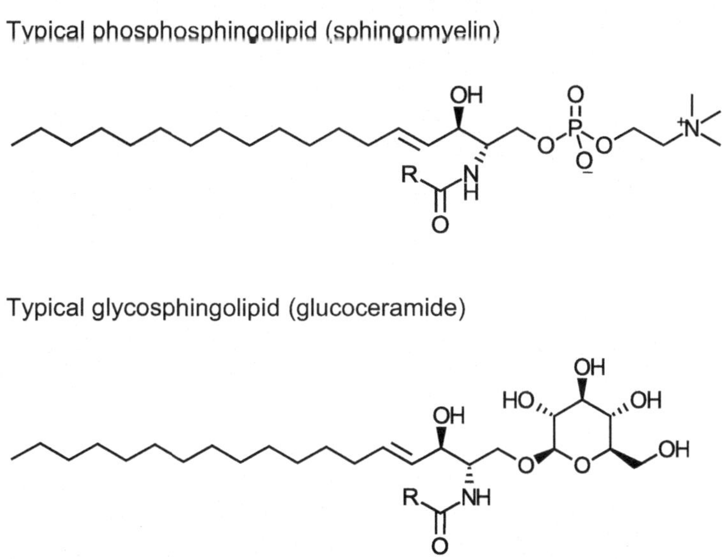

Figure 1.6. Typical structures of sphingolipids (phosphosphingolipid and glycosphingolipid classes), based on a d18:1 ceramide (R = fatty acid group).

Table 1.11. Proportions of individual phospholipids and ceramides in bovine milk (as a percentage of total polar lipids w/w)

	Garcia *et al.* (2012)[a]	Zuo *et al.* (2013)	Lopez *et al.* (2014)[b]	Liu *et al.* (2015)[b]	Barry *et al.* (2016)
Phosphatidylethanolamine	32.9	30.2	29.3	22.5	23.2
Phosphatidylinositol	4.2	9.9	10.5	11.2	7.8
Phosphatidylserine	10.8	7.3	15.0	11.5	9.6
Phosphatidylcholine	29.6	25.2	21.8	22.1	36.0
Sphingomyelin	22.5	27.4	23.0	28.4	23.4
Glucoceramide				0.9	
Lactoceramide				3.5	

[a]Converted from mol% by typical MW.
[b]Calculated from graphs.

Table 1.12. Fatty acid composition of phospholipids from bovine milk

Fatty acid composition (%, w/w)						
Fatty acid	PE[a]	PI[a]	PS[a]	PC[a]	Sph[a]	Sph (2)[b]
14:0	1.0	1.2	1.0	7.1	4.1	1.8
15:0	.5	.5	.4	1.5	1.1	0.3
16:0	11.4	7.6	8.4	32.2	36.1	18.0
16:1	2.7	2.1	2.2	3.4	0.6	0.6
17:0	.9	.9	1.0	1.0	1.5	0.5
18:0	10.3	22.7	25.7	7.5	8.7	4.1
18:1	47.0	35.9	35.5	30.1	1.5	4.5
18:2	13.5	9.5	9.7	8.9	0.2	0.8
18:3	2.3	2.2	2.1	1.4		
20:0	.6	1.5	1.7	.4	1.2	0.6
20:3	1.7	4.1	3.1	1.0		
20:4	2.7	5.0	1.8	1.2		
20:5	1.0	1.0	.8	.5		
21:0	.5	1.1	1.2	.3	1.5	0.8
22:0					14.4	15.9
22:4	.8	1.1	1.1	.4		
22:5 n-6	.6	.9	.9	.4		
22:5 n-3	1.0	1.3	1.7	.3		
22:6	.1	.2	.2	.0		
23:0					17.4	30.4
24:0					11.3	17.3
PUFA	23.7	25.3	21.4	14.1	0.2	0.8
Saturated	26.4	36.7	40.6	51.6	97.6	90.6
% of total PL	31.1	5.2	8.5	26.4	28.7	-

[a]Bitman and Wood (1990), 12 cows, 42 days of lactation.
[b]Morrison *et al.* (1965) example of the variation.

Table 1.11. These changes have been reviewed by Contarini and Povolo (2013) and Verardo *et al.* (2017) suggesting changes due to natural variation between breeds, feeds, seasons and stage of lactation but also analysis. Gallier *et al.* (2010), Le *et al.* (2011) and Barry *et al.* (2016) warned about variations in levels reported in the literature due to different extraction methods as the polar lipids are concentrated for analysis. ^{31}P NMR methods (Mackenzie *et al.* 2009; Diehl 2015) have become popular because of easier calibration, though only applicable to phosphorus-containing polar lipids. There are also quantitative HPLC mass spectroscopy methods (Fong *et al.* 2007; Liu *et al.* 2015) which are able to differentiate the phospholipid groups by means of the signature headgroups (though phosphatidylcholine and sphingomyelin have to be separated by HPLC as they have the same headgroup). Wang and Zhou (2017) have reviewed the methods of phospholipid quantitation that are available.

Table 1.12 shows the fatty acid composition of the phospholipids of milk samples from 12 cows (Bitman and Wood 1990). The major fatty acids of the glycerophospholipids are 18:1, 16:0, 18:0 and 18:2. None of the phospholipids have significant amounts of fatty acids below 14:0 (unlike the triacylglycerols). For identification of glycerophospholipids, electrospray ionisation mass spectrometry (ESI-MS) has been described by Kerwin *et al.* (1994), Myher and Kuksis (1995) and Fong *et al.* (2007). Such methods have enabled the identification of the headgroups; hence, the identification of phospholipid species and the fatty acid constituents. Furthermore, these methods can identify the two fatty acids on the same phospholipid. This has been carried out by Fong *et al.* (2007) and Liu *et al.* (2015) who

Table 1.13. Distribution of fatty acids at positions *sn*-1 and *sn*-2 of phosphatidylcholine and phosphatidylethanolamine from bovine and human milks[a]

	Fatty acid composition (mol%)							
	Bovine[b]				Human[c]			
	Fatty acid phosphatidylcholine		Phosphatidylethanolamine		Phosphatidylcholine		Phosphatidylethanolamine	
	sn-1	*sn*-2	*sn*-1	*sn*-2	*sn*-1	*sn*-2	*sn*-1	*sn*-2
14:0	5.6	10.8	1.9	1.3	3.4	4.9	1.0	1.0
16:0	41.9	30.6	19.7	4.7	34.2	32.3	9.3	8.2
16:1	0.6	1.2	1.2	2.2	1.5	2.2	1.8	3.3
18:0	17.5	2.4	19.0	1.3	34.9	2.1	65.4	1.3
18:1	20.3	27.8	45.8	47.8	14.3	13.7	18.1	15.3
18:2	2.7	9.2	2.9	21.4	2.7	30.9	4.4	30.2
18:3	0.8	1.8	1.1	4.5	–	2.0	–	5.1
20:3	–	1.6	0.2	2.2	–	3.9	–	5.4
20:4	0.2	1.2	0.2	3.0	–	6.6	–	20.9
22:6	–	–	–	–	–	0.8	–	5.2

[a]From Christie (1995).
[b]Morrison *et al.* (1965).
[c]Morrison and Smith (1967).

listed the phospholipid species observed with the likely fatty acids.

Sphingomyelin has a unique composition of fatty acids compared to the other phospholipids (Table 1.12) as the fatty acids are mainly long-chain and saturated (i.e. 16:0, 22:0, 23:0 and 24:0). To illustrate the differences in fatty acid composition of sphingomyelin, the fatty acid analysis from another study is also listed. Furthermore, Bitman and Wood (1990) and Nyberg (1995) found that the ratio of 16:0 to very long-chain fatty acids is relatively high, while Morrison *et al.* (1965) and Ramstedt *et al.* (1999) found a lower ratio. The major sphingoloid base of sphingomyelin is reported to be sphingosine, a dihydroxy 18C amino alcohol (d18:1) (which introduces unsaturation into the molecule) while d16:1, d16:0 and d18:0 are found to be the minor sphingoloid bases (Ramstedt *et al.* 1999). Karlsson *et al.* (1998), using HPLC/MS with electrospray and atmospheric pressure chemical ionisation on sphingomyelin, found the same sphingoloid bases as Ramstedt *et al.* (1999). Fong *et al.* (2007) and Liu *et al.* (2015) included sphingomyelin in their phospholipid work referenced above.

The specific phospholipids differ in the degree of unsaturation (Table 1.12). Phosphatidylethanolamine has a low content (26%) of saturated fatty acids and has especially a high content of linoleic acid, far higher than found in triacylglycerols of milk fat. Phosphatidylinositol and phosphatidylserine have 37–40% saturated fatty acids, whereas phosphatidylcholine has over 50%. The fatty acids in sphingomyelin are almost all completely saturated.

The positional distribution of fatty acids in phosphatidylcholine and phosphatidylethanolamine, the major glycerophospholipids of bovine milk, were investigated by Morrison *et al.* (1965) and are shown in Table 1.13. Unlike triacylglycerols, phospholipids do not contain short-chain fatty acids (14:0 being the shortest chain fatty acid present at a significant level). This is probably due to differences in the route of synthesis, as most short-chain fatty acids are found at the *sn*-3 position of triacylglycerols, which in phospholipids is occupied by the phosphate moiety. In phospholipids, the polyunsaturated fatty acids tend to be esterified preferentially at the *sn*-2 position, while the saturated fatty acids show a preference for the *sn*-1 position (Table 1.13). For phosphatidylethanolamine, 18:2 and 18:3 are found predominantly at the *sn*-2 position, while 18:1 is fairly evenly distributed and 16:0 and 18:0 are predominantly at the *sn*-1 position. In phosphatidylcholine, which is more saturated than phosphatidylethanolamine, the distribution of saturated and unsaturated fatty acids is less distinct between the *sn*-1 and *sn*-2 positions, though 18:0 and the polyunsaturated fatty acids still show the preference described above.

1.4.2 Ceramides and Gangliosides

The glycoceramides (glycosphingolipids) have one or more hexose sugar units attached at position 1 of the ceramide (Figure 1.6), rather than the phosphorylcholine group which is present in sphingomyelin. The concentration of hexose ceramides in polar lipids, as determined by a number of authors, is about 5% of the polar lipids (Table 1.11). The fatty acid composition of the ceramides has been reported by Christie (1995) and tends to follow the fatty acid trends found in sphingomyelin.

The gangliosides are complex ceramide polyhexosides, which contain one or more acidic sugars [known as sialic acid or *N*-acetylneuramic acid (NANA)]. The specific names of the gangliosides identify their structure. The letter G followed by M, D, T or Q designates mono-, di-, tri- or quatra- sialic acid groups. The carbohydrate sequence that is attached to the ceramide is designated by a number (5-n), where n is the number of neutral sugar residues. For example, GD3 is a ganglioside with two NANA units and two neutral sugar residues.

A number of gangliosides, namely GM3, GM2, GM1, GD3, GD2 and GD1, have been isolated from bovine milk (Keenan and Dylewski 1995), the major ones being GD3 (50%) and GM3 (20%). Martin *et al.* (2001) investigated the seasonal variation in the total gangliosides [measured as lipid-bound sialic acid (LBSA)]. They found that the level of gangliosides was high in the colostrum, which decreased through transitional and mature milk, before rising a little in late lactation milk; mean values were 3.5, 1.3, 0.9 and 1.9 mg/g milk, respectively. Puente *et al.* (1992), who measured the seasonal trend in individual gangliosides, showed that GD3 was slightly higher in colostrum and late-lactation milk compared to mid-lactation milk, while GM3 showed the reverse trend.

Table 1.14. Fatty acid composition of bovine and human milk gangliosides GD3 and GM3[a]

	Fatty acid composition (%, w/w)			
	Bovine milk		Human milk	
Fatty acid	GD3	GM3	GD3	GM3
12:0	0.4	0.5	2.7	1.1
14:0	1.0	1.6	3.8	1.8
15:0	0.3	0.5	1.0	0.6
16:0	10.3	11.9	18.4	11.7
16:1	0.1	0.4	1.4	1.3
17:0	0.3	0.3	0.3	0.4
18:0	4.6	5.7	18.7	16.6
18:1	2.4	4.1	3.8	6.0
18:2	1.2	0.8	2.6	4.2
18:3	0.3	0.4	1.2	0.8
19:0	0.1	0.1	11.1	3.1
20:0	1.5	1.6	5.6	9.5
21:0	1.4	1.3	0.2	0.6
22:0	25.6	23.9	10.9	18.6
22:1	0.9	1.1	1.1	0.8
23:0	26.7	24.1	1.8	2.9
24:0	18.9	17.2	7.7	11.1
24:1	2.8	3.1	6.8	7.6
25:0	1.1	1.0	0	0

[a]Adapted from Bode *et al.* (2004).

The major fatty acids of GD3 and GM3 (the major gangliosides of bovine milk) are 22:0, 23:0, 24:0 and 16:0, with both species of ganglioside having similar fatty acid compositions (Table 1.14). These fatty acids are also a feature of the sphingomyelin structure. Martin *et al.* (2001), who investigated the seasonal variation in the fatty acids of the gangliosides, found that the fatty acids were very highly saturated (97%) in colostrum (2 days post partum), while they were much less saturated (68%) in transitional (15 day) milk. Subsequently, the level of saturation increased (78%) in mature (90 day) milk and a further increase was observed (83%) in late lactation milk (10th month).

The introduction of mass spectroscopy has greatly improved the specificity of the analysis of gangliosides in bovine milk. Fong *et al.* (2011) measured GD3 and GM3 in a range of dairy products, the major GD3 peaks being D34:1, d40:1, d41:1, d42:1. Le *et al.* (2011) using mass

Table 1.15. Cholesterol content of different dairy products[a]

	Whole milk	Cream	Skim milk	Butter	Buttermilk
Fat content (%)	4.7	43.0	0.06	82.5	1.3
Cholesterol (mg/g fat)	3.3	3.1	44	2.8	8.5

[a]From Russell and Gray (1979).

spectroscopy found that for days 2, 15 and 90, the GD3 concentration was 15.2, 3.3 and 2.4 mg/L, respectively, and GM3 concentration was 0.98, 0.15 and 0.15 mg/L, respectively. Rivas-Serna *et al.* (2015) found that colostrum and whole milk contained 4.54 and 3.17 mg/L GD3 and 0.35 and 0.1 mg/L GM3, respectively.

1.4.3 Health Issues

In addition to their importance in cell membranes and in cell signalling, specific polar lipids are recognised to have a number of positive health effects relating to immune function, heart health, brain health and cancer. These effects are related to either the polar lipids themselves or to their metabolites (Vesper *et al.* 1999). Sphingomyelin, plasmalogen and ceramides have shown strong anti-tumour activity. Sphingomyelin can influence cholesterol metabolism and coronary heart disease. Sphingomyelin and gangliosides exhibit anti-infection activity. Phospholipids may also protect against mucosal damage. Kullenberg *et al.* (2012), Contarini and Povolo (2013) and Chapter 10 describe the health effects of phospholipids in more detail.

1.5 Minor Constituents

1.5.1 Sterols

Sterols are minor components of milk lipids, which make up just 0.3% of the total fat (Table 1.1). The principal component is cholesterol, which accounts for over 95% of the total; about 10% of the cholesterol is present in the esterified form. Small amounts of other sterols, namely campesterol, stigmasterol and β-sitosterol, have also been identified in milk fat (Mincione *et al.* 1977).

The cholesterol content, expressed as mg/g fat, of dairy products varies considerably (Table 1.15). Cholesterol values for skim milk (~44 mg/g fat) and buttermilk (~8.5 mg/g fat) are much higher than the normal value of cholesterol (3 mg/g fat) found in milk. As a result of churning, buttermilk contains a considerable quantity of MFGM material and a relatively high concentration of small fat globules. Similarly, during separation, small fat globules and membrane material are concentrated in skim milk. Since small fat globules have a greater amount of membrane material relative to the amount of fat in the core as compared to large fat globules, it can be concluded from the above data that the amount of cholesterol (mg/g fat) is greater in the membrane material than in the fat core of the globule.

1.5.2 Carotenoids

The principal pigment in milk fat is β-carotene, accounting for about 95% of the total carotenoids present. In milk, β-carotene is found in the core of the milk fat globules and is absent from the membrane (Jensen and Nielsen 1996).

The concentration of β-carotene in milk depends on the level of β-carotene in feed and on the breed of cow. Carotenoid pigments are particularly high in fresh grass, but substantially lower in a mix of concentrate and hay, a normal winter feed. Channel Island cows, Jerseys and Guernseys have a higher level of β-carotene and a lower level of vitamin A in their milk fat than other breeds, such as Friesians. In New Zealand, cows fed fresh spring grass can have β-carotene up to 13 mg/g fat (Jersey) or 8 mg/g fat (Friesian) in their milk. However, when cows are grazed on mature summer pasture, these values decrease substantially to about 7 and 4 mg/g fat for Jersey and Friesian, respectively (Winkelman *et al.* 1999). In view of these variations, it is not surprising that value for β-carotene in commercial butters ranges from 2.5 to 12:5 mg/g fat (Buss *et al.* 1984). Gentili *et al.* (2013) in a study of milk from various animal species found that cow's milk had 243 μg/L of β-carotene.

1.5.3 Fat-Soluble Vitamins

A significant nutritional aspect of milk lipids is that vitamins A, D, E and K are dissolved in the fat phase. Milk fat is considered to be a significant source of vitamin A, but a poor source of vitamins D and K (McBean and Speckmann 1988).

Calderon *et al.* (2007a, b) studied variation in carotenoids, vitamin A and vitamin E which respect to lactation and diet. They found that carotenoid and vitamin concentrations in milk drastically decreased during the first week of lactation, then did not vary significantly throughout the remainder of the experiment (3 months). With regard to diet, in going to a higher total carotenoid diet, the total carotenoids increased in the milk, but the vitamin A levels stayed constant. The change took about 8–40 days to plateau.

Vitamin A is a fat-soluble vitamin involved in critical biological functions, such as embryonic development, growth and vision. It has three primary forms; retinol, retinal and retinoic acid. In addition, β-carotene can be converted to some extent in the body into retinol and is therefore called provitamin A. The bioactivity of these vitamin A compounds varies considerably, ranging from 100% for all-*trans* retinol, 75% for 13-*cis* retinol and to just 17% for β-carotene. All-trans retinol is the major form of vitamin A in milk fat, with values ranging from 8.0 to 12:0 mg/g fat in samples of commercial butter. In contrast, 13-*cis* retinol is present at a very low concentration, 0.5–1:5 mg/g fat (Buss *et al.* 1984). The total vitamin A bioactivity can be obtained by summation of the concentrations and activities of the different forms of vitamin A. In milk fat, this gives an average value of approximately 12 retinol equivalents/g fat or 40 IU/g fat.

Vitamin E is an effective scavenger of lipid peroxy radicals and is efficient at protecting unsaturated fatty acids against lipid peroxidation. The chemistry of vitamin E is rather complex as there are eight compounds, four tocopherols and four tocotrienols, which exhibit vitamin E activity. The relative bioactivity of the various compounds varies considerably, from 1.0 for α-tocopherol to 0.03 for δ-tocopherol. In milk, α-tocopherol accounts for virtually all of vitamin E, although very small amounts of β-tocopherols and γ-tocopherols are present.

The level of vitamin D in milk fat is very low, about 0.01–0:02 mg/g fat, which equates to 0.4–0.8 IU/g fat. Gill *et al.* (2016) found in a seasonal study of New Zealand milk that vitamin D_3 levels ranged from 167 ng/L in winter to 615 ng/L in the summer whereas 25OH-D_3 showed little variation at <50 ng/L. Interestingly, the vitamin D3 levels correlated with the exposure to sunlight (solar radiation) as the seasons changed.

Vitamin K is also present in milk at a very low concentration. The level of phylloquinone (vitamin K1) is reported to be 0.1–0:2 mg/g fat (Haroon *et al.* 1982). Using mass spectroscopy, Gentili *et al.* (2013) found that vitamin K1 to be 8.8 μg/L in the milk and menaquinone-4 (vitamin K2) to be 8.6 μg/L.

1.5.4 Flavour Compounds

The chemistry of the flavour of milk fat and butter is very complex, involving multiple compounds that each contribute to the overall aroma and taste. Whilst more than 200 volatile compounds have been identified in milk fat, the majority of those compounds are present at concentrations below their individual flavour threshold level, which means that they will not contribute to the overall flavour profile. The principal factor that can change the concentration of the volatile compounds in milk, and thus all subsequent dairy products, is the feeding regime of the cow (Bendall, 2001). However, the differences in flavour are primarily caused by concentration differences of a common set of flavour compounds, rather than the occurrence of compounds uniquely associated with a particular feed.

Schieberle *et al.* (1993) established which compounds make a significant contribution to the flavour of butter through a four-stage process. Firstly, of the many volatile compounds in butter, those which served as potent odorants of an intensely flavoured sour cream butter were identified by using gas chromatography–olfactometry (GC–O) on extracts from the butter. In the second stage, selected potent odorants with the highest flavour dilution (FD) values from GC–O, and which were therefore most likely to be important to the overall flavour, were accurately quantified for a range of different butter types. In a third stage, those measured concentrations were compared as a ratio to the sensory thresholds for the pure compounds measured in a bland fat matrix of sunflower oil. Each ratio is the odour activity values (OAVs) for each potent odorant. A compound with a concentration in the butter that exceeds its sensory threshold (i.e. having an OAV >1) can potentially contribute to the overall flavour impression of the butter; while any compound having an OAV of <1 is unlikely to contribute to the flavour (although for odorants with similar characters that can be additive in nature, contribution may still occur). Different OAVs of a shared set of compounds for different sample types are what give rise to the flavour variation of those different sample types. Finally, the importance of the presumptive key odorants of butter was assessed by sensory evaluation of the original butter for similarity with a model mixture created by the combination of the complete set of quantified compounds in a bland fat matrix of sunflower oil. Additionally, sensory evaluation of a series of 'omission models', generated by combining all (except one) of the presumptive key odorants quantified compounds, allowed for both the importance of a specific individual compound to the overall flavour of butter, as well that compound's contribution to the sensory character of the butter, to be established.

For butter, the most important flavour compounds are diacetyl, δ-decalactone and butyric acid — with the relative importance of those compounds differing for sour cream butter, sweet cream butter, Irish sour cream butter, cultured butter and a traditional farmhouse style of sour cream butter. Other compounds that may each contribute a lesser role to the overall flavour of butter include: skatole, γ-dodec-*cis*-6-enolactone, caprylic acid, non-*trans*-2-enal, nona-*cis*-3, *cis*-6-dienal, non-*cis*-2-enal, γ-octalactone, pent-1-en-3-one, deca-*trans*-2,*trans*-4-dienal, *trans*-4, 5-epoxydec-*trans*-2-enal, hexanal, oct-1-en-3-one and nona-*trans*-2 *trans*-4-dienal (Schieberle *et al.* 1993).

The potent odorants of fresh butter oil are similar to those for butter. Widder *et al.* (1991) used GC–O to identify the potent odorants as: δ-decalactone, diacetyl, butyric acid, vanillin, γ-dodec-*cis*-6-enolactone, acetic acid, hex-1-en-3-one, oct-1-en-3-one, hex-*cis*-3-enal, guaiacol, non-*cis*-2-enal, octa-1, *cis*-5-dien-3-one, non-*trans*-2-enal, deca-*trans*-2,trans-4-dienal, δ-octalactone and skatole.

During the storage of butter oil, lipid oxidation occurs to give rise to oxidative off-flavours. When GC–O was used to compare fresh butter oil with stored butter oil, the latter had markedly increased FD values for aldehydes and vinyl ketones. The highest FD values arising were found from: oct-1-en-3-one, octa-1,*cis*-5-dien-3-one, non-*trans*-2-enal, hex-*cis*-3-enal, non-*cis*-2-enal, nona-*trans*-2,*trans*-4-dienal, hexanal, nonanal and deca-*trans*-2,*trans*-4-dienal (Widder *et al.* 1991). As noted earlier, milk fat contains very small amounts of hydroxy acids, esterified to triacylglycerols. These act as precursors of the flavoursome γ-lactones and δ-lactones (Kinsella *et al.* 1967).

Fresh milk fat contains virtually no methyl ketones. However, on heating, the trace amounts of β-keto acids present in milk fat are decarboxylated to yield methyl ketones (Parks *et al.* 1964). Not only does the methyl ketone, nonan-2-one, contribute to the rich flavour associated with baked foods that are made with butter; but through its absence from baked goods that are made with margarine, nonan-2-one also serves to differentiate between butter- and margarine-containing bakery foods. δ-Decalactone and γ-dodec-*cis*-6-enolactone are also absent from margarine-containing bakery foods (Gassenmeier and Schieberle, 1994).

Table 1.16. Average value of fat content of milk from various species

	Fat content (%)	References
Cow	3.5	Elagamy (2003)
Sheep	7.8	Elagamy (2003)
Goat	4.2	Elagamy (2003)
Buffalo	6.8	Elagamy (2003)
Camel	4.0	Elagamy (2003)
Rabbit	18.3	Jenness and Sloan (1970)
Seal	42.1	Iverson *et al.* (1997)
Pig	6.8	Mellies *et al.* (1978)
Human	3.8	Jensen (1988)

1.6 Milk Fat from Different Animal Species

A wide range of animals, large or small, living on land or in water, surviving in hot or cold climates produce milk. Inevitably, there are variations in the composition of milk from these animals. Some of these variations are within species caused by factors, such as region, stage of lactation and season, as discussed earlier. However, when comparing different species, additional factors need to be considered that relate to the physiology of the animals, such as whether the animal is a ruminant or a non-ruminant; suckling frequency; development factors, such as growth rate and environmental factors, such as temperature.

The detailed composition of milk from a wide range of mammals has been reviewed by IDF (1986), Alston-Mills (1995), Iverson and Oftedal (1995), Oftedal and Iverson (1995), IDF (1995), Elagamy (2003) and Oftedal (2011). In the current discussion, the focus is on milks that are consumed by humans, though other examples have been included by way of comparison. Obviously, the variations within a species mean that quoted values are intended only to be indicative of the typical concentrations rather than definitive values.

1.6.1 Gross Composition

The range of fat content in the milk of different animal species varies widely (Table 1.16). The milk of the cow, human, goat and camel have a low fat content. The milk of the buffalo and sheep have values that are a little higher, whereas the fat content of the milk of the rabbit and seal is much higher. No distinct relationship is observed between the fat content of milk of ruminants (cow, sheep, goat, buffalo and camel) and that of non-ruminants (human, rabbit, seal and pig). However, domesticated animals tend to produce milk that is lower in fat content than non-domesticated animals (Oftedal and Iverson 1995).

Low suckling frequency and a high metabolic rate have been associated with a high fat content. It has been suggested that the high fat content of rabbit's milk is due to the infrequent suckling of the young, about once a day (Oftedal and Iverson 1995). Marine mammals may have a high fat content in their milk to facilitate fat deposition in the suckling young and, thus, to reduce heat loss from the neonate. In addition, high solid content reduces the water requirement in the milk, which could be advantageous when the water has to come from the body reserves of the mother. The major lipid class in the milk of animals is triacylglycerols, and this usually accounts for 97–98% of the total lipids (Christie 1995).

1.6.2 Fatty Acids

The fatty acids in milk fat are derived from two sources, *de novo* synthesis of fatty acids in the mammary gland and plasma lipids (see Chapter 2). *De novo* synthesis generally involves short-chain and medium-chain fatty acids and some

Table 1.17. Major fatty acids in milk triacylglycerols from various species

Fatty acid composition (%, w/w)									
Fatty acid	Cow	Sheep	Goat	Buffalo	Camel	Rabbit	Fur Seal	Pig	Human
4:0	3.9	4.0	3.1	3.6	1.0				0.2
6:0	2.5	2.8	2.2	1.6					0.2
8:0	1.5	2.7	2.4	1.1	0.5	22.4			0.5
10:0	3.2	9.0	6.3	1.9	0.1	20.1		0.7	1.0
12:0	3.6	5.4	2.9	2.0	0.5	2.9	0.2	0.5	4.4
14:0	11.1	11.8	7.7	8.7	10.0	1.7	8.7	4.0	6.3
14:1	0.8				1.5		0.4		
15:0	1.2						0.3		0.4
16:0	27.9	25.4	22.0	30.4	31.5	14.2	18.5	32.9	22.0
16:1	1.5	3.4	1.9	3.4	9.0	2.0	10.0	11.3	3.7
18:0	12.2	9.0	10.6	10.1	14.0	3.8	1.8	3.5	8.1
18:1	21.1	20.0	23.7	28.7	25.0	13.6	22.5	35.2	34.0
18:2	1.4	2.1	2.7	2.5	3.0	14.0	1.4	11.9	10.9
18:3	1.0	1.4	1.0	2.5		4.4	0.8	0.7	0.3
>C_{19}							29.4		

Cow, Creamer and MacGibbon (1996); Sheep, Glass *et al.* (1967); Goat, Posati and Orr (1976); Buffalo, Glass *et al.* (1967); Camel, Gnan and Sheriha (1981); Rabbit, Glass *et al.* (1967); Antarctic fur seal, Iverson *et al.* (1997); Pig, Glass *et al.* (1967); Human, Jensen *et al.* (1995).

16:0. The proportions of various fatty acids depend on the specific balance between enzymatic chain elongation and chain termination. The plasma lipids are derived from the diet and also from storage in the body tissues. For non-ruminants, the diet has a large influence on the fatty acid composition but for ruminants, biohydrogenation in the rumen results in much less impact of diet on the final fatty acids absorbed into the bloodstream.

One distinct difference between the fatty acid composition of the various species is the proportion of short-chain fatty acids in the milk of cow, sheep, goat and buffalo compared to the milk of camel, rabbit, fur seal, pig and human (Table 1.17). Interestingly, the milk of the camel (a ruminant) contains little C4 to C12 fatty acids, whereas the rabbit (a non-ruminant) has over 40% of the fatty acids as the medium-chain length fatty acids, C8 to C10.

Generally, the milk fat of non-ruminants has a higher level of polyunsaturated fatty acids than the milk fat of ruminants, due to the direct absorption of these fatty acids from the diet, without the biohydrogenation in the rumen (Table 1.17). Marine mammals, such as the fur seal, have also high levels of long-chain polyunsaturated fatty acids, 20:5 and 22:6, due to the presence of these fatty acids in the diet (Iverson *et al.* 1997).

Precht *et al.* (2001) found that the proportion of *trans*-18:1 fatty acid isomers was similar in the milk fat of the cow, goat and sheep: vaccenic acid (11t-18:1) was the major isomer in all these milks, accounting for 51, 37 and 47% of the total, respectively. The isomers between 9t and 16t, excluding the 11t, made up most of the rest (4–10% each).

Methods for distinguishing cow, sheep and goat milk, using the differences in fatty acid composition, have been reviewed by Ramos and Juarez (1986). These methods involve comparing the ratios of fatty acids. For example, the 12:0/10:0 fatty acid ratio is consistently higher in cow's milk (0.9–1.3) than in sheep's milk (0.4–0.8) or goat's milk (0.3–0.4) (Bernassi 1990; Ramos and Juarez 1986). Other ratios that have been found include 14:0/12:0 and 14:0/8:0 and more complex combinations, such as 10:0/(12:0 + 16:0 + 18:1). More recent methods have used principal component analysis (Smiddy *et al.* 2012).

Table 1.18. Composition of the fatty acids esterified at each position of the triacylglycerols in the milks from various species

Fatty acid composition (mol% of total fatty acids)															
	Cow			Sheep			Goat			Pig			Human		
	sn-1	*sn*-2	*sn*-3	*sn*-1	*sn*-2	*sn-3*	*sn*-1	*sn*-2	*sn*-3	*sn*-1	*sn*-2	*sn*-3	*sn*-1	*sn*-2	*sn*-3
4:0		0.4	30.6			10.8			13.2						
6:0		0.7	13.8			10.4			10.6						
8:0	0.3	3.5	4.2	0.3	2.0	4.4	1.7	1.2	4.6						
10:0	1.4	8.1	7.5	1.4	5.2	10.3	3.3	6.9	12.2				0.2	0.2	1.1
12:0	3.5	9.5	4.5	2.2	4.7	3.5	4.0	4.6	1.2				1.3	2.1	5.6
14:0	13.1	25.6	6.9	8.2	17.6	5.3	8.4	20.3	2.7	2.4	6.8	3.7	3.2	7.3	6.9
16:0	43.8	38.9	9.3	38.0	23.8	2.5	43.6	33.9	3.4	21.8	57.6	15.4	16.1	58.2	5.5
18:0	17.6	4.6	6.0	19.1	12.6	9.1	15.3	6.3	7.7	6.9	1.1	1.4	15.0	3.3	1.8
18:1	19.7	8.4	17.1	18.7	19.3	27.2	16.1	16.1	30.2	49.6	13.9	51.7	46.1	12.7	50.4
18:2				2.7	4.2	6.0	0.3	2.5	4.5	11.3	8.4	11.5	11.0	7.3	15.0
18:3				2.2	1.7	4.4				1.4	1.0	1.8	0.4	0.6	1.3

Cow, Parodi (1979); Sheep and Goat, Kuksis *et al.* (1973); Pig, Christie and Moore (1970); Human, Breckenridge *et al.* (1969).

1.6.3 Triacylglycerols

Triacylglycerols are synthesised in the mammary gland, and the enzyme specificity and substrate availability play a role in the final structure of the molecule. Table 1.18 shows the fatty acids esterified at each position of the triacylglycerols in the milk of various species. The distinctive feature of the triacylglycerols of ruminants (cow, sheep and goat) is the presence of the short-chain fatty acids, 4:0 and 6:0, which are esterified almost exclusively at the *sn*-3 position. In human and porcine milk fat more than 50% of the 16:0 is at the *sn*-2 position; however, in the milk fat of cow, sheep and goat, 16:0 is distributed more evenly between the *sn*-1 and *sn*-2 positions. 18:1 is esterified preferentially at the *sn*-1 and *sn*-3 positions in human, porcine and bovine milk fat, although for the cow, the total amount of 18:1 is less. In ovine and caprine milk fat, 18:1 is favoured at the *sn*-3 position and there are smaller but similar amounts at the other two positions.

Ruiz-Sala *et al.* (1996) compared the triacylglycerol composition of the milk fat of sheep, cow and goat by HPLC, using ELSD detection. They found significant differences in the proportions of very low molecular weight triacylglycerols (carbon number ≤34) in the three milk fats, with sheep having the highest proportion (18.2%), followed by goat (15.2%) and cow (10.8%). This supports the earlier work of the same group (Barron *et al.* 1990) using HPLC and UV detection. Andreotti *et al.* (2002) used ^{13}C nuclear magnetic resonance (NMR) to study the low molecular weight triacylglycerols in cow, sheep and goat milk fats. Principal component analysis of the signals from multiple samples showed distinct clusters relating to each of the three sources of milk fat, providing a possible method for distinguishing between the milks of these three species.

Precht (1992) developed a method for determining foreign fats in cow's milk using a mathematical combination for the levels of triacylglycerols grouped by even carbon number (C24 … C54). This formed the basis of the ISO/IDF standard for detecting foreign fats in bovine milk fat (IDF 2019). The limit of detection of adulteration depends on the foreign fat, but is typically 2–7%. Goudjil *et al.* (2003) and Fontecha *et al.* (1998) used a modification of the Precht method to detect non-milk fats in sheep and goat milks. While effective for non-milk fats, the methods are less sensitive in distinguishing adulteration of one milk fat with another. For instance, they were able to detect only 30% of cow's milk in sheep's milk.

Table 1.19. Phospholipid classes in the milk from various animal species

		Cow	Buffalo	Sheep	Donkey	Camel	Human	Human[a]
PE	%	30	31	36	31	36	12	35
PI	%	10	4	3	4	6	8	4
PS	%	7	4	3	4	5	14	9
PC	%	25	30	30	25	26	25	13
SM	%	27	32	28	36	28	40	37
Total PL	mg/g lipid	4.8	3.2	4.3	4.0	4.7	5.1	

Phospholipid classes expressed as % of total phospholipid. Total phospholipid expressed as mg polar lipid per g of total lipid.
Adapted from Zou *et al.* (2013).
[a]Data from Ma *et al.* (2017).
PE phosphatidylethanolamine, *PI* phosphatidylinositol, *PS* phosphatidylserine, *PC* phosphatidylcholine, *SM* sphingomyelin, *PL* phospholipid.

When comparing triacylglycerols of milks from a variety of species, Smiddy *et al.* (2012) found and separated into three general groups. The first group was cow, water buffalo, sheep and goat's milk, the second was horse and donkey milk and the final group was camel milk. A combination of triacylglycerol stereo-specific analysis by Grignard reagent, separation by TLC and fatty acid methyl ester (FAME) analysis of the three fractions allowed Blasi *et al.* (2013) to observe 5% cow's milk contamination in ewe's milk.

Ten-Domenech *et al.* (2015) used reverse-phase HPLC with ELSD detection to separate milk fat triacylglycerols into about 50 peaks of different chain lengths and degrees of saturation. These peaks were identified by ACPI MS fragmentation in the positive mode, using the parent ion and the fatty acid fragments. Linear discriminant analysis was carried out on a training set of samples of cow, goat, sheep and human milk from the HPLC–ELSD data and the result was used to evaluate further samples correctly. Thus, there is the possibility to classify milks according to their mammalian origin by using triacylglycerol (TAG) profiles obtained by HPLC–ELSD.

1.6.4 Polar Lipids

The proportions of corresponding phospholipid classes in the milk of various animal species are remarkably similar (Table 1.19). In each case, phosphatidylethanolamine, phosphatidylcholine and sphingomyelin are the major phospholipids, with smaller amounts of phosphatidylinositol and phosphatidylserine. It is possible that the plasmalogen in human milk affects that PE value depending on whether plasmalogen is included in the PE measurement or not. Garcia *et al.* (2012) reported about 10% plasmalogen in PE in cow's milk and 40% in human milk. Morrison (1968) reiterated the fact that the total proportion of choline-containing phospholipids (phosphatidylcholine and sphingomyelin) is remarkably constant, suggesting that these phospholipids perform a similar function in all milks.

The positional distribution of the fatty acids in two phospholipids, phosphatidylcholine and phosphatidylethanolamine, in bovine and human milk is described in Table 1.13 (Christie 1995). The fatty acids in bovine milk phospholipids have a lower degree of unsaturation than those in human milk phospholipids. However, the distribution of fatty acids between the two positions is similar. Polyunsaturated fatty acids are attached preferentially at the *sn*-2 position, the monounsaturates are distributed similarly between the two positions, and 18:0 is esterified preferentially at the *sn*-1 position; while the other saturated fatty acids (14:0, 16:0) are distributed more evenly (bovine phosphatidylethanolamine being an exception). The phospholipids in human milk show remarkable selectivity for 18:0 at the *sn*-1 and 18:2 at the *sn*-2 positions.

Pan and Izumi (2000) found that the concentration of gangliosides in human milk (9.5 mg LBSA/L) is higher than in cows' milk (4.0 mg LBSA/L). Fong *et al.* (2011) found milk at 13.7 mg/L total gangliosides while using a similar mass spectroscopy procedure. Ma *et al.* (2015) found that human breast milk content varied but had an average of 18.5 mg/L total gangliosides. The gangliosides in human and bovine milk also differ in their fatty acid composition (Bode *et al.* 2004). GD3 and GM3 for each species had similar fatty acid composition, but differences existed between the species (Table 1.14). Human milk gangliosides contain less very long-chain saturated fatty acids, especially 23:0, than bovine milk gangliosides, and more of 16:0 and 18:0. It is not known whether these differences have any influence on the physiological effects of the gangliosides. While there has been recent mass spectrometry data for human and bovine milk gangliosides individually, there has been no new comparison with the same analytical procedure.

Pereira (2014) reviewed the vitamins in the milk of a range of species (cow, goat, sheep and human) and found that the levels were not dissimilar.

References

Alston-Mills, B. P. (1995). Comparative analysis of non-human milks. C. Comparative analysis of milks used for human consumption. In R. G. Jensen (Ed.), *Handbook of milk composition* (pp. 828–855). San Diego: Academic.

Amores, G. & Virto, M. (2019). Total and free fatty acids analysis in milk and dairy fat. *Separations, 6*, 14.

Anankanbil, S., Larsen, M. K., Weisbjerg, M. R. & Wiking, L. (2018). Effects of variation in fatty acids and triglyceride composition on melting behavior in milk fat. *Milchwissenschaft, 71*, 4–9.

Andreotti, G., Lamanna, R., Trivellone, E. & Motta, A. (2002). 13C NMR spectra of TAG: An easy way to distinguish milks from different animal species. *Journal of the American Chemical Society, 79*, 123–127.

Bainbridge, M. L., Cersosimo, L. M., Wright, A. D. & Kraft, J. (2016). Content and composition of branched-chain fatty acids in bovine milk are affected by lactation stage and breed of dairy cow. *PLoS One, 11*, e0150386.

Barron, L. J. R., Hierro, M. T. G. & Santa-Maria, G. (1990). HPLC and GC analysis of the triglyceride composition of bovine, ovine and caprine milk fat. *Journal of Dairy Research, 57*, 517–526.

Barry, K. M., Dinan, T. G., Murray, B. A. & Kelly, P. M. (2016). Comparison of dairy phospholipid preparative extraction protocols in combination with analysis by high performance liquid chromatography coupled to a charged aerosol detector. *International Dairy Journal, 56*, 179–185.

Bendall, J. G. (2001). Aroma compounds of fresh milk from New Zealand cows fed different diets. *Journal of Agricultural and Food Chemistry, 49*, 4825–4832.

Bernassi, R. (1990). Sulla composizione del grasso de lette di Pecora. *Il. Latte, 36*, 468–470.

Bitman, J. & Wood, D. L. (1990). Changes in milk fat phospholipids during lactation. *Journal of Dairy Science, 73*, 1208–1216.

Blasi, F., Lombardi, G., Damiani, P., Simonetti, M. S., Giua, L. & Cossignani, L. (2013). Triacylglycerol stereospecific analysis and linear discriminant analysis for milk speciation. *Journal of Dairy Research, 80*, 144–151.

Bode, L., Beermann, C., Mank, M., Kohn, G. & Boehm, G. (2004). Human and bovine gangliosides differ in their fatty acid composition. *Journal of Nutrition, 134*, 3016–3020.

Brechany, E. Y. & Christie, W. W. (1992). Identification of the saturated oxo fatty acids in cheese. *Journal of Dairy Research, 59*, 57–64.

Breckenridge, W. C. & Kuksis, A. (1968). Structure of bovine milk fat triglycerides, I Short and medium chain length. *Lipids, 3*, 291–300.

Breckenridge, W. C. & Kuksis, A. (1969). Structure of bovine milk fat triglycerides. II. Long chain lengths. *Lipids, 4*, 197–204.

Breckenridge, W. C., Marai, L. & Kuksis, A. (1969). Triglyceride structure of human milk fat. *Canadian Journal of Biochemistry, 47*, 761–769.

Brouwer, I. A. (2016). *Effect of trans-fatty acid intake on blood lipids and lipoproteins: a systematic review and meta-regression analysis*. Geneva: World Health Organization.

Buss, D. H., Jackson, P. A. & Scuffam, D. (1984). Composition of butters on sale in Britain. *Journal of Dairy Research, 51*, 637–641.

Byrdwell, W. C. (2001). Atmospheric pressure chemical ionization mass spectrometry for analysis of lipids. *Lipids, 36*, 327–346.

Calderon, F., Chauveau-Duriot, B., Martin, B., Graulet, B., Doreau, M. & Noziere, P. (2007a). Variations in carotenoids, vitamins A and E, and color in cow's plasma and milk during late pregnancy and the first three months of lactation. *Journal of Dairy Science, 90*, 2335–2346.

Calderon, F., Chauveau-Duriot, B., Pradel, P., Martin, B., Graulet, B., Doreau, M. & Noziere, P. (2007b). Variations in carotenoids, vitamins A and E, and color in cow's plasma and milk following a shift from hay diet to diets containing increasing levels of carotenoids and vitamin E. *Journal of Dairy Science, 90*, 5651–5664.

Chen, Z. Y. & Nawar, W. W. (1991). Prooxidation and antioxidative effects of phospholipids in milk fat. *Journal of the American Oil Chemists' Society, 68*, 938–940.

Chen, S., Bobe, G., Zimmerman, S., Hammond, E. G., Luhman, C. M., Boylston, T. D., Freeman, A. E. & Beitz, D. C. (2004). Physical and sensory properties of dairy products from cows with various milk fatty acid compositions. *Journal of Agricultural and Food Chemistry, 52*, 3422–3428.

Chen, B., Grandison, A. S. & Lewis, M. J. (2017a). Best use for milk – A review. I – Effect of breed variations on the physicochemical properties of bovine milk. *International Journal of Dairy Technology, 70*, 3–15.

Chen, B., Grandison, A. S. & Lewis, M. J. (2017b). Best use for milk – A review. II – Effect of physiological, husbandry and seasonal factors on the physicochemical properties of bovine milk. *International Journal of Dairy Technology, 70*, 155–164.

Christie, W. W. (1978). The composition, structure and function of lipids in the tissues of ruminant animals. *Progress in Lipid Research, 17*, 111–205.

Christie, W. W. (1995). Composition and structure of milk lipids. In P. F. Fox (Ed.), *Advanced dairy chemistry. 2: Lipids* (2nd ed., pp. 1–36). London: Chapman and Hall.

Christie, W. W. & Clapperton, J. L. (1982). Structures of the triglycerides of cows' milk, fortified milks (including infant formulae), and human milk. *Journal of the Society of Dairy Technology, 35*, 22–24.

Christie, W. W. & Han, X. (2010). *Lipid Analysis – Isolation, separation, identification and lipidomic analysis*. Bridgwater: The Oily Press.

Christie, W. W. & Moore, J. H. (1970). A comparison of the structures of triglycerides from various pig tissues. *Biochimica et Biophysica Acta (BBA) - Lipids and Lipid Metabolism, 210*, 46–56.

Contarini, G. & Povolo, M. (2013). Phospholipids in milk fat: composition, biological and technological significance, and analytical strategies. *International Journal of Molecular Sciences, 14*, 2808–2831.

Cossignani, L. Pollini, L. & Blasi, F. (2019). Invited review: Authentication of milk by direct and indirect analysis of triacylglycerol molecular species. *Journal of Dairy Science, 102*, 5871–5882.

Creamer, L. K. & MacGibbon, A. K. H. (1996). Some recent advances in the basic chemistry of milk proteins and lipids. *International Dairy Journal, 6*, 539–568.

Currie, G. J. & Kallio, H. (1993). Triacylglycerols of human milk: rapid analysis by ammonia negative ion tandem mass spectrometry. *Lipids, 28*, 217–222.

Deeth, H. C. (1997). The role of phospholipids in the stability of milk fat globules. *Australian Journal of Dairy Technology, 52*, 44–46.

Diehl, B. (2015). NMR of polar lipids. In M. Ahmad & X. Xu (Eds.), *Polar lipids; biology, chemistry, and technology* (pp. 391–438). Urbana: AOCS Press.

Dunshea, F. R., Walker, G. P., Ostrowska, E. & Doyle, P. T. (2008). Seasonal variation in the concentrations of conjugated linoleic and *trans* fatty acids in milk fat from commercial dairy farms is associated with pasture and grazing management and supplementary feeding practices. *Australian Journal of Experimental Agriculture, 48*, 1062–1075.

Elagamy, E. I. (2003). Properties of milk and milk fats from species other than cow. In B. Rossell (Ed.), *Oils and Fats Volume 3 – Dairy Fats* (pp. 27–65). Leatherhead: Leatherhead Publishing.

Fong, B. Y., Norris, C. S. & MacGibbon, A. K. H. (2007). Protein and lipid composition of bovine milk-fat-globule membrane. *International Dairy Journal, 17*, 275–288.

Fong, B., Norris, C. & McJarrow, P. (2011). Liquid chromatography–high-resolution electrostatic ion-trap mass spectrometric analysis of GD3 ganglioside in dairy products. *International Dairy Journal, 21*, 42–47.

Fontecha, J. D., Fraga, V. M. J. & Juarez, M. (1998). Triglyceride analysis by gas chromatography in assessment of authentic goat milk fat. *Journal of the American Oil Chemists' Society, 75*, 1893–1896.

Gallier, S., Gragson, D., Cabral, C., Jimenez-Flores, R. & Everett, D. W. (2010). Composition and fatty acid distribution of bovine milk phospholipids from processed milk products. *Journal of Agricultural and Food Chemistry, 58*, 10503–10511.

Garcia, C., Lutz, N. W., Confort-Gouny, S., Cozzone, P. J., Armand, M. & Bernard, M. (2012). Phospholipid fingerprints of milk from different mammalians determined by 31P NMR: towards specific interest in human health. *Food Chemistry, 135*, 1777–1783.

Gassenmeier, K. & Schieberle, P. (1994). Comparison of important odorants in puff-pastries prepared with butter or margarine. *LWT- Food Science and Technology, 27*, 282–288.

Gentili, A., Caretti, F., Bellante, S., Ventura, S., Canepari, S. & Curini, R. (2013). Comprehensive profiling of carotenoids and fat-soluble vitamins in milk from different animal species by LC-DAD-MS/MS hyphenation. *Journal of Agricultural and Food Chemistry, 61*, 1628–1639.

Glass, R. L., Troolin, H. A. & Jenness, R. (1967). Comparative biochemical studies of milks. IV. Constituent fatty acids of milk fats. *Comparative Biochemistry and Physiology, 22*, 415–25.

Gill, B. D., Zhu, X. & Indyk, H. E. (2016). The determination of vitamin D3 and 25-hydroxyvitamin D3 in early lactation and seasonal bovine milk. *International Dairy Journal, 63*, 29–34.

Gnan, S. O. & Sheriha, A. M. (1981). Composition of Libyan camel's milk. *Australian Journal of Dairy Technology, 41*, 33–35.

Goudjil, H., Fontecha, J., Fraga, M. J. & Juarez, M. (2003). TAG composition of ewe's milk fat. Detection of foreign fats. *Journal of the American Oil Chemists' Society, 80*, 219–222.

Gresti, J., Bugaut, M., Maniogui, C. & Bezard, J. (1993). Composition of molecular species of triacylglycerols in bovine milk fat. *Journal of Dairy Science, 76*, 1850–1869.

Grummer, R. R. (1991). Effect of feed on the composition of milk fat. *Journal of Dairy Science, 74*, 3244–3257.

Ha, Y. L., Grimm, N. K. & Pariza, M. W. (1987). Anticarcinogens from fried ground beef: heat-altered derivatives of linoleic acid. *Carcinogenesis, 8*, 1881–1887.

Hakova, E., Vrkoslav, V., Mikova, R., Schwarzova-Peckova, K., Bosakova, Z. & Cvacka, J. (2015). Localization of double bonds in triacylglycerols using high-performance liquid chromatography/atmospheric pressure chemical ionization ion-trap mass spectrometry. *Analytical and Bioanalytical Chemistry, 407*, 5175–5188.

Hanus, O., Samkova, E., Krizova, L., Hasonova, L. & Kala, R. (2018). Role of fatty acids in milk fat and the influence of selected factors on their variability – a review. *Molecules*. https://doi.org/10.3390/molecules23071636.

Haroon, Y., Shearer, M. J., Rahim, S., Gunn, W. G., McEnery, G. & Barkhan, P. (1982). The content of phylloquinone (vitamin K1) in human milk, cow's milk and infant formula foods determined by high-performance liquid chromatography. *Journal of Nutrition, 112*, 1105–1117.

Harvatine, K. J., Boisclair, Y. R. & Bauman, D. E. (2009). Recent advances in the regulation of milk fat synthesis. *Animal: An International Journal of Animal Bioscience, 3*, 40–54.

Hebeisen, D. F., Hoeflin, F., Reusch, H. P., Junker, E. & Lauterburg, B. H. (1993). Increased concentrations of omega-3 fatty acids in milk and platelet rich plasma of grass-fed cows. *International Journal for Vitamin and Nutrition Research, 63*, 229–233.

Heck, J. M., van Valenberg, H. J., Dijkstra, J. & van Hooijdonk, A. C. (2009). Seasonal variation in the Dutch bovine raw milk composition. *Journal of Dairy Science, 92*, 4745–4755.

Huang, T. C. & Kuksis, A. (1967). A comparative study of the lipids of globular membrane and fat core and of the milk serum of cows. *Lipids, 2*, 453–470.

Huyghebaert, A. & Hendrickx, H. (1971). Studies on Belgian butterfat 3 The fatty acid composition. *Milchwissenschaft, 26*, 613–617.

IDF. (1986). *Production and utilisation of ewe's and goat's milk*. In *Bulletin 202*. International Dairy Federation: Brussels.

IDF. (1995). Production and utilisation of ewe's and goat's milk. In *IDF/CIRVAL Seminar*. Crete: International Dairy Federation. Brussels.

IDF. (2002). *IDF Standard 184 Milk fat – determination of the fatty acid composition by gas-liquid chromatography*. Brussels: International Dairy Federation.

IDF. (2019). *IDF Standard 202 Milk and milk products – Determination of milk fat purity by gas chromatographic analysis of triglycerides*. Brussels: International Dairy Federation.

Iverson, S. J. & Oftedal, O. T. (1995). Comparative analysis of non-human milks B. Philogenetic and ecological variation in the fatty acid composition of milks. In R. G. Jensen (Ed.), *Handbook of milk composition* (pp. 789–827). San Diego: Academic.

Iverson, S. J., Arnould, J. P. Y. & Boyd, I. L. (1997). Milk fatty acid signatures indicate both major and minor shifts in the diet of lactating Antarctic fur seals. *Canadian Journal of Zoology, 75*, 188–197.

Iverson, J. L. & Sheppard, A. L. (1986). Determination of fatty acids in butter fat using temperature-programmed gas chromatography of the butyl esters. *Food Chemistry, 21*(3), 223–234.

Jenkins, T. C. (1993). Lipid metabolism in the rumen. *Journal of Dairy Science, 76*, 3851–3863.

Jenness, R. (1974). *The composition of milk*. New York: Academic Press.

Jenness, R. & Sloan, R. E. (1970). The composition of milks of various species: a review. *Dairy Science Abstracts, 32*, 599–612.

Jensen, R. G. (1988). *The lipids of human milk*. Boca Raton: CRC Press.

Jensen, R. G. (2000). Fatty acids in milk and dairy products. In C. K. Chow (Ed.), *Fatty acids in foods and their health implications* (2nd ed., pp. 109–123). New York: Marcel Dekker Inc.

Jensen, R. G. (2002). The composition of bovine milk lipids: January 1995 to December 2000. *Journal of Dairy Science, 85*, 295–350.

Jensen, R. G. & Clark, R. W. (1988). Lipid composition and properties. In N. P. Wong, R. Jenness, M. Keeney, & E. H. Marth (Eds.), *Fundamentals of dairy chemistry* (pp. 171–213). New York: Van Nostrand.

Jensen, R. G. & Newberg, D. S. (1995). Milk lipids B. Bovine milk lipids. In R. G. Jensen (Ed.), *Handbook of milk composition* (pp. 543–575). San Diego: Academic.

Jensen, S. K. & Nielsen, K. N. (1996). Tocopherols, retinol, β-carotene, and fatty acids in the globule membrane and fat globule core in cows' milk. *Journal of Dairy Research, 63*, 566–574.

Jensen, R. G., Bitman, J., Carlson, S. E., Couch, S. C., Hamosh, M. & Newberg, D. S. (1995). Milk lipids A. Human milk lipids. In R. G. Jensen (Ed.), *Handbook of milk composition* (pp. 495–542). San Diego: Academic.

Kalo, P., Kemppinen, A., Ollilainen, V. & Kuksis, A. (2004). Regiospecific determination of short-chain triacylglycerols in butterfat by normal-phase HPLC with on-line electrospray-tandem mass spectrometry. *Lipids, 39*, 915–928.

Kalo, P., Kemppinen, A. & Ollilainen, V. (2009). Determination of triacylglycerols in butterfat by normal-phase HPLC and electrospray-tandem mass spectrometry. *Lipids, 44*, 169–195.

Karlsson, A. A., Mitchelsen, P. & Odham, G. (1998). Molecular species of sphingomyelin: determination by HPLC/MS with electrospray and HPLC/MS/MS with atmospheric pressure chemical ionization. *Journal of Mass Spectrometry, 33*, 1192–1198.

Kataoka, K. & Nakae, T. (1973). Comparative studies of milk constituents of various mammals in Japan. VIII.

Comparison in phospholipids composition of the milk from various mammals. *Japanese Journal of Dairy Science, 22*, 137–142.

Keenan, T. W. & Dylewski, D. P. (1995). Intracellular origin of milk lipid globules and the nature and structure of the milk lipid globule membrane. In P. F. Fox (Ed.), *Advanced dairy chemistry. 2: Lipids* (2nd ed., pp. 89–130). London: Chapman and Hall.

Kerwin, J. L., Tuininga, A. R. & Ericsson, L. H. (1994). Identification of molecular-species of glycerophospholipids and sphingomyelin using electrospray mass-spectrometry. *Journal of Lipid Research, 35*, 1102–1114.

Kinsella, J. E. & Houghton, G. (1975). Phospholipids and fat secretion by cows on normal and low fiber diets: lactational trends. *Journal of Dairy Science, 58*, 1288–1293.

Kinsella, J. E., Patton, S. & Dimick, P. S. (1967). The flavor potential of milk fat. A review of its chemical nature and biochemical origin. *Journal of the American Oil Chemists' Society, 44*, 449–454.

Kuksis, A., Marai, L. & Myher, J. J. (1973). Triglyceride structure of milk fats. *Journal of the American Oil Chemists' Society, 50*, 193–201.

Kullenberg, D., Taylor, L. A., Schneider, M. & Massing, U. (2012). Health effects of dietary phospholipids. *Lipids in Health and Disease, 11*, 3.

Le, T. T., Miocinovic, J., Nguyen, T. M., Rombaut, R., van Camp, J. & Dewettinck, K. (2011). Improved solvent extraction procedure and high-performance liquid chromatography-evaporative light-scattering detector method for analysis of polar lipids from dairy materials. *Journal of Agricultural and Food Chemistry, 59*, 10407–10413.

Lee, Y. J. & Jenkins, T. C. (2011). Biohydrogenation of linolenic acid to stearic acid by the rumen microbial population yields multiple intermediate conjugated diene isomers. *Journal of Nutrition, 141*, 1445–1450.

Lindmark-Mansson, H. (2008). Fatty acids in bovine milk fat. *Food & Nutrition Research, 52*. https://doi.org/10.3402/fnr.v52i0.1821

Lindmark-Mansson, H., Fonden, R. & Pettersson, H. E. (2003). Composition of Swedish dairy milk. *International Dairy Journal, 13*, 409–425.

Liu, Z., Moate, P., Ezerniks, V., Cocks, B. G. & Rochfort, S. (2015). Identification and quantification of triacylglycerols containing n-3 long-chain polyunsaturated fatty acids in bovine milk. *Journal of Dairy Science, 98*, 8473–8485.

Liu, Z., Wang, J., Cocks, B. G. & Rochfort, S. (2017). Seasonal variation of triacylglycerol profile of bovine milk. *Metabolites, 7*. https://doi.org/10.3390/metabo7020024.

Lok, C. M. (1979). Identification of chiral diacylglycerols in fresh milk fat. *Receuil. Journal of Royal Netherlands Chemical Society, 98*, 92–95.

Lopez, C. (2011). Milk fat globules enveloped by their biological membrane: Unique colloidal assemblies with a specific composition and structure. *Current Opinion in Colloid & Interface Science, 16*, 391–404.

Lopez, C., Briard-Bion, V. & Ménard, O. (2014). Polar lipids, sphingomyelin and long-chain unsaturated fatty acids from the milk fat globule membrane are increased in milks produced by cows fed fresh pasture based diet during spring. *Food Research International, 58*, 59–68.

MacGibbon, A. K. H. (1988). Thermal analysis of milkfat and butter. *Chemistry in New Zealand, 52*, 59.

MacGibbon, A. K. H. & Mclennan, W. D. (1987). Hardness of New-Zealand patted butter – Seasonal and regional variations. *New Zealand Journal of Dairy Science, 22*, 143–156.

MacKenzie, A., Vyssotski, M. & Nekrasov, E. (2009). Quantitative analysis of dairy phospholipids by 31P NMR. *Journal of the American Oil Chemists' Society, 86*, 757–763.

Ma, L., MacGibbon, A. K. H., Jan Mohamed, H. J. B., Loy, S., Rowan, A., McJarrow, P. & Fong, B. Y. (2015). Determination of ganglioside concentrations in breast milk and serum from Malaysian mothers using a high performance liquid chromatography-mass spectrometry-multiple reaction monitoring method. *International Dairy Journal, 49*, 62–71.

Ma, L., MacGibbon, A. K. H., Jan Mohamed, H. J. B., Loy, S., Rowan, A., McJarrow, P. & Fong, B. Y. (2017). Determination of phospholipid concentrations in breast milk and serum using a high performance liquid chromatography–mass spectrometry–multiple reaction monitoring method. *International Dairy Journal, 71*, 50–59.

Martin, M. J., Martin-Sosa, S. & Hueso, P. (2001). Bovine milk gangliosides: Changes in ceramide moiety with stage of lactation. *Lipids, 36*, 291–298.

McBean, L. D. & Speckmann, E. W. (1988). Nutritive value of dairy foods. In N. P. Wong, R. Jenness, M. Keeney, & E. H. Marth (Eds.), *Fundamentals of dairy chemistry* (pp. 343–407). New York: Van Nostrand.

Mellies, M. J., Ishikawa, T. T., Gartside, P., Burton, K., MacGee, J., Allen, K., Steiner, P. M., Brady, D. & Glueck, C. J. (1978). Effects of varying maternal dietary cholesterol and phytosterol in lactating women and their infants. *American Journal of Clinical Nutrition, 31*, 1347–1354.

Mincione, B., Musso, S., De Franciscus, S. & G. (1977). Studies on the different species. Sterol content of cow's milk. *Milchwissenschaft, 32*, 599–603.

Moate, P. J., Chalupa, W., Boston, R. C. & Lean, I. J. (2008). Milk fatty acids II: Prediction of the production of individual fatty acids in bovine milk. *Journal of Dairy Science, 91*, 1175–1188.

Moore, J. H. & Christie, W. W. (1979). Lipid metabolism in the mammary gland of ruminant animals. *Progress in Lipid Research, 17*, 347–395.

Morrison, W. R. (1968). The distribution of phospholipids in some mammalian milks. *Lipids, 3*, 101–103.

Morrison, W. R. (1970). Milk lipids. In F. D. Gunstone (Ed.), *Topics in lipid chemistry- I* (pp. 51–106).

Morrison, W. R., Jack, E. L. & Smith, L. M. (1965). Fatty acids of bovine milk glycolipids and phospholipids and their specific distribution in the diacylglycerophospholipids. *Journal of the American Oil Chemists' Society, 42*, 1142–1147.

Mottram, H. R. & Evershed, R. P. (2001). Elucidation of the composition of bovine milk fat triacylglycerols using high-performance liquid chromatography-atmospheric pressure chemical ionisation mass spectroscopy. *Journal of Chromatography. A, 926*, 239–253.

Muuse, B. G., Werdmuller, G. A., Geerts, J. P. & Deknegt, R. J. (1986). Fatty-acid profile of Dutch butterfat. *Netherlands Milk and Dairy Journal, 40*, 189–201.

Myher, J. J. & Kuksis, A. (1995). Electrospray-MS for lipid indentification. *Inform, 6*, 1068–1071.

Myher, J. J., Kuksis, A., Marai, L. & Sandra, P. (1988). Identification of the more complex triacylglycerols in bovine milk fat by GC-MS using polar capillary columns. *Journal of Chromatography, 452*, 93–118.

Myher, J. J., Kuksis, A. & Marai, L. (1993). Indentification of the less common isologous short-chain triacylglycerols in the most volatile 2.5% molecular distillate of butter oil. *Journal of the American Oil Chemists' Society, 70*, 1183–1191.

Nagy, K., Sandoz, L., Destaillats, F. & Schafer, O. (2013). Mapping the regioisomeric distribution of fatty acids in triacylglycerols by hybrid mass spectrometry. *Journal of Lipid Research, 54*, 290–305.

Noble, R. C. (1978). Digestion, absorption and transport of lipids in ruminant animals. *Progress in Lipid Research, 17*, 55–91.

Norris, C. S., Fong, B. Y., de Bueger, N. & MacGibbon, A. K. H. (2003). Phospholipids in dairy products: milkfat globule membrane from milk and colostrum. *Australian Journal of Dairy Technology, 58*, 200.

Nyberg, L. (1995). Sphingomyelin from bovine milk. In G. Cevec & F. Paltauf (Eds.), *Phospholipids: Characterization, metabolism, and novel biological applications* (pp. 120–125). Champaign: American Oil Chemists Society Press.

O'Connor, T. P. & O'Brien, N. M. (1995). Lipid oxidation. In P. F. Fox (Ed.), *Advanced dairy chemistry. 2: Lipids* (2nd ed., pp. 309–347). London: Chapman and Hall.

O'Donnell-Megaro, A. M., Barbano, D. M. & Bauman, D. E. (2011). Survey of the fatty acid composition of retail milk in the United States including regional and seasonal variations. *Journal of Dairy Science, 94*, 59–65.

Oftedal, O. T. (2011). Milk composition: Species comparison. In W. G. Pond, D. E. Ullrey, & C. K. Baer (Eds.), *Encyclopedia of animal science* (pp. 764–767). Boca Raton: CRC Press.

Oftedal, O. T. & Iverson, S. J. (1995). Comparative analysis of non-human milks. A. phytogenetic variation in gross composition of milks. In R. G. Jensen (Ed.), *Handbook of milk composition* (pp. 749–789). San Diego: Academic.

Osorio, J. S., Lohakare, J. & Bionaz, M. (2016). Biosynthesis of milk fat, protein, and lactose: roles of transcriptional and posttranscriptional regulation. *Physiological Genomics, 48*, 231–256.

Pan, X. L. & Izumi, T. (2000). Variations in the ganglioside compositions of human milk, cow's milk and infant formula. *Early Human Development, 57*, 25–31.

Parks, O. W., Keeney, M., Katz, I. & Schwartz, D. P. (1964). Isolation and characterization of the methyl ketone precursor in butter fat. *Journal of Lipid Research, 5*, 232–235.

Parodi, P. W. (1977). Conjugated octadecadienoic acids of milk fat. *Journal of Dairy Science, 60*, 1150–1153.

Parodi, P. W. (1979). Stereospecific distribution of fatty acids in bovine milk fat triglycerides. *Journal of Dairy Research, 46*, 75–81.

Parodi, P. W. (1980). Separation of milk fat triglycerides into classes by silver ion adsorption thin-layer chromatography. *Australian Journal of Dairy Technology, 35*(35), 17–22.

Parodi, P. W. (1982). Positional distribution of fatty acids in the triglyceride classes of milk fat. *Journal of Dairy Research, 49*, 73–80.

Parodi, P. W. (1999). Conjugated linoleic acid and other anticarcinogenic agents of bovine milk fat. *Journal of Dairy Science, 82*, 1339–1349.

Parodi, P. W. (2004). Milk fat in human nutrition. *Australian Journal of Dairy Technology, 59*, 3–59.

Patton, S. & Jensen, R. G. (1976). *Biomedical aspects of lactation*. New York: Pergamon.

Patton, S. & Keenan, T. W. (1971). The relationship of milk phospholipids to membranes of the secretory cell. *Lipids, 6*, 58–61.

Pereira, P. C. (2014). Milk nutritional composition and its role in human health. *Nutrition, 30*, 619–627.

Petersen, M. B. & Jensen, S. K. (2014). Biohydrogenation of fatty acids is dependent on plant species and feeding regimen of dairy cows. *Journal of Agricultural and Food Chemistry, 62*, 3570–3576.

Pimentel, L., Gomes, A., Pintado, M. & Rodriguez-Alcala, L. M. (2016). Isolation and analysis of phospholipids in dairy foods. *Journal of Analytical Methods in Chemistry, 2016*, 9827369.

Pitas, R. E., Sampugna, J. & Jensen, R. G. (1967). Triglyceride structure of cows' milk fat. I. Preliminary observations on the fatty acid composition of positions 1, 2 and 3. *Journal of Dairy Science, 50*, 1332–1336.

Posati, L. P. & Orr, M. L. (1976). Composition of foods, dairy and egg products. In *Agricultural Handbook* (pp. 77–109). Washington, DC: USDA-ARS, Consumer and Food Economics Inst.

Precht, D. (1992). Detection of foreign fat in milkfat. 2. Quantitative-evaluation of foreign fat mixtures. *Zeitschrift für Lebensmittel-Untersuchung und-Forschung, 194*, 107–114.

Precht, D. & Frede, E. (1994). Determination of the solid fat-content in milk-fat by gas-chromatographic triglyceride analysis. *Fett Wiss Technology, 96*, 324–330.

Precht, D. & Molkentin, J. (1996). Rapid analysis of the isomers of trans-octadecenoic acid in milk fat. *International Dairy Journal, 6*, 791–809.

Precht, D. & Molkentin, J. (1997). Trans-geometrical and positional isomers of linoleic acid including conjugated linoleic acid (CLA) in German milk and vegetable fats. *Fett-Lipid, 99*, 319–326.

Precht, D. & Molkentin, J. (2000). Trans unsaturated fatty acids in bovine milk fat and dairy products. *European Journal of Lipid Science and Technology, 102*, 635–640.

Precht, D., Molkentin, J., Destaillats, F. & Wolff, R. L. (2001). Comparative studies on individual isomeric 18:1 acids in cow, goat, and ewe milk fats by low-temperature high-resolution capillary gas-liquid chromatography. *Lipids, 36*, 827–832.

Puente, R., Garcia-Pardo, L. A. & Hueso, P. (1992). Gangliosides in bovine milk. Changes in content and distribution of individual ganglioside levels during lactation. *Biological Chemistry Hoppe-Seyler, 373*, 283–288.

Ramos, M. & Juarez, M. (1986). Chromatographic, electrophoretic and immunological methods for detecting mixtures of milks from different species. In *Production and utilisation of ewe's and goat's milk* (Bulletin 202) (pp. 175–187). Brussels: International Dairy Federation.

Ramstedt, B., Leppimaki, P., Axberg, M. & Slotte, J. P. (1999). Analysis of natural and synthetic sphingomyelins using high-performance thin-layer chromatography. *European Journal of Biochemistry, 266*, 997–1002.

Ratnayake, W. M. N. & Pelletier, G. (1992). Positional and geometric isomers of linoleic acid in partially hydrogenated oils. *Journal of the American Oil Chemists' Society, 69*, 95–105.

Rego, O. A., Cabrita, A. R. J., Rosa, H. J. D., Alves, S. P., Duarte, V., Fonseca, A. J. M., Vouzela, C. F. M., Pires, F. R. & Bessa, R. J. B. (2016). Changes in milk production and milk fatty acid composition of cows switched from pasture to a total mixed ration diet and back to pasture. *Italian Journal of Animal Science, 15*, 76–86.

Rivas-Serna, I. M., Polakowski, R., Shoemaker, G. K., Mazurak, V. C. & Clandinin, M. T. (2015). Profiling gangliosides from milk products and other biological membranes using LC/MS. *Journal of Food Composition and Analysis, 44*, 45–55.

Robinson, N. P. & MacGibbon, A. K. H. (1998). The composition of New Zealand milk fat triacylglycerols by reversed-phase high-performance liquid chromatography. *Journal of the American Oil Chemists' Society, 75*, 993–999.

Robinson, N. P. & MacGibbon, A. K. (2000). Determination of the conjugated linoleic acid-containing triacylglycerols in New Zealand bovine milk fat. *Lipids, 35*, 789–796.

Rombaut, R., Camp, J. V. & Dewettinck, K. (2005). Analysis of phospho- and sphingolipids in dairy products by a new HPLC method. *Journal of Dairy Science, 88*, 482–488.

Rombaut, R., Van Camp, J. V. & Dewettinck, K. (2006). Phospho- and sphingolipid distribution during processing of milk, butter and whey. *International Journal of Food Science and Technology, 41*, 435–443.

Ruiz-Sala, P., Hierro, M. T. G., Martinez-Castro, I. & Santa-Maria, G. (1996). Triglyceride composition of ewe, cow and goat milk fat. *Journal of the American Oil Chemists' Society, 73*, 283–293.

Russell, C. E. & Gray, I. K. (1979). The cholesterol content of dairy products. *New Zealand Journal of Dairy Science and Technology, 14*, 281–289.

Schieberle, P., Gassenmeier, K., Guth, H., Sen, A. & Grosch, W. (1993). Character impact odor compounds of different kinds of butter. *LWT - Food Science and Technology, 26*, 347–356.

Schwendel, B. H., Wester, T. J., Morel, P. C., Fong, B., Tavendale, M. H., Deadman, C., Shadbolt, N. M. & Otter, D. E. (2017). Pasture feeding conventional cows removes differences between organic and conventionally produced milk. *Food Chemistry, 229*, 805–813.

Shingfield, K. J., Chilliard, Y., Toivonen, V., Kairenius, P. & Givens, D. I. (2008). *Trans* fatty acids and bioactive lipids in ruminant milk. In Z. Bosze (Ed.), *Bioactive components of milk* (pp. 3–65). New York: Springer.

Smiddy, M. A., Huppertz, T. & van Ruth, S. M. (2012). Triacylglycerol and melting profiles of milk fat from several species. *International Dairy Journal, 24*, 64–69.

Spanos, G. A., Schwartz, S. J., van Breemen, R. B. & Huang, C. H. (1995). High-performance liquid chromatography with light-scattering detection and desorption chemical-ionization tandem mass spectrometry of milk fat triacylglycerols. *Lipids, 30*, 85–90.

Taylor, M. W. & Hawke, J. C. (1975a). Structural analysis of the triacylglycerols of bovine milkfats. *New Zealand Journal of Dairy Science and Technology, 10*, 49–57.

Taylor, M. W. & Hawke, J. C. (1975b). The triacylglycerol compositions of bovine milkfats. *New Zealand Journal of Dairy Science and Technology, 10*, 40–48.

Ten-Domenech, I., Beltran-Iturat, E., Herrero-Martinez, J. M., Sancho-Llopis, J. V. & Simo-Alfonso, E. F. (2015). Triacylglycerol analysis in human milk and other mammalian species: Small-scale sample preparation, characterization, and statistical classification using HPLC-ELSD profiles. *Journal of Agricultural and Food Chemistry, 63*, 5761–5770.

Toth, T., Viszket, E., Csavajda, E., Tanai, A. & Varga, L. (2012). Seasonal changes in the fatty acid composition of raw bovine milk produced in the western part of Hungary. *Milchwissenschaft-Milk Science International, 67*, 261–264.

Tzompa-Sosa, D. A., van Aken, G. A., van Hooijdonk, A. C. & van Valenberg, H. J. (2014). Influence of C16:0 and long-chain saturated fatty acids on normal variation of bovine milk fat triacylglycerol structure. *Journal of Dairy Science, 97*, 4542–4551.

Valeille, K. & Martin, J. C. (2004). Complete stereospecific determination of conjugated linoleic acids in triacylglycerol of milk-fat. *Reproduction, Nutrition, Development, 44*, 459–464.

Vanhoutte, B. & Huyghebaert, A. (2003). Chemical properties of milk fat. In B. Rossell (Ed.), *Oils and fats, volume 3 – Dairy fats* (pp. 67–98). Leatherhead: Leatherhead International Ltd.

Verardo, V., Gomez-Caravaca, A. M., Arraez-Roman, D. & Hettinga, K. (2017). Recent advances in phospholipids from colostrum, milk and dairy by-products. *International Journal of Molecular Sciences, 18*. https://doi.org/10.3390/ijms18010173.

Vesper, H., Schmelz, E. M., Nikolova-Karakashian, M. N., Dillehay, D. L., Lynch, D. V. & Merrill, A. H., Jr. (1999). Sphingolipids in food and the emerging importance of sphingolipids to nutrition. *Journal of Nutrition, 129*, 1239–1250.

Walstra, P. & Jenness, R. (1974). *Dairy Chemistry and Physics*. New York: Wiley.

Wang, T. & Zhou, D. (2017). Advances in phospholipid quantification methods. *Current Opinion in Food Science, 16*, 15–20.

Weihrauch, J. L., Brewington, C. R. & Schwartz, D. P. (1974). Trace constituents in milk fat: isolation and identification of oxo fatty acids. *Lipids, 9*, 883–890.

Widder, S., Sen, A. & Grosch, W. (1991). Changes in the flavour of butter oil during storage – Identification of potent odorants. *Zeitschrift für Lebensmittel-Untersuchung und -Forschung, 193*, 32–35.

Winkelman, A. M., Johnson, D. L. & MacGibbon, A. K. (1999). Estimation of heritabilities and correlations associated with milk color traits. *Journal of Dairy Science, 82*, 215–224.

Wolff, R. L., Bayard, C. C. & Fabien, R. J. (1995). Evaluation of sequential methods for the determination of butterfat fatty acid composition with emphasis on trans-18:1 acids. Application to the study of seasonal variations in French butters. *Journal of the American Oil Chemists' Society, 72*, 1471–1483.

Xu, S. L., Wei, F., Xie, Y., Lv, X., Dong, X. Y. & Chen, H. (2018). Research advances based on mass spectrometry for profiling of triacylglycerols in oils and fats and their applications. *Electrophoresis, 39*, 1558–1568.

Yang, B., Chen, H., Stanton, C., Ross, R. P., Zhang, H., Chen, Y. Q. & Chen, W. (2015). Review of the roles of conjugated linoleic acid in health and disease. *Journal of Functional Foods, 15*, 314–325.

Yoshinaga, K., Obi, J., Tago, A., Kato, Y., Nagai, T., Yoshida, A. & Gotoh, N. (2019). Analysis of hydroxy triacylglycerol as a lactone precursor in milk fat using liquid chromatography electrospray ionization tandem mass spectrometry. *Food Chemistry, 274*, 298–304.

Zegarska, Z. (2003). Milk lipids. In E. Sikorski & A. Kolakowska (Eds.), *Chemical and functional properties of food lipids* (pp. 265–277). Boca Raton: CRC Press.

Zegarska, Z., Jaworski, J., Paszczyk, B., Charkiewicz, J. & Borejszo, Z. (2001). Fatty acid composition with emphasis on *trans* C18:1 isomers of milk fat from lowland black-and-white and polish red cows. *Polish Journal of Food and Nutrition Sciences, 10/51*, 41–44.

Zou, X., Huang, J., Jin, Q., Guo, Z., Liu, Y., Cheong, L., Xu, X. & Wang, X. (2013). Lipid composition analysis of milk fats from different mammalian species: potential for use as human milk fat substitutes. *Journal of Agricultural and Food Chemistry, 61*(29), 7070–7080.

Origin of Fatty Acids and Influence of Nutritional Factors on Milk Fat

2

D. L. Palmquist and K. J. Harvatine

Abbreviations

ACBP	Acyl-CoA-binding protein
ACC	Acetyl-CoA carboxylase
ACP	Acyl carrier protein
Akt	Serine/threonine protein kinase
AMP	Adenosine monophosphate
ATP	Adenosine triphosphate
cAMP	Cyclic AMP
CDP	Cytidine diphosphate
CLA	Conjugated linoleic acid
CMP	Cytidine monophosphate
CoASH	Coenzyme A
CTP	Cytidine Triphosphate
DGAT	Diacylglycerol acyltransferase
DHA	Docosahexaenoic acid
DM	Dry matter
EPA	Eicosapentaenoic acid
FABP	Fatty acid-binding protein
FAT	Fatty acid translocator
GLUT	Glucose transporter
GPAT	Glycerol phosphate acyl transferase
HMP	Hexose monophosphate pathway
IGF	Insulin-like growth factor
LPL	Lipoprotein lipase
NADH + H^+	Reduced nicotine adenine dinucleotide
NADPH + H^+	Reduced nicotine adenine dinucleotide phosphate
NEFA	Non-esterified fatty acids (also called "free" fatty acids, FFA)
OAA	Oxaloacetic acid
PFK	Phosphofructokinase
PPi	Inorganic Pyrophosphate
SCD	Stearoyl-CoA desaturase
sn	Stereospecific numbering (of positions on asymmetric glycerol)
SREBPs	Sterol regulatory element-binding proteins
TAG	Triacylglycerol
VLDL	Very low-density lipoproteins

D. L. Palmquist (✉)
Department of Animal Sciences, Ohio Agricultural Research and Development Center/The Ohio State University, Wooster, OH, USA
e-mail: palmquist.1@osu.edu

K. J. Harvatine
Department of Animal Sciences, Pennsylvania State University, University Park, PA, USA

Summary
Ruminant milk fat is of unique composition among terrestrial mammals, due to its great diversity of component fatty acids. The diversity arises from the effects of ruminal microbial biohydrogenation on dietary unsaturated fat and synthesis of odd- and branched-chain fatty acids, and the range of fatty acids synthesized *de novo* in the mammary gland.

P. L. H. McSweeney et al. (eds.), *Advanced Dairy Chemistry, Volume 2*,
https://doi.org/10.1007/978-3-030-48686-0_2

Forty to 60% of milk fatty acids are long-chain (predominantly C_{18}) fatty acids derived from the diet, dependent on the amount of fat in the diet. Fatty acids from C_4 to C_{14} are synthesized *de novo* in the mammary gland, whereas C_{16} arises from both diet and *de novo* synthesis.

Milk fat is the most variable component of milk, both in concentration and composition. In dairy cattle, both the concentration and composition of milk fat are influenced by the diet. Concentration is reduced by feeding diets that contain large proportions of readily fermentable carbohydrates (starch) and unsaturated fat, among other factors, that result in altered rumen fermentation and the formation of numerous *trans* fatty acids. Conversely, the percentage of fat in milk can be increased by feeding rumen-inert fats. Nevertheless, subtle changes in composition and manufacturing functionality can be achieved by feeding different fats. Those fatty acids synthesized *de novo*, especially C_{12} to C_{16}, palmitic acid, and oleic acid, show greatest variation when supplemental fats are fed depending on the profile of the fat source and feeding level.

Modern developments in the manufacture of rumen-protected and rumen-inert fats and availability of oils enriched in specific fatty acids, together with increased understanding of ruminal and animal lipid metabolism, provide considerable flexibility in manipulation of the composition of milk fat for specific nutritional and manufacturing needs.

Future advances in the science of milk fat and nutrition will come from focusing on the unique biological properties of minor milk fatty acids arising from ruminal biohydrogenation and possibly some of *de novo* mammary origin.

2.1 Introduction

Few natural biological lipids exceed bovine milk fat in complexity of fatty acids and triacylglycerol (TAG) structure. This, together with its importance commercially as a human food, has generated considerable knowledge on the synthesis and composition of milk fat.

Garton (1963) marvelled at the "bewildering complexity of unsaturated fatty acids present in the milk fat of ruminants," noting that these were apparently associated with ruminal microbial metabolism, in particular, ruminal biohydrogenation. Jensen (2002) conducted the most comprehensive review of the diversity of lipids in bovine milk including 416 fatty acids and lamented the paucity of new information on the content of trace fatty acids and complex lipids. Little work has been done in this area since this review, but modern lipidomics techniques may advance it in the near future.

Sixty years ago the dual sources of origin of milk fatty acids were also recognized, i.e., *de novo* synthesis of the short-chain fatty acids in the mammary glands and longer-chain fatty acids of dietary origin. However, little was known of the quantitative significance of each. Rapid progress in studies of fatty acid synthesis in the mammary glands occurred between1960 and 1980. More recent efforts have been directed toward characterizing genetic determinants of, and the physiological regulation of, lipid synthesis and evaluating the effects of milk fatty acids on human health and as unique regulators of cell and gene function. Supplements enriched in palmitic acid have garnered recent applied interest due to availability as a by-product of palm oil manufacturing and their ability to increase milk fat yield (Loften *et al.* 2014).

As knowledge of the biological functions of minor milk fatty acids has increased, a new appreciation for studies of effects of nutritional and animal management on milk fat composition has been gained. Because animal nutrition and management are key factors in the manipulation of milk fat composition, this chapter will address the synthesis of milk fatty acids and glycerides from the standpoint of regulation and opportunities for manipulation by various feeding strategies.

2.2 Origin of the Fatty Acids in Milk Fat

2.2.1 Overview

The question of the origins of milk fat, whether wholly from the diet or synthesized by the animal, was an early topic of debate (Jordan and Jenter 1897). The evolution of the field ranged from a leading theory put forth by the eminent lipid chemist, Hilditch (1947), that proposed that the short-chain fatty acids arise from the degradation of oleic acid, which was entirely wrong, to key insights provided by isotope approaches in the 1950s through the 1970s.

2.2.2 Mammary Uptake of Fatty Acids

Modern concepts of milk fat synthesis developed rapidly with the carefully designed studies in lactating goats by Popják *et al.* (1951) which showed unequivocally the *de novo* synthesis of short-chain fatty acids in the mammary gland from ^{14}C-labelled acetate. Also, the incorporation of tritium-labelled stearic acid into milk fat was demonstrated by Glascock *et al.* (1956). From further studies Glascock *et al.* (1966) concluded that up to 48% of milk fatty acids were derived from β-lipoprotein TAGs. Because the long-chain fatty acids comprise approximately 50% of milk fat, this observation implied that more than 90% of these fatty acids were of plasma origin, consistent with the demonstration of little fatty acid elongation in the mammary gland (Palmquist *et al.* 1969).

Palmquist and Conrad (1971) fed or intravenously infused 1-^{14}C palmitic acid into lactating cows and mathematically partitioned ^{14}C secretion in milk fat as originating from two pools, diluted due to turnover of the dietary and endogenous fatty acids. The proportions of the long-chain fatty acids in milk derived from the diet were influenced by the composition of the diet. In further studies, Palmquist and Mattos (1978) injected 1-^{14}C-linoleic acid as labelled chylomicron TAGs, VLDL, or albumin-bound fatty acids. From curve analysis of labelled milk fat secretion and reanalysis of earlier data, they concluded that 88% of long-chain fatty acids in milk were derived directly from TAGs of intestinal lipoproteins, while 12% were derived from TAGs of endogenous origin. A model was developed, showing that 76% of absorbed TAG linoleic acid was taken up directly by the mammary glands (Figure 2.1). Using a more direct approach, Glascock *et al.* (1983) injected ^{3}H TAGs into sheep at various stages of lactation and measured the partition of label to adipose TAG, milk lipids, or oxidization, as measured by appearance of ^{3}H in the body water. Partitioning of blood TAGs to milk clearly declined with increasing days in lactation (Figure 2.2), with maximal incorporation (46%) at 18 days of lactation. This demonstrated modification of the partitioning of plasma TAGs in different physiological states, consistent with the principles of homeorhesis that prioritize milk synthesis over body fat described by Bauman and Currie (1980).

Detailed studies by Annison and colleagues (Annison *et al.* 1967; *West et al.* 1972) showed that both plasma TAG fatty acids and non-esterified fatty acids (NEFAs) were taken up by the mammary glands of lactating goats. No net uptake of NEFAs occurred, although the specific activity of radiolabelled NEFAs decreased across the gland, indicating labelled NEFA uptake and replacement by fatty acids arising from the hydrolysis of plasma TAGs in the capillaries by mammary lipoprotein lipase (LPL) and subsequent mixing with plasma NEFAs. Net uptake of plasma NEFAs by the mammary glands occurs at high NEFA concentrations (Miller *et al.* 1991; Nielsen and Jakobsen 1994) observed immediately *postpartum* or in subclinical ketosis. Usually, in established or mid-lactation, net uptake is variable (+ and −), averaging about zero, so that uptake of fatty acids from the NEFA pool is not evident unless labelled tracers are used. However, milk fat concentration is positively correlated ($r = 0.76$, $P < 0.05$) with plasma NEFA concentration (Pullen *et al.* 1989).

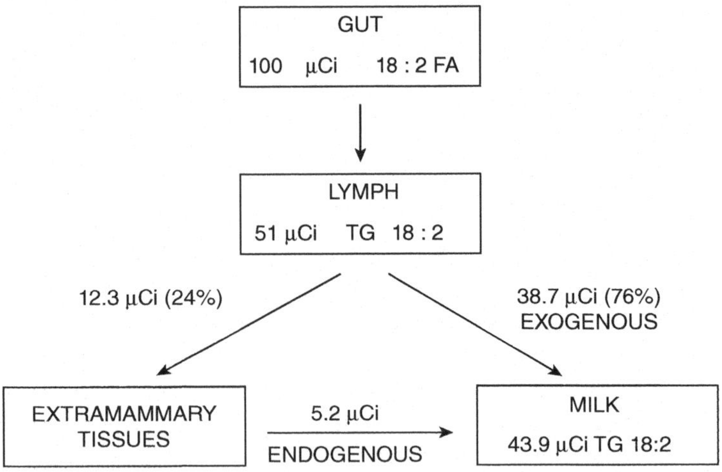

Figure 2.1. Model of linoleic acid transport from intestine to milk. Incorporation into intestinal lipoproteins accounted for 51% of absorbed ^{14}C linoleic acid; of this 76% was incorporated directly into milk, whereas 10% was incorporated into milk after cycling through other metabolic pool(s). Adapted from Palmquist and Mattos (1978).

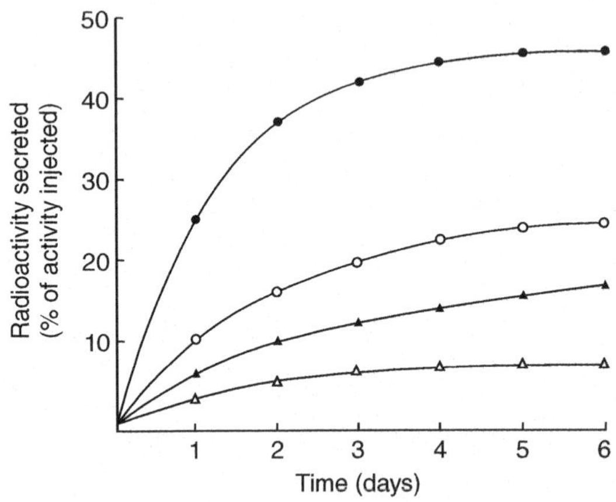

Figure 2.2. Cumulative secretion of radioactivity in milk fat of sheep after intravenous injection of an emulsion of [^{3}H] triacylglycerol at (●) 18 days, (○) 35 days, (▲) 58 days, and (Δ) 73 days after lambing. (From Glascock *et al.* 1983; reprinted with the permission of Cambridge University Press).

2.2.3 Lipoprotein Lipase

The TAG-rich chylomicra from the intestine and very low-density lipoproteins (VLDL) from the liver are the primary sources of long-chain fatty acids taken up by the mammary glands. Uptake is mediated by lipoprotein lipase (LPL), an enzyme that hydrolyzes TAGs to form fatty acids, glycerol, and perhaps 2-monoacylglycerol. Mammary gland LPL was described first by Korn (1962) who showed its presence in cow's milk where it can increase free fatty acid concentration in milk. The enzyme in tissue and milk is identical (Castberg *et al.* 1975). Characteristics and regulation of mammary LPL were discussed in detail by Barber *et al.* (1997). LPL is associated with vascular endothelial surfaces, bound by heparin sulfate chains; it is released rapidly by intravenous injection of heparin, which competes with endothelial binding sites of the enzyme, and by the presence of TAG-rich lipoproteins. Mammary LPL activity increases markedly immediately prior to parturition and remains high throughout lactation (Liesman *et al.* 1988), whereas it is simultaneously downregulated in adipose tissue (Hamosh *et al.* 1970; Shirley *et al.* 1973) also under apparent control of homeorhetic regulation (Bauman and Currie 1980).

It is important to note that lipoprotein lipase is unable to release fatty acids from phospholipid and cholesterol ester fractions in the blood. The majority of plasma polyunsaturated fatty acids are found in these fractions, limiting their incorporation into milk fat.

2.2.4 Transport of Long-Chain Fatty Acids into Mammary Cells

The mechanism(s) for the transfer of fatty acids from the capillaries into mammary cells is not well understood. Enjalbert *et al.* (1998) observed a nearly linear correlation between mammary arteriovenous difference and arterial concentration of NEFAs + TAGs in the range of 400–750 μmolar (μM), or below saturation for uptake, as described by Baldwin *et al.* (1980). Veerkamp *et al.* (1991) suggested that saturation of fatty acid uptake may result from subsequent intracellular processes (see below), from limitations of intracellular diffusion or from diffusive equilibration between extracellular albumin and intracellular fatty acid-binding protein (FABP). Although fatty acids can move through the phospholipid membrane, this mechanism may be too slow to account for fatty acid uptake (Hajri and Abumrad 2002). Although not directly investigated in the mammary gland, the current model for fatty acid uptake in other tissues includes the fatty acid transport proteins (FATP) and a membrane bound glycoprotein (FAT/CD36) that assist movement of the fatty acid into the cytoplasm where it associates with a fatty acid-binding protein (Schwenk *et al.* 2010). Rates of diffusion may differ between albumin-bound NEFAs and free FA from TAG-rich lipoproteins tightly associated with the capillary endothelium (Hajri and Abumrad 2002). Because the concentration of coenzyme A is very low and well below saturation in cytosol, it could well be that fatty acyl-CoA incorporation into TAGs which frees up coenzyme A for acyl-CoA synthetase is the limiting steps in the rate of fatty acid uptake.

The FATPs and FABPs have been implicated in transmembrane and intracellular transport of fatty acids (Veerkamp *et al.* 1991; Storch and Thumser 2000). The FATPs and FABPs are groups of tissue-specific proteins that bind long-chain fatty acids (C_{16}–C_{20}) with high affinity and a molar stoichiometry of 1:1. Most FABPs bind unsaturated fatty acids with higher affinity than saturated fatty acids. In addition to transport functions, it has been proposed that they modulate specific enzymes of lipid metabolism, regulate expression of fatty acid-responsive genes, maintain cellular membrane fatty acid levels, and reduce the concentration of fatty acids in the cell, thereby removing their inhibitory effect on metabolic processes.

More recently, a greater role in regulating concentrations and transport of fatty acids in the cytosol has been proposed for the acyl-CoA-binding proteins (ACBP; Knudsen *et al.* 2000). ACBP bind long-chain acyl-CoA with a tenfold higher affinity than does FABP (Rasmussen *et al.* 1990), thus being much more effective in protecting membranes from damaging effects of long-chain acyl-CoA esters. Many factors regulate the concentration of unbound acyl-CoA in cells (Figure 2.3). Under conditions in which the concentration of ACBP is not adequate to bind long-chain acyl-CoA, FABP takes over the buffering function, thus protecting cellular membranes from damage (Knudsen *et al.* 2000). The concentration of ACBP is highest in liver cytosol and, interestingly, was found to be lowest in the mammary glands of the cow (Mikkelsen and Knudsen 1987). One might expect, based on the presumption of an association of ACBP concentration with tissue lipid metabolic rates, that the concentration of ACBP might be high in mammary tissue.

2.2.5 Summary of the Supply of Long-Chain Fatty Acids to the Mammary Gland

More than 95% of C_{18} and longer-chain fatty acids in milk fat are derived from the blood TAG-rich lipoproteins and predominantly originate from dietary absorption. Non-esterified fatty acids are taken up also, but net uptake is measurable only when the concentrations of NEFAs are high, notably in early lactation. When cows are in a positive energy balance, adipose tissues contribute less than 15% of long-chain fatty acid uptake. Efficient uptake from plasma TG and little uptake of phospholipids and cholesterol esters decrease polyunsaturated fatty acid available for milk fat synthesis. The proportion of the palmitic acid in milk fat derived from blood lipids is variable, as the proportion synthesized *de novo* in mammary tissue can decrease to 30% or less of the total with increasing dietary fat.

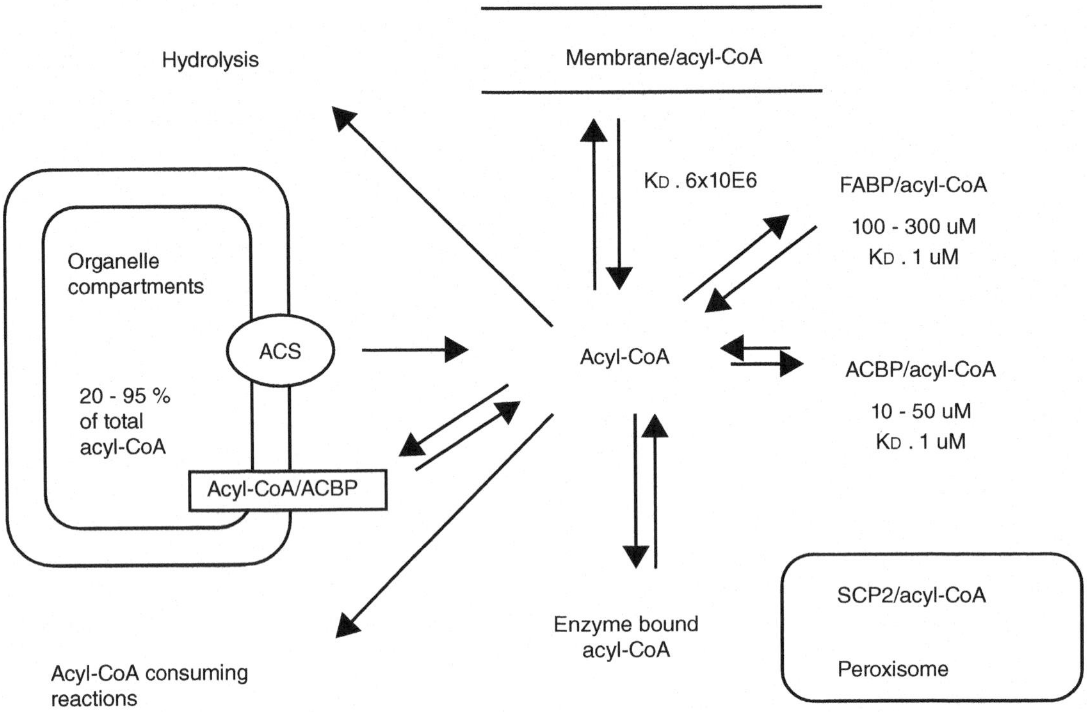

Figure 2.3. Regulation of cytosolic-free acyl-CoA concentration. The major factors affecting the cytosolic-free long-chain acyl-CoA ester (LCA) concentration are the rates of LCA synthesis, uptake and consumption, the concentration of acyl-CoA-binding protein (ACBP), fatty acid-binding protein (FABP), and the acyl-CoA hydrolase activity. The two binding proteins buffer large fluctuations in free LCA concentration. The acyl-CoA hydrolases are suggested to function as a "scavenger" system to prevent the accumulation of free unprotected LCA and to ensure sufficient-free CoA to support β-oxidation and other CoA-dependent enzymes. Sterol carrier protein 2 (SCP2), which binds LCA and very long chain-LCA, is localized in the peroxisomes. It is suggested that SCP2 acts as a peroxisomal pool former for LCA destined for β-oxidation in this organelle. (Adapted from Knudsen *et al.*, *J. Nutrition* 130:294S–298S, 2000, with permission).

2.3 Uptake of Non-lipid Metabolites by Lactating Mammary Glands

The major nutrients utilized for *de novo* milk fat synthesis are glucose, acetate, and β-hydroxybutyrate. Kinetics for the uptake of these from blood was reported by Miller *et al.* (1991).

In fasting animals, the arterial supply of glucose is limiting for milk synthesis (Chaiyabutr *et al.* 1980), whereas in fed animals, glucose uptake is independent of arterial concentration (Miller *et al.* 1991). In the lactating mammary tissues of both ruminants and non-ruminants, glucose is taken up via facilitative transport systems, namely, the insulin-independent GLUT1 and probably a Na^+-dependent glucose transporter (Zhao *et al.* 1996; Shennan and Peaker 2000; Nielsen *et al.* 2001). There is no evidence for the insulin-sensitive GLUT 4 transporter in lactating mammary tissues of either rats or cattle. Baldwin *et al.* (1980) suggested, based on the limited arteriovenous difference data available in the literature at that time, that Michaelis-Menten type equations could be used to describe glucose uptake across the bovine udder, but the highly scattered data fit only slightly better than a linear equation. Uptake of β-hydroxybutyrate and TAGs by the lactating mammary glands also was well described by Michaelis-Menten relationships, whereas the uptake of acetate was strongly linear

(Baldwin *et al.* 1980). These authors cautioned that the uptake of metabolites by lactating mammary glands is interrelated, and factors other than arterial blood glucose concentration govern glucose uptake and utilization (Miller *et al.* 1991; Baldwin 1995).

A serine/threonine protein kinase, Akt, is thought to regulate the expression or function of glucose transport proteins in adipose tissue and also may represent a central signalling molecule in mammary gland development and function. Expression of constitutively activated Akt in mammary glands of transgenic mice resulted in precocious lipid accumulation; fat content of the milk was 65–70% by volume, compared to 25–30% in wild-type mice (Schwertfeger *et al.* 2003).

2.4 Fatty Acid Synthesis in Mammary Glands

Fatty acid synthesis in lactating mammary glands was discussed in detail by Hillgartner *et al.* (1995) and Barber *et al.* (1997). Comparative aspects of fatty acid synthesis in mammary glands of ruminants and non-ruminants were reviewed by Dils (1986), and an in-depth review of lipid metabolism in ruminant mammary glands was provided by Moore and Christie (1981). Requirements for fatty acid synthesis are a carbon source and reducing equivalents in the form of NADPH + H^+. In ruminants, acetate and β-hydroxybutyrate are the primary carbon sources utilized, and glucose and acetate are the primary sources of reducing equivalents (Bauman and Davis 1974). In non-ruminants, glucose is the primary source of both carbon and reducing equivalents (Figure 2.4). Enzyme activities and utilization of substrates by mammary tissues are increased dramatically relative to those of non-mammary tissues at the onset of lactation (Bauman and Currie 1980; Vernon *et al.* 1987).

2.4.1 Sources of Carbon and Reducing Equivalents for Fatty Acid Synthesis

The basic starting substrate for fatty acid synthesis is acetyl-CoA (see below). In ruminants, the provision of this substrate is straightforward. Acetate from blood (+ CoA + ATP) is converted by the cytosolic acetyl-CoA synthetase (EC 6.2.1.1) to AMP and acetyl-CoA, which can then be used for fatty acid synthesis.

The reductive steps of FAS have a specific requirement for NADPH, whereas glycolysis, also a cytosolic process, generates NADH. In non-ruminants, about half of the required NADPH is generated during the transport of C_2 from the mitochondrion to the cytosol through the activity of citrate lyase and malic enzyme (Figure 2.4; Bauman and Davis 1974).

The low activity of ATP-citrate lyase in ruminants was originally proposed to block the incorporation of glucose C into fatty acid carbon in ruminants (Ballard *et al.* 1969). However, physiologically, the negative feedback of acetyl-CoA formed from acetate on pyruvate dehydrogenase probably results in essentially complete inhibition of the enzyme activity in mammary as well as other tissues, thereby yielding a net sparing of glucose which is appropriate to the limited availability of glucose in the ruminant. Inhibition of pyruvate dehydrogenase by acetyl-CoA *in vivo* is consistent with the low incorporation of pyruvate and lactate into fatty acids *in vivo* and the incorporation of significant amounts of lactate and pyruvate carbon into fatty acids by bovine mammary tissue at high substrate concentrations *in vitro* in the absence of acetate (Forsberg *et al.* 1985; Torok *et al.* 1986). Conversion of glucose to fatty acids is low even in the absence of acetate, indicating a block somewhere between fructose-6-phosphate and pyruvate, possibly at phosphofructokinase (PFK) or in triose-phosphate metabolism, but this issue has not received any significant attention.

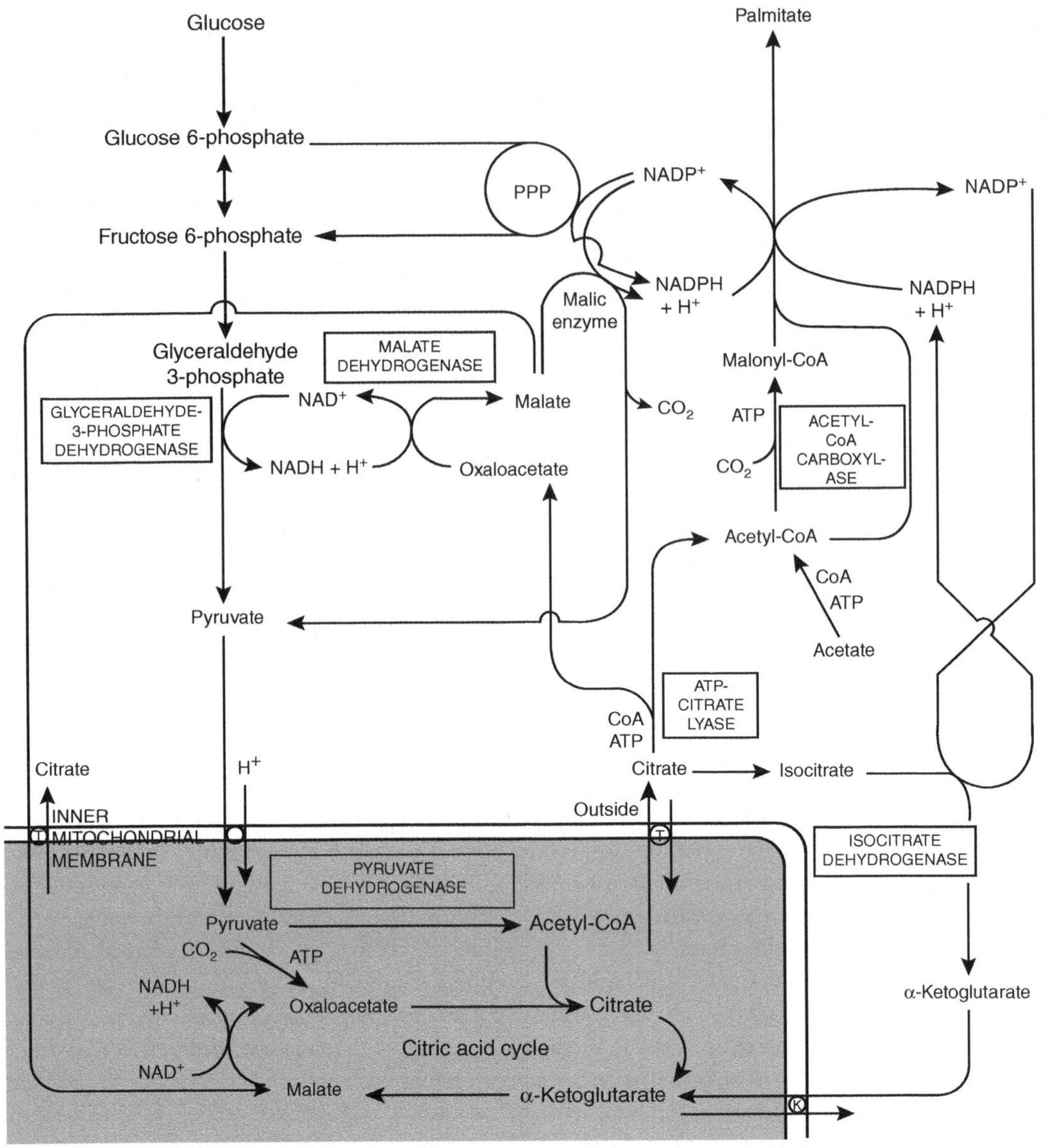

Figure 2.4. The provision of acetyl-CoA and NADPH for lipogenesis. PPP, pentose phosphate pathway; T, tricarboxylate transporter; K, α-ketoglutarate transporter. In ruminants, pyruvate dehydrogenase, ATP-citrate lyase, and malic enzyme activities are low and perhaps non-functional. (From Murray *et al.*, *Harper's Biochemistry*, 21st ed., 1988, p. 207, Appleton and Lange, Norwalk, CT; reproduced with permission of The McGraw-Hill Companies).

Citrate occurs at a rather high concentration in milk, being generated in the citric acid cycle from acetate, and the carboxylation of pyruvate and propionate. Acetyl-CoA is a strong positive effector of pyruvate carboxylase. Because acetyl-CoA transport from the mitochondrion to the cytosol is not necessary in ruminants, cytosolic oxaloacetic acid (OAA) and malate are not generated, so that NADPH + H^+ from the malic enzyme is not available to support fatty acid synthesis. Instead, isocitrate diffuses from the mitochondrion to the cytosol and is used to generate the NADPH + H^+ required via the cytosolic NADP-isocitrate dehydrogenase (EC 1.1.1.42), accounting for as much as 50–60% of total NADPH used by ruminant fatty acid synthase (Mellenberger *et al.* 1973; Mellenberger and Bauman 1974). The α-ketoglutarate generated in this reaction is transported back into the mitochondrion (Figure 2.4). A unique aspect of this pathway is that NADPH is generated by oxidation of acetate, thus sparing glucose (Bauman *et al.* 1970). High-fat diets reduce *de novo* fatty acid synthesis, thus decreasing demand for reducing equivalents. This spares mammary glucose, freeing it for lactose synthesis (Cant *et al.* 1993). Isocitrate oxidation is decreased, causing citrate concentration in milk to increase (Faulkner and Pollock 1989).

Glucose oxidation via the hexose monophosphate pathway (HMP) is an equally important source of NADPH + H^+ in ruminants and nonruminants, and it also provides glycerol-3-phosphate for fatty acid esterification as an alternative to the glycolytic pathway. Glucose oxidation via the HMP accounts for 10–40% of ruminant mammary glucose utilization and 25–40% of mammary CO_2 output (Bickerstaffe and Annison 1974; Cant *et al.* 1993). Wood *et al.* (1965) examined glucose utilization in perfused cow udders. Twenty-three to 30% of glucose was metabolized via the HMP, 10% was utilized in the Embden-Meyerhof pathway, and 60–70% was converted to lactose. A similar proportion was oxidized in the HMP in rats, but only 10–20% was utilized for lactose and up to 50% underwent glycolysis to provide C_2 units for lipogenesis (Katz and Wals 1972).

2.4.2 Acetyl-CoA Carboxylase

The first committed step for the incorporation of acetate carbon into fatty acids is mediated by acetyl-CoA carboxylase (ACC; E.C.6.4.1.2) in two steps, as follows (Allred and Reilly 1997):

$$\text{Enzyme} + \text{ATP} + \text{HCO}_3^- \rightarrow \text{Enz} - \text{biotin} - \text{CO}_2^- + \text{ADP} + \text{Pi}$$

$$\text{Enz} - \text{biotin} - \text{CO}_2^- + \text{acetyl} - \text{CoA} \rightarrow \text{Enzyme} + \text{malonyl} - \text{CoA}$$

The transcriptional regulation of ACC1 is complex with multiple promoters and splice variants that are expressed in a tissue- and signal-specific manner [reviewed by (Travers and Barber 2001; Barber *et al.* 2005)]. During lactation expression from promoter III (PIII) is increased 15- to 30-fold in mammary tissue from cows, sheep, and humans but has little or no expression in other tissues (Barber and Travers 1998; Mao and Seyfert 2002; Barber *et al.* 2003; Molenaar *et al.* 2003). However, PIII accounts for only ~15% of the total ACC transcripts in the bovine lactating mammary gland (Molenaar *et al.* 2003), while PI and PII each account for over 30% (Mao *et al.* 2001; Mao and Seyfert 2002; Molenaar *et al.* 2003), and the remainder is uncharacterized isoforms. Regulation by Sterol regulatory element-binding protein 1 (SREBP1), a key lipogenic transcription factor in the mammary gland, can occur at the PII and PIII promoters [reviewed by (Barber *et al.* 2005; Tong 2005)]. PI and PII result in transcripts that differ in their 5′ non-translated regions but produce the same protein (Barber *et al.* 2005). The protein translated from PIII transcripts replaces the first 110 AA of the PI- and PII-derived protein with 17 hydrophilic AA and eliminates three serine (Ser) residues that are phosphorylated (**Phos**) *in vivo*, although these three sites have no clear effect on ACC activity (Barber and Travers 1998). The PIII product does modify the AMPK consensus motif for Phos of Ser-79, a Phos site that greatly reduces ACC activity, and the hydrophilic N-terminal of the PIII protein also has been proposed to allow unique interactions with cellular components (Barber *et al.* 2005). Lastly, alternate exon splicing results in the presence (+24 nt) or absence (Δ24 nt) of 24 nucleotides coding for 8 AA acids proximal to the Ser-1200 Phos motif (Kong *et al.* 1990). The +24 nt transcript is observed predominantly in the liver, adipose tissue, and non-lactating mammary tissue of sheep and rats, while the Δ24 nt transcript is increased over tenfold at the initiation of lactation and accounts for up to 80% of the ACC1 transcript in the mammary gland (Kong *et al.* 1990; Barber *et al.* 2001). The Δ24 nt may be more susceptible to Phos, although the role of Ser-1200 Phos is disputed and may function to modify protein structure rather than inhibit enzyme activity (Barber *et al.* 2005). Clearly, ACC1 transcription is complex but is highly regulated by physiological state.

The ACC1 protein has a long half-life [48–58 h in well-fed state (Volpe and Vagelos 1973)] and is regulated by multiple post-translational mechanisms. Post-translational mechanisms include allosteric activation and inhibition, protein polymerization, serine Phos, and protein degradation [see reviews (Kim 1997; Barber *et al.* 2005)]. Most of the post-translational regulatory mechanisms of ACC1 were characterized 40–50 years ago (Volpe and Vagelos 1973).

In the liver, adipose, and possibly mammary tissues, ACC is regulated acutely by phosphorylation/dephosphorylation and by allosteric mechanisms involving fatty acids (fatty acyl-CoA) and citrate, whereas the amount of protein is regulated by several hormones, including insulin, growth hormone, possibly via IGF-1, and prolactin (Allred and Reilly 1997; Barber *et al.* 1997; Kim 1997). Short-term regulation of ACC in liver and adipose tissues is well characterized. Briefly, Lee and Kim (1979) reported that incubation of rat adipocytes with epinephrine phosphorylated ACC and reduced enzyme activity by 61% within 30 min. Witters *et al.* (1979) reported similar effect of glucagon on rat hepatocyte ACC. Similar regulation is expected in ruminant mammary tissue but has not been well studied, possibly because acute hormonal actions require the use of cells with intact membranes. The high level of connective tissue in ruminant mammary tissue requires extensive incubation with collagenase to permit the isolation of secretory cells, and these, most often, have severely damaged cell membranes, which are then unable to respond to hormonal signals. Activity of ACC is increased by citrate, and the K_m for the binding of citrate to phosphorylated ACC of 2.4 mM was reduced to 0.2 mM by dephosphorylation (Carlson and Kim 1974), and the dephosphorylated form of ACC is more susceptible to inhibition by palmitoyl-CoA. Phosphorylation of ACC is expected to be a key short to intermediate regulator of activity of the lipogenic pathway. ACC1 is allosterically inhibited by malonyl-CoA and long-chain acyl-CoA and is activated by citrate. Inactive ACC1 is found as a protein-dimer, while active ACC1 has a polymer (filament) structure of 10–20 dimers (Gregolin *et al.* 1966; Ryder *et al.* 1967). Polymer formation is nutritionally responsive and mostly studied in rodent models (Clarke and Clarke 1982; Borthwick *et al.* 1987).

Phosphorylation of multiple serine residues of ACC1 by AMPK and PKA is well described and results in reduced Vmax and responsiveness to allosteric activators [(Ha *et al.* 1994); see review (Barber *et al.* 2005)]. The biological importance of many of the Phos events is highly contested, but strong evidence supports the importance of Phos of Ser-79. As previously mentioned, the PIII ACC1 isoform results in the loss of the AMPK consensus motif near Ser-79 and may result in Phos resistance.

In contrast to the acute inactivation of ACC by cAMP-elevating hormones, it is well established that insulin stimulates fatty acid synthesis in several tissues, including lactating rat mammary acini within a few minutes of treatment (Williamson *et al.* 1983). Responses of ruminant lipogenic tissues to insulin are generally far less dramatic than in rodents. For example, Yang and Baldwin (1973) found only a doubling of fatty acid synthesis in bovine adipocytes treated with insulin during a 3-h incubation. Lactating cow mammary tissue does not appear to exhibit acute responses to insulin *in vitro* (Forsberg *et al.* 1985; Laarveld *et al.* 1985) although the tissue does have insulin receptors (Oscar *et al.* 1986). It has been reported also that epidermal growth factor stimulates fatty acid synthesis and the phosphorylation of ACC in rat liver and adipose tissues (Holland and Hardie 1985), suggesting that several peptide hormones sharing homology with insulin, such as IGF-1, could enhance lipogenesis similarly.

2.4.3 Fatty Acid Synthase

Mammalian fatty acid synthase (FAS; EC 6.2.1.3) is one of the most complex multifunctional enzymes that have been characterized, as this single polypeptide contains all the catalytic components required for a series of 37 sequential transactions (Smith 1994). It consists of two identical polypeptides of approximately 2500 amino acid residues (MW, ca. 270 kDa), each containing seven catalytic subunits: (1) ketoacyl-synthase, (2) malonyl/acetyl transferase, (3) dehydrase, (4) enoyl reductase, (5) β-keto reductase, (6) acyl carrier protein (ACP), and (7) thioesterase. Although some components of the complex can carry out their respective catalytic steps in the monomeric form, only in the FAS dimer do the subunits attain conformations that facilitate coupling of the individual reactions of fatty acid synthesis to occur (Smith *et al.* 2003).

$$
\begin{aligned}
&\text{Acetyl} - \text{CoA} + 7\,\text{Malonyl} - \text{CoA} + \\
&14\,\text{NADPH} + 14\text{H}^+ \rightarrow \\
&\text{Palmitic acid} + 7\text{CO}_2 + 8\,\text{CoA} + \\
&14\,\text{NADP}^+ + 6\,\text{H}_2\text{O}.
\end{aligned}
$$

The entire sequence is described in exquisite detail by Smith *et al.* (2003, see Figure 2.5). Importantly, the primer is usually acetyl-CoA, and the chain extender substrate is usually

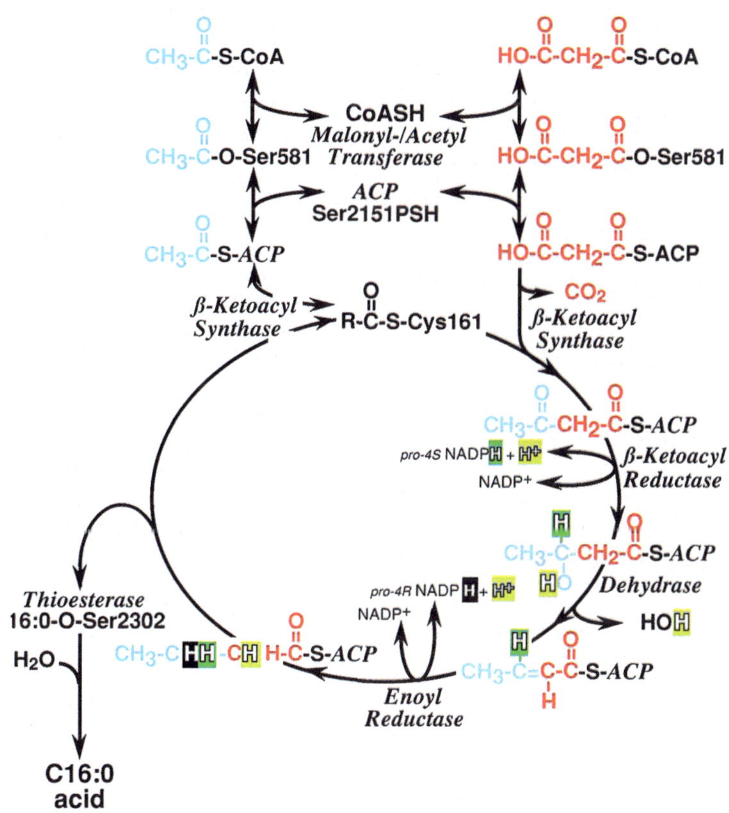

Figure 2.5. Reaction sequence for the biosynthesis of fatty acids *de novo* by the animal FAS. The condensation reaction proceeds with stereochemical inversion of the malonyl C-2; the β-ketoacyl moiety is reduced by NADPH to D-β hydroxyacyl moiety, which then is dehydrated to a *trans*-enoyl moiety; finally, the enoyl moiety is reduced to a saturated acyl moiety by NADPH, with the simultaneous addition of a solvent proton. The two C atoms at the methyl end of the fatty acid are derived from acetyl-CoA, the remainder from malonyl-CoA. The entire series of reactions takes approximately 1 second. PSH, phosphopantetheine. (Reprinted from *Prog. in Lipid Res.*, vol. 42, S. Smith, A. Witkowski and A.K. Joshi, Structural and functional organization of the animal fatty acid synthase, pp. 289–317, copyright (2003), with permission from Elsevier).

malonyl-CoA. The original acetyl moiety becomes the terminal methyl of the growing acyl chain. Cycling continues, with the addition of C_2 units from malonyl-CoA until the chain length of the nascent fatty acid reaches up to C_{16} when a thioesterase specific for that chain length releases the fatty acid and terminates the cycle.

2.4.4 Regulation of Acyl Chain Length

The chain length of the fatty acid product is influenced by numerous factors. The same acyl transferase accomplishes loading of both the acetyl and malonyl substrates. The catalytic rate of the acyl transferase increases as chain length progresses from C_2 to C_{12}; the total concentration of covalently bound saturated intermediates decreases with increasing chain length up to C_{14}, implying that the early β-ketoacylsynthase reactions are slower than the later ones (Smith *et al.* 2003). Mass action coefficients for chains longer than C_{12} become progressively smaller, with little accumulation of $C_{18.}$ Acyltransferase is not entirely specific for the acetyl and malonyl moieties. β-Ketobutyryl, β-hydroxybutyryl, and crotonyl residues all are incorporated (Dodds *et al.* 1981; Joshi *et al.* 1997). Indeed, Kumar and associates (Nandedkar *et al.* 1969; Lin and Kumar 1972) have shown that lactating mammary glands utilize butyryl-CoA more efficiently than acetyl-CoA as a "primer" for FAS. Further, these authors showed that butyryl-CoA is synthesized from acetyl-CoA by, essentially, a reversal of β-oxidation in both the liver and mammary glands of rabbits, rats, and cows. The acyl-CoA dehydrogenase associated with β-oxidation is a FAD-linked dehydrogenase, which renders β-oxidation essentially irreversible on thermodynamic grounds. The enzyme that catalyzes the reverse reaction in the Kumar scheme is linked via NADPH + H, which renders the reduction of crotonyl-CoA thermodynamically favorable. Thus, this enzyme, unique to the reversal of β-oxidation, could be called "crotonyl-CoA reductase." This activity was much lower in rat adipose and pigeon liver tissues. This synthetic pathway is independent of malonyl-CoA and thus is not subject to regulation by acetyl-CoA carboxylase. These observations explain the incorporation of β-OH butyrate as the methyl terminal C_4 moiety of up to 50–60% of fatty acids synthesized *de novo* by lactating mammary glands (Palmquist *et al.* 1969; Smith *et al.* 1974). Also, small quantities of C_6, C_8, and C_{10} in milk fat may be synthesized via this pathway. Propionic acid is incorporated as a primer leading to the synthesis of an odd number of carbon atoms in the acyl chain. The branched-chain volatile acids, *iso*-valeric, *iso*-butyric, and 2-methyl butyric, also can serve as primers, giving rise to *iso*- and *anteiso* acyl chains (Massart-Leen *et al.* 1981; Ha and Lindsay 1990). Methymalonyl-CoA also may be incorporated, substituting for malonyl-CoA, leading to multi-branched acyl products; the synthesis of these is terminated with chain lengths less than 16 carbons, and the rate of synthesis is one-tenth that for straight-chain products (Smith 1994). Some branched-chain products, in particular 13-methyl tetradecanoic acid, have been shown to have regulatory effects on cell function (Yang *et al.* 2000; Parodi 2003). The branch chain fatty acids are uniquely found in ruminant meat and milk and may have bioactive properties

Because the acyl transferase releases intermediates, some of the intermediates may escape transfer to the β-ketoacylsynthase, resulting in short-chain fatty acids being incorporated into milk fat. In non-ruminants, this is limited mainly to butyrate, whereas the more relaxed specificity of the acyl transferase in ruminants allows medium-chain acyl-CoA to be released in significant amounts (Hansen and Knudsen 1980). An increased concentration of malonyl-CoA in the medium increases the relative proportions of longer-chained fatty acids (Knudsen 1979; Knudsen and Grunnet 1982). Indeed, this can be inferred from studies in which intestinal glucose supply (Hurtaud *et al.* 2000) or systemic insulin concentrations (Griinari *et al.* 1997) were increased experimentally. Removal of intermediate-chain length acyl-CoA is dependent on the presence of an acceptor, such as albumin (Knudsen and Grunnet 1982), α-glycerol-phosphate, or diacylglycerol. The incorporation of fatty acids into TAGs is greatly enhanced by the presence of non-limiting concentrations of α-glycerol-phosphate or diacylglycerol (Grunnet and Knudsen 1981; Hansen *et al.* 1984b).

Regulation of chain termination reactions in ruminant and non-ruminant mammary glands was discussed in detail by Smith (1994; see also Barber *et al.* 1997). In non-ruminant mammary tissue, medium-chain fatty acids are generated by a novel chain-terminating enzyme that is not part of the FAS complex, known as thioesterase II to distinguish it from the chain-terminating enzyme (thioesterase I) associated with FAS (Smith 1994). Thioesterase II, though a 29 kDa protein independent of FAS, appears to function identically to thioesterase I. However, its specificity differs, producing C_8 to C_{14} fatty acids.

2.4.4.1 Elongation of C_{16} Acyl Chains

Bishop *et al.* (1969) reported that 4.6% of intravenously infused methyl [^{3}H] palmitate was incorporated into longer-chained FA of milk fat when a cow was fed a low-fat diet, whereas Palmquist *et al.* (1969) reported no significant labelling of C_{18} acids in milk fat after intramammary infusions of 1-^{14}C-acetate. Chain lengthening of long-chain CoA esters occurs in the microsomes, using malonyl-CoA as the acetyl donor (Bernert and Sprecher 1979). This is not a significant pathway for the supply of C_{18} fatty acids for milk fat synthesis.

2.5 Stearoyl CoA Desaturase

Stearoyl-CoA desaturase (SCD; EC 1.14.99.5) is the rate-limiting enzyme for the conversion of saturated to monounsaturated fatty acids. As such, it plays a major role in regulating the unsaturation of membranes and TAG composition, and evidence indicates that SCD plays an important regulatory role in lipid metabolism. SCD-1-deficient mice exhibit increased fatty acid oxidation and reduced lipid synthesis (Ntambi and Miyazaki 2004; ALJohani *et al.* 2017). Its role is especially important in ruminants, in which the majority of absorbed C_{18} fatty acids are in the form of stearic acid, the major product of ruminal biohydrogenation. The activity of SCD in ruminants is high in lactating mammary gland and adipose tissue and somewhat lower in intestinal tissue. SCD activity is low in non-lactating mammary tissue, being induced by lactation (Kinsella 1972); though normally inactive in ruminant liver, SCD may be induced in liver and muscle by high-fat diets (Chang *et al.* 1992).

SCD is located in the endoplasmic reticulum; its primary substrates are stearoyl-CoA and 18 carbon *trans* monoene-CoAs, whereas considerably lower activity is observed for palmitoyl-CoA and myristoyl-CoA (Bickerstaffe and Annison 1970). Desaturation of a range of *trans* monoenes yields *trans*-x, *cis*-9 octadecadienoic acids as products and is particularly important considering the large amounts of *trans* octadecenoic acids that are formed in the rumen by biohydrogenation (Shingfield *et al.* 2003; Jenkins *et al.* 2008). The quantitatively most important is desaturation of vaccenic acid (*trans*-11 $C_{18:1}$) to rumenic acid (*cis*-9, *trans*-11 $C_{18:2}$), commonly called conjugated linoleic acid (CLA).

Considerable variation among cows (Kinsella 1972; Kelsey *et al.* 2003) and breeds (DePeters *et al.* 1995) in the extent of stearic acid desaturation and in mRNA expression for SCD (Taniguchi *et al.* 2004) has been observed. SCD mRNA expression is downregulated by *trans*-10, *cis*-12 $C_{18:2}$, which also causes milk fat depression, and other *trans* FA that do not reduce milk fat synthesis (Harvatine *et al.* 2009; Bauman *et al.* 2011). Experimentally, SCD can also be inhibited with the cyclopropenyl fatty acid sterculic acid and cobalt-EDTA that is sometimes used as a liquid flow marker, as discussed below (e.g., Griinari *et al.* 2000; Shingfield *et al.* 2008).

2.6 Triacylglycerol Synthesis

Enzymes for triacylglycerol synthesis are associated with the endoplasmic reticulum and the inner and outer mitochondrial membranes (Coleman and Mashek 2011). Glycerol-3-

phosphate required for esterification of fatty acids is generated by glycolysis or by phosphorylation of free glycerol by glycerol kinase (Kinsella 1968; Bickerstaffe and Annison 1971). Although some free glycerol may be taken up from blood during lipolysis of lipoprotein TAGs, most glycerol-3-phosphate is probably derived from glycolysis (Bickerstaffe and Annison 1971).

Enzymes and regulation of TAG synthesis have been reviewed in depth by Coleman and Mashek (2011).

For metabolism to occur, fatty acids must be activated to their acyl-CoA esters:

$$\text{Fatty acid} \underset{\text{CoASH,Mg}^{++}}{\overset{\text{acyl CoA synthetase}}{\rightarrow}} \text{Acyl CoA} + \text{AMP} + \text{PPi}$$

Acyl-CoA synthetase is expressed in several isoforms and is distributed on the cytosolic surfaces of the endoplasmic reticulum, peroxisomal, and outer mitochondrial membranes. Activity of this enzyme responds to changes in physiological state, suggesting that it may play a role in regulating the entry of fatty acids into synthetic or oxidative pathways (Grevengoed *et al.* 2014).

Acylation of glycerol-3-phosphate is the first committed step in TAG synthesis and the activity of acyl-CoA: glycerol-*sn*-3-phosphate acyl transferase (GPAT) is the lowest of the transacylation enzymes in this pathway, suggesting a regulatory role in TAG synthesis (Coleman and Mashek 2011). Although microsomal GPAT constitutes the majority of total GPAT activity in most tissues, its role in TAG synthesis is uncertain (Coleman and Mashek 2011), and mitochondrial GPAT, located on the mitochondrial outer membrane, is characterized more completely. Its activity is influenced by nutritional and hormonal changes in adipose tissue (Coleman and Mashek 2011).

Cooper and Grigor (1980) reported that acylation at *sn*-1 in rat mammary TAGs favors oleic acid, whereas palmitic acid predominates at *sn*-2. This pattern is true in most species, although in the cow, palmitic acid is found nearly equally at the *sn*-1 and *sn*-2 positions (Christie 1985; Parodi 1982; Jensen and Newburg 1995). The product, 1-acyl-lysophosphatidate, is acylated by acyl-CoA:1-acylglycerol-*sn*-3-phosphate acyltransferase. The microsomal fraction from lactating cow mammary gland transfers acyl chains C_8 to C_{18}, but not C_4 or C_6, to the *sn*-2 position of 1-acyl-lysophosphatidate. The chain length specificity is $C_{16} > C_{14} > C_{12} > C_{10} > C_8$, similar to the pattern found at the *sn*-2 position of bovine milk fat (Marshall and Knudsen 1977).

Phosphatidic acid (1,2 diacylglycerol-*sn*-3-phosphate) occupies a central point in lipid biosynthesis. It can be converted to CDP-diacylglycerol, a precursor for the biosynthesis of acidic phospholipids, or dephosphorylated to produce a diacylglycerol, the precursor of TAGs, phosphatidylserine, phosphatidylcholine, and phosphatidylethanolamine (Coleman and Mashek 2011). In TAG synthesis, dephosphorylation of phosphatidic acid is mediated by phosphatidic acid phosphatase-1, a Mg^{2+}-requiring enzyme that is transferred from the cytosol to the endoplasmic reticulum in the presence of fatty acids or acyl-CoA. Activity in liver is stimulated by glucagon, glucocorticoids, cAMP, and growth hormone and inhibited by insulin (Coleman and Mashek 2011).

Diacylglycerol acyltransferase (*sn*-1,2-diacylglycerol transacylase) esterifies both long- and short-chain fatty acids at the *sn*-3 position. Mice that lack both copies of the gene for diacylglycerol acyltransferase 1 (DGAT 1) are unable to secrete milk (Smith *et al.* 2000). This gene was mapped to a region close to the quantitative trait locus on bovine chromosome 14 for variation in milk fat content (Grisart *et al.* 2002). The functional SNP was mapped to a substitution of alanine for lysine in the enzyme (Winter *et al.* 2002). The wild-type allele ($DGAT1^K$) exceeds the $DGAT1^A$ allele by +0.34% units in fat and + 0.08% units in protein, whereas milk and protein yields are reduced (Grisart *et al.* 2004). A genotyping study of 38 *Bos indicus* and *Bos taurus* breeds from five continents showed that most domesticated dairy breeds have predominantly the $DGAT1^A$ allele, whereas $DGAT1^K$ was the major allele (69%) in the Jersey breed (Kaupe *et al.* 2004).

In nearly all species studied, oleic acid is the fatty acid preferably esterified at the *sn*-3 of TAGs; $C_{18:3\ n\text{-}3}$ predominates in the koala and horse. In ruminant species, butyric and caproic acids are esterified exclusively at *sn*-3 and slightly exceed the molar percentage of oleic acid esterified at that position (Parodi 1982). Lin *et al.* (1976) examined the acyl specificity for TAG synthesis in lactating rat mammary glands, particularly with respect to the unique positioning of short- and medium-chain length fatty acids at *sn*-3. Whereas the acyl transferases for *sn*-1 and *sn*-2 showed high specificity for long-chain fatty acids (C_{16} and C_{18}), no such specificity was observed for acylation of diacylglycerol. They concluded that a lack of acyl chain specificity for this position caused accumulation of shorter acyl chains at *sn*-3.

Knudsen and colleagues (Hansen and Knudsen 1980; Marshall and Knudsen 1980; Grunnet and Knudsen 1981; Hansen *et al.* 1984a, b) examined in detail TAG synthesis and the specific incorporation of the short- and medium-chain fatty acids into TAGs in goat mammary glands. The synthesis of medium-chain fatty acids is dependent on simultaneous removal of the acyl-CoA produced, whereas long-chain fatty acids are released as free acids by thioesterase I. As described above, long-chain fatty acids are esterified preferentially at positions *sn*-1 and *sn*-2; the ready supply of diacylglycerols allows short- and medium-chain fatty acids to be esterified rapidly, facilitating their removal from FAS. Their studies showed the importance of the rate of activation of fatty acids in the mammary gland relative to the rate of *de novo* synthesis and the supply of α-glycerol phosphate for milk fat synthesis. If the supply of exogenous fatty acids is low, the relative concentration of short- and medium-chain fatty acids could be increased, even though total synthesis (yield) was not increased. Conversely, with an increasing supply of exogenous long-chain fatty acids, *de novo* synthesis may be reduced because they compete for the diacylglycerol transferase. Limiting the supply of α-glycerol phosphate similarly would limit diacylglycerol supply, also causing *de novo* synthesis to be reduced. These observations demonstrate also that regulation of the relative proportions of short-, medium-, and long-chain fatty acids is much more complex than simply by regulation of ACC. Finally, these authors concluded that a simple explanation for the unique occurrence of short- and medium-chain fatty acids in milk fat could be that α-glycerol phosphate is rate-limiting for TAG synthesis in all ruminant tissues except mammary glands (Hansen *et al.* 1984b).

Interestingly, for maximum TAG synthesis, a preference was shown for palmitoyl-CoA as substrate for the initial acylation of glycerol-3-phosphate (Kinsella and Gross 1973), apparently accelerating the rate of supply of substrate as acceptor for *de novo*-synthesized fatty acids, whereas oleic acid reduced total *de novo* synthesis, apparently by competing with butyryl CoA for the esterification of diacylglycerol (Hansen and Knudsen 1987).

2.6.1 Fatty Acid Esterification by the Monoacylglycerol Pathway

Evidence for the esterification of fatty acids via the 2-monoacylglycerol pathway was shown for mammary glands of guinea pigs by McBride and Korn (1964) and of goats by Bickerstaffe and Annison (1971). However, even though this is an important pathway for the esterification of fatty acids in the intestine, later studies have established that this pathway is not functional in mammary tissue (Christie 1985; Hansen *et al.* 1986).

2.7 Synthesis of Complex Lipids

The complex lipids in milk fat are comprised of the phosphoglycerides, phosphatidylcholine, phosphatidylserine, phosphatidylethanolamine, phosphatidylinositol, and plasmalogens. Also, the non-glyceride phospholipid, sphingomyelin, occurs in important amounts (Jensen 2002). Bitman and Wood (1990) described the distribution of phospholipid classes in bovine milk and their fatty acid composition. The phospholipids comprise about 1% and cholesterol 0.4–0.5% of the total milk fat. These occur almost completely in the milk fat globule membrane. The fatty acid composition of phosphoglycerides and sphingolipids and their incorporation into membranes with cholesterol are coordinately regulated processes that maintain membrane integrity and function. Coordination is mediated by sterol regulatory genes and sterol regulatory element-binding proteins (SREBPs; Ridgway *et al.* 1999), which are transcription factors that are central regulators of lipid synthesis and have a key role in milk fat depression (see below). Glycosphingolipids are highly enriched in the outer leaflet of the apical plasma membrane domain of polarized epithelial cells. They are a key component of membrane lipid rafts and thus are involved in exocytosis of milk components. Their metabolites act as second messengers in regulating the expression of cell receptors (Hoekstra *et al.* 2003). Sphingolipids are enriched in the milk fat globule membrane and as a food component have been implicated in cell regulation and anti-cancer activity (Vesper *et al.* 1999).

2.7.1 Synthesis of Phospholipids

Synthesis of phosphatidylcholine and phosphatidylethanolamine begins with activation of choline (or ethanolamine) with ATP via choline kinase to yield phosphocholine (phosphoethanolamine) + ADP; the activated base is transferred via CTP and phosphocholine cytidyltransferase to form CDP-choline (CDP-ethanolamine) and PPi. The base is then transferred to the *sn*-3 of diacylglycerol via phosphocholine diacylglycerol transferase to yield phosphatidylcholine (phosphatidylethanolamine) + CMP. The cytidyltransferase is believed to be the rate-limiting or regulatory step in the pathway. Phosphatidylserine is formed by a direct transfer and substitution of serine for ethanolamine in phosphatidylethanolamine. Phosphatidylserine can be decarboxylated to form phosphatidylethanolamine.

Phosphatidylcholine is preferentially synthesized in lactating mammary tissue (Kinsella 1973), possibly regulated by the differential activities of choline kinase and ethanolamine kinase. Choline kinase has a lower K_m and a higher V_{max} with its substrate than does ethanolamine kinase. Also, choline kinase is inhibited slightly by ethanolamine, whereas choline is a potent competitive inhibitor of ethanolamine kinase. Thus, the intracellular concentration of choline probably regulates the synthesis of these two phosphoglycerides (Infante and Kinsella 1976). Choline supplementation has been extensively studied for prevention of fatty liver diseases in early lactation but has not been specifically investigated for milk fat.

The *sn*-1 and *sn*-2 acyl groups of phosphoglycerides differ among phospholipid classes (Bitman and Wood 1990) and from the patterns of TAGs. However, the CDP-choline: diacylglycerol transferase has little specificity for the molecular species of fatty acids in the diacylglycerol moiety. Therefore, molecular remodelling via specific phospholipases, followed by reacylation via specific acyl-CoA transferases, may account for the unique fatty acid profiles of the various phosphoglycerols. Kinsella and Infante (1974) showed that the acyl-CoA:1-acyl-*sn*-glycerol-3-phosphorylcholine acyltransferase preferentially esterified oleic acid at the *sn*-2 position and that the predominant molecular species of phosphatidylcholine in mammary cells is diunsaturated.

Synthesis of phosphatidylinositol follows a slightly different pathway from the other phosphoglycerides. Phosphatidate is activated by CTP-phosphatidate cytidyltransferase to form CDP-diacylglycerol. Free inositol is then incorporated by CDP-diacylglycerol inositol transferase, with the release of CMP. Phosphatidylinositol usually constitutes less than 5% of total milk phospholipids (Bitman and Wood 1990).

The glycerol ethers and plasmalogens are a unique class of phosphoglycerides that contain an ether-linked chain at the *sn*-1 position. These are formed by the incorporation of acyl-CoA at the *sn*-1 position of dihydroxyacetone phosphate. The acyl group is then exchanged for a long-chain alcohol to yield 1-alkyl dihydroxyacetone phosphate, which is reduced to 1-alkylglycerophosphate. This product is acylated to 1-alkyl, 2-acylglycerophosphate which then may be incorporated into various phosphoglycerols or acylated to form a neutral alkylglycerol. The alkyl residue of the phosphatidylethanolamine typically is desaturated at the 1,2 position, involving O_2 and NADPH to yield 1-alkenyl, 2-acyl-glycerol-3-phosphatidylethanolamine (plasmalogen), found at high concentrations in the mitochondrial membrane. The alkylglycerols occur at very low concentrations in milk fat (~ 0.01% of the fat) but are

5- to 20-fold higher in the fat of colostrum (Ahrné *et al.* 1980).

2.7.2 Sphingolipids

The sphingolipids are very complex molecules, occurring in numerous molecular forms (Jensen 2002). The basic structure, sphingomyelin, consists of a complex unsaturated amino alcohol (sphingosine) ether-linked to phosphocholine and to an acyl chain by an amide bond. It contains no glycerol (Vesper *et al.* 1999). In sphingomyelin synthesis, sphingosine is acylated by acyl-CoA to form a ceramide; phosphocholine is then transferred to a ceramide from phosphatidylcholine or from CDP-choline to yield sphingomyelin. Sphingomyelin constitutes about 25% of the total phospholipids in dairy products, which are the most abundant source of sphingolipids in the human diet. The sphingolipids are highly bioactive and, as such, are considered to be functional in foods (Vesper *et al.* 1999; Parodi 2006: McFadden and Rico 2019). The neutral sphingolipids contain no phosphocholine, and therefore are not phospholipids. These are formed by addition of various sugar residues to ceramide. These can be quite complex (Jensen 2002). Sphingolipids often contain very long-chain (C_{24}) fatty acids, especially those found in the brain.

2.7.3 Cholesterol

Cholesterol makes up 95% of the total milk sterols, and because it is associated with the milk fat globule membrane, its content is highly correlated with the total fat content (Jensen 2002). Only about 10% of the cholesterol is esterified.

Cholesterol is taken up rapidly from plasma lipoproteins and is synthesized also by the mammary glands (Clarenburg and Chaikoff 1966). Quantitative contributions of each, and whether all lipoproteins or only chylomicra and VLDL contribute cholesterol, are less certain (Clarenburg and Chaikoff 1966; Raphael *et al.* 1975a, b). From steady-state estimates of labelled cholesterol in lipoproteins, plasma contributed 83% of milk cholesterol in lactating rats (Clarenburg and Chaikoff 1966) and 45–50% in a lactating goat (Raphael *et al.* 1975b). The mechanism of cholesterol uptake also is not well defined; however, an active cholesterol esterase is present in mammary tissue (Ross and Rowe 1984; Shand and West 1991; Small *et al.* 1991), suggesting that both free and esterified lipoprotein cholesterol may be taken up and utilized. The cholesterol transporters ABCA1 and ABCG1 are differentially expressed across the lactation cycle and inversely related to plasma cholesterol concentration and may play functional roles in cholesterol transport (Mani *et al.* 2010).

2.8 Milk Fat Globular Membrane and Fat Secretion

Fatty acids to be secreted in milk fat are esterified in the endoplasmic reticulum (**ER**) and assembled into lipid droplets that move to the apex of the cell. Due to the hydrophobic nature of esterified FAs, milk fat is secreted from the mammary epithelial cell as a lipid droplet surrounded by a protein-rich polar lipid coat called the "milk fat globule membrane" (**MFGM**; see reviews by Patton and Keenan (1975), Mather and Keenan (1998) and McManaman 2012); see Chap. 7.

The origin of the MFGM and the mechanism of cellular milk fat secretion continue to be areas of intense investigation. Proteomic approaches have identified many of the associated proteins (Cavaletto *et al.* 2008), and knockout mouse models have demonstrated the essential role for some of these proteins in milk fat secretion including butyrophilin (Ogg *et al.* 2004) and xanthine oxidoreductase (McManaman *et al.* 2002). More recently, the MFGM has been proposed to be important to milk fat absorption in humans and has been proposed to have health-promoting effect (Beals *et al.* 2019).

2.9 Physiological Factors that Influence Milk Fat Composition

2.9.1 Genetics

Milk fat concentration and yield are highly heritable [0.45 and 0.29, respectively; (Welper and Freeman 1992)] in the dairy cow, and milk fat is unique in that the genetic variation is due to a limited number of SNPs with large individual effects (Hayes *et al.* 2010). The largest effect is a K232A SNP in diacylglycerol acyltransferase [DGAT1; (Grisart *et al.* 2002)] followed by the F279Y SNP in the growth hormone receptor [GHR; milk fat allele substitution effect 0.46% units; (Signorelli *et al.* 2009)]. Wang *et al.* (2012) identified four QTLs that explained over 46% of the genetic variation in milk fat concentration. The DGAT1 SNP also has an impact on milk fatty acid profile, including *de novo* synthesized FA (Schennink *et al.* 2007). Other SNPs associated with milk fatty acid profile have also been identified including a QTL on chromosome 17 that is related to *de novo* FA (Duchemin *et al.* 2014).

In addition to the well-characterized differences in milk fat content among and within breeds of dairy cattle, differences occur also in fatty acid composition. Numerous studies indicate that SCD activity, and thus 18:1/18:0 ratio, varies among breeds (Beaulieu and Palmquist 1995; DePeters *et al.* 1995) and within a breed (Kelsey *et al.* 2003). Genetic polymorphisms in sterol CoA desaturase 1 and sterol response element binding protein 1, which regulates many lipogenic enzymes, have also been identified.

2.9.2 Stage of Lactation

Milk fat is highest at the initiation of lactation coinciding with rapid mobilization from body stores. At parturition, the proportions of C_{12} to C_{16} fatty acids are relatively high in bovine colostrum. Proportions of short-chain fatty acids and stearic and oleic acids increase rapidly as the mobilization of adipose tissue commences, becoming relatively stable by 1 week after parturition (Laakso *et al.* 1996).

Proportions and yields of fatty acids that synthesized *de novo* in the mammary gland (C_6 to C_{16}) increase during the early weeks of lactation, with compensating decreases in the proportions of all C_{18} fatty acids (Figure 2.6; Karijord *et al.* 1982; Lynch *et al.* 1992). The proportion of butyric acid is high at parturition and does not increase with advancing lactation (Lynch *et al.* 1992), consistent with its synthesis being independent of malonyl-CoA. As adipose tissue mobilization declines, due to increasingly positive energy balance and depletion of stored tissue, the proportions of short- and medium-chain fatty acids in milk fat increase. The time required for stabilization of milk fatty acid composition depends on the amount of stored fat, milk fat yield, energy balance, and the quantity of fat in the diet. The effect on the proportion of $C_{14:0}$ in milk fat of supplementing dietary fat to lactating cows at parturition or delayed to the sixth week of lactation is shown in Figure 2.7. Without fat supplementation, the proportion of $C_{14:0}$ increased until 8 weeks of lactation; with fat supplementation, the maximum proportion of $C_{14:0}$ was lower.

2.9.3 Daily Patterns and Circadian Rhythms

There is a well-recognized daily pattern to milk composition with higher milk yield and lower milk fat concentration in the morning compared to the evening milkings. It is not uncommon for milk fat concentration to differ more than 0.5% units between milkings across the day and is especially seen in large dairies shipping multiple tankers of milk each day. The cow has a daily pattern of intake with a majority of feed consumed over half of the day. The daily pattern of milk synthesis may be driven by temporal changes in nutrient absorption over the day or may be due to endogenous timekeepers in the mammary gland that modify metabolic capacity across the day. The daily pattern of milk synthesis can be modified by the timing of feed intake, and increasing feeding times per day decreased the amplitude of

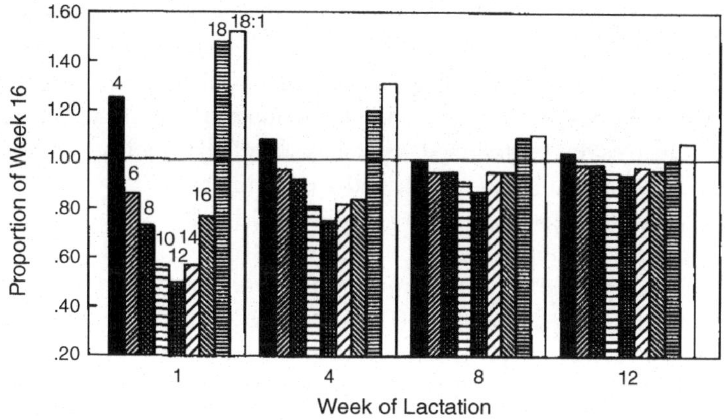

Figure 2.6. Proportions of individual milk fatty acids in milk fat at 1, 4, 8, and 12 weeks of lactation relative to their proportions at 16 weeks. (From Palmquist *et al.* 1993, *Journal of Dairy Science*.76:1753–1771).

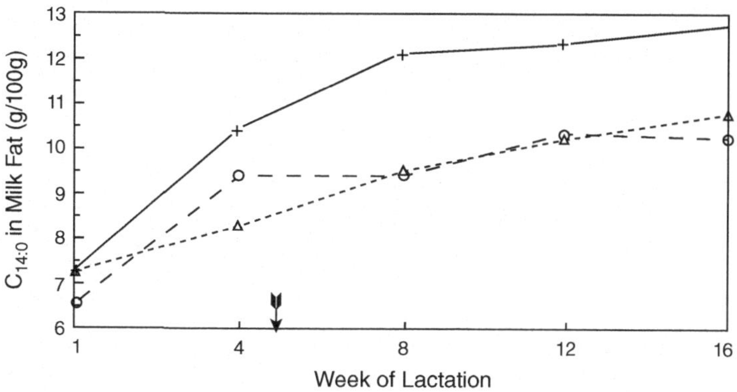

Figure 2.7. Changes in the proportion of $C_{14:0}$ in milk fat of cows fed a control diet (—+—) or added fat beginning at parturition (–Δ–) or sixth week of lactation (–O–). (From Palmquist *et al.* 1993, *Journal of Dairy Science*.76:1753–1771).

the daily rhythm (Rottman *et al.* 2015). Temporal changes in nutrient absorption can entrain cellular timekeeping mechanisms, and nutritional entrainment of the biological clock that regulates metabolism has been demonstrated in the mouse. Regardless of the mechanism, spreading feed intake across the day is expected to stabilize rumen fermentation and availability of nutrients for milk synthesis and thus reduce variation in milk fat across the day.

2.9.4 Seasonal Rhythms

Circannual or seasonal rhythms in milk fat concentration and profile occur over the course of a year. These patterns may be driven by changes in diet and stage of lactation, especially in grazing and seasonal calving herds and by endogenous timekeeping mechanisms responding to day length and changes in day length. The predominance of total mixed rations (TMR) feeding in the USA reduces the effect of diet composition on the annual rhythm. In the USA, milk fat concentration is highest near the winter solstice and lowest near the summer solstice (Salfer *et al.* 2019). Milk yield is highest near the spring equinox resulting in peak milk fat yield between the winter solstice and spring equinox. The amplitude of the annual rhythm in milk fat concentration is larger in the northern regions of the USA, but the amplitude of the annual rhythm of milk fat yield is higher in the southern regions. On average milk fat concentration varies approximately 0.25% units across the year.

Changes in fatty acid profile across the year under pasture grazing changes as expected with increased unsaturated fatty acids and vaccenic acid during the grazing months. Seasonal changes across the year under TMR feeding is not as well characterized, although recent work from the Miner Institute indicated that a majority of the annual rhythm was explained by short- and medium-chain fatty acids (Woolpert *et al.* 2016). The consistency of the annual rhythms across the year and from year to year supports photoperiod or changes in photoperiod as a key regulator rather than heat stress. Importantly, experimentally induced heat stress decreases milk yield and increases milk fat concentration. Constant long-day photoperiod increases milk yield without changing milk composition (Dahl *et al.* 2000), and the ability of photoperiod management to modify seasonal rhythms is not clear.

2.10 Effects of Dietary Fat on Composition of Milk Fat

2.10.1 Effects of Low-Fat Diets

Very low-fat diets reduce milk and fat yields (Maynard and McCay 1929; Banks *et al.* 1976) and greatly reduce the proportions and yields of the C_{18} fatty acids, with the proportions of $C_{16:0}$ approaching 50% of the total fat yield (Virtanen 1966; Banks *et al.* 1976).

Banks *et al.* (1976) reported that increasing the C_{18} content of low-fat diets resulted in a linear increase of C_{18} fatty acids in the milk fat with C_{18} fatty acids transferred to milk fat with 54% efficiency. This is consistent with other estimates of a maximum transfer of 60%, assuming 80% digestibility of dietary fat (Palmquist 1991) and 75% uptake of absorbed lipoprotein TAGs by the mammary glands (Palmquist and Mattos 1978). A meta-analysis conducted by Glasser *et al.* (2008) reported a quadratic relationship between duodenal flow of C_{18} fatty acids and milk C_{18} fatty acid yield concurrent with a linear decrease in *de novo* synthesized fatty acids that limited the increase in milk fat yield. Transfer of dietary C_{16} is heavily confounded by effects of changing dietary fat intake on *de novo* synthesis of C16, but recent work with high palmitic acid supplements generally reports a 15–20 g increase in milk fat yield for every 100 g of palmitic acid fed. Using arterial-venous difference, Enjalbert *et al.* (1998) reported that mammary C16:0 extraction increased from 21.8 to 67.1% with supplementation of palmitic acid, while extraction efficiency of stearic and oleic acids increased to a lesser extent with supplementation and may explain differences in milk fat response to C_{16} and C_{18} fatty acids.

2.10.2 Effects of Specific Fatty Acids

Incorporation of dietary unsaturated fat into milk fat by ruminants is low because of the efficient ruminal biohydrogenation process (Jenkins *et al.* 2008) and their rapid incorporation into phospholipid and cholesterol ester fractions that are unavailable to the mammary gland. Nevertheless, dietary fatty acids have profound effects on milk fat composition that have led to a prodigious amount of literature in the late twentieth century (for reviews see: Grummer 1991; Palmquist *et al.* 1993; Shingfield *et al.* 2013).

Acyl chain length (C_{16} vs C_{18}) influences the proportions of these in milk fat; the apparent effects of C_{16} are more subtle because of compensation by reduced *de novo* $C_{16:0}$ synthesis when long-chain fatty acids are supplemented in the diet. Palmitic acid was increased up to 53% of fatty acids when a high (68%) $C_{16:0}$ supplement was added to a low-fat diet, while palmitic is normally 20–30% of fatty acids. Similarly, supplementing soy oil (90% C_{18}) increased the total C_{18} of milk fat from 25% to 60% of milk fatty acids. Yields of C_6 to C_{14} were reduced by both supplements, whereas the yield of $C_{16:0}$ was increased by palm oil and reduced by soya oil (Banks *et al.* 1976). Similarly, supplementing increasing amounts of coconut oil (high in $C_{12:0}$ and $C_{14:0}$) increased the proportions of these in milk fat and reduced the proportions and yields of short-chain fatty acids and $C_{16:0}$ (Storry *et al.* 1971). However, abomasal infusion of butter oil increases the short- and medium-chain concentration in milk fat and increases milk fat yield indicating less inhibition of *de novo* lipogenesis (Kadegowda *et al.* 2008)

Oleic acid was increased to 48% of total milk fatty acids by feeding oleamide as a rumen-protected source of oleic acid (Jenkins 1998), and abomasal infusion of high oleic sunflower oil linearly increases oleic to 57% of fatty acids at 1000 g/d (Drackley *et al.* 2007). Proportions of all *de novo*-synthesized milk fatty acids, except butyric, were reduced. LaCount *et al.* (1994) also reported linear transfer of abomasally infused oleic acid to milk fat (slope = 0.541; 0–350 g infused/day). Linoleic acid from canola also was transferred linearly (slope = 0.527; 0–90 g infused/day). These transfers from the intestine are nearly identical to that reported by Banks *et al.* (1976). Hagemeister *et al.* (1991) reported 42–57% transfer of abomasally infused linolenic acid to milk fat. Similar transfer efficiencies have been reported in recent work summarized by Shingfield *et al.* (2013)

Mammary uptake of individual fatty acids from plasma was explored by Enjalbert *et al.* (1998). Uptakes of palmitic, stearic, and oleic acids were similar and nearly linear in the range of 400–750 μmolar in the plasma. As uptake increased, mammary balance (milk content minus uptake) of butyric acid increased linearly, and the mean fatty acid chain length of synthesized fatty acids decreased linearly. Desaturation of stearic acid increased linearly as stearic acid uptake increased ($y = 0.52x$, $r^2 = 0.75$, $P < 0.001$). Decreasing mean chain length of synthesized fatty acids and increasing desaturation with increasing long-chain fatty acid uptake were interpreted as compensating responses to maintain the fluidity of milk fat at body temperature. Maintenance of milk fat fluidity has been suggested as a basic physiological requirement in the regulation of milk fat synthesis (Timmen and Patton 1988). An interesting exception to this is the report of Emanuelson *et al.* (1991) who fed heat-treated rapeseed to cows in late lactation. The stearic acid content of the milk fat was 28.8% of the total weight of fatty acids, and the milk fat coalesced into floating butter globules immediately upon milking. The levels of palmitic (21.6%) and oleic (28.8%) acids were within normal ranges, but the proportions of short- and medium-chain fatty acids were rather low. The phenomenon was explained later by the discovery that cobalt inhibits SCD activity (see below). A number of studies have directly altered the stearic acid desaturation through inhibition of SCD1 enzyme including abomasal infusions of sterculic oil (e.g., Griinari *et al.* 2000), bioactive CLA isomers that do not reduce milk fat (e.g., Perfield *et al.* 2006), and dietary administration of the rumen marker CoEDTA (Shingfield *et al.* 2008). Thus, inhibition of SCD impacts melting temperature, but it appears that the mammary gland is able to compensate for this change in fluitidy by other mechanisms.

2.10.3 Feeding for Specific Milk Fatty Acid Profiles

A very large body of literature that focused on feeding effects on the composition of milk fat has been published in recent years, driven by the increasing public interest and concern for the role of fat in the human diet. Most of the feeding studies have used manipulation of the type and amount of fat in the ration of lactating cows as the experimental approach. In many studies, the role of ruminal biohydrogenation or its manipulation has been a major focus (e.g., Toral *et al.* 2018). Therefore, the effects, but not the regulation, of ruminal biohydrogenation will be addressed. Issues in these studies have included (1) increased unsaturation/polyunsaturation; (2) reduced saturation; (3) increased n-3/fish oil fatty acids; (4) increased CLA content of milk fat; and (5) increased odd- and branched-chain fatty acids.

Early efforts to modify milk fatty acid profile were successful using an insoluble formaldehyde-crosslinked protein to encapsulate unsaturated vegetable oils. In numerous studies, linoleic acid was increased to as high as 35%, w/w, of the total milk fatty acid (reviewed by McDonald and Scott 1977). A typical milk fatty acid profile from cows fed a protected sunflower/soybean (70/30) supplement is shown in Table 2.1. Though feeding protected polyunsaturated fats furthered understanding of the regulation of milk fat synthesis, it has not found practical application. In addition to cost, it has been difficult to assure product quality (consistency of protection); government and the public have been reluctant to approve formaldehyde as a component of feed ingredients because

Table 2.1. Fatty acid composition (weight % of total fatty acids) of milk fat from cows fed a standard diet or supplemented with protected supplements

Fatty acid	Control[a]	Sunflower/soybean[a]	Canola[b]	Oleamide[c]
C_4	3.2	2.8	3.2	1.6
C_6	2.2	1.4	2.4	1.1
C_8	1.1	0.8	1.9	0.6
$C_{10:0}$	3.6	1.7	3.2	1.5
$C_{12:0}$	3.9	1.7	3.6	1.9
$C_{14:0}$	11.4	5.9	9.5	8.2
$C_{15:0}$	2.1	0.9	–	–
$C_{16:0}$		15.2	19.9	22.5
$C_{16:1}$	2.7	0.5	3.3	2.9
$C_{18:0}$	10.9	14.0	9.2	12.5
$C_{18:1}$	28.6	37.6	29.2	43.4
$C_{18:2}$	3.0	16.6	4.9	2.3
C18:3	1.0	0.9	2.6	–

[a]Calculated from Barbano and Sherbon (1980). Formaldehyde-protected sunflower/soybean (70/30); 1250 g oil/day.
[b]Ashes *et al.* (1992). Formaldehyde-protected canola; 520 g oil/day.
[c]Jenkins, T.C. (1999) JDS 82:1525–1531. Oleamide at 5% of the diet.

some amino acids may be transformed to potential carcinogens; and, importantly, highly polyunsaturated milk fat has very poor oxidative stability, and its physical properties are not well suited for processed products (McDonald and Scott 1977; Fauteux *et al.* 2016). Formaldehyde treatment of high oleic acid feeds was also explored, and Ashes *et al.* (1992) reported that feeding 0.52 kg/d of protected canola seeds increased milk fat percentage and yield and decreased the proportions of $C_{14:0}$ and $C_{16:0}$ 20 and 25%, respectively, whereas proportions of $C_{18:0}$, $C_{18:1}$, $C_{18:2}$, and $C_{18:3}$ were increased by 30, 22, 122, and 62%, respectively.

Several other procedures have been developed to protect unsaturated fatty acids from ruminal biohydrogenation. Of these, only the amide derivative has extensive research documentation (e.g., Jenkins 1998) but has not been applied commercially. Often, calcium soaps of palm oil or canola fatty acids are referred to as "protected." These are not protected from ruminal biohydrogenation (Table 2.2) but rather are ruminally inert, having relatively small effects on the rumen microbial population and biohydrogenation pathways. The commercial interest to increase milk fat unsaturation has declined with more recent convincing studies that saturated milk fatty acids have minimal effects on cardiovascular health (Lordan *et al.* 2018). However, there is interest in supplements to provide essential fatty acids to the cow to improve reproduction and immunity.

2.10.4 Supplementation with Oilseeds and Commercial Fats

Numerous types of fat are available commercially as supplemental energy sources. Many are products of the rendering industry and include tallow, lard (pork), and poultry fats. By-products from palm and soybean oil processing and corn distilleries are also commonly available along with recycled cooking oils from the restaurant industry. Blending of these oils is also common. Generally, unsaturated oils are undesirable as energy supplements for lactating cows (Palmquist and Jenkins 1980; Jenkins and Harvatine 2014). More recently, high-fat by-products such as corn distillers' grains with solubles have been a considerable source of dietary fat. Plant and animal oils and many high-fat by-products are economical sources of energy, but care must be used as they contribute to diet-induced milk fat depression when altered rumen biohydrogenation of their unsaturated fatty acids occurs.

Table 2.2. Fatty acid composition of milk fat from cows fed various fat supplements

	Fat supplement										
		Calcium salts									
	None	Palm oil	Palm oil	Canola	Soy	Linseed	Whole roasted soybeans	Tallow	Fish oil	High C16:0 prill	C16 and C18 prill
% of feed DM	–	3.0	2.3	5.0	4.9	4.3	4.6	5.4	2.0	1.9	3.0
Reference	a	a	b	c	c	C	d	d	e	b	f
Fatty acid	Weight % of reported fatty acids										
C_4	3.25	3.15	4.41	4.52	4.65	4.92	3.65	2.65	3.88	4.26	4.06
C_6	2.28	1.92	2.24	2.33	2.39	2.89	2.34	1.59	2.66	2.27	2.56
C_8	1.57	1.24	1.10	1.16	1.14	1.52	1.31	0.83	1.30	1.14	1.24
$C_{10:0}$	3.36	2.31	2.28	2.59	2.34	3.09	2.77	1.75	2.83	2.44	2.70
$C_{12:0}$	4.11	2.92	2.53	2.53	2.21	2.71	3.03	2.23	3.16	2.73	3.04
$C_{14:0}$	13.13	10.35	9.01	9.39	8.10	9.20	9.75	10.70	11.40	9.61	11.09
$C_{14:1\ c\text{-}9}$	0.95	0.74	0.71	1.10	0.74	0.71	1.05	1.73	0.77	0.79	1.15
$C_{15:0}$	1.10	0.82	0.74	1.06	0.84	0.96	–	–	0.98	0.80	1.38
$C_{16:0}$	32.58	34.75	28.7	19.26	19.25	19.09	25.44	31.68	27.56	33.7	33.85
$C_{16:1\ c\text{-}9}$	1.83	2.18	1.22	1.20	0.90	0.84	1.33	3.08	1.40	1.55	2.44
$C_{18:0}$	10.74	10.50	10.10	15.04	14.76	14.99	12.79	9.40	8.11	8.25	8.18
$C_{18:1\ c\text{-}9}$	20.23	24.60	20.7	26.95	25.99	22.92	21.88	23.73	15.08	17.8	19.16
$C_{18:1\ t\text{-}11}$	–	–	1.38	8.42	12.59	10.18	3.95	4.96	2.34	1.14	0.66
$C_{18:2}$	2.70	2.76	3.14	2.32	2.45	3.30	5.27	2.36	2.20	2.99	2.26
$C_{18:3}$	1.28	1.13	0.49	0.20	0.19	0.31	1.00	0.43	0.85	0.50	0.43
$C_{18:2\ c\text{-}9,\ t\text{-}11}$	–	–	0.72	–	–	–	1.03	0.81	0.88	0.64	0.38
$C_{18:2\ t\text{-}10,\ c\text{-}12}$	–	–	–	–	–	–	–	–	0.04	–	–
$C_{20:5\ n\text{-}3}$	–	–	–	–	–	–	–	–	0.24	–	–
$C_{22:5\ n\text{-}3}$	–	–	–	–	–	–	–	–	0.28	–	–
$C_{22:6\ n\text{-}3}$	–	–	–	–	–	–	–	–	0.26	–	–

[a]Schauff and Clark (1992).
[b]Rico *et al.* (2014) Journal of Dairy Science 97:5637–5644.
[c]Chouinard *et al.* (1998).
[d]Morales *et al.* (2000).
[e]AbuGhazaleh *et al.* (2002).
[f]Weiss *et al.* (2011) *Journal of Dairy Science*. 94:931–939.

Calcium salts of palm oil fatty acid distillate are used widely as an energy supplement. With a content of 45–50% $C_{16:0}$, these calcium salts increase palmitic acid in milk fat compared to oilseeds (Table 2.2). Dry fat prills highly enriched in palmitic acid (>85%) are a by-products of palm oil manufacturing that have increased recently in the market as they increase milk fat around 20 g for every 100 g fed. Whole oilseeds (cottonseed, canola, soybeans) also are used widely as energy supplements in dairy diets. All except cottonseed oil (25% $C_{16:0}$) contain predominantly C_{18} fatty acids. The unsaturated fatty acids in oilseeds are nearly completely biohydrogenated even when fed whole. This results in increased absorption of stearic acid, which is converted to oleic by the SCD enzyme resulting in an increase in the proportion of $C_{18:1}$ in milk fat and a reduction in the proportions of C_6 to C_{16} fatty acids, but especially $C_{14:0}$ and $C_{16:0}$. Feeding whole soybeans can modestly increase the proportions of $C_{18:2}$ and $C_{18:3}$ in milk fat; apparently, some portions of the soybeans pass from the rumen undegraded

(Morales *et al.* 2000) and thereby increase the susceptibility of the milk fat to oxidative rancidity (Timmons *et al.* 2001). Tallow has long been a staple energy source for dairy diets; however, Onetti *et al.* (2002) showed that supplemental tallow may reduce milk fat percent modestly, which was attributed to increased concentrations of *trans*-10 C18:1 in the milk fat (see milk fat depression). Increasing the proportion of corn silage in dietary forage (reducing alfalfa silage) also increased the content of *trans*-10 C18:1 and lowered milk fat percent (Onetti *et al.* 2002) (Table 2.3).

Increasing public concern related to the composition of dietary fats has increased interest in the n-3 fatty acid content of milk fat (Palmquist 2009). The content of linolenic acid in milk fat from cows grazing pasture may be more than double that from cows fed preserved feeds (Dhiman *et al.* 1999; Moallem 2018) owing to the high content and rumen escape of $C_{18:3}$ n-3 from forages. Others have investigated the transfer of the long-chain (C_{20} and C_{22}) n-3 fatty acids of fish oil to milk fat. Adding 2% menhaden fish oil to the diet of lactating cows increased the content of $C_{20:5}$ n-3 and $C_{22:6}$ n-3 (EPA and DHA, respectively) from 0.05 and 0.04% in the control to 0.24 and 0.26% in milk fat of supplemented cows (AbuGhazaleh *et al.* 2002). Feeding fish oil fatty acids in combination with vegetable oils synergistically increases the CLA content of milk fat (see Chap. 3).

Changes in the proportions of fatty acids in milk fat by supplementation of various oils and oilseeds are summarized in Figure 2.8 (Grummer 1991). Hermansen (1995) and Moate *et al.* (2008) developed regression equations to predict the composition of milk fat based on diet fatty acid profile. The models generally effectively predicted milk fatty acid profile but will require more refined modelling in the future.

2.10.5 Low Milk Fat Syndrome

Low milk fat syndrome, now more commonly called diet-induced milk fat depression, is up to a 50% reduction in milk fat and has been investigated intensively since the mid-twentieth century. Early investigators pursued the link between changes in the ruminal acetate:propionate ratio and the percentage of fat in milk (van Soest 1963). A glucogenic response, whereby increased production of ruminal propionate would increase blood glucose and insulin concentrations, with decreased fatty acid release from adipose tissue, was proposed by McClymont and Vallance (1962) and developed further by van Soest (1963). Whereas intravenous infusion of glucose or glycerol (Vallance and McClymont 1959), or duodenal infusion of glucose (Hurtaud *et al.* 2000), has been shown to cause small reductions in milk fat percentage, these increase the relative proportion of *de novo*-synthesized fatty acids in milk fat (Hurtaud *et al.* 2000), contrary to the consistent decrease in these in classical low milk fat syndrome (Bauman and Griinari 2001). Thus, it became apparent that other aspects of lipid metabolism were involved (Davis and Brown 1970). Over the past two decades, it has become clear that diet-induced milk fat depression is mediated by bioactive *trans* fatty acids including *trans*-10, *cis*-12 $C_{18:2}$ (Bauman and Griinari 2001, 2003; Bauman *et al.* 2011) that are products of changes in ruminal biohydrogenation in the presence of unsaturated fatty acids and an altered microbial populations caused by low ruminal pH or other factors. However *trans*-10, *cis*-12 $C_{18:2}$ explains only a portion of the decrease in milk fat during milk fat depression, and other bioactive fatty acid isomer(s) or other metabolites that are highly correlated with the *trans*-10 biohydrogenation pathway are yet to be identified (Bauman and Griinari 2003).

Depressed milk fat synthesis is associated with decreased enzyme mRNA abundance for acetyl-CoA carboxylase, fatty acid synthase, stearoyl-CoA desaturase, lipoprotein lipase, and glycerol phosphate acyl transferase consistent with a coordinated decrease in the lipid synthesis pathway and also decreases in protein abundance and/or activity of lipid synthesis enzymes. Specifics of regulation during milk fat depression have been previously reviewed (Bauman *et al.* 2011; Harvatine *et al.* 2009). Briefly, milk fat depression is characterized by a specific reduc-

Table 2.3. Least square means for fatty acid composition of milk fat when decreasing proportions of corn silage were fed without (0%) or with (2%) tallow[a]

	0% Tallow			2% Tallow			Significance (P<)[b]	
Corn silage[c]	50%	37.5%	25%	50%	37.5%	25%	F	L
Fatty acid	(g/100 g of fatty acids)							
C_4 to C_{14}	25.3	24.6	24.7	21.4	21.1	21.4	0.001	NS
$C_{16:0}$	28.9	29.1	29.1	28.1	27.9	27.8	0.01	NS
$C_{18:0}$	8.1	8.7	8.7	8.2	9.1	9.2	NS	0.01
$C_{18:1}$	23.2	23.3	23.0	26.3	26.1	25.5	0.001	NS
$C_{18:1}$ isomers								
trans-6/8	0.37	0.39	0.37	0.56	0.54	0.50	0.001	NS
trans-9	0.51	0.56	0.49	0.72	0.65	0.71	0.001	NS
trans-10	1.3	1.0	0.9	2.2	1.8	1.4	0.001	0.001
trans-11	1.1	1.1	1.0	0.94	1.1	1.0	NS	NS
trans-12	0.41	0.41	0.42	0.42	0.50	0.52	0.01	0.04
trans-16	0.06	0.05	0.05	0.03	0.04	0.05	NS	NS
cis-9	17.8	18.2	17.8	20.0	19.8	19.4	0.001	NS
cis-11	0.87	0.92	0.93	0.94	0.94	0.95	0.04	NS
cis-12	0.41	0.44	0.52	0.22	0.33	0.44	0.001	0.001
$C_{18:2\ c9t11}$	0.60	0.57	0.58	0.61	0.63	0.62	0.06	NS
$C_{18:2\ t10c12}$	0.01	0.01	0.02	0.01	0.02	0.02	NS	NS
$C_{18:2}$	4.5	4.5	4.5	4.2	4.2	4.5	NS	NS
$C_{18:3}$	0.33	0.40	0.50	0.28	0.39	0.45	0.01	0.001
Other	8.8	8.2	8.4	10.3	10.1	9.5	0.01	NS

[a]Onetti *et al.* (2002).
[b]F = Main effect of fat; L = linear effect of forage.
[c]Diets: 50% forage and 50% concentrate (DM). Forages were (1) 50% of diet DM as corn silage, (2) 37.5% corn silage and 12.5% alfalfa silage, and (3) 25% corn silage and 25% alfalfa silage.

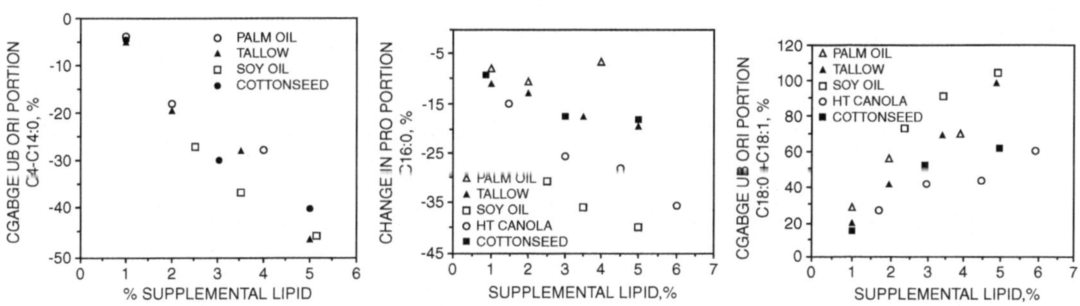

Figure 2.8. Changes in proportions of milk fatty acids relative to control treatments with increasing supplementation of fats or oilseeds A. C_4 to $C_{14:0}$; B. $C_{16:0}$; C. $C_{18:0}$ + $C_{18:1}$. (From Grummer 1991, *Journal of Dairy Science* 74:3244–3257).

tion in mammary lipid synthesis with little phenotype observed outside of the mammary gland. Sterol response element binding protein 1 is a central transcriptional regulator of lipid synthesis that has been consistently observed to be decreased during milk fat depression. Thyroid hormone-responsive spot 14 is a second lipogenic factor that is consistently decreased, but its function is not clear. Robust support for other regulatory systems is lacking, but other signalling systems are likely involved (Table 2.3).

2.10.6 Milk Fat Composition and Quality

The uniqueness of milk fat is not limited to its fatty acid profile. If the 400 fatty acids of milk fat were distributed randomly in the milk fat TAGs, the total theoretical number of glycerides would be 64×10^6 (Jensen 2002); however, distribution is not random, as noted above (see triacylglycerol synthesis). The predominant locations of fatty acids in TAGs are shown in Table 2.4; however, distributions in high- and low-molecular-weight TAGs can differ widely from the mean (Morrison and Hawke 1977b). Also, 36% of TAGs were found to contain $C_{4:0}$ or $C_{6:0}$ and two long-chain fatty acids (Jensen 2002). Jensen (2002) listed 22 TAG structures that were found at >1 mol % in the milk fat; these totaled 42.7 mol % of the total.

The acyl carbon number (CN; total carbon in the acyl chains) of milk fat TAGs typically ranges from 26 to 54. This distribution contributes significantly to the physical characteristics of plasticity and spreadability (functionality or rheological properties) of milk fat. This property is caused by a large proportion of the fat occurring in the molten state at room temperature, supported in a matrix of solid fat that makes up only a small percentage of the total fat (German *et al.* 1997). Changing the average fatty acid chain length in milk fat will change the acyl chain number; an extreme example is shown in Figure 2.9. In this case most of the CN54 acyl chains were linoleic acid from feeding a protected lipid supplement (Morrison and Hawke 1977a). Butter made from high linoleic (>20%) milk fats is slower to churn, more susceptible to oxidation on storage, and breaks down with oiling off at a temperature above 10 °C. Cheeses made from milk with

Table 2.4. Predominant distribution (mole percent) of fatty acids in milk fat triacylglycerols[a]

	TAG position		
Fatty acid	*sn*-1	*sn*-2	*sn*-3
C_4	1.6	0.3	98.1
C_6	3.1	3.9	93.0
C_8	10.3	55.2	34.5
$C_{10:0}$	15.2	56.6	28.2
$C_{12:0}$	23.7	62.9	13.4
$C_{14:0}$	27.3	65.6	7.1
$C_{16:0}$	44.1	45.4	10.5
$C_{18:0}$	54.0	16.2	29.8
$C_{18:1}$	37.3	21.2	41.5

[a]Adapted from Jensen (2002).

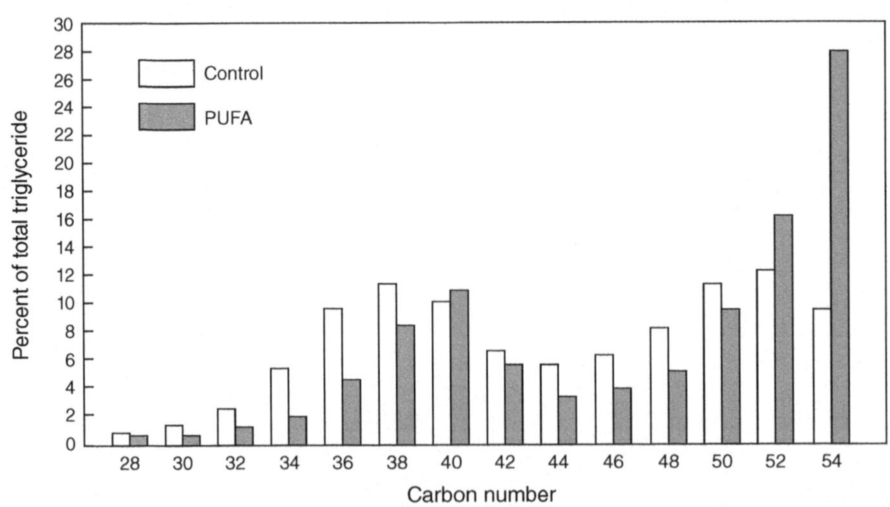

Figure 2.9. Distribution of TAG by acyl carbon number in milk fat from cows fed diets with no added fat (control) or high in polyunsaturated fats. See Table 2.1 for fatty acid profile. (From Palmquist *et al.* 1993, *Journal of Dairy Science* 76:1753–1771).

10–12% $C_{18:2}$ were acceptable, whereas cheeses with a higher linoleic acid content had off-flavor, a soft body, and a mealy texture (McDonald and Scott 1977). Most recently, feeding palmitic acid supplements increased butter melting temperature compared to control and stearic acid because it is highly incorporated into milk fat and is a poor substrate for the desaturase enzymes (Enjalbert *et al.* 2000; Chamberlain *et al.* 2016).

References

AbuGhazaleh, A. A., Schingoethe, D. J., Hippen, A. R., Kalscheur, K. F. & Whitlock, L. A. (2002). Fatty acid profiles of milk and rumen digesta from cows fed fish oil, extruded soybeans or their blend. *Journal of Dairy Science, 85*, 2266–2276.

Ahrné, L., Björck, L., Raznikiewicz, T. & Claesson, O. (1980). Glycerol ether in colostrum and milk from cow, goat, pig and sheep. *Journal of Dairy Science, 63*, 741–745.

ALJohani, A. M., Syed, D. N. & Ntambi, J. M. (2017). Insights into stearoyl-CoA desaturase-1 regulation of systemic metabolism. *Trends in Endocrinology and Metabolism, 28*, 831–842.

Allred, J. B. & Reilly, K. E. (1997). Short-term regulation of acetyl CoA carboxylase in tissues of higher animals. *Progress in Lipid Research, 35*, 371–385.

Annison, E. F., Linzell, J. L., Fazakerley, S. & Nichols, B. W. (1967). The oxidation and utilization of palmitate, stearate, oleate and acetate by the mammary gland of the fed goat in relation to their overall metabolism, and the role of plasma phospholipids and neutral lipids in milk-fat synthesis. *Biochemical Journal, 102*, 637–647.

Ashes, J. R., St. Vincent Welch, P., Gulati, S. K., Scott, T. W., Brown, G. H. & Blakeley, S. (1992). Manipulation of the fatty acid composition of milk by feeding protected canola seeds. *Journal of Dairy Science, 75*, 1090–1096.

Baldwin, R. L. (1995). *Modeling ruminant digestion and metabolism* (pp. 370–387). London: Chapman and Hall.

Baldwin, R. L., Smith, N. E., Taylor, J. & Sharp, M. (1980). Manipulating metabolic parameters to improve growth rate and milk secretion. *Journal of Animal Science, 51*, 1416–1428.

Ballard, F. J., Hanson, R. W. & Kronfeld, D. S. (1969). Gluconeogenesis and lipogenesis in tissue from ruminant and nonruminant animals. *Federation Proceedings, 28*, 218–231.

Banks, W., Clapperton, J. L. & Ferrie, M. E. (1976). Effect of feeding fat to dairy cows receiving a fat-deficient basal diet. II. Fatty acid composition of the milk fat. *Journal of Dairy Research, 43*, 219–227.

Barbano, D. M. & Sherbon, J. W. (1980). Polyunsaturated protected lipid: Effect on triglyceride molecular weight distribution. *Journal of Dairy Science, 63*, 731–740.

Barber, M. C. & Travers, M. T. (1998). Elucidation of a promoter activity that directs the expression of acetyl-coa carboxylase alpha with an alternative n-terminus in a tissue-restricted fashion. *Biochemical Journal, 333*, 17–25.

Barber, M. C., Clegg, R. A., Travers, M. T. & Vernon, R. G. (1997). Review. Lipid metabolism in the lactating mammary gland. *Biochimica et Biophysica Acta, 1347*, 101–126.

Barber, M. C., Pooley, L. & Travers, M. T. (2001). Developmental regulation of alternatively spliced acetyl-coa carboxylase-alphamrnas encoding isozymes with or without an eight amino acid domain upstream of the ser-1200 phosphorylation motif in the mammary gland. *Journal of Molecular Endocrinology, 27*, 349–356.

Barber, M. C., Vallance, A. J., Kennedy, H. T. & Travers, M. T. (2003). Induction of transcripts derived from promoter iii of the acetyl-coa carboxylase-alpha gene in mammary gland is associated with recruitment of srebp-1 to a region of the proximal promoter defined by a dnase i hypersensitive site. *Biochemical Journal, 375*, 489–501.

Barber, M., Price, N. & Travers, M. (2005). Structure and regulation of acetyl-CoA carboxylase genes of metazoa. *Biochimica et Biophysica Acta, 1733*, 1–28.

Bauman, D. E. & Currie, W. B. (1980). Partitioning of nutrients during pregnancy and lactation: A review of mechanisms involving homeostasis and homeorhesis. *Journal of Dairy Science, 63*, 1514–1529.

Bauman, D. E. & Davis, C. L. (1974). Biosynthesis of milk fat. In B. L. Larson & V. R. Smith (Eds.), *Lactation* (Vol. II, pp. 31–75). NewYork: Academic Press.

Bauman, D. E. & Griinari, J. M. (2001). Regulation and nutritional manipulation of milk fat: Low-fat milk syndrome. *Livestock Production Science, 70*, 15–29.

Bauman, D. E. & Griinari, J. M. (2003). Nutritional regulation of milk fat synthesis. *Annual Review of Nutrition, 23*, 203–227.

Bauman, D. E., Brown, R. E. & Davis, C. L. (1970). Pathways of fatty acid synthesis and reducing equivalent generation in mammary gland of rat, sow, and cow. *Archives of Biochemistry and Biophysics, 140*, 237–244.

Bauman, D. E., Harvatine, K. J. & Lock, A. L. (2011). Nutrigenomics, rumen-derived bioactive fatty acids, and the regulation of milk fat synthesis. *Annual Review of Nutrition, 31*, 299–319.

Beals, E., Kamita, S. G., Sacchi, R., Demmer, E., Rivera, N., Rogers-Soeder, T. S., Gertz, E. R., Van Loan, M. D., German, J. B., Hammock, B. D., Smilowitz, J. T. & Zivkovic, A. M. (2019). Addition of milk fat globule membrane-enriched supplement to a high-fat meal attenuates insulin secretion and induction of soluble epoxide hydrolase gene expression in the postprandial state in overweight and obese subjects. *Journal of Nutritional Science, 8*, e16.

Beaulieu, A. D. & Palmquist, D. L. (1995). Differential effects of high fat diets on fatty acid composition in milk of Jersey and Holstein cows. *Journal of Dairy Science, 78*, 1336–1344.

Bernert, J. T., Jr. & Sprecher, H. (1979). Factors regulating the elongation of palmitic and stearic acid by rat liver microsomes. *Biochimica et Biophysica Acta, 574*, 18–24.

Bickerstaffe, R. & Annison, E. F. (1970). The desaturase activity of goat and sow mammary tissue. *Comparative Biochemistry and Physiology, 35*, 653–665.

Bickerstaffe, R. & Annison, E. F. (1971). Triglyceride synthesis in goat and sow mammary tissue. *International Journal of Biochemistry, 2*, 153–162.

Bickerstaffe, R. & Annison, E. F. (1974). The metabolism of glucose, acetate, lipids and amino acids in lactating dairy cows. *Journal of Agriculture Science (Cambridge), 82*, 71–85.

Bishop, C., Davies, T., Glascock, R. F. & Welch, V. A. (1969). A further study of bovine serum lipoproteins and an estimation of their contribution to milk fat. *Biochemical Journal, 113*, 629–633.

Bitman, J. & Wood, D. L. (1990). Changes in milk fat phospholipids during lactation. *Journal of Dairy Science, 73*, 1208–1216.

Borthwick, A. C., Edgell, N. J. & Denton, R. M. (1987). Use of rapid gel-permeation chromatography to explore the inter-relationships between polymerization, phosphorylation and activity of acetyl-CoA carboxylase. Effects of insulin and phosphorylation by cyclic AMP-dependent protein kinase. *Biochemical Journal, 241*, 773–782.

Cant, J. P., DePeters, E. J. & Baldwin, R. L. (1993). Mammary uptake of energy metabolites in dairy cows fed fat and its relationship to milk protein depression. *Journal of Dairy Science, 76*, 2254–2265.

Carlson, C. A. & Kim, K. H. (1974). Regulation of hepatic acetyl coenzyme A carboxylase by phosphorylation and dephosphorylation. *Archives of Biochemistry and Biophysics, 164*, 478–489.

Castberg, H. B., Egelrud, T., Solberg, P. & Olivecrona, T. (1975). Lipases in bovine milk and the relationship between the lipoprotein lipase and tributyrate hydrolyzing activities in cream and skim-milk. *Journal of Dairy Research, 42*, 255–266.

Cavaletto, M., Giuffrida, M. G. & Conti, A. (2008). Milk fat globule membrane components – A proteomic approach. *Advances in Experimental Medicine and Biology, 606*, 129–141.

Chaiyabutr, N., Faulkner, A. & Peaker, M. (1980). The utilization of glucose for the synthesis of milk components in the fed and starved lactating goat *in vivo*. *Biochemical Journal, 186*, 301–308.

Chamberlain, M. B., Pareas, J. W., Juchem, S. O., DePeters, E. J., Getachew, G., Taylor, S. J., et al. (2016). Feeding lactating Holstein cows a lipid source high in palmitic acid changes the fatty acid composition and thermal properties of lipids in milk and butter. *Professional Animal Scientists, 32*, 672–680.

Chang, J. H. P., Lunt, D. K. & Smith, S. B. (1992). Fatty acid composition and fatty acid elongase and stearoyl-CoA desaturase activities in tissues of steers fed high oleate sunflower seed. *Journal of Nutrition, 122*, 2074–2080.

Chouinard, P. W., Girard, V. & Brisson, G. J. (1998). Fatty acid profile and physical properties of milk fat from cows fed calcium salts of fatty acids with varying unsaturation. *Journal of Dairy Science, 81*, 471–481.

Christie, W. W. (1985). Structure of the triacyl-*sn*-glycerols in the plasma and milk of the rat and rabbit. *Journal of Dairy Research, 52*, 219–222.

Clarenburg, R. & Chaikoff, I. L. (1966). Origin of milk cholesterol in the rat: Dietary versus endogenous sources. *Journal of Lipid Research, 7*, 27–37.

Clarke, B. A. & Clarke, S. D. (1982). Polymer-protomer transition of acetyl-CoA carboxylase as a regulator of lipogenesis in rat liver. *Archives of Biochemistry and Biophysics, 218*, 92–100.

Coleman, R. A. & Mashek, D. G. (2011). Mammalian triacylglycerol metabolism: Synthesis, lipolysis, and signaling. *Chemical Reviews, 111*, 6359–6386.

Cooper, S. M. & Grigor, M. R. (1980). Fatty acid specificities of microsomal acyltransferases esterifying positions-1 and -2 of acylglycerols in mammary glands from lactating rats. *Biochemical Journal, 187*, 289–295.

Dahl, G. E., Buchanan, B. A. & Tucker, H. A. (2000). Photoperiodic effects on dairy cattle: A review. *Journal of Dairy Science, 83*, 885–893.

Davis, C. L. & Brown, R. E. (1970). Low-fat milk syndrome. In A. T. Phillipson (Ed.), *Physiology of digestion and metabolism in the ruminant* (pp. 545–565). Newcastle-upon-Tyne: Oriel Press.

DePeters, E. J., Medrano, J. F. & Reed, B. A. (1995). Fatty acid composition of milk fat from three breeds of dairy cattle. *Canadian Journal of Animal Science, 75*, 267–269.

Dhiman, T. R., Anand, G. R., Satter, L. D. & Pariza, M. W. (1999). Conjugated linoleic acid content of milk from cows fed different diets. *Journal of Dairy Science, 82*, 2146–2156.

Dils, R. R. (1986). Comparative aspects of milk fat synthesis. *Journal of Dairy Science, 69*, 904–910.

Dodds, P. F., Guzman, M. G. F., Chalberg, S. C., Anderson, G. J. & Kumar, S. (1981). Acetoacetyl-CoA reductase activity of lactating bovine mammary fatty acid synthase. *Journal of Biological Chemistry, 256*, 6282–6290.

Drackley, J. K., Overton, T. R., Ortiz-Gonzalez, G., Beaulieu, A. D., Barbano, D. M., Lynch, J. M. & Perkins, E. G. (2007). Responses to increasing amounts of high-oleic sunflower fatty acids infused into the abomasum of lactating dairy cows. *Journal of Dairy Science, 90*, 5165–5175.

Duchemin, S. I., Visker, M. H., Van Arendonk, J. A. & Bovenhuis, H. (2014). A quantitative trait locus on bos taurus autosome 17 explains a large proportion of the genetic variation in de novo synthesized milk fatty acids. *Journal of Dairy Science, 97*, 7276–7285.

Emanuelson, M., Murphy, M. & Lindberg, J.-E. (1991). Effects of heat-treated and untreated full-fat rapeseed and tallow on rumen metabolism, digestibility, milk composition and milk yield in lactating cows. *Animal Feed Science and Technology, 34*, 291–309.

Enjalbert, F., Nicot, M.-C., Bayourthe, C. & Moncoulon, R. (1998). Duodenal infusions of palmitic, stearic or oleic acids differently affect mammary gland metabolism of fatty acids in lactating dairy cows. *Journal of Nutrition, 128*, 1525–1532.

Enjalbert, F., Nicot, M. C., Bayourthe, C. & Moncoulon, R. (2000). Effects of duodenal infusions of palmitic, stearic, or oleic acids on milk composition and physical properties of butter. *Journal of Dairy Science, 83*, 1428–1433.

Faulkner, A. & Pollock, H. T. (1989). Changes in the concentration of metabolites in milk from cows fed on diets supplemented with soyabean oil or fatty acids. *Journal of Dairy Research, 56*, 179–183.

Fauteux, M. C., Gervais, R., Rico, D. E., Lebeuf, Y. & Chouinard, P. Y. (2016). Production, composition, and oxidative stability of milk highly enriched in polyunsaturated fatty acids from dairy cows fed alfalfa protein concentrate or supplemental vitamin E. *Journal of Dairy Science, 99*, 4411–4426.

Forsberg, N. E., Baldwin, R. L. & Smith, N. E. (1985). Roles of lactate and its interactions with acetate in maintenance and biosynthesis in bovine mammary tissue. *Journal of Dairy Science, 68*, 2550–2556.

Garton, G. A. (1963). The composition and biosynthesis of milk lipids. *Journal of Lipid Research, 4*, 237–254.

German, J. B., Morand, L., Dillard, C. J. & Xu, R. (1997). Milk fat composition: Targets for alteration of function and nutrition. In R. A. S. Welch, D. J. W. Burns, S. R. Davis, A. I. Popay & C. G. Prosser (Eds.), *Milk Composition, Production and Biotechnology* (pp. 35–72). Wallingford: CAB International.

Glascock, R. F., Duncombe, W. G. & Reinius, L. R. (1956). Studies on the origin of milk fat. 2. The secretion of dietary long-chain fatty acids in milk fat by ruminants. *Biochemical Journal, 62*, 535–541.

Glascock, R. F., Welch, V. A., Bishop, C., Davies, T., Wright, E. W. & Noble, R. C. (1966). An investigation of serum lipoproteins and of their contribution to milk fat in the dairy cow. *Biochemical Journal, 98*, 149–156.

Glascock, R. F., Smith, R. W. & Walsh, A. (1983). Partition of circulating triglycerides between formation of milk fat and other metabolic pathways in sheep. *Journal of Agricultural Sciences (Cambridge), 101*, 33–38.

Glasser, F., Schmidely, P., Sauvant, D. & Doreau, M. (2008). Digestion of fatty acids in ruminants: A meta-analysis of flows and variation factors: 2. C18 fatty acids. *Animal, 2*, 691–704.

Gregolin, C., Ryder, E., Warner, R. C., Kleinschmidt, A. K. & Lane, M. (1966). Liver acetyl CoA carboxylase: The dissociation-reassociation process and its relation to catalytic activity. *Proceedings of the National Academy of Sciences, 56*, 1751–1758.

Grevengoed, T. J., Klett, E. L. & Coleman, R. A. (2014). Acyl-CoA metabolism and partitioning. *Annual Review of Nutrition, 34*, 1–30.

Griinari, J. M., McGuire, M. A., Dwyer, D. A., Bauman, D. E. & Palmquist, D. L. (1997). Role of insulin in the regulation of milk fat synthesis in dairy cows. *Journal of Dairy Science, 80*, 176–1084.

Griinari, J. M., Corl, B. A., Lacy, S. H., Chouinard, P. Y., Nurmela, K. V. & Bauman, D. E. (2000). Conjugated linoleic acid is synthesized endogenously in lactating dairy cows by delta(9)-desaturase. *Journal of Nutrition, 130*, 2285–2291.

Grisart, B., Coppieters, W., Farnir, F., Karim, L., Ford, C., Berzi, P., Cambisano, N., Mni, M., Reid, S., Simon, P., Spelman, R., Georges, M. & Snell, R. (2002). Positional candidate cloning of a qtl in dairy cattle: Identification of a missense mutation in the bovine dgat1 gene with major effect on milk yield and composition. *Genome Research, 12*, 222–231.

Grisart, B., Farnir, F., Karim, L., Cambisano, N., Kim, J. J., Kvasz, A., Mni, M., Simon, P., Frere, J. M., Coppieters, W. & Georges M. (2004). Genetic and functional confirmation of the causality of the dgat1 k232a quantitative trait nucleotide in affecting milk yield and composition. *Proceedings of the National Academy of Sciences of the United States of America 101*(8):2398–2403.

Grummer, R. R. (1991). Effect of feed on the composition of milk fat. *Journal of Dairy Science, 74*, 3244–3257.

Grunnet, I. & Knudsen, J. (1981). Direct transfer of fatty acids synthesized 'de novo' from fatty acid synthetase into triacylglycerols without activation. *Biochemical and Biophysical Research Communications, 100*, 629–636.

Ha, J. K. & Lindsay, R. C. (1990). Method for the quantitative analysis of volatile free and total branched-chain fatty acids in cheese and milk fat. *Journal of Dairy Science, 73*, 1988–1999.

Ha, J., Daniel, S., Broyles, S. S. & Kim, K. H. (1994). Critical phosphorylation sites for acetyl-CoA carboxylase activity. *Journal of Biological Chemistry, 269*, 22162–22168.

Hagemeister, H., Precht, D., Franzen, M. & Barth, C. A. (1991). α-Linolenic acid transfer into milk fat and its elongation by cows. *Fett Wissenschaft und Technologie, 93*, 387–391.

Hajri, T. & Abumrad, N. A. (2002). Fatty acid transport across membranes: Relevance to nutrition and metabolic pathology. *Annual Review of Nutrition, 22*, 383–415.

Hamosh, M., Clary, T. R., Chernick, W. W. & Scow, R. O. (1970). Lipoprotein lipase activity of adipose and mammary tissue and plasma triglyceride in pregnant and lactating rats. *Biochimica et Biophysica Acta, 210*, 473–482.

Hansen, J. K. & Knudsen, J. (1980). Transacylation as a chain-termination mechanism in fatty acid synthesis by mammalian fatty acid synthetase. Synthesis of butyrate and hexanoate by lactating cow mammary gland fatty acid synthetase. *Biochemical Journal, 186*, 287–294.

Hansen, H. O. & Knudsen, J. (1987). Effect of exogenous long-chain fatty acids on individual fatty acid synthesis by dispersed ruminant mammary gland cells. *Journal of Dairy Science, 70*, 1350–1354.

Hansen, H. O., Grunnet, I. & Knudsen, J. (1984a). Triacylglycerol synthesis in goat mammary gland. The effect of ATP, Mg^{2+} and glycerol 3-phosphate on the esterification of fatty acids synthesized *de novo*. *Biochemical Journal, 220*, 513–519.

Hansen, H. O., Grunnet, I. & Knudsen, J. (1984b). Triacylglycerol synthesis in goat mammary gland. Factors influencing the esterification of fatty acids synthesized *de novo*. *Biochemical Journal, 220*, 521–527.

Hansen, H. O., Jensen, S. S. & Knudsen, J. (1986). Absence of monoacylglycerol pathway for triacylglycerol synthesis in goat mammary gland. *Biochemical Journal, 238*, 173–176.

Harvatine, K. J., Boisclair, Y. R. & Bauman, D. E. (2009). Recent advances in the regulation of milk fat synthesis. *Animal, 3*, 40–54.

Hayes, B. J., Pryce, J., Chamberlain, A. J., Bowman, P. J. & Goddard, M. E. (2010). Genetic architecture of complex traits and accuracy of genomic prediction: Coat colour, milk-fat percentage, and type in holstein cattle as contrasting model traits. *PLoS Genetics, 6*, e1001139.

Hermansen, J. E. (1995). Prediction of milk fatty acid profile in dairy cows fed dietary fat differing in fatty acid composition. *Journal of Dairy Science, 78*, 872–879.

Hilditch, T. P. (1947). *The chemical constitution of natural fats* (2nd ed.). London: Wiley.

Hillgartner, F. B., Salati, L. M. & Goodridge, A. G. (1995). Physiological and molecular mechanisms involved in nutritional regulation of fatty acid synthesis. *Physiological Reviews, 75*, 47–76.

Hoekstra, D., Maier, O., van der Wouden, J. M., Slimane, T. A. & van Ijzendoorn, C. D. (2003). Membrane dynamics and cell polarity: The role of sphingolipids. *Journal of Lipid Research, 44*, 869–877.

Holland, R. & Hardie, D. G. (1985). Both insulin and epidermal growth factor stimulate fatty acid synthesis and increase phosphorylation of acetyl-CoA carboxylase and ATP-citrate lyase inisolated hepatocytes. *FEBS Letters, 181*, 308–312.

Hurtaud, C., Lemosquet, S. & Rulquin, H. (2000). Effect of graded duodenal infusions of glucose on yield and composition of milk from dairy cows. 2. Diets base on grass silage. *Journal of Dairy Science, 83*, 2952–2962.

Infante, J. P. & Kinsella, J. E. (1976). Phospholipid synthesis in mammary tissue. Choline and ethanolamine kinases: Kinetic evidence for two discrete active sites. *Lipids, 11*, 727–735.

Jenkins, T. C. (1998). Fatty acid composition of milk from Holstein cows fed oleamide or canola oil. *Journal of Dairy Science, 81*, 794–800.

Jenkins, T. C. & Harvatine, K. J. (2014). Lipid feeding and milk fat depression. *Veterinary Clinics of North America. Food Animal Practice, 30*, 623–642.

Jenkins, T. C., Wallace, R. J., Moate, P. J. & Mosley, E. E. (2008). Board-invited review: Recent advances in biohydrogenation of unsaturated fatty acids within the rumen microbial ecosystem. *Journal of Animal Science, 86*, 397–412.

Jensen, R. G. (2002). Invited review: The composition of bovine milk lipids: January 1995 to December 2000. *Journal of Dairy Science, 85*, 295–350.

Jensen, R. G. & Newburg, D. S. (1995). Milk lipids. B. Bovine milk lipids. In R. G. Jensen (Ed.), *Handbook of Milk Composition* (pp. 543–575). San Diego: Academic Press.

Jordan, W. H. & Jenter, C. G. (1897). *The source of milk fat* (Bulletin 132) (pp. 455–488). Geneva: New York Agricultural Experiment Station.

Joshi, A. K., Witkowski, A. & Smith, S. (1997). Mapping of functional interactions between domains of the animal fatty acid synthase by mutant complementation *in vitro*. *Biochemistry, 36*, 2316–2322.

Kadegowda, A., Piperova, L., Delmonte, P. & Erdman, R. (2008). Abomasal infusion of butterfat increases milk fat in lactating dairy cows. *Journal of Dairy Science, 91*, 2370–2379.

Karijord, Ø., Standal, N. & Syrstad, O. (1982). Sources of variation in composition of milk fat. *Zeitschrift für Tierzüchtung und Züchtungsbiologie, 99*, 81–93.

Katz, J. & Wals, P. A. (1972). Pentose cycle and reducing equivalents in rat mammary-gland slices. *Biochemical Journal, 128*, 879–899.

Kaupe, B., Winter, A., Fries, R. & Erhardt, G. (2004). DGAT1 polymorphism in *Bos indicus* and *Bos taurus* cattle breeds. *Journal of Dairy Research, 71*, 182–187.

Kelsey, J. A., Corl, B. A., Collier, R. J. & Bauman, D. E. (2003). The effect of breed, parity, and stage of lactation on conjugated linoleic acid (CLA) in milk fat from dairy cows. *Journal of Dairy Science, 86*, 2588–2597.

Kim, K.-H. (1997). Regulation of mammalian acetyl-coenzyme A carboxylase. *Annual Review of Nutrition, 17*, 77–99.

Kinsella, J. E. (1968). The incorporation of [$^{14}C_3$] glycerol into lipids by dispersed bovine mammary cells. *Biochimica et Biophysica Acta, 164*, 540–549.

Kinsella, J. E. (1972). Stearyl Co-A as a precursor of oleic acid and glycerolipids in mammary microsomes from lactating bovine: Possible regulatory step in milk triglyceride synthesis. *Lipids, 7*, 349–355.

Kinsella, J. E. (1973). Preferential labeling of phosphatidylcholine during phospholipids synthesis by bovine mammary tissue. *Lipids, 8*, 393–400.

Kinsella, J. E. & Gross, M. (1973). Palmitic acid and initiation of mammary glyceride synthesis *via* phosphatidic acid. *Biochimica et Biophysica Acta, 316*, 109–113.

Kinsella, J. E. & Infante, J. P. (1974). Acyl-CoA acyl-sn glycerol-3 phosphorylcholine acyl transferase of bovine mammary tissue. *Lipids, 9*, 748–751.

Knudsen, J. (1979). Medium-chain fatty acid synthesis in lactating-rabbit mammary gland. *Biochemical Journal, 181*, 267–274.

Knudsen, J. & Grunnet, I. (1982). Transacylation as a chain-termination mechanism in fatty acid synthesis by mammalian fatty acid synthetase. Synthesis of medium-chain-length (C_8-C_{12}) acyl-CoA esters by goat mammary gland fatty acid synthetase. *Biochemical Journal, 202*, 139–143.

Knudsen, J., Neergaard, T. B. F., Gaigg, B., Jensen, M. V. & Hansen, J. K. (2000). Role of acyl-CoA binding protein in acyl-CoA metabolism and acyl-CoA-mediated cell signaling. *Journal of Nutrition, 130*, 294S–298S.

Kong, I., Lopez-Casillas, F. & Kim, K. (1990). Acetyl-CoA carboxylase mrna species with or without inhibitory coding sequence for ser-1200 phosphorylation. *Journal of Biological Chemistry, 265*, 13695–13701.

Korn, E. D. (1962). The lipoprotein lipase of cow's milk. *Journal of Lipid Research, 3*, 246–249.

Laakso, P., Manninen, P., Mäkinen, J. & Kallio, H. (1996). Postparturition changes in the triacylglycerols of cow colostrums. *Lipids, 31*, 937–943.

Laarveld, B., Chaplin, R. K. & Brockman, R. P. (1985). Effects of insulin on the metabolism of acetate, beta-hydroxybutyrate and triglycerides by the bovine mammary gland. *Comparative Biochemistry and Physiology. B, 82*, 265–267.

LaCount, D. W., Drackley, J. K., Laesch, S. O. & Clark, J. H. (1994). Secretion of oleic acid in milk fat in response to abomasal infusions of canola or high oleic sunflower fatty acids. *Journal of Dairy Science, 77*, 1372–1385.

Lee, K. H. & Kim, K.-H. (1979). Stimulation by epinephrine of in vivo phosphorylation and inactivation of acetyl coenzyme A carboxylase of rat epididymal adipose tissue. *Journal of Biological Chemistry, 254*, 1450–1453.

Liesman, J. S., Emery, R. S., Akers, R. M. & Tucker, H. A. (1988). Mammary lipoprotein lipase in plasma of cows after parturition or prolactin infusion. *Lipids, 23*, 504–507.

Lin, C. Y. & Kumar, S. (1972). Pathway for the synthesis of fatty acids in mammalian tissues. *Journal of Biological Chemistry, 247*, 604–606.

Lin, C. Y., Smith, S. & Abraham, S. (1976). Acyl specificity in triglyceride synthesis by lactating rat mammary gland. *Journal of Lipid Research, 17*, 647–656.

Loften, J. R., Linn, J. G., Drackley, J. K., Jenkins, T. C., Soderholm, C. G. & Kertz, A. F. (2014). Invited review: Palmitic and stearic acid metabolism in lactating dairy cows. *Journal of Dairy Science, 97*, 4661–4674.

Lordan, R., Tsoupras, A., Mitra, B. & Zabetakis, I. (2018). Dairy fats and cardiovascular disease: Do we really need to be concerned? *Food, 7*, E29.

Lynch, J. M., Barbano, D. M., Bauman, D. E., Hartnell, G. F. & Nemeth, M. A. (1992). Effect of a prolonged-release formulation of N-methionyl bovine somatotropin (Sometribove) on milk fat. *Journal of Dairy Science, 75*, 1794–1809.

Mani, O., Korner, M., Sorensen, M. T., Sejrsen, K., Wotzkow, C., Ontsouka, C. E., Friis, R. R., Bruckmaier, R. M. & Albrecht, C. (2010). Expression, localization, and functional model of cholesterol transporters in lactating and nonlactating mammary tissues of murine, bovine, and human origin. *American Journal of Physiology. Regulatory, Integrative and Comparative Physiology, 299*, R642–R654.

Mao, J. & Seyfert, H.-M. (2002). Promoter II of the bovine acetyl-coenzyme A carboxylase – alpha-encoding gene is widely expressed and strongly active in different cells. *Biochimica et Biophysica Acta, 1576*, 324–329.

Mao, J., Marcos, S., Davis, S. K., Burzlaff, J. & Seyfert, H.-M. (2001). Genomic distribution of three promoters of the bovine gene encoding aceyl-CoA carboxylase alpha and evidence that the nutritionally regulated promoter I contains a repressive element different from that in the rat. *Biochemical Journal, 358*, 127–135.

Marshall, M. O. & Knudsen, J. (1977). The specificity of 1-acyl-*sn*-glycerol 3-phosphate acyltransferase in microsomal fractions from lactating cow mammary gland towards short, medium and long chain acyl-CoA esters. *Biochimica et Biophysica Acta, 489*, 236–241.

Marshall, M. O. & Knudsen, J. (1980). Factors influencing the in vitro activity of diacylglycerol acyltransferase from bovine mammary gland and liver towards butyryl-CoA and palmitoyl-CoA. *Biochimica et Biophysica Acta, 617*, 393–397.

Massart-Leen, A. M., DePooter, H., Decloedt, M. & Schamp, N. (1981). Composition and variability of the branched-chain fatty acid fraction in the milk of goats and cows. *Lipids, 16*, 286–292.

Mather, I. H. & Keenan, T. W. (1998). Origin and secretion of milk lipids. *Journal of Mammary Gland Biology and Neoplasia, 3*, 259–273.

Maynard, L. A. & McCay, C. M. (1929). The influence of a low-fat diet upon fat metabolism during lactation. *Journal of Nutrition, 2*, 67–81.

McBride, O. W. & Korn, E. D. (1964). Presence of glycerokinase in guinea pig mammary gland and the incorporation of glycerol into glycerides. *Journal of Lipid Research, 5*, 442–417.

McClymont, G. L. & Vallance, S. (1962). Depression of blood glycerides and milk fat synthesis by glucose infusion. *Proceedings of the Nutrition Society, 21*, xli.

McDonald, I. W. & Scott, T. W. (1977). Foods of ruminant origin with elevated content of polyunsaturated fatty acids. *World Review of Nutrition and Dietetics, 26*, 144–207.

McFadden, J. W. & Rico, J. E. (2019). Invited review: Sphingolipid biology in the dairy cow: The emerging role of ceramide. *Journal of Dairy Science, **102***, 7919–7919.

McManaman, J. L. (2012). Milk lipid secretion: Recent biomolecular aspects. *Biomolecular Concepts, 3*, 581–591.

McManaman, J. L., Palmer, C. A., Wright, R. M. & Neville, M. C. (2002). Functional regulation of xanthine oxidoreductase expression and localization in the mouse mammary gland: Evidence of a role in lipid secretion. *Journal of Physiology, 545*, 567–579.

Mellenberger, R. W. & Bauman, D. E. (1974). Fatty acid synthesis in rabbit mammary tissue during pregnancy and lactation. *Biochemical Journal, 138*, 373–379.

Mellenberger, R. W., Bauman, D. E. & Nelson, D. R. (1973). Fatty acid and lactose synthesis in cow mammary tissue. *Biochemical Journal, 136*, 741–748.

Mikkelsen, J. & Knudsen, J. (1987). Acyl-CoA-binding protein from cow. Binding characteristics and cellular and tissue distribution. *Biochemical Journal, 248*, 709–714.

Miller, P. S., Reis, B. L., Calvert, C. C., DePeters, E. J. & Baldwin, R. L. (1991). Patterns of nutrient uptake by the mammary glands of lactating dairy cows. *Journal of Dairy Science, 74*, 3791–3799.

Moallem, U. (2018). Invited review: Roles of dietary n-3 fatty acids in performance, milk fat composition, and reproductive and immune systems in dairy cattle. *Journal of Dairy Science, 101*, 8641–8661.

Moate, P. J., Chalupa, W., Boston, R. C. & Lean, I. J. (2008). Milk fatty acids ii: Prediction of the production of individual fatty acids in bovine milk. *Journal of Dairy Science, 91*, 1175–1188.

Molenaar, A., Mao, J., Oden, K. & Seyfert, H.-M. (2003). All three promoters of the actetyl-coenzyme A-carboxylase α-encoding gene are expressed in mammary cells of ruminants. *Journal of Histochemistry and Cytochemistry, 51*, 1073–1081.

Moore, J. H. & Christie, W. W. (1981). Lipid metabolism in the mammary gland of ruminants. In W. W. Christie (Ed.), *Lipid metabolism in ruminant animals* (pp. 227–277). Oxford: Pergamon Press.

Morales, M. S., Palmquist, D. L. & Weiss, W. P. (2000). Effects of fat source and copper on unsaturation of blood and milk triacylglycerol fatty acids in Holstein and Jersey cows. *Journal of Dairy Science, 83*, 2105–2111.

Morrison, I. M. & Hawke, J. C. (1977a). Triglyceride composition of bovine milk fat with elevated levels of linoleic acid. *Lipids, 12*, 994–1004.

Morrison, I. M. & Hawke, J. C. (1977b). Positional distribution of fatty acids in the triglycerides of bovine milk fat with elevated levels of linoleic acid. *Lipids, 12*, 1005–1011.

Murray, R. K., Granner, D. K., Mayes, P. A. & Rodwell, V. W. (1988). *Harper's biochemistry* (21st ed.). Norwalk/San Mateo: Appleton & Lange.

Nandedkar, A. K. N., Schirmer, E. W., Pynadath, T. I. & Kumar, S. (1969). Biosynthesis of fatty acid in mammary tissue. I. Purification and properties of fatty acid synthetase from lactating-goat mammary tissue. *Archives of Biochemistry and Biophysics, 134*, 554–562.

Nielsen, M. O. & Jakobsen, K. (1994). Changes in mammary uptake of free fatty acids, triglyceride, cholesterol and phospholipid in relation to milk synthesis during lactation in goats. *Comparative Biochemistry and Physiology, 109A*, 857–867.

Nielsen, M. O., Madsen, T. G. & Hedeboe, A. M. (2001). Regulation of mammary glucose uptake in goats: Role of mammary gland supply, insulin, IGF-1 and synthetic capacity. *Journal of Dairy Research, 68*, 337–349.

Ntambi, J. M. & Miyazaki, M. (2004). Regulation of stearoyl-CoA desaturases and role in metabolism. *Progress in Lipid Research, 43*, 91–104.

Ogg, S. L., Weldon, A. K., Dobbie, L., Smith, A. J. & Mather, I. H. (2004). Expression of butyrophilin (btn1a1) in lactating mammary gland is essential for the regulated secretion of milk-lipid droplets. *Proceedings of the National Academy of Sciences of the United States of America, 101*, 10084–10089.

Onetti, S. G., Shaver, R. D., McGuire, M. A., Palmquist, D. L. & Grummer, R. R. (2002). Effect of supplemental tallow on performance of dairy cows fed diets with different corn silage: Alfalfa silage ratios. *Journal of Dairy Science, 85*, 632–641.

Oscar, T. P., Baumrucker, C. R. & Etherton, T. D. (1986). Insulin binding to bovine mammary membranes: Comparison of microsomes versus smooth membranes. *Journal of Animal Science, 62*, 179–186.

Palmquist, D. L. (1991). Influence of source and amount of dietary fat on digestibility in lactating cows. *Journal of Dairy Science, 74*, 1354–1360.

Palmquist, D. L. (2009). Omega-3 fatty acids in metabolism, health, and nutrition for modified animal product foods. *Professional Animal Scientists, 25*, 207–249.

Palmquist, D. L. & Conrad, H. R. (1971). Origin of plasma fatty acids in lactating cows fed high grain or high fat diets. *Journal of Dairy Science, 54*, 1025–1033.

Palmquist, D. L. & Jenkins, T. C. (1980). Fat in lactation rations: Review. *Journal of Dairy Science, 63*, 1–14.

Palmquist, D. L. & Mattos, W. (1978). Turnover of lipoproteins and transfer to milk fat of dietary (1-Carbon-14) linoleic acid in lactating cows. *Journal of Dairy Science, 61*, 561–565.

Palmquist, D. L., Davis, C. L., Brown, R. E. & Sachan, D. S. (1969). Availability and metabolism of various substrates in ruminants. V. Entry rate into the body and incorporation into milk fat of D(-)β-hydroxybutyrate. *Journal of Dairy Science, 52*, 633–638.

Palmquist, D. L., Beaulieu, A. D. & Barbano, D. M. (1993). Feed and animal factors influencing milk fat composition. *Journal of Dairy Science, 76*, 1753–1771.

Parodi, P. W. (1982). Positional distribution of fatty acids in triglycerides from milk of several species of mammals. *Lipids, 17*, 437–442.

Parodi, P. W. (2003). Anti-cancer agents in milk fat. *Australian Journal of Dairy Technology, 58*, 114–118.

Parodi, P.W. (2006). Nutritional significance of milk lipids. In P. F. Fox & P. L. H. McSweeny (Eds.), *Advanced dairy chemistry* (Vol. 2, 3rd ed., pp. 601–639). London: Springer.

Patton, S. & Keenan, T. W. (1975). The milk fat globule membrane. *Biochimica et Biophysica Acta, 415*, 273–309.

Perfield, J. W., II, Delmonte, P., Lock, A. L., Yurawecz, M. P. & Bauman, D. E. (2006). *Trans*-10, *trans*-12 conjugated linoleic acid does not affect milk fat yield but reduces delta9-desaturase index in dairy cows. *Journal of Dairy Science, 89*, 2559–2566.

Popják, G., French, T. H. & Folley, S. J. (1951). Utilization of acetate for milk-fat synthesis in the lactating goat. *Biochemical Journal, 48*, 411–416.

Pullen, D. L., Palmquist, D. L. & Emery, R. S. (1989). Effect of days of lactation and methionine hydroxy analog on incorporation of plasma fatty acids into plasma triglycerides. *Journal of Dairy Science, 72*, 49–58.

Raphael, B. C., Patton, S. & McCarthy, R. D. (1975a). The serum lipoproteins as a source of milk cholesterol. *FEBS Letters, 58*, 47–49.

Raphael, B. C., Patton, S. & McCarthy, R. D. (1975b). Transport of dietary cholesterol into blood and milk of the goat. *Journal of Dairy Science, 58*, 971–976.

Rasmussen, J. T., Börchers, T. & Knudsen, J. (1990). Comparison of the binding affinities of acyl-CoA-binding protein and fatty-acid-binding protein for long-chain acyl-CoA esters. *Biochemical Journal, 265*, 849–855.

Ridgway, N. D., Byers, D. M., Cook, H. W. & Storey, M. K. (1999). Integration of phospholipids and sterol metabolism in mammalian cells. *Progress in Lipid Research, 38*, 337–360.

Ross, A. C. & Rowe, J. F. (1984). Cholesterol esterification by mammary gland microsomes from the lactating rat. *Proceedings of the Society for Experimental Biology and Medicine, 176*, 42–47.

Rottman, L. W., Ying, Y., Zhou, K., Bartell, P. A. & Harvatine, K. J. (2015). The effects of feeding rations that differ in neutral detergent fiber and starch concentration within a day on production, feeding behavior, total-tract digestibility, and plasma metabolites and hormones in dairy cows. *Journal of Dairy Science, 97*, 4673–4684.

Ryder, E., Gregolin, C., Chang, H. C. & Lane, M. D. (1967). Liver acetyl CoA carboxylase: Insight into the mechanism of activation by tricarboxylic acids and acetyl coa. *Proceedings of the National Academy of Sciences, 57*, 1455–1462.

Salfer, I. J., Dechow, C. D. & Harvatine, K. J. (2019). Annual rhythms of milk and milk fat and protein production in dairy cattle in the United States. *Journal of Dairy Science, 102*, 742–753.

Schauff, D. J. & Clark, J. H. (1992). Effects of feeding diets containing calcium salts of long chain fatty acids to lactating dairy cows. *Journal of Dairy Science, 75*, 2990–3002.

Schennink, A., Stoop, W. M., Visker, M. H., Heck, J. M., Bovenhuis, H., van der Poel, J. J., van Valenberg, H. J. & van Arendonk, J. A. (2007). Dgatl underlies large genetic variation in milk-fat composition of dairy cows. *Animal Genetics, 38*, 467–473.

Schwenk, R. W., Holloway, G. P., Luiken, J. J. F. P., Bonen, A. & Glatz, J. F. C. (2010). Fatty acid transport across the cell membrane: Regulation by fatty acid transporters. *Prostoglandins, Leukotrienes, and Essential Fatty Acids,* ***82***, 149–154.

Schwertfeger, K. L., McManaman, J. L., Palmer, C. A., Neville, M. C. & Anderson, S. M. (2003). Expression of constitutively activated Akt in the mammary gland leads to excess lipid synthesis during pregnancy and lactation. *Journal of Lipid Research, 44*, 1100–1112.

Shand, J. H. & West, D. W. (1991). Acyl-CoA: Cholesterol acyltransferase activity in the rat mammary gland: Variation during pregnancy and lactation. *Lipids, 26*, 150–154.

Shennan, D. B. & Peaker, M. (2000). Transport of milk constituents by the mammary gland. *Physiological Reviews, 80*, 925–951.

Shingfield, K. J., Ahvenjärvi, S., Toivonen, V., Arölä, A., Nurmela, K. V. V., Huhtanen, P. & Griinari, J. M. (2003). Effect of dietary fish oil on biohydrogenation of fatty acids and milk fatty acid content in cows. *Animal Science, 77*, 165–179.

Shingfield, K. J., Arola, A., Ahvenjarvi, S., Vanhatalo, A., Toivonen, V., Griinari, J. M. & Huhtanen, P. (2008). Ruminal infusion of cobalt-edta reduce mammary delta9-desaturase index and alter milk fatty acid composition in lactating cows. *Journal of Nutrition, 138*, 710–717.

Shingfield, K. J., Bonnet, M. & Scollan, N. D. (2013). Recent developments in altering the fatty acid composition of ruminant-derived foods. *Animal, 7*(Suppl 1), 132–162.

Shirley, J. F., Emery, R. S., Convey, E. M. & Oxender, W. D. (1973). Enzymic changes in bovine adipose and mammary tissue serum and mammary tissue hormonal changes with initiation of lactation. *Journal of Dairy Science, 56*, 569–574.

Signorelli, F., Orru, L., Napolitano, F., De Matteis, G., Scata, M. C., Catillo, G., Marchitelli, C. & Moioli, B. (2009). Exploring polymorphisms and effects on milk traits of the dgat1, scd1 and ghr genes in four cattle breeds. *Livestock Science, 125*, 74–79.

Small, C. A., Yeaman, S. J., West, D. W. & Clegg, R. A. (1991). Cholesterol ester hydrolysis and hormone-sensitive lipase in lactating rat mammary tissue. *Biochimica et Biophysica Acta, 1082*, 251–254.

Smith, S. (1994). The animal fatty acid synthase: One gene, one polypeptide, seven enzymes. *The FASEB Journal, 8*, 1248–1259.

Smith, G. H., McCarthy, S. & Rook, J. A. F. (1974). Synthesis of milk fat from β-hydroxybutyrate and acetate in lactating goats. *Journal of Dairy Research, 41*, 175–191.

Smith, S. J., Cases, S., Jensen, D. R., Chen, H. C., Sande, E., Tow, B., Sanan, D. A., Raber, J., Eckel, R. H. & Farese, R. V., Jr. (2000). Obesity resistance and multiple mechanisms of triglyceride synthesis in mice lacking DGAT 1. *Nature Genetics, 25*, 87–90.

Smith, S., Witkowski, A. & Joshi, A. K. (2003). Structural and functional organization of the animal fatty acid synthase. *Progress in Lipid Research, 42*, 289–317.

Storch, J. & Thumser, A. E. A. (2000). The fatty acid transport function of fatty acid-binding proteins. *Biochimica et Biophysica Acta, 1486*, 28–44.

Storry, J. E., Hall, A. J. & Johnson, V. W. (1971). The effects of increasing amounts of dietary coconut oil on milk-fat secretion in the cow. *Journal of Dairy Research, 38*, 73–77.

Taniguchi, M., Mannen, H., Oyama, K., Shimakura, Y., Oka, A., Watanabe, H., Kojima, T., Komatsu, M., Harper, G. S. & Tsuji, S. (2004). Differences in stearoyl-CoA desaturase mRNA levels between Japanese Black and Holstein cattle. *Livestock Production Science, 87*, 215–220.

Timmen, H. & Patton, S. (1988). Milk fat globules: Fatty acid composition, size and in vivo regulation of fat liquidity. *Lipids, 23*, 685–689.

Timmons, J. S., Weiss, W. P., Palmquist, D. L. & Harper, W. J. (2001). Relationships among dietary roasted soybeans, milk components, and spontaneous oxidized flavor of milk. *Journal of Dairy Science, 84*, 2440–2449.

Tong, L. (2005). Acetyl-coenzyme a carboxylase: Crucial metabolic enzyme and attractive target for drug discovery. *Cellular and Molecular Life Sciences, 62*, 1784–1803.

Toral, P. G., Monahan, F. J., Hervas, G., Frutos, P. & Moloney, A. P. (2018). Review: Modulating ruminal lipid metabolism to improve the fatty acid composition of meat and milk. Challenges and opportunities. *Animal, 12*, s272–s281.

Torok, E. D., Beitz, D. C., Johnson, D. C., Baldner-Shank, G. L. & McGilliard, A. D. (1986). Use of different precursors for lipogenesis in ruminant mammary tissue. *Nutrition Research, 6*, 1211–1218.

Travers, M. T. & Barber, M. C. (2001). Acetyl-CoA carboxylase-{alpha}: Gene structure-function relationships. *Journal of Animal Science, 79*, E136–E143.

van Soest, P. J. (1963). Ruminant fat metabolism with particular reference to factors affecting low milk fat and feed efficiency. A review. *Journal of Dairy Science, 46*, 204–216.

Vallance, W. S. & McClymont, G. L. (1959). Depression in percentage of milk fat by parenteral glucose infusion and glycerol feeding. *Nature, 183*, 466–467.

Veerkamp, J. H., Peeters, R. A. & Maatman, R. G. H. J. (1991). Structural and functional features of different types of cytoplasmic fatty acid-binding proteins. *Biochimica et Biophysica Acta, 1081*, 1–24.

Vernon, R. G., Faulkner, A., Finley, E., Pollock, H. & Taylor, E. (1987). Enzymes of glucose and fatty acid metabolism of liver, kidney, skeletal muscle, adipose tissue and mammary gland of lactating and non-lactating sheep. *Journal of Animal Science, 64*, 1395–1411.

Vesper, H., Schmelz, E. M., Nikolova-Karakashian, M. N., Dillehay, D. L., Lynch, D. V. & Merrill, A. H., Jr. (1999). Sphingolipids in food and the emerging importance of sphingolipids to nutrition. *Journal of Nutrition, 129*, 1239–1250.

Virtanen, A. I. (1966). Milk production of cows on protein-free feed. Studies of the use of urea and ammonium salts as the sole nitrogen source open new important perspectives. *Science, 153*, 1603–1614.

Volpe, J. J. & Vagelos, P. R. (1973). Saturated fatty acid biosynthesis and its regulation. *Annual Review of Biochemistry, 42*, 21–60.

Wang, X., Wurmser, C., Pausch, H., Jung, S., Reinhardt, F., Tetens, J., Thaller, G. & Fries, R. (2012). Identification and dissection of four major qtl affecting milk fat content in the german holstein-friesian population. *PLoS One, 7*, e40711.

Welper, R. D. & Freeman, A. E. (1992). Genetic parameters for yield traits of holsteins, including lactose and somatic cell score. *Journal of Dairy Science, 75*, 1342–1348.

West, C. E., Bickerstaffe, R., Annison, E. F. & Linzell, J. L. (1972). Studies on the mode of uptake of blood triglycerides by the mammary gland of the lactating goat. The uptake and incorporation into milk fat and mammary lymph of labeled glycerol, fatty acids and triglycerides. *Biochemical Journal, 126*, 477–490.

Williamson, D. H., Munday, M. R., Jones, R. G., Roberts, A. F. & Ramsey, A. J. (1983). Short-term dietary regulation of lipogenesis in the lactating mammary gland of the rat. *Advances in Enzyme Regulation, 21*, 135–145.

Winter, A., Krämer, W., Werner, F. A. O., Kollers, S., Kata, S., Durstewitz, G., Buitkamp, J., Womack, J. E., Thaller, G. & Fries, R. (2002). Association of a lysine-232/alanine polymorphism in a bovine gene encoding acyl-CoA: Diacylglycerol acyltransferase (DGAT1) with variation at a quantitative trait locus for milk fat content. *Proceedings of the National Academy of Sciences, 99*, 9300–9305.

Witters, L. A., Moriarity, D. & Martin, D. B. (1979). Regulation of hepatic acetyl coenzyme A carboxylase by insulin and glucagon. *Journal of Biological Chemistry, 254*, 6644–6649.

Wood, H. G., Peeters, G. J., Verbeke, R., Lauryssens, M. & Jacobson, B. (1965). Estimation of the pentose cycle in the perfused cow's udder. *Biochemical Journal, 96*, 607–615.

Woolpert, M. E., Dann, H. M., Cotanch, K. W., Melilli, C., Chase, L. E., Grant, R. J. & Barbano, D. M. (2016). Management, nutrition, and lactation performance are related to bulk tank milk de novo fatty acid concentration on northeastern US dairy farms. *Journal of Dairy Science, 99*, 8486–8497.

Yang, Y. T. & Baldwin, R. L. (1973). Preparation and metabolism of isolated cells from bovine adipose tissue. *Journal of Dairy Science, 56*, 350–365.

Yang, Z., Liu, S., Chen, X., Chen, H., Huang, M. & Zheng, J. (2000). Induction of apoptotic cell death and *in vivo* growth inhibition of human cancer cells by a saturated branched-chain fatty acid, 13-methyltetradecanoic acid. *Cancer Research, 60*, 505–509.

Zhao, F.-Q., Dixon, W. T. & Kennelly, J. J. (1996). Localization and gene expression of glucose transporters in bovine mammary gland. *Comparative Biochemistry and Physiology, 115B*, 127–134.

Conjugated Linoleic Acid: Biosynthesis and Nutritional Significance

3

D. E. Bauman, A. L. Lock, R. Conboy Stephenson, K. Linehan, R. P. Ross, and C. Stanton

Biomedical studies with animal models have shown that RA and VA have anticarcinogenic and antiatherogenic properties, with the effects of VA being related to its conversion to RA. The anticarcinogenic effects have been observed for a wide range of cancer types, but the most impressive results have been reported in relation to mammary cancer. Of special importance, RA and VA are potent anticarcinogens when supplied as natural food components in the form of VA/RA-enriched butter. The functional food considerations of CLA isomers in dairy products realistically relate only to RA as the major isomer, although this should include VA because of its use in humans for the endogenous synthesis of RA. The RA and VA content in milk fat are directly related and they can be markedly enhanced through the use of diet formulation and nutritional management of dairy cows.

Trans-10, *cis*-12 CLA is another CLA isomer in milk fat which can affect lipid metabolism. It is generally present at low concentrations in milk fat (typically <0.2% of CLA); under some dietary conditions, a portion of the rumen biohydrogenation shifts to produce more of this isomer, although it is still only a minor portion of total CLA. These dietary conditions are associated with milk fat depression and as little as 2 g/d of *trans*-10, *cis*-12 CLA leaving the rumen will reduce milk fat synthesis by 20%. Because of the potency and specificity of this CLA isomer, it is being developed as a dairy management tool to allow for a controlled reduction in milk fat output.

CLA isomers in milk fat and how they relate to both animal agriculture and human health are rapidly expanding fields. Milk and dairy products offer exciting opportunities in the area of functional foods, and the functional properties of VA and RA in milk further serve to illustrate the value of dairy products in the human diet.

D. E. Bauman
Department of Animal Science, Cornell University, Ithaca, NY, USA

A. L. Lock
Department of Animal Science, Michigan State University, East Lansing, MI, USA

R. Conboy Stephenson · K. Linehan
Teagasc, Moorepark, Fermoy, Co., Cork, Ireland

R. P. Ross
APC Microbiome Ireland, University College Cork, Cork, Ireland

C. Stanton (✉)
Teagasc, Moorepark, Fermoy, Co., Cork, Ireland

APC Microbiome Ireland, University College Cork, Cork, Ireland
e-mail: Catherine.stanton@teagasc.ie

3.1 Introduction

An adequate supply of good-quality food is essential for human health and well-being. Milk and meat products derived from ruminants represent important sources of nutrients in human diets, providing energy, high-quality protein, and

P. L. H. McSweeney et al. (eds.), *Advanced Dairy Chemistry, Volume 2*,
https://doi.org/10.1007/978-3-030-48686-0_3

essential minerals and vitamins (National Research Council Committee on Technological Options to Improve the Nutritional Attributes of Animal 1988; Murphy and Allen 2003). Nutritional quality is increasingly an important consideration in food choices because of the growing consumer awareness of the link between diet and health. Many foods contain microcomponents that have beneficial effects beyond those associated with their traditional nutrient content, and these are often referred to as "functional food" components. One such component in foods derived from ruminants is conjugated linoleic acid (CLA).

CLA refers to a mixture of positional and geometric isomers of linoleic acid (*cis*-9, *cis*-12 octadecadienoic acid) with a conjugated double-bond system. The structure of two CLA isomers is contrasted with linoleic and vaccenic acids in Figure 3.1. The presence of CLA isomers in ruminant fat is related to the biohydrogenation of polyunsaturated fatty acids (PUFA) in the rumen. Ruminant fats are relatively more saturated than most plant oils and this is also a consequence of biohydrogenation of dietary PUFA by rumen bacteria. Increases in saturated fatty acids are considered undesirable; however, CLA was given GRAS designation by the FDA in 2008, exempting it from classification as a *trans*-fat on food labels. Furthermore, consumption of CLA has been shown to be associated with many health benefits (den Hartigh 2019), and food products derived from ruminants are the major dietary source of CLA for humans.

The anticarcinogenic activity of CLA has been clearly established, but biomedical studies with animal models have identified an impressive range of additional positive health effects for CLA as summarized in Chapter 13. Particularly noteworthy is the fact that CLA is a potent anticarcinogen when supplied as a natural food component in the form of CLA-enriched butter as discussed later in this review.

The presence of CLA in ruminant milk has been known for more than 80 years, and in this chapter, we first review the dietary sources of CLA and provide an overview of the analytical challenges associated with quantifying CLA in foods and biological samples. Second, we review the origin of the different CLA isomers present in milk fat, developing the interrelationships between biohydrogenation intermediates produced in the rumen, synthesis of CLA in the tissues, and the presence of these isomers in milk fat. Third, we highlight the nutritional and physiological factors that affect the level of CLA in milk fat and discuss milk quality considerations for dairy products that have a naturally enhanced content of CLA. Finally, we review the biological effects of CLA related to the dairy cow and dairy products. This includes its effects on milk fat synthesis in dairy cows, an area that has progressed rapidly and promises to contribute to our general understanding of the regulation of lipid metabolism. Our review of the biological effects also includes the significance of CLA in dairy products and its potential as a functional food component that benefits human health.

3.2 Dietary Sources

The predominant source of CLA in human diets is ruminant-derived food products. In the USA, dairy products provide about 70% of the intake of CLA and beef products account for another 25% (Ritzenthaler *et al.* 2001). Similar values for the contribution of different food classes have been reported for other countries (Parodi *et al.* 2003).

Scientists at the University of Reading, UK, first demonstrated that fatty acids obtained from summer butter differed from those obtained from winter butter by exhibiting a much stronger spectrophotometric absorption at 230 µm (Booth *et al.* 1933). It was subsequently concluded that the adsorption at this wavelength was due to a conjugated double-bond pair (Moore 1939). Parodi (1977) was the first to identify *cis*-9, *trans*-11 octadecadienoic acid as a fatty acid in milk fat that contained the conjugated double-bond pair. As analytical techniques improved, it was discovered that milk fat and body fat from ruminants contained many isomers of CLA that differ by position (e.g., 7–9, 8–10, 9–11, 10–12, 11–13) or geometric orientation (*cis-trans*, *trans-cis*, *cis-cis*, and *trans-trans*) of the double-bond

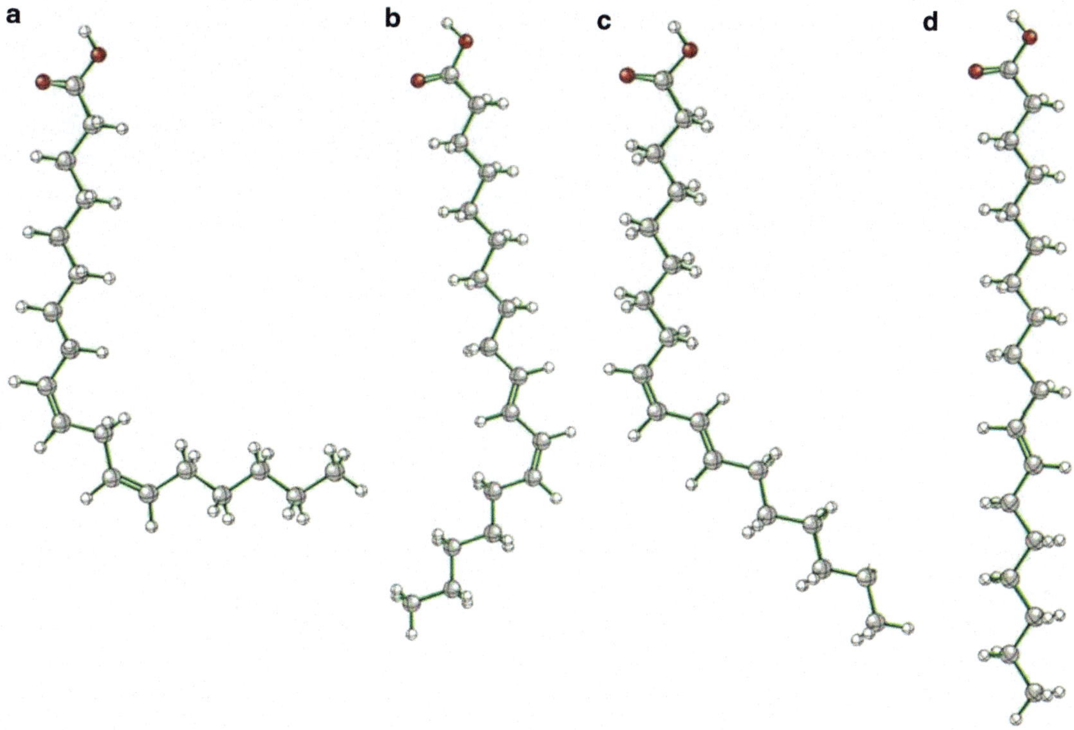

Figure 3.1. Chemical structures (from left to right): linoleic acid (*cis*-9, *cis*-12 18:2) (**a**); *trans*-10, *cis*-12 conjugated linoleic acid (**b**); rumenic acid (*cis*-9, *trans*-11 conjugated linoleic acid) (**c**) and vaccenic acid (*trans*-11 18:1) (**d**).

pair. The range of CLA isomers and their levels in milk and dairy products are summarized in Table 3.1. *Cis*-9, *trans*-11 is the major CLA isomer in ruminant fat, representing about 75–90% of the total CLA, and the common name of "rumenic acid" (RA) has been proposed for this isomer because of its unique relationship to ruminants (Kramer *et al.* 1998). The second most common isomer is *trans*-7, *cis*-9 CLA, representing about 10% of total CLA. Each of the other CLA isomers is at a low concentration when present, generally representing less than 0.5% of the total CLA in ruminant fat. Daily CLA intake can be increased by taking commercially produced CLA isomer supplements, produced by heating linoleic acid in the presence of alkali or by partial hydrogenation of linoleic acid (Banni 2002). The synthetic derivation of CLA is often termed "mixed" CLA because of the approximately 1:1 ratio of *cis*-9, *trans*-11 and *trans*-10, *cis*-12 CLA. The fortification of dairy products such as yogurt, milk, and cheese with CLA (usu-

Table 3.1. Range of positional and geometric isomers of conjugated $C_{18:2}$ fatty acids in milk and dairy products. (Adapted from Lock and Bauman 2004)[a]

Isomer	% of total CLA isomers
trans-7, *cis*-9	1.2–8.9
trans-7, *trans*-9	<0.1–2.4
trans-8, *cis*-10	<0.1–1.5
trans-8, *trans*-10	0.2–0.4
cis-9, *trans*-11	72.6–91.2
trans-9, *trans*-11	0.8–2.9
trans-10, *cis*-12	<0.1–1.5
trans-10, *trans*-12	0.3–1.3
cis-11, *trans*-13	0.2–4.7
trans-11, *cis*-13	0.1–8.0
trans-11, *trans*-13	0.3–4.2
cis-12, *trans*-14	<0.01–0.8
trans-12, *trans*-14	0.3–2.8
cis-*cis* isomers	0.1–4.8

[a]Data derived from seven studies where fatty acid analysis was carried out on milk samples (Precht and Molkentin 1997; Piperova *et al.* 2002; Kraft *et al.* 2003; Shingfield *et al.* 2003; Kay *et al.* 2004), butter (Bauman *et al.* 2000) or cheese (Rickert *et al.* 1999).

ally mixed CLA) is another method by which humans can increase their CLA intake (Rodriguez-Alcala and Fontecha, 2007). A number of studies have examined the impact of such CLA-fortified foods on human health (Derakhshande-Rishehri *et al.* 2015; Haghighatdoost and Hariri 2018; Mirzaii *et al.* 2016; den Hartigh 2019).

3.3 Analytical Challenges

Biological samples generally contain multiple isomers of CLA, many at very low concentrations, and each may differ in their biological effects. Milk fat alone consists of over 400 fatty acids (Månsson 2008). Thus, the ability to determine the concentration of specific isomers is becoming increasingly important, and frequently a combination of analytical methods is required to quantify fully CLA isomers and related fatty acids (Kramer *et al.* 2004; Christie *et al.* 2007). The analysis of CLA typically requires their conversion to derivatives that can be separated from other fatty acids in the sample, usually by either gas chromatography (GC) or high-performance liquid chromatography (HPLC). Early investigations generally used high temperature, acid-catalyzed methylation to prepare fatty acid methyl esters (FAME), but subsequent work established that this procedure causes extensive isomerization, producing mainly *trans*/*trans* isomers, so alkali-catalyzed methods are most suitable (Amores and Virto 2019). Therefore, data on the distribution of CLA isomers from early investigations are highly questionable (Yurawecz *et al.* 1998).

GC provides the basis for most analytical approaches reported in the literature, and the use of alkali-catalyzed methylation has proven to be the most accurate method for the analysis of CLA (Yurawecz *et al.* 1998). Sodium methoxide is the catalyst used most widely and has the advantage that it does not isomerize conjugated double bonds or form methoxy artefacts. Kramer *et al.* (1997) demonstrated that this procedure completely methylated test samples containing mainly triglycerides, but it did not methylate free fatty acids. Since milk fat consists of ~98% triglycerides, the contribution made by other lipids can often be disregarded so that only a base-catalyzed methylation is necessary (Yurawecz *et al.* 1999). The transmethylation procedure described by Christie (1982) with modifications by Chouinard *et al.* (1999b) to minimize the loss of highly volatile short chain fatty acids is used widely and recommended for the analysis of fatty acids in milk by the authors. In the GC analysis, the type of column is another important consideration. Long (100–120 m) highly polar columns are used typically and offer a reasonable degree of separation for CLA isomers, and especially *trans* 18:1 fatty acids (Christie *et al.* 2007); 100 meter polar columns coated with 100% cyanopropyl polysiloxane give optimum CLA isomer resolution (Christie *et al.* 2007; Amores and Virto 2019). However, with typical GC procedures, *trans*-7, *cis*-9 CLA and RA co-elute, and other CLA isomers are often not separable, especially if there is a low concentration of one isomer relative to another. Therefore, CLA standards are essential for GC; a commercial CLA mixture is useful but does not contain all isomers found in natural samples. Alternative internal standards such as sorbic acid have been employed to good effect (Czauderna *et al.* 2011). Analysis of the CLA content and profile of animal tissues or biological fluids containing a mixture of lipid classes is more difficult. In order for all of the fatty acids to be methylated, a two-stage methylation procedure is recommended. Kramer *et al.* (1997) evaluated many different combinations of acid/base catalysts and concluded that the best compromise was the use of sodium methoxide followed by a mild acidic methylation, which resulted in the methylation of the majority of the fatty acids with minimal isomerization of the CLA isomers. However, mild boron triflouride or 1% methanolic sulfuric acid with a minimal temperature and reaction time are often used with good success (Christie *et al.* 2007).

Additional analytical methods are appropriate when a more complete characterization of the CLA isomers in biological samples is required. Most often, a combination of GC and silver ion HPLC is used and permits excellent separation and identification of positional and geometrical isomers of CLA (Momchilova and Nikolova-Damyanova 2012); see Adlof (2003) and (Kramer *et al.* 2004) for detailed reviews of this approach. The use of gas chromatography-mass spectrometry (GC-MS) has become increasingly popular and represents a very powerful technique for identification of the position of double bonds in fatty acids (see Dobson 2003) and the orientation of those bonds in CLA isomers (Michaud *et al.* 2003). ^{13}C-NMR spectroscopy has proven to be the single most comprehensive method for the identification and quantification of all the positional and geometrical isomers present in commercial CLA preparations (Davis *et al.* 1999). However, the methodology requires substantial amounts of sample and is not likely to be applicable to tissue extracts at natural levels. A powerful tandem mass spectrometry (MS/MS) technique using acetonitrile/chemical ionization has been developed for determining the double-bond position and geometry in CLA isomers from natural samples (Alves *et al.* 2011).

Novel optical-based methods have been reported as the most promising techniques because they have a great potential for real-time applications (Tao and Ngadi 2017). Coppa *et al.* (2010) used NIRS to predict the CLA composition of bovine milk. Meurens *et al.* (2005) and Bernuy *et al.* (2008) demonstrated that Fourier transform-RS (FT-RS) technique has a great potential in predicting total CLA in bovine milk. More recently, Prema *et al.* (2013) analyzed total CLA in the lipid fraction of cheeses by ^{1}H NMR spectroscopy. Some authors analyze CLA by NMR as part of a more general metabolomics analysis, without previous lipid extraction (Tsiafoulis *et al.* 2019). For example, O'Callaghan *et al.* (2018) identified and quantified 49 metabolites in cow milk samples, including total CLA. Sun *et al.* (2013) developed a novel method for the *de novo* identification of specific CLA isomers in which silver ion liquid chromatography is coupled to in-line ozonolysis/mass spectrometry (Ag^+-LC/O_3-MS). They also reported an approach for direct determination of double-bond positions in FAME by coupling ozonolysis in-line with mass spectrometry (in-line O_3-MS) (Sun *et al.* 2013).

The increasing knowledge of CLA-producing bacteria has seen the development of rapid screening techniques to identify potential CLA producers. Barrett *et al.* (2007) reported a spectrophotometric method for screening a large number of culture supernatants for CLA production spectrophotometrically at a wavelength of 233 nm for CLA production following incubation in a medium containing linoleic acid; this method eliminates the need for GC during the screening process. Liu *et al.* (2012) reported a method to screen bacterial strains with the ability to produce specific CLA isomers. An ultraviolet spectral scan method was used for preliminary assessment of production of CLA, and the follow-up capillary electrophoresis analysis was implemented to detect the composition of CLA isomers.

The particular objectives and the anticipated use of the analytical data will determine the extent to which individual CLA isomers need to be separated, identified, and quantified (Christie *et al.* 2007). The methodology for the analysis of CLA and related fatty acids continues to evolve and it is recommended that the reader consult recent reviews and publications in this area before undertaking such analysis for the first time. We recommend Christie *et al.* (2007) and Kramer *et al.* (2004) as excellent practical guides on the analysis of CLA

3.4 Origin of CLA in Milk Fat

3.4.1 Lipid Metabolism in the Rumen

The diet of lactating dairy cows typically contains 4–5% fat, with the major PUFAs being linoleic and linolenic acids that are supplied mainly from dietary concentrates and forages, respectively. When dietary lipids enter the rumen, they undergo two major processes, lipolysis and biohydrogenation. Lipolysis is the initial step resulting in the hydrolysis of the ester linkages in the triglycerides, phospholipids, and glycolipids.

Hydrolysis of dietary lipids in the rumen involves extracellular lipases that are produced by the rumen bacteria; there is little evidence to support significant roles for rumen protozoa and fungi, or salivary or plant lipases in rumen hydrolysis. The extent of hydrolysis of dietary lipids in the rumen is generally high (>85%), and a number of factors that affect the rate and extent of hydrolysis have been identified, including the pH of the rumen and the type/quantity of fat entering the rumen (see Doreau *et al.* 1997; Doreau and Ferlay 1994; Harfoot 1981; Harfoot and Hazlewood 1997; Lourenço *et al.* 2010 for reviews).

Biohydrogenation of PUFAs is the second major transformation that dietary lipids undergo in the rumen and this process first requires that the fatty acid is free. As a consequence, rates are always less than those of hydrolysis and factors that affect hydrolysis also affect rates of biohydrogenation; 92% of α-linolenic acid and 86% linoleic acid that are ingested by the cow undergo biohydrogenation in the rumen (Privé *et al.* 2015). In the 1960s and 1970s, an extensive series of *in vitro* and *in vivo* studies examined rumen biohydrogenation (Dawson and Kemp 1970; Harfoot 1981; Harfoot and Hazlewood 1997; Keeney 1970). Most biohydrogenation (>80%) occurs in association with the small dense (fine) food particles and this has been attributed to extracellular enzymes of bacteria either associated with the feed or free in suspension. A few species of rumen bacteria capable of carrying out the biohydrogenation reactions have been identified and predominant pathways have been elucidated. Biohydrogenation is now recognized as a protective mechanism by the bacterial community to reduce the toxic effects of unsaturated fatty acids on microbial growth. Kemp and Lander (1984) classified rumen bacteria involved in biohydrogenation into two groups based on their metabolic pathways. Group A included bacteria that could hydrogenate 18 carbon PUFAs to *trans* 18:1 fatty acids, whereas only a few species, characterized as Group B, could hydrogenate *trans* 18:1 fatty acids to stearic acid (Harfoot and Hazlewood 1997). Strains of *Butyrivibrio fibrisolvens*, in addition to a few strains of *Eubacterium ruminantium* and *Pseudobutyrivibrio*, have been identified as members of Group A. Bacteria capable of carrying out stearate formation (Group B) include *Clostridium proteoclasticum*, the genus *Fuscillus*, and *Butyrivibrio hungatei* strain SU6. However, subsequent 16s rRNA gene sequencing suggests that *B. hungatei* SU6 clusters more closely with *C. proteoclasticum*, compromising its taxonomic validity. Moreover, the genus *'Fusocillus'* is also in doubt as cultures deposited in culture collections were nonviable, preventing modern molecular phylogenetic analysis. Consequently, Jenkins *et al.* (2008) suggested that strains of *C. proteoclasticum* are the sole members of Group B. Furthermore, Moon *et al.* (2008) proposed the reclassification of *C. proteoclasticum* as *Butyrivibrio proteoclasticus*, due to its 16s rRNA gene sequencing. Therefore, this reclassification suggests that *Butyrivibrio proteoclasticus* is the only member of Group B (See Jarvis and Moore 2010; Jenkins *et al.* 2008, for reviews). Thus, complete biohydrogenation of unsaturated fatty acids generally requires bacteria from both groups. However, research has identified an occasional exception where a specific bacterial strain can carry out the complete biohydrogenation of PUFAs to C18:0 (see Palmquist *et al.* 2005).

The initial step in rumen biohydrogenation of linoleic and linolenic acids involves an isomerization of the *cis*-12 double bond to a *trans*-11 configuration, resulting in a conjugated dieonic or trienoic fatty acid (Figure 3.2). Next is a reduction of the *cis*-9 double bond resulting in a *trans*-11 octadecenoic acid. Therefore, RA is an intermediate formed only during the biohydrogenation of linoleic acid. The conversion of RA to vaccenic acid (*trans*-11 18:1; VA) is catalyzed by a reductase. The structure of VA is given in Figure 3.1 and it is noteworthy that it is an intermediate in the biohydrogenation of both linoleic and linolenic acids (Figure 3.2). VA is a unique and valuable intermediate of biohydrogenation due to its ability to act as a substrate for the formation of *cis*-9, *trans*-11 CLA in mammary tissue (Griinari *et al.* 2000). The final step is a further reduction of the *trans*-monoenes, producing stearic acid. Reduction of *trans*-octadecenoic fatty acids to stearic acid is generally the rate-

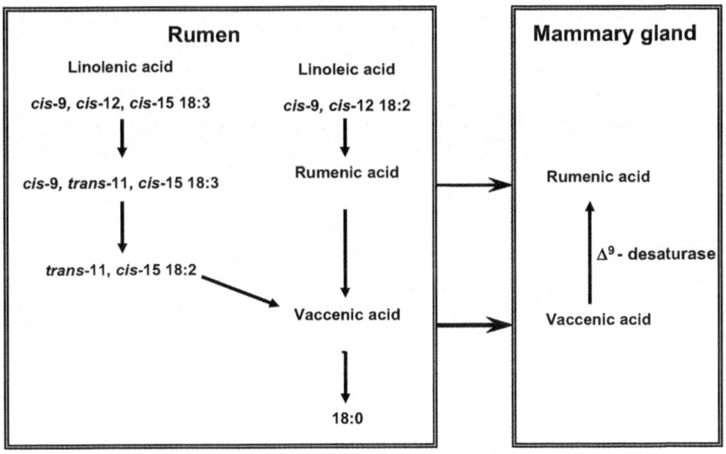

Figure 3.2. Pathways for ruminal and endogenous synthesis of rumenic acid (*cis*-9, *trans*-11 CLA) in the lactating dairy cow. Pathways for biohydrogenation of linoleic and linolenic acids yielding vaccenic acid (*trans*-11 18:1) are shown in the rumen box and endogenous synthesis by Δ^9-desaturase is shown in the mammary gland box. (Adapted from Bauman *et al.* 2003).

limiting step and, as a consequence, there is often an accumulation of *trans* fatty acids in the rumen (Keeney 1970).

As analytical techniques improved, we have gained an appreciation of the complexity of the biohydrogenation processes occurring in the rumen. In addition to the major pathways involving RA and VA as intermediates, there must be many additional pathways. A remarkable range of *trans*-18:1 and CLA isomers are produced during biohydrogenation and their outflows from the rumen based on limited data from growing cattle and lactating cows are shown in Table 3.2. This range of CLA and *trans*-18:1 isomers is not accounted for by known pathways of rumen biohydrogenation. Isomerase is the enzyme that catalyzes the key step that introduces the conjugated double-bond system and, unfortunately, this enzyme has been studied in only a few species of rumen bacteria (Kepler and Tove 1967; Kepler *et al.* 1970; Yokoyama and Davis 1971). The isomerase from *Butyrivibrio fibrisolvens* is a particulate enzyme bound to the bacterial cell membrane and it has an absolute substrate requirement for a *cis*-9, *cis*-12 diene system and a free carboxyl group (Kepler and Tove 1967; Kepler *et al.* 1970). If the initial isomerization involves the *cis*-12 double bond, then a *cis*-9, *trans*-11 conjugated diene is produced, whereas if the initial double bond isomerized is the *cis*-9, then *trans*-10, *cis*-12 conjugated diene is produced. Most rumen bacteria capable of carrying out this isomerization produce mainly RA from linoleic acid. However, Kim *et al.* (2002) demonstrated that the rumen bacterium, *Megasphaera elsdenii* YJ-4, produced predominately *trans*-10, *cis*-12 CLA and only a minor quantity of RA when incubated with linoleic acid. Nevertheless, the extent to which the various pathways of biohydrogenation are associated with specific enzymes and species of bacteria or reflect a general lack of specificity of the bacteria and their enzymes is not known.

It appears that the type of diet, rather than level of intake, is a major factor affecting biohydrogenation, and diet-induced changes in the rumen environment can shift the biohydrogenation pathways, resulting in dramatic changes in the fatty acid intermediates.

In addition, studies have established that the metabolism of radiolabeled oleic and eladic acids by mixed ruminal microorganisms results in extensive labeling of a wide range of *trans* octadecenoic fatty acids (*trans*-6 to *trans*-16) as well as of stearic acid (Mosley *et al.* 2002; Proell *et al.* 2002). In a few cases, the biological implications of the changes

Table 3.2. Range of double bond positions in *trans* $C_{18:1}$ and conjugated $C_{18:2}$ fatty acids and their ruminal outflow in growing and lactating cattle[a]

Trans 18:1		Conjugated 18:2	
Isomer	Ruminal outflow (g/day)	Isomer	Ruminal outflow (g/day)
trans-4	0.5–0.7	*trans*-7, *cis*-9	<0.01
trans-5	0.4–0.6	*trans*-7, *trans*-9	<0.01–0.05
trans-6-8	0.4–6.7	*trans*-8, *cis*-10	0.01–0.02
trans-9	0.8–6.2	*trans*-8, *trans*-10	<0.01–0.10
trans-10	1.7–29.1	*cis*-9, *cis*-11	<0.01–0.01
trans-11	5.0–121.0	*cis*-9, *trans*-11	0.19–2.86
trans-12	0.5–9.5	*trans*-9, *trans*-11	0.22–0.55
trans-13 + 14	6.5–22.9	*trans*-10, *cis*-12	0.02–0.32
trans-15	3.2–8.5	*trans*-10, *trans*-12	0.05–0.06
trans-16	3.1–8.0	*cis*-11, *trans*-13	0.01–0.10
		trans-11, *cis*-13	0.01–0.46
		trans-11, *trans*-13	0.09–0.40
		cis-12, *trans*-14	<0.01–0.05
		trans-12, *trans*-14	0.08–0.19

[a]Data derived from three studies where samples were collected from the duodenum (Duckett *et al.* 2002; Piperova *et al.* 2002) or omasum (Shingfield *et al.* 2003).

in the pathways of rumen biohydrogenation have been established and these will be discussed in later sections. A more comprehensive discussion of lipid metabolism in the rumen and its effects on the production of CLA and *trans* 18:1 isomers is provided in a review by Palmquist *et al.* (2005).

3.4.2 *cis*-9, *trans*-11 CLA (Rumenic Acid)

Initially, it was assumed that the RA in milk fat and body fat of ruminants originated from incomplete biohydrogenation in the rumen. This hypothesis was based on the fact that RA was the major CLA isomer in ruminant fat and the first intermediate in the major biohydrogenation pathway for linoleic acid (Figure 3.2). A close linear relationship was also observed between the levels of VA and RA in milk fat (Griinari and Bauman 1999; Jahreis *et al.* 1997; Jiang *et al.* 1996; Lawless *et al.* 1998), consistent with the concept that these two fatty acid intermediates had escaped complete biohydrogenation in the rumen and were subsequently absorbed from the digestive tract and used for milk fat synthesis. However, there were a number of inconsistencies with this idea. First, the kinetics of rumen biohydrogenation are such that CLA represents only a transitory product, and VA is the major biohydrogenation intermediate that accumulates in the rumen (Harfoot and Hazlewood 1997; Keeney 1970). Second, nutrition studies demonstrated that increases in the milk fat content of CLA occurred when linseed oil and other dietary sources of linolenic acid were fed (e.g., Dhiman *et al.* 2000; Kelly *et al.* 1998; Lock and Garnsworthy 2002). As previously discussed, RA is not an intermediate in the biohydrogenation of linolenic acid, but the biohydrogenation of both linoleic and linolenic acids produces VA as an intermediate. Third, the ratio of VA to RA is >50:1 in rumen fluid but only about 3:1 in milk fat. Based on these considerations, Griinari and Bauman (1999) proposed that endogenous synthesis could be an important source of the RA found in milk fat, with the synthesis involving the enzyme Δ^9-desaturase and VA as the substrate (see Figure 3.2). Previous investigations with Δ^9-desaturase from rat liver established that while the preferred reaction was the conversion of stearic acid to oleic acid, this enzyme could also desaturate positional isomers of *trans*-octadecenoic acids (Mahfouz *et al.* 1980; Pollard *et al.* 1980).

The first study to show directly that milk fat CLA could originate via endogenous synthesis was Griinari *et al.* (2000); these authors infused 12.5 g/d of VA into the abomasum of dairy cows and observed a 31% increase in the concentration of RA in milk fat. This investigation clearly demonstrated the potential for endogenous synthesis, but additional studies were needed to determine its actual importance. To address this, two approaches have been used; one approach was to inhibit Δ^9-desaturase directly, and this has involved the use of sterculic oil which contains two cyclopropene fatty acids, sterculic acid and malvalic acid, which are specific inhibitors of Δ^9-desaturase (Bickerstaffe and Johnson 1972; Jeffcoat and Pollard 1977; Phelps *et al.* 1965). To account for the fact that inhibition of Δ^9-desaturase may not be complete, these investigations employed a correction factor based on the ratio in milk fat of *cis*-9 C14:1 to C14:0 (another product: substrate pair of Δ^9-desaturase that is almost exclusively synthesized in the mammary gland). Griinari *et al.* (2000), using this approach, estimated that 64% of the RA in milk fat was of endogenous origin in cows fed an alfalfa hay/corn grain-based diet. This represented the first direct demonstration that endogenous synthesis was the major source of RA in milk fat. Subsequent investigations using the same approach extended the results to other dietary situations (total mixed diets with or without plant oils and pasture) and in all cases endogenous synthesis was the predominant source of the RA in milk fat (Corl *et al.* 2001; Kay *et al.* 2004). Results obtained with grazing cows are of special note because pasture is high in linolenic acid and endogenous synthesis accounted for >91% of the total RA in milk fat (Kay *et al.* 2004).

The second approach to quantify the contribution of endogenous synthesis to milk fat CLA was indirect and involved a comparison of ruminal outflow with secretion in milk; rumen output of RA would represent the maximum proportion of RA secreted in milk fat and the remainder would have to be derived from endogenous synthesis. For this approach, representative samples of digesta were obtained and data for CLA content were combined with marker-derived estimates of flow rates of digesta. Lock and Garnsworthy (2002), who conducted the first such investigation, estimated rumen output of CLA in nonlactating cows and then extrapolated results to lactating cattle on the basis of feed intake. Their estimates indicated that endogenous synthesis accounted for over 80% of the RA in milk fat in cows fed a grass silage/concentrate diet supplemented with various plant oils. Comparable results were obtained in subsequent studies using this approach with cows fed corn-silage diets containing either a high or low level of forage (Piperova *et al.* 2002), and grass silage/concentrate diets including fish oil supplements (Shingfield *et al.* 2003). Palmquist *et al.* (2004) used a mathematical modeling approach to quantify the importance of endogenous synthesis of CLA in adipose tissue of lambs, but it has not yet been applied to lactating cows.

Overall, investigators using different diets and experimental approaches have found similar results; the major source of RA in milk fat is endogenous synthesis (Figure 3.2). Thus, endogenous synthesis is the basis for *cis*-9, *trans*-11 being the predominant CLA isomer in milk fat and the relatively constant ratio between VA and RA observed in milk fat reflects the substrate:product relationship for Δ^9-desaturase.

3.4.3 *trans*-7, *cis*-9 CLA

Yurawecz *et al.* (1998) were the first to identify the presence of *trans*-7, *cis*-9 CLA in milk fat and to do so they used combinations of silver nitrate-mHPLC, GC-MS, and Fourier transform infrared spectroscopy. This CLA isomer had not been detected previously because it coeluted with RA in GC methods that were in routine use. Thus, concentrations of RA reported in the scientific literature typically include *trans*-7, *cis*-9 CLA as a component. Across studies, the level of *trans*-7, *cis*-9 CLA in milk fat has generally been on the order of 10% of RA and several fold greater than any of the other CLA isomers (Bauman *et al.* 2000; Corl *et al.* 2002; Piperova *et al.* 2002; Sehat *et al.* 1998; Yurawecz *et al.* 1998). Early investigations with Δ^9-desaturase from rat liver had established that *trans*-7 18:1 could serve as a substrate for this enzyme (Mahfouz *et al.* 1980; Pollard *et al.* 1980). Furthermore, *trans*-7 18:1 is present in rumen outflow, albeit at low concentrations (Table 3.2), being produced as an intermediate in the biohydrogenation of oleic acid and 18-carbon PUFAs, as discussed earlier. Based on this, when

Yurawecz *et al.* (1998) initially discovered *trans*-7, *cis*-9 CLA in milk fat, they speculated that it might originate from endogenous synthesis.

A number of the investigators who determined endogenous synthesis of RA also examined the source of *trans*-7, *cis*-9 CLA in milk fat. Corl *et al.* (2002) inhibited the activity of Δ^9-desaturase in lactating dairy cows with sterculic oil as a source of cyclopropene fatty acids and with *trans*-10, *cis*-12 CLA, a specific inhibitor of both activity and gene expression for Δ^9-desaturase (Baumgard *et al.* 2000; Bretillon *et al.* 1999; Lee *et al.* 1998; Park *et al.* 2000). Their data indicated that the *trans*-7, *cis*-9 CLA in milk fat was "derived almost exclusively from endogenous synthesis via Δ^9-desaturase" (Corl *et al.* 2002). Consistent with this, they also observed that there was no detectable *trans*-7, *cis*-9 CLA in rumen fluid. Piperova *et al.* (2002) used the indirect approach to calculate rumen outflow by combining duodenal content of *trans*-7, *cis*-9 CLA with an estimate of digesta flow. When estimates of rumen output of this CLA isomer were compared to that secreted in milk, they concluded that "almost the entire [amount of] *trans*-7, *cis*-9 CLA" found in milk fat must be produced postruminally. Thus, results from these two very different approaches were in agreement that the source of *trans*-7, *cis*-9 CLA in milk fat was endogenous synthesis via Δ^9-desaturase from ruminally produced *trans*-7 18:1.

3.4.4 The Δ^9-Desaturase Enzyme System

The predominance of endogenous synthesis as the source of RA and *trans*-7, *cis*-9 CLA in milk fat highlights the critical role of Δ^9-desaturase in the biology of CLA. Although referred to as Δ^9-desaturase in this review, this enzyme is also known as stearoyl-CoA desaturase (EC 1.14.19.1) in biochemistry texts because stearic acid is its most common substrate. The oxidative reaction catalyzed by Δ^9-desaturase involves cytochrome b5, NAD(P)-cytochrome b5 reductase, and molecular oxygen (Figure 3.3). The CoA ester of VA is the substrate for the formation of RA, but the preferred substrates for Δ^9-desaturase are stearoyl-CoA and palmitoyl-CoA, which are converted to oleoyl-CoA and palmitoleoyl-CoA, respectively (Ntambi 1999). For ruminants, a substantial activity of Δ^9-desaturase has been reported in mammary tissue (Bickerstaffe and Annison 1970; Kinsella 1972; McDonald and Kinsella 1973; Wahle 1974), adipose tissue (Barber *et al.* 2000; Cameron *et al.* 1994; Chang *et al.* 1992; Wahle 1974), and in intestinal epithelium (Bickerstaffe and Annison 1969). In contrast to rodents, the ruminant liver has only negligible activity. Both bovine (Chung *et al.* 2000; Cooney and Headon 1989) and ovine (Ward *et al.* 1998) Δ^9-desaturase genes have been cloned and only one gene has been found. This is similar to humans, but differs from rodents where two isoforms of the gene have been identified in rats and three isoforms of the gene have been characterized in mice (Ntambi and Miyazaki 2004).

Our understanding of the regulation of Δ^9-desaturase in ruminants is limited, with current knowledge coming mainly from investigations on rodents. Δ^9-Desaturase has no known allosteric or feedback inhibition involving its substrates or products. However, it is regulated by dietary factors such as glucose and PUFAs, and by hormones such as insulin and glucagon (Ntambi and Miyazaki 2004). The enzyme protein has a relatively short half-life (~4 h) and thus gene transcription is its major point of regulation (Ozols 1997). Both PUFAs and *trans*-10, *cis*-12 CLA downregulate gene expression, but RA has no effect (Choi *et al.* 2000; Lee *et al.* 1998; Ntambi and Miyazaki 2004). Interestingly, the cyclopropene fatty acids in sterculic oil do not affect expression of the Δ^9-desaturase gene or protein, but they directly inhibit the activity of the enzyme (Gomez *et al.* 2003).

At a cellular level, regulation of Δ^9-desaturase in mammary tissue appears to involve the sterol-response-element-binding-protein (SREBP) family of transcription factors (Peterson *et al.* 2004). PUFAs inhibit the processing of SREBP 1 and

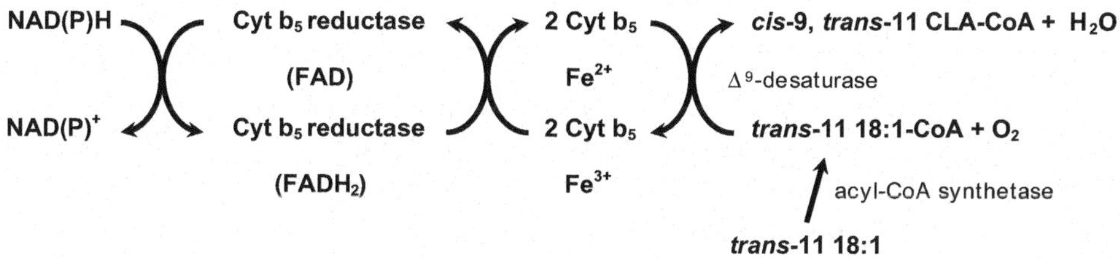

Figure 3.3. The Δ^9-desaturase enzyme system showing the conversion of vaccenic acid (*trans*-11 18:1) to rumenic acid (*cis*-9, *trans*-11 CLA).

may decrease the abundance of the precursor protein, leading to reduction in transcription of many genes in the lipogenic pathways, including Δ^9-desaturase (Horton *et al.* 2002; Shimano 2001). Ward *et al.* (1998) reported high expression of Δ^9-desaturase mRNA in adipose tissues and mammary glands of lactating sheep, and the expression was decreased by 80% in adipose tissues of animals during pregnancy and lactation, a time when lipogenic activity is increased in mammary glands and decreased in adipose tissues (Bauman and Currie 1980). The relationship between substrate and product for Δ^9-desaturase is reflected by the desaturase index, defined as [RA ÷ (RA + VA)] (Kelsey *et al.* 2003). The desaturase index in milk fat represents a proxy for Δ^9-desaturase and a several-fold range is observed among individuals. This is discussed in Section 3.5.2, but provides a strong indication that there are genetic differences among individuals with respect to this enzyme.

3.4.5 Other CLA Isomers

In contrast to *cis*-9, *trans*-11 and *trans*-7, *cis*-9, the other isomers of CLA found in the milk fat of ruminants appear to originate exclusively from rumen output. This conclusion is based, in large part, on the fact that these minor *cis*-*trans*, *trans*-*cis*, *cis*-*cis*, and *trans*-*trans* isomers are detected in rumen fluid (Corl *et al.* 2002) and duodenal fluid (Piperova *et al.* 2002; Shingfield *et al.* 2003), and estimates of digesta flow indicate that rumen output is more than adequate to account for the trace amounts secreted in milk fat (Piperova *et al.* 2002; Shingfield *et al.* 2003). Furthermore, there has been no demonstration that other mammalian desaturases act in a manner analogous to Δ^9-desaturase to synthesize CLA endogenously from *trans* octadecenoic fatty acids. Thus, these CLA isomers found at trace levels in milk fat are logically of rumen origin and represent intermediates formed in the biohydrogenation of PUFAs.

Information on the effect of diet on the production of minor isomers of CLA in the rumen and alterations in their content in milk fat is limited. Diet-induced changes in *trans*-10, *cis*-12 CLA have been best described, and their biological effects in the dairy cow will be discussed in Section 3.6.1. Griinari and Bauman (1999) presented a putative pathway for the biohydrogenation of linoleic acid where the initial isomerization involved the *cis*-9 double bond, thereby resulting in the production of *trans*-10, *cis*-12 CLA and *trans*-10 18:1 as intermediates. As discussed earlier, rumen bacteria have been identified that produce *trans*-10, *cis*-12 CLA when incubated with linoleic acid (Kim *et al.* 2002; Verhulst *et al.* 1987), and the addition of *trans*-10, *cis*-12 CLA to the rumen results in the increased formation of *trans*-10 18:1 (Loor and Herbein 2001).

Diet also influences the rumen output and milk fat content of other minor CLA isomers, although these isomers always remain a small portion of the total CLA in milk fat. For example, dietary supplements rich in linolenic acid increased the relative proportions of *trans*-11, *cis*-13 CLA, *trans*-11, *trans*-13 CLA, *cis*-12, *trans*-14 CLA, and *trans*-12, *trans*-14 CLA (Shingfield *et al.* 2006). Ruminal output of *trans*-

trans CLA isomers with double bonds at positions 9, 11 and 10, 12 was enhanced when diets contained high amounts of concentrates (Piperova *et al.* 2002) or were supplemented with fish oil (Shingfield *et al.* 2003). Kraft *et al.* (2003) reported that *trans*-11, *cis*-13 CLA represented 2–8% of total CLA in milk fat from cows grazing in the Alps. Kramer *et al.* (2004) verified this when they found relatively high concentrations of this isomer in a cheese sample produced from the milk of Alpine cows and *trans*-11, *cis*-13 CLA was also observed at significant concentrations in a sample of Yak milk fat (Cruz-Hernandez *et al.* 2004). Kraft *et al.* (2003) suggested that the *trans*-11, *cis*-13 CLA might be produced via rumen biohydrogenation of linolenic acid based on the lipid content of Alpine pasture, but this has not been established directly. More recent studies have demonstrated the presence of *trans*-11, *cis*-13 CLA in the digesta (Shen *et al.* 2011; Shingfield *et al.* 2008) and subcutaneous adipose tissue (Dugan *et al.* 2007; Petri *et al.* 2014) of ruminants. Recent work by Garcia *et al.* (2017) investigated the presence and origin of *trans*-11, *cis*-13 CLA. It was proposed that *trans*-11, *cis*-13 may originate in a similar dual origin mechanism to RA. The study suggests that levels of *trans*-13, *cis*-11 CLA in ruminant milk fat was due to an alternative and direct host FADS3-catalyzed Δ^{13} desaturation of VA, in addition to incomplete biohydrogenation of linolenic acid. Furthermore, the same group tested this hypothesis *in vivo* using lactating female rats. The rats were placed on either a high or low VA diet. Rats consuming the high VA diet produced *trans*-11, *cis*-13 CLA. Garcia *et al.* (2017) concluded that despite low metabolic conversion, results demonstrated that lactating mammary tissues can produce *trans*-11, *cis*-13 CLA from dietary VA. In addition, they suggest FADS3 enzyme as a *trans* vaccinate Δ^{13} desaturase *in vivo* (Garcia *et al.* 2018).

Overall, most investigations of the effects of diet on CLA have not used the detailed analytical methods required to resolve the full range of minor CLA isomers. Thus, more completely identifying rumen biohydrogenation pathways and establishing their relationship to specific rumen bacteria and diets are important areas for future research.

3.5 Modification of CLA in Milk Fat

The discovery of health benefits of CLA and a recognition of the potential of RA as a functional food component in dairy products has stimulated research to identify factors that affect the CLA content of milk fat. These efforts have focused on enhancing the CLA content per unit of fat and centered on RA as the predominant CLA isomer. From the preceding discussion on the origin of RA (Section 3.4.2), there are four possibilities to consider: (i) increase the 18-carbon PUFA precursors in the diet (linoleic and linolenic acids); (ii) maintain rumen biohydrogenation pathways that result in the production of VA as an intermediate; (iii) inhibit the final step in the biohydrogenation of 18-carbon PUFA so that VA accumulates; and (iv) increase Δ^9-desaturase and the desaturation of VA to RA in the mammary gland. In the following sections, we will discuss results of investigations designed to establish dietary and nutritional conditions that maximize rumen outflow of VA and RA, optimize the amount and activity of Δ^9-desaturase in mammary tissue, and identify the physiological basis for the large differences among individuals in terms of the production of CLA. Obviously, before CLA-enriched foods are widely marketed, effects on quality and consumer acceptability of the dairy products also need to be examined, and the limited research on this will be summarized also.

3.5.1 Dietary and Nutritional Effects

Numerous studies have shown that diet is the most significant factor affecting the CLA content of milk fat, and its concentration can be increased several fold by dietary means (see reviews by

Shingfield *et al.* 2013; Chilliard *et al.* 2000, 2001; Bauman *et al.* 2001; Stanton *et al.* 2003; Lock and Bauman 2004). As cited above, one key to increasing milk CLA is to increase the dietary intake of 18-carbon PUFAs, thereby providing more substrate for rumen biohydrogenation. The dietary supply of linoleic and linolenic acids is most easily increased by the addition of plant oils rich in these fatty acids, and a range of plant oils have been investigated and shown to be effective in increasing the level of CLA in milk fat. For example, dietary supplements of soybean, sunflower, rapeseed, or linseed oils have been used successfully to increase the level of CLA in milk fat (Kelly *et al.* 1998; Dhiman *et al.* 2000; Chouinard *et al.* 2001; Lock and Garnsworthy 2002). For a detailed review of the effects of dietary supplementation on milk fat CLA, see Kliem and Shingfield (2016).

The slow release of PUFAs in the rumen typically creates favorable conditions for the accumulation of *trans*-18:1 fatty acids, thereby increasing rumen output of VA (Bauman and Griinari 2001). The coat of oil seeds offers some protection against rumen biohydrogenation, and thus the use of different oil seeds and processing techniques has been investigated also; correspondingly, a range of oilseeds containing both linoleic and linolenic acids have been shown to be affective in increasing the CLA content of milk fat (AbuGhazaleh and Buckles 2007; Bu *et al.* 2007; Stanton *et al.* 1997; Dhiman *et al.* 2005). In general, oil seeds that are rich in 18-carbon PUFAs and processed so that the oil is accessible to the bacteria involved in biohydrogenation result in greater increases in milk CLA compared with whole oil seeds, but are not as efficient as using the pure oil (Lock *et al.* 2004). The use of calcium salts of fatty acids derived from plant oils has also been investigated because of the partial protection that the calcium-fatty acid complex offers from rumen biohydrogenation. Chouinard *et al.* (2001) fed calcium salts of fatty acids derived from rape, soybean, and linseed oils; all three increased the CLA content of milk fat, with the largest increases occurring for those containing the greatest amounts of linoleic and linolenic acids (soybean and linseed, respectively). Furthermore, Theurer *et al.* (2009) fed calcium salts and PUFA, which resulted in a 19% increase of CLA and a 14% increase in vaccenic acid in milk fat. They attributed the increase in milk CLA to incomplete biohydrogenation of linoleic and linolenic acids to VA and synthesis of CLA from VA in the mammary gland.

The amount of 18-carbon PUFAs that can be added to the diets of dairy cows is limited due to the adverse effects these PUFAs can have on the metabolism of rumen bacteria, thereby impairing rumen fermentation and animal performance (Kliem and Shingfield 2016; Muruz and Çetinkaya 2019; Jenkins 1993). Thus, dairy cattle diets are generally restricted to less than 7% total lipid, and this provides an upper limit in the use of lipid supplements. Oilseeds and chemical protection of oils offer some benefit as they often allow for greater amounts of the oil to be fed before negative effects on microbial growth and metabolism are realized. When a high level of oil is added, up to tenfold increases in the CLA content of milk fat are observed, but because of negative effects on rumen bacteria discussed above, these levels are often transient and decline within a few weeks to stabilize at ~four- to fivefold increases (e.g., Bauman *et al.* 2000). In addition, there is often a fine line between supplying additional lipid supplements to increase milk fat CLA content and causing changes in the rumen environment; for example, under some conditions, the rumen environment may be modified to produce more *trans*-10 18:1 and *trans*-10, *cis*-12 CLA as intermediates and this results in a dramatic reduction in milk fat synthesis (Kliem and Shingfield 2016).

Another means through which dietary and nutritional factors can increase the CLA content of milk fat is by inhibiting the terminal step in biohydrogenation (Figure 3.2). This typically occurs either directly or indirectly *via* changes in the rumen environment; the net result is an accumulation of VA, thereby increasing the rumen outflow of this precursor for the endogenous synthesis of CLA (see Chilliard *et al.* 2007; Frutos *et al.* 2017). Downregulation of the SCD gene by FA in marine lipids or BH intermediates might also contribute to this response (Carreño *et al.* 2016) although data on *in vivo* SCD activity are inconclusive (Faulconnier *et al.* 2018). A limited number of bacterial species have been shown to carry out the final biohydrogenation step and, presumably, changes in the rumen environment lead to a reduction in these species and/or a reduction in their capacity to reduce VA to stearic acid. Several dietary situations also have these effects and they include alterations in the forage: concentrate ratio, dietary supplements of fish oil, and restricted feeding (see Bauman and Griinari 2001). The most consistently effective of these is the use of fish oils. Fish oils themselves provide very little 18-carbon PUFAs precursors to allow for increased rumen VA output, indicating that this increase occurs through an inhibition of the biohydrogenation of VA; indeed, C22:6 *n*-3 (docosahexaenoic acid, DHA), a major *n*-3 fatty acid in fish oil, has been shown to promote accumulation of VA in mixed ruminal cultures when incubated with linoleic acid (AbuGhazaleh and Jenkins 2004). Both linoleic and linolenic acids are plentiful in forages and concentrates which provide sufficient 18-carbon PUFA precursors. A range of fish and marine oils have been used with success, and similar to the supply of 18-carbon PUFAs discussed above, both lipid supplements and fish by-products (fish meal) have been shown to be effective (AbuGhazaleh *et al.* 2003; Offer *et al.* 1999; Shingfield *et al.* 2003; Whitlock *et al.* 2002). Marine algae also contain long-chain PUFAs and have also been effective (Franklin *et al.* 1999).

The most effective dietary treatments for increasing the CLA content of milk fat are those that both increase the supply of 18-carbon PUFAs and modify the rumen environment. The most widely studied of these is the use of fresh pasture, with numerous studies indicating that fresh pasture results in a two- to threefold increase in the CLA content of milk fat (AbuGhazaleh *et al.* 2007; Gulati *et al.* 2018; Stanton *et al.* 1997; O'Callaghan *et al.* 2016a, b; Slots *et al.* 2009). Fresh-pasture-associated increases in milk CLA content have been attributed to increased α-linolenic acid content in grasses, which is extensively biohydrogenated in the rumen (Kay *et al.* 2004). The degree of response, however, decreases as the pasture matures and the proportion in the diet decreases. Correspondingly, seasonal effects on milk CLA content have been reported, with the trend that content is greatest when fresh pasture is plentiful, and decreases throughout the growing season (Riel 1963; Auldist *et al.* 1999; Banni *et al.* 1996; Lock and Garnsworthy 2003). These results cannot be explained fully in terms of the fatty acid composition and supply of PUFAs that grass provides; therefore, there must be additional factors or components of grass that promote the production of VA in the rumen, and these lessen in effect as the pasture matures (Lock and Bauman 2004). Presumably, these factors inhibit the conversion of VA to stearic acid, as discussed previously. The effect of different farming systems has also been investigated, with systems differentiated by the amount and type of forages typically fed to cows. In general, production systems with the greatest proportion of fresh forage in the diet give the highest level of CLA in milk fat. For example, Jahreis *et al.* (1997) reported that cows that grazed during the summer months had a higher level of CLA in milk than cows housed all-year round and fed conserved forage. More recently, O'Callaghan *et al.* (2016a, b) showed greater than twofold increase in the conjugated linoleic acid C18:2 *cis*-9, *trans*-11 content of raw milk compared with that of the total mixed ration (TMR) feeding system over an entire lactation.

Research on the effects of direct-fed microbials (probiotics) on ruminal BH and the FA profile of milk and meat are an area of great interest. Information is still very limited, but encouraging results (i.e., increased milk *trans*-11-C18:1 and *cis*-9-*trans*-11-CLA concentration) have been shown in goats fed strains of *Butyrivibrio fibrisolvens* (Shivani *et al.* 2015) *Lactobacillus plantarum* (Maragkoudakis *et al.* 2010), and a mixture of *Lactobacillus reuteri*, *Lactobacillus alimentarius*, *Enterococcus faecium*, and *Bifidobacterium bifidum* (Apas *et al.* 2015). Greater milk 18:2n-6 and 18:3n-3 proportions were also found in the two latter studies.

In terms of the persistency of changes, the response of rumen microbiota to biohydrogenation modulators may vary over time, with some inconsistent effects most likely being due to interspecies differences. For instance, in cows fed a ration based on corn silage, fish oil induced decreases in milk 18:0 and *cis*-9-C18:1 concentrations which were transient, and the large enhancements in milk *trans*-11-C18:1 and *cis*-9-*trans*-11-CLA found during the first days on treatment subsequently declined (Shingfield *et al.* 2006). However, the effects of marine lipids in grazing cows (AbuGhazaleh *et al.* 2009) were similar to those in ewes (Bichi *et al.* 2013), with rapid and persistent enrichments in milk *trans*-11-C18:1 and *cis*-9-*trans*-11-CLA that suggest major changes in the rumen microbiota at the beginning of treatment and relative stability afterward. Nevertheless, slower responses in milk *trans*-10-C18:1 suggest that changes in microorganisms involved in alternative BH pathways may take longer (Shingfield *et al.* 2006; Boeckaert *et al.* 2008; Carreño *et al.* 2016).

Although the use of fresh pasture has striking effects on enhancing the CLA content of milk fat, a similar increase is possible using standard dietary ingredients such as plant oils/oilseeds and fish oil/fish meal supplements. Further, there is some indication that dietary regimes involving a combination of supplements can have an additive effect on increasing the level of CLA in milk; for example, Whitlock *et al.* (2002) observed higher levels with a combination of plant oil and fish oil than when either was fed alone. In all of the dietary situations designed to enhance the level of CLA in milk fat, it is vital that the normal VA pathway of biohydrogenation is maintained. If shifts in biohydrogenation occur, then the pattern of *trans* fatty acids changes and there will be a reduction in the rumen output of VA, and as a consequence a reduction in the level of CLA in milk fat. This shift in the pathways of biohydrogenation is also associated with an increased risk of depression of milk fat synthesis (see Section 3.6.1).

3.5.2 Physiological Factors

Physiological factors also have an impact on the content of CLA in milk fat. Surveys have shown an eightfold range in the milk fat content of CLA among herds (Riel 1963; Kelly and Bauman 1996) and these differences in large part reflect diet and nutritional effects as discussed above. However, substantial differences are observed among cows within a herd consuming the same diet. Investigations involving diets ranging from corn-based total mixed rations to pasture have all shown a two- to threefold range in the milk fat content of CLA among individual cows (e.g., Kelly *et al.* 1998; Lawless *et al.* 1998; Lock and Garnsworthy 2002; Lock and Garnsworthy 2003; Peterson *et al.* 2002b). Thus, across diets that result in substantial differences in the average milk fat content of CLA, a similar two- to threefold range is observed among cows consuming the same diet. This variation would in large part be related to individual differences in, (1) rumen output of VA and to a lesser extent CLA, and (2) the amount and activity of Δ^9-desaturase.

The final method to enhance the level of CLA in milk is to increase endogenous synthesis and this must be the basis for the variation among cows in a herd fed the same diet. Undoubtedly, the variation in Δ^9-desaturase among individuals has a genetic basis (Bauman and Griinari 2003; Garnsworthy *et al.* 2010). It has been demonstrated that milk fat CLA can differ two- to threefold in individuals fed the same diet (Kelsey *et al.* 2003). Furthermore, individual cows maintain their relative ranking for Δ^9-desaturase index irrespective of dietary FA manipulations (Lock *et al.* 2006), suggesting that Δ^9-desaturase activity may have a genetic component. Two Δ^9-desaturase isoforms have been identified in cattle, Δ^9-desaturase 1 is located on chromosome 26 and expressed in a variety of tissues among which are adipose and mammary tissues (Taniguchi *et al.* 2004). Δ^9-desaturase 5 is located on chromosome 6 and expressed primarily in the brain (Lengi and Corl 2007) and it is also expressed in the bovine mammary tissue (Jacobs *et al.* 2011). The exact number of Δ^9-desaturase genes in bovine is unknown (Campbell *et al.* 2001). The single nucleotide polymorphism (SNP) valine (V) for alanine (A) at position 293 of Δ^9-desaturase 1 (Taniguchi *et al.* 2004) has been suggested as a candidate gene to explain the variability of the FA profile in bovine milk (Schennink *et al.* 2008). More recently, SNP g.133A>C was reported as a marker with a dual impact on milk yield and FA desaturation level, making it an interesting candidate for assisted selection programs of the species (Gu *et al.* 2019)

The genetic components underlying Δ^9-desaturase activity among individuals may also be indirectly evaluated. This is possible because milk fat contains four major fatty acid pairs that represent a product/substrate relationship for Δ^9-desaturase, myristoleic/myristic acid, Palmitoleic/palmitic acid, oleic/stearic acid, and RA/VA. Ratios for these pairs of fatty acids, referred to as a desaturase index, represent a proxy for Δ^9-desaturase activity. In the largest study to examine this, Kelsey *et al.* (2003) found that the variation in milk fat content of RA and the desaturase index was about threefold among individuals consuming the same diet. Other investigations have also observed a two- to threefold range in desaturase index among cows in the same herd (Lock and Garnsworthy 2002; Lock and Garnsworthy 2003; Peterson *et al.* 2002b). Peterson *et al.* (2002b) also demonstrated a consistency in the individual hierarchy in desaturase index over time when cows were fed the same diet and a consistency in the individual hierarchy when cows were switched between diets. Presumably, this variation reflects individual differences in the activity of Δ^9-desaturase involving the regulation of gene expression, primary or tertiary structure of the enzyme due to gene polymorphisms, post-translational modifications, or other factors affecting the interaction between the enzyme and the substrate or product.

Based on the desaturase index, studies (Garnsworthy *et al.* 2010; Krag *et al.* 2013; Schennink *et al.* 2007) provide evidence that additive genetic effects are responsible for a significant proportion of the phenotypic variation in Δ^9-desaturase activity in dairy cows. They also reported that the heritability of Δ^9-desaturase activity is comparable to the heritability of milk yield. Desaturase activity could, therefore, be used in future breeding programs to improve the FA profile of milk fat by increasing MUFA and RA concentrations and decreasing SFA concentrations. Increasing the Δ^9-desaturase activity would not only impact the level of CLA in milk fat but also increase other unsaturated fatty acids that are products of this enzyme. As a consequence of these changes, the saturated: unsaturated content of milk fat would be altered, resulting in an improvement in the "human health" characteristics of milk fat. Thus, establishing the heritability of individual differences in the desaturase index and the extent to which this could be used in genetic selection programs is of interest. This potential to improve the fatty acid composition of milk fat is also the basis for work to produce transgenic goats that have greater expression of Δ^9-desaturase in the mammary gland (Reh *et al.* 2004).

Several specific physiological factors have been examined for effects on the level of CLA in milk fat, but because of the large impact of diet and the wide range among individuals, it is important that these comparisons involve a reasonable number of cows fed a common diet. These conditions were met in the studies by Kelsey *et al.* (2003) and Lock *et al.* (2003), who both found that the CLA content of milk fat and the desaturase index had no relationship to parity or stage of lactation (days in milk). Likewise, they observed that milk fat content of CLA and desaturase index also had no relationship to milk yield, milk fat percent, or yield of milk fat. The study by Kelsey *et al.* (2003) involved over 200 cows fed the same diet and found no difference between Holstein and Brown Swiss breeds. In contrast, several studies have reported breed differences in the CLA content of milk fat (Lawless *et al.* 1999; White *et al.* 2001; Whitlock *et al.* 2002), which may reflect differences in desaturase index among breeds. However, these studies often involved very few animals or were confounded by diet, or both. Using a larger data set, DePeters *et al.* (1995) reported breed differences in the desaturase index in milk fat of dairy cows, consistent with the suggestion that the activity of Δ^9-desaturase is higher in Holstein than in Jersey mammary tissue (Beaulieu and Palmquist 1995). However, if breed differences exist, they would appear to be minor compared with the magnitude of dietary effects and variation among cows in terms of both the CLA content of milk fat and desaturase index (Hanuš *et al.* 2016; Samková *et al.* 2018).

3.5.3 Manufacturing and Product Quality Considerations

Consumer surveys indicate an interest in dairy products that are enriched in CLA (Ramaswamy *et al.* 2001). As outlined in preceding sections, the level of CLA in milk fat can be enhanced several fold naturally by diet formulation and selection of individual cows with elevated milk fat CLA. But central to marketing and consumer acceptance of CLA-enriched foods is a consideration of the effects of processing and storage, and the final sensory characteristics of CLA-enriched products. Many dairy products undergo a microbial fermentation during processing and the effects of these on the CLA content have been of special interest. Several studies have investigated this and found that food processing and manufacturing have little or no effect on CLA content (Luna *et al.* 2005; Rodriguez-Alcala and Fontecha 2007; Shantha *et al.* 1992; Gnädig *et al.* 2004; Jiang *et al.* 1996; Lin *et al.* 1999; Shantha and Decker 1995; Werner *et al.* 1992). As emphasized in the review by Sébédio *et al.* (2003), any changes in the CLA content related to processing or to storage are minimal when compared to the variations associated with diet formulations and differences between individual cows. Thus, the final concentration of CLA in dairy products is, in large part, related to the CLA concentration in the raw milk fat and the fat content of the final product.

Consumer acceptability of CLA-enriched dairy products is also dependent on their organoleptic properties. Off-flavors due to fatty acid oxidation are of prime concern because diet formulation methods used to enhance naturally milk fat with respect to CLA generally cause an increase in the proportion of unsaturated fatty acids in the milk fat (Lock and Bauman 2004). Reports on sensory characteristics and quality of naturally enriched dairy products, typically having a two- to threefold increase in milk fat CLA, have generally indicated no differences from unenriched dairy products (Ramaswamy *et al.* 2001; Baer *et al.* 2001; Avramis *et al.* 2003; Gonzalez *et al.* 2003). For instance, O'Callaghan *et al.* (2016a, b) reported an increased sensory score for butter made from pasture-derived milk which had a twofold higher CLA content than butter made from milk of cows fed TMR. Similar findings have been reported by other authors (Faulkner *et al.* 2018; Fretin *et al.* 2019; Khanal *et al.* 2005; O'Callaghan *et al.* 2017). CLA enrichment through dairy products made from cows fed on pasture systems appears more efficient compared in terms of both CLA enrichment and improving consumer taste perception versus confined systems (Kilcawley *et al.* 2018).

In contrast, Lacasse *et al.* (2002) found that 2.7%-fat milk from cows fed either protected (3% of dry matter) or unprotected fish oil (3.7% of dry matter) scored significantly lower in flavor and taste. However, the levels of fish oil used in this study were significantly greater than used by others.

Lynch *et al.* (2005) compared the flavor, organoleptic, and storage characteristics of standard 2%-fat milk with 2%-fat milk that had an approximately tenfold higher level of CLA. The naturally enhanced milk (CLA and VA equaled 47 and 121 mg/g fatty acids, respectively) was produced through individual selection and nutritional management of the cows. Initial evaluation of the milk and evaluation over a 14-day post-pasteurization period indicated no flavor differences as determined by triangle taste tests. Similarly, sensory results indicated no differences in susceptibility to the development of oxidized off-flavors between the control and CLA-enhanced milks, even when milk was stored under light (Lynch *et al.* 2005). Thus, flavor and consumer acceptability were maintained in a dairy product with substantially enhanced levels of CLA and VA.

The research discussed above involved dairy products that were naturally enriched with CLA through formulation of diets known to increase the level of CLA in milk fat and selection of individual cows with a higher level of CLA in their milk fat. Campbell *et al.* (2003) used an alternative approach involving fortification of milk fat with synthetic CLA during the manufacturing process. They added 1 or 2% CLA-containing triglycerides to skim milk together with vitamin E and rosemary extract to retard lipid oxidation. Descriptive sensory analysis revealed the fortified milk had a "grassy/vegetable oil" flavor and consumer acceptability scores were low, although acceptability was improved when a chocolate flavor was added.

Overall, results to date indicate that manufacturing and quality characteristics were normal for dairy products naturally enriched with CLA and consumer acceptability was comparable to unenriched dairy products. However, the single study examining fortification of skim milk with synthetic CLA during the manufacturing process had poor consumer acceptability.

3.6 Biological Effects of CLA Isomers

A broad overview of the biological effects of CLA is presented elsewhere in this volume (Chapter 10), so the emphasis in the following section will be twofold. First, the biology of *trans*-10, *cis*-12 CLA in the dairy cow will be summarized because under certain dietary conditions, production of this isomer in the rumen can profoundly affect milk fat synthesis. Second, the biological effects of RA when supplied as a natural component of the diet will be reviewed because this CLA isomer represents a functional component of milk fat that has potential health benefits; this will also briefly include discussion of the *trans*-10 *cis*-12 isomer as it is often included in studies as part of mixed CLA isomer supplementation. Although many other CLA isomers are present in milk fat, they are present at concentrations much too low to have a significant effect.

3.6.1 *trans*-10, *cis*-12 CLA and Lipid Metabolism

3.6.1.1 Inhibition of Milk Fat Synthesis

Investigations in which the transfer of CLA to milk fat in dairy cows was examined showed that supplementation of mixed isomers of CLA resulted in a dramatic reduction in milk fat secretion (Chouinard *et al.* 1999a, b; Loor and Herbein 1998). Decreases of up to 50% in milk fat yield occurred and the effects were reversed when supplementation was terminated. Furthermore, effects were specific for milk fat with the yield of milk and other milk components being relatively unaffected. Initial investigations were of short duration (<7 days) and the CLA supplement was infused abomasally as a convenient experimental method to avoid possible alterations during rumen fermentation. However, subsequent long-term studies (20 weeks) demonstrated that the reduction in milk fat synthesis was maintained when a rumen-protected formulation of CLA was used (Bernal-Santos *et al.* 2003; Perfield *et al.* 2002).

Early investigations utilized CLA supplements that were composed of a mixture of, generally, four or more isomers. Baumgard *et al.* (2000) reported the first evidence of the differential effect of specific CLA isomers on milk fat synthesis; they demonstrated that abomasal infusion of *trans*-10, *cis*-12 CLA resulted in an immediate decrease in milk fat synthesis whereas RA had no effect. However, two subsequent studies have demonstrated the potential role of two additional CLA isomers in reducing milk fat synthesis: *cis*-10, *trans*-12 CLA and *trans*-9, *cis*-11 CLA. The first study (Sæbø *et al.* 2005b) examined the effects of geometric isomers of *trans*-10, *cis*-12 CLA. This study demonstrated that a mixed isomer CLA supplementation (which included *trans*-10-*cis*-12, *cis*-10 *trans*-12, *trans*-10 *trans*-12 and *cis*-10-*cis*-12) effectively reduced milk fat synthesis. Due to the abundance of *cis*-10, *trans*-12 CLA in milk and low transfer efficiency, authors suggested that *cis*-10, *trans*-12 was the isomer responsible for the decreased milk fat synthesis. Similarly, a study by Perfield and colleagues also used a mixed isomer CLA supplementation that successfully reduced milk fat synthesis. The supplementation was composed of three CLA isomers; *trans*-9, *cis*-11 *trans*-9, *trans*-11 and *cis*-9, *trans*-11. *Cis*-9, *trans*-11 CLA has previously been determined to be unsuccessful in reducing milk fat synthesis. The study also examined *trans*-9, *trans*-11 CLA as an isolated isomer, which proved ineffective. Thus, by deduction, Perfield *et al.* (2007) concluded that *trans*-9, *cis*-11 CLA was the active component responsible for reduced milk fat synthesis. As both studies were carried out using mixtures of isomers, studies investigating the potential effects of these isomers in isolation are required. In a number of other studies, additional CLA isomers have been examined via abomasal infusion and these have included *trans*-8, *cis*-10 CLA, *cis*-11, *trans*- 13 CLA, *cis*-9, *trans*-11 CLA, *trans*-9, *trans*-11 CLA, and *trans*-10, *trans*-12 CLA (Perfield *et al.* 2004a, b, 2006, 2007; Sæbø *et al.* 2005b). Although all of these isomers were taken up by the mammary gland and incorporated into milk fat, none affected the rate of milk fat synthesis. In addition, Kadegowda *et al.* (2008) suggested the involvement of *trans*-7, *cis*-9 CLA in reduced milk fat synthesis. However, this isomer has only been studied indirectly via abomasal infusion of stearic acid and in all studies, no effect was seen on milk fat output (Bauman *et al.* 2011). Thus, *trans*-10, *cis*-12 CLA remains the most important isomer when investigating reduced milk fat synthesis.

The initial step in the metabolism of linoleic and linolenic acids to form eicosanoids is catalysis by Δ^6-desaturase. The metabolite is formed by the action of Δ^6-desaturase on *trans*-10, *cis*-12 CLA is *cis*-6, *trans*-10, *cis*-12 18:3. Investigations of this fatty acid, as well as *cis*-6, *trans*-8, *cis*-12 18:3, have established that neither of these conjugated trienoic 18:3 fatty acids affects milk fat synthesis or any other lactational variable (Sæbø *et al.* 2005a).

Relationships between *trans*-10, *cis*-12 CLA and milk fat synthesis have been examined. There is a curvilinear relationship between the reduction in milk fat yield and the abomasal infusion dose of *trans*-10, *cis*-12 CLA (Figure 3.4). *Trans*-10, *cis*-12 CLA is a very potent inhibitor of milk fat synthesis in dairy cows; a dose of 2.0 g/d (<0.01% of dry matter intake) reduced milk fat synthesis by 20%. *Trans*-10, *cis*-12 CLA is also incorporated into milk fat and in this case, the relationship is linear (Figure 3.4); a summary of seven studies showed that the transfer efficiency

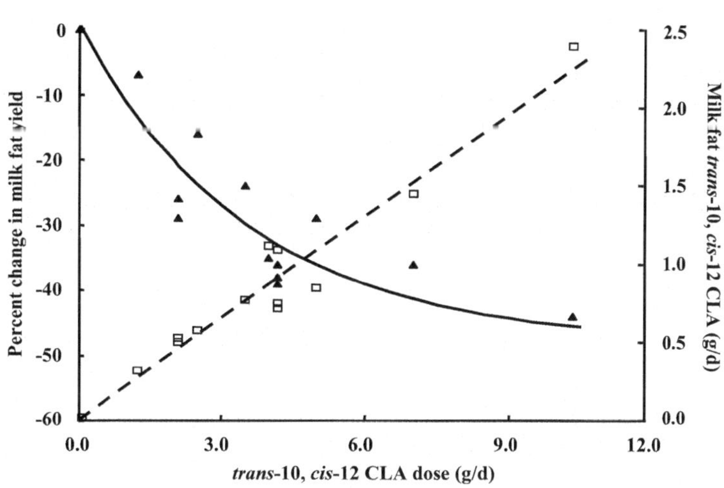

Figure 3.4. Relationships between the dose of *trans*-10, *cis*-12 CLA infused into the abomasum and (i) change in milk fat yield (▲; $y = -48.26 + 49.03exp^{-0.2782x}$; $R^2 = 0.86$), and (ii) secretion of *trans*-10, *cis*-12 CLA into milk fat (□; $y = 0.2175x + 0.0111$; $R^2 = 0.94$). Adapted from a summary by de Veth *et al.* (2004) using data from Baumgard *et al.* (2000, 2001, 2002), Peterson *et al.* (2002a), Sæbø *et al.* (2005a, b), de Veth *et al.* (2004) and Perfield *et al.* (2004c).

of abomasally infused *trans*-10, *cis*-12 CLA into milk fat averaged 22% (de Veth *et al.* 2004). The linear relationship in transfer to milk fat is remarkable when one considers that the yield of milk fat is simultaneously decreased as the abomasal dose of *trans*-10, *cis*-12 CLA is increased. This suggests that the mechanisms which coordinate the CLA-induced decrease in the use of preformed fatty acids for milk fat synthesis have a less pronounced effect on the mammary uptake and incorporation of *trans*-10, *cis*-12 CLA into milk fat, but the basis for this difference is unknown.

Initial studies on CLA showed that the reduction in milk fat secretion reflected decreases in fatty acid levels of all chain lengths, but effects were most pronounced for those synthesized *de novo* (Chouinard *et al.* 1999a, b; Loor and Herbein 1998). As investigations focused on *trans*-10, *cis*-12 CLA and expanded to include a range of doses, it was discovered that at lower doses, the reduction in milk fat was distributed more uniformly among the fatty acids synthesized *de novo* (short-chain and medium-chain lengths) and the longer-chain fatty acids taken up from the blood (Baumgard *et al.* 2001; Peterson *et al.* 2002a). Likewise, an inhibition of Δ^9-desaturase which resulted in a marked shift in the fatty acid composition of milk fat was observed only at doses of *trans*-10, *cis*-12 CLA where milk fat production was reduced by >20%. At lower doses of *trans*-10, *cis*-12 CLA, the ratio of fatty acids representing product/substrate for Δ^9-desaturase was unaffected (Baumgard *et al.* 2001; de Veth *et al.* 2004; Peterson *et al.* 2002a).

The changes observed in the fatty acid composition of milk in CLA-supplemented cows suggest that many of the processes involved in milk fat synthesis must be affected. Baumgard *et al.* (2002) conducted the first investigation of this by quantifying the abundance of mRNA for several lipogenic enzymes in mammary tissue obtained 5 days after treatment with *trans*-10, *cis*-12 CLA. They found that the 48% reduction in milk fat yield corresponded to a reduction of similar magnitude in the abundance of mRNA for genes that encoded for enzymes involved in the uptake and transport of circulating fatty acids (lipoprotein lipase and fatty acid-binding protein), *de novo* fatty acid synthesis (acetyl CoA carboxylase and fatty acid synthase), desaturation of fatty acids (Δ^9-desaturase), and triglyceride synthesis (glycerol phosphate acyltransferase and acylglycerol phosphate acyltransferase) (Gervais *et al.* 2009; Harvatine and Bauman 2006; Peterson *et al.* 2003; Piperova *et al.* 2000). Subsequent work using a bovine mammary epithelial cell line has given similar results when cells were incubated with *trans*-10, *cis*-12 CLA (Peterson *et al.* 2004).

The biochemical responses described above support the hypothesis that the reduction in the production of milk fat involves a coordinated regulation of key lipogenic enzymes in the mammary gland, and logical candidates as a central regulator of lipid synthesis are the sterol-response-element-binding-proteins (SREBP). The role of SREBP in the regulation of lipid metabolism has been characterized elegantly in rodents where promoters for sterol response elements have been identified in genes for key enzymes in the pathways of fatty acid synthesis and metabolism (see reviews by Horton *et al.* 2002; Shimano 2001). Recently, Peterson *et al.* (2004) found that bovine mammary epithelial cells also contain SREBP. They demonstrated further that *trans*-10, *cis*-12 CLA reduced lipid synthesis in these cells through inhibition of the proteolytic activation of SREBP-1 and subsequent reduction in translational activation of lipogenic genes. Thus, the mechanism whereby *trans*-10, *cis*-12 CLA affects milk fat synthesis appears to involve alterations in the activation of this transcription factor. Moreover, a study by Harvatine and Bauman (2006) investigated the SREBP transcription factor system in the mammary tissue of milk fat depression (MFD); their study supports the findings of Peterson *et al.* (2004). They demonstrated that cows with MFD due to *trans*-10, *cis*-12 CLA infusion resulted in a decrease in mammary expression of SREBP-1, SREBP-1 activation protein, and a coordinated reduction in SREBP-1-responsive lipogenic enzymes. In addition, a 59% reduction in SREBP-1 as a result of MFD was seen by Gervais *et al.* (2009).

Therefore, it is apparent from these findings that SREBP-1 plays a key role in the regulation of milk fat synthesis. Harvatine and colleagues also identified the thyroid hormone-responsive spot 14 (S14) as a *trans*-10 *cis*-12 CLA-responsive gene. They used multivariate analysis to demonstrate a significant relationship between expression of S14 and lipogenic enzymes (FAS and lipoprotein lipase), suggesting the involvement of S14 in mammary regulation of milk fat synthesis (Harvatine and Bauman 2006).

3.6.1.2 Relationship to Diet-Induced Milk Fat Depression

Under particular dietary situations, a reduction in the content and yield of milk fat occurs in dairy cows. This has commonly been referred to as milk fat depression (MFD) and recent investigations indicate that this metabolic syndrome is related, at least in part, to effects of specific CLA isomers on rates of milk fat synthesis. First described over a century ago, MFD has most often been observed with diets that are low in roughage and high in starch, diets containing plant or fish oil supplements, and diets where effective fiber is reduced by processing the forage (e.g., grinding or pelleting). Effects are specific for milk fat and decreases of up to 50% have been observed.

Diet-induced MFD has been the subject of extensive research, especially over the last 60 years (see reviews by Bauman and Griinari 2001; Bauman *et al.* 2011; Davis and Brown 1970; Jenkins and Harvatine 2014; Palmquist *et al.* 1993). Many theories have been advanced to explain diet-induced MFD. However, most have been found inadequate to explain the cause and mechanism of this phenomenon (Bauman and Griinari 2001, 2003; Doreau *et al.* 1999). A shift in rumen fermentation is clearly involved and the occurrence corresponds to a marked increase in the *trans*-C18:1 content of milk fat (Davis and Brown 1970; Griinari *et al.* 1998). While VA is generally the principal *trans*-C18:1 isomer in milk fat, a key development was the discovery by Griinari *et al.* (1998) that the change with diet-induced MFD specifically involved an increase in the *trans*-10-C18:1 isomer. Subsequently, this was verified for other dietary conditions (Offer *et al.* 2001; Peterson *et al.* 2003; Piperova *et al.* 2000), and it established that diet-induced MFD involved a shift in the rumen pathways of biohydrogenation, as indicated in Figure 3.5.

Bauman and Griinari (2001) recognized the central role of rumen biohydrogenation in MFD and proposed that "under certain dietary conditions, the pathways of rumen biohydrogenation are altered to produce unique fatty acid interme-

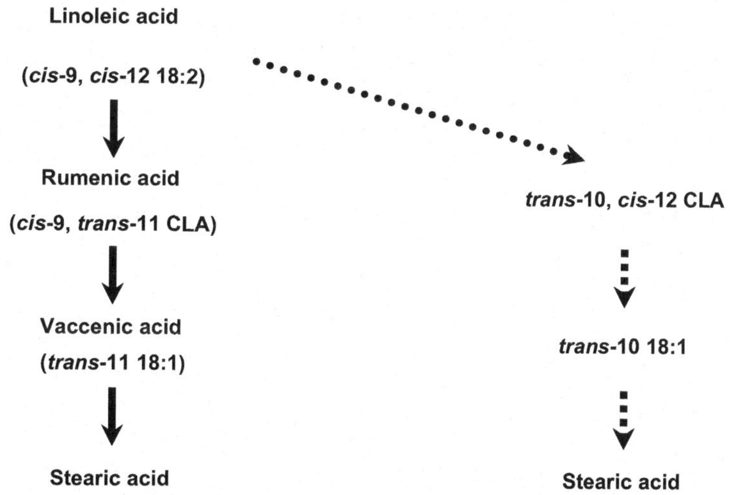

Figure 3.5. Generalized scheme of ruminal biohydrogenation of linoleic acid under normal conditions (solid line) and during diet-induced milk fat depression (dotted line). (Adapted from Griinari and Bauman 1999).

diates which are potent inhibitors of milk fat synthesis." This is referred to as the "biohydrogenation theory" and results have demonstrated that diet-induced MFD is generally correlated with the level of *trans*-10, *cis*-12 CLA in milk fat (Bauman and Griinari 2001; Peterson *et al.* 2003; Piperova *et al.* 2004). Further, Bauman and Griinari (2001) suggested that additional unique biohydrogenation intermediates that inhibit fat synthesis may be produced under dietary conditions causing MFD and recent work has offered further support for this idea (Perfield *et al.* 2002; Peterson *et al.* 2003; Piperova *et al.* 2004). The level of *trans*-10 C18:1 in milk fat is also highly correlated with the onset of diet-induced MFD, but to date there have been no investigations establishing a direct effect of this fatty acid (see the discussion by Bauman and Griinari 2003).

Investigations of the mechanism of diet-induced MFD indicate that it involves a coordinated reduction in the abundance of mRNA for key enzymes involved in the pathways of milk fat synthesis (Ahnadi *et al.* 2002; Peterson *et al.* 2003; Piperova *et al.* 2000). Thus, mechanisms appear to be identical to those discussed earlier to explain the reduction in the production of milk fat observed with dietary supplementation with *trans*-10, *cis*-12 CLA. Overall, diet-induced MFD represents a natural situation where the production of *trans*-10, *cis*-12 CLA, and probably other unique biohydrogenation intermediates in the rumen, results in a decrease in mammary synthesis of fatty acids and a reduction in milk fat secretion. As knowledge of the biology of CLA increases, comparisons with the physiology of diet-induced MFD will continue to be of interest.

3.6.1.3 Use as a Management Tool

Dietary supplementation with CLA to reduce milk fat yield has potential use as a management tool in milk production. Milk fat is the major "cost" of milk synthesis accounting for over one-half of the energy needed for milk synthesis; consequently, a reduction in milk fat output will result in a sparing of energy that can be used for other purposes. Commercial situations where this could have application include markets where production is regulated by a quota system based on milk fat, and nutritional situations where cows cannot consume sufficient energy to meet their requirements. Examples of the latter include the onset of lactation and the early lactation period, and under adverse environmental conditions such as heat stress or weather-related feed shortages (Griinari and Bauman 2003).

Commercial application of *trans*-10, *cis*-12 CLA as a management tool requires a CLA formulation that must have two characteristics; it must offer protection of the CLA from alterations by rumen bacteria and the CLA must subsequently become available for absorption in the small intestine. The majority of protection methods are pH-dependent and take advantage of the transition occurring between rumen pH (5.8–6.7) and pH of the abomasum (2–4). To date, the majority of research on rumen-protected CLA has used supplements consisting of calcium salts of free fatty acids. Perfield *et al.* (2002) used this formulation in the first long-term investigation using cows in late lactation; they observed that the reduction in the production of milk fat (23% decrease) was maintained over the 20 week treatment period, whereas yields of milk and other milk components, maintenance of pregnancy, and cow well-being were unaffected. A consistent reduction in the level of milk fat has also been observed in subsequent studies using calcium salts of CLA over treatment periods ranging from 3 to 20 weeks involving primiparous and multiparous cows at different stages of lactation and under different dietary and management practices (Bernal-Santos *et al.* 2003; Giesy *et al.* 2002; Moore *et al.* 2004; Piperova *et al.* 2004; Selberg *et al.* 2004).

The preparation of dietary supplements containing CLA using other methods of rumen-protection has been investigated less extensively compared to calcium salts of CLA, but have included formulations where the protection was by treatment with formaldehyde, the formation of amide bonds, and lipid encapsulation (de Veth *et al.* 2003; Perfield *et al.* 2004a). The transfer of *trans*-10, *cis*-12 CLA to milk fat offers a convenient method to evaluate the effectiveness of rumen protection methods. While all methods resulted in a reduction in the production of milk fat, the transfer efficiency of *trans*-10, *cis*-12 CLA from rumen-protected supplements was much lower than the 22% transfer efficiency reported for investigations involving abomasal infusions (de Veth *et al.* 2004). Thus, the CLA in these formulations is protected only partially from rumen biohydrogenation and there is some indication that the rumen metabolism of a portion of the dietary supplement of CLA may result in the production of other fatty acids that are also able to inhibit milk fat synthesis (Perfield *et al.* 2002; Piperova *et al.* 2004). Additional aspects of the potential application of dietary supplements of *trans*-10, *cis*-12 CLA as a management tool are discussed by Griinari and Bauman (2006).

3.6.2 Rumenic Acid and Human Health

3.6.2.1 Cancer

Since the original discovery of the antimutagen properties of CLA (Ha *et al.* 1987; Pariza *et al.* 1979), its anti-carcinogenic effects have received widespread interest. There are biomedical models for most types of cancer and many of these have been used to investigate the role of CLA as an anticarcinogen (see reviews by Banni 2003; Belury 2002; Parodi 2004; Scimeca 1999). These include the use of human cancer cell lines, transplanted cell lines, and *in situ* organ site carcinogenesis models. The latter are of particular value in cancer investigations and dietary supplements of CLA have been shown to be effective in inhibiting chemically induced skin papillomas, forestomach neoplasia, and preneoplastic lesions and tumors in the colon and mammary gland (see Bauman *et al.* 2006; Parodi 2004). The majority of studies have used a mixture of CLA isomers produced synthetically from vegetable oil, typically containing two or four predominant isomers; the two-isomer mix contains almost equal proportions of RA and *trans*-10, *cis*-12 CLA, whereas the four-isomer mixture also includes the *trans*-8, *cis*-10 and *cis*-11, *trans*-13 isomers.

Following the work of Pariza and colleagues, a number of studies using murine models have demonstrated the potential anticarcinogenic effects of CLA supplementation in a broad range of cancers including colon (Kohno *et al.* 2004a, b; Liew *et al.* 1995; Shiraishi *et al.* 2010; Soel *et al.* 2007), stomach (Ha *et al.* 1990), prostate (Cesano *et al.* 1998), and liver (Yamasaki *et al.* 2005). The anticarcinogenic effect of CLA is particularly impressive in studies on chemically induced mammary cancer; dietary intake of CLA gives a dose-dependent reduction in the incidence and number of tumors (Ip *et al.* 1991) and is independent of the type or level of fat in the diet (Ip *et al.* 1991, 1994). Most impressive is the fact that feeding CLA during the peripubertal period provided protection against mammary tumor development even when the carcinogen was administered at a later time (Thompson *et al.* 1997). Conversely, when rats received no CLA supplementation until they were older and had mature mammary glands, the protective effect was achieved only when CLA was fed continuously during the tumor promotion period following administration of the carcinogen (Ip *et al.* 1995).

However, in contrast, a number of studies investigating the potential anticarcinogenic effects of CLA supplementation have found no beneficial effects (Petrik *et al.* 2000; Wong *et al.* 1997). Furthermore, a more recent pilot study conducted in PyMT transgenic mice found that CLA mixed isomer supplementation resulted in an increased incidence of tumor growth, suggesting that CLA supplementation can potentially promote tumor growth and progression in breast cancer (Flowers *et al.* 2010). Despite these findings, the majority of studies support the hypothesis that CLA has potentially anticarcinogenic effects in modulating tumor development. The discrepancy in findings between studies can perhaps be explained by differences in dose, experimental designs, and, perhaps most importantly, timing. As outlined above, the study by Thompson *et al.* (1997) suggested that CLA supplementation during mammary gland maturation is critical in altering the development of specific target cells that may be vulnerable to carcinogen-induced transformation.

The use of a functional food approach would have many advantages as a strategy to prevent cancer. Since CLA is found predominately in dairy fats in human diets, a series of studies have used the rat prepubertal mammary cancer model to investigate the anticarcinogenic potential of CLA when supplied as a naturally enriched butter that was produced using dietary regimes described in Section 3.5. As discussed in Section 3.4, the majority of the RA in milk fat is synthesized endogenously from VA, and, as a consequence, the levels of VA and CLA in milk fat generally approximate a 3:1 ratio and change in concert (Bauman *et al.* 2003; Palmquist *et al.* 2005). Thus, the enriched butter is higher in both RA and VA. The initial investigation established that RA was an effective anticarcinogen when it was supplied as a dietary food in a natural form (esterified in triglycerides; Table 3.3; Ip *et al.* 1999). Importantly, tissue concentrations of RA were greater in rats fed the VA/RA-enriched butter than in rats fed a comparable amount of chemically synthesized RA, suggesting the possibility of endogenous synthesis from VA. In addition, two studies investigated the effect of CLA-enriched milk fat on MCF-7, a widely used breast cancer cell line. Both studies demonstrated a significant reduction in cell number for MCF-7 (Miller *et al.* 2003; O'Shea *et al.* 2000). As discussed in Section 3.4.4, mammals possess a Δ^9-desaturase, and the ability to convert VA to RA has been demonstrated for several species, including humans (Turpeinen *et al.* 2002; see also review by Palmquist *et al.* 2005). In addition, Banni *et al.* (2001) observed that feeding rats increasing amounts of pure VA resulted in a progressive increase in tissue concentration of RA, and a corresponding reduction in the number of premalignant mammary lesions, an early marker for mammary tumors. Subsequent investigations established that dietary VA derived from VA/RA-enriched butter also resulted in a dose-dependent increase in the accumulation of CLA in the mammary fat pad, which was accompanied by a parallel decrease in tumor incidence and tumor number (Corl *et al.* 2003), and that the anticarcinogenic effects of VA were predominately, perhaps exclusively, mediated through its conversion to RA via Δ^9-desaturase (Lock *et al.* 2004). Therefore, VA and RA derived from milk fat are both anticarcinogenic and this series of preclinical investigations clearly demonstrate the feasibility of a functional food approach using dairy products enriched in VA and RA in the prevention of mammary cancer. Furthermore, Koronowicz and colleagues highlighted further the benefits of naturally derived CLA over chemically synthesized CLA in MCF-7 cell lines. In two separate studies, they demonstrated that fatty acid extracts from CLA-enriched egg yolks were more efficient in reducing MCF-7 cancer cell proliferation than synthetic CLA isomers (Koronowicz *et al.* 2016, 2017).

Premalignant lesions and tumors grow when the rate of cell proliferation exceeds cell death, and investigations to date suggest that the anticarcinogenic effects of CLA involve a multitude of mechanisms. These include a decrease in cell proliferation, an increased rate of apoptosis, inhibition of angiogenesis, modulation of the immune

Table 3.3. Bioassay of mammary cancer prevention in rats fed different sources of conjugated linoleic acids*

		CLA content (μg/mg lipid)		Mammary tumors	
Dietary treatment	Total CLA in diet (%)	Plasma	Mammary fat	Incidence	Total No.
Control butter	0.1	5.4^a	7.2^a	28/30^a (93%)	92^a
High CLA butter	0.8	23.3^c	36.5^c	15/30^b (50%)	43^b
Control butter and synthetic CLA	0.8	18.4^c	26.2^b	16/30^b (53%)	46^b

Adapted from Ip *et al.* (1999).
*Dietary treatments were initiated at weaning and continued for 30 days. All animals were then injected with methylnitrosurea (MNU) to induce mammary tumors and switched to a 5% corn oil diet with no CLA. They remained on this diet for 24 weeks and were then sacrificed for tissue analysis. Values with unlike superscripts in the same column (a, b, c) differ ($P < 0.05$).

cell environment, alteration in the eicosanoid signaling pathways, and a possible antioxidant role (see Banni 2003; Belury 2002; Ip *et al.* 2003). Particular mechanisms may vary in importance depending on the tissue-specific process being regulated, and the opportunity to exploit the diversity in the mechanism of action of CLA may form the basis for the range in tissues and cancer types in which CLA is effective. Pierre *et al.* (2013) demonstrated this specificity in a study investigating the effects of *cis*-9, *trans*-11, CLA and *trans*-10, *cis*-12, CLA separately on apoptosis in colon cancer cells. Their study revealed that only *trans*-10, *cis*-12, CLA treatment was effective in triggering cell death in colon cancer cells. It is apparent that CLA acts in a dose-dependent and isomer-specific manner.

Evaluating the specific role of CLA in health maintenance and the prevention of cancer in humans is difficult. Since cancer takes many years to develop, documenting that dietary CLA is beneficial in health maintenance and the prevention of this disease is a major challenge. A limited number of epidemiological studies gave conflicting results (Aro *et al.* 2000; Chajès *et al.* 2003; McCann *et al.* 2004; Voorrips *et al.* 2002). This inconsistency is not surprising. Dairy products are used in recipes for many manufactured food products, and estimating CLA intake is further complicated by the fact that CLA is a fatty acid and dairy products vary widely in fat content; milk fat varies widely in CLA content, and analysis of RA is difficult and reported values are often inaccurate (see reviews by Bauman *et al.* 2006; Parodi 2004). A clear example of the challenges associated with epidemiological studies is the study of Larsson *et al.* (2005) who investigated the association of CLA intake from high-fat dairy food with colorectal cancer in women. This cohort study with a 14.8-year follow-up suggested that high consumption of high-fat dairy foods may lower the risk of colorectal cancer. However, no significant effect of CLA was observed in the same study when a multivariate model was used to mimic intake of high-fat dairy foods and CLA. Furthermore, a study by the same group examining the associations of CLA intake and mammary cancer in a Swedish mammography cohort found no significant effect of dietary CLA consumption and risk of breast cancer (Larsson *et al.* 2009). Overall, findings from epidemiological studies are sparse and inconsistent.

Another approach would be dietary interventions using biomarkers as end points to predict reduced cancer risk, but to date there are no consensus biomarkers for breast cancer and many other cancer types. However, a clinical trial by Mohammadzadeh *et al.* (2013) investigated the link between CLA administration and specific biomarkers of angiogenesis and tumor invasion (TNF-A, IL-1B, IL-6, MMP-2, MMP-9, and hsCRP) in patients with rectal cancer, who were undergoing chemotherapy. Serum levels were measured before and after intervention and demonstrated that in comparison with the placebo (sunflower oil), CLA supplementation resulted in significant reductions in serum levels of TNF-α, hsCRP, and MMP-9. Furthermore, a separate clinical trial by McGowan *et al.* (2013) investigated the effects of CLA supplementation in women with stage I-III breast cancer prior to tumor removal. Spot 14 (S14), as previously noted in Section 3.6.1, is a regulator of fatty acid synthesis, and hypothesized to augment breast cancer cell growth (Martel *et al.* 2006). McGowan *et al.* (2013) demonstrated that CLA supplementation caused a reduction in S14. These findings support previous work by Donnelly *et al.* (2008), who demonstrated that exposure to CLA caused a reduction in S14 expression in T47D human breast cancer cell line. To date, the number of studies investigating the effects of CLA supplementation in humans, specifically those with cancer, is limited. Clearly, assessing the role of dietary CLA in functional foods for the prevention of cancer presents some unusual difficulties, and thus many of the traditional approaches to evaluate human health effects have substantial limitations (see Koronowicz and Banks 2018, for review). Evidently, a greater understanding of the precise mechanism of action responsible for these anticancer effects is required for consensus and to ensure transferability of the positive results seen in murine models to humans.

3.6.2.2 Atherosclerosis

Although initial interest focused mainly on the anticarcinogenic effects of CLA, interest has now increased in the antiatherogenic potential of CLA. However, investigations of the effects of CLA on atherosclerosis are limited compared with anticarcinogenic studies.

Early studies investigating the antiatherogenic properties of CLA supplements were often carried out in rabbits due to their ability to develop human-like atherosclerosis lesions under certain dietary conditions. A number of animal studies have demonstrated that dietary supplementation with mixtures of CLA isomers can reduce the development of atherosclerotic lesions (Kritchevsky *et al.* 2000, 2002; Lee *et al.* 1994; Wilson *et al.* 2000) and even induce the regression of pre-existing lesions in rabbits (Kritchevsky *et al.* 2000, 2004). Kritchevsky *et al.* (2004) demonstrated that pure isomers of RA and *trans*-10, *cis*-12 CLA were equally effective in reducing cholesterol-induced atherogenesis in rabbits.

Since these initial animal studies, which mainly involved rabbits and hamsters, research has moved to more cost-effective alternatives of murine models. A number of studies have demonstrated that CLA is effective at combating the development of atherosclerosis in genetically modified mice. ApoE-deficient (apoE-/-) and Ldlr-/- mice are the most commonly used models. For example, in ApoE-/- mice, RA not only prevented the progression, but also induced the regression of atherosclerotic lesions (Toomey *et al.* 2003). This model has been used widely in studies of atherosclerosis because it spontaneously develops lesions on a regular low-fat, low-cholesterol diet with a histopathology similar to lesions that develop in humans (Meir and Leitersdorf 2004).

However, a study by Cooper *et al.* (2008) revealed that CLA supplementation was not effective in preventing the development or progression of atherosclerosis in ApoE-/- mice. Furthermore, one study, in the atherosclerosis-susceptible C57BL/6 mouse, showed that CLA had no effect on atherosclerotic lesions, and could even promote their development (Munday *et al.* 1999). Additionally, the effects of individual isomers have been investigated and in contrast to a study by Kanter *et al.* (2018) which found the *cis*-10, *trans*-12 CLA isomer effective in treating atherosclerosis in Ldlr-/- mice, Arbonés-Mainar *et al.* (2006) found that *cis*-10, *trans*-12 CLA promotes atherosclerosis in ApoE-/- mice. This same study revealed that *cis*-9, *trans*-11, CLA was effective in impeding atherosclerosis development. Thus, an isomer mix is potentially crucial in evaluating the anti-atherogenic effects of CLA.

Changes in both total plasma cholesterol and individual lipoprotein cholesterol concentrations have been implicated as major determinants of the risk of atherosclerosis and this has led to a number of studies which specifically investigated the effects of CLA on cholesterol and lipoprotein metabolism in animal models. Most have used a synthetic source composed of a mixture of CLA isomers, and results have been inconsistent, with some showing beneficial changes in blood lipid variables while others have shown no effect (Bauman *et al.* 2006). Lock *et al.* (2005) completed a study using the Golden Syrian hamster to examine the potential of CLA when fed as a component of a functional food (VA/RA-enriched butter) as part of a diet that was high in cholesterol (0.2%) and fat (20%). Compared with the control animals, those fed the VA/RA-enriched butter showed a number of beneficial effects, including reduced total plasma cholesterol and VLDL and LDL cholesterol lipoproteins, suggesting that CLA may modify the production of atherogenic lipoproteins by the liver. In addition, the VA/RA-enriched butter produced a less atherogenic profile than an equivalent diet in which the VA/RA-enriched butter was replaced by *trans* fatty acids from partially hydrogenated vegetable oil (Lock *et al.* 2005). Consistent with these findings, it has been proposed (McLeod *et al.* 2004) that RA could, in the absence of other CLA isomers, improve hepatic lipid metabolism. This may explain why VA/CLA-enriched butter elicited such impressive effects compared to studies in which synthetic CLA isomer mixtures are used, since naturally derived sources of CLA provide essentially only RA. Furthermore, a more recent study investigated the effects of CLA-enriched hen eggs (both naturally enriched and supplemented) on pre-established atherosclerosis in KO mice. Although both egg types had some effect in reducing aortic plaques, the naturally enriched eggs significantly reduced total plasma cholesterol, and the number of atherogenic macrophages in plaques and an increased surface area occupied with smooth muscles cells in atherosclerosis lesions was observed (Franczyk-Żarów *et al.* 2008). Conversely, Kostogrys *et al.* (2012) investigated the effects of margarine supplemented with CLA isomers and found no significant effect of *trans*-10, *cis*-12 or *cis*-9, *trans*-11 CLA in double KO mice.

It is important to note that elevated and/or altered plasma lipid levels is only one of a wide range of risk factors that contribute to the clinical manifestations of cardiovascular disease in humans (Lusis 2000). Consequently, in some studies, the reduced incidence of atherosclerosis in animals fed CLA was not accompanied by an improvement in the plasma lipid profile during the CLA feeding phase (Wilson *et al.* 2000). Reasons for these effects are not understood fully. However, atherosclerosis can also be considered as a chronic inflammatory disease (Libby 2002) and several important antiinflammatory effects have been associated with the use of RA; these include a reduction in the expression of COX-2, PGE2, reduced release of nitric oxide, a decreased production of proinflammatory cytokines, and PPARg activation (Toomey *et al.* 2003; Urquhart *et al.* 2002, Yu *et al.* 2002). In addition, the effects of CLA supplementation on inflammation (hs-CRP, IL-6) and oxidative stress markers (GPx, MDA) in atherosclerosis were evaluated in a clinical trial by Eftekhari *et al.* (2013). The participants in this study suffered from atherosclerosis and the study found a significant reduction in hs-CRP, GPx, and MDA in comparison with baseline levels in the CLA-treated group; the reduction in IL-6 was not significant. Inflammation and oxidative stress play a critical role in atherosclerosis. The findings of this study suggest that CLA supplementation may possess anti-inflammatory and antioxidant properties that can be exploited in atherosclerosis treatment. In contrast, a subsequent study found that 12 weeks of CLA supplementation had no significant effect on hs-CRP in addition to ADMA in obese and overweight women (Dus-Zuchowska *et al.* 2016).

Since results from studies with biomedical models indicate potential, there is obvious interest in the effects of RA consumption in foods on the risk of atherogenesis in humans. The use of surrogate biomarkers for disease risk is more readily achievable for atherosclerosis than for cancer in humans and a number of genetic and environmental risk factors have been identified, with the relative abundance of the different lipoproteins being of primary importance (Lusis 2000). To date, there have been no epidemiological studies that have examined the intake of CLA derived from foods with the risk of atherosclerosis. However, as discussed in Section 3.6.2.1, the challenge of adequately evaluating the effect of dietary intake of CLA from different food sources presents some special limitations.

Several human intervention studies involving dietary supplements of CLA in the form of capsules have shown plasma lipid variables as secondary observations, but most utilized mixed isomers of CLA and gave variable results (Bauman *et al.* 2006). However, two studies examined the specific effects of RA on blood lipids in healthy subjects; a CLA supplement containing a 50:50 mixture of RA and *trans*-10, *cis*-12 CLA significantly improved plasma triacylglycerol and VLDL metabolism, with an 80:20 CLA isomer blend significantly reducing the concentration of VLDL-cholesterol, providing further evidence for the role of RA in altering the hepatic lipid metabolism (Noone *et al.* 2002). Conversely, a subsequent study assessing the effect of CLA supplementation on endothelial function and markers of cardiovascular disease including LDL cholesterol, HDL cholesterol, and triglycerides found that CLA supplementation had no significant effect (Pfeuffer *et al.* 2011). In this study, the participants were overweight men who received 3.4 g of CLA mixture (50:50 *cis*-9, *trans*-11 and *trans*-10, *cis*-12 CLA) over the course of 4 weeks. Utilizing pure CLA isomers, it was observed that RA and *trans*-10, *cis*-12 CLA had opposing effects on blood lipids in healthy humans; plasma triacylglycerol, total plasma cholesterol, LDL-cholesterol, and LDL:HDL-cholesterol were all lower during supplementation with RA compared to *trans*-10, *cis*-12 CLA (Tricon *et al.* 2004). A companion study showed that CLA isomers were readily incorporated into plasma and cellular lipids to a similar extent and in a dose-dependent manner (Burdge *et al.* 2004). Although these data are limited, they provide support that some of the antiatherosclerosis effects of CLA reported in animal models will extend to humans. However, a subsequent study investigating the effects of *cis*-9, *trans*-11 CLA supplementation in overweight and obese but otherwise healthy participants did not support any antiatherosclerotic effects of RA (Sluijs *et al.* 2009). In this study, aortic pulse wave velocity (a measure of aortic stiffness and early marker of atherosclerosis) and other cardiovascular disease and atherosclerosis risk factors were assessed. Over a 6-month period, participants received 4 g of CLA or a placebo per day. The study demonstrated no change in pulse wave velocity and other atherosclerosis indicators including triglycerides, LDL, HDL, and total cholesterol. Perhaps, most importantly are the results of a 2-month clinical trial investigating the effects of CLA supplementation on the lipid profile of patients with atherosclerosis. Unfortunately, no antiatherogenic effects were seen. CLA supplementation resulted in no significant reduction of LDL cholesterol, HDL cholesterol, triglycerides, and total cholesterol (Eftekhari *et al.* 2014).

Evidently, the effect of CLA is dose-, isomer-, tissue-, and species-specific; further studies are required to substantiate the antiatherogenic properties of CLA. As previously mentioned, often in atherosclerosis studies utilizing animal models, the beneficial effects did not involve alterations in plasma lipids. Thus, CLA studies in humans that focus on the possible role of eicosanoid- and cytokine-related effects could also be of importance.

Bibliography

AbuGhazaleh, A. A., Schingoethe, D. J., Hippen, A. R. & Kalscheur, K. F. (2003). Milk conjugated linoleic acid response to fish oil supplementation of diets differing in fatty acid profiles. *Journal of Dairy Science, 86*, 944–953.

AbuGhazaleh, A. A. & Jenkins, T. C. (2004). Docosahexaenoic acid promotes vaccenic acid accumulation in mixed ruminal cultures when incubated with linoleic acid. *Journal of Dairy Science, 87*, 1047–1050.

AbuGhazaleh, A. A. & Buckles, W. R. (2007). The effect of solids dilution rate and oil source on trans C18:1 and conjugated linoleic acid production by ruminal microbes in continuous culture. *Journal of Dairy Science, 90*, 963–969.

AbuGhazaleh, A. A., Felton, D. O. & Ibrahim, S. A. (2007). Milk conjugated linoleic acid response to fish oil and sunflower oil supplementation to dairy cows managed under two feeding systems. *Journal of Dairy Science, 90*, 4763–4769.

AbuGhazaleh, A. A., Potu, R. B. & Ibrahim, S. (2009). Short communication: The effect of substituting fish oil in dairy cow diets with docosahexaenoic acid-micro algae on milk composition and fatty acids profile. *Journal of Dairy Science, 92*, 6156–6159.

Adlof, R. O. (2003). Application of silver-ion chromatography to the separation of conjugated linoleic acid isomers. In J. Sebedio, W. W. Christie, & R. Adlof (Eds.), *Advances in conjugated linoleic acid research* (pp. 43–61). Champaign: AOCS Publishing.

Ahnadi, C. E., Beswick, N., Delbecchi, L., Kennelly, J. J. & Lacasse, P. (2002). Addition of fish oil to diets for dairy cows. II. Effects on milk fat and gene expression of mammary lipogenic enzymes. *Journal of Dairy Research, 69*, 521–531.

Alves, S. P., Tyburczy, C., Lawrence, P., Bessa, R. J. & Brenna, J. T. (2011). Acetonitrile covalent adduct chemical ionization tandem mass spectrometry of non-methylene-interrupted pentaene fatty acid methyl esters. *Rapid Communications in Mass Spectrometry, 25*, 1933–1941.

Amores, G. & Virto, M. (2019). Total and free fatty acids analysis in milk and dairy fat. *Separations, 6*(1), 14.

Apas, A. L., Arena, M. E., Colombo, S. & Gonzalez, S. N. (2015). Probiotic administration modifies the milk fatty acid profile, intestinal morphology, and intestinal fatty acid profile of goats. *Journal of Dairy Science, 98*, 47–54.

Arbonés-Mainar, J. M., Navarro, M. A., Guzmán, M. A., Arnal, C., Surra, J. C., et al. (2006). Selective effect of conjugated linoleic acid isomers on atherosclerotic lesion development in apolipoprotein E knockout mice. *Atherosclerosis, 189*, 318–327.

Aro, A., Männistö, S., Salminen, I., Ovaskainen, M.-L., Kataja, V. & Uusitupa, M. (2000). Inverse association between dietary and serum conjugated linoleic acid and risk of breast cancer in postmenopausal women. *Nutrition and Cancer, 38*, 151–157.

Auldist, M. J., Kay, J. K., Thomson, N. A., Napper, A. R. & Kolver, E. S. (1999). *New Zealand Society of Animal Production, 62*: 240–241.

Avramis, C. A., Wang, H., McBride, B. W., Wright, T. C. & Hill, A. R. (2003). Physical and processing properties of milk, butter, and *Cheddar* cheeses from cows fed supplemental fish meal. *Journal of Dairy Science, 86*, 2568–2576.

Baer, R. J., Ryali, J., Schingoethe, D. J., Kasperson, K. M., Donovan, D. C., Hippen, A. R. & Franklin, S. T. (2001). Composition and properties of milk and butter from cows fed fish oil. *Journal of Dairy Science, 84*, 345–353.

Banni, S. (2002). Conjugated linoleic acid metabolism. *Current Opinion in Lipidology, 13*, 261–266.

Banni, S. (2003). Conjugated linoleic acids as anticancer nutrients: Studies *in vivo* and cellular mechanisms. Sebedio, J. Christie, W. W. Adolf, R. (Eds.). *Advances in conjugated linoleic acid research, 2*, 267–282.

Banni, S., Carta, G., Contini, M. S., Angioni, E., Deiana, M., et al. (1996). Characterization of conjugated diene fatty acids in milk, dairy products, and lamb tissues. *Journal of Nutritional Biochemistry, 7*, 150–155.

Banni, S., Angioni, E., Murru, E., Carta, G., Paola, M. M., et al. (2001). Vaccenic acid feeding increases tissue levels of conjugated linoleic acid and suppresses development of premalignant lesions in rat mammary gland. *Nutrition and Cancer, 41*, 91–97.

Barber, M., Ward, R., Richards, S., Salter, A., Buttery, P., et al. (2000). Ovine adipose tissue monounsaturated fat content is correlated to depot-specific expression of the stearoyl-CoA desaturase gene. *Journal of Animal Science, 78*, 62–68.

Barrett, E., Ross, R. P., Fitzgerald, G. F. & Stanton, C. (2007). Rapid screening method for analyzing the conjugated linoleic acid production capabilities of bacterial cultures. *Applied and Environmental Microbiology, 73*, 2333–2337.

Bauman, D. E. & Currie, W. B. (1980). Partitioning of nutrients during pregnancy and lactation: A review of mechanisms involving homeostasis and homeorhesis. *Journal of Dairy Science, 63*, 1514–1529.

Bauman, D. & Griinari, J. (2001). Regulation and nutritional manipulation of milk fat: Low-fat milk syndrome. *Livestock Production Science, 70*, 15–29.

Bauman, D. E. & Griinari, J. M. (2003). Nutritional regulation of milk fat synthesis. *Annual Review of Nutrition, 23*, 203–227.

Bauman, D., Barbano, D., Dwyer, D. & Griinari, J. (2000). Production of butter with enhanced conjugated linoleic acid for use in biomedical studies with animal models. *Journal of Dairy Science, 83*, 2422–2425.

Bauman, D. E., Corl, B. A. & Peterson, D. G. (2003). The biology of conjugated linoleic acids in ruminants. In Sebedio, J. Christie, W. W. Adolf, R. (Eds.), *Advances in conjugated linoleic acid research* (pp. 152–179). Champaign: AOCS Publishing.

Bauman, D., Lock, A., Corl, B., Ip, C., Salter, A. & Parodi, P. (2006). Milk fatty acids and human health: Potential role of conjugated linoleic acid and *trans* fatty acids. In K. Sejrsen, T. Hvelplund & M. O. Nielsen (Eds.), *Ruminant physiology: Digestion, metabolism and impact of nutrition on gene expression, immunology and stress* (pp. 523–555). Wageningen: Wageningen Academic Publishers.

Bauman, D. E., Harvatine, K. J. & Lock, A. L. (2011). Nutrigenomics, rumen-derived bioactive fatty acids, and the regulation of milk fat synthesis. *Annual Review of Nutrition, 31*, 299–319.

Baumgard, L. H., Corl, B. A., Dwyer, D. A., Sæbø, A. & Bauman, D. E. (2000). Identification of the conjugated linoleic acid isomer that inhibits milk fat synthesis. *American Journal of Physiology – Regulatory, Integrative and Comparative Physiology, 278*, R179–RR84.

Baumgard, L. H., Sangster, J. K. & Bauman, D. E. (2001). Milk fat synthesis in dairy cows is progressively reduced by increasing supplemental amounts of *trans*-10, *cis*-12 conjugated linoleic acid (CLA). *Journal of Nutrition, 131*, 1764–1769.

Baumgard, L., Matitashvili, E., Corl, B., Dwyer, D. & Bauman, D. (2002). *Trans*-10, *cis*-12 conjugated linoleic acid decreases lipogenic rates and expression of genes involved in milk lipid synthesis in dairy cows. *Journal of Dairy Science, 85*, 2155–2163.

Beaulieu, A. D. & Palmquist, D. L. (1995). Differential-effects of high-fat diets on fatty-acid composition in milk of Jersey and Holstein cows. *Journal of Dairy Science, 78*, 1336–1344.

Belury, M. A. (2002). Dietary conjugated linoleic acid in health: Physiological effects and mechanisms of action. *Annual Review of Nutrition, 22*, 505–531.

Bernal-Santos, G., Perfield, J., II, Barbano, D., Bauman, D. & Overton, T. (2003). Production responses of dairy cows to dietary supplementation with conjugated linoleic acid (CLA) during the transition period and early lactation. *Journal of Dairy Science, 86*, 3218–3228.

Bernuy, B., Meurens, M., Mignolet, E. & Larondelle, Y. (2008). Performance comparison of UV and FT-Raman spectroscopy in the determination of conjugated linoleic acids in cow milk fat. *Journal of Agricultural and Food Chemistry, 56*, 1159–1163.

Bichi, E., Frutos, P., Hervás, G., Keisler, D., Loor, J. J. & Toral, P. G. (2013). Dietary marine algae and its influence on tissue gene network expression during milk fat depression in dairy ewes. *Animal Feed Science and Technology, 186*, 36–44.

Bickerstaffe, R. & Annison, E. (1969). Glycerokinase and desaturase activity in pig, chicken and sheep intestinal epithelium. *Comparative Biochemistry and Physiology, 31*, 47–54.

Bickerstaffe, R. & Annison, E. (1970). The desaturase activity of goat and sow mammary tissue. *Comparative Biochemistry and Physiology, 35*, 653–665.

Bickerstaffe, R. & Johnson, A. (1972). The effect of intravenous infusions of sterculic acid on milk fat synthesis. *British Journal of Nutrition, 27*, 561–570.

Boeckaert, C., Vlaeminck, B., Fievez, V., Maignien, L., Dijkstra, J. & Boon, N. (2008). Accumulation of *trans* C18:1 fatty acids in the rumen after dietary algal supplementation is associated with changes in the *Butyrivibrio* community. *Applied and Environmental Microbiology, 74*, 6923–6930.

Booth, R. G., Dann, W. J., Kon, S. K. & Moore, T. (1933). A new variable factor in butter fat. *Chemistry & Industry, 52*, 270.

Bretillon, L., Chardigny, J., Gregoire, S., Berdeaux, O. & Sebedio, J. (1999). Effects of conjugated linoleic acid isomers on the hepatic microsomal desaturation activities in vitro. *Lipids, 34*, 965–969.

Burdge, G. C., Lupoli, B., Russell, J. J., Tricon, S., Kew, S., et al. (2004). Incorporation of *cis*-9, *trans*-11 or *trans*-10, *cis*-12 conjugated linoleic acid into plasma and cellular lipids in healthy men. *Journal of Lipid Research, 45*, 736–741.

Bu, D. P., Wang, J. Q., Dhiman, T. R. & Liu, S. J. (2007). Effectiveness of oils rich in linoleic and linolenic acids to enhance conjugated linoleic acid in milk from dairy cows. *Journal of Dairy Science, 90*, 998–1007.

Cameron, P., Rogers, M., Oman, J., May, S., Lunt, D. & Smith, S. (1994). Stearoyl coenzyme A desaturase enzyme activity and mRNA levels are not different in subcutaneous adipose tissue from Angus and American Wagyu steers. *Journal of Animal Science, 72*, 2624–2628.

Campbell, E. M., Gallagher, D. S., Davis, S. K., Taylor, J. F. & Smith, S. B. (2001). Rapid communication: mapping of the bovine stearoyl-coenzyme A desaturase (SCD) gene to BTA261. *Journal of Animal Science, 79*, 1954–1955.

Campbell, W., Drake, M. A. & Larick, D. K. (2003). The impact of fortification with conjugated linoleic acid (CLA) on the quality of fluid milk. *Journal of Dairy Science, 86*, 43–51.

Carreño, D., Hervás, G., Toral, P. G., Castro-Carrera, T. & Frutos, P. (2016). Fish oil-induced milk fat depression and associated downregulation of mammary lipogenic genes in dairy ewes. *Journal of Dairy Science, 99*, 7971–7981.

Cesano, A., Visonneau, S., Scimeca, J., Kritchevsky, D. & Santoli, D. (1998). Opposite effects of linoleic acid and conjugated linoleic acid on human prostatic cancer in SCID mice. *Anticancer Research, 18*, 1429–1434.

Chajès, V., Lavillonnière, F., Maillard, V., Giraudeau, B., Jourdan, M.-L., et al. (2003). Conjugated linoleic acid content in breast adipose tissue of breast cancer patients and the risk of metastasis. *Nutrition and Cancer, 45*, 17–23.

Chang, J. H., Lunt, D. K. & Smith, S. B. (1992). Fatty acid composition and fatty acid elongase and stearoyl-CoA desaturase activities in tissues of steers fed high oleate sunflower seed. *Journal of Nutrition, 122*, 2074–2080.

Chilliard, Y., Ferlay, A., Mansbridge, R. M. & Doreau, M. (2000). Ruminant milk fat plasticity: Nutritional control of saturated, polyunsaturated, trans and conjugated fatty acids. *Annales De Zootechnie, 49*, 181–205.

Chilliard, Y., Ferlay, A. & Doreau, M. (2001). Effect of different types of forages, animal fat or marine oils in cow's diet on milk fat secretion and composition, especially conjugated linoleic acid (CLA) and polyunsaturated fatty acids. *Livestock Production Science, 70*, 31–48.

Chilliard, Y., Glasser, F., Ferlay, A., Bernard, L., Rouel, J. & Doreau, M. (2007). Diet, rumen biohydrogenation and nutritional quality of cow and goat milk fat. *European Journal of Lipid Science and Technology, 109*, 828–855.

Choi, Y., Kim, Y.-C., Han, Y.-B., Park, Y., Pariza, M. W. & Ntambi, J. M. (2000). The *trans*-10, *cis*-12 isomer of conjugated linoleic acid downregulates stearoyl-CoA desaturase 1 gene expression in 3T3-L1 adipocytes. *Journal of Nutrition, 130*, 1920–1924.

Chouinard, P. Y., Corneau, L., Barbano, D. M., Metzger, L. E. & Bauman, D. E. (1999a). Conjugated linoleic acids alter milk fatty acid composition and inhibit milk fat secretion in dairy cows. *Journal of Nutrition, 129*, 1579–1584.

Chouinard, P., Corneau, L., Sæbø, A. & Bauman, D. (1999b). Milk yield and composition during abomasal infusion of conjugated linoleic acids in dairy cows. *Journal of Dairy Science, 82*, 2737–2745.

Chouinard, P. Y., Corneau, L., Butler, W. R., Bauman, D. E., Chilliard, Y. & Drackley, J. K. (2001). Effect of dietary lipid source on conjugated linoleic acid concentrations in milk fat. *Journal of Dairy Science, 84*, 680–690.

Christie, W. W. (1982). A simple procedure for rapid transmethylation of glycerolipids and cholesteryl esters. *Journal of Lipid Research, 23*, 1072–1075.

Christie, W. W., Dobson, G. & Adlof, R. O. (2007). A practical guide to the isolation, analysis and identification of conjugated linoleic acid. *Lipids, 42*, 1073–1084.

Chung, M., Ha, S., Jeong, S., BOK, J., CHO, K., et al. (2000). Cloning and characterization of bovine stearoyl CoA desaturasel cDNA from adipose tissues. *Bioscience, Biotechnology, and Biochemistry, 64*, 1526–1530.

Cooney, A. & Headon, D. (1989). *Molecular cloning of the bovine gene encoding the stearyl coenzyme A desaturase*. London: Portland Press Limited.

Cooper, M. H., Miller, J. R., Mitchell, P. L., Currie, D. L. & McLeod, R. S. (2008). Conjugated linoleic acid isomers have no effect on atherosclerosis and adverse effects on lipoprotein and liver lipid metabolism in apoE−/− mice fed a high-cholesterol diet. *Atherosclerosis, 200*, 294–302.

Coppa, M., Ferlay, A., Leroux, C., Jestin, M., Chilliard, Y., et al. (2010). Prediction of milk fatty acid composition by near infrared reflectance spectroscopy. *International Dairy Journal, 20*, 182–189.

Corl, B. A., Baumgard, L. H., Dwyer, D. A., Griinari, J. M., Phillips, B. S. & Bauman, D. E. (2001). The role of Δ9-desaturase in the production of *cis*-9, *trans*-11 CLA. *Journal of Nutritional Biochemistry, 12*, 622–630.

Corl, B. A., Barbano, D. M., Bauman, D. E. & Ip, C. (2003). *cis*-9, *trans*-11 CLA derived endogenously from *trans*-11 18:1 reduces cancer risk in rats. *Journal of Nutrition, 133*, 2893–2900.

Corl, B. A., Baumgard, L. H., Griinari, J. M., Delmonte, P., Morehouse, K. M., et al. (2002). *Trans*-7, *cis*-9 CLA is synthesized endogenously by Δ9-desaturase in dairy cowsin dairy cows. *Lipids, 37*, 681–688.

Cruz-Hernandez, C., Deng, Z., Zhou, J., Hill, A. R., Yurawecz, M. P., et al. (2004). Methods for analysis of conjugated linoleic acids and *trans*-18: 1 isomers in dairy fats by using a combination of gas chromatography, silver-ion thin-layer chromatography/gas chromatography, and silver-ion liquid chromatography. *Journal of AOAC International, 87*, 545–562.

Czauderna, M., Kowalczyk, J., Marounek, M., Mlchalski, J. P., Rozbicka-Wieczorek, A. J. & Krajewska, K. A. (2011). A new internal standard for HPLC assay of conjugated linoleic acid in animal tissues and milk. *Czech Journal of Animal Science, 56*(1), 23–29.

de Veth, M. J., Griinari, J. M., Pfeiffer, A.-M. & Bauman, D. E. (2004). Effect of CLA on milk fat synthesis in dairy cows: Comparison of inhibition by methyl esters and free fatty acids, and relationships among studies. *Lipids, 39*, 365–372.

den Hartigh, L. J. (2019). Conjugated linoleic acid effects on cancer, obesity, and atherosclerosis: A review of pre-clinical and human trials with current perspectives. *Nutrients, 11*, 370.

Davis, C. & Brown, R. (1970). Low-fat milk syndrome. In A. T. Phillipson (Ed.), *Physiology of Digestion and Metabolism in the Ruminant* (pp. 545–565). Newcastle upon Tyne, UK: Oriel Press Limited.

Davis, A. L., Mc Neill, G. P. & Caswell, D. C. (1999). Analysis of conjugated linoleic acid isomers by 13C NMR spectroscopy. *Chemistry and Physics of Lipids, 97*, 155–165.

Dawson, R. & Kemp, P. (1970). Biohydrogenation of dietary fats in ruminants. *Physiology of Digestion and Metabolism in the Ruminant*, 504–518.

de Veth, M., McFadden, J., Griinari, J., Gulati, S., Luchini, N. & Bauman, D. (2003). Comparison of the effect of different rumen protected forms of CLA on milk fat synthesis. *Journal of Dairy Science, 86*, 146–147.

DePeters, E. J., Medrano, J. F. & Reed, B. A. (1995). Fatty acid composition of milk fat from three breeds of dairy cattle. *Canadian Journal of Animal Science, 75*, 267–269.

Derakhshande Rishehri, S. M., Mansourian, M., Kelishadi, R. & Heidari-Beni, M. (2015). Association of foods enriched in conjugated linoleic acid (CLA) and CLA supplements with lipid profile in human studies: A systematic review and meta-analysis. *Public Health Nutrition, 18*, 2041–2054.

Dhiman, T., Satter, L., Pariza, M., Galli, M., Albright, K. & Tolosa, M. (2000). Conjugated linoleic acid (CLA) content of milk from cows offered diets rich in linoleic and linolenic acid. *Journal of Dairy Science, 83*, 1016–1027.

Dhiman, T. R., Nam, S. H. & Ure, A. L. (2005). Factors affecting conjugated linoleic acid content in milk and meat. *Critical Reviews in Food Science and Nutrition, 45*, 463–482.

Dobson, G. (2003). Gas chromatography-mass spectrometry of conjugated linoleic acids and metabolites. In Sebedio, J. Christie, W. W. Adlof, R. (Eds.). *Advances in conjugated linoleic acid research* (pp. 19–42). Champaign: AOCS Publishing.

Donnelly, C., Olsen, A. M., Lewis, L. D., Eisenberg, B. L., Eastman, A. & Kinlaw, W. B. (2008). Conjugated linoleic acid (CLA) inhibits expression of the Spot 14 (THRSP) and fatty acid synthase genes and impairs the growth of human breast cancer and liposarcoma cells. *Nutrition and Cancer, 61*, 114–122.

Doreau, M. & Ferlay, A. (1994). Digestion and utilisation of fatty acids by ruminants. *Animal Feed Science and Technology, 45*, 379–396.

Doreau, M., Demeyer, D. & Van Nevel, C. (1997). Transformation and effects of unsaturated fatty acids in the rumen. Consequences on milk fat secretion. In R. A. S. Welch (Ed.), *Milk composition, production and biotechnology* (pp. 73–92). Wallingford: CAB International.

Doreau, M., Chilliard, Y., Rulquin, H. & Demeyer, D. (1999). Manipulation of milk fat in dairy cows. *Recent Advances in Animal Nutrition*, 81–110.

Duckett, S. K., Andrae, J. G. & Owens, F. N. (2002). Effect of high-oil corn or added corn oil on ruminal biohydrogenation of fatty acids and conjugated linoleic acid formation in beef steers fed finishing diets. *Journal of Animal Science, 80*, 3353–3360.

Dugan, M. E., Kramer, J. K., Robertson, W. M., Meadus, W. J., Aldai, N. & Rolland, D. C. (2007). Comparing subcutaneous adipose tissue in beef and muskox with emphasis on trans 18:1 and conjugated linoleic acids. *Lipids, 42*, 509–518.

Dus-Zuchowska, M., Madry, E., Krzyzanowska, P., Bogdanski, P. & Walkowiak, J. (2016). Twelve-week-conjugated linoleic acid supplementation has no effects on the selected markers of atherosclerosis in obese and overweight women. *Food and Nutrition Research, 60*, 32776.

Eftekhari, M. H., Aliasghari, F., Babaei-Beigi, M. A. & Hasanzadeh, J. (2013). Effect of conjugated linoleic acid and omega-3 fatty acid supplementation on inflammatory and oxidative stress markers in atherosclerotic patients. *ARYA Atherosclerosis, 9*, 311–318

Eftekhari, M. H., Aliasghari, F., Beigi, M. A. B. & Hasanzadeh, J. (2014). The effect of conjugated linoleic acids and omega-3 fatty acids supplementation on lipid profile in atherosclerosis. *Advanced Biomedical Research, 3*, 15.

Faulconnier, Y., Bernard, L., Boby, C., Domagalski, J., Chilliard, Y. & Leroux, C. (2018). Extruded linseed alone or in combination with fish oil modifies mammary gene expression profiles in lactating goats. *Animal, 12*, 1564–1575.

Faulkner, H., O'Callaghan, T. F., McAuliffe, S., Hennessy, D., Stanton, C., et al. (2018). Effect of different forage types on the volatile and sensory properties of bovine milk. *Journal of Dairy Science, 101*, 1034–1047.

Flowers, M., Schroeder, J. A., Borowsky, A. D., Besselsen, D. G., Thomson, C. A., et al. (2010). Pilot study on the effects of dietary conjugated linoleic acid on tumorigenesis and gene expression in PyMT transgenic mice. *Carcinogenesis, 31*, 1642–1649.

Franczyk-Żarów, M., Kostogrys, R. B., Szymczyk, B., Jawień, J., Gajda, M., et al. (2008). Functional effects of eggs, naturally enriched with conjugated linoleic acid, on the blood lipid profile, development of atherosclerosis and composition of atherosclerotic plaque in apolipoprotein E and low-density lipoprotein receptor double-knockout mice (apoE/LDLR−/−). *British Journal of Nutrition, 99*, 49–58.

Franklin, S. T., Martin, K. R., Baer, R. J., Schingoethe, D. J. & Hippen, A. R. (1999). Dietary marine algae (*Schizochytrium* sp.) increases concentrations of conjugated linoleic, docosahexaenoic and *trans* vaccenic acids in milk of dairy cows. *Journal of Nutrition, 129*, 2048–2054.

Fretin, M., Martin, B., Buchin, S., Desserre, B., Lavigne, R., et al. (2019). Milk fat composition modifies the texture and appearance of Cantal-type cheeses but not their flavor. *Journal of Dairy Science, 102*, 1131–1143.

Frutos, P., Toral, P. G. & Hervás, G. (2017). Individual variation of the extent of milk fat depression in dairy ewes fed fish oil: Milk fatty acid profile and mRNA abundance of candidate genes involved in mammary lipogenesis. *Journal of Dairy Science, 100*, 9611–9622.

Garcia, C., Duby, C., Catheline, D., Toral, P. G., Bernard, L., et al. (2017). Synthesis of the suspected *trans*-11, *cis*-13 conjugated linoleic acid isomer in ruminant mammary tissue by FADS3-catalyzed Δ13-desaturation of vaccenic acid. *Journal of Dairy Science, 100*, 783–796.

Garcia, C., Guillocheau, E., Richard, L., Drouin, G., Catheline, D., et al. (2018). Conversion of dietary *trans*-vaccenic acid to *trans*11, *cis*13-conjugated linoleic acid in the rat lactating mammary gland by Fatty Acid Desaturase 3-catalyzed methyl-end Δ13-desaturation. *Biochemical and Biophysical Research Communications, 505*, 385–391.

Garnsworthy, P. C., Feng, S., Lock, A. L. & Royal, M. D. (2010). Heritability of milk fatty acid composition and stearoyl-CoA desaturase indices in dairy cows. *Journal of Dairy Science, 93*, 1743–1748.

Gervais, R., McFadden, J., Lengi, A., Corl, B. & Chouinard, P. (2009). Effects of intravenous infusion of *trans*-10, *cis*-12 18: 2 on mammary lipid metabolism in lactating dairy cows. *Journal of Dairy Science, 92*, 5167–5177.

Giesy, J., McGuire, M., Shafii, B. & Hanson, T. (2002). Effect of dose of calcium salts of conjugated linoleic acid (CLA) on percentage and fatty acid content of milk fat in midlactation Holstein cows. *Journal of Dairy Science, 85*, 2023–2029.

Gnädig, S., Chamba, J.-F., Perreard, E., Chappaz, S., Chardigny, J.-M., et al. (2004). Influence of manufacturing conditions on the conjugated linoleic acid content and the isomer composition in ripened French Emmental cheese. *Journal of Dairy Research, 71*, 367–371.

Gomez, F. E., Bauman, D. E., Ntambi, J. M. & Fox, B. G. (2003). Effects of sterculic acid on stearoyl-CoA desaturase in differentiating 3T3-L1 adipocytes. *Biochemical and Biophysical Research Communications, 300*, 316–326.

Gonzalez, S., Duncan, S. E., O'Keefe, S. F., Sumner, S. S. & Herbein, J. H. (2003). Oxidation and textural characteristics of butter and ice cream with modified fatty acid profiles. *Journal of Dairy Science, 86*, 70–77.

Griinari, J. & Bauman, D. (1999). Biosynthesis of conjugated linoleic acid and its incorporation into meat and milk in ruminants. *Advances in Conjugated Linoleic Acid Research, 1*, 180–200.

Griinari, J. & Bauman, D. (2003). Update on theories of diet-induced milk fat depression and potential applications. *Recent Advances in Animal Nutrition*, 115–156.

Griinari, J. & Bauman, D. (2006). Milk fat depression: Concepts, mechanisms and management applications. *Ruminant Physiology: Digestion, Metabolism and Impact of Nutrition on Gene Expression, Immunology and Stress*, 389–417.

Griinari, J., Bauman, D., Chilliard, Y., Perajoki, P. & Nurmela, K. (2000). Dietary influences on conjugated linoleic acids (CLA) in bovine milk fat. 3rd Meeting of the European Section of AOCS. Oils and Fats, from Basic Research to Industrial Applications. June 18-21, 2000. Helsinki, Finland. (Abs).

Griinari, J., Dwyer, D., McGuire, M., Bauman, D., Palmquist, D. & Nurmela, K. (1998). *Trans*-octadecenoic acids and milk fat depression in lactating dairy cows. *Journal of Dairy Science, 81*, 1251–1261.

Griinari, J., Corl, B., Lacy, S., Chouinard, P., Nurmela, K. & Bauman, D. (2000). Conjugated linoleic acid is synthesized endogenously in lactating dairy cows by Δ9-desaturase. *Journal of Nutrition, 130*, 2285–2291.

Gu, M., Cosenza, G., Iannaccone, M., Macciotta, N. P. P., Guo, Y., et al. (2019). The single nucleotide polymorphism g. 133A> C in the stearoyl CoA desaturase gene (SCD) promoter affects gene expression and quali-quantitative properties of river buffalo milk. *Journal of Dairy Science, 102*, 442–451.

Gulati, A., Galvin, N., Lewis, E., Hennessy, D., O'Donovan, M., et al. (2018). Outdoor grazing of dairy cows on pasture versus indoor feeding on total mixed ration: Effects on gross composition and mineral content of milk during lactation. *Journal of Dairy Science, 101*, 2710–2723.

Ha, Y., Grimm, N. & Pariza, M. (1987). Anticarcinogens from fried ground beef: Heat-altered derivatives of linoleic acid. *Carcinogenesis, 8*, 1881–1887.

Ha, Y. L., Storkson, J. & Pariza, M. W. (1990). Inhibition of benzo (a) pyrene-induced mouse forestomach neoplasia by conjugated dienoic derivatives of linoleic acid. *Cancer Research, 50*, 1097–1101.

Haghighatdoost, F. & Hariri, M. (2018). Effect of conjugated linoleic acid supplementation on serum leptin concentration: A systematic review and meta-analysis. *Endocrine, Metabolic & Immune Disorders Drug Targets, 18*, 185–193.

Hanuš, O., Křížová, L., Samková, E., Špička, J., Kučera, J., et al. (2016). The effect of cattle breed, season

and type of diet on the fatty acid profile of raw milk. *Archives Animal Breeding, 59*, 373–380.
Harfoot, C. (1981). Lipid metabolism in the rumen. In *Lipid metabolism in ruminant animals* (pp. 21–55). New York: Elsevier.
Harfoot, C. & Hazlewood, G. (1997). Lipid metabolism in the rumen. In P. N. Hobson (Ed.), *The rumen microbial ecosystem* (pp. 382–426). Elsevier: London.
Harvatine, K. J. & Bauman, D. E. (2006). SREBP1 and thyroid hormone responsive spot 14 (S14) are involved in the regulation of bovine mammary lipid synthesis during diet-induced milk fat depression and treatment with CLA. *Journal of Nutrition, 136*, 2468–2474.
Horton, J. D., Goldstein, J. L. & Brown, M. S. (2002). SREBPs: activators of the complete program of cholesterol and fatty acid synthesis in the liver. *Journal of Clinical Investigation, 109*, 1125–1131.
Ip, C., Chin, S. F., Scimeca, J. A. & Pariza, M. W. (1991). Mammary cancer prevention by conjugated dienoic derivative of linoleic acid. *Cancer Research, 51*, 6118–6124.
Ip, C., Singh, M., Thompson, H. J. & Scimeca, J. A. (1994). Conjugated linoleic acid suppresses mammary carcinogenesis and proliferative activity of the mammary gland in the rat. *Cancer Research, 54*, 1212–1215.
Ip, C., Scimeca, J. A. & Thompson, H. (1995). Effect of timing and duration of dietary conjugated linoleic acid on mammary cancer prevention. *Nutrition and Cancer, 24*(3), 241–247.
Ip, C., Banni, S., Angioni, E., Carta, G., McGinley, J., et al. (1999). Conjugated linoleic acid–enriched butter fat alters mammary gland morphogenesis and reduces cancer risk in rats. *Journal of Nutrition, 129*, 2135–2142.
Ip, M. M., Masso-Welch, P. A. & Ip, C. (2003). Prevention of mammary cancer with conjugated linoleic acid: Role of the stroma and the epithelium. *Journal of Mammary Gland Biology and Neoplasia, 8*, 103–118.
Jacobs, A. A. A., Van Baal, J., Smits, M. A., Taweel, H. Z. H., Hendriks, W. H., Van Vuuren, A. M. & Dijkstra, J., (2011). Effects of feeding rapeseed oil, soybean oil, or linseed oil on stearoyl-CoA desaturase expression in the mammary gland of dairy cows. *Journal of Dairy Science, 94*, 874-887.
Jahreis, G., Fritsche, J. & Steinhart, H. (1997). Conjugated linoleic acid in milk fat: High variation depending on production system. *Nutrition Research, 17*, 1479–1484.
Jarvis, G. & Moore, E. (2010). Lipid metabolism and the rumen microbial ecosystem. *Handbook of Hydrocarbon and Lipid Microbiology*, 2245–2257.
Jeffcoat, R. & Pollard, M. (1977). Studies on the inhibition of the desaturases by cyclopropenoid fatty acids. *Lipids, 12*, 480–485.
Jenkins, T. C. (1993). Lipid metabolism in the rumen. *Journal of Dairy Science, 76*, 3851–3863.
Jenkins, T. C. & Harvatine, K. J. (2014). Lipid feeding and milk fat depression. *Veterinary Clinics: Food Animal Practice, 30*, 623–642.
Jenkins, T., Wallace, R., Moate, P. & Mosley, E. (2008). Board-invited review: Recent advances in biohydrogenation of unsaturated fatty acids within the rumen microbial ecosystem. *Journal of Animal Science, 86*, 397–412.
Jiang, J., Bjoerck, L., Fondén, R. & Emanuelson, M. (1996). Occurrence of conjugated *cis*-9, *trans*-11-octadecadienoic acid in bovine milk: Effects of feed and dietary regimen. *Journal of Dairy Science, 79*, 438–445.
Kadegowda, A., Piperova, L. & Erdman, R. (2008). Principal component and multivariate analysis of milk long-chain fatty acid composition during diet-induced milk fat depression. *Journal of Dairy Science, 91*, 749–759.
Kanter, J., Goodspeed, L., Wang, S., Kramer, F., Wietecha, T., et al. (2018). 10, 12 conjugated linoleic acid-driven weight loss is protective against atherosclerosis in mice and is associated with alternative macrophage enrichment in perivascular adipose tissue. *Nutrients, 10*, 1416.
Kay, J., Mackle, T., Auldist, M., Thomson, N. & Bauman, D. (2004). Endogenous synthesis of *cis*-9, *trans*-11 conjugated linoleic acid in dairy cows fed fresh pasture. *Journal of Dairy Science, 87*, 369–378.
Keeney, M. (1970). Lipid metabolism in the rumen. In A. T. Phillipson (Ed.), *Physiology of digestion and metabolism in the ruminant*. Newcastle-upon-Tyne: Oriel Press.
Kelly, M. L. & Bauman, D. E. (1996). Conjugated linoleic acid: A potent anticarcinogen found inmilk fat. *Proceedings of Cornell Nutrition Conference for Feed Manufacturers, Cornell University, Ithaca, NY*, 68–74.
Kelly, M. L., Berry, J. R., Dwyer, D. A., Griinari, J., Chouinard, P. Y., et al. (1998). Dietary fatty acid sources affect conjugated linoleic acid concentrations in milk from lactating dairy cows. *Journal of Nutrition, 128*, 881–885.
Kelsey, J. A., Corl, B. A., Collier, R. J. & Bauman, D. E. (2003). The effect of breed, parity, and stage of lactation on conjugated linoleic acid (CLA) in milk fat from dairy cows. *Journal of Dairy Science, 86*, 2588–2597.
Kemp P. & Lander, D. J. (1984). Hydrogenation in vitro of α-linolenic acid to stearic acid by mixed cultures of pure strains of rumen bacteria. *Microbiology, 130*, 527–533.
Kepler, C. R. & Tove, S. (1967). Biohydrogenation of unsaturated fatty acids III. Purification and properties of a linoleate Δ12-cis, Δ11-*trans*-isomerase from *Butryrivibrio fibrisolvens*. *Journal of Biological Chemistry, 242*, 5686–5692.
Kepler, C. R., Tucker, W. & Tove, S. (1970). Biohydrogenation of unsaturated fatty acids IV. Substrate specificity and inhibition of linoleate Δ12-CIS, Δ11-*trans*-isomerase from *butyrivibrio fibrisolvens*. *Journal of Biological Chemistry, 245*, 3612–3620.
Khanal, R. C., Dhiman, T. R., Ure, A. L., Brennand, C. P., Boman, R. L. & McMahon, D. J. (2005). Consumer acceptability of conjugated linoleic acid-enriched

milk and cheddar cheese from cows grazing on pasture. *Journal of Dairy Science, 88*, 1837–1847.

Kilcawley, K. N., Faulkner, H., Clarke, H. J., O'Sullivan, M. G. & Kerry, J. P. (2018). Factors influencing the flavour of bovine milk and cheese from grass based versus non-grass based milk production systems. *Foods, 7*, 37.

Kim, Y., Liu, R., Rychlik, J. & Russell, J. (2002). The enrichment of a ruminal bacterium (*Megasphaera elsdenii* YJ-4) that produces the *trans*-10, *cis*-12 isomer of conjugated linoleic acid. *Journal of Applied Microbiology, 92*, 976–982.

Kinsella, J. (1972). Stearyl CoA as a precursor of oleic acid and glycerolipids in mammary microsomes from lactating bovine: Possible regulatory step in milk triglyceride synthesis. *Lipids, 7*, 349–355.

Kliem, K. E. & Shingfield, K. J. (2016). Manipulation of milk fatty acid composition in lactating cows: Opportunities and challenges. *European Journal of Lipid Science and Technology, 118*, 1661–1683.

Kohno, H., Suzuki, R., Yasui, Y., Hosokawa, M., Miyashita, K. & Tanaka, T. (2004a). Pomegranate seed oil rich in conjugated linolenic acid suppresses chemically induced colon carcinogenesis in rats. *Cancer Science, 95*, 481–486.

Kohno, H., Yasui, Y., Suzuki, R., Hosokawa, M., Miyashita, K. & Tanaka, T. (2004b). Dietary seed oil rich in conjugated linolenic acid from bitter melon inhibits azoxymethane-induced rat colon carcinogenesis through elevation of colonic PPARγ expression and alteration of lipid composition. *International Journal of Cancer, 110*, 896–901.

Koronowicz, A. A. & Banks, P. (2018). Antitumor properties of CLA-enriched food products. *Nutrition and Cancer, 70*, 529–545.

Koronowicz, A. A., Banks, P., Domagała, D., Master, A., Leszczyńska, T., et al. (2016). Fatty acid extract from CLA-enriched egg yolks can mediate transcriptome reprogramming of MCF-7 cancer cells to prevent their growth and proliferation. *Genes and Nutrition, 11*, 22.

Koronowicz, A. A., Banks, P., Master, A., Domagała, D., Piasna-Słupecka, E., et al. (2017). Fatty acids of CLA-enriched egg yolks can induce transcriptional activation of peroxisome proliferator-activated receptors in MCF-7 breast cancer cells. *PPAR Research, 2017*.

Kostogrys, R. B., Franczyk-Żarów, M., Maslak, E., Gajda, M. & Chłopicki, S. (2012). Effects of margarine supplemented with t10c12 and C9T11 CLA on atherosclerosis and steatosis in apoE/LDLR-/-mice. *Journal of Nutrition, Health and Aging, 16*, 482–490.

Kraft, J., Collomb, M., Möckel, P., Sieber, R. & Jahreis, G. (2003). Differences in CLA isomer distribution of cow's milk lipids. *Lipids, 38*, 657–664.

Krag, K., Poulsen, N. A., Larsen, M. K., Larsen, L. B., Janss, L. L. & Buitenhuis, B. (2013). Genetic parameters for milk fatty acids in Danish Holstein cattle based on SNP markers using a Bayesian approach. *BMC Genetics, 14*, 79.

Kramer, J. K. G., Fellner, V., Dugan, M. E. R., Sauer, F. D., Mossoba, M. M. & Yurawecz, M. P. (1997). Evaluating acid and base catalysts in the methylation of milk and rumen fatty acids with special emphasis on conjugated dienes and total *trans* fatty acids. *Lipids, 32*, 1219–1228.

Kramer, J. K., Parodi, P. W., Jensen, R. G., Mossoba, M. M., Yurawecz, M. P. & Adlof, R. O. (1998). Rumenic acid: A proposed common name for the major conjugated linoleic acid isomer found in natural products. *Lipids, 33*, 835.

Kramer, J. K., Cruz-Hernandez, C., Deng, Z., Zhou, J., Jahreis, G. & Dugan, M. E. (2004). Analysis of conjugated linoleic acid and *trans* 18:1 isomers in synthetic and animal products. *American Journal of Clinical Nutrition, 79*, 1137S–1145S.

Kritchevsky, D., Tepper, S. A., Wright, S., Tso, P. & Czarnecki, S. K. (2000). Influence of conjugated linoleic acid (CLA) on establishment and progression of atherosclerosis in rabbits. *Journal of the American College of Nutrition, 19*, 472S–477S.

Kritchevsky, D., Tepper, S. A., Wright, S. & Czarnecki, S. K. (2002). Influence of graded levels of conjugated linoleic acid (CLA) on experimental atherosclerosis in rabbits. *Nutrition Research, 22*, 1275–1279.

Kritchevsky, D., Tepper, S. A., Wright, S., Czarnecki, S. K., Wilson, T. A. & Nicolosi, R. J. (2004). Conjugated linoleic acid isomer effects in atherosclerosis: Growth and regression of lesions. *Lipids, 39*, 611–616.

Lacasse, P., Kennelly, J. J., Delbecchi, L. & Ahnadi, C. E. (2002). Addition of protected and unprotected fish oil to diets for dairy cows. I. Effects on the yield, composition and taste of milk. *Journal of Dairy Research, 69*, 511–520.

Larsson, S. C., Bergkvist, L. & Wolk, A. (2005). High-fat dairy food and conjugated linoleic acid intakes in relation to colorectal cancer incidence in the Swedish Mammography Cohort. *American Journal of Clinical Nutrition, 82*, 894–900.

Larsson, S. C., Bergkvist, L. & Wolk, A. (2009). Conjugated linoleic acid intake and breast cancer risk in a prospective cohort of Swedish women. *American Journal of Clinical Nutrition, 90*, 556–560.

Lawless, F., Murphy, J., Harrington, D., Devery, R. & Stanton, C. (1998). Elevation of conjugated *cis*-9, *trans*-11-octadecadienoic acid in bovine milk because of dietary supplementation. *Journal of Dairy Science, 81*, 3259–3267.

Lawless, F., Stanton, C., L'escop, P., Devery, R., Dillon, P. & Murphy, J. J. (1999). Influence of breed on bovine milk *cis*-9, *trans*-11-conjugated linoleic acid content. *Livestock Production Science, 62*, 43–49.

Lee, K. N., Kritchevsky, D. & Parizaa, M. W. (1994). Conjugated linoleic acid and atherosclerosis in rabbits. *Atherosclerosis, 108*, 19–25.

Lee, K. N., Pariza, M. W. & Ntambi, J. M. (1998). Conjugated linoleic acid decreases hepatic stearoyl-CoA desaturase mRNA expression. *Biochemical and Biophysical Research Communications, 248*, 817–821.

Lengi, A. & Corl, B. (2007). Identification and characterization of a novel bovine stearoyl-CoA desaturase isoform with homology to human SCD5. *Lipids, 42*, 499–508.

Libby, P. (2002). Inflammation in atherosclerosis. *Nature, 420*, 868–874.

Liew, C., Schut, H., Chin, S., Pariza, M. & Dashwood, R. (1995). Protection of conjugated linoleic acids against 2-amino-3-methylimidazo [4, 5-f] quinoline-induced colon carcinogenesis in the F344 rat: A study of inhibitory mechanisms. *Carcinogenesis, 16*, 3037–3043.

Lin, H., Boylston, T. D., Luedecke, L. O. & Shultz, T. D. (1999). Conjugated linoleic acid content of Cheddar-type cheeses as affected by processing. *Journal of Food Science, 64*, 874–878.

Liu, X., Li, H., Chen, Y. & Cao, Y. (2012). Method for screening of bacterial strains biosynthesizing specific conjugated linoleic acid isomers. *Journal of Agricultural and Food Chemistry, 60*, 9705–9710.

Lock, A. & Garnsworthy, P. (2002). Independent effects of dietary linoleic and linolenic fatty acids on the conjugated linoleic acid content of cows' milk. *Animal Science, 74*, 163–176.

Lock, A. L. & Garnsworthy, P. C. (2003). Seasonal variation in milk conjugated linoleic acid and Δ9-desaturase activity in dairy cows. *Livestock Production Science, 79*, 47–59.

Lock, A. L., Bauman, D. E. (2004). Modifying milk fat composition of dairy cows to enhance fatty acids beneficial to human health. *Lipids, 39*, 1197–1206.

Lock, A. L., Bauman, D. E. & Garnsworthy, P. C. (2003). Effects of milk yield and milk fat production on milk *cis*-9, *trans*-11 CLA and delta-9 desaturase activity. *Journal of Dairy Science, 86*, 245.

Lock, A. L., Corl, B. A., Barbano, D. M., Bauman, D. E. & Ip, C. (2004). The anticarcinogenic effect of *trans*-11 18: 1 is dependent on its conversion to *cis*-9, *trans*-11 CLA by Δ9-desaturase in rats. *Journal of Nutrition, 134*, 2698–2704.

Lock, A. L., Horne, C. A., Bauman, D. E. & Salter, A. M. (2005). Butter naturally enriched in conjugated linoleic acid and vaccenic acid alters tissue fatty acids and improves the plasma lipoprotein profile in cholesterol-fed hamsters. *Journal of Nutrition, 135*, 1934–1939.

Lock, A., Harvatine, K., Drackley, J. & Bauman, D. E. (2006). *Concepts of fat and fatty acid digestion in ruminants*.

Loor, J. J. & Herbein, J. H. (1998). Exogenous conjugated linoleic acid isomers reduce bovine milk fat concentration and yield by inhibiting de novo fatty acid synthesis. *Journal of Nutrition, 128*, 2411–2419.

Loor, J. J. & Herbein, J. H. (2001). Alterations in blood plasma and milk fatty acid profiles of lactating Holstein cows in response to ruminal infusion of a conjugated linoleic acid mixture. *Animal Research, 50*, 463–476.

Lourenço, M., Ramos-Morales, E. & Wallace, R. (2010). The role of microbes in rumen lipolysis and biohydrogenation and their manipulation. *Animal, 4*, 1008–1023.

Luna, P., de la Fuente, M. A. & Juarez, M. (2005). Conjugated linoleic acid in processed cheeses during the manufacturing stages. *Journal of Agricultural and Food Chemistry, 53*, 2690–2695.

Lusis, A. J. (2000). Atherosclerosis. *Nature, 407*, 233–241.

Lynch, J. M., Lock, A. L., Dwyer, D. A., Noorbakhsh, R., Barbano, D. M. & Bauman, D. E. (2005). Flavor and stability of pasteurized milk with elevated levels of conjugated linoleic acid and vaccenic acid. *Journal of Dairy Science, 88*, 489–498.

Mahfouz, M., Valicenti, A. & Holman, R. (1980). Desaturation of isomeric *trans*-octadecenoic acids by rat liver microsomes. *Biochimica et Biophysica Acta (BBA)-Lipids and Lipid Metabolism, 618*, 1–12.

Månsson, H. L. (2008). Fatty acids in bovine milk fat. *Food and Nutrition Research, 52*. https://doi.org/10.3402/fnr.v52i0.1821.

Maragkoudakis, P. A., Mountzouris, K. C., Rosu, C., Zoumpopoulou, G., Papadimitriou, K., et al. (2010). Feed supplementation of *Lactobacillus plantarum* PCA 236 modulates gut microbiota and milk fatty acid composition in dairy goats – A preliminary study. *International Journal of Food Microbiology, 141*, S109–SS16.

Martel, P. M., Bingham, C. M., McGraw, C. J., Baker, C. L., Morganelli, P. M., et al. (2006). S14 protein in breast cancer cells: Direct evidence of regulation by SREBP-1c, superinduction with progestin, and effects on cell growth. *Experimental Cell Research, 312*, 278–288.

McCann, S. E., Ip, C., Ip, M. M., McGuire, M. K., Muti, P., et al. (2004). Dietary intake of conjugated linoleic acids and risk of premenopausal and postmenopausal breast cancer, Western New York Exposures and Breast Cancer Study (WEB Study). *Cancer Epidemiology and Prevention Biomarkers, 13*, 1480–1484.

McDonald, T. & Kinsella, J. (1973). Stearyl-CoA desaturase of bovine mammary microsomes. *Archives of Biochemistry and Biophysics, 156*, 223–231.

McGowan, M. M., Eisenberg, B. L., Lewis, L. D., Froehlich, H. M., Wells, W. A., et al. (2013). A proof of principle clinical trial to determine whether conjugated linoleic acid modulates the lipogenic pathway in human breast cancer tissue. *Breast Cancer Research and Treatment, 138*, 175–183.

McLeod, R. S., LeBlanc, A. M., Langille, M. A., Mitchell, P. L. & Currie, D. L. (2004). Conjugated linoleic acids, atherosclerosis, and hepatic very-low-density lipoprotein metabolism. *American Journal of Clinical Nutrition, 79*, 1169S–1174S.

Meir, K. S. & Leitersdorf, E. (2004). Atherosclerosis in the apolipoprotein E–deficient mouse: A decade of progress. *Arteriosclerosis, Thrombosis, and Vascular Biology, 24*, 1006–1014.

Meurens, M., Baeten, V., Yan, S. H., Mignolet, E. & Larondelle, Y. (2005). Determination of the conjugated linoleic acids in cow's milk fat by Fourier transform Raman spectroscopy. *Journal of Agricultural and Food Chemistry, 53*, 5831–5835.

Michaud, A. L., Yurawecz, M. P., Delmonte, P., Corl, B. A., Bauman, D. E. & Brenna, J. T. (2003).

Identification and characterization of conjugated fatty acid methyl esters of mixed double bond geometry by acetonitrile chemical ionization tandem mass spectrometry. *Analytical Chemistry, 75*, 4925–4930.

Miller, A., Stanton, C., Murphy, J. & Devery, R. (2003). Conjugated linoleic acid (CLA)-enriched milk fat inhibits growth and modulates CLA-responsive biomarkers in MCF-7 and SW480 human cancer cell lines. *British Journal of Nutrition, 90*, 877–885.

Mirzaii, S., Mansourian, M., Derakhshandeh-Rishehri, S.-M., Kelishadi, R. & Heidari-Beni, M. (2016). Association of conjugated linoleic acid consumption and liver enzymes in human studies: A systematic review and meta-analysis of randomized controlled clinical trials. *Nutrition, 32*, 166–173.

Mohammadzadeh, M., Faramarzi, E., Mahdavi, R., Nasirimotlagh, B. & Asghari, J. M. (2013). Effect of conjugated linoleic acid supplementation on inflammatory factors and matrix metalloproteinase enzymes in rectal cancer patients undergoing chemoradiotherapy. *Integrative Cancer Therapies, 12*, 496–502.

Momchilova, S. M. & Nikolova-Damyanova, B. M. (2012). Advances in silver ion chromatography for the analysis of fatty acids and triacylglycerols-2001 to 2011. *Analytical Sciences, 28*, 837–844.

Moon, C. D., Pacheco, D. M., Kelly, W. J., Leahy, S. C., Li, D., et al. (2008). Reclassification of *Clostridium proteoclasticum* as *Butyrivibrio proteoclasticus comb. nov.*, a butyrate-producing ruminal bacterium. *International Journal of Systematic and Evolutionary Microbiology, 58*, 2041–2045.

Moore, T. (1939). Spectroscopic changes in fatty acids: General. *Biochemical Journal, 33*, 1635.

Moore, C., Hafliger, H., III, Mendivil, O., Sanders, S., Bauman, D. & Baumgard, L. (2004). Increasing amounts of conjugated linoleic acid (CLA) progressively reduces milk fat synthesis immediately postpartum. *Journal of Dairy Science, 87*, 1886–1895.

Mosley, E. E., Powell, G. L., Riley, M. B. & Jenkins, T. C. (2002). Microbial biohydrogenation of oleic acid to *trans* isomers *in vitro*. *Journal of Lipid Research, 43*, 290–296.

Munday, J. S., Thompson, K. G. & James, K. A. (1999). Dietary conjugated linoleic acids promote fatty streak formation in the C57BL/6 mouse atherosclerosis model. *British Journal of Nutrition, 81*, 251–255.

Murphy, S. P. & Allen, L. H. (2003). Nutritional importance of animal source foods. *Journal of Nutrition, 133*, 3932S–3935S.

Muruz, H. & Çetinkaya, N. (2019). The effect of dairy cow feeding regime on functional milk production. *International Advanced Researches and Engineering Journal, 3*, 1–6.

National Research Council Committee on Technological Options to Improve the Nutritional Attributes of Animal Products. (1988). *Designing foods: Animal product options in the marketplace*. Washington, DC: National Academies Press (US).

Noone, E. J., Roche, H. M., Nugent, A. P. & Gibney, M. J. (2002). The effect of dietary supplementation using isomeric blends of conjugated linoleic acid on lipid metabolism in healthy human subjects. *British Journal of Nutrition, 88*, 243–251.

Ntambi, J. M. (1999). Regulation of stearoyl-CoA desaturase by polyunsaturated fatty acids and cholesterol. *Journal of Lipid Research, 40*, 1549–1558.

Ntambi, J. M. & Miyazaki, M. (2004). Regulation of stearoyl-CoA desaturases and role in metabolism. *Progress in Lipid Research, 43*, 91–104.

O'Callaghan, T. F., Faulkner, H., McAuliffe, S., O'Sullivan, M. G., Hennessy, D., et al. (2016a). Quality characteristics, chemical composition, and sensory properties of butter from cows on pasture versus indoor feeding systems. *Journal of Dairy Science, 99*, 9441–9460.

O'Callaghan, T. F., Hennessy, D., McAuliffe, S., Kilcawley, K. N., O'Donovan, M., et al. (2016b). Effect of pasture versus indoor feeding systems on raw milk composition and quality over an entire lactation. *Journal of Dairy Science, 99*, 9424–9440.

O'Callaghan, T. F., Mannion, D. T., Hennessy, D., McAuliffe, S., O'Sullivan, M. G., et al. (2017). Effect of pasture versus indoor feeding systems on quality characteristics, nutritional composition, and sensory and volatile properties of full-fat Cheddar cheese. *Journal of Dairy Science, 100*, 6053–6073.

O'Callaghan, T. F., Vazquez-Fresno, R., Serra-Cayuela, A., Dong, E., Mandal, R., et al. (2018). Pasture feeding changes the bovine rumen and milk metabolome. *Metabolites, 8*.

O'Shea, M., Devery, R., Lawless, F., Murphy, J. & Stanton, C. (2000). Milk fat conjugated linoleic acid (CLA) inhibits growth of human mammary MCF-7 cancer cells. *Anticancer Research, 20*, 3591–3601.

Offer, N. W., Marsden, M., Dixon, J., Speake, B. K. & Thacker, F. E. (1999). Effect of dietary fat supplements on levels of n-3 polyunsaturated fatty acids, *trans* acids and conjugated linoleic acid in bovine milk. *Journal of Animal Science, 69*, 613–625.

Offer, N., Marsden, M. & Phipps, R. (2001). Effect of oil supplementation of a diet containing a high concentration of starch on levels of *trans* fatty acids and conjugated linoleic acids in bovine milk. *Animal Science, 73*, 533–540.

Ozols, J. (1997). Degradation of hepatic stearyl CoA Δ9-desaturase. *Molecular Biology of the Cell, 8*, 2281–2290.

Palmquist, D., Beaulieu, A. D. & Barbano, D. (1993). Feed and animal factors influencing milk fat composition. *Journal of Dairy Science, 76*, 1753–1771.

Palmquist, D., St-Pierre, N. & McClure, K. (2004). Tissue fatty acid profiles can be used to quantify endogenous rumenic acid synthesis in lambs. *Journal of Nutrition, 134*, 2407–2414.

Palmquist, D. L., Lock, A. L., Shingfield, K. J. & Bauman, D. E. (2005). Biosynthesis of conjugated linoleic acid in ruminants and humans. *Advances in Food and Nutrition Research, 50*, 179–217.

Pariza, M. W., Ashoor, S. H., Chu, F. S. & Lund, D. B. (1979). Effects of temperature and time on mutagen formation in pan-fried hamburger. *Cancer Letters, 7*, 63–69.

Park, Y., Storkson, J. M., Ntambi, J. M., Cook, M. E., Sih, C. J. & Pariza, M. W. (2000). Inhibition of hepatic stearoyl-CoA desaturase activity by *trans*-10, *cis*-12 conjugated linoleic acid and its derivatives. *Biochimica et Biophysica Acta (BBA)-Molecular and Cell Biology of Lipids, 1486*, 285–292.

Parodi, P. W. (1977). Conjugated octadecadienoic acids of milk fat. *Journal of Dairy Science, 60*, 1550–1553.

Parodi, P. W. (2003). Conjugated linoleic acid in food. In Sebedio, J. Christie, W. W. Adlof, R., (Eds.). *Advances in conjugated linoleic acid research*, (pp. 152–179). Champaign: AOCS Publishing.

Parodi, P. W. (2004). Milk fat in human nutrition. *Australian Journal of Dairy Technology, 59*, 3–59.

Perfield, J., II, Bernal-Santos, G., Overton, T. & Bauman, D. (2002). Effects of dietary supplementation of rumen-protected conjugated linoleic acid in dairy cows during established lactation. *Journal of Dairy Science, 85*, 2609–2617.

Perfield, J., II, Sæbø, A. & Bauman, D. (2004a). Use of conjugated linoleic acid (CLA) enrichments to examine the effects of *trans*-8, *cis*-10 CLA, and *cis*-11, *trans*-13 CLA on milk-fat synthesis. *Journal of Dairy Science, 87*, 1196–1202.

Perfield, J., II, Lock, A., Pfeiffer, A. & Bauman, D. (2004b). Effects of amide-protected and lipid-encapsulated conjugated linoleic acid (CLA) supplements on milk fat synthesis. *Journal of Dairy Science, 87*, 3010–3016.

Perfield, J. W., Sæbø, A. & Bauman, D. E. (2004c). Use of conjugated linoleic acid (CLA) enrichment to examine the effects of *trans*-8, *cis*-10 CLA, and *cis*-11, *trans*-13 CLA on milk-fat synthesis. *Journal of Animal Science, 87*, 1196–1202.

Perfield, J., II, Delmonte, P., Lock, A., Yurawecz, M. & Bauman, D. (2006). *Trans*-10, *trans*-12 conjugated linoleic acid does not affect milk fat yield but reduces Δ9-desaturase index in dairy cows. *Journal of Dairy Science, 89*, 2559–2566.

Perfield Ii, J., Lock, A., Griinari, J., Sæbø, A., Delmonte, P., et al. (2007). *Trans*-9, *cis*-11 conjugated linoleic acid reduces milk fat synthesis in lactating dairy cows. *Journal of Dairy Science, 90*, 2211–2218.

Peterson, D., Baumgard, L. & Bauman, D. (2002a). Milk fat response to low doses of *trans*-10, *cis*-12 conjugated linoleic acid (CLA). *Journal of Dairy Science, 85*, 1764–1766.

Peterson, D. G., Kelsey, J. A. & Bauman, D. E. (2002b). Analysis of variation in *cis*-9, *trans*-11 conjugated linoleic acid (CLA) in milk fat of dairy cows. *Journal of Dairy Science, 85*, 2164–2172.

Peterson, D. G., Matitashvili, E. A. & Bauman, D. E. (2003). Diet-induced milk fat depression in dairy cows results in increased *trans*-10, *cis*-12 CLA in milk fat and coordinate suppression of mRNA abundance for mammary enzymes involved in milk fat synthesis. *Journal of Nutrition, 133*, 3098–3102.

Peterson, D. G., Matitashvili, E. A. & Bauman, D. E. (2004). The inhibitory effect of *trans*-10, *cis*-12 CLA on lipid synthesis in bovine mammary epithelial cells involves reduced proteolytic activation of the transcription factor SREBP-1. *Journal of Nutrition, 134*, 2523–2527.

Petri, R. M., Mapiye, C., Dugan, M. E. & McAllister, T. A. (2014). Subcutaneous adipose fatty acid profiles and related rumen bacterial populations of steers fed red clover or grass hay diets containing flax or sunflower-seed. *PLoS One, 9*, e104167.

Petrik, M. B. H., McEntee, M. F., Johnson, B. T., Obukowicz, M. G. & Whelan, J. (2000). Highly unsaturated (n-3) fatty acids, but not α-linolenic, conjugated linoleic or γ-linolenic acids, reduce tumorigenesis in Apc Min/+ mice. *Journal of Nutrition, 130*, 2434–2443.

Pfeuffer, M., Fielitz, K., Laue, C., Winkler, P., Rubin, D., et al. (2011). CLA does not impair endothelial function and decreases body weight as compared with safflower oil in overweight and obese male subjects. *Journal of the American College of Nutrition, 30*, 19–28.

Phelps, R. A., Shenstone, F., Kemmerer, A. & Evans, R. (1965). A review of cyclopropenoid compounds: Biological effects of some derivatives. *Poultry Science, 44*, 358–394.

Pierre, A.-S., Minville-Walz, M., Fèvre, C., Hichami, A., Gresti, J., et al. (2013). *Trans*-10, *cis*-12 conjugated linoleic acid induced cell death in human colon cancer cells through reactive oxygen species-mediated ER stress. *Biochimica et Biophysica Acta (BBA)-Molecular and Cell Biology of Lipids, 1831*, 759–768.

Piperova, L. S., Teter, B. B., Bruckental, I., Sampugna, J., Mills, S. E., et al. (2000). Mammary lipogenic enzyme activity, *trans* fatty acids and conjugated linoleic acids are altered in lactating dairy cows fed a milk fat–depressing diet. *Journal of Nutrition, 130*, 2568–2574.

Piperova, L. S., Sampugna, J., Teter, B. B., Kalscheur, K. F., Yurawecz, M. P., et al. (2002). Duodenal and milk *trans* octadecenoic acid and conjugated linoleic acid (CLA) isomers indicate that postabsorptive synthesis is the predominant source of *cis*-9-containing CLA in lactating dairy cows. *Journal of Nutrition, 132*, 1235–1241.

Piperova, L., Moallem, U., Teter, B., Sampugna, J., Yurawecz, M., et al. (2004). Changes in milk fat in response to dietary supplementation with calcium salts of *trans*-18: 1 or conjugated linoleic fatty acids in lactating dairy cows. *Journal of Dairy Science, 87*, 3836–3844.

Pollard, M. R., Gunstone, F. D., James, A. T. & Morris, L. J. (1980). Desaturation of positional and geometric isomers of monoenoic fatty acids by microsomal preparations from rat liver. *Lipids, 15*, 306–314.

Precht, D. & Molkentin, J. (1997). *Trans*-geometrical and positional isomers of linoleic acid including conjugated linoleic acid (CLA) in German milk and vegetable fats. *European Journal of Lipid Science and Technology, 99*, 319–326.

Prema, D., Pilfold, J., Krauchi, J., Church, J., Donkor, K. & Cinel, B. (2013). Rapid determination of total conjugated linoleic acid content in select Canadian cheeses by H-1 NMR spectroscopy. *Journal of Agricultural and Food Chemistry, 61*(41), 9915–9921.

Privé, F., Newbold, C. J., Kaderbhai, N. N., Girdwood, S. G., Golyshina, O. V., et al. (2015). Isolation and characterization of novel lipases/esterases from a bovine rumen metagenome. *Applied Microbiology and Biotechnology, 99*, 5475–5485.

Proell, J. M., Mosley, E. E., Powell, G. L. & Jenkins, T. C. (2002). Isomerization of stable isotopically labeled elaidic acid to *cis* and *trans* monoenes by ruminal microbes. *Journal of Lipid Research, 43*, 2072–2076.

Ramaswamy, N., Baer, R. J., Schingoethe, D. J., Hippen, A. R., Kasperson, K. M. & Whitlock, L. A. (2001). Composition and flavor of milk and butter from cows fed fish oil, extruded soybeans, or their combination. *Journal of Dairy Science, 84*, 2144–2151.

Reh, W. A., Maga, E. A., Collette, N. M., Moyer, A., Conrad-Brink, J. S., et al. (2004). Hot topic: Using a stearoyl-CoA desaturase transgene to alter milk fatty acid composition. *Journal of Dairy Science, 87*, 3510–3514.

Rickert, R., Steinhart, H., Fritsche, J., Sehat, N., Yurawecz, M. P., Mossoba, M. M., Roach, J.A.G., Eulitz, K., Ku, Y. & Kramer, J.K.G. (1999). Enhanced resolution of conjugated linoleic acid isomers by tandem-column silver-ion high performance liquid chromatography. *Journal of High Resolution Chromatography, 22*, 144–148.

Riel, R. R. (1963). Physico-chemical characteristics of Canadian milk fat. Unsaturated fatty acids. *Journal of Dairy Science, 46*, 102–106.

Ritzenthaler, K. L., McGuire, M. K., Falen, R., Shultz, T. D., Dasgupta, N. & McGuire, M. A. (2001). Estimation of conjugated linoleic acid intake by written dietary assessment methodologies underestimates actual intake evaluated by food duplicate methodology. *Journal of Nutrition, 131*, 1548–1554.

Rodriguez-Alcala, L. M. & Fontecha, J. (2007). Hot topic: Fatty acid and conjugated linoleic acid (CLA) isomer composition of commercial CLA-fortified dairy products: Evaluation after processing and storage. *Journal of Dairy Science, 90*, 2083–2090.

Sæbø, A., Perfield, J., Delmonte, P., Yurawecz, M., Lawrence, P., et al. (2005a). Milk fat synthesis is unaffected by abomasal infusion of the conjugated diene 18: 3 isomers *cis*-6, *trans*-10, *cis*-12 and *cis*-6, *trans*-8, *cis*-12. *Lipids, 40*, 89–95.

Sæbø, A., Sæbø, P. C., Griinari, J. M. & Shingfield, K. J. (2005b). Effect of abomasal infusions of geometric isomers of 10, 12 conjugated linoleic acid on milk fat synthesis in dairy cows. *Lipids, 40*, 823–832.

Samková, E., Koubová, J., Hasoňová, L., Hanuš, O., Kala, R., et al. (2018). Joint effects of breed, parity, month of lactation, and cow individuality on the milk fatty acids composition. *Mljekarstvo: časopis za unaprjeđenje proizvodnje i prerade mlijeka, 68*, 98–107.

Schennink, A., Stoop, W. M., Visker, M. W., Heck, J. M. L., Bovenhuis, H., et al. (2007). DGAT1 underlies large genetic variation in milk-fat composition of dairy cows. *Animal Genetics, 38*, 467–473.

Schennink, A., Heck, J. M., Bovenhuis, H., Visker, M. H., van Valenberg, H. J. & van Arendonk, J. A. (2008). Milk fatty acid unsaturation: Genetic parameters and effects of stearoyl-CoA desaturase (SCD1) and acyl CoA: Diacylglycerol acyltransferase 1 (DGAT1). *Journal of Dairy Science, 91*, 2135–2143.

Scimeca, J. A. (1999). Cancer inhibition in animals. *Advances in Conjugated Linoleic Acid Research, 1*, 420–443.

Sébédio, J.-L., Christie, W. W. & Adlof, R. (2003). In Sebedio, J. Christie, W. W. Adolf, R. (Eds.). *Advances in conjugated linoleic acid research.* Champaign: AOCS Publishing.

Sehat, N., Kramer, J. K., Mossoba, M. M., Yurawecz, M. P., Roach, J. A., et al. (1998). Identification of conjugated linoleic acid isomers in cheese by gas chromatography, silver ion high performance liquid chromatography and mass spectral reconstructed ion profiles. Comparison of chromatographic elution sequences. *Lipids, 33*, 963–971.

Selberg, K., Lowe, A., Staples, C., Luchini, N. & Badinga, L. (2004). Production and metabolic responses of periparturient Holstein cows to dietary conjugated linoleic acid and *trans*-octadecenoic acids. *Journal of Dairy Science, 87*, 158–168.

Shantha, N. C. & Decker, E. A. (1995). Conjugated linoleic acid concentrations in cooked beef containing antioxidants and hydrogen donors. *Journal of Food Lipids, 2*, 57–64.

Shantha, N. C., Decker, E. A. & Ustunol, Z. (1992). Conjugated linoleic acid concentration in processed cheese. *Journal of the American Oil Chemists' Society, 69*, 425–428.

Shen, X., Dannenberger, D., Nuernberg, K., Nuernberg, G. & Zhao, R. (2011). *Trans*-18: 1 and CLA isomers in rumen and duodenal digesta of bulls fed *n*-3 and *n*-6 PUFA-based diets. *Lipids, 46*, 831–841.

Shimano, H. (2001). Sterol regulatory element-binding proteins (SREBPs): Transcriptional regulators of lipid synthetic genes. *Progress in Lipid Research, 40*, 439–452.

Shingfield, K., Ahvenjärvi, S., Toivonen, V., Ärölä, A., Nurmela, K., et al. (2003). Effect of dietary fish oil on biohydrogenation of fatty acids and milk fatty acid content in cows. *Animal Science, 77*, 165–179.

Shingfield, K. J., Reynolds, C. K., Hervas, G., Griinari, J. M., Grandison, A. S. & Beever, D. E. (2006). Examination of the persistency of milk fatty acid composition responses to fish oil and sunflower oil in the diet of dairy cows. *Journal of Dairy Science, 89*, 714–732.

Shingfield, K. J., Ahvenjärvi, S., Toivonen, V., Vanhatalo, A., Huhtanen, P. & Griinari, J. M. (2008). Effect of incremental levels of sunflower-seed oil in the diet on ruminal lipid metabolism in lactating cows. *British Journal of Nutrition, 99*, 971–983.

Shingfield, K. J., Bonnet, M. & Scollan, N. D. (2013). Recent developments in altering the fatty acid composition of ruminant-derived foods. *Animal, 7*(Suppl 1), 132–162.

Shiraishi, R., Iwakiri, R., Fujise, T., Kuroki, T., Kakimoto, T., et al. (2010). Conjugated linoleic acid suppresses colon carcinogenesis in azoxymethane-pretreated rats with long-term feeding of diet containing beef tallow. *Journal of Gastroenterology, 45*, 625–635.

Shivani, S., Srivastava, A., Shandilya, U., Kale, V. & Tyagi, A. (2015). Dietary supplementation of *Butyrivibrio fibrisolvens* alters fatty acids of milk and rumen fluid in lactating goats: Alteration of fatty acid profile of goat milk. *Journal of the Science of Food and Agriculture, 96*(5), 1716–1722.

Slots, T., Butler, G., Leifert, C., Kristensen, T., Skibsted, L. H. & Nielsen, J. H. (2009). Potentials to differentiate milk composition by different feeding strategies. *Journal of Dairy Science, 92*, 2057–2066.

Sluijs, I., Plantinga, Y., De Roos, B., Mennen, L. I. & Bots, M. L. (2009). Dietary supplementation with *cis*-9, *trans*-11 conjugated linoleic acid and aortic stiffness in overweight and obese adults. *American Journal of Clinical Nutrition, 91*, 175–183.

Soel, S. M., Choi, O. S., Bang, M. H., Park, J. H. Y. & Kim, W. K. (2007). Influence of conjugated linoleic acid isomers on the metastasis of colon cancer cells in vitro and in vivo. *Journal of Nutritional Biochemistry, 18*, 650–657.

Stanton, C., Lawless, F., Kjellmer, G., Harrington, D., Devery, R., et al. (1997). Dietary influences on bovine milk *cis*- 9, *trans*-11-conjugated linoleic acid content. *Journal of Food Science, 62*, 1083–1086.

Stanton, C., Murphy, J., McGrath, E., Devery, R. (2003). Animal feeding strategies for conjugated linoleic acid enrichment of milk. In Sebedio, J. Christie, W. W. Adlof, R. (Eds.). *Advances in conjugated linoleic acid research* (pp. 123–145), Champaign: AOCS Publishing.

Sun, C., Black, B. A., Zhao, Y. Y., Ganzle, M. G. & Curtis, J. M. (2013). Identification of conjugated linoleic acid (CLA) isomers by silver ion-liquid chromatography/in-line ozonolysis/mass spectrometry (Ag$^+$-LC/O_3-MS). *Analytical Chemistry, 85*, 7345–7352.

Taniguchi, M., Utsugi, T., Oyama, K., Mannen, H., Kobayashi, M., et al. (2004). Genotype of stearoyl-CoA desaturase is associated with fatty acid composition in Japanese Black cattle. *Mammalian Genome: Official Journal of the International Mammalian Genome Society, 15*, 142–148.

Tao, F. & Ngadi, M. (2017). Applications of spectroscopic techniques for fat and fatty acids analysis of dairy foods. *Current Opinion in Food Science, 17*, 100–112.

Theurer, M. L., Block, E., Sanchez, W. K. & McGuire, M. A. (2009). Calcium salts of polyunsaturated fatty acids deliver more essential fatty acids to the lactating dairy cow. *Journal of Dairy Science, 92*, 2051–2056.

Thompson, H., Zhu, Z., Banni, S., Darcy, K., Loftus, T. & Ip, C. (1997). Morphological and biochemical status of the mammary gland as influenced by conjugated linoleic acid: Implication for a reduction in mammary cancer risk. *Cancer Research, 57*, 5067–5072.

Toomey, S., Roche, H., Fitzgerald, D. & Belton, O. (2003). *Regression of pre-established atherosclerosis in the apoE−/− mouse by conjugated linoleic acid.* London: Portland Press Limited.

Tricon, S., Burdge, G. C., Kew, S., Banerjee, T., Russell, J. J., et al. (2004). Opposing effects of *cis*-9, *trans*-11 and *trans*-10, *cis*-12 conjugated linoleic acid on blood lipids in healthy humans. *American Journal of Clinical Nutrition, 80*, 614–620.

Tsiafoulis, C. G., Papaemmanouil, C., Alivertis, D., Tzamaloukas, O., Miltiadou, D., et al. (2019). NMR-based muetabolomics of the lipid fraction of organic and conventional bovine milk. *Molecules, 24*.

Turpeinen, A. M., Mutanen, M., Aro, A., Salminen, I., Basu, S., et al. (2002). Bioconversion of vaccenic acid to conjugated linoleic acid in humans. *American Journal of Clinical Nutrition, 76*, 504–510.

Urquhart, P., Parkin, S. M., Rogers, J. S., Bosley, J. A. & Nicolaou, A. (2002). The effect of conjugated linoleic acid on arachidonic acid metabolism and eicosanoid production in human saphenous vein endothelial cells. *Biochimica et Biophysica Acta (BBA)-Molecular and Cell Biology of Lipids, 1580*, 150–160.

Verhulst, A., Janssen, G., Parmentier, G. & Eyssen, H. (1987). Isomerization of polyunsaturated long chain fatty acids by propionibacteria. *Systematic and Applied Microbiology, 9*, 12–15.

Voorrips, L. E., Brants, H. A., Kardinaal, A. F., Hiddink, G. J., van den Brandt, P. A. & Goldbohm, R. A. (2002). Intake of conjugated linoleic acid, fat, and other fatty acids in relation to postmenopausal breast cancer: The Netherlands Cohort Study on Diet and Cancer. *American Journal of Clinical Nutrition, 76*, 873–882.

Wahle, K. (1974). Desaturation of long-chain fatty acids by tissue preparations of the sheep, rat and chicken. *Comparative Biochemistry and Physiology Part B: Comparative Biochemistry, 48*, 87–105.

Ward, R. J., Travers, M. T., Richards, S. E., Vernon, R. G., Salter, A. M., et al. (1998). Stearoyl-CoA desaturase mRNA is transcribed from a single gene in the ovine genome. *Biochimica et Biophysica Acta (BBA)-Lipids and Lipid Metabolism, 1391*, 145–156.

Werner, S. A., Luedecke, L. O. & Shultz, T. D. (1992). Determination of conjugated linoleic acid content and isomer distribution in three Cheddar-type cheeses: Effects of cheese cultures, processing, and aging. *Journal of Agricultural and Food Chemistry, 40*, 1817–1821.

White, S. L., Bertrand, J. A., Wade, M. R., Washburn, S. P., Green, J. T., Jr. & Jenkins, T. C. (2001). Comparison of fatty acid content of milk from Jersey and Holstein cows consuming pasture or a total mixed ration. *Journal of Dairy Science, 84*, 2295–2301.

Whitlock, L. A., Schingoethe, D. J., Hippen, A. R., Kalscheur, K. F., Baer, R. J., et al. (2002). Fish oil and extruded soybeans fed in combination increase conjugated linoleic acids in milk of dairy cows more than when fed separately. *Journal of Dairy Science, 85*, 234–243.

Wilson, T., Nicolosi, R., Chrysam, M. & Kritchevsky, D. (2000). Conjugated linoleic acid reduces early aortic atherosclerosis greater than linoleic acid in hypercholesterolemic hamsters. *Nutrition Research, 20*, 1795–1805.

Wong, M. W., Chew, B., Wong, T., Hosick, H., Boylston, T. & Shultz, T. (1997). Effects of dietary conjugated linoleic acid on lymphocyte function and growth of mammary tumors in mice. *Anticancer Research, 17*, 987–993.

Yamasaki, M., Nou, S., Tachibana, H. & Yamada, K. (2005). Cytotoxity of the *trans*10, *cis*12 isomer of conjugated linoleic acid on rat hepatoma and its modulation by other fatty acids, tocopherol, and tocotrienol. *In Vitro Cellular & Developmental Biology. Animal, 41*, 239.

Yokoyama, M. T. & Davis, C. (1971). Hydrogenation of unsaturated fatty acids by *Treponema* (*Borrelia*) strain B25, a rumen spirochete. *Journal of Bacteriology, 107*, 519–527.

Yu, Y., Correll, P. & Heuvel, J. V. (2002). Conjugated linoleic acid decreases production of pro-inflammatory products in macrophages: Evidence for a PPARγ-dependent mechanism. *Biochimica et Biophysica Acta (BBA)-Molecular and Cell Biology of Lipids, 1581*, 89–99.

Yurawecz, M. P., Roach, J. A., Sehat, N., Mossoba, M. M., Kramer, J. K., et al. (1998). A new conjugated linoleic acid isomer, 7 *trans*, 9 *cis*-octadecadienoic acid, in cow milk, cheese, beef and human milk and adipose tissue. *Lipids, 33*, 803–809.

Yyrawecz, M. P., Mossoba, M. M., Kramer, J. K. G., Pariza, M. W. & Nelson, G. (1999). Methylation procedures for conjugated linoleic acid. *Advances in Conjugated Linoleic Acid Research, 1*, 64–82.

4 Intracellular Origin of Milk Fat Globules, Composition and Structure of the Milk Fat Globule Membrane Highlighting the Specific Role of Sphingomyelin

C. Lopez

4.1 Introduction

Milk secretion is fundamental to mammalian life since this biological fluid is the exclusive feeding of newborns providing energy and bioactive molecules essential for their development and health. Milk secretion is also of major commercial importance for the manufacture of many dairy products that are consumed by more than 6 billion people worldwide (FAO 2018). Among milk components, lipids are of particular interest. The mechanisms involved in their secretion by the mammary gland lead to the formation of lipid droplets, called the milk fat globules, which are unique in their composition and structure and specific to milk. Milk fat globules are enveloped by a biological membrane, the milk fat globule membrane (MFGM), that is rich in polar lipids, cholesterol, and membrane-specific proteins. The composition of the MFGM has gained much interest due to its natural variety of polar lipids, with for example a large proportion of sphingomyelin (>25% polar lipids) compared to other sources of dietary polar lipids, and the presence of glycoproteins. The structure of the MFGM has also gained much attention in recent years since it is the interface between the triacylglycerol core of milk fat globules and the surrounding aqueous phase and is therefore involved in many interfacial mechanisms, such as the adsorption of bacteria and the digestion of milk lipids by enzymes. An increased understanding of the relationship between the composition, the structure, and the functions of the MFGM is essential in the development of MFGM-rich ingredients (through the valorization of by-products of the dairy industry, e.g. buttermilk) for the preparation of processed emulsions bioinspired by milk fat globules coated by the MFGM, for example in infant milk formulae.

In writing this chapter, I have emphasized new findings since the previous edition of this book dedicated to dairy lipids was published in 2006. Much of the new information gained in the structure of the MFGM concerns the lateral phase separation of polar lipids in the outer bilayer with formation of ordered microdomains rich in sphingomyelin and cholesterol, and a heterogeneous distribution of the proteins. The concept of homogenous fluid lipid matrix no longer exists for the structure of the MFGM. The concept of a patchwork in the MFGM has emerged since 2008. This concept is based on studies highlighting that all MFGM polar lipids are not the same from chemical and biophysical points of view, which induces separation of saturated high-phase-transition temperature (T_m) polar lipids (mainly milk sphingomyelin) in ordered phases from unsaturated low-T_m polar lipids

C. Lopez (✉)
INRAE, UMR STLO, Rennes, France

INRAE, UR BIA, Nantes, France
e-mail: christelle.lopez@inrae.fr

P. L. H. McSweeney et al. (eds.), *Advanced Dairy Chemistry, Volume 2*,
https://doi.org/10.1007/978-3-030-48686-0_4

in the fluid phase. The patchwork concept also includes the attractive interactions between milk sphingomyelin and cholesterol that are responsible for the formation of complexes. The impact of processing, i.e. mechanical and thermal treatments, on the composition and structure of the MFGM are out of the scope of this chapter. For those interested in earlier literature and coverage of historical aspects of research on the intracellular origin, growth, transit, and secretion of milk fat globules, the comprehensive book chapter by Keenan and Mather (2006) is recommended.

4.2 Intracellular Origin and Secretion of Milk Fat Globules

Milk fat globules originate from the mammary epithelial cells of lactating females. The formation and intracellular transit of lipid droplets in the cells, as well as the secretory mechanisms of milk fat globules by the cells during lactation, have been studied by a number of investigators since the pioneering works in the 1950s (Bargmann and Knoop 1959). They have been the focus of many papers, book chapters (Keenan and Mather 2006), and reviews (Heid and Keenan 2005a). While we currently have a general overview of the steps leading to the secretion of fat globules in milk, knowledge about the molecular mechanisms involved in milk fat globule formation, intracellular transit, and secretion is still lacking.

It has been assumed from ultrastructural similarities obtained from electron microscopy that the same or closely similar mechanisms occur in many mammalian species. In the milk-secreting epithelial cells, the triacylglycerols (TAGs) are secreted from the endoplasmic reticulum after their accumulation between the outer and inner halves of the bilayer. TAGs are then released into the cytosol in the form of droplets of various sizes that are the intracellular precursors of milk fat globules (Zaczek and Keenan 1990). The diameters of the intracellular lipid droplets range from less than 0.5 μm for the microlipid droplets to more than 4 μm. The intracellular microlipid droplets can grow in volume by fusing with each other to form large droplets termed cytoplasmic lipid droplets. Whatever their size, the intracellular lipid droplets have a TAG-rich core and they are surrounded by a coat material composed of proteins (such as adipophilin) and polar lipids originating from the endoplasmic reticulum membrane (Keenan and Mather 2006). The intracellular lipid droplets transit from their sites of origin, mostly in basal cell regions, to the apical region of the cell where they will be secreted from the cell. The mechanisms involved in this apical migration of the intracellular lipid droplets are not known with certainty, but may involve cytoskeletal elements (Heid and Keenan 2005b).

The widely accepted mechanism of milk fat globule secretion involves the contact of intracellular lipid droplets with the apical surface of the cell and their progressive envelopment in differentiated regions of apical plasma membrane up to the point where they are dissociated from the epithelial cells into the gland lumen (Figure 4.1). The extracellular lipid droplets, that are called milk fat globules, are surrounded by an apical plasma membrane. However, an alternative mechanism based on an observed association between lipid droplets and secretory vesicles in apical cell regions could occur under certain conditions. Wooding (1971a) proposed that progressive fusion of neighboring vesicles on the surfaces of lipid droplets may lead to the formation of intracytoplasmic vacuoles containing both casein micelles and lipid droplets enveloped with secretory vesicle membrane. The content of such vacuoles could be released from the cell by exocytosis. To date, no definite conclusion about the plasma membrane or secretory vesicle membrane origin of the MFGM can be made from the extant biochemical data. However, available cytochemical evidence favors plasma membrane envelopment as the principal mechanism for the secretion of milk fat globules. For example, butyrophilin, a major integral membrane protein of the MFGM, is expressed on the apical plasma membrane of the epithelial cells. The glycoprotein mucin 1 found in the membrane surrounding fat globules in milk is also concentrated in apical regions of the plasma membrane of the lactating

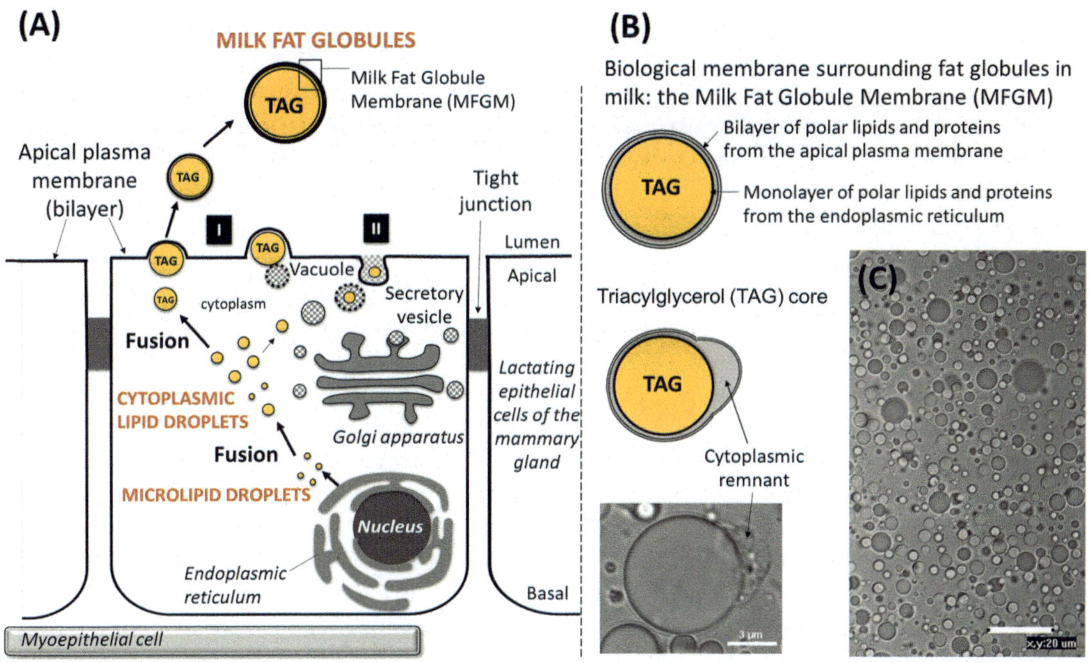

Figure 4.1. (**A**) Intracellular origin, growth, transit, and secretion of milk fat globules. (I) Secretion of milk fat globules by envelopment in apical plasma membrane. (II) Secretory vesicles may surround intracellular lipid droplets and fuse together to form a vacuole containing the fat droplet. The vacuole content may be released by exocytosis. (Adapted from Heid and Keenan (2005b)). (**B**) Schematic representation of milk fat globules enveloped by their biological membrane and showing the presence of cytoplasmic remnants (not to scale), bottom: microscopy image of a cytoplasmic crescent attached to a human milk fat globule. Adapted from Lopez and Ménard (2011). (**C**) Microscopy image showing the size distribution of milk fat globules in milk.

cells. Furthermore, immuno-gold labeling experiments make it unlikely that substantial portions of the membrane surrounding fat globules originate from secretory vesicle membranes (Keenan and Mather 2006). It is well accepted by the scientific community that immediately after release from the secretory cell, milk fat globules are surrounded by a biological membrane called the milk fat globule membrane (MFGM). The MFGM is structured as a trilayer of polar lipids with a monolayer of polar lipids and proteins from the endoplasmic reticulum in contact with the TAG core and an outer bilayer membrane from the apical plasma membrane of the epithelial cells (Figure 4.1).

During the secretion process, milk fat globules are frequently enveloped compactly by apical plasma membrane. However, in some instances, closure of the membrane around the fat droplet can entrain some cytoplasm between the monolayer and the bilayer membrane (Figure 4.1). The result is a milk fat globule with a remnant of cytoplasm attached that can vary from thin slivers of cellular material to cases where the volume of cytoplasm exceeds the fat globule core volume. Figure 4.1 shows a cytoplasmic remnant attached to a fat globule in human breast milk (Lopez and Ménard 2011). According to Wooding, the remnants of cytoplasm could also originate from trapping of cytoplasm by fusing secretory vesicles during formation of vacuoles containing a droplet. An abnormality in the protein coat along the cytoplasmic face of the apical plasma membrane may interfere with adhesion of the membrane to the intracellular lipid droplet and may be responsible for the phenomenon of crescent formation (Huston and Patton 1990). Remnants of cytoplasm attached to milk fat globules have been observed in the milks of numerous species, and the proportion of such fat globules appears to vary both between species and in individuals within a species, from about 1 to 5% (Huston and Patton 1990; Janssen and Walstra 1982). Electron microscopy observations showed that the remnants of cytoplasm attached to milk fat globules contain nearly all intracellular membranes and organelles of the milk-secreting cells. They are, therefore, a route by which cellular constituents can enter milk and could represent a means for providing bioactive molecules, such as hormones or growth factors, to the nursing young.

Milk fat globule size ranges over three orders of magnitude, from less than 200 nm to greater than 15 μm. The mechanisms controlling the size distribution of fat globules in milk integrate the intracellular formation and transit as well as the secretion. Three distinct mechanisms regulating intracellular lipid droplet size in mammary epithelial cells have been reported, that is, co-regulation of fat content and TAG-synthesis capacity of the cells, membrane material availability (polar lipids in the cell), and fusion between intracellular lipid droplets (Argov-Argaman 2019). The latter is controlled by the membrane's polar lipid composition and involves mitochondrial enzymes. The size distribution of fat globules secreted in milk can vary within a herd, as a function of the stage of lactation, diet, and season (Argov-Argaman *et al.* 2014, 2016; Logan *et al.* 2014; Lopez *et al.* 2008a, 2014; Mesilati-Stahy *et al.* 2015). The mean diameter of fat globules has been positively related to the total fat content of milk, and depends on the mammalian species that produce the milk. For example, in mammalian species such as buffalo that produce a high amount of milk fat (7% *versus* 4% for bovine milk), the size of fat globules is large, that is, 5 μm *versus* 3.5 μm in bovine milk (Ménard *et al.* 2010). Therefore, the control of milk fat globule size involves various points of regulation at the molecular, genetic, and metabolic levels. Furthermore, the availability of membrane material with the ability to stabilize fat globules upon their secretion governs their size in milk.

4.3 Composition of the Milk Fat Globule Membrane and Biophysical Properties of Polar Lipids

4.3.1 Composition

The membrane that surrounds milk fat globules, the MFGM, is derived mainly from the apical plasma membrane of mammary secretory cells at the time of milk fat secretion (Heid and Keenan 2005b). The MFGM is therefore chemically very similar to the plasma membrane (Keenan *et al.* 1970). However, some parts of the MFGM that initially surround fat globules after their secretion, or some MFGM components, can be lost following secretion either within the mammary gland or in expressed milk (Evers 2004). The manner in which milk is collected, handled, and stored may also cause loss and compositional alteration of the MFGM. Finally, the protocols used for the isolation of MFGM to perform chemical analysis (e.g., centrifugation, churning, cooling or freezing, treatment with detergents or solvents) can also affect the composition of the MFGM. For further information on this aspect, the reader is referred to the review by Dewettinck *et al.* (2008).

The MFGM is composed of TAGs, membrane-specific proteins (mainly glycoproteins), enzymes, a variety of polar lipids including the sphingolipids (mainly sphingomyelin and glycosphingolipids), cholesterol, and other minor components. TAGs are the most abundant class of lipid found in MFGM preparations, but the sample preparation method used has a major impact on their presence because these neutral lipids originating from the core of milk fat globules remain associated with the membrane during isolation. The quantitative data reported in the literature for the composition of MFGM material are highly variable due to differences in isolation, purification, and analytical techniques used.

4.3.1.1 Protein Composition

The proteins represent 25–70% of the total mass of the MFGM, and modern proteomic techniques have revealed that there are more than 100 different MFGM proteins (Fong *et al.* 2007; Reinhardt and Lippolis 2006). The proteome of the human MFGM has also been characterized over 12 months of lactation (Liao *et al.* 2011). Among well-described proteins present in high concentrations in the MFGM are the butyrophilin, the mucins (mucin 1, mucin 15 previously called PAS III), the cluster of differentiation CD36 (also called PAS IV), lactadherin (also called PAS-6/7 or MFG-E8), adipophilin, fatty acid-binding protein (FABP), and the enzyme xanthine dehydrogenase/oxidase (Keenan and Mather 2006; Mather 2000). The structures, amino acid sequences, and properties of major MFGM proteins are reviewed and discussed in detail elsewhere (Mather 2000; Singh 2006). However, little is known about their specific composition in the MFGM and functions. About 30 different enzymes or enzymatic activities have been detected in MFGM preparations from bovine milk (Fong *et al.* 2007; Keenan and Mather 2006). Several of the enzymes associated with the MFGM have been purified and at least partially characterized, and it is known that some MFGM-associated enzymes are involved in degradative events that can produce flavor defects in dairy products.

4.3.1.2 Lipid Composition

Cholesterol represents about 90% of the sterols identified in the MFGM, with the cholesterol content being about 300 mg/100 g fat, and represents ~30 wt % of the MFGM lipid fraction (Et-Thakafy *et al.* 2017; Mesilati-Stahy and Argov-Argaman 2014; Yao *et al.* 2016).

Milk polar lipids account for about 0.5–1% of milk lipids and are mainly located in the MFGM. The amount of polar lipids found in milk depends on the size of fat globules and on the amount of surface covered by the MFGM. This means that for a similar amount of total fat, small fat globules are more numerous and contain a higher amount of MFGM and polar lipids compared to large fat globules (Lopez *et al.* 2011). Studies reported a positive relationship between the amount of fat in the milk and the size of fat globules. This means that milks with a small amount of fat contain small fat globules and a large amount of MFGM and polar lipids (Lopez *et al.* 2014).

The MFGM contains two main classes of polar lipids, that is, the glycerophospholipids and the sphingolipids. The glycerophospholipids have fatty acids at positions *sn*-1 and *sn*-2 and a phosphate linked to a polar headgroup on the *sn*-3 position. The major milk glycerophospholipids are phosphatidylcholine (PC) and phosphatidylethanolamine (PE), while phosphatidylserine (PS) and phosphatidylinositol (PI) are quantitatively more minor (Figure 4.2) (Dewettinck *et al.* 2008; Fong *et al.* 2007; Lopez 2011; Yao *et al.* 2016). About one-third to one-half of the polar lipids in the MFGM are sphingolipids. Sphingolipids are based on a sphingosine backbone and, according to the headgroup, can be differentiated into sphingomyelin (phosphocholine), ceramide (H), glycosylceramides (glucose; galactose for cerebrosides), lactosylceramides (lactose), and with more complex glycosyl residues are the gangliosides (combination of glucose, monosaccharides, N-acetylgalactosamine, sialic acid, and others) (Nilsson 2016). The main sphingolipids are the milk sphingomyelin (milk-SM; Figure 4.2) and the glycosphingolipids. The relative proportion of the individual classes of MFGM polar lipids depends on the size of milk fat globules (Lopez *et al.* 2011) and on mammal species (Garcia *et al.* 2012). For example, milk-SM represents 20–45 wt % of milk polar lipids depending on the mammalian species, and milk-SM is the most abundant polar lipid species in human milk (Figure 4.2) (Giuffrida *et al.* 2013; Lopez and Ménard 2011; Yao *et al.* 2016; Zou *et al.* 2012).

Among the dietary sources of polar lipids that are commercially available, bovine milk polar lipids naturally contain a high proportion of SM (25–27% of total polar lipids) compared to eggs (egg lecithin is rich in PC and contains about 1.5% SM) or polar lipids from vegetable origin such as soybean lecithin that are devoid of SM (Burling and Graverholt 2008) (Figure 4.2). The MFGM surrounding fat globules in milk and dairy products, and by-products of the butter industry (i.e., buttermilk and butter serum) are major and convenient sources of exogenous PS and sphingolipids in the human diet, particularly milk-SM and ganglio-

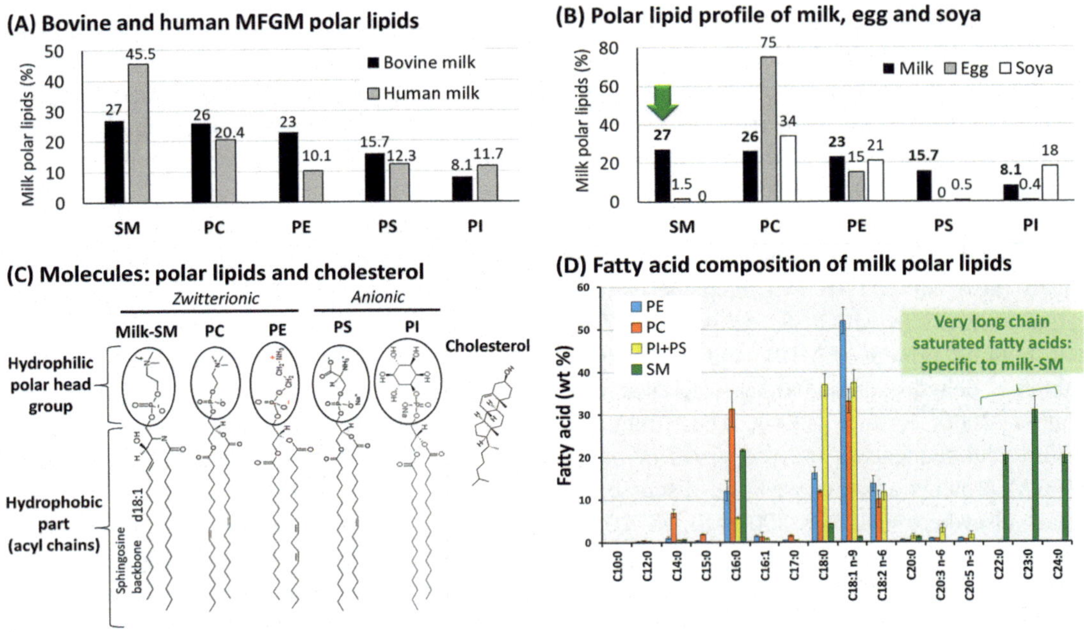

Figure 4.2. (**A**) Main polar lipids located in the milk fat globule membrane (MFGM) from bovine and human milks. Adapted from Lopez and Ménard (2011) and Yao *et al.* (2016). (**B**) Comparison of the relative proportion of polar lipid classes in milk, eggs, and soybean lecithin. Adapted from Burling and Graverholt (2008). (**C**) Molecules of polar lipids showing their polar headgroup and the hydrophobic tail. (**D**) Acyl chain composition of polar lipids from bovine milk. Adapted from Sanchez-Juanes *et al.* (2009).

sides (Bourlieu *et al.* 2018; Dewettinck *et al.* 2008; Lopez *et al.* 2017a, 2019; Vanderghem *et al.* 2010). Scientific evidence of the nutritional and health benefits of these sphingolipids is accumulating. Milk-SM is involved in decreasing of intestinal cholesterol absorption and contributes to improving lipid metabolism and cardio-metabolic health (Eckhardt *et al.* 2002; Noh and Koo 2004a; Norris *et al.* 2016; Nyberg *et al.* 2000). Glycosphingolipids are quantitatively minor constituents of MFGM polar lipids, but they have been studied widely because of their role in a number of biological phenomena and the role of their breakdown products. For further details on this topic, the reader is referred to a review by Nilsson (2016).

While polar lipids constitute a very small quantitative proportion of the total milk lipids, they play an important structural role because they are amphiphilic molecules with a hydrophobic tail and a hydrophilic headgroup able to form lamellar structures such as those in membranes. This is the reason why polar lipids are the key structural elements of the MFGM. Milk polar lipids are also efficient in forming liposomes and stabilizing emulsions (Lopez *et al.* 2017b; Singh 2006). Milk-SM, PC, and PE are zwitterionic polar lipids, while PI and PS are anionic polar lipids (Figure 4.2). Milk polar lipids are involved in electrostatic interactions, and their interactions are influenced by changes in pH and ionic strength. The isoelectric point of the MFGM is pI ~4.2 (Lopez *et al.* 2017b). In their hydrophobic part, milk polar lipids are characterized by a wide diversity of saturated and unsaturated acyl residues (Sanchez-Juanes *et al.* 2009; Yao *et al.* 2016) (Figure 4.2). The short- and medium-chain length fatty acids with less than 14 atoms of carbon are absent from MFGM polar lipids, while they are present in milk TAGs. Each class of glycerophospholipid (PC, PE, PI, PS) corresponds to several molecular species combining fatty acids onto the *sn*-1 and *sn*-2 positions of the glycerol backbone. In milk-SM, a single fatty acid is bound via an amide linkage. The fatty acid composition of milk-SM is completely different from the fatty acid composition of the other polar lipids found in the MFGM. Milk-SM contains several molecular species with long-chain saturated fatty acids (C16:0, C18:0, C22:0, C23:0, C24:0)

(Figure 4.2) (Fong *et al.* 2007; Sanchez-Juanes *et al.* 2009). This unique composition of milk-SM leads to specific biophysical properties (e.g., high phase-transition temperature T_m, formation of ordered domains with a high degree of packing, and attractive interactions with cholesterol) (Lopez *et al.* 2010, 2018; Malmsten *et al.* 1994); each of these points is developed further in Section 4.3.2 of this chapter. In addition, the complex chemical composition of milk-SM molecular species in comparison to other dietary sources of SM such as brain-SM (~50% C18:0-SM) or egg-SM (~85% C16:0-SM), which leads to specific physical properties (Et-Thakafy *et al.* 2018), is involved in biological functions and health benefits (e.g., cholesterol-lowering effect) (Noh and Koo 2004b). The fatty acid composition of milk polar lipids is not constant. Changes in the fatty acid composition of MFGM polar lipids in response to cow diet have been reported (Lopez *et al.* 2008b, 2014). A diet rich in unsaturated fatty acids (e.g., grass, addition of polyunsaturated fatty acid-rich seeds such as linseeds) leads to an increased amount of unsaturated fatty acids in MFGM glycerophospholipids. But the fatty acid composition of milk-SM is not affected by diet, and this major milk sphingolipid remains rich in long-chain saturated fatty acids. It has also been reported that the relative proportion of MFGM polar lipid classes (i.e., milk-SM, PC, PE, PI, and PS) is not affected by diet.

The MFGM is a dietary source of many molecules with structural properties as well as bioactive compounds of nutritional and health interest, for example, the unsaturated fatty acids provided by the polar lipids and the high amount of sphingolipids, mainly milk-SM.

4.3.2 Biophysical Properties of Milk Polar Lipids: Specific Roles of Sphingomyelin and Cholesterol

Lipid species with different chemical compositions have individual intrinsic properties, such as the melting transition temperature (T_m), and may therefore exhibit different phases depending on the temperature T (i.e., $T < T_m$ or $T > T_m$). As a result, lipids experience different types of interactions with each other, which in turn dictate the membrane fluidity, lateral phase segregation, and the membrane mechanical properties. Gathering information about milk polar lipid properties and interactions as a function of temperature is essential in the understanding of their functions in the MFGM, in food-grade ingredients and in products such as emulsions (creams, spreads, infant milk formulae) containing milk polar lipids and MFGM fragments.

4.3.2.1 Thermotropic and Structural Behavior of Saturated MFGM Polar Lipids

The polar lipids located in the MFGM correspond to a wide range of lipid species, including various polar heads and acyl chains (Figure 4.2). Under fully hydrated conditions, MFGM polar lipids organize as lamellar structures (bilayers such as in biological membranes where the hydrophobic tails are in the inner part and the polar heads are in the outer part) and can exhibit different packing of the acyl chains as a function of temperature (ordered packing for $T < T_m$ or fluid state for $T > T_m$).

The thermotropic phase behavior of milk-SM has been the focus of interest of several studies since it is present at high concentrations (20–45% of milk polar lipids; Figure 4.2) and plays major structural and metabolic roles. Thermograms recorded by differential scanning calorimetry (DSC) on heating of milk-SM bilayers exhibit a broad asymmetric multicomponent endothermic peak composed by a sharp peak that determines the melting transition temperature (T_m), with broad components on both sides of the main peak (Figure 4.3). For fully hydrated milk-SM bilayers (i.e., less than 50 mol% milk-SM), the phase-transition temperature, T_m, is 34.3 ± 0.1 °C (Figure 4.3) (Cheng *et al.* 2017; Malmsten *et al.* 1994; Murthy *et al.* 2015). Interestingly, temperature influences the physical state of milk-SM molecules, and close to the physiological temperature of 37 °C in the human body, the fluid state of milk-SM may facilitate it being involved in certain biological functions. The width of the endothermic event recorded between the onset and completion temperatures spans from about 30 to 40 °C. This complex thermal behavior of milk-SM results

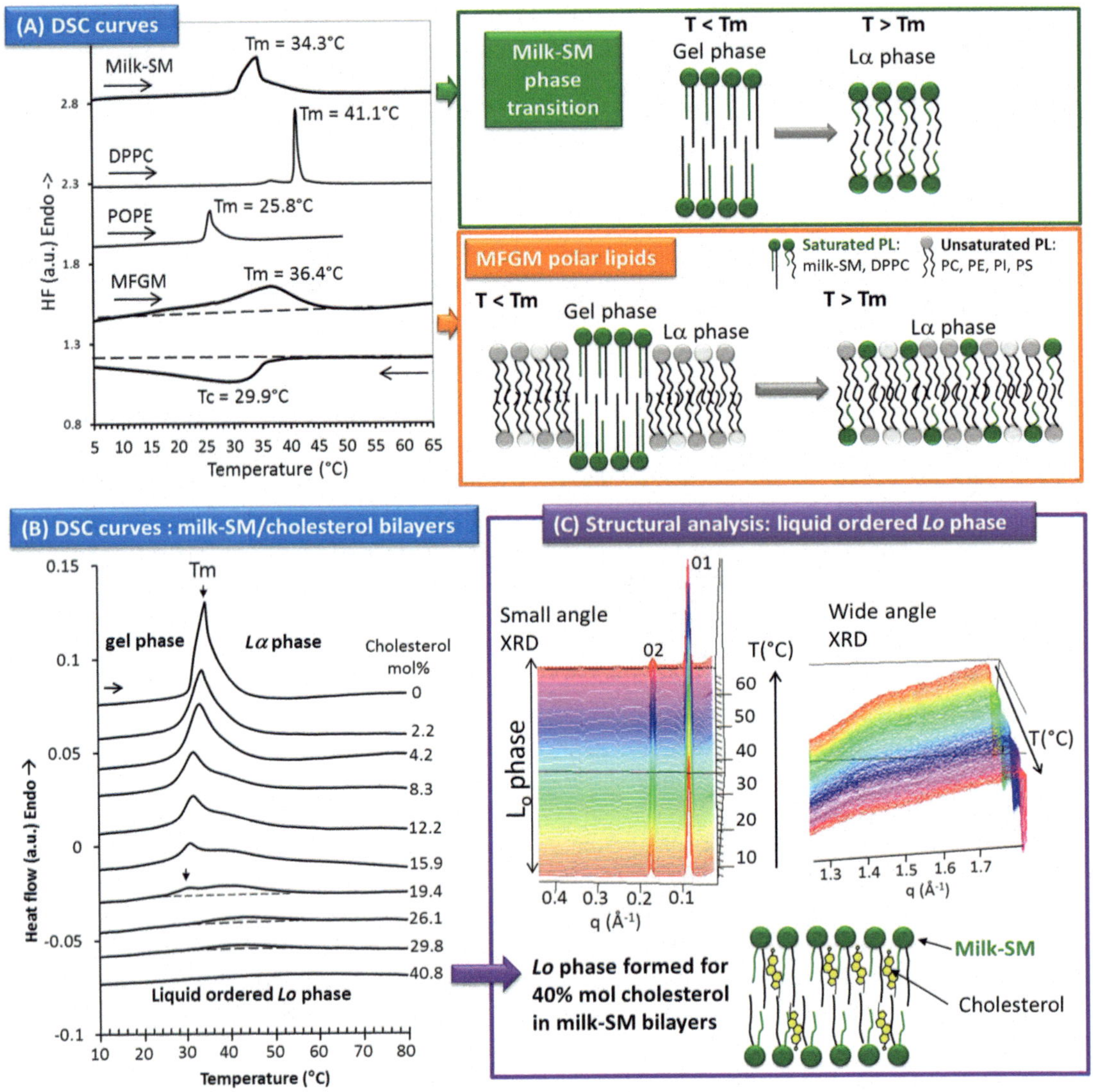

Figure 4.3. (**A**) Thermotropic phase behavior of polar lipids recorded by differential scanning calorimetry (DSC) showing the gel to fluid phase transition temperature and schematic representations. Adapted from Murthy *et al.* (2016b). (**B**) Role of cholesterol on the thermotropic phase behavior of milk sphingomyelin. Adapted from Lopez *et al.* (2018). (**C**) X-ray diffraction (XRD) patterns showing the non temperature-dependent structural behavior of milk sphingomyelin organized in the liquid-ordered Lo phase in the presence of cholesterol. Adapted from Lopez *et al.* (2018). Abreviations: *SM* sphingomyelin, *DPPC* dipalmitoyl-phosphatidylcholine, *POPE sn*-1 palmitoyl *sn*-2 oleoyl phosphatidylethanolamine, *MFGM* milk fat globule membrane, T_m phase transition temperature.

from the large heterogeneity in the hydrocarbon chains (from C16:0 to C24:0; Figure 4.2), the successive melting of the individual SM species, and to structural reorganizations occurring on heating of the bilayers. Structural analysis of fully hydrated milk-SM bilayers performed by X-ray diffraction showed the formation of a gel phase at temperature less than T_m and a fluid liquid-crystalline phase (noted Lα) at temperature greater than T_m (Figure 4.3) (Cheng *et al.* 2017; Malmsten *et al.* 1994; Murthy *et al.* 2015).

The polar lipids found in the MFGM have a mixture of saturated and unsaturated acyl chains (Figure 4.2). The unsaturated milk polar lipids are fluid at temperature greater than 0 °C (e.g., T_m-DOPC = −20 °C; where DOPC signifies dioleoyl-PC), while some polar lipids containing at least one saturated acyl chain exhibit high gel to fluid phase-transition temperatures (e.g., T_m-POPE = 25.8 °C with P = palmitic acid, O = oleic acid, PE = phosphatidylethanolamine; T_m-DPPC = 41.1 °C; where DPPC signifies dipalmitoyl-PC) (Figure 4.3). It is important to note that major PC species composed of a saturated acyl chain (e.g., P: palmitic acid C16:0) and an unsaturated acyl chain (e.g., O: oleic acid C18:1 n-9) are in the fluid state at temperature greater than 0 °C, such as POPC which exhibits a phase-transition temperature at T_m = −4.7 °C.

Figure 4.3 shows that the biophysical properties of MFGM polar lipids are influenced by temperature. High-T_m saturated polar lipids (mainly milk-SM, but also DPPC and POPE) are involved in the complex thermotropic phase behavior of the complex mixture of MFGM polar lipids, which exhibits a main phase transition at $T_m = 36.4 \pm 0.2$ °C (Murthy *et al.* 2015). Structural analysis of fully hydrated mixtures of polar lipids extracted from the MFGM showed (1) at temperature less than T_m, the coexistence of a fluid liquid-crystalline Lα phase, together with a gel phase composed, respectively, of the unsaturated low-T_m polar lipids and the saturated high-T_m polar lipids (milk-SM, DPPC), and (2) at temperature greater than T_m, a fluid liquid-crystalline Lα phase that includes the melted saturated and unsaturated polar lipids (Figure 4.3) (Murthy *et al.* 2015).

4.3.2.2 Attractive Interactions Between Milk-Sphingomyelin and Cholesterol

In recent years, information on the phase state of milk-SM as a function of the molar ratio of cholesterol and temperature has been provided by biophysical techniques such as DSC, X-ray diffraction (XRD), and Langmuir isotherms. The addition of cholesterol to milk-SM bilayers or MFGM polar lipid bilayers changes their phase state from the gel phase (for $T < T_m$) or the Lα phase (for $T > T_m$) to the liquid-ordered Lo phase whatever T (Lopez *et al.* 2018). Figure 4.3 shows that increasing cholesterol concentration in milk-SM bilayers decreases the endothermic gel to Lα phase transition recorded by DSC, until its disappearance. This means that the gel to Lα phase transition is abolished by the presence of cholesterol molecules and that milk-SM bilayers are in the liquid-ordered Lo phase, as shown by XRD. In the binary mixture milk-SM/cholesterol, the liquid-ordered Lo phase, is completed for 40 mol% cholesterol in the bilayers (Figure 4.3). In the complex MFGM polar lipid extract, the liquid-ordered Lo phase is completed for 27 mol% cholesterol, which corresponds to milk-SM/cholesterol molar ratio of 50/50 (Murthy *et al.* 2016a). The XRD signature of the liquid-ordered Lo phase is a lamellar structure with a lateral disorder in the acyl chains organization (Figure 4.3). The structural parameters of the lamellar liquid-ordered Lo phase do not evolve as a function of temperature, in accordance with the absence of phase transition (Figure 4.3). Furthermore, the presence of cholesterol induces attractive interactions with milk-SM molecules, with a decrease in the area occupied per milk-SM molecule leading to a condensing effect of cholesterol, as demonstrated using Langmuir isotherms (Cheng *et al.* 2017; Murthy *et al.* 2015). This means that ordered lipid domains composed of milk-SM and cholesterol in the Lo phase can be present in the MFGM, whatever the temperature (see Section 4.4 of this chapter about the structure of the MFGM; presence of ordered lipid domains). Furthermore, the attractive interactions between milk-SM and cholesterol provide scientific explanation for the formation of complexes that may be involved in decreased cholesterol absorption in the intestine as reported in a number of *in vivo* studies (Noh and Koo 2004b; Nyberg *et al.* 2000).

4.4 Structure of the Milk Fat Globule Membrane: New Scientific Advances Highlighting the Role of Sphingomyelin in the Formation of Ordered Domains

For more than 50 years, the structure of the MFGM has been the focus of attention of several research groups. The first morphological studies performed using electron microscopy with negatively stained preparations provided details about the structure of the MFGM and remain very informative (Wooding 1971b; Wooding and Mather 2017). Researchers agree on the fact that the MFGM is organized as a trilayer of polar lipids, where the internal monolayer originates from the endoplasmic reticulum of the epithelial cells while the external bilayer of the MFGM derives from the apical plasma membrane during secretion of the fat globule by the mammary cell (Figure 4.1).

4.4.1 Membrane-Specific Proteins Inserted in a Trilayer of Polar Lipids

For many years, authors were mainly interested in the proteins associated with the MFGM, in terms of localization and bioactive properties (Dewettinck *et al.* 2008; Rasmussen 2009). Some of the MFGM proteins are integral (e.g., butyrophilin), or transmembrane proteins integrated in the outer bilayer of the MFGM (e.g., mucin 1, mucin 15, and CD36), while others are reported to be more loosely attached in the outer part of the membrane (e.g., lactadherin and PP3). The intermembrane space between the monolayer and the bilayer of polar lipids is occupied with protein coat and the cytosolic parts of the transmembrane proteins (e.g., butyrophilin, xanthine dehydrogenase/oxidoreductase, and adipophilin) (Figure 4.4). There is substantial evidence that butyrophilin and xanthine dehydrogenase/oxidoreductase interact directly with each other by disulfide bonds. The MFGM glycoproteins

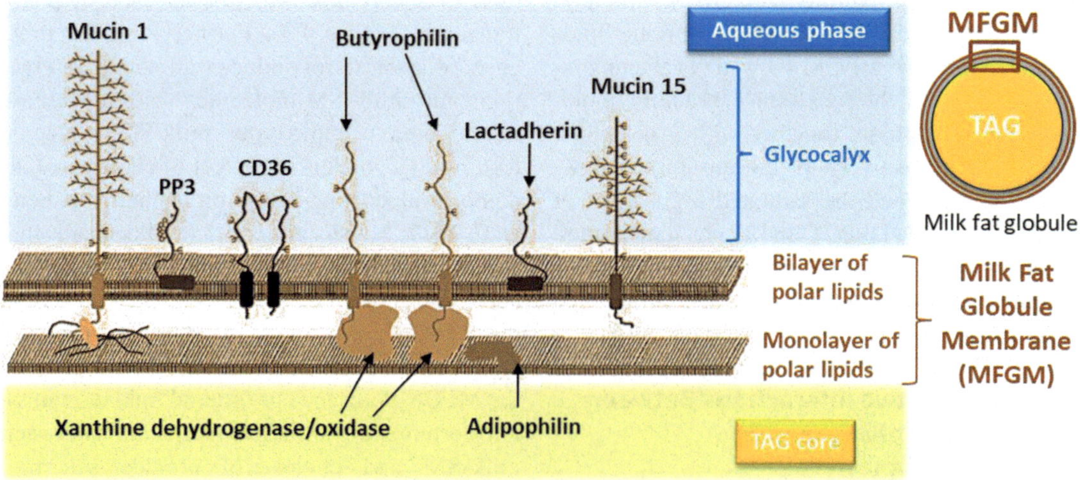

Figure 4.4. Schematic representation of the milk fat globule membrane (MFGM) showing the insertion of proteins within the trilayer of polar lipids. The transmembrane proteins (mucin 1, mucin 15, butyrophilin, and CD 36) and peripheral bound proteins (lactadherin and PP3) are embedded in the outer bilayer of polar lipids. The glycoproteins protrude in the aqueous phase and form the glycocalyx. The cytosolic parts of the transmembrane proteins and "coat"-proteins (xanthine dehydrogenase/oxidase and adipophilin) are located in the space between the bilayer and the monolayer of polar lipids. In this representation, the polar lipids are organized as a homogeneous fluid matrix. Adapted with permission from J.T. Rasmussen (Aarhus University, DK).

(mainly the mucins and the butyrophilin), together with the MFGM glycolipids, form a glycocalyx that protrudes in the aqueous phase surrounding fat globules in milk. The glycocalyx is involved in the physical stability of fat globules in milk and in interactions with bacteria, toxins, and viruses. These glycosylated components from the MFGM have protective functions against infections in the human body by preventing pathogen adhesion to the intestinal epithelium (Douellou *et al.* 2017; Sprong *et al.* 2012). Glycosylated components from the human MFGM are considered as beneficial in the health of breast-fed neonates (Hamosh *et al.* 1999; Peterson *et al.* 1998).

The localization of polar lipids within the three layers of the MFGM is the subject of debate. As for other biological membranes, authors reported an asymmetry in the localization of polar lipids in the MFGM. Milk-SM, PC, and cholesterol have been reported to be preferably located in the outer bilayer of the MFGM, while PE, PI, and PS could be mainly concentrated in the inner monolayer of the MFGM (Deeth 1997). Other studies reported the localization of PS in the outer bilayer of the MFGM (Zheng *et al.* 2014). The negative zeta potential values of bovine and breast milk fat globules (−12 and −7 mV, respectively) and the high sensitivity of the MFGM to cations such as calcium may correspond to the localization of the anionic polar lipids PS and PI at the surface of the MFGM in contact with the aqueous environment.

Structural observations performed by electron microscopy showed that the portion of the MFGM originating from the apical plasma membrane has a typical bilayer appearance (Heid and Keenan 2005b). This outer bilayer of the MFGM has long been described as having a homogeneous organization of polar lipids in the plane of the membrane according to the fluid mosaïc membrane, as reviewed in Dewettinck *et al.* (2008). As the concept of lipid lateral phase separation and presence of domains called "rafts" started to emerge in cellular biology (Simons and Ikonen 1997), questions were raised as to whether the MFGM obeyed similar rules as other biological membranes. Moreover, the different physical properties of milk polar lipids (saturated high-T_m polar lipids, such as milk-SM *versus* unsaturated low-T_m polar lipids) raised questions about their homogeneous accommodation in the outer bilayer of the MFGM.

The development of methods, innovative protocols, and the combination of microscopy techniques (electron microscopy, confocal microscopy, and atomic force microscopy) made possible an increased knowledge on the structure of the MFGM and evidenced lateral topographical and mechanical heterogeneities due to the lateral phase separation of milk-SM. This part of the chapter summarizes the main recent research advances in the structure of the MFGM and polar lipid assemblies and highlights the specific role of milk-SM and milk-SM/cholesterol complexes.

4.4.2 Ordered Lipid Domains Rich in Sphingomyelin and the Heterogeneous Distribution of Proteins Observed *In Situ* in the MFGM Around Fat Globules

New scientific information about the organization of MFGM emerged in 2008 thanks to the potential and application of confocal laser scanning microscopy (CLSM) combined with the specific labeling of membrane components (e.g., lipids, proteins, and sugars) that permit structural observations of the MFGM *in situ* in milk. Using lipophilic and lectin fluorescent dyes for CLSM observations, Evers *et al.* (2008) showed the heterogeneous distribution of lipids and proteins in the MFGM. The presence of nonfluorescent areas around fat globules in milk raised the question of the possible absence of a bilayer membrane at these locations. Using the exogenous phospholipid DOPE, head-labeled with the fluorescent dye rhodamine (i.e., Rhodamine-DOPE), that selectively partitions in fluid phospholipids, Christelle Lopez's group (INRAE, France) revealed the presence of nonfluorescent ordered lipid domains on the surface of milk fat globules (Figure 4.5) (Lopez *et al.* 2008c, 2010). These ordered lipid domains, first observed in bovine milk at room temperature, are circular in shape and their size is in the range of micrometers. It is also possible that such domains with a smaller

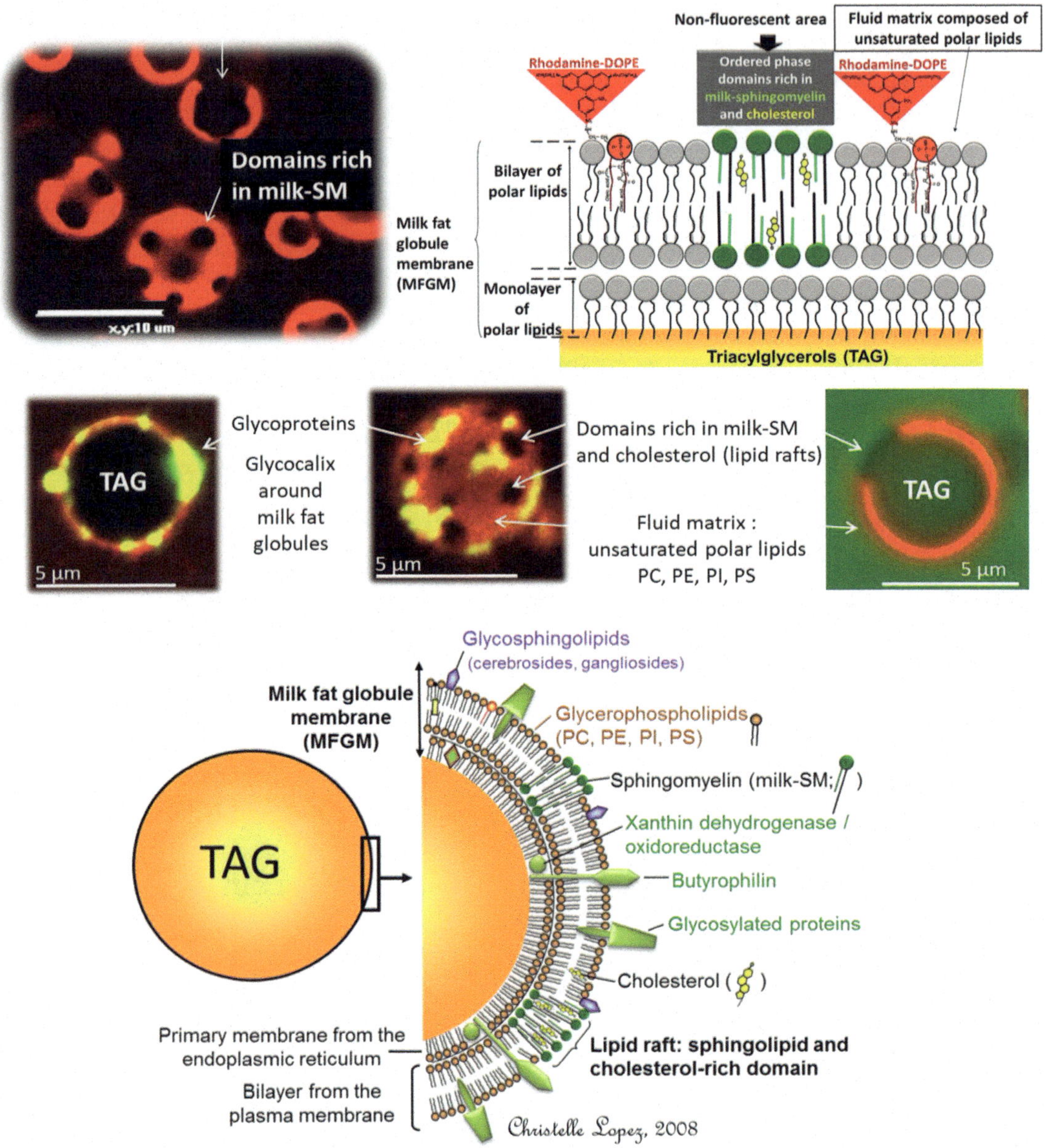

Figure 4.5. Confocal laser scanning microscopy (CLSM) images showing the heterogeneous distribution of proteins and polar lipids in the milk fat globule membrane as observed *in situ* in bovine milk. Adapted from Lopez *et al.* (2010). Schematic representations of the organization of polar lipids and proteins in the MFGM, showing the formation of ordered domains rich in sphingomyelin and cholesterol. Adapted from Lopez *et al.* (2008b).

size exist in the MFGM, but they are not detected by CLSM. These microdomains were interpreted as the lateral segregation of milk-SM in the gel phase. Complexes of milk-SM together with cholesterol in the liquid-ordered Lo phase could also explain these domains. This interpretation was formulated on the basis that (i) milk-SM and cholesterol are major lipid components present in the outer bilayer of the MFGM and (ii) milk-SM and cholesterol exhibit attractive interactions involved in the formation of lipid domains (Cheng *et al.* 2017). The double labeling of fluid phospholipids (with Rh-DOPE) and membrane proteins (with fast green FCF, and various lectins such as WGA-

488) confirmed that the nonfluorescent domains do not correspond to the localization of proteins in the MFGM (Lopez *et al.* 2010) (Figure 4.5). The lectin WGA-488 showed that the glycosylated molecules (i.e., glycoproteins and glycolipids) are heterogeneously distributed in the MFGM, protrude in the surrounding aqueous phase to form the glycocalyx, and are not located in the ordered lipid domains (Figure 4.5). The presence of ordered lipid domains enriched in milk-SM and cholesterol in the MFGM has been reported in the MFGM whatever the size of milk fat globules (i.e., 1.6 versus 6.3 μm) even if the amount of milk-SM is lower in small fat globules (Lopez *et al.* 2011). Moreover, CLSM observations revealed the diffusion of the ordered lipid domains in the plane of the MFGM, without any coalescence (Et-Thakafy *et al.* 2017; Lopez *et al.* 2010; Nguyen *et al.* 2016) (Figure 4.6).

Based on these new scientific findings, obtained from CLSM observations, an updated model of the structure of the MFGM was proposed by Christelle Lopez's group in 2008, with additional information about the lateral phase separation of milk-SM and cholesterol to form ordered domains in the outer bilayer of the MFGM and the heterogeneous distribution of proteins in the fluid matrix of the MFGM (Figures 4.5 and 4.10; Lopez 2011; Lopez *et al.* 2008b, 2010). These ordered lipid domains found in the outer bilayer of the MFGM were called lipid rafts (Lopez *et al.* 2010), analogous to rafts in mammalian plasma membranes (Simons and Ikonen 1997). The presence of ordered lipid domains in the MFGM was confirmed by other groups, using CLSM and the fluorescent dye Rh-DOPE to label the fluid matrix of the MFGM. These ordered lipid domains have been observed *in situ* in milks from various species: bovine milk (Et-Thakafy *et al.* 2017; Gallier *et al.* 2010a; Lopez *et al.* 2010; Zou *et al.* 2015), human breast milk (Gallier *et al.* 2015; Lopez and Ménard 2011; Zou *et al.* 2012), buffalo milk (Nguyen *et al.* 2015), goat and sheep milks (Et-Thakafy *et al.* 2017), and yak milk (Luo *et al.* 2018). Differences in the number, size, and shape of the ordered lipid domains have been characterized at the surface of milk fat globules and could be related to specific lipid composition of the MFGM from various mammalian species, for example, the ratio between milk-SM and cholesterol.

Most of the CLSM observations of the MFGM have been performed at room temperature after cooling the milk from the physiological temperature or after storage at 4 °C. However, investigations of the MFGM structure as a function of temperature are of primary importance to better understand its functions at the physiological temperature of 37 °C upon digestion of milk fat globules, and in the range spanning from 4 to 60 °C for technological applications. The ordered lipid domains formed in the MFGM have been observed over a wide range of temperature, that is, from 4 to 60 °C (Figure 4.6), including the physiological temperature of 37 °C (Et-Thakafy *et al.* 2017; Lopez and Ménard 2011; Nguyen *et al.* 2016; Zou *et al.* 2015). This means that these ordered lipid domains are present at the surface of milk fat globules at the physiological temperature and then upon digestion in the gastrointestinal tract. The ordered lipid domains rich in milk-SM and cholesterol and/or the difference in composition, height, and mechanical properties between the domains and the surrounding fluid phase could therefore be involved in the mechanisms of milk fat globule digestion by the digestive enzymes. On heating, some lipid domains melt at temperatures greater than 34 °C that corresponds to the gel/fluid Lα phase-transition temperature of milk-SM, which represents more than 25% of MFGM polar lipids. Other domains remain present in the MFGM at temperatures greater than 50 °C. The presence of the ordered domains above the phase-transition temperature of the saturated MFGM polar lipids (T_m of MFGM and SM = 36.4 and 34.0 °C, respectively) suggests that these domains should be in the nontransitioning liquid-ordered Lo phase, that corresponds to interactions between milk-SM and cholesterol (Figure 4.3). These results show that, as a function of temperature, the coexistence of ordered domains rich in milk-SM with or without cholesterol is possible (Et-Thakafy *et al.* 2017).

The dynamics of the lipid domains formed in the MFGM were investigated as a function of

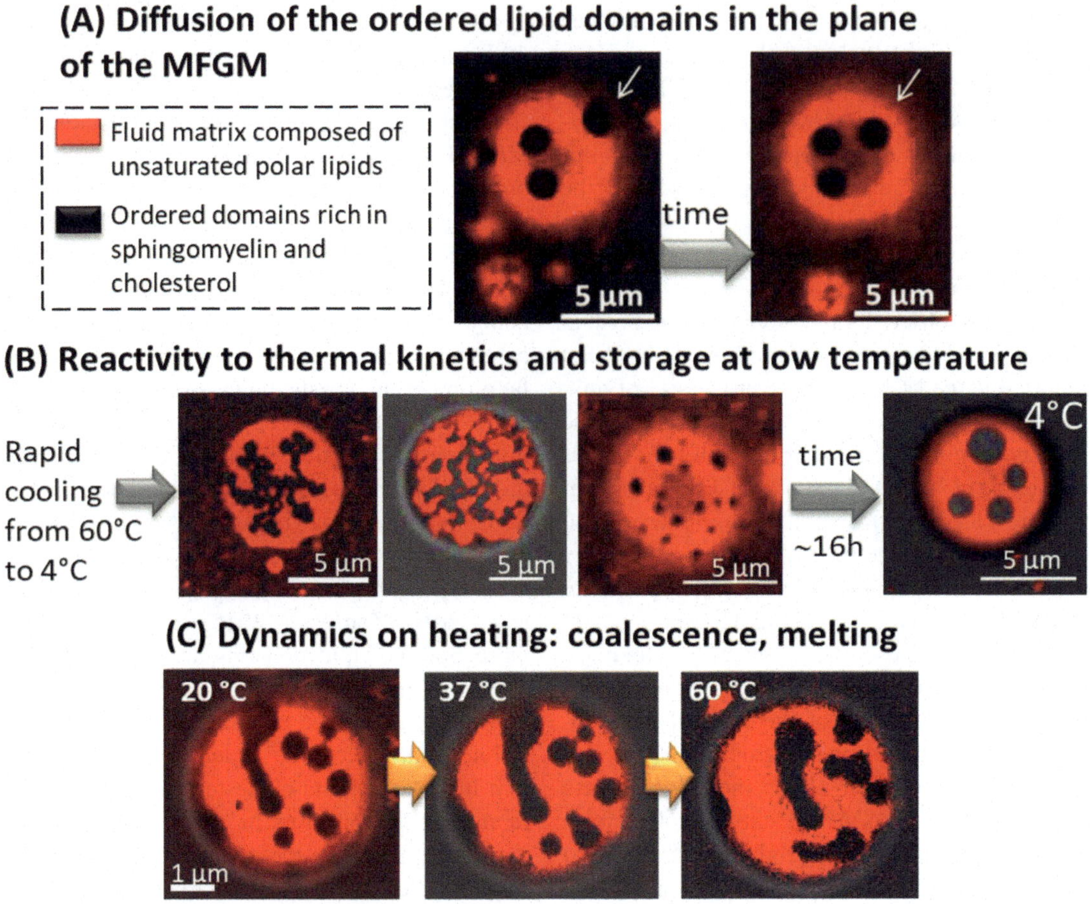

Figure 4.6. Confocal laser scanning microscopy (CLSM) images showing the dynamics of the polar lipids in the milk fat globule membrane (MFGM) *in situ* in milk. (**A**) Diffusion of the ordered lipid domains in the plane of the MFGM. (**B**) Reactivity of the lipid domains to thermal kinetics and storage of milk at low temperature. (**C**) Dynamics of the lipid domains on heating of milk from 20 to 60° C. The polar lipids were labeled with the fluorescent dye rhodamine-DOPE that integrates the fluid matrix of polar lipids and reveal the nonfluorescent areas that correspond to ordered lipid domains rich in milk sphingomyelin and cholesterol. Adapted from Et-Thakafy *et al.* (2017).

temperature (Figure 4.6). Cooling of milk fat globules from 60 to 4 °C enhances the mechanisms of nucleation of saturated high T_m polar lipids in the MFGM, which leads to the formation of small numerous domains or elongated domains (Figure 4.6). Extended storage at low temperature after rapid cooling induces lipid reorganization within the MFGM, with growth, leading to the formation of microdomains with a circular shape. All these structural investigations performed using CLSM showed that the organization of polar lipids and proteins is heterogeneous and that the MFGM is a highly dynamic system.

4.4.3 Topography and Mechanical Properties of MFGM Model Membranes Examined with Nanoscale Resolution

The observation of ordered lipid domains *in situ* in the MFGM surrounding fat globules in milk raised questions about their physical properties and the consequences on the topography and the mechanical properties of the MFGM. The high spatial resolution of atomic force microscopy (AFM) has permitted investigations of model lipid membranes at the nanometer scale.

4.4.3.1 Milk-Sphingomyelin Molecules Are Responsible for Topographical and Mechanical Heterogeneities in Membranes

The main role played by milk-SM on the topography and mechanics of the MFGM was first evidenced using AFM with the observation of hydrated planar supported lipid bilayers (SLBs) composed of an equimolar mixture of milk-SM and dioleoylphosphatidylcholine (DOPC; unsaturated polar lipid mimicking the fluid phase of the MFGM). AFM images recorded at 20 °C (i.e., for $T < T_m$ of milk-SM) show the lateral segregation of milk-SM-rich domains in the gel phase surrounded by a DOPC-rich fluid phase (Figure 4.7). In the bilayer, the domains are thicker than the surrounding fluid phase, due to elongation of the acyl chains of milk-SM in the ordered gel phase. The difference in height between the milk-SM domains and the fluid phase is ~1 nm. This height mismatch between polar lipids in the gel and polar lipids in the fluid states creates disorder that is minimized through phase separation and the formation of domains. At complete equilibrium, the domains tend to circularity to minimize the phase boundary.

The gel phase domains are more rigid than the fluid phase (Figure 4.7). The milk-SM-rich domains exhibited breakthrough forces (F_B) of 31.6 ± 1.9 nN compared with 5.7 ± 2.3 nN for the continuous DOPC-rich phase (Murthy *et al.* 2018). The higher F_B values exhibited for the milk-SM domains reflect the tight acyl chain packing of high-T_m polar lipids in the gel phase, compared to the continuous phase where the acyl chains are in disordered state.

Interestingly, the presence of subdomains in the gel phase was occasionally reported in bilayers containing milk-SM (Guyomarc'h *et al.* 2014, 2017) (Figure 4.7). The two levels characterized by AFM demonstrated the existence of gel–gel phase separation between the SM molecular species present in milk-SM. These two levels of gel phase domains were interpreted as the segregation of individual species composed by different *N*-acyl chain lengths and/or the co-existence of two interdigitated organizations of the long SM molecules (C atoms >20). By using the ternary system, milk-SM/DOPC/cholesterol (3:7:1), Bhojoo *et al.* (2018) also observed subdomains in the bilayers, while they did not observe these subdomains with egg-SM/DOPC/cholesterol (2:2:1). These different levels in the ordered phase domains could be specific to milk-SM that is naturally composed of a mixture of molecular species with chain lengths varying from C16:0 to C24:0. This natural complexity of milk-SM has also been shown to alter the elasticity and the mechanical stability of the ordered lamellar structures (Et-Thakafy *et al.* 2018). Thus, membranes with composition as complex as that of the MFGM are expected to be more compliant to mechanical stress than membranes composed by saturated polar lipids with homogeneous acyl chain length such as DPPC.

4.4.3.2 Milk Polar Lipid Membranes Exhibit Topographical and Mechanical Heterogeneity due to the Formation of Milk-Sphingomyelin-Rich Domains

Using the Langmuir-Blodgett technique, the group of Prof Jiménez-Flores (CalPoly, USA) first pioneered the deposition of MFGM components isolated from buttermilk powder to form a monolayer on a mica surface and performed observations using AFM imaging in air (Jiménez-Flores and Brisson 2008). The same group used this approach to show that polar lipids isolated from milk fractions, or processed milk, phase separate and form domains at various temperatures and lateral surface pressures (Gallier *et al.* 2010b, 2012). Murthy *et al.* (2015) also reported the formation of domains in MFGM polar lipid monolayers, protruding by about 1.5 nm above the continuous phase at a surface pressure of 30 $mN.m^{-1}$ and 20 °C (i.e., below the phase-transition temperature of MFGM polar lipids which have T_m of 36.4 °C). Figure 4.8 shows the lateral phase separation of polar lipids and the formation of circular milk-SM domains in the liquid-condensed phase protruding from the surrounding liquid-expanded phase composed of unsaturated MFGM polar lipids.

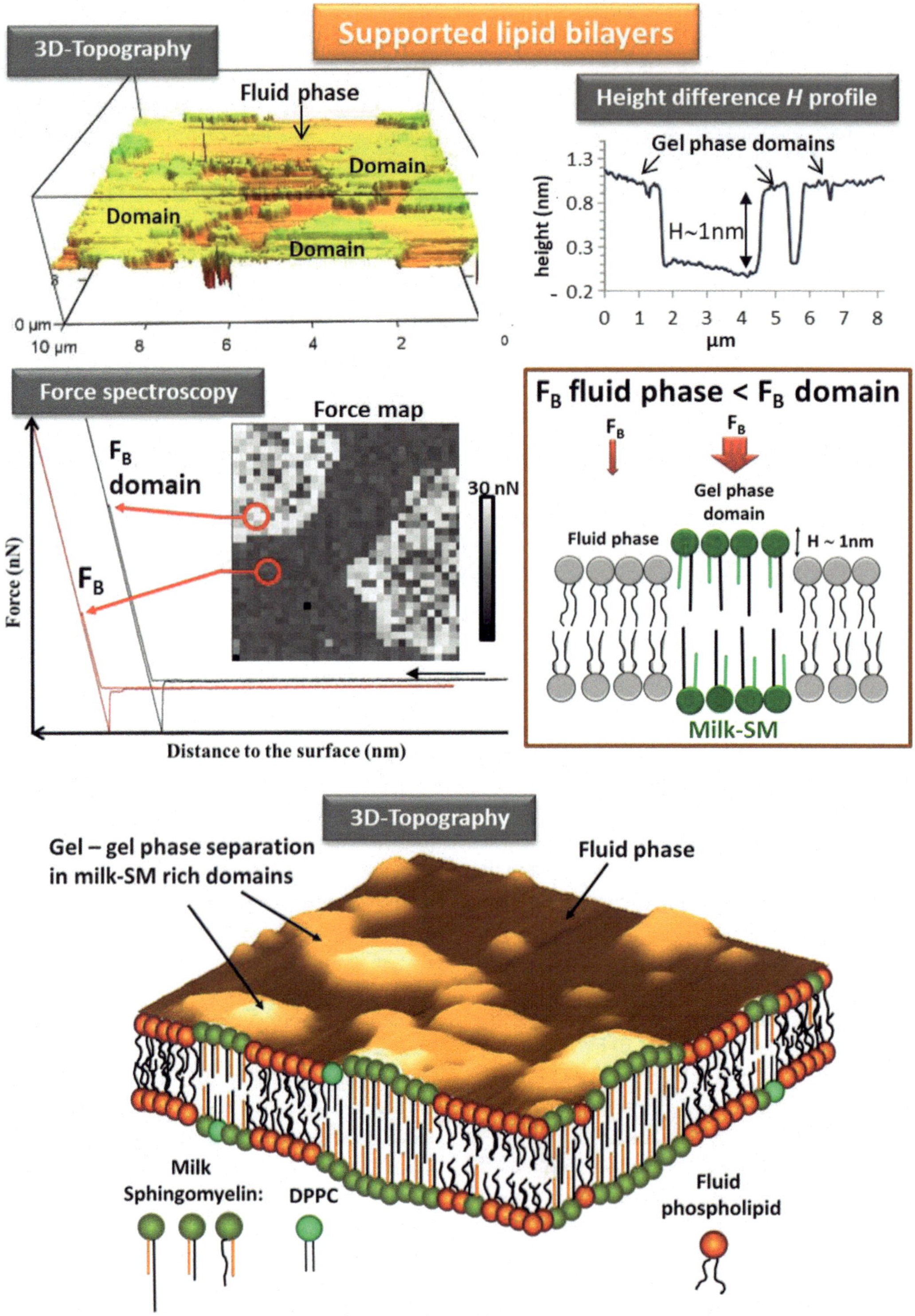

Figure 4.7. Membranes of polar lipids investigated by atomic force microscopy (AFM) imaging and force spectroscopy. The bilayers were composed of an equimolar proportion of milk sphingomyelin and dioleoylphosphatidylcholine to mimic the fluid phase of polar lipids in the milk fat globule membrane (MFGM). AFM images show the lateral phase separation of milk-sphingomyelin in ordered gel phase domains that are higher and more rigid than the surrounding fluid phase (F_B represents breakthrough forces, with the size of the arrows being directly proportional to rigidity). Bottom: 3D topography AFM image combined with a schematic representation of polar lipids in the MFGM, showing the lateral phase segregation of milk-sphingomyelin in domains and the existence of a gel–gel phase separation. Adapted with permission from Guyomarc'h *et al.* (2017) and Lopez *et al.* (2015, 2019).

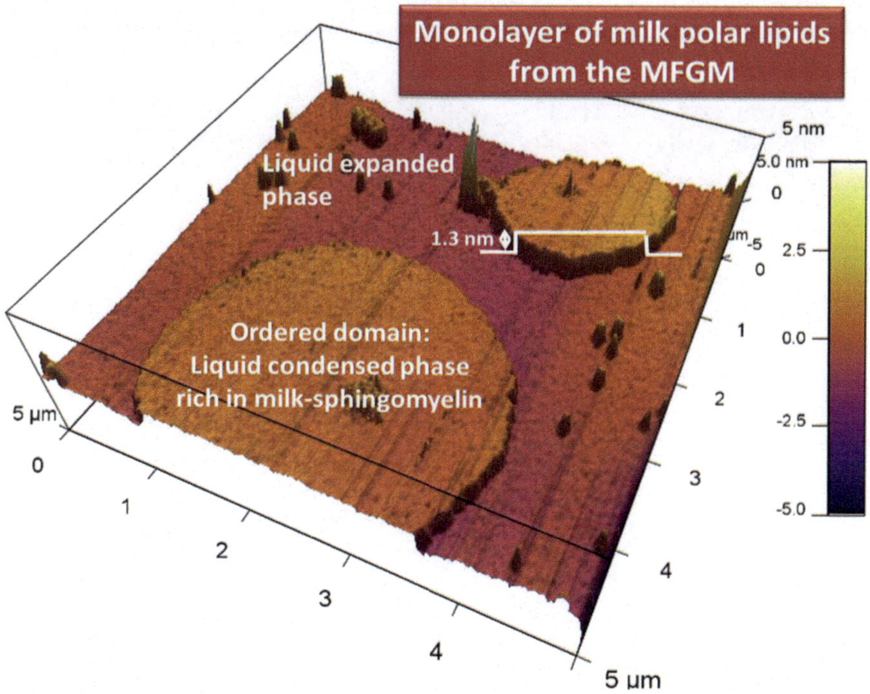

Figure 4.8. Monolayer of polar lipids extracted from the milk fat globule membrane (MFGM) investigated using atomic force microscopy (AFM). The AFM image shows the formation of circular ordered lipid domains that are rich in milk sphingomyelin. These domains are higher than the surrounding fluid phase.

To take account of the natural complexity of the MFGM, the topography and mechanical properties of hydrated SLBs composed of polar lipids extracted from the MFGM were examined (Murthy *et al.* 2016b). As in milk-SM/DOPC bilayers, SLBs made of MFGM polar lipids exhibit, at 20 °C, a lateral phase separation with the formation of milk-SM-rich domains protruding by 1 nm from the surrounding fluid phase composed of the unsaturated MFGM polar lipids PE, PC, PI, and PS. The lipid domains exhibited a higher F_B than the surrounding fluid phase, that is, 30 versus 15 nN (Figure 4.9).

4.4.3.3 Temperature Governs the Physical State of Polar Lipids with Consequences on the Topography and Mechanical Properties of the Membrane

In multicomponent lipid bilayers, such as the outer bilayer of the MFGM, the presence of a variety of saturated and unsaturated polar lipids with distinct T_m makes these systems more prone to temperature variations. The impact of temperature on the topography and nanomechanics of membranes was studied using AFM temperature-controlled imaging and force mapping in the range from 60 to 6 °C upon cooling of SLB composed of milk-SM/DOPC (50/50 mol%) or the complex blend of milk polar lipids extracted from the MFGM (Guyomarc'h *et al.* 2014; Murthy *et al.* 2016b). These studies showed that milk-SM and MFGM polar lipid bilayers are dynamic systems exhibiting changes in their topography and mechanical properties as a function of temperature (Figure 4.9). At temperature above 35 °C (i.e., above T_m of milk-SM), and in the absence of cholesterol, the polar lipid molecules are homogeneously distributed in the membrane and the rupture force of the fluid membrane (F_B) was 2 nN (Figure 4.9). The segregation of the milk-SM-rich domains into salient features protruding by about 1 nm above the fluid phase was detected upon phase transition at 35 °C (Murthy *et al.* 2015) and below that temperature (Guyomarc'h *et al.* 2014; Murthy *et al.*

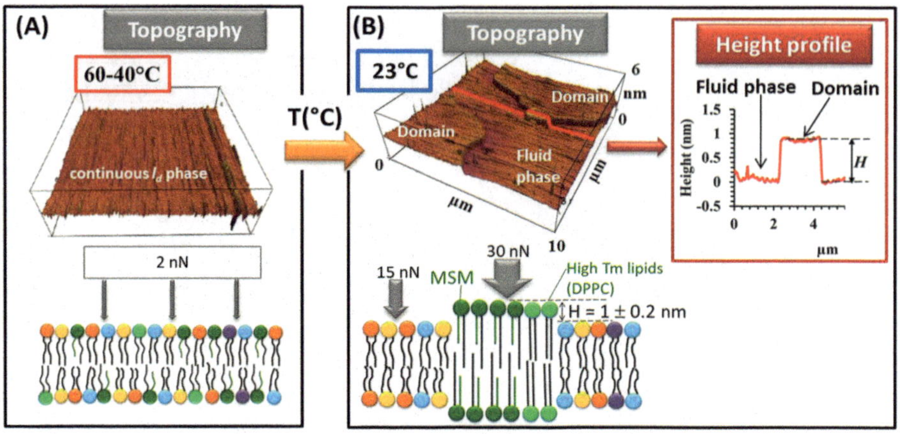

Figure 4.9. Impact of temperature on the topography and mechanical properties of membranes composed of polar lipids from the milk fat globule membrane (MFGM). Atomic force microscopy (AFM) topographical images and force maps showing, (**A**) above 40° C, a flat and fluid membrane and (**B**) at temperature less than 35 °C, the formation of ordered lipid domains rich in milk-sphingomyelin (MSM) that are higher and more rigid than the surrounding fluid phase. Adapted with permission from Murthy *et al.* (2016b).

2016b) (Figure 4.9). At 35 °C, the formation of lipid domains was associated with heterogeneities in the mechanical properties since the F_B values were 23 nN in the gel phase milk-SM-rich domains and 6 nN in the fluid phase. The higher rupture force recorded for the domains is due to the very tight packing of milk-SM molecules in the gel phase, as verified using X-ray diffraction (Murthy *et al.* 2016b). Decreasing the temperature leads to an increase in the rupture forces both in the domains and in the fluid phase, for example, F_B of 30 nN for the domains versus 15 nN for the fluid phase at 25 and 6 °C (Murthy *et al.* 2016b).

The heterogeneities in the structure and the mechanical properties of the membrane were induced by the temperature-dependent gel to fluid phase immiscibility between high-T_m polar lipids (mainly milk-SM) and the unsaturated fluid polar lipids. Undoubtedly, milk-SM molecules are key structural components of the MFGM.

4.4.3.4 Cholesterol Molecules Affect the Topography and the Mechanical Properties of Membranes: Condensing and Fluidizing Effects

The functional role of cholesterol in the MFGM was poorly understood until recently, despite the fact that cholesterol is known to play a crucial role in the lateral organization of lipids and formation of lipid rafts in mammalian cell membranes. Research studies performed over the last 5 years have revealed the main role played by cholesterol in milk-SM containing membranes, that is, changes in the topography of the ordered lipid domains and in the mechanical properties of the membranes.

In the ternary system composed of milk-SM/DOPC/cholesterol (40/40/20 mol%), the AFM height images showed the coexistence of the milk-SM/cholesterol-enriched liquid-ordered Lo phase and DOPC-enriched Lα phase on both monolayers or bilayers (Guyomarc'h *et al.* 2014). The presence of cholesterol yielded narrow and more elongated milk-SM/cholesterol Lo phase domains compared to those formed by the gel phase milk-SM in the binary milk-SM/DOPC (50/50 mol%) system, as revealed in both Langmuir-Blodgett monolayers and hydrated SLBs (Guyomarc'h *et al.* 2014). Introduction of 20 mol% cholesterol in the binary system milk-SM/DOPC induced a significant decrease in the height difference (H) between the ordered domains and the Lα phase. In hydrated bilayers, this decrease was from H_{gel} of 0.8–1.1 nm to H_{Lo} of 0.4–0.6 nm. In Langmuir-Blodgett monolayers, the decrease was from 1.4 ± 0.2 nm for the gel phase milk-SM domains down to 1.1 ± 0.1 nm for the Lo phase domains in the presence of 20 mol% cholesterol. Preferred partitioning of cholesterol into milk-SM domains resulted in the gel to Lo phase transition of these domains, associated with a reduction of their thickness (thinning of the membrane). As the proportion of cholesterol increased, the DOPC Lα phase could also progressively increase in thickness (thickening of the membrane) and could further contribute to reduce H values, that is, the height mismatch between the ordered domains and the surrounding fluid phase. AFM experiments performed on hydrated SLB showed that cholesterol decreases the F_B value of the membrane. In SLB composed of milk-SM/DOPC/cholesterol (50/50/0 and 40/40/20 mol%), the addition of cholesterol significantly decreased the F_B value, corresponding to a decreased resistance of the bilayer to perforation, and revealed the fluidizing effect of cholesterol on milk-SM/DOPC bilayers (Guyomarc'h *et al.* 2014). The addition of cholesterol in a complex mixture of polar lipids extracted from the MFGM has also been demonstrated to affect the topography of the membranes, both in Langmuir-Blodgett monolayers (Murthy *et al.* 2015) and in hydrated bilayers (Murthy *et al.* 2016a).

In monolayers composed of the MFGM polar lipid extract, Murthy *et al.* (2015) revealed that cholesterol plays two major roles on the lateral packing of the polar lipids, first through reducing their molecular area (attractive interactions leading to a condensing effect) and second through dispersing the domains into scattered subunits in the continuous phase (fluidizing effect on the domains). The increased addition of cholesterol up to 30 mol% of polar lipids (milk-SM/cholesterol of 50/50 mol%) induces a decrease in the area of the gel phase (liquid-condensed) milk-SM-rich domains, as a result of the condensing effect of the cholesterol, and the formation of a number of small domains as a result of the fluidizing and therefore dispersing effect of the cholesterol. Moreover, the height difference between the domains and the continuous fluid (liquid-expanded) phase decreased. These results were interpreted in terms of nucleation effects of cholesterol and decrease of the line tension between the domains and the fluid phase, hence increasing the number of the domains and decreasing their size, as revealed both in monolayers and in bilayers (Murthy *et al.* 2015, 2016a). As a consequence of the fluidizing effect of cholesterol on the milk-SM-rich domains, the addition of cholesterol in a complex mixture of polar lipids extracted from the MFGM affects the mechanical properties of the membranes, by decreasing the resistance of the membrane (and especially the domains) to rupture (Murthy *et al.* 2016b).

4.5 Updated Model of the MFGM Organization Highlighting the Main Role of Sphingomyelin in Topographical and Mechanical Heterogeneities

The mechanisms involved in the formation and secretion of milk fat globules are complex and involve the step-by-step building of the MFGM. As a result, the MFGM is a trilayer of polar lipids, with a monolayer originating from the endoplasmic reticulum and a bilayer from the apical plasma membrane of the mammary epithelial cell. New scientific information results from a better knowledge of the biophysical properties of MFGM polar lipids obtained thanks to the development of microscopy techniques, such as confocal microscopy and atomic force microscopy.

Figure 4.10 shows an updated model for the organization of polar lipids and proteins in the MFGM as adapted from Lopez *et al.* (2015). The complex composition of MFGM polar lipids associated with their different biophysical properties (saturated high-T_m polar lipids such as sphingomyelin *versus* unsaturated low-T_m polar lipids) leads to a patchwork, with the formation of ordered lipid microdomains. The lipid phase segregation that occurs in the outer bilayer of the membrane is a consequence of the physical state of the polar lipids and their attractive interactions, that is, ordered lipid domains rich in milk-SM and cholesterol, surrounded by a fluid phase of the unsaturated polar lipids PC, PE, PI, and PS. These ordered lipid domains are higher and more rigid than the surrounding fluid matrix, which induces topographical and mechanical heterogeneities. The phase separation of polar lipids that is due to different physical properties of the polar lipids induces heterogeneities in the charge. The microdomains are rich in milk-SM that is zwitterionic, while the unsaturated polar lipids integrate PS and PI that are anionic (Figure 4.2). The proteins are embedded in the fluid matrix composed of the unsaturated polar lipids and are then heterogeneously distributed in the MFGM.

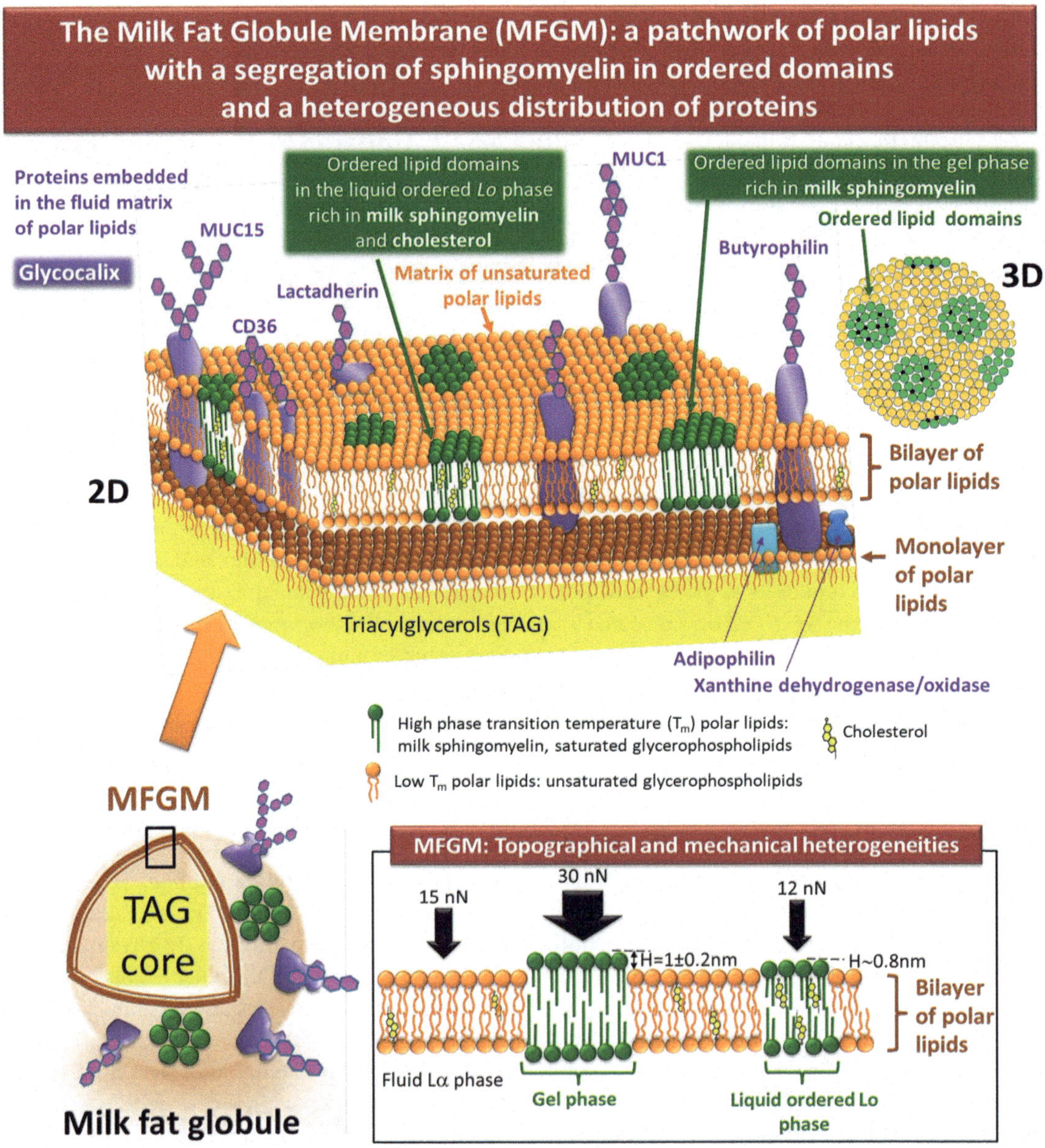

Figure 4.10. Updated model of the milk fat globule membrane (MFGM), highlighting the lateral phase separation of polar lipids in the outer bilayer with the formation of ordered lipid domains rich in milk-sphingomyelin and cholesterol. The proteins are embedded in the fluid matrix composed of the unsaturated polar lipids. Adapted from Lopez *et al.* (2015).

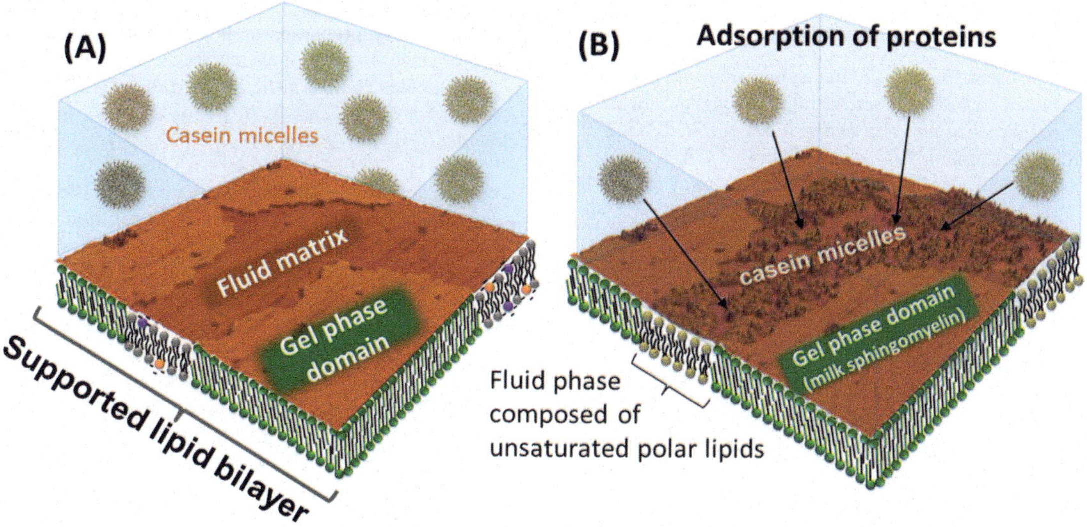

Figure 4.11. Impact of milk polar lipid phase state and charge on the adsorption of proteins. The casein micelles preferentially adsorb on the fluid phase of polar lipids devoid of anionic lipids. The casein micelles do not adsorb on the ordered microdomains rich in sphingomyelin. The experiments were performed on supported lipid bilayers in hydrated conditions. Adapted from Obeid *et al.* (2019).

These heterogeneities in the phase state and charge of polar lipids in the outer bilayer of the MFGM can modulate the interactions with proteins such as the milk proteins (caseins and whey proteins) or enzymes, as already reported (Bourlieu *et al.* 2016; Crespo-Villanueva *et al.* 2018; Obeid *et al.* 2019). Figure 4.11 shows that the casein micelles do not adsorb on the ordered microdomains rich in sphingomyelin, but adsorb on the fluid phase of polar lipids.

4.6 Perspectives

Over the next 5–10 years, milk polar lipids will undoubtedly be the focus of much attention in order to understand the functional role of phase separation in the MFGM that leads to a heterogeneous distribution of proteins. Milk sphingomyelin will also continue to receive attention for a better understanding of its role in the MFGM but also in the mechanism of milk fat globule digestion and its role in the intestine, for example, toward the absorption of cholesterol. I am convinced that the milk fat globules will be the gold standard for the composition and structure of lipids in infant milk formulae, for the nutritional and health benefits of infants. Milk polar lipids and MFGM-enriched ingredients prepared from cream or by-products of the dairy industry will also be further used as dietary emulsifiers to benefit their specific functional properties.

Acknowledgments The scientist colleagues whose past work has contributed to present knowledge about the intracellular origin and secretion of milk fat globules, as well as on the composition and structure of the MFGM, but whose publications could not be cited in this chapter, are warmly acknowledged. I express my gratitude to all the colleagues, PhD students, and postdoctoral researchers who have worked with me on the research area of the MFGM. The research performed by my group on the structure of the MFGM has been supported by the French National Research Institute for Agriculture, Food and Environment (INRAE) and by CNIEL (French Dairy Interbranch Organization, Paris, France).

References

Argov-Argaman, N. (2019). Symposium review: Milk fat globule size: Practical implications and metabolic regulation. *Journal of Dairy Science, 102*, 2783–2795.

Argov-Argaman, N., Mesilati-Stahy, R., Magen, Y. & Moallem, U. (2014). Elevated concentrate-to-forage ratio in dairy cow rations is associated with a shift in the diameter of milk fat globules and remodeling of their membranes. *Journal of Dairy Science, 97*, 6286–6295.

Argov-Argaman, N., Hadaya, O., Glasser, T., Muklada, H., Dvash, L., Mesilati-Stahy, R. & Landau, S. Y. (2016). Milk fat globule size, phospholipid contents and composition of milk from purebred and Alpine-crossbred Mid-Eastern goats under confinement or grazing condition. *International Dairy Journal, 58*, 2–8.

Bargmann, W. & Knoop, A. (1959). Über die Morphologie der Milchsekretion Licht- und elektronenmikroskopische Studien an der Milchdrüse der Ratte. *Zeitschrift für Zellforschung, 49*, 344–388.

Bhojoo, U., Chen, M. & Zou, S. (2018). Temperature induced lipid membrane restructuring and changes in nanomechanics. *Biochimica et Biophysica Acta (BBA) – Biomembranes, 1860*, 700–709.

Bourlieu, C., Paboeuf, G., Chever, S., Pezennec, S., Cavalier, J.-F., Guyomarc'h, F., Deglaire, A., Bouhallab, S., Dupont, D., Carrière, F., et al. (2016). Adsorption of gastric lipase onto multicomponent model lipid monolayers with phase separation. *Colloids and Surfaces B: Biointerfaces, 143*, 97–106.

Bourlieu, C., Cheillan, D., Blot, M., Daira, P., Trauchessec, M., Ruet, S., Gassi, J.-Y., Beaucher, E., Robert, B., Leconte, N., et al. (2018). Polar lipid composition of bioactive dairy co-products buttermilk and butterserum: Emphasis on sphingolipid and ceramide isoforms. *Food Chemistry, 240*, 67–74.

Burling, H. & Graverholt, G. (2008). Milk – A new source for bioactive phospholipids for use in food formulations. *Lipid Technology, 20*, 229–231.

Cheng, K., Ropers, M.-H. & Lopez, C. (2017). The miscibility of milk sphingomyelin and cholesterol is affected by temperature and surface pressure in mixed Langmuir monolayers. *Food Chemistry, 224*, 114–123.

Crespo-Villanueva, A., Gumí-Audenis, B., Sanz, F., Artzner, F., Mériadec, C., Rousseau, F., Lopez, C., Giannotti, M. I. & Guyomarc'h, F. (2018). Casein interaction with lipid membranes: Are the phase state or charge density of the phospholipids affecting protein adsorption? *Biochimica et Biophysica Acta (BBA) – Biomembranes, 1860*, 2588–2598.

Deeth, H. C. (1997). The role of phospholipids in the stability of milk fat globules. *Australian Journal of Dairy Technology, 52*, 44–46.

Dewettinck, K., Rombaut, R., Thienpont, N., Le, T. T., Messens, K. & Van Camp, J. (2008). Nutritional and technological aspects of milk fat globule membrane material. *International Dairy Journal, 18*, 436–457.

Douellou, T., Montel, M. C. & Sergentet, D. T. (2017). Anti-adhesive properties of bovine oligosaccharides and bovine milk fat globule membrane-associated glycoconjugates against bacterial food enteropathogens. *Journal of Dairy Science, 100*, 3348–3359.

Eckhardt, E. R. M., Wang, D. Q.-H., Donovan, J. M. & Carey, M. C. (2002). Dietary sphingomyelin suppresses intestinal cholesterol absorption by decreasing thermodynamic activity of cholesterol monomers. *Gastroenterology, 122*, 948–956.

Et-Thakafy, O., Guyomarc'h, F. & Lopez, C. (2017). Lipid domains in the milk fat globule membrane: Dynamics investigated *in situ* in milk in relation to temperature and time. *Food Chemistry, 220*, 352–361.

Et-Thakafy, O., Delorme, N., Guyomarc'h, F. & Lopez, C. (2018). Mechanical properties of milk sphingomyelin bilayer membranes in the gel phase: Effects of naturally complex heterogeneity, saturation and acyl chain length investigated on liposomes using AFM. *Chemistry and Physics of Lipids, 210*, 47–59.

Evers, J. M. (2004). The milkfat globule membrane—Compositional and structural changes post secretion by the mammary secretory cell. *International Dairy Journal, 14*, 661–674.

Evers, J. M., Haverkamp, R. G., Holroyd, S. E., Jameson, G. B., Mackenzie, D. D. S. & McCarthy, O. J. (2008). Heterogeneity of milk fat globule membrane structure and composition as observed using fluorescence microscopy techniques. *International Dairy Journal, 18*, 1081–1089.

FAO. (2018). *Food and Agriculture Organization of the United Nations Gateway to dairy production and products: Milk and milk products*.

Fong, B. Y., Norris, C. S. & MacGibbon, A. K. H. (2007). Protein and lipid composition of bovine milk-fat-globule membrane. *International Dairy Journal, 17*, 275–288.

Gallier, S., Gragson, D., Jiménez-Flores, R. & Everett, D. (2010a). Using confocal laser scanning microscopy to probe the milk fat globule membrane and associated proteins. *Journal of Agricultural and Food Chemistry, 58*, 4250–4257.

Gallier, S., Gragson, D., Jiménez-Flores, R. & Everett, D. W. (2010b). Surface characterization of bovine milk phospholipid monolayers by Langmuir isotherms and microscopic techniques. *Journal of Agricultural and Food Chemistry, 58*, 12275–12285.

Gallier, S., Gragson, D., Jiménez-Flores, R. & Everett, D. W. (2012). β-Casein–phospholipid monolayers as model systems to understand lipid–protein interactions in the milk fat globule membrane. *International Dairy Journal, 22*, 58–65.

Gallier, S., Vocking, K., Post, J. A., Van de Heijning, B., Acton, D., Van der Beek, E. M. & Van Baalen, T. (2015). A novel infant milk formula concept: Mimicking the human milk fat globule structure. *Colloid Surfaces B-Biointerfaces, 136*, 329–339.

Garcia, C., Lutz, N. W., Confort-Gouny, S., Cozzone, P. J., Armand, M. & Bernard, M. (2012). Phospholipid fingerprints of milk from different mammalians determined by 31P NMR: Towards specific interest in human health. *Food Chemistry, 135*, 1777–1783.

Giuffrida, F., Cruz-Hernandez, C., Flück, B., Tavazzi, I., Thakkar, S. K., Destaillats, F. & Braun, M. (2013). Quantification of phospholipids classes in human milk. *Lipids, 48*, 1051–1058.

Guyomarc'h, F., Zou, S., Chen, M., Milhiet, P.-E., Godefroy, C., Vie, V. & Lopez, C. (2014). Milk sphingomyelin domains in biomimetic membranes and the role of cholesterol: Morphology and nanomechanical properties investigated using AFM and force spectroscopy. *Langmuir, 30*, 6516–6524.

Guyomarc'h, F., Chen, M., Et-Thakafy, O., Zou, S. & Lopez, C. (2017). Gel-gel phase separation within milk sphingomyelin domains revealed at the nanoscale using atomic force microscopy. *Biochimica et Biophysica Acta (BBA) – Biomembranes, 1859*, 949–958.

Hamosh, M., Peterson, J. A., Henderson, T. R., Scallan, C. D., Kiwan, R., Ceriani, R. L., Armand, M., Mehta, N. R. & Hamosh, P. (1999). Protective function of human milk: The milk fat globule. *Seminars in Perinatology, 23*, 242–249.

Heid, H. W. & Keenan, T. W. (2005a). Intracellular origin and secretion of milk fat globules. *European Journal of Cell Biology, 84*, 245–258.

Heid, H. W. & Keenan, T. W. (2005b). Intracellular origin and secretion of milk fat globules. *European Journal of Cell Biology, 84*, 245–258.

Huston, G. & Patton, S. (1990). Factors related to the formation of cytoplasmic crescents on milk-fat globules. *Journal of Dairy Science, 73*, 2061–2066.

Janssen, M. & Walstra, P. (1982). Cytoplasmic remnants in milk of certain species. *Netherlands Milk and Dairy Journal, 36*, 365–368.

Jiménez-Flores, R. & Brisson, G. (2008). The milk fat globule membrane as an ingredient: Why, how, when? *Dairy Science Technology, 88*, 5–18.

Keenan, T. W. & Mather, I. H. (2006). Intracellular origin of Milk fat globules and the nature of the milk fat globule membrane. In P. F. Fox & P. L. H. McSweeney (Eds.), *Advanced dairy chemistry volume 2 lipids* (pp. 137–171). Springer: Boston.

Keenan, T. W., Morré, D. J., Olson, D. E., Yunghans, W. N. & Patton, S. (1970). Biochemical and morphological comparison of plasma membrane and milk fat globule membrane from bovine mammary gland. *Journal of Cell Biology, 44*, 80–93.

Liao, Y., Alvarado, R., Phinney, B. & Lönnerdal, B. (2011). Proteomic characterization of human milk fat globule membrane proteins during a 12 month lactation period. *Journal of Proteome Research, 10*, 3530–3541.

Logan, A., Auldist, M., Greenwood, J. & Day, L. (2014). Natural variation of bovine milk fat globule size within a herd. *Journal of Dairy Science, 97*, 4072–4082.

Lopez, C. (2011). Milk fat globules enveloped by their biological membrane: Unique colloidal assemblies with a specific composition and structure. *Current Opinion in Colloid & Interface Science, 16*, 391–404.

Lopez, C. & Ménard, O. (2011). Human milk fat globules: Polar lipid composition and in situ structural investigations revealing the heterogeneous distribution of proteins and the lateral segregation of sphingomyelin in the biological membrane. *Colloids and Surfaces B: Biointerfaces, 83*, 29–41.

Lopez, C., Briard-Bion, V., Menard, O., Rousseau, F., Pradel, P. & Besle, J.-M. (2008a). Phospholipid, sphingolipid, and fatty acid compositions of the milk fat globule membrane are modified by diet. *Journal of Agricultural and Food Chemistry, 56*, 5226–5236.

Lopez, C., Briard-Bion, V., Menard, O., Rousseau, F., Pradel, P. & Besle, J.-M. (2008b). Phospholipid, sphingolipid, and fatty acid compositions of the milk fat globule membrane are modified by diet. *Journal of Agricultural and Food Chemistry, 56*, 5226–5236.

Lopez, C., Briard-Bion, V., Beaucher, E. & Ollivon, M. (2008c). Multiscale characterization of the organization of triglycerides and phospholipids in emmental cheese: From the microscopic to the molecular level. *Journal of Agricultural and Food Chemistry, 56*, 2406–2414.

Lopez, C., Madec, M.-N. & Jimenez-Flores, R. (2010). Lipid rafts in the bovine milk fat globule membrane revealed by the lateral segregation of phospholipids and heterogeneous distribution of glycoproteins. *Food Chemistry, 120*, 22–33.

Lopez, C., Briard-Bion, V., Ménard, O., Beaucher, E., Rousseau, F., Fauquant, J., Leconte, N. & Robert, B. (2011). Fat globules selected from whole milk according to their size: Different compositions and structure of the biomembrane, revealing sphingomyelin-rich domains. *Food Chemistry, 125*, 355–368.

Lopez, C., Briard-Bion, V. & Ménard, O. (2014). Polar lipids, sphingomyelin and long-chain unsaturated fatty acids from the milk fat globule membrane are increased in milks produced by cows fed fresh pasture based diet during spring. *Food Research International, 58*, 59–68.

Lopez, C., Cauty, C. & Guyomarc'h, F. (2015). Organization of lipids in milks, infant milk formulas and various dairy products: Role of technological processes and potential impacts. *Dairy Science & Technology, 95*, 863–893.

Lopez, C., Blot, M., Briard-Bion, V., Cirié, C. & Graulet, B. (2017a). Butter serums and buttermilks as sources of bioactive lipids from the milk fat globule membrane: Differences in their lipid composition and potentialities of cow diet to increase n-3 PUFA. *Food Research International, 100*, 864–872.

Lopez, C., Cauty, C., Rousseau, F., Blot, M., Margolis, A. & Famelart, M.-H. (2017b). Lipid droplets coated with milk fat globule membrane fragments: Microstructure and functional properties as a function of pH. *Food Research International, 91*, 26–37.

Lopez, C., Cheng, K. & Perez, J. (2018). Thermotropic phase behavior of milk sphingomyelin and role of cholesterol in the formation of the liquid ordered phase examined using SR-XRD and DSC. *Chemistry and Physics of Lipids, 215*, 46–55.

Lopez, C., Cauty, C. & Guyomarc'h, F. (2019). Unraveling the complexity of milk fat globules to tailor bioinspired emulsions providing health benefits: The key role played by the biological membrane. *European Journal of Lipid Science and Technology, 121*, 1800201.

Luo, J., Huang, Z., Liu, H., Zhang, Y. & Ren, F. (2018). Yak milk fat globules from the Qinghai-Tibetan Plateau: Membrane lipid composition and morphological properties. *Food Chemistry*, 731–737.

Malmsten, M., Bergenståhl, B., Nyberg, L. & Odham, G. (1994). Sphingomyelin from milk-characterization of liquid crystalline, liposome and emulsion properties. *Journal of the American Oil Chemists' Society, 71*, 1021–1026.

Mather, I. H. (2000). A review and proposed nomenclature for major proteins of the milk-fat globule membrane1,2. *Journal of Dairy Science, 83*, 203–247.

Ménard, O., Ahmad, S., Rousseau, F., Briard-Bion, V., Gaucheron, F. & Lopez, C. (2010). Buffalo vs. cow milk fat globules: Size distribution, zeta-potential compositions in total fatty acids and in polar lipids from the milk fat globule membrane. *Food Chemistry, 120*, 544–551.

Mesilati-Stahy, R. & Argov-Argaman, N. (2014). The relationship between size and lipid composition of the bovine milk fat globule is modulated by lactation stage. *Food Chemistry, 145*, 562–570.

Mesilati-Stahy, R., Moallem, U., Magen, Y. & Argov-Argaman, N. (2015). Altered concentrate to forage ratio in cows ration enhanced bioproduction of specific size subpopulation of milk fat globules. *Food Chemistry, 179*, 199–205.

Murthy, A. V. R., Guyomarc'h, F., Paboeuf, G., Vié, V. & Lopez, C. (2015). Cholesterol strongly affects the organization of lipid monolayers studied as models of the milk fat globule membrane: Condensing effect and change in the lipid domain morphology. *Biochimica et Biophysica Acta (BBA) – Biomembranes, 1848*, 2308–2316.

Murthy, A. V. R., Guyomarc'h, F. & Lopez, C. (2016a). Cholesterol decreases the size and the mechanical resistance to rupture of sphingomyelin rich domains, in lipid bilayers studied as a model of the milk fat globule membrane. *Langmuir, 32*, 6757–6765.

Murthy, A. V. R., Guyomarc'h, F. & Lopez, C. (2016b). The temperature-dependent physical state of polar lipids and their miscibility impact the topography and mechanical properties of bilayer models of the milk fat globule membrane. *Biochimica et Biophysica Acta – Biomembranes, 1858*, 2181–2190.

Murthy, A. V. R., Guyomarc'h, F. & Lopez, C. (2018). Palmitoyl ceramide promotes milk sphingomyelin gel phase domains formation and affects the mechanical properties of the fluid phase in milk-SM/DOPC supported membranes. *Biochimica et Biophysica Acta (BBA) – Biomembranes, 1860*, 635–644.

Nguyen, H. T. H., Ong, L., Beaucher, E., Madec, M.-N., Kentish, S. E., Gras, S. L. & Lopez, C. (2015). Buffalo milk fat globules and their biological membrane: In situ structural investigations. *Food Research International, 67*, 35–43.

Nguyen, H. T. H., Madec, M.-N., Ong, L., Kentish, S. E., Gras, S. L. & Lopez, C. (2016). The dynamics of the biological membrane surrounding the buffalo milk fat globule investigated as a function of temperature. *Food Chemistry*, 343–351.

Nilsson, A. (2016). Role of sphingolipids in infant gut health and immunity. *Journal of Pediatrics, 173*, S53–S59.

Noh, S. K. & Koo, S. I. (2004a). Milk sphingomyelin is more effective than egg sphingomyelin in inhibiting intestinal absorption of cholesterol and fat in rats. *Journal of Nutrition, 134*, 2611–2616.

Noh, S. K. & Koo, S. I. (2004b). Milk sphingomyelin is more effective than egg sphingomyelin in inhibiting intestinal absorption of cholesterol and fat in rats. *Journal of Nutrition, 134*, 2611–2616.

Norris, G. H., Jiang, C., Ryan, J., Porter, C. M. & Blesso, C. N. (2016). Milk sphingomyelin improves lipid metabolism and alters gut microbiota in high fat diet-fed mice. *Journal of Nutritional Biochemistry, 30*, 93–101.

Nyberg, L., Duan, R.-D. & Nilsson, Å. (2000). A mutual inhibitory effect on absorption of sphingomyelin and cholesterol. *Journal of Nutritional Biochemistry, 11*, 244–249.

Obeid, S., Guyomarc'h, F., David-Briand, E., Gaucheron, F., Riaublanc, A. & Lopez, C. (2019). The phase and charge of milk polar lipid membrane bilayers govern their selective interactions with proteins as demonstrated with casein micelles. *Journal of Colloid and Interface Science, 534*, 279–290.

Peterson, J. A., Patton, S. & Hamosh, M. (1998). Glycoproteins of the human milk fat globule in the protection of the breast-fed infant against infections. *NEO, 74*, 143–162.

Rasmussen, J. T. (2009). Bioactivity of milk fat globule membrane proteins. *Australian Journal of Dairy Technology, 64*, 63–67.

Reinhardt, T. A. & Lippolis, J. D. (2006). Bovine milk fat globule membrane proteome. *Journal of Dairy Research, 73*, 406–416.

Sanchez-Juanes, F., Alonso, J. M., Zancada, L. & Hueso, P. (2009). Distribution and fatty acid content of phospholipids from bovine milk and bovine milk fat globule membranes. *International Dairy Journal, 19*, 273–278.

Simons, K. & Ikonen, E. (1997). Functional rafts in cell membranes. *Nature, 387*, 569–572.

Singh, H. (2006). The milk fat globule membrane—A biophysical system for food applications. *Current Opinion in Colloid & Interface Science, 11*, 154–163.

Sprong, R. C., Hulstein, M. F. E., Lambers, T. T. & van der Meer, R. (2012). Sweet buttermilk intake reduces colonisation and translocation of Listeria monocytogenes in rats by inhibiting mucosal pathogen adherence. *British Journal of Nutrition, 108*, 2026–2033.

Vanderghem, C., Bodson, P., Danthine, S., Paquot, M., Deroanne, C. & Blecker, C. (2010). Milk fat globule membrane and buttermilks: From composition to valorization. *Biotechnologie, Agronomie, Société et Environnement, 14*, 485–500.

Wooding, F. B. P. (1971a). The mechanism of secretion of the milk fat globule. *Journal of Cell Science, 9*, 805–821.

Wooding, F. B. P. (1971b). The structure of the milk fat globule membrane. *Journal of Ultrastructure Research, 37*, 388–400.

Wooding, F. B. P. & Mather, I. H. (2017). Ultrastructural and immunocytochemical evidence for the reorganisation of the milk fat globule membrane after secretion. *Cell and Tissue Research, 367*, 283–295.

Yao, Y., Zhao, G., Xiang, J., Zou, X., Jin, Q. & Wang, X. (2016). Lipid composition and structural characteristics of bovine, caprine and human milk fat globules. *International Dairy Journal, 56*, 64–73.

Zaczek, M. & Keenan, T. (1990). Morphological evidence for an endoplasmic-reticulum origin of milk lipid globules obtained using lipid-selective staining procedures. *Protoplasma, 159*, 179–182.

Zheng, H., Jimenez-Flores, R. & Everett, D. W. (2014). Lateral lipid organization of the bovine milk fat globule membrane is revealed by washing processes. *Journal of Dairy Science, 97*, 5964–5974.

Zou, X., Guo, Z., Jin, Q., Huang, J., Cheong, L., Xu, X. & Wang, X. (2015). Composition and microstructure of colostrum and mature bovine milk fat globule membrane. *Food Chemistry, 185*, 362–370.

Zou, X.-Q., Guo, Z., Huang, J.-H., Jin, Q.-Z., Cheong, L.-Z., Wang, X.-G. & Xu, X.-B. (2012). Human milk fat globules from different stages of lactation: A lipid composition analysis and microstructure characterization. *Journal of Agricultural and Food Chemistry, 60*, 7158–7167.

Physical Chemistry of Milk Fat Globules

5

T. Huppertz, T. Uniacke-Lowe, and A. L. Kelly

5.1 Introduction

The presence of fat globules in milk was first reported by Van Leeuwenhoek in 1674, based on microscopic analysis of milk placed in a fine capillary tube; since then, the physical and colloidal properties of milk fat globules and their size distribution have been the subject of considerable study. These properties of milk fat globules are responsible for some of the properties and phenomena observed in liquid dairy products, e.g. the colour and creaming of milk, and are integral to the manufacture and characteristics of many dairy products, e.g. butter and ice cream. Furthermore, the properties of milk fat globules are also influenced by enzymatic processes, such as lipolysis, with implications for processes such as the ripening of cream and the flavour of some cheese varieties. Finally, milk fat globules can be affected greatly by processes applied to the milk, particularly homogenization, which has significant implications not only for the properties of milk fat globules but also of casein micelles in milk.

This chapter describes important aspects of the physical and colloidal chemistry of milk fat globules, and in particular recent research in the area, which underpins many of the phenomena described in other chapters of this book. The relevant aspects of processes that affect the stability of fat globules, including storage, homogenization and heating, and the resulting interactions with other milk constituents, including caseins and whey proteins, will be reviewed also.

5.2 The Nature and Size Distribution of Milk Fat Globules

Fat in milk is present predominantly in spherical droplets ranging from <0.2 to >15 μm in diameter; bovine milk typically contains $>10^{10}$ globules per mL. The composition of the fat in the milk fat in globules has been discussed in detail in Chapters 1 and 2. Fat globules are dispersed in the continuous phase of milk plasma, which contains casein micelles, serum proteins, sugars and minerals, and can be considered both a colloidal suspension and an oil-in-water emulsion. Fat globules in milk are naturally emulsified by a complex layer of surface material, the milk fat

T. Huppertz
Wageningen University and Research, Wageningen, The Netherlands

FrieslandCampina, Amersfoort, The Netherlands

T. Uniacke-Lowe · A. L. Kelly (✉)
School of Food and Nutritional Sciences, University College Cork, Cork, Ireland
e-mail: a.kelly@ucc.ie

P. L. H. McSweeney et al. (eds.), *Advanced Dairy Chemistry, Volume 2*,
https://doi.org/10.1007/978-3-030-48686-0_5

Table 5.1. Estimated average composition of milk fat globule membranes

Component	mg/100 g fat globules	mg/m^2 fat surface	% of membrane material
Protein	1800	9.0	70
Phospholipids	650	3.2	25
Cerebrosides	80	0.4	3
Cholesterol	40	0.2	2
Neutral glycerides	+	+	?
Water	+	+	?
Carotenoids and vitamin A	0.04	2×10^{-4}	–
Iron	0.3	1.5×10^{-3}	–
Copper	0.01	5×10^{-5}	–
Total	*> 2570*	*> 12.8*	*100*

Adapted from Walstra *et al.* (1999).

globule membrane (MFGM), which accounts for 2–6% of the mass of the fat globules (Keenan and Mather 2002) and maintains the integrity of the lipid droplets and helps to protect them from destabilization (Walstra *et al.* 1999). The composition of the MFGM (Table 5.1) is closer to that of a cell membrane, from which it is largely derived, than to either milk fat or milk serum. Widely differing compositions of the MFGM, particularly in terms of triglyceride profile, have been reported. Several enzymes are found in the MFGM, including alkaline phosphatase and xanthine oxidoreductase, which make up a significant portion of the membrane protein content, as well as monoglycerides and free fatty acids. For a more detailed description of the secretion of milk fat globules and the exact structure of the MFGM, the reader is referred to Chapter 4.

One of the most important properties of fat globules in milk is their size, both in terms of mean (average) size, but also the range or distribution of sizes and the effects of processes and treatments thereon. A review of the early studies on milk fat globule size was published by Campbell (1932). The size distribution of milk fat globules may vary greatly with the analytical method used, giving a certain degree of unreliability to results obtained using some older methods (Walstra *et al.* 1969).

Various microscopic techniques such as optical microscopy (Tatar *et al.* 2015), fluorescence microscopy (Martini *et al.* 2006; Evers *et al.* 2008), confocal laser scanning microscopy (Ong *et al.* 2010), transmission electron microscopy (Goff *et al.* 1987), scanning electron microscopy (Bermúdez-Aguirre *et al.* 2008) and holographic video microscopy (Cheong *et al.* 2009) can be used to measure milk fat globule size, although these techniques have drawbacks including processing time required to get sufficient data for determination of size distribution (Truong *et al.* 2016). The advantage of microscopic techniques, however, is that they allow direct visualization of fat globule shape, distribution and microstructure of milk fat globules, MFGM and fat crystals.

Methods such as dynamic light-scattering (Robin and Paquin 1991; Dalgleish and Hallett 1995), small-angle laser light scattering (Muir *et al.* 1991; Michalski *et al.* 2001a), Coulter counting (Hillbrick *et al.* 1998), ultrasound (Miles *et al.* 1990) and electroacoustics (Wade *et al.* 1996; Wade and Beattie 1997) provide more accurate and reproducible results. Laser light scattering methods, where particles scatter light from one or two laser beams with an angular pattern directly related to their size, are now favoured for measuring fat globule size. Many of these measurement techniques generate complex primary data, which must be processed using specific algorithms or programmes to yield useful data for milk fat globule size.

Control of interference by other milk constituents with measurements is key to measurement of milk fat globule sizes; for example, dissociation of casein micelles by calcium-chelating agents, such as trisodium citrate or ethylenediamine tetra-acetic acid (EDTA), may be used to avoid interference by the micelles in particle size measurement, while clusters of fat globules can be disrupted by adding a low level of anionic sodium dodecyl sulphate (SDS) or the non-ionic surfactant Tween 20 can be used in combination with EDTA (McCrae and Lepoetre 1996). SDS and Tween 20 act by removing absorbed materials from the oil–water interface and disrupt the agglomerates of fat globules (De Feijter *et al.* 1987; Thiebaud *et al.* 2003) and maintaining the globules in a dispersed form, enabling size measurement of individual globules.

Table 5.2. Parameters describing the milk fat globule size distribution in unhomogenized bovine milk

	Range (μm)	Average (μm)
Number mean diameter: $d_n = d_{1,0} = S_1/S_0$	0.67–1.0	0.81
Volume mean diameter: $d_v = d_{3,0} = (S_3/S_0)^{1/3}$	1.5–2.1	1.8
Volume surface-weighted mean diameter: $d_{vs} = d_{3,2} = S_3/S_2$	2.5–4.6	3.34
Volume moment-weighted mean diameter $d_{vm} = d_{4,3} = S_4/S_3$		3.53[1]

Values from Walstra (1969a).

Table 5.3. Fat globule size in milk from various species

Species	D_{vs} (μm)	Reference
Cow	3.9	Walstra (1969a)
	4.0	Rüegg and Blanc (1981)
	~3.5	Van Boekel and Folkerts (1991)
	5.32	Mehaia (1995)
	3.51	Attaie and Richter (2000)
Reindeer	3.95	Uniacke-Lowe and Fox (2011)
Goat	4.89	Mehaia (1995)
	2.76	Attaie and Richter (2000)
	3.20	El-Zeini (2006)
	2.22	Tatar *et al.* (2015)
Camel	4.40	Farah and Rüegg (1991)
	4.40	Mehaia (1995)
	2.99	El-Zeini (2006)
Human milk colostrum	1.74	Rüegg and Blanc (1981)
	8.9	Michalski *et al.* (2005)
	3.51	Zou *et al.* (2012)
Human milk transitional	1.84	Rüegg and Blanc (1981)
	2.8	Michalski *et al.* (2005)
	3.14	Zou *et al.* (2012)
Human milk mature	4.10	Rüegg and Blanc (1981)
	4.00	Michalski *et al.* (2005)
	3.25	Zou *et al.* (2012)
Mare	1.05	Uniacke-Lowe (2011)
	1.30	Devle *et al.* (2012)
Donkey	1.92	Martini *et al.* (2014)
Ewe	4.95	Mehaia (1995)
	5.00	Gervilla *et al.* (2001)
	3.76	El-Zeini (2006)
Yak	6.09	Luo *et al.* (2018)
Buffalo	8.7	El-Zeini (2006)
	5.0	Ménard *et al.* (2010) and Ahmad *et al.* (2013)

A plot of the number distribution, i.e. the number of globules per unit volume, N, in a certain size class, divided by the width of the size class, Δd, as a function of size, d, shows three sub-distributions (Walstra 1969a, 1995): a subclass of 'small particles', comprising ~80% of the number of particles but only ~3% of the mass of fat, the main fractions, comprising ~95% of fat, and a subclass of large globules, comprising ~2% of the fat. Besides the number distribution, distributions of mass, volume or surface area can also be calculated by multiplying the number frequency by mass, volume or surface area, respectively, for each size class. Plotting volume frequency versus particle diameter is the most common method of presentation of globule size data (Walstra 2003).

Several different parameters can be used to express the mean size of the milk fat globules. These parameters are derived from the so-called moments of the size distribution function; the nth moment of the distribution function is equal to:

$$S_n = \quad d_i^n N_i$$

where N_i is the number particles present and d_i is the particle diameter in size class i. These moments have no physical meaning but are particularly useful as auxiliary parameters in the calculation of characteristic numbers of size distribution. Some common parameters characterizing mean globule size are given in Table 5.2, as are means and ranges of such values for bovine milk. The specific surface area of the fat globules, A, can be derived from the volume surface-weighted mean diameter:

$$A = 6\varphi / d_{3,2}$$

where φ is the volume fraction of milk fat. A typical mean value for A is ~2.2 m^2 g^{-1} fat in unhomogenized bovine milk (range 1.9–2.5 m^2 g^{-1} fat; Walstra 1969b).

As illustrated in Table 5.3, considerable interspecies differences in milk fat globule size have been reported. The smallest fat globules are found in equine milk (Uniacke-Lowe 2011) and the largest are found in buffalo milk (El-Zeini 2006). Compared to bovine milk (4.0 μm, Walstra 1969a and Rüegg and Blanc 1981), $d_{3,2}$ values are lower in caprine (3.2 µm, El-Zeini 2006; 2.76 µm, Attaie and Richter 2000; 2.22 µm Tatar *et al.* 2015), equine milk (1.05 µm, Uniacke-Lowe 2011; 1.3 µm, Devle *et al.* 2012) and asinine milk (1.92 µm, Martini *et al.* 2014) but are higher in ovine milk (4.95 µm, Mehaia 1995 and Gervilla *et al.* 2001), camel milk (4.4 µm, Mehaia 1995 and Farah and Rüegg 1991), yak milk (6.09 µm, Luo *et al.* 2018) and buffalo milk (8.7 µm, El-Zeini 2006). In contrast, $d_{3,2}$ values for mature human milk (4.0 µm, Rüegg and Blanc 1981 and Michalski *et al.* 2005) and reindeer milk (3.95 µm, Uniacke-Lowe and Fox 2011) are similar to those for bovine milk. The mean diameter of human milk fat globule is largest in colostrum, followed by mature milk and smallest in transitional milk (Rüegg and Blanc 1981; Michalski *et al.* 2005; Zou *et al.* 2012).

Milk fat globule size greatly impacts on the formation of milk fat clusters, as in creaming and cold agglutination (see Sections 5.7 and 5.8, respectively, below), which in turn affects the physical stability of milk and dairy products. In general, the smaller the fat globule, the more stable the milk is (Truong *et al.* 2016).

In bovine milk, the average milk fat globule size decreases with advancing lactation (Walstra 1969a) and is positively correlated with the fat content of the milk (Wiking *et al.* 2003) and daily fat yield (Wiking *et al.* 2004). For other species, a positive correlation between fat content and fat globule size has also been reported (El-Zeini 2006), with some exceptions. Equine milk has only 1.4% fat and small fat globules of ~1 µm (Uniacke-Lowe 2011). Similarly, asinine milk is low in fat (1.5%, Malacarne *et al.* 2002) with fat globules of ~1.9 µm (Martini *et al.* 2014). On the other hand, buffalo milk has ~7.5% fat (Ménard *et al.* 2010), and this can be as high as 15% fat depending on lactation stage (Abd El-Salam and El-Shibiny 2011; Ahmad 2013), with fat globules of up to 8.7 µm in diameter (El-Zeini 2006). Reindeer milk has 21–22% fat at peak lactation (Gjøstein *et al.* 2004; Luick *et al.* 1974); but in this case, the fat globules are similar in size to those of bovine milk (Table 5.3). The reason for the formation of larger fat globules during the synthesis of milk fat is important and may be related to limitations in the production of MFGM when fat globules are enveloped during their secretion from the epithelial cells of the mammary gland (Wiking *et al.* 2004). Thus, the level of available MFGM may be a limiting factor in the formation of small fat globules in some high-fat content milks such as buffalo milk.

According to the Laplace equation, large globules have lower stability against rupture and a lower resistance to deformation and coalescence under mechanical pressure than the small globules (Walstra *et al.* 2006). Therefore, the larger fat globules in buffalo milk indicate that the globules are more prone to disruption during churning (Hammad 1993). Similarly, yak milk is reported to be suitable for making butter and ghee products due to its large fat globules (Luo *et al.* 2018). The Laplace pressure of buffalo milk is reported to be 0.8–1.6 kPa, compared to 1.1–2.2 kPa for bovine milk (Ménard *et al.* 2010).

Milk fat globule size can also be influenced by several treatments applied to milk. Homogenization, as discussed in Section 5.12, is a mechanical treatment which has long been applied by milk processors to reduce fat globule size and prevent creaming during storage of liquid milk. Van Boekel and Folkerts (1991) reported that batch heating or indirect ultra-high temperature (UHT) heating of milk at 90–150 °C did not influence d_{vs}, whereas direct UHT heating reduced d_{vs} progressively with increasing temperature. Treatment of milk or cream at a high hydrostatic pressure (up to 600 MPa) has little effect on milk fat globule size (Dumay *et al.* 1996; Gervilla *et al.* 2001; Huppertz *et al.* 2003).

5.3 Differences in the Composition of Milk Fat Globules

Walstra and Borggreve (1966) reported that in milk from a single milking of a single cow, considerable differences in refractive index existed between milk fat globules of similar diameter, indicating differences in the composition of the globules. Furthermore, the observation by Walstra (1967) that the fat in a small proportion of the fat globules in milk melts at a temperature considerably higher than the average melting point (37 °C) also indicates compositional differences between fat globules.

It is now accepted that the composition (i.e. fatty acid composition of the triacylglycerides) of fat globules also varies with globule size. Timmen and Patton (1988) found less $C_{4:0} - C_{10:0}$ and $C_{18:0}$ fatty acids and more $C_{18:1}$ in smaller than in larger fat globules. The fatty acid composition of globules also differs with season; the $C_{18:1}$ and $C_{18:2}$ acid content of milk obtained in winter increases with fat globule size, but a reverse effect is observed in spring milk; in winter, the content of $C_{14:0}$ and $C_{16:0}$ fatty acids decreased with fat globule size (Briard *et al.* 2003). In both spring and winter, there was significantly more $C_{14:0}$, $C_{16:1}$ and less $C_{18:0}$ fatty acids in small fat globules compared to large globules (Briard *et al.* 2003). Higher levels of $C_{18:1}$ and $C_{18:2}$ in small compared to large globules in spring milk can be explained partially by the fact that the fat globule membrane, which represents a larger proportion of the mass of smaller globules, contains a higher proportion of these fatty acids than bulk fat (Jensen and Nielsen 1996); however, the higher proportion of $C_{18:1}$ and $C_{18:2}$ in the membrane alone cannot fully explain compositional differences between globules of different size; thus, it may be assumed that their level in the fat core is also higher (Briard *et al.* 2003). Wiking *et al.* (2004) reported a positive correlation between average milk fat globule size in milk and the concentration of C_{16}, $C_{16:1}$, C_{18} and $C_{18:1}$ fatty acids.

5.4 Fat Crystals in Globules

Crystallization of fats (triglycerides) is a complex phenomenon, especially for milk fat, due to its very broad fatty acid composition (see Chapter 1). Principles of crystallization of milk fat have been reviewed extensively elsewhere (Mulder and Walstra 1974; Walstra *et al.* 1995; Chapters 7 and 8). Whether a quantity of milk fat is present as a continuous mass (e.g. anhydrous milk fat or butter oil) or in numerous small globules (e.g. as in milk or cream) has a considerable influence on its crystallization behaviour. The crystallization state affects many properties of the milk fat globules, e.g. their susceptibility to partial coalescence and their resistance against disruption. Some reasons why crystallization of fat in globules may differ from that in bulk milk fat are (Mulder and Walstra 1974):

1. Heat dissipation in bulk fat is considerably slower than in milk or cream; this is related to the lower thermal conductivity of bulk fat and, in particular, the fact that bulk fat cannot be agitated efficiently.
2. Not all fat globules contain the catalytic impurities required to start heterogeneous nucleation, so that nuclei would have to form spontaneously in those globules. Söderberg *et al.* (1989) observed that deeper supercooling was necessary to induce crystallization in milk fat globules than in bulk fat, whereas Lopez *et al.* (2002a) observed that, with decreasing globule size in cream, a deeper super-cooling was required for crystallization of milk fat inside the globules.
3. The surface layer of the fat globule may act as a catalytic impurity, e.g. when it contains mono- or diglycerides with long-chain fatty acid residues; however, there is still some uncertainty as to whether this process actually occurs (see Walstra 1995). Although concentric layers of apparently crystalline fat have been observed in electron micrographs of freeze-etched or freeze-fractured milk or cream samples (Buchheim 1970; Henson *et al.* 1971), these observations could not be confirmed using other microscopy techniques. Noda and Yamamoto (1994) reported that it is thermodynamically favourable for fat crystals to be located at the oil/water interface, rather than in the interior of the droplet, which may explain the presence of fat crystals at the membrane.
4. The composition of bulk fat is uniform, but differences from globule to globule are known to occur (see Section 5.3); consequently, considerable differences may occur in the final melting point of the fat between different globules. Shi *et al.* (2001) proposed that a higher ratio of higher-melting triglycerides to lower-melting ones in bulk milk fat was associated with faster crystallization rates.

The dispersed state has a considerable effect on fat crystal polymorphism. Lopez *et al.* (2000, 2001c) observed that fat crystallization in milk fat globules is more disordered than in bulk fat. On slow cooling, milk fat crystallizes in the α form in cream (Lopez *et al.* 2001a), whereas, in anhydrous milk fat, it crystallizes in the β′ form and then in the α form (Lopez *et al.* 2001b). Rapid cooling of cream or anhydrous milk fat from 60 to 4 °C led to the formation of α crystalline structures which transformed into β′ structures rapidly in anhydrous milk fat, and more slowly in cream. On prolonged storage, these crystal structures evolve further, leading to the co-existence of α, β′ and β structures (Lopez *et al.* 2002b). Furthermore, Lopez *et al.* (2002a) observed a greater disorder and smaller size of the fat crystals in milk fat globules with reduced fat globule size.

Crystallization of milk fat in globules is also influenced by exposure to high hydrostatic pressure. High pressure (HP) treatment at 100–500 MPa at 23 °C induces crystallization of milk fat within the globules, and crystallization proceeds further during storage at 23 °C (Buchheim and Abou El-Nour 1992; Buchheim *et al.* 1996). Acceleration of crystallization of milk fat by HP treatment is due to a shift of the solid/liquid transition temperature towards a higher value (Frede and Buchheim 2000). HP-induced crystallization of milk fat was strongly delayed by a reduction in fat globule size (Buchheim *et al.* 1996). These HP-induced changes in crystallization behaviour of globular milk fat may offer opportunities to overcome the necessity for super-cooling to obtain particular levels of crystalline fat.

5.5 Colloidal Interactions

Colloidal interactions form the basis of several of many of the properties of emulsions, as well as the changes observed in emulsions over time; such interactions govern whether droplets remain as separate entities or aggregate. In this section, a brief overview of the predominant colloidal interactions of importance for the stability of emulsions of milk fat globules is given.

The interactions between two emulsion droplets can be described in terms of the interaction energy, or inter-droplet pair potential $w(h)$, which is the energy required to bring two emulsion droplets from an infinite distance apart to a surface-to-surface separation difference, h (McClements 1999):

$$w(h) = w_{\text{attractive}}(h) + w_{\text{repulsive}}(h)$$

If attractive forces dominate at all separations, $w(h)$ is always positive, and the interaction energy, i.e. the free energy needed to bring two droplets from an infinite distance closer together, will be negative, and the droplets will tend to aggregate. Conversely, if repulsive forces dominate at all separations, and the positive interaction energy is several times larger than the average kinetic energy involved in the encounter of two particles by Brownian motion, droplets tend to remain as individual entities. In many cases, however, $w(h)$ is neither positive nor negative over the entire distance h.

The classical DLVO (Derjaguin-Landau-Verwey-Overbeek) theory (Derjaguin and Landau 1941; Verwey and Overbeek 1948) states that the stability of a colloidal suspension essentially depends on two independent interactions between colloidal particles: van der Waals attractions and electrostatic repulsion:

$$w(h) = w_{\text{van der Waals}}(h) + w_{\text{electrostatic}}(h)$$

Van der Waals forces are attractive forces which act between all molecules; they arise from the attraction between electrostatically or orientationally polarized molecules. Their strength decreases with droplet separation, increases with droplet size and depends on the physical properties of the droplets and the surrounding medium and on the thickness and composition of the absorbed emulsifier layer (Bergenståhl and Claesson 1997; Friberg 1997; McClements 1999).

Electrostatic interactions occur between molecules that contain a permanent electrical charge, such as ions and polar molecules. The approach of two identically charged surfaces leads to increases in the counter-ion concentration between the surfaces, which generate a repulsive force as a result of increased osmotic pressure (Dickinson and Stainsby 1988; Bergenståhl and Claesson 1990, 1997).

The surface charge of milk fat globules cannot be measured directly, but the zeta potential (ζ), which approximates the potential at a certain distance from the particle surface, can be measured electrokinetically (Tuinier and De Kruif 2002). The ζ of milk fat globules can be defined as the potential at the shear layer, indicating the degree of globule surface coverage by plasma proteins (Michalski *et al.* 2002). The surface charge, as estimated by ζ, is ~ −13 to −14 mV for unhomogenized bovine milk fat globules (Jack and Dahle 1937; Payens 1963, 1964; Michalski *et al.* 2001b) and ~ −20 mV after homogenization (Wade and Beattie 1997; Michalski *et al.* 2001b). Dalgleish (1984) reported slightly lower values for ζ, i.e. −10 mV for unhomogenized and − 13 to −17 mV for homogenized milk fat globules. The overlap of electric double layers will cause a local increase in potential, implying that work must be performed to bring particles closer together.

Thus, according to the DLVO theory, aggregation of milk fat globules should occur if the van der Waals attraction is larger than the electrostatic repulsion. However, calculation of these forces for milk and application of the data to the DLVO theory results in a negative interaction energy at all distances (Walstra 1995), so that immediate aggregation of milk fat globules should be observed. Aggregation of fat globules, however, does not occur, even when electrostatic interactions are minimal. Thus, there must be a second repulsive force acting, i.e. steric repulsion; the DLVO theory may then be extended to:

$$w(h) = w_{\text{van der Waals}}(h) + w_{\text{electrostatic}}(h) + w_{\text{steric}}(h)$$

Repulsive steric forces are encountered when the outer segments of two polymer-covered surfaces begin to overlap. These interactions usually lead to a repulsive force due to the unfavourable reduction in entropy associated with confining the chains between surfaces (Tadros and Vincent 1983; Israelachvili 1992; Walstra 1996). In the case of milk fat globules, steric repulsion is provided by glycoproteins of the milk fat globule membrane, which have highly hydrophilic moieties protruding from the globule surface. Hydrolysis of these glycoproteins by papain causes aggregation of milk fat globules (Shimizu *et al.* 1980).

From a colloidal stability point of view, milk fat globules with high ζ potential are electrically stabilized, while globules with low ζ potentials tend to coagulate or flocculate (Zou *et al.* 2012). Human milk fat globules have a lower ζ potential (−7.8 mV, Michalski *et al.* 2001b; −7.6 mV, Lopez and Menard 2011 or −7.25 mV, Zou *et al.* 2012;) than the ζ potential of milk fat globules from other mammals such as equine (−10.3 mV, Uniacke-Lowe 2011), bovine (−13.5 mV, Michalski *et al.* 2001b), yak (−11.9 mV, Luo *et al.* 2018), buffalo (−11.0 mV, Ménard *et al.* 2010), and reindeer (−12.5 mV, Uniacke-Lowe and Fox 2011) milk fat globules. The ζ potentials for human colostrum, transitional and mature milk fat globules are −5.60, −6.72 and − 7.25 mV, respectively, indicating an increasing trend as lactation progresses (Zou *et al.* 2012).

The differences in ζ potential of human milk fat globules at different stages of lactation or from different species are largely due to differences in the composition of polar lipids and proteins in the MFGM, and minerals present in the aqueous environment. However, the low ζ potential of human milk fat globules may be of special benefit to the digestion and metabolism of human milk fat by the infant (Zou *et al.* 2012). The fat globules of reindeer milk, which are similar in size to those in bovine milk (Table 5.3), form an exceptionally stable emulsion and show little or no tendency to flocculate compared to those of bovine milk, which is attributed to the higher zeta potential of reindeer fat globules, reflecting significant differences in the electrical character of the surface layers of the fat globules of reindeer and bovine milk (Aikio *et al.* 2002).

5.6 Separation of Milk

Because milk fat globules have a lower density than milk plasma, they tend to rise under the influence of a gravitational force. For perfect spheres, the rate of rise, v, is given by Stokes' Law:

$$v = a\left(\rho_p - \rho_f\right)d^2 / 18\eta_p$$

where a is the acceleration due to the gravitational force, ρ_p is the mass density of the plasma, ρ_f is the mass density of the fat, d is the diameter of the fat globule and η_p is the viscosity of the plasma. For gravity creaming, $a = g \approx 9.8$ m s^{-2}. For creaming in a centrifugal field, $a = R^2\ \omega$, where R is the effective centrifugal radius, and ω is the angular velocity (= $2\pi n/60$, where n is the number of revolutions per minute).

To predict v correctly, several prerequisites must be met (Mulder and Walstra 1974; Walstra and Oortwijn 1975; Walstra 1995), most notably: (1) globules must be perfect and homogeneous spheres; (2) other particles in the plasma must be considerably smaller than the fat globules; (3) Brownian motion must be small compared to the rate of rise; (4) counter-flow of liquid due to globule movement must be negligible; and (5) mutual interaction between globules must be absent. Troy and Sharp (1928) found that in milk highly diluted with milk plasma, the rate of rise of individual milk fat globules, as well as roughly spherical clusters of milk fat globules, correlated well with Stokes' law. However, Walstra and Oortwijn (1975) observed that the rate of rise of fat globules in undiluted milk systems under the influence of gravity was lower than predicted by Stokes' law, in particular for milk of high fat content or containing small fat globules.

The creaming rate (defined as the proportion of the fat arriving in the cream layer per unit time) is proportional to the creaming parameter, H (Walstra and Oortwijn 1975):

$$H = S_5 / S_3 = \ N_i d_i^5 / \ N_i d_i^3$$

This parameter shows a linear relation to the creaming rate if the effect of aggregation of the globules is excluded (Rüegg and Blanc 1981); it can be seen that larger globules, in particular, affect H, and thus the creaming rate.

The presence of clusters of fat globules affects creaming considerably. Such clusters will rise faster than the individual globules because of their larger size. Clusters may be formed due to cold agglutination (see Section 5.8) or due to inefficient homogenization (i.e. formation of homogenization clusters, see Section 5.13). Also, small clusters of fat globules may be formed during sterilization of heat-evaporated milk at the onset of heat-induced coagulation (Schmidt *et al.* 1971).

The separation of milk can be significantly accelerated by application of a centrifugal force, which is the principle of separation (skimming) of milk in industrial practice; the design of a separator is depicted in Figure 5.1. The objective of centrifugal separation is to achieve the lowest possible fat content in the skimmed milk while removing the fat greatly (~tenfold) concentrated in a cream phase. With the exception of high-fat products, Stokes' law can be applied rather accurately to the rate of rise of milk fat globules in a centrifugal field. Centrifugal separation is more efficient at elevated temperatures, as the factor $(\rho_p - \rho_f)/\eta_p$ increases more than tenfold in a linear fashion over the temperature range 0–80 °C (Mulder and Walstra 1974). The fat content of the skimmed milk depends on the proportion of the fat in very small globules (e.g. <1 μm), which are the most difficult to separate, and the level of non-globular fat.

5.7 Physical Instability of Emulsions

The stability of an emulsion denotes its ability to resist changes in its properties over time, i.e. a higher emulsion stability implies slower change in emulsion properties. When considering the stability of an emulsion, it is of major importance to distinguish between thermodynamic stability and kinetic stability. Thermodynamics predict whether or not a process will occur, whereas kinetics predict the rate of the process,

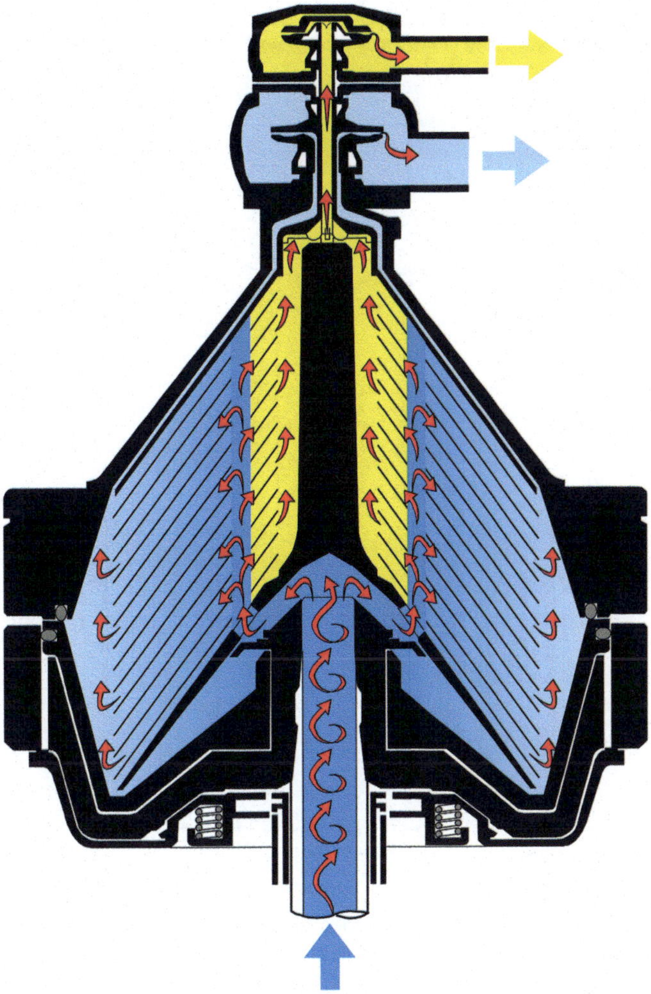

Figure 5.1. Principle of operation of a centrifugal milk separator. (Reproduced with permission from *Dairy Processing Handbook*, Tetra Pak Processing Systems AB, Lund, Sweden, 1995).

if it does occur. All food emulsions are thermodynamically unstable and will break down if left long enough; thus, it is kinetic stability which is responsible for the occurrence of the different types of instability that are observed in food emulsions.

Instability of an emulsion may be physical or chemical in nature. Chemical instability, which results in an alteration in the chemical structure of the lipid molecules due to oxidation or hydrolysis (McClements 1999), will not be covered in this chapter; for more information, the reader is referred to Chapter 11. Physical instability results in an alteration in the spatial distribution or structural organization of the globules (i.e. the dispersed phase of the emulsion). The key mechanisms responsible for the physical instability of emulsions, as depicted in Figure 5.2, can be divided into two categories: gravitational separation and droplet aggregation.

Gravitational separation involves the movement of emulsion droplets due to the fact that they differ in density from the surrounding liquid. If the droplets have a lower density than the surrounding medium, they tend to move upwards, a process referred to as *creaming*. Conversely, droplets or particles that have a density higher than the surrounding medium tend to move downwards under the influence of a gravitational force, i.e. *sedimentation*.

Droplet aggregation is said to occur when droplets stay together for a time much longer than they would in the absence of colloidal interactions (Walstra 2003), i.e. than can be accounted

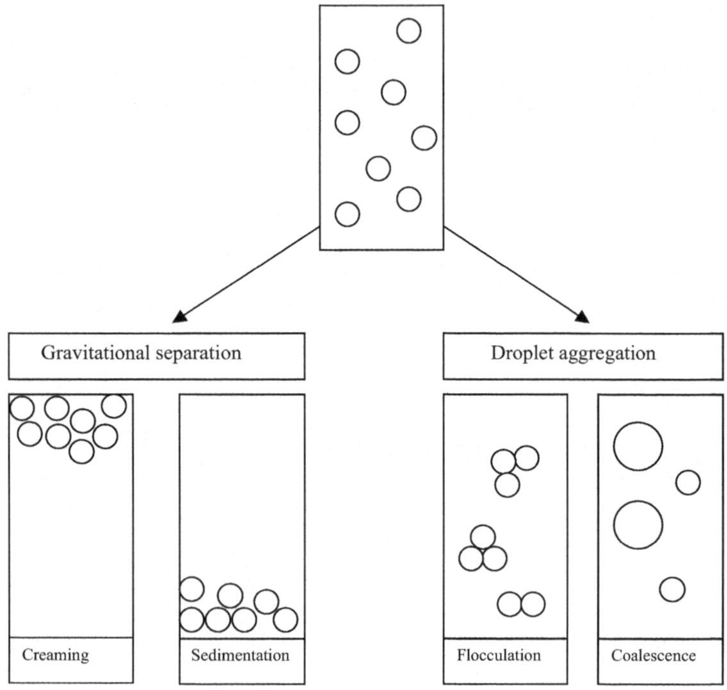

Figure 5.2. Schematic overview of types of instability of emulsions.

for by collisions due to Brownian motion. Mechanisms responsible for the physical instability of droplets through aggregation are flocculation, coalescence or partial coalescence.

- *Flocculation* of droplets is defined as the aggregation of droplets to give three-dimensional floccules, wherein the droplets remain as individual entities (Tadros and Vincent 1983). Flocculation can be distinguished from coagulation by the fact that the former denotes weak, reversible interactions, whereas the latter denotes strong, and often irreversible, interactions (Walstra 2003). Flocculation occurs as a result of collision; the extent of flocculation is determined by both the total number of droplet collisions per unit time per unit emulsion volume and the likelihood that an encounter between droplets will lead to aggregation. The most effective way to control the rate and extent of flocculation is by regulating the colloidal interactions between the droplets.
- *Depletion flocculation* of droplets can occur in samples containing different-sized colloids, wherein it is energetically favourable for the droplets to be surrounded by particles of the same size. This phenomenon is known to exist in systems containing both emulsion droplets and polymers and is known to occur for mixtures of milk and casein-stabilized emulsions (Ten Grotenhuis *et al.* 2003). Depletion flocculation becomes particularly important at small interparticle distances, i.e. when volume fractions of particles are high (e.g. >0.3).
- *Coalescence* is the process by which two or more fat globules merge to form one larger spherical fat globule through the rupture of the liquid film between emulsion droplets. It is the principal mechanism by which an emulsion moves towards its thermodynamically stable state, through a decrease in free energy as a result of the decrease in contact area between the oil and water phases (Tadros and Vincent 1983; McClements 1999; Walstra 1996, 2003). Coalescence of milk fat globules will be discussed in more detail in Section 5.9.
- *Partial coalescence* involves the formation of anisometrically shaped conglomerates of droplets due to the fact that true coalescence is prevented, e.g. because the globules contain a network of crystalline fat (Walstra 1996, 2003;

McClements 1999). The ultimate driving force behind partial coalescence is a decrease in interfacial free energy, although other processes are also involved (Walstra 2003). Partial coalescence of milk fat globules will be discussed in more detail in Section 5.9.

The primary form of destabilization relevant to bovine milk is creaming. The creaming of bovine milk has been the subject of research since, at least, the work of Babcock (1889). The relevant literature has been reviewed by Hammer (1916), Dunkley and Sommer (1944), Mulder and Walstra (1974), Euber and Brunner (1984), Brunner (1978), Walstra and Jenness (1984) and Keenan *et al.* (1988). However, the creaming of only a few other species has been studied, and then only superficially.

5.8 Analytical Methods for Evaluating Creaming of Milk

From an analytical perspective, the simplest method of following gravitational separation and creaming in milk is visual observation. One of the earliest methods of measuring the volume of cream produced by a milk sample consisted of standing bottles of identical size and shape beside each other and comparing the depth of the cream layer or measuring the distance from the top of the bottle to the line dividing the cream from the milk. Harding *et al.* (1922) developed a relatively precise method which is still in use today. Milk samples are poured into round-bottomed test tubes, of precise dimensions, to a specific height. The tubes are cooled in ice water and held at 4 °C for ~20 h, and the depth of the resulting cream layer is measured in mm, where each mm of cream represents a certain proportion of cream by volume (Harding *et al.* 1922).

Creaming can alternatively be defined as the volume of cream produced from a specified volume of milk in a glass tube, of various dimensions, at a stated temperature after certain time intervals, usually up to 24 h. The rate of creaming can be derived from the data obtained from this type of experiment if a sufficient number of sampling times is used. The creaming index can be calculated from the ratio of cream height over the total height of the emulsion upon standing (see Hammer 1916; Dunkley and Sommer 1944; Keynon *et al.* 1966; Euber and Brunner 1984).

Farah and Rüegg (1991) adapted the methods above and added 2 drops of nigrosine solution per 100 mL milk in a creaming assay on camel milk, carried out in graduated measuring cylinders, which allowed clear optical distinction between the aqueous (blue) and fat (white) phases. The relatively simple method of Speroni and Bertoni (1984) to measure natural creaming in milk is frequently used prior to the production of Gruyère, Parmigiano-Reggiano and Grana Padano cheeses when milk is allowed to cream naturally before cheese production (e.g. Abeni *et al.* 2005). Milk (14 mL) is placed in plastic tubes for 3 h in flowing water at 15 °C. Fat on the surface is removed by aspiration, and then 2 mL of milk is aspirated from the bottom of the tube using a syringe and analysed for milk fat content. Natural creaming is calculated as:

$$\%\text{natural creaming} = (\text{MF1} - \text{MF2}) / \text{MF1} \times 100$$

where MF1 = milk fat content before natural creaming and MF2 = milk fat content after natural creaming.

In the last decade, dispersion stability analysers such as the LUMiSizer (LUM, Berlin, Germany) and the Turbiscan (Formulaction, Toulouse, France) have been used to assess and quantify creaming in milk samples. Both instruments use a near-infrared (NIR) light source to detect instabilities such as creaming or sedimentation in samples by scanning the entire length of a sample over a defined time period, at a defined temperature, which can typically be varied (in a range from 4 to 60 °C); however, the principles of how they operate differ, with the LUMiSizer accelerating physical instability by application of centrifugal force (200 to 4000 rpm), whereas the Turbiscan studies separation under gravity.

The LUMiSizer can also perform high-resolution particle size analysis in the range of 10 nm to 300 μm (depending on physical properties), while particle sizes of 1 mm to 10 nm can be measured by the Turbiscan. The migration rate of particles is dependent on Stokes Law, and hydrodynamic diameter can thus be calculated from the migration rate.

The LUMiSizer uses artificial conditions to monitor sample stability, especially for forms of instability such as creaming where fat globules are in weak equilibrium within a liquid phase, and can predict broad stability under accelerated conditions, where instability is not dependent on longer time-scale changes in a sample. The Turbiscan, on the other hand, does not use external stress and evaluates stability as the sample is in a natural state. Increasing temperature rather than using an external force is regarded by some researchers as being preferable for acceleration of destabilization processes while maintaining realistic testing conditions.

5.9 Cold Agglutination

When bovine milk is stored in the cold under quiescent conditions, a cream layer will form rapidly due to the rise of milk fat globules under the influence of gravity, and the rate of separation is faster and the mobility of large fat globules faster at 4 °C than at higher temperatures (Ma and Barbano 2000). The rate of rise of milk fat globules during creaming is considerably faster than can be accounted for by Stokes' law for individual globules (Troy and Sharp 1928). A cream layer may be evident in bovine milk within 20 min of milking. However, the appearance of a cream layer, if formed as a result of the rise of individual fat globules of ~4 μm diameter, according to Stokes' equation, would take approximately 50 h (Fox *et al.* 2015). This is due to the fact that milk fat globules tend to rise in large clusters, which rise at a considerably faster rate than individual globules. Merthens (1933b) reported that addition of colostrum to milk enhanced creaming considerably, suggesting that one or more agents enriched in colostrum promoted creaming. The clustering of milk fat globules during cold storage markedly resembles the agglutination of bacteria or red blood cells, due to the action of the immunoglobulin IgM, in terms of dependence on pH, concentration and valency of cations. Hence, the clustering of milk fat globules in the cold is referred to as cold agglutination.

In terms of understanding the mechanism for cold agglutination of milk fat globules, two of the most important phenomena are the 'Merthens effect' and the 'Samuelson effect'. Merthens (1933a) observed that milk recombined from homogenized skim milk and unhomogenized cream has poor creaming ability ('Merthens effect'). It was proposed initially that this is due to denaturation of the agglutinin on homogenization, but Koops *et al.* (1966) showed that this was not the case. Samuelson *et al.* (1954) showed that two components are required for cold agglutination: a homogenization-labile component and a heat-labile component ('Samuelsson effect'). Homogenization at a pressure as low as 1 MPa, or even mild shearing, impairs the agglutinating tendency of skimmed milk (Walstra 1980). The euglobulin fraction of milk, implicated by many early investigators as the agglutinin, associates with the milk fat globules, particularly in the cold, and is not homogenization-labile (Payens 1964). Subsequent studies (Payens 1964, 1968; Payens *et al.* 1965; Gammack and Gupta 1967; Payens and Both 1970; Stadhouders and Hup 1970) identified immunoglobulin M (IgM) as the heat-labile agglutinin in the euglobulin fraction of milk. Gammack and Gupta (1970) showed that lipoprotein particles in the aqueous phase are a prerequisite for the rapid creaming of milk, which supports the earlier observations by Hansson (1949) that creaming of milk is enhanced by the addition of phospholipids.

Euber and Brunner (1984) proposed a mechanism for cold agglutination which involves three components: (1) the milk fat globules, (2) IgM, the heat-labile component, which functions as a cold agglutinin; and (3) the so-called skim milk membrane (SMM), the homogenization-labile component, consisting of lipoprotein particles present in the aqueous phase of milk. Euber and Brunner (1984) suggested that these components interact through specific carbohydrate moieties. IgM can interact with both SMM and the fat globules, whereas SMM interacts with IgM only. Fat globules can be clustered to a limited extent by IgM alone, but clustering is considerably more extensive in the presence of SMM, which acts as a cross-linking agent. Environmental factors that affect the uptake of IgM by fat globules or SMM include ionic strength, dielectric constant, pH and the temperature of the suspending medium (Euber and Brunner 1984). Based on studies using confocal microscopy and fluorescent labelling of immunoglobulins, Hansen *et al.* (2019) reported that immunoglobulin-driven agglutination of milk fat globules was more extensive at 5 °C than at 15 °C and hardly occurred at 20 or 37 °C. The key mechanism involved appeared to involve the presence of an immunoglobulin receptor on the MFGM, which was not found at warm temperatures.

Cold agglutination is influenced also by processing conditions. Agitation of milk during cold storage impairs creaming, but heating milk to 40–50 °C normally restores the creaming capacity of the milk on cold storage (Merthens 1933a). Heating milk at a higher temperature, up to ~62 °C, improves the creaming capacity, relative to that of fresh milk (Rowland 1937). A similar increase in creaming capacity was observed after high-pressure treatment at 100–250 MPa by Huppertz *et al.* (2003), who showed that clusters of milk fat globules formed on cold storage of milk treated at 200 MPa were larger than those formed in unpressurized milk. However, the exact mechanism for heat or HP-induced increases in creaming of milk has not yet been described.

Heating milk at >62 °C (Orla-Jensen *et al.* 1929; Rowland 1937), or treating it at a pressure ≥ 400 MPa (Huppertz *et al.* 2003), impairs the rate of creaming of milk fat globules. Huppertz *et al.* (2003) showed that clustering of milk fat globules on cold storage did not occur in milk treated at 600 MPa. Thermal or high-pressure-induced inhibition of cold agglutination is probably the result of denaturation of IgM; heat-induced denaturation of immunoglobulins reduced agglutination in a manner proportional to the level of denaturation (Hansen *et al.* 2019). Heat-induced interactions of caseins or whey proteins with the MFGM may also prevent cold agglutination (Van Boekel and Walstra 1995). Addition of colostral euglobulin to heated milk restores its creaming capacity (Keynon and Jenness 1958).

The mechanism of gravity separation of cream in milk remains an active topic of research, particularly in the context of pre-treatment of milk for some cheese types (such as Parmigianno Reggiano). D'Inecco *et al.* (2018) reported that partial coalescence occurred in milk creamed at a temperature higher than 22 °C, and that warming milk to 37 °C followed by separation at cold temperatures altered the fat globule size distribution, potentially by affecting IgA- and IgM-mediated interactions of fat globules with clostridial spores; adding purified immunoglobulins to raw milk increased their partition into the cream layer. The role of IgM in the separation of spores through their interaction with fat globules was also demonstrated by Geer *et al.* (2014a), while Caplan *et al.* (2013) proposed that gravity separation of milk could result in reduced-fat milk with decreased bacterial and somatic cell counts.

Interestingly, Geer *et al.* (2014b) reported that during gravity separation, somatic cells in milk were critical for the upward movement of fat globules (and also bacterial spores), and that immunoglobulins alone were not sufficient to cause such separation. They proposed that somatic cells may interact with immunoglobulins and that these then aggregate with fat, with the cells providing buoyancy for the upward separation due to entrapped air.

Clustering of milk fat globules in the cold, followed by rapid creaming, is not a universal phenomenon. As described earlier, it occurs in bovine milk, but not, or to a considerably lower extent, in caprine (Jenness and Parkash 1971), buffalo (Fahmi 1951; Abo-Elnaga 1966; Wahba *et al.* 1977; Ismail *et al.* 1972), camel (Farah and Rüegg 1991) and reindeer (Uniacke-Lowe and Fox 2011; Gonzales-Janolino 1968a) milk. This has been related to the fact that clustering of milk fat globules does not occur in the milks from these species. The clustering of fat globules has been monitored in a thin (1 mm) body of milk between glass plates (Dunkley and Sommer

1944; Keynon *et al.* 1966); the clusters can be seen with the naked eye or photographed. Dispersion analysers such as the Turbiscan (as described above) can be used to accurately monitor clustering of fat globules during creaming of milk. Changes in the particle size due to coalescence can be detected on the basis of the back-scattering signal in the middle zone of the sample cell. If the fat globules grow with time due to agglomeration or coalescence, the number of scattering centres decreases in the middle zone of the sample cell, decreasing the back-scattering signal.

Jenness and Parkash (1971) showed that milk reconstituted from caprine cream and bovine skim milk creams rapidly, whereas milk reconstituted from bovine cream and caprine skim milk shows a very low level of creaming. Similar results were observed on reconstituting cream and skimmed milk from bovine and camel milk (Farah and Rüegg 1991). The poor creaming properties of buffalo milk were attributed to its poor clustering ability (Abo-Elnaga *et al.* 1966; Ahmad 2013). Addition of euglobulin, isolated from buffalo colostrum, considerably increased the creaming capacity of buffalo milk (Wahba *et al.* 1977). Gonzales-Janolino (1968b) observed poor creaming of mixtures of cow's cream and caribou' skimmed milk, whereas a mixture of agglutinin-rich bovine-skimmed milk and caribou' cream creamed extensively. Further experiments showed that caribou milk lacks the homogenization-labile component (Gonzales-Janolino 1968b). Thus, it is apparent that cold agglutination of fat globules in milk is highly dependent on the species of origin.

Ongoing research (authors' unpublished data) has confirmed the significant differences found in the creaming behaviours of fresh milk samples from a variety of species. Milk from seven species (porcine, bovine, human, reindeer, buffalo, goat and camel) was assayed in a LUMiSizer at 14 °C at 500 rpm (~ 36 g). The intensity of transmitted NIR light, measured over the entire length of the samples, was recorded every 120 s for 2.5 h (75 profiles recorded). The slope (%/h) of the integral transmission versus time curves was calculated as an indicator of the initial creaming rate for each sample. The higher the slope value, the more creaming that has occurred under these experimental conditions. Slopes calculated were 0.1147 for porcine milk, 4.6551 for bovine milk, 14.0912 for human milk, 0.2107 for reindeer milk, 0.8383 for buffalo milk, 0.2409 for goat milk and 0.8840 for camel milk. Human milk, under these conditions, creamed significantly more than any of the other samples, while goat and reindeer milk did not cream (Figure 5.3). Furthermore, human milk became more translucent as it creamed, due to the small size and low concentration of human casein micelles rendering the contrast in colour between skim milk and cream phases much greater than in milk of other species. Buffalo and camel milk creamed very poorly, as reported earlier. Freezing milk for periods from 1 day to 2 months had little effect on the creaming of the milk of all species studied once the milk was thawed slowly overnight at 4 °C.

5.10 Coalescence and Partial Coalescence

When two or more emulsion droplets come into contact, a thin film of the liquid continuous phase forms between them. Coalescence is the process whereby liquid droplets merge to form a single larger droplet as a result of the rupture of both this liquid film and the interfacial membranes of the droplets. Coalescence moves an emulsion towards a thermodynamically stable state, because it involves a decrease in the contact area between the phases (Tadros and Vincent 1983; Walstra 1996; McClements 1999).

The current state of understanding of coalescence is unsatisfactory because of the number of variables involved and the fact that some fundamental problems have not been resolved fully (Walstra 2003), but some general understanding has been obtained. The susceptibility of droplets to coalescence is determined by the nature of the forces that act on and between the droplets and the resistance of the droplet membranes to rupture. Coalescence may be induced by collisions or by prolonged contact between the emulsion droplets. Collision-induced coalescence can be

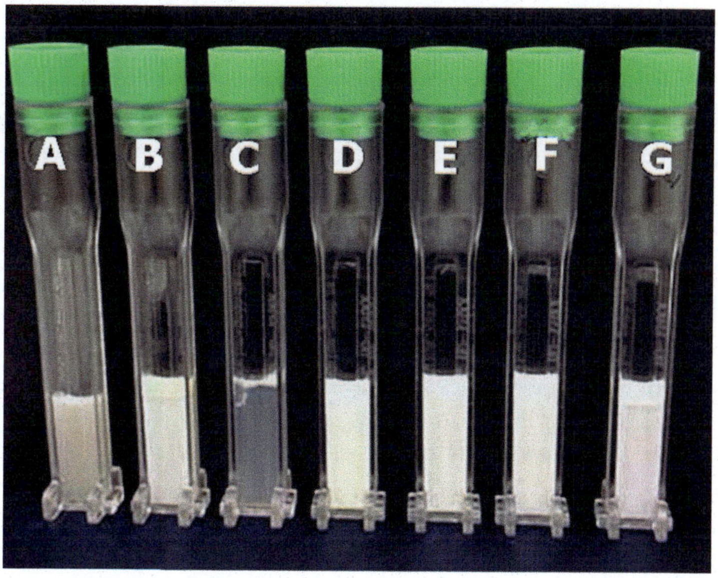

Figure 5.3. Images showing the creaming ability of the milk from seven species after analysis using a LUMiSizer at 14 °C for 2.5 h at 500 rpm. *From left*: A, porcine; B, bovine; C, human; D, reindeer; E, buffalo; F, goat and G, camel milk.

due to movement of the droplets by Brownian motion, gravity or applied mechanical forces. Coalescence induced by prolonged contact occurs spontaneously after the droplets have been in contact for a sufficient period, e.g. in emulsions which contain flocculated droplets or droplets that have accumulated at the top or bottom of the emulsion due to gravitational separation, respectively. The probability of film rupture will be larger if the interfacial tension, γ, is small, and if the colloidal repulsion between the droplets is stronger (Walstra 2003). Furthermore, susceptibility to coalescence increases with droplet size.

Shimizu *et al.* (1980) reported that removal of the polar head of phospholipids in the milk fat globule membrane by phospholipase C results in oiling-off; thus, it appears that the polar head of the phospholipids plays an important role in the stability of milk fat globules against coalescence. It is notable that several species of psychrotrophic bacteria isolated from raw milk can produce phospholipases, which may contribute to spoilage phenomena in milk (Sadiq *et al.* 2016; Vithanage *et al.* 2016).

Indirect UHT treatment can cause aggregation of fat globules due to partial heat-induced coagulation; direct UHT treatment, which involves greater turbulence and flash boiling, does not cause aggregation (Melsen and Walstra 1989). Mulder and Walstra (1974) reported that coalescence of fat globules in cream may occur during treatment in a heat exchanger, but Van Boekel and Folkerts (1991) could not confirm this for direct or indirect UHT treatment of unhomogenized milk. Streuper and Van Hooijdonk (1986) observed coalescence on UHT treatment of milk, but only if back-pressure in the apparatus allowed some boiling of the liquid.

Whereas true coalescence is of limited importance in the case of milk and dairy products, partial coalescence is of far greater importance, in particular in the preparation of products such as whipped cream, butter and ice cream. Partial coalescence occurs when two or more partially crystalline emulsion droplets come into contact. A fat crystal protruding from a globule may pierce the film between close globules, which leads to conjunction of the globules, resulting in the formation of an irregularly shaped aggregate (Van Boekel and Walstra 1981). The aggregate partially retains the shape of the globules from which it was formed, because the fat crystal network within the droplets prevents a complete merger. Partial coalescence differs from true coalescence in that it tends to be much faster and that, due to the formation of irregular aggregates or clumps, it increases the effective volume fraction of the dispersed phase. Prerequisites for partial coalescence include the presence of a network of fat crystals in the globules (Boode *et al.* 1993) and that the fat crystals are located at the oil–water interface; if fat crystals are totally wetted by either the oil or water phase, they do not affect emulsion stability (Boode and Walstra 1993).

In the case of milk fat globules, partial coalescence can lead to the formation of irregularly shaped granules, e.g. butter clumps, or the formation of a continuous network, e.g. in whipped cream or ice cream. Walstra *et al.* (1999) reported that the following factors affect the rate of partial coalescence in milk:

1. *Application of a velocity gradient* of shear rate increases the rate of collision between fat globules and presses globules closer together, thus enhancing the possibility of a protruding crystal bridging the gap between globules. However, above a certain velocity, the rate of partial coalescence decreases (Boode *et al.* 1993).
2. An increased *fat content* increases the rate of clumping.
3. The *proportion of solid fat* is crucial. Partial coalescence cannot occur if there are no fat crystals, but if there is too much solid fat, there may not be enough liquid fat to hold globules together (Boode *et al.* 1993).
4. *Fat globule size* also influences the extent of partial coalescence. Larger globules are less stable against partial coalescence, due to the fact that they may have larger fat crystals and the probability of a crystal sticking out far enough is thus higher;
5. The *surface layer on the globules* plays an important role. Natural fat globules are reasonably stable, but the presence of a surface layer of protein, e.g. after homogenization or recombination, increases the stability of the globules considerably through colloidal repulsion.

Partial coalescence is probably also involved in a defect in unhomogenized milk that is not kept at a sufficiently low temperature, referred to as 'bitty cream' or 'broken cream'. Bacterial phospholipases can hydrolyse up to 60% of the phospholipids on the milk fat globule membrane (O'Mahony and Shipe 1972), making the globules more susceptible to partial coalescence, which leads to the formation of large particles of cream floating in the milk (Stone 1952a, b; Stone and Rowlands 1952; Labots and Galesloot 1959). In market milk, the bitty cream defect has been largely eliminated by homogenization.

5.11 Rebodying

The term rebodying is often used for the phenomenon whereby cooled cream, when warmed (e.g. to 30 °C) and subsequently re-cooled, becomes more viscous, or even (at a sufficiently high fat content) solid-like (Hoffmann 1999). Rebodying is caused by partial coalescence (Oortwijn and Walstra 1982b) and occurs in unhomogenized or weakly homogenized systems, in which much of the natural MFGM is retained. The extent and nature of the rebodying phenomenon depends on the rate of cooling (which determines whether the number [fast cooling] or size [slow cooling] of the fat crystals increases) and the temperature history of the cream, particularly cycling through higher and lower temperatures. Repeated rebodying can cause fat separation. Warming cream that has undergone rebodying to 30 °C increases fat globule size, probably as a result of full coalescence of globules that are already in the partially coalesced state (Oortwijn and Walstra 1982b).

In UHT-treated whipping cream, rebodying results in clumping of fat or the formation of cream plugs (small lumps of partially solidified fatty material), which cannot be redispersed in the product by gentle shaking. The fat droplets in the cream plug are aggregated and partially coalesced. Streuper and Van Hooijdonk (1986) reported that the firmness of the plug in UHT-treated cream increased with increasing rate of cooling of the cream. The formation of a cream plug in heat-treated unhomogenized cream can be prevented completely by addition of carrageenan, in combination with protein-fat powder (Precht *et al.* 1987). Dickinson *et al.* (1989) introduced the term 'cohesive cream' to describe a concentrated emulsion layer in which the flocculated oil or fat droplets have become compressed into a coherent structure that cannot be redispersed by mild agitation; these authors also reported that the formation of cohesive cream in liqueurs is enhanced at a low pH, a high calcium content or a low level of caseinate emulsifier, as well as by temperature fluctuations during storage.

Studies on model creams have further clarified the possible mechanism for rebodying. For an increase in viscosity to occur on re-cooling, it is necessary that, after the warming step, <10% of the fat remains solid; if all fat is melted, rebodying does not occur (Boode *et al.* 1991; Noda and Yamamoto 1994; Mutoh *et al.* 2001). Sugimoto *et al.* (2001) observed that the increase in viscosity on re-cooling warmed cream is accompanied by a substantial increase in the concentration of protein on the fat globule surface and proposed the following mechanism for an increase in viscosity on re-cooling of warmed cream. At a critical level of solid fat (<10%) in cream, fat crystals approach the oil droplet surface, which causes conformational changes in the proteins absorbed at the oil-droplet surface. A rapid decrease in the fluidity of triacylglycerides on cooling causes further changes in the conformation and charge of the surface proteins, which leads to attraction between serum proteins and those on the surface; this results in an increase in viscosity and solidification of the cream. However, further studies may be necessary to establish if this mechanism also applies to dairy cream.

5.12 Factors That Affect the Surface Layers of Fat Globules in Milk and Cream

The surface layers of the fat globules in milk are affected by various treatments. Effects of homogenization are described in Sections 5.13 and 5.14; however, other treatments such as cooling and heating, as well as environmental conditions, also influence the surface layers of milk fat globules and are described in this section.

Cooling of milk induces the release of up to 15% of phospholipids from the MFGM (Koops and Tarassuk 1959; Baumrucker and Keenan 1973), resulting in an increase in the phospholipid content of the milk plasma (Patton *et al.* 1980). Cooling also causes a shift in xanthine oxidase activity from the fat to the plasma phase and results in the reversible adsorption of the cryoglobulins onto the fat globules (Mulder and Walstra 1974). Furthermore, cooling induced the migration of copper from the milk fat globules to the milk plasma (Mulder and Walstra 1974). Freezing and subsequent thawing cause considerable clumping of milk fat globules, particularly in cream, primarily caused by mechanical pressure differences in the frozen products (Mulder and Walstra 1974).

Heat treatment can also affect the composition of the MFGM. The amount of protein associated with the fat globules increases on heating; the newly bound protein is largely denatured whey protein, particularly β-lactoglobulin (Dalgleish and Banks 1991; Corredig and Dalgleish 1998). Interactions of whey proteins with the MFGM probably occur primarily via sulphydryl-disulphide interchange reactions (Kim and Jimenez-Flores 1995; Lee and Sherbon 2002). Heating can also result in the formation of high molecular weight protein complexes between xanthine oxidase, butyrophilin and denatured whey proteins (Ye *et al.* 2002); the kinetics of such reactions were reported by Ye *et al.* (2004).

If milk is heated prior to homogenization, less whey protein is incorporated into the MFGM than would be the case if the order of these steps is reversed (Sharma and Dalgleish 1994). The association of denatured whey protein, principally β-lactoglobulin, with the MFGM increases its protein content and reduces that of lipids proportionately. Membrane glycoproteins, such as PAS-6 and 7, become less evident on electrophoretic (SDS-PAGE) analysis following heat treatment (Houlihan *et al.* 1992; Iametti *et al.* 1997; Lee and Sherbon 2002), which may be due to mechanical damage to the globules caused by pumping and circulation through the pasteurizer (Iametti *et al.* 1997) or due to their displacement by denatured whey proteins during heating (Houlihan *et al.*

1992), although the exact mechanism is not clear (Lee and Sherbon 2002). The effects of heat treatment and homogenization on fat globules are compared schematically in Figure 5.4.

Heat treatment can also result in the release of H_2S from the globules and transfer of copper from plasma to globules (Mulder and Walstra 1974). Heating can also reduce the triacylglyceride content of the MFMG (Houlihan *et al.* 1992), but conflicting results have been published on the effect of heat treatment on the phospholipid content of the MFGM. Koops and Tarassuk (1959) and Greenbank and Pallansch (1961) observed a reduction in the phospholipid content of the MFGM on heating, but Houlihan *et al.* (1992) reported that heat treatment did not influence this parameter. In a system comprising milk fat globules in simulated milk ultrafiltrate, warming to 45–50 °C for 10 min resulted in the loss of up to 50% of total protein from the MFGM, perhaps due to the melting of the lipid phase and subsequent rearrangement of the globule surface (Ye *et al.* 2002).

Because concentration of milk by thermal evaporation can also damage the MFGM, and drying can damage the MFGM considerably, milk to be used for evaporation or drying is usually homogenized to strengthen the globule mem-

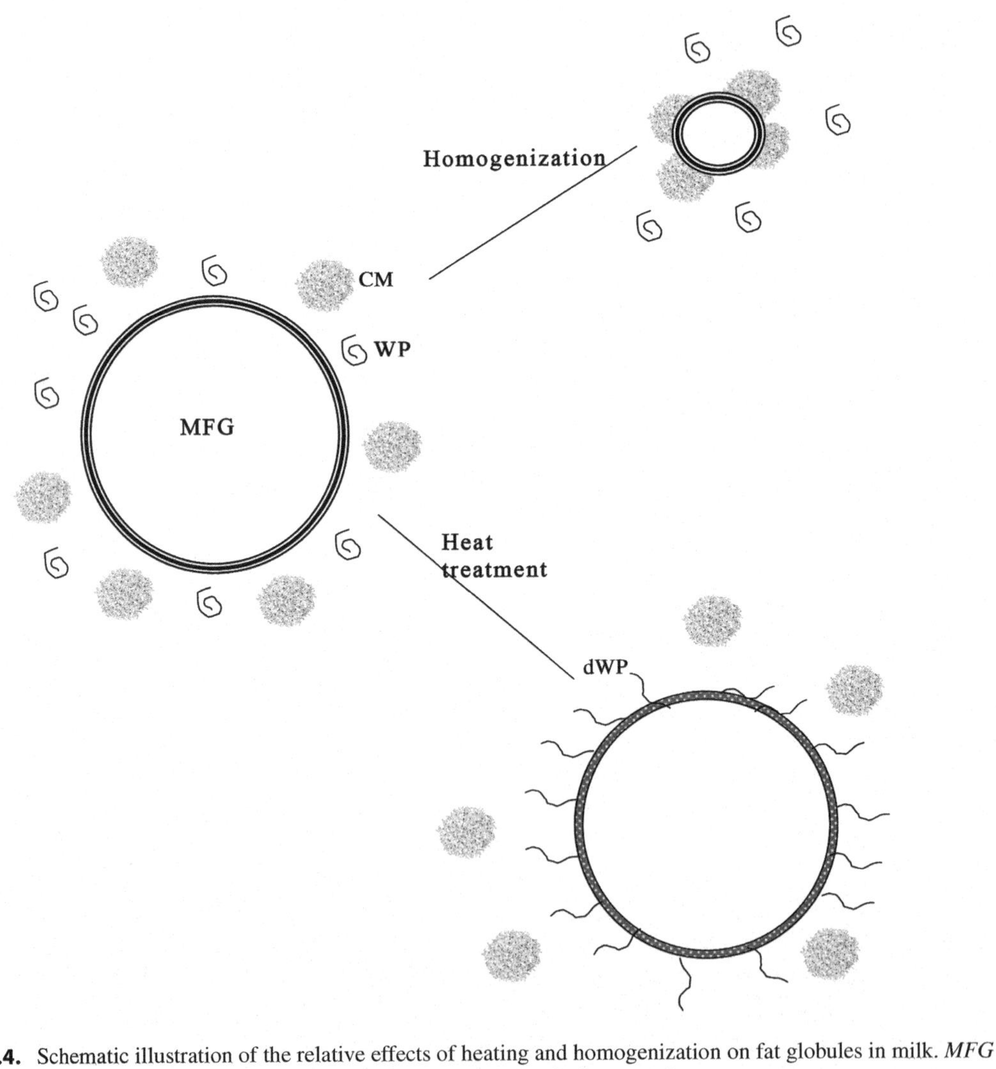

Figure 5.4. Schematic illustration of the relative effects of heating and homogenization on fat globules in milk. *MFG* milk fat globule, *CM* casein micelle, *WP* whey protein, *dWP* denatured whey protein.

branes by binding of caseins. Furthermore, contact with air bubbles can change the MFGM, which has important implications for products such as ice cream and whipping cream (see Chapter 9).

5.13 Disruption of Globules

Fat globules are relatively fragile, particularly when the fat is liquid, and can be disrupted readily by a number of conditions experienced in dairy processing operations. Viscous (shearing) or inertial (cavitation, turbulence) forces, in particular, can damage the MFGM and cause physical rupture and sub-division of the globules. Rupture occurs when droplets are deformed beyond a critical value for longer than a critical time. Resistance to deformation is related to the Laplace pressure and the ratio of the viscosity of the fat to that of the plasma. The Laplace pressure refers to the difference between the pressures at the concave and convex sides of a curved interface of two fluids. For a spherical droplet, the Laplace pressure, P_L, can be expressed as follows (Walstra 2003):

$$P_L = 2\gamma / R$$

where γ is the interfacial tension and R is the radius of the droplet. The value for γ for native milk fat globules is very small, but increases as globules are deformed. To disrupt a droplet, it is necessary to apply an external force which is considerably larger than P_L, and the duration of application must be longer than the time required to deform and disrupt the droplet (McClements 1999).

Globules can be deformed by the shearing action of liquid; if the viscous stress, ηS (where η is the viscosity of milk serum and S is the velocity gradient), equals or exceeds P_L, the globule may be disrupted; this typically requires very high velocity gradients, and, even then, the ratio of the viscosity of the milk fat to plasma protects all but the largest fat globules from shear-induced disruption, in cases of non-turbulent flow (Walstra 1995). Disruption occurs more readily under turbulent flow conditions, depending on the amount of turbulent energy dissipated per unit time and per unit volume of liquid. Such conditions are encountered only very transiently, e.g. in the valve of homogenizers or at the top of a rapidly rotating stirrer blade.

Homogenization is a process designed to reduce the size of the milk fat globules and thus retard separation of fat globules to such an extent that a cream layer does not form in homogenized milk products during their shelf-life. During homogenization, pre-warmed (~ 40 °C) milk (in which the fat is in a liquid state; homogenization is less effective when the fat is partially solid) at a pressure of 10–20 MPa is passed through a small orifice. Shearing, impact and distortion effects combine to stress the fat globules to such an extent that they split into a greater number of smaller globules (usually <2 μm diameter; Figure 5.5). The principle of operation of a homogenizer is shown in Figure 5.6. The extent of the reduction depends on a number of factors, including the valve geometry of the homogenizer used, the number of passes through the valve and, in particular, the homogenization pressure (Walstra 1975). The relationship between d_{vs} and homogenization pressure (p_h) is given by:

$$\log d_{vs} = \text{constant} - 0.6 \log p_h$$

where the constant varies between −2 and − 2.5, depending on the type of homogenizer and other processing conditions (Walstra 1975). The principal mechanism responsible for globule disruption during homogenization is probably pressure fluctuations under turbulent flow conditions (Walstra 1995). Creaming of fat globules in homogenized milk is considerably slower than that of the original globules, due to the reduction in fat globule size and the adsorption of milk proteins onto the fat globules, which increases their density and therefore reduces the rate of creaming, as well as causing inactivation of the homogenization-labile component involved in cold agglutination.

Michalski *et al.* (2002) studied the types of particles in homogenized milk and identified three classes: (1) disrupted globules covered mainly by caseins and some of the original MFGM (the surface layer of fat globules in

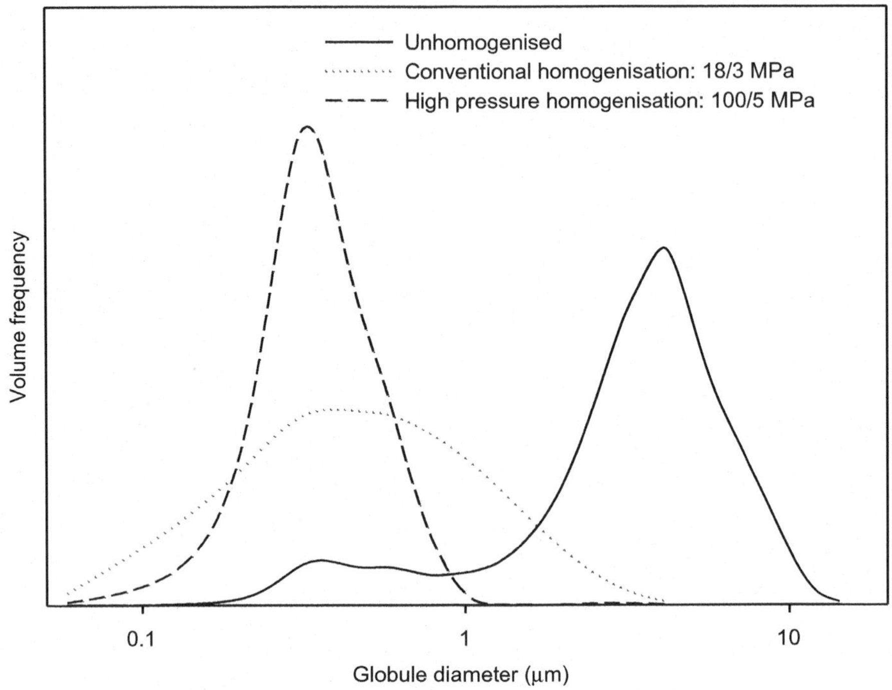

Figure 5.5. Effect of conventional (18/3 MPa) and high pressure (100/5 MPa) homogenization on the volume frequency distribution of fat globules in bovine milk.

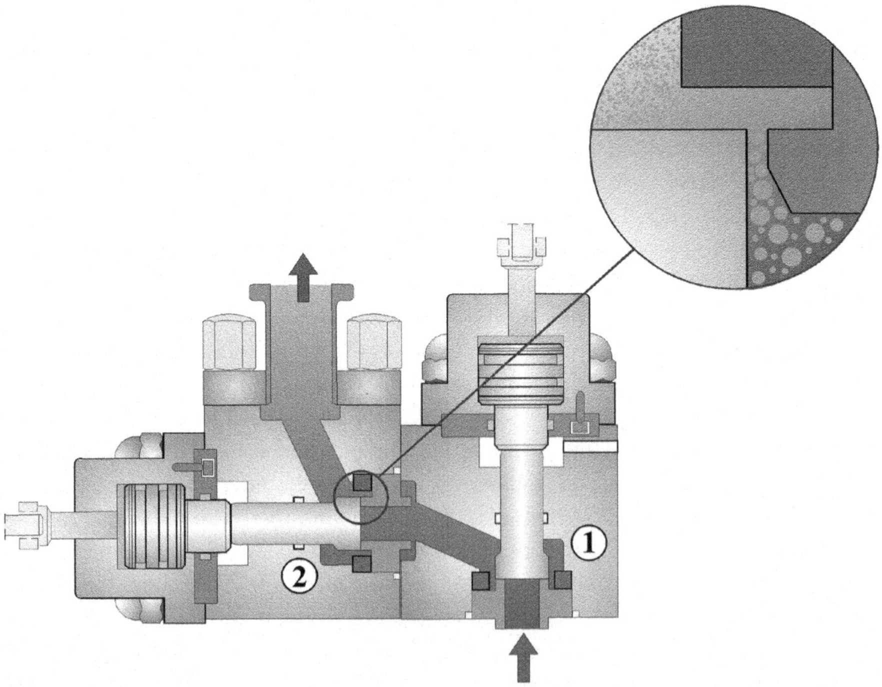

Figure 5.6. Principle of operation of a typical two-stage homogenizer for liquid milk, indicating the first (1) and second stages (2). (Reproduced with permission from *Dairy Processing Handbook*, Tetra Pak Processing Systems AB, Lund, Sweden, 1995).

homogenized milk is discussed in Section 5.13); (2) a population of very small native fat globules of around 100 nm in diameter, which appeared unaffected by homogenization, and (3) small newly formed lipid-protein complexes with a new casein-rich membrane, a type of particle also produced by ultrasonication or pumping. The presence of the population of small fat globules that are apparently unaffected by homogenization may be explained by the fact that their size is smaller than the Kolmogorov scale (Michalski *et al.* 2002), which approximates the minimum size of particles that can be affected at a given homogenization pressure, i.e. 320, 240, 180 or 140 nm at 5, 10, 20 or 40 MPa, respectively (Mulder and Walstra 1974).

In recent years, novel homogenizing devices that reduce milk fat globule size considerably more than the traditional homogenizers by operating at a higher pressure (100–300 MPa; Figure 5.5), such as high-pressure homogenizers (Hayes and Kelly 2003a; Thiebaud *et al.* 2003) and microfluidizers (McCrae 1994; Strawbridge *et al.* 1995; Hardham *et al.* 2000) have been used. The operating principle of a high-pressure homogenizer is generally similar to that of a conventional two-stage mechanical homogenizer, but it operates at a significantly higher pressure. The forces exerted by high-pressure homogenization, including shear, cavitation, impacts, turbulence and frictional heating, can also kill bacteria, inactivate enzymes, denature whey proteins and alter several properties of milk (Hayes and Kelly 2003a, b; Thiebaud *et al.* 2003; Hayes *et al.* 2005).

In a microfluidizer, forces are generated by impinging high-velocity fluid jets (Paquin 1999). Bucci *et al.* (2018) reported formation of a stable emulsion on microfluidization of milk at pressures of 75–170 MPa, leading to particle sizes substantially lower than those achieved by conventional homogenization. Microfluidization has been found to improve the textural properties of low-fat yoghurt through modification of the acid gel structure (Ciron *et al.* 2011, 2012).

Ultrasonic treatment can also disrupt milk fat globules, probably through cavitation and other shear and shock effects (Villamiel *et al.* 1999). Wu *et al.* (2001) reported that high-amplitude ultrasound homogenization of milk for yoghurt manufacture achieved similar effects as conventional homogenization.

High-pressure jet processing, in which milk is subjected to a high pressure (up to 600 MPa) and then pushed through a very narrow orifice (1–10 μm), subjects fat globules to forces including shear, pressure, turbulence, impingement and cavitation. There have been a number of reports of the impact of such processing on milk proteins, but fewer on whole milk; Tran *et al.* (2018) reported a bimodal particle size distribution in such milk, with the formation of stable casein-fat particles of relatively high density.

Heffernan *et al.* (2011) compared high-pressure homogenization, microfluidization, ultrasound homogenization and homogenization using orifice nozzles and radial diffusion principles for the emulsification of cream liqueurs. Microfluidization produced the smallest globule sizes, while some of the processes evaluated yielded a wide range of particle sizes. For all technologies, processing pressure was inversely correlated with droplet size, and small droplet sizes were generally associated with increased physical storage stability (except in the case of ultrasound homogenization). The applications of a range of homogenization technologies for milk systems were reviewed by Tobin *et al.* (2015).

5.14 Milk Fat Globules in Homogenized Milk and Cream

The decrease in fat globule size on homogenization results in a significant increase in globule surface area. This new surface area is far too large to be covered and stabilized by the original amount of MFGM material; therefore, the surface becomes covered by casein micelles, or fragments thereof, with some whey protein also becoming attached when homogenization is combined with heat treatment (Darling and Butcher 1978; García-Risco *et al.* 2002; Lee and

Sherbon 2002). The protein load per unit surface area increases as homogenization pressure is increased, but homogenization pressure has little effect on the composition of surface protein (Cano-Ruiz and Richter 1997). The membrane in homogenized milk is thicker than that in fresh raw milk, and the proportion of the casein in milk that becomes associated with the fat globules has been estimated to be ~6–8% (Fox 2002). In extreme cases, a high protein load on the globules may lead to gravitational or centrifugal sedimentation instead of upward separation in evaporated milk (Fox *et al.* 1960) or homogenized raw milk (Michalski *et al.* 2002), respectively.

Fat globules in homogenized milk, due to their small size and the presence of a high level of casein on their surface, can, in effect, behave like casein micelles, which has significant implications for the heat-, acid- or rennet-coagulation properties of milk and hence for the properties of resulting products. These effects can be positive in the case of acidified milk products, such as yoghurt, but they are generally undesirable in the case of cheese. Fat-casein complexes may accelerate the kinetics of coagulation, particularly heat coagulation; homogenization reduces the heat stability of whole milk (Sweetsur and Muir 1983).

As discussed in Section 5.9, milk fat globules are naturally susceptible to partial coalescence or clumping. Freshly homogenized fat globules are particularly unstable and tend to coalesce into clusters or clumps, i.e. homogenization clusters; these clusters are formed as a result of the sharing of casein micelles between globules (even a single casein micelle can form a bridge between two neighbouring globules), because the amount of surface-active material is insufficient to cover the newly formed interface (Ogden *et al.* 1976; Darling and Butcher 1978). The formation of homogenization clusters is usually prevented by use of a second, or even third, homogenization stage, usually at a lower pressure than the first stage (Kiesner *et al.* 1997). The tendency to form homogenization clusters is enhanced by a high fat content, small fat globule size and a high surface protein load. Intense heat treatment before homogenization can also increase the tendency of globules to form clusters (Walstra 1995).

The presence of homogenization clusters increases the viscosity of cream (Niar *et al.* 2000), particularly during product storage; high-fat cream can acquire the consistency of a thick paste. This has implications for processes where only the cream is homogenized, followed by recombination with skim milk (i.e. liquid milk processing); if clusters form, products will cream readily. When fat is emulsified directly in skim milk, clusters are formed if the fat concentration exceeds 12% (w/w) (Oortwijn and Walstra 1982b).

At a given homogenization pressure, average fat globule size decreases with increasing fat volume fraction up to a certain level, above which an increase is observed (Phipps 1983), indicating the formation of homogenization clusters. In a conventional valve homogenizer, fat clustering in milk or cream occurred at a pressure > ~20 MPa (McCrae 1994; Noda and Yamamoto 1996; McCrae and Lepoetre 1996), whereas on high-pressure homogenization, milk fat globule size decreased up to 250 MPa, but increased at 300 MPa, due to the formation of homogenization clusters (Thiebaud *et al.* 2003). In a microfluidizer, some fat clustering occurred following treatment at 35 MPa, but none was reported after treatment at 103 MPa (McCrae 1994).

Homogenized milk is very susceptible to hydrolytic rancidity, as the protective function of the MFGM has been compromised (Iametti *et al.* 1997); for this reason, homogenization should be combined with pasteurization to inactivate lipases in milk. Several methods have been developed to evaluate damage to the MFGM (Iametti *et al.* 1997; Evers 2004). Homogenized milk is also more susceptible to the so-called sunlight flavour or light-activated flavour defect (Dunkley *et al.* 1962). This flavour defect results from conversion of methionine to methional catalysed by riboflavin activated by photo-oxidation; however, the exact mechanism through which homogenization influences this process has thus far not been described. Furthermore, homogenized milk is less prone to copper-catalysed lipid oxidation than unhomogenized milk (Tarassuk and Koops

1960; Dunkley *et al.* 1962), which is probably due to the fact that oxidation-susceptible phospholipids are more uniformly distributed following homogenization and are less likely to propagate lipid oxidation (Tarassuk and Koops 1960).

5.15 Milk Fat Globules in Recombined Milk

A class of products in which the altered fat globule surface layers is of considerable importance is that of recombined milk. Typically, in the manufacture of recombined milk, skim milk powder, water and a source of milk fat (e.g. anhydrous milk fat; AMF) are mixed and homogenized to emulsify the fat and yield a stable product. Since AMF contains little or no MFGM material, the membrane surrounding fat globules in recombined milk does not contain any original MFGM material; the nature of the new membrane is highly influenced by adsorption conditions (e.g. composition of the continuous phase, agitation, temperature, heat treatment and fat: protein ratio). Both caseins and whey proteins are present in the membrane of recombined milk (Oortwijn *et al.* 1977), but the proportion of whey proteins on the membrane is lower than that in milk (Walstra and Oortwijn 1982; Sharma *et al.* 1996a, b). In addition to recombined milk, various other products are also prepared by recombining a fat source and milk protein sources, e.g. infant formula products as well as clinical nutrition products.

As the ratio of protein to fat in recombined milk is increased, the surface protein load on the fat globules increases; at higher ratios, little further effect is observed (Sharma *et al.* 1996a). The protein load per unit surface area on the fat globules in recombined milk is influenced markedly by the form of protein present in the liquid phase, i.e. it is markedly higher when casein micelles are present, compared to when only whey protein or sodium caseinate is present (Oortwijn and Walstra 1979; Sharma and Singh 1998, 1999). Furthermore, the surface protein load decreases with increasing temperature during emulsification (Oortwijn and Walstra 1979; Sharma *et al.* 1996a) and with increasing homogenization pressure (Sharma *et al.* 1996a) and is increased by heat treatment of milk prior to emulsification (Oortwijn and Walstra 1979; Sharma *et al.* 1996a). Heat treatment prior to emulsification also increases the level of β-lactoglobulin in the membrane (Sharma *et al.* 1996a).

The casein micelles on the globule surface in recombined milk often appear to be disrupted, which may be either due to homogenization or to the process of adsorption (Sharma *et al.* 1996b). The extent of disruption of micelles in recombined milk is greater than that in freshly homogenized milk (Sharma *et al.* 1996b) and increases with the temperature of homogenization (40–70 °C; Oortwijn and Walstra 1982a). Disruption of casein micelles in recombined milk was not observed after fixation of casein micelles with glutaraldehyde or addition of the surfactant Tween 20 prior to homogenization (Oortwijn *et al.* 1977); the latter effect is probably due to the preferential absorption of Tween 20 compared to casein micelles on the micelle surface. The addition of surfactants before or after recombination also decreases the protein surface load (Oortwijn and Walstra 1979). Destabilization of casein micelles by reducing the colloidal calcium phosphate content reduces the protein load on fat globules in recombined milk and alters the proportions of individual caseins at the globule surface (Sharma *et al.* 1996a). Sharma *et al.* (1996b) reported that it is far more difficult to remove κ-casein than α_s-caseins or β-casein from the fat globule surface in recombined milk and suggested that a portion of the κ-casein is associated directly with the globule surface, which was confirmed by Su and Everett (2003).

Inclusion of certain emulsifiers prevents fat separation in UHT-processed recombined milk, with Tween 20 being most effective; refined monoglyceride actually enhanced creaming slightly, perhaps due to protein displacement from surface layers, thus reducing the effective density of the globules (Mayhill and Newstead 1992). Addition of soy lecithin may reduce the stability of fat globules in recombined cream against coalescence (Melsen and Walstra 1989).

Heating of recombined milk products at 130 °C at pH 6.7 leads to the formation of chains of fat globules and casein particles, linked via the latter. Furthermore, at pH values <6.7, the surface of the casein micelles on the fat globule surface develops appendages on heating, possibly whey protein aggregates, while those in milk of pH >6.7 remained free of whey protein on heating (Singh *et al.* 1996). Addition of AMF (without homogenization) or homogenization (in the absence of AMF) did not influence the heat coagulation time of the skim milk, but the heat stability of milk recombined from AMF and skim milk is considerably lower than that of the skim milk (Sharma and Singh 1999). Furthermore, the heat stability of recombined milk decreased with decreasing fat globule size, which may be linked to a higher surface protein concentration and a lower proportion of κ-casein on smaller fat globules than larger ones (Sharma and Singh 1999). McCrae *et al.* (1994) reported that heat-induced interactions between whey protein and casein adsorbed at the fat globule surface promote the heat coagulation of recombined milk. However, the relatively low heat stability of recombined milk can be increased considerably by addition of soy lecithin, either pre- or post-homogenization (McCrae and Muir 1992).

5.16 Free Fat

'Free fat' is a term used in the literature to denote a particular parameter that has been claimed to correlate with the degree of damage to, or stability of, fat globules; various definitions have been given for this ambiguous term (Evers 2004), e.g. 'fat inside damaged globules' (Fink and Kessler 1983), 'fat that is enclosed insufficiently by an undamaged membrane' (Kessler and Fink 1992) or 'fat that has leaked out of globules' (Fink and Kessler 1983). Other authors have defined free fat as a method-dependent parameter, e.g. the proportion of fat extracted by centrifugation at 60 °C (Te Whaiti and Fryer 1975) or solvent-extractable fat (Deeth and Fitz-Gerald 1978).

The question of whether free fat actually occurs in milk or cream is, in fact, controversial, with frequent suggestions that it is an artefact of the method used for its measurement (e.g. that organic solvents can damage the MFGM and extract some fat; Evers *et al.* 2001). Walstra and co-workers (Van Boekel and Walstra 1989, 1995; Van Boekel and Folkerts 1991; Walstra 1995; Walstra *et al.* 1999) suggested that there is more than sufficient protein in milk to cover any uncovered fat very rapidly, e.g. in ~10 ms. The efficiency of methods to quantify the level of free fat was recently reviewed by Evers (2004), who, in agreement with Walstra *et al.* (1995), concluded that these methods have poor repeatability due to their poor robustness, i.e. very precise experimental control is required to obtain repeatable results. In some cases, the extraction method used may damage the fat globules, thereby magnifying, or even generating, the extractable fat level (Evers *et al.* 2001).

While free fat remains a controversial concept in liquid dairy products, it undoubtedly has an important role in whole milk powder (WMP); for certain applications (e.g. chocolate manufacture) high free fat WMP is favoured. Keogh *et al.* (2003) reported that the particle size in chocolate mixes after refining and the viscosity of the molten chocolate decreases as the free fat content of the WMP increased; such changes have significant implications for the mouthfeel and smoothness of chocolate. Schmidmeier *et al.* (2019) found high levels of free fat in a coarse (large particle size) fraction of a fat-filled milk powder which was prone to the formation of white flecks on reconstitution, and linked this to poor emulsion stability of such powder on reconstitution. The topic of free fat in dairy powders was reviewed by Vignolles *et al.* (2007), who concluded that both processing conditions and composition strongly affect free fat in dairy powders.

When considering so-called free fat in dairy powders, care should be taken to distinguish between values based on traditional solvent extraction methods and those obtained by methods such as X-ray photoemission spectroscopy (XPS). The latter focusses only on the actual composition of the surface layer of the particle, and values are typically expressed as % of particle surface covered by fat. For solvent extraction, however, not only surface fat but also some accessible fat further into the particle is extracted. Values for this method are typically expressed as g extractable fat per 100 g total fat or powder (Vignolles *et al.* 2007).

For more information on the role of fat in milk powder and chocolate, see Chapter 9.

5.17 Influence of Fat Globules on Rheological Properties of Milk and Cream

Rheological properties of emulsions are of importance in food science for various reasons. Some sensory attributes (e.g. creaminess, smoothness, thickness and flowability) of food emulsions are directly related to their rheological properties. Furthermore, the shelf-life of many food emulsions depends on rheological characteristics of the phases; for example, the rate of creaming of milk depends on the viscosity of the milk plasma (McClements 1999). The content of lactose or whey proteins in milk has little influence on the viscosity of milk; fat content has a major influence, although by far the greatest influence is that of the casein content (McCarthy 2003). The influence of milk fat globules on the rheological properties of milk and cream, in particular the viscosity, will be discussed in this section.

If fat globules are present as separate particles, the fat content is $\leq$40% and the milk fat completely molten, milk and cream behave as Newtonian fluids at intermediate and at high shear rates (Phipps 1969; McCarthy 2003), i.e. its viscosity is not influenced by shear rate ($\tau = \eta \times \gamma$, where τ is the shear stress [*Pa*], η is the viscosity [*Pa s*] and γ is the shear rate [s^{-1}]). For a Newtonian fluid, the Eilers equation (Eilers 1941) is generally obeyed (Walstra 1995):

$$\eta = \eta_0 \left[1 + \frac{1.25\varphi}{1 - \varphi / \varphi_{\max}} \right]^2$$

where η is the viscosity of the product, η_0 is the viscosity of the continuous phase, ϕ is the volume fraction of spherical particles and ϕ_{max} is the hypothetical volume fraction when the particles are in the closest possible packing arrangement. Van Vliet and Walstra (1980) showed that, if ϕ is taken as ϕ_{fat} + 0.16 (ϕ for skim milk $\approx$ 0.16), $\eta_0 = 1.02\,\eta_{water}$ and $\phi_{max} = 0.88$, values calculated from the Eilers equation are in good agreement with the experimental data from Phipps (1969).

For studying the viscosity of particle suspensions, a number of other models (e.g. the Einstein model, the Batchelor model and the Krieger-Dougherty model) have also been proposed and applied over the years (e.g. Walstra 2003; Willenbacher and Georgieva 2013). However, the Einstein model is relevant only for infinite dilutions and assumes a linear relationship between concentration and viscosity. The Batchelor model addresses some issues of the Einstein model by including a factor accounting for particle interactions, but still applies only to rather dilute systems. A more useful model is the Krieger-Dougherty model, which is not only applicable to dilute but also to concentrated particle suspensions; it states that

$$\eta = \eta_{serum} \left(1 - \frac{\varphi}{\varphi_{\max}} \right)^{-[\eta]\varphi_{\max}}$$

where η and η_{serum} are the viscosity of the sample and of the serum phase of the sample, φ is the volume fraction of particles, φ_{max} is the maximum volume fraction of particles on random packing and [η] is the intrinsic viscosity. For monodisperse spherical particles, φ_{max} = 0.64, and this value increases with polydispersity of the particle size distribution. In addition, for systems with bi- or multi-modal particle size distributions, φ_{max} may become as high as 0.85. For solid hard spheres, [η] = 2.5, but on swelling of particles and on deviations from spherical shape, [η] increases. The aspects are of particular importance when flocculation or aggregation of emulsion droplets occurs, as the fractal geometry of the flocs or aggregates formed essentially determines [η].

At a temperature < 40 °C, milk does not behave as a Newtonion fluid; the deviation from Newtonian flow is larger at a temperature further below 40 °C (Randhahn 1973; Wayne and

Shoemaker 1988; Kristensen *et al.* 1997). Rather, milk viscosity decreases with increasing shear rate at a temperature below 40 °C (Randhahn 1973), which Mulder and Walstra (1974) suggested may be due to the disruption of clusters of coalesced milk fat globules which were formed as a result of cold agglutination.

Rheological properties of milk and cream are influenced by various processes, e.g. heat treatment, cooling or homogenization. McClements (1999) reported that the main factors which determine the rheological properties of emulsions can be divided into five groups:

5.17.1 Dispersed Phase Volume Fraction

With an increase in dispersed-phase volume fraction, the viscosity of an emulsion increases. This increase in viscosity is linear at a low droplet concentration (McClements 1999); the viscosity of an emulsion of milk fat globules in milk plasma increases linearly with fat content up to 30% (Bakshi and Smith 1984; Kyazze and Starov 2004), whereas the viscosity of low-fat milk (≤2.0% fat) increases in a near-linear fashion with fat content (Phillips *et al.* 1995). However, above a certain volume fraction of the dispersed phase, the droplets in emulsions are packed so closely that flow is impaired, giving the emulsion a gel-like character (McClements 1999). For instance, the viscosity of cream increases very significantly with increasing fat content when the fat content is >50% (Prentice 1968; Mulder and Walstra 1974).

5.17.2 Rheology of the Component Phases

The viscosity of an emulsion is directly proportional to the viscosity of the continuous phase; any alteration in the rheological properties of the continuous phase results in a corresponding alteration in the rheology of the whole emulsion (McClements 1999). The rheological properties of the dispersed phase, i.e. the milk fat globules in the case of milk and cream, have only a minor influence on the rheology of the emulsion (Walstra 1996; McClements 1999). This is illustrated well by the influence of temperature on milk viscosity. A decrease in temperature, particularly below ambient temperature, results in an increase in milk viscosity (Randhahn 1973; Prentice 1992; Kristensen *et al.* 1997); even though considerable crystallization of fat occurs in the globules on cooling, changing the rheological properties of the fat, increases in milk viscosity are dominated by changes in the milk serum, primarily increases in hydration of casein micelles (Prentice 1992).

5.17.3 Droplet Size

The viscosity of dilute emulsions is independent of the size of its droplets when long-range attractive and repulsive colloidal interactions between droplets are negligible, and the thickness of the surface layer is small compared to the droplet diameter; however, when long-range colloidal interactions are present and/or the thickness of the surface layer is a significant proportion of particle diameter, particle size has a considerable effect on emulsion viscosity. The viscosity of an emulsion increases with increasing thickness of the surface layer, due to an increase in the effective volume fraction of the dispersed phase (Pal 1996). Homogenization of milk leads to an increase in milk viscosity (Whitnah *et al.* 1956; Randhahn 1973), which Prentice (1992) suggested is due to the adsorption of casein particles on the fat globule surface, thereby increasing the effective volume fraction of the dispersed phase. The formation of homogenization clusters also leads to increases in viscosity of the product, as discussed in Section 5.13.

5.17.4 Colloidal Interactions

Colloidal interactions between emulsion droplets play a primary role in determining emulsion rheology. If attractive interactions predominate over repulsive interactions flocculation can occur, which leads to an increase in the effective volume fraction of the dispersed phase and thus increases viscosity. As discussed in Sections 5.5 and 5.6, flocculation of natural milk fat globules does not occur. However, clustering of milk fat globules due to cold agglutination increases the effective volume fraction of the milk fat globules, thereby increasing viscosity (Prentice 1992).

5.17.5 Particle Charge Interactions

As discussed earlier (Section 5.5), the charge on an emulsion droplet can influence the rheological properties of the emulsion. First, the charge determines whether droplets tend to aggregate (see Section 5.6). Furthermore, droplet charge can also influence the rheological properties of the emulsion through the primary electroviscous effect (Pal 1996); movement of a charged droplet through a fluid results in distortion of the surrounding cloud of counter-ions, which causes an attraction between the charge on the droplet and the charge associated with the 'cloud' of counter-ions that lags slightly behind it. This attraction opposes the movement of droplets and thus increases the viscosity of the emulsion because more energy is needed to cause droplets to move (Pal 1996; McClements 1999). This mechanism may be involved in the increase in viscosity observed on homogenization of milk (Whitnah *et al.* 1956; Randhahn 1973). The fat globules in homogenized milk have a higher net-negative surface charge than those in unhomogenized milk (Dalgleish 1984; Michalski *et al.* 2001b), which may thus, through the primary electroviscous effect, result in increased milk viscosity, although this has to be confirmed with experimental data.

Overall, it is apparent that although fat globules are not the predominant milk constituent affecting the rheological properties of milk and cream, they still exhibit a considerable influence.

5.18 Conclusions

Whole milk or cream are emulsions of milk fat globules in milk plasma. The physico-chemical properties of the milk fat globules affect many properties of liquid dairy products such as milk and cream, and as such should always be considered when studying the stability of liquid dairy products. The physico-chemical properties of the milk fat globules can also be influenced through a wide variety of common industrial processes, as described in this chapter, and once chosen and controlled carefully, these processes can be efficiently used to give products desired characteristics, e.g. in terms of storage stability or rheological properties. Although much is known concerning physico-chemical properties of the milk fat globules, and instability of dairy emulsions can be controlled well with the current state of knowledge, gathering further information concerning the physical chemistry of milk fat globules, and the underlying fundamental problems, remains crucial. Pursuit of fundamental problems often leads to good results, sometimes in unexpected ways. Thus, it is important to continue the constant enhancement of our understanding of areas such as those described in this chapter.

References

Abd El-Salam, M. & El-Shibiny, S. (2011). A comprehensive review on the composition and properties of buffalo milk. *Dairy Science & Technology, 91*, 663–699.

Abeni, F., Degano, L., Calza, F., Giangiacomo, R. & Pirlo, G. (2005). Milk quality and automatic milking: Fat globule size, natural creaming, and lipolysis. *Journal of Dairy Science, 88*, 3519–3529.

Abo-Elnaga, I. G. (1966). Factors affecting creaming of cows' and buffaloes' milk. *Milchwissenschaft, 21*, 429–433.

Abo-Elnaga, I. G., El-Sadek, G. M. & El-Sokary, A. M. (1966). Clustering of fat globules in cows' and buffaloes' milk, creaming mechanism and the physical arrangement of globules in gravity cream. *Milchwissenschaft, 21*, 210–215.

Ahmad, S. (2013). Buffalo milk. In Y. W. Park & G. F. H. Haenlein (Eds.), *Milk and dairy products in human nutrition: Production, composition and health* (pp. 519–553). Chichester: Wiley.

Ahmad, S., Anjum, F. M., Huma, N., Sameen, A. & Zahoor, T. (2013). Composition and physico-chemical characteristics of buffalo milk with particular emphasis on lipids, proteins, minerals, enzymes and vitamins. *Journal of Animal and Plant Science, 23*, 62–74.

Aikio, P., Nieminen, M., Holand, Ø., Mossing, T., Alatossava, T. & Malinen, H. L. (2002). *Försök med mjölkning av vajor. Prosjektrapport Interreg. 54.*

Report from the Sapmi Interrieg Project (in Swedish and partly in English).

Attaie, R. & Richter, R. L. (2000). Size distributions of fat globules in goat milk. *Journal of Dairy Science, 83*, 940–944.

Babcock, S. M. (1889). *The constitution of milk, and some of the conditions that affect the separation of cream* (Wisconsin Agricultural Experimentation Station Bulletin, Vol. 18). Madison: University of Wisconsin, Agricultural Experiment Station.

Bakshi, A. S. & Smith, D. E. (1984). Effect of fat content and temperature on viscosity in relation to pumping requirements of fluid milk products. *Journal of Dairy Science, 67*, 1157–1160.

Baumrucker, C. R. & Keenan, T. W. (1973). Membranes of mammary gland. VIII. Stability of milkfat globule membrane in secreted milk. *Journal of Dairy Science, 56*, 1092–1094.

Bergenståhl, B. A. & Claesson, P. M. (1990). Surface forces in emulsions. In K. Larsson & S. E. Friberg (Eds.), *Food emulsions* (2nd ed., pp. 41–96). New York: Marcel Dekker Inc.

Bergenståhl, B. A. & Claesson, P. M. (1997). Surface forces in emulsions. In K. Larsson & S. E. Friberg (Eds.), *Food emulsions* (3rd ed., pp. 57–109). New York: Marcel Dekker Inc.

Bermúdez-Aguirre, D., Mawson, R. & Barbosa-Cánovas, G. V. (2008). Microstructure of fat globules in whole milk after thermosonication treatment. *Journal of Food Science, 73*, 325–332.

Boode, K. & Walstra, P. (1993). Partial coalescence of oil-in-water emulsions. 1. Nature of aggregation. *Colloids and Surfaces A, 81*, 121–137.

Boode, K., Bisperink, C. & Walstra, P. (1991). Destabilization of O/W emulsions containing fat crystals by temperature cycling. *Colloids and Surfaces, 61*, 55–74.

Boode, K., Walstra, P. & De Groot-Mostert, A. E. A. (1993). Partial coalescence in oil-in-water emulsions. 2. Influence of properties of the fat. *Colloids and Surfaces A, 81*, 139–151.

Briard, V., Leconte, N., Michel, F. & Michalski, M. C. (2003). The fatty acid composition of small and large naturally occurring milk fat globules. *European Journal of Lipid Science and Technology, 105*, 677–682.

Brunner, J. R. (1978). Physical equilibria in milk: The lipid phase. In B. H. Webb, A. H. Johnson & J. A. Johnson (Eds.), *Fundamentals of dairy chemistry* (2nd ed., pp. 474–602). Westport: The AVI Publishing Company, Inc.

Buchheim, W. (1970). Die submikroskopische struktur der milchfettes und ihre bedeutung fur die hertellung von butter. *Proceedings of XVIII International Dairy Congress (Sidney), 1D*, 73.

Buchheim, W. & Abou El-Nour, A. M. (1992). Induction of milkfat crystallization in the emulsified state by high hydrostatic pressure. *Fat Science Technology, 94*, 369–373.

Buchheim, W., Schütt, M. & Frede, E. (1996). High pressure effects on emulsified fats. In R. Hayashi & C. Balny (Eds.), *High pressure bioscience and biotechnology* (pp. 331–336). Amsterdam: Elsevier Science B.V.

Bucci, A. J., Van Hekken, D. L., Tunick, M. H., Reyne, J. A. & Tomasula, P. M. (2018). The effects of microfluidization on the physical, microbial, chemical and coagulation properties of milk. *Journal of Food Engineering, 247*, 160–167.

Campbell, M. H. (1932). *A study of the fat-globule size in milk* (Bulletin 341). Burlington: Vermont Agricultural Experiment Station.

Cano-Ruiz, M. E. & Richter, R. L. (1997). Effect of homogenization pressure on the milk fat globule membrane proteins. *Journal of Dairy Science, 80*, 2732–2739.

Caplan, Z., Melilli, C. & Barbano, D. M. (2013). Gravity separation of fat, somatic cells, and bacteria in raw and pasteurized milks. *Journal of Dairy Science, 96*, 2011–2019.

Cheong, F. C., Xiao, K. & Grier, D. G. (2009). Technical note: Characterising individual milk fat globules with holographic video microscopy. *Journal of Dairy Science, 92*, 95–99.

Ciron, C. I. E., Gee, V. L., Kelly, A. L. & Mauty, M. A. E. (2011). Effect of microfluidisation of heat-treated on rheology and sensory properties of reduced-fat yoghurt. *Food Hydrocolloids, 25*, 1470–1476.

Ciron, C. I. E., Gee, V. L., Kelly, A. L., Mauty, M. A. E. (2012). Modifying the microstructure of low-fat yoghurt by microfluidisation of milk at different pressures to enhance rheological and sensory properties. *Food Chemistry, 130*, 510–519.

Corredig, M. & Dalgleish, D. G. (1998). Effect of heating of cream on the properties of milk fat globule membrane isolates. *Journal of Agricultural and Food Chemistry, 46*, 2533–2540.

Dalgleish, D. G. (1984). Measurement of electrophoretic mobilities and zeta-potentials of particles from milk using laser Doppler electrophoresis. *Journal of Dairy Research, 51*, 425–438.

Dalgleish, D. G. & Banks, J. M. (1991). The formation of complexes between serum proteins and fat globules during heating of whole milk. *Milchwissenschaft, 46*, 75–78.

Dalgleish, D. G. & Hallett, F. R. (1995). Dynamic light scattering: Applications to food systems. *Food Research International, 28*, 181–193.

Darling, D. F. & Butcher, D. V. (1978). Milk-fat globule membrane in homogenized cream. *Journal of Dairy Research, 45*, 197–208.

Deeth, H. C. & Fitz-Gerald, C. H. (1978). Effect of mechanical agitation of raw milk on the milk-fat globule in relation to the level of induced lipolysis. *Journal of Dairy Research, 45*, 373–380.

De Feijter, J. A., Benjamins, J. & Tamboer, M. (1987). Adsorption displacement of proteins by surfactants in oil-in-water emulsions. *Colloids and Surfaces, 27*, 243–266.

Derjaguin, B. & Landau, L. (1941). Theory of the stability of strongly charged lyophobic sols and of the adhesion of strongly charged particles in solution of electrolytes. *Acta Physicochimica USSR, 14*, 633–662.

Devle, H., Vetti, I., Naess-Andresen, C. F., Rukke, E. O., Vegarud, G. & Ekeberg, D. (2012). A comparative study of fatty acid profiles in ruminant and nonruminant milk. *European Journal of Lipid Science and Technology, 114*, 1036–1043.

Dickinson, E. & Stainsby, G. (1988). Emulsion stability. In E. Dickinson & G. Stainsby (Eds.), *Advances in food emulsions and foams* (pp. 1–44). London: Elsevier Applied Science.

Dickinson, E., Narhan, S. K. & Stainsby, G. (1989). Factors affecting the properties of cohesive creams formed from cream liqueurs. *Journal of the Science of Food and Agriculture, 48*, 225–234.

Dumay, E., Lambert, C., Funtenberger, S. & Cheftel, J. C. (1996). Effects of high pressure on the physico-chemical characteristics of dairy creams and model oil/water emulsions. *Lebensmittel Wissenschaft und-Technologie, 29*, 606–625.

Dunkley, W. L. & Sommer, H. H. (1944). *The creaming of milk* (Research Bulletin 151). Madison: Agricultural Experiment Station, University of Wisconsin.

Dunkley, W. L., Pangborn, R. M. & Franklin, J. D. (1962). Influence of homogenization, copper, and ascorbic acid on light-activated flavor in milk. *Journal of Dairy Science, 45*, 1040–1044.

D'Inecco, P., Ong, L., Pellegrino, L., Faoro, F., Barbiroli, A., & Gras, S. (2018). Effect of temperature on the microstructure of fat globules and the immunoglobulin-mediated interactions between fat and bacteria in natural raw milk creaming. *Journal of Dairy Science, 101*, 2984–2997.

Eilers, H. (1941). Die viscosität von Emulsionen hochviskoser Stoffe als Funktion de Konzentration. *Kolloid Zeitschrift, 97*, 313–321.

El-Zeini, H. M. (2006). Microstructure, rheological and geometrical properties of fat globules of milk from different animal species. *Polish Journal of Food and Nutrition Science, 15*, 147–154.

Euber, J. R. & Brunner, J. R. (1984). Reexamination of fat globule clustering and creaming in cow milk. *Journal of Dairy Science, 67*, 2821–2832.

Evers, J. M. (2004). The milk fat globule membrane – Methodologies for measuring milk fat globule (membrane) damage. *International Dairy Journal, 14*, 747–760.

Evers, J. M., Crawford, R. A., Robinson, N. P., Harding, D. R. K. & McCarthy, O. J. (2001). "Free fat" solvent extraction methods damage milkfat globules. *Milchwissenschaft, 56*, 504–508.

Evers, J. M., Haverkamp, R., Holroyd, S. & McCarthy, O. J. (2008). Heterogeneity of milk fat globule membrane structure and composition as observed using fluorescence microscopy techniques. *International Dairy Journal, 18*, 1081–1089.

Fahmi, A. H. (1951). The rate of creaming of raw cows' and buffaloes' milk. *Journal of Dairy Research, 18*, 106–112.

Farah, Z. & Rüegg, M. (1991). The creaming properties and size distribution of fat globules in camel milk. *Journal of Dairy Science, 74*, 2901–2904.

Fink, A. & Kessler, H. G. (1983). Determination of free fat in cream for the assessment of damage caused to fat globules by processing. *Milchwissenschaft, 38*, 330–334.

Fox, P. F. (2002). Fat globules in milk. In H. Roginski, J. Fuquay & P. F. Fox (Eds.), *Encyclopedia of dairy sciences* (pp. 1564–1568). London: Academic.

Fox, K. K., Holsinger, V. H., Caha, J. & Pallansch, M. J. (1960). Formation of a fat protein complex in milk by homogenization. *Journal of Dairy Science, 43*, 1396–1406.

Fox, P. F., Uniacke-Lowe, T., McSweeney, P. L. H. & O'Mahony, J. A. (2015). Dairy Chemistry and Biochemistry, 2nd Edn., Springer, Switzerland.

Frede, E. & Buchheim, W. (2000). The influence of high pressure upon the phase transition behaviour of milk-fat and milk-fat fractions. *Milchwissenschaft, 55*, 683–686.

Friberg, S. E. (1997). Emulsion stability. In S. E. Friberg & K. Larsson (Eds.), *Food emulsions* (3rd ed., pp. 1–55). New York: Marcel Dekker.

Gammack, D. B. & Gupta, B. B. (1967). Isolation of lipoprotein particles from bovine milk by a gel-filtration procedure. *Biochemical Journal, 103*, 72.

Gammack, D. B. & Gupta, B. B. (1970). Clustering of milk fat globules. *Proceedings of XVIII International Dairy Congress (Sidney), 1E*, 20.

García-Risco, M. R., Villamiel, M. & López-Fandiño, R. (2002). Effect of homogenization on protein distribution during storage of indirectly heated UHT milk. *Le Lait, 82*, 589–599.

Geer, S. R. & Barbano, D. M. (2014a). The effect of immunoglobulins and somatic cells on the gravity separation of fat, bacteria and spores in pasteurized whole milk. *Journal of Dairy Science, 97*, 2027–2038.

Geer, S. R. & Barbano, D. M. (2014b). Effect of colostrum on gravity separation of milk somatic cells in skim milk. *Journal of Dairy Science, 97*, 687–693.

Gervilla, R., Ferragut, V. & Guamis, B. (2001). High hydrostatic pressure effects on colour and milk-fat globule of ewe's milk. *Journal of Food Science, 66*, 880–885.

Gjøstein, H., Holand, Ø. & Weladji, R. B. (2004). Milk production and composition in reindeer (*Rangifer tarandus*): Effect of lactational stage. *Comparative Biochemistry and Physiology. Part A, 137*, 649–656.

Goff, H. D., Liboff, M., Jordan, W. K. & Kinsella, J. E. (1987). The effects of polysorbate 80 on the fat emulsion in ice cream mix: Evidence from transmission electron microscopy studies. *Food Structure, 6*, 193–198.

Gonzales-Janolino, V. T. (1968a). The creaming properties of caraboas' milk. Part I. A comparison of the creaming properties of cows' milk and caraboas' milk. *Milchwissenschaft, 23*, 21–23.

Gonzales-Janolino, V. T. (1968b). The creaming of caraboas' milk. Part II. A study of the factors influencing the creaming ability of caraboas' milk. *Milchwissenschaft, 23*, 204–207.

Greenbank, G. R. & Pallansch, M. J. (1961). Migration of phosphatides in processing dairy products. *Journal of Dairy Science, 44*, 1597–1602.

Hammad, Y. A. (1993). Instability of buffalo milk fat globules. *Annals of Agricultural Science, 38*, 111–125.

Hammer, B. W. (1916). *Studies on the creaming ability of milk* (Iowa Agricultural Experiment Station Bulletin, Vol. 31). Ames: Agricultural Experiment Station, Iowa State College of Agriculture and Mechanic Arts

Hansen, S. F., Larson, L. B. & Wiking, L. (2019). Thermal effects on IgM-milk fat globule-mediated agglutination. *Journal of Dairy Research, 86*, 108–113.

Hansson, E. (1949). The creaming of milk. *Proceedings of XII International Dairy Congress (Stockholm), 2*, 30–35.

Hardham, J. F., Imison, B. W. & French, H. M. (2000). Effect of homogenization and microfluidisation on the extent of fat separation during storage of UHT milk. *Australian Journal of Dairy Technology, 55*, 16–22.

Harding, H. A., Kelly, F. W. & Chrisler, E. S. (1922). Methods of measuring the volume of cream on milk. *Journal of Dairy Science, 5*, 468–478.

Hayes, M. G. & Kelly, A. L. (2003a). High pressure homogenization of raw whole bovine milk (a) effects on fat globule size and other properties. *Journal of Dairy Research, 70*, 297–305.

Hayes, M. G. & Kelly, A. L. (2003b). High pressure homogenization of raw whole bovine milk (b) effects on indigenous enzyme activity. *Journal of Dairy Research, 70*, 307–313.

Hayes, M. G., Fox, P. F. & Kelly, A. L. (2005). Potential applications of high pressure homogenization in processing of liquid milk. *Journal of Dairy Research, 72*, 25–33.

Heffernan, S. P., Kelly, A. L., Mulvihill, D. M., Lambrich, U. & Schuchmann, H. P. (2011). Efficiency of a range of homogenisation technologies in the emulsification and stabilization of cream liqueurs. *Innovative Food Science and Emerging Technologies, 12*, 628–634.

Henson, A. F., Holdsworth, G. & Chandan, R. C. (1971). Physicochemical analyses of bovine milk fat globule membrane. II. Electron microscopy. *Journal of Dairy Science, 54*, 1752–1763.

Hillbrick, G. C., McMahon, D. J. & Deeth, H. C. (1998). Electrical impedance particle size method (Coulter Counter) detects the large fat globules in poorly homogenized UHT processed milk. *Australian Journal of Dairy Technology, 53*, 17–21.

Hoffmann, W. (1999). Lagerstabilität von H-Schlagsahne. *Kieler Milchwirtschaftliche Forschungsberichte, 51*, 125–136.

Houlihan, A. V., Goddard, P. A., Kitchen, B. J. & Masters, C. J. (1992). Changes in structure of the bovine milk fat globule membrane on heating whole milk. *Journal of Dairy Research, 59*, 321–329.

Huppertz, T., Fox, P. F. & Kelly, A. L. (2003). High pressure-induced changes in the creaming properties of bovine milk. *Innovative Food Science and Emerging Technologies, 4*, 349–359.

Iametti, S., Versuraro, L., Tragna, S., Giangiacomo, R. & Bonomi, F. (1997). Surface properties of the fat globule in treated creams. *International Dairy Journal, 7*, 375–380.

Ismail, A. A., El-Ganam & Sirry, I. (1972). Ability of buffalo and cow milk to form cream layer. *Zeitschrift für Lebensmittel-Untersuchung und -Forschung, 145*, 275–279.

Israelachvili, J. N. (1992). *Intermolecular and surface forces*. London: Academic.

Jack, E. L. & Dahle, C. D. (1937). The electrokinetic potential of milk fat. I. General electrophoretic studies. *Journal of Dairy Science, 20*, 551–556.

Jenness, R. & Parkash, S. (1971). Lack of a fat globule clustering agent in goats' milk. *Journal of Dairy Science, 54*, 123–126.

Jensen, S. K. & Nielsen, K. N. (1996). Tocopherols, retinol, beta-carotene and fatty acids in fat globule membrane and fat globule core in cows' milk. *Journal of Dairy Research, 63*, 565–574.

Keenan, T. W. & Mather, I. H. (2002). Milk fat globule membrane. In H. Roginski, J. Fuquay & P. F. Fox (Eds.), *Encyclopedia of dairy sciences* (pp. 1568–1576). London: Academic.

Keenan, T. W., Mather, I. H. & Dylewski, D. P. (1988). Physical equilibria in milk: The lipid phase. In N. P. Wong, R. Jenness, M. Kenney & E. M. Marth (Eds.), *Fundamentals of dairy chemistry* (3rd ed., pp. 511–582). New York: Van Nostrand Reinhold Company.

Keogh, M. K., Murray, C. A. & O'Kennedy, B. T. (2003). Effect of selected properties of ultrafiltered spray-dried milk powders on some properties of chocolate. *International Dairy Journal, 13*, 719–726.

Kessler, H. G. & Fink, A. (1992). *Physico-chemical effect of pasteurization on cream properties* (Bulletin 271) (pp. 11–17). Brussels: International Dairy Federation.

Keynon, A. & Jenness, R. (1958). Effects of proteins on agglutination of fat globules. *Journal of Dairy Science, 41*, 716.

Keynon, A. J., Jenness, R. & Anderson, R. K. (1966). Role of milk immunoglobulins in fat globule clustering and the ring test phenomenon. *Journal of Dairy Science, 49*, 1144–1148.

Kiesner, C., Hinrichsen, M. & Jahnke, S. (1997). Mehrstufiger Druckabbau beim Homogenisieren von Rahm. *Kieler Milchwirtschaftliche Forschungsberichte, 49*, 197–206.

Kim, H.-H. Y. & Jimenez-Flores, R. (1995). Heat-induced interactions between the proteins of milk fat globule membrane and skim milk. *Journal of Dairy Science, 78*, 24–35.

Koops, J. & Tarassuk, N. P. (1959). The effect of various processing treatments on the partition of phosphatides between the fat phase and the milk plasma. *Netherlands Milk and Dairy Journal, 13*, 180–189.

Koops, J., Payens, T. A. J. & Kerkhof-Mogot, M. F. (1966). The effect of homogenization on the spontaneous creaming of milk. *Netherlands Milk and Dairy Journal, 20*, 296–299.

Kristensen, D., Jensen, P. Y., Madsen, F. & Birdi, K. S. (1997). Rheology and surface tension of selected processed dairy fluids: Influence of temperature. *Journal of Dairy Science, 80*, 2282–2290.

Kyazze, G. & Starov, V. (2004). Viscosity of milk: Influence of cluster formation. *Colloid Journal, 3*, 316–321.

Labots, H. & Galesloot, T. E. (1959). Vlokvorming in de roomlaag van gepasteuriseerde melk. *Netherlands Milk and Dairy Journal, 13*, 79–99.

Lee, S. J. & Sherbon, J. W. (2002). Chemical changes in bovine milk fat globule membrane caused by heat treatment and homogenization of whole milk. *Journal of Dairy Research, 69*, 555–567.

Lopez, C. & Menard, O. (2011). Human milk fat globules: Polar lipid′ composition and in situ structural investigations revealing the heterogeneous distribution of proteins and the lateral segregation of sphingomyelin in the biological membrane. *Colloids and Surfaces, B: Biointerfaces, 83*, 29–41.

Lopez, C., Lesieur, P., Keller, G. & Ollivon, M. (2000). Thermal and structural behavior of milk fat. 1. Unstable species of cream. *Journal of Colloid and Interface Science, 229*, 62–71.

Lopez, C., Lavigne, F., Lesieur, P., Bourgaux, C. & Ollivon, M. (2001a). Thermal and structural behavior of milk fat. 1. Unstable species of anhydrous milk fat. *Journal of Dairy Science, 84*, 756–766.

Lopez, C., Lavigne, F., Lesieur, P., Keller, G. & Ollivon, M. (2001b). Thermal and structural behavior of milk fat. 2. Crystalline forms obtained by slow cooling. *Journal of Dairy Science, 84*, 2404–2412.

Lopez, C., Lesieur, P., Bourgaux, C., Keller, G. & Ollivon, M. (2001c). Thermal and structural behavior of milk fat. 2. Crystalline forms obtained by slow cooling of cream. *Journal of Colloid and Interface Science, 240*, 150 161.

Lopez, C., Bourgaux, C., Bernadou, S., Keller, G. & Ollivon, M. (2002a). Thermal and structural behavior of milk fat. 3. Influence of cooling rate and droplet size on cream crystallization. *Journal of Colloid and Interface Science, 254*, 64–78.

Lopez, C., Bourgaux, C., Lesieur, P. & Ollivon, M. (2002b). Crystalline structures formed in cream and anhydrous milk fat at 4°C. *Le Lait, 82*, 317–335.

Luick, J. R., White, R. G., Gau, A. M. & Jenness, R. (1974). Compositional changes in the milk secreted by grazing reindeer. I. Gross composition and ash. *Journal of Dairy Science, 57*, 1325–1333.

Luo, J., Huang, Z., Liu, H., Zhang, Y. & Ren, F. (2018). Yak milk fat globules from the Qinghai-Tibetan Plateau: Membrane lipid composition and morphological properties. *Food Chemistry, 245*, 731–737.

Ma, Y. & Barbano, D. M. (2000). Gravity separation of raw bovine milk: fat globule size distribution and fat content of milk fractions. *Journal of Dairy Science, 83*, 1719–1727.

Malacarne, M., Martuzzi, F., Summer, A. & Mariani, P. (2002). Review: Protein and fat composition of mare's milk: Some nutritional remarks with reference to human and cow's milk. *International Dairy Journal, 12*, 869–877.

Martini, M., Cecchi, F. & Scolozzi, C. (2006). Relationship between fat globule size and chemical and fatty acid composition of cow's milk in mid lactation. *Italian Journal of Animal Science, 5*, 349–358.

Martini, M., Altomonte, I. & Salari, F. (2014). Amiata donkeys: Fat globule characteristics, milk gross composition and fatty acids. *Italian Journal of Animal Science, 13*, 123–126.

Mayhill, P. G. & Newstead, D. F. (1992). The effect of milkfat fractions and emulsified type on creaming in normal-solids UHT recombined milk. *Milchwissenschaft, 47*, 75–79.

McCarthy, O. J. (2003). Liquid products and semi-solid products. In H. Roginski, J. W. Fuquay & P. F. Fox (Eds.), *Encyclopedia of dairy sciences* (pp. 2445–2456). Amsterdam: Academic.

McClements, D. J. (1999). *Food emulsions, principles, practice and techniques*. Boca Raton: CRC Press.

McCrae, C. H. (1994). Homogenization of milk emulsions: Use of microfluidizer. *Journal of the Society of Dairy Technology, 47*, 28–31.

McCrae, C. H. & Muir, D. D. (1992). Heat stability of recombined milk; influence of lecithins on the heat coagulation time-pH profile. *Journal of Dairy Research, 59*, 177–185.

McCrae, C. H. & Lepoetre, A. (1996). Characterization of dairy emulsions by forward lobe laser light scattering – Application to milk and cream. *International Dairy Journal, 6*, 247–256.

McCrae, C. H., Hirst, D., Law, A. J. R. & Muir, D. (1994). Heat stability of homogenized milk; role of interfacial protein. *Journal of Dairy Research, 61*, 507–516.

Mehaia, M. A. (1995). The fat globule size distribution in camel, goat, ewe and cow milk. *Milchwissenschaft, 50*, 260–263.

Melsen, J. P. & Walstra, P. (1989). Stability of recombined milk fat globules. *Netherlands Milk and Dairy Journal, 43*, 63–78.

Ménard, O., Ahmad, S., Rousseau, F., Briard-Bion, V., Gaucheron, F. & Lopez, C. (2010). Buffalo vs. cow milk fat globules: Size distribution, zeta-potential, compositions in total fatty acids and in polar lipids from the milk fat globule membrane. *Food Chemistry, 120*, 544–551.

Merthens, E. (1933a). Beitrage zum Aufrahmungsproblem. I. Praktische Versuche zum Problem der Aufrahmung. *Milchwirtschaftliche Forschungsberichte, 14*, 1–20.

Merthens, E. (1933b). Beitrage zum Aufrahmungsproblem. II. Anreicheung von capillaraktiven natürlichen Milchbestandteilen in Milch zur Verbesserung der Aufrahmung. *Milchwirtschaftliche Forschungsberichte, 14*, 37–42.

Michalski, M. C., Briard, V. & Michel, F. (2001a). Optical parameters of fat globules for laser light scattering measurements. *Le Lait, 81*, 787–796.
Michalski, M. C., Michel, F., Sainmont, D. & Briard, V. (2001b). Apparent ζ-potential as a tool to assess mechanical damages to the milk fat globule membrane. *Colloids and Surfaces B: Biointerfaces, 23*, 23–30.
Michalski, M. C., Michel, F. & Geneste, C. (2002). Appearance of submicronic particles in the milk fat globule size distribution upon mechanical treatments. *Le Lait, 82*, 193–208.
Michalski, M. C., Briard, V., Michel, F., Tasson, F. & Poulain, P. (2005). Size distribution of fat globules in human colostrum, breast milk, and infant formula. *Journal of Dairy Science, 88*, 1927–1940.
Miles, C. A., Shore, D. & Langley, K. R. (1990). Attenuation of ultrasound in milks and creams. *Ultrasonics, 28*, 394–400.
Muir, D. D., McCrae-Homsma, C. H. & Sweetsur, A. W. M. (1991). Characterisation of dairy emulsions by forward lobe laser light scattering – Application to cream liqueurs. *Milchwissenschaft, 46*, 691–694.
Mulder, H. & Walstra, P. (1974). *The Milk fat globule: Emulsion science as applied to milk products and comparable foods*. Wageningen: PUDOC.
Mutoh, T. A., Nakagawa, S., Noda, M., Shiinoki, Y. & Matsumura, Y. (2001). Relationship between characteristics of oil droplets and solidification of thermally treated creams. *Journal of the American Oil Chemists' Society, 78*, 177–182.
Niar, M. G., Mistry, V. V. & Oommen, B. S. (2000). Yield and functionality of Cheddar cheese as influenced by homogenization of cream. *International Dairy Journal, 10*, 647–657.
Noda, M. & Yamamoto, H. (1994). Effects of tempering on physical properties of cream. *Nippon Shokuhin Kogyo Gakkaishi, 41*, 327–334.
Noda, M. & Yamamoto, H. (1996). The effect of homogenization on the physical properties of whip cream. *Nippon Shokuhin Kagaku Kogaku Kiashi, 43*, 896–903.
Ogden, L. V., Walstra, P. & Morris, H. A. (1976). Homogenization-induced clustering of fat globules in cream and model systems. *Journal of Dairy Science, 59*, 1727–1737.
O'Mahony, J. P. & Shipe, W. F. (1972). Hydrolysis of the lipoprotein fractions of milk by phospholipase C. *Journal of Dairy Science, 55*, 408–412.
Ong, L., Dagastine, R. R., Kentish, S. E. & Gras, S. L. (2010). The effect of milk processing on the microstructure of the milk fat globule and rennet induced gel observed using confocal laser scanning microscopy. *Journal of Food Science, 75*, 135–145.
Oortwijn, H. & Walstra, P. (1979). The membranes of recombined fat globules. 2. Composition. *Netherlands Milk and Dairy Journal, 33*, 134–154.
Oortwijn, H. & Walstra, P. (1982a). The membranes of recombined fat globules. 3. Mode of formation. *Netherlands Milk and Dairy Journal, 36*, 103–113.
Oortwijn, H. & Walstra, P. (1982b). The membranes of recombined fat globules. 4. Effects on properties of recombined milks. *Netherlands Milk and Dairy Journal, 36*, 279–290.
Oortwijn, H., Walstra, P. & Mulder, H. (1977). The membranes of recombined fat globules. 1. Electron microscopy. *Netherlands Milk and Dairy Journal, 31*, 134–147.
Orla-Jensen, S., Le Dous, C., Jacobsen, J. & Otte, N. C. (1929). Recherches sur la pasteurisation basse du lait. Une nouvelle reaction pour controller la temperature de chauffage. *Le Lait, 9*, 622–648, 724–740, 816–831, 914–927 and 1032–1046.
Pal, R. (1996). Rheology of emulsions containing polymeric liquids. In P. Becher (Ed.), *Encyclopedia of emulsion technology* (Vol. 4, pp. 93–264). New York: Marcel Dekker Inc.
Paquin, P. (1999). Technological properties of high pressure homogenizers: The effect on fat globules, milk proteins, and polysaccharides. *International Dairy Journal, 9*, 329–335.
Patton, S., Long, C. & Sokka, T. (1980). Effect of storing milk on cholesterol and phospholipid content of skim milk. *Journal of Dairy Science, 63*, 697–700.
Payens, T. A. J. (1963). Influence of particle-size on the agglutination of the milk-fat globules; a preliminary account. *Netherlands Milk and Dairy Journal, 17*, 150–153.
Payens, T. A. J. (1964). Das Problem der Kolloidchemischen Stabilitat der Fettkugelchen in der Milch. *Kieler Milchwirtschaftliche Forschungsberichte, 16*, 457–472.
Payens, T. A. J. (1968). Neuere Forschungen auf dem Gebiet der Milcheiweisse. *Milchwissenschaft, 23*, 325–330.
Payens, T. A. J. & Both, P. (1970). Cryoglobulins from milk and a mechanism for cold agglutination of milk fat globules. *Immunochemistry, 7*, 869.
Payens, T. A. J., Koops, J. & Kerkhof Mogot, M. F. (1965). Adsorption of euglobulin on agglutinating milk fat globules. *Biochimica et Biophysica Acta, 94*, 576–578.
Phillips, L., McGiff, M. L., Barbano, D. M. & Lawless, H. T. (1995). The influence of fat on the sensory properties, viscosity, and color of lowfat milk. *Journal of Dairy Science, 78*, 1258–1266.
Phipps, L. W. (1969). The interrelationship of the viscosity, fat content and temperature of cream between 40 and 80°C. *Journal of Dairy Research, 36*, 417–426.
Phipps, L. W. (1983). Effects of fat concentration on homogenization of cream. *Journal of Dairy Research, 50*, 91–96.
Precht, D., Peters, K. H. & Petersen, J. (1987). Qualitätsverbesserung von Schlagsahne durch Zusatz von Carrageen under Milchinhaltstoffen. I. Einfluss von Zusätzen auf die Pfropfenbildung bei pasteurisiertem, unhomogenisiertem Rahm. *Milchwissenschaft, 42*, 635–639.
Prentice, J. H. (1968). Measurement of some flow properties of market cream. *S.C.I. Monograph, 27*, 265–279.

Prentice, J. H. (1992). *Dairy rheology: A concise guide*. New York: VCH Publishers, Inc.

Randhahn, H. (1973). Beitrag zum Fliessverhalten von Milch- und Milchkonzentraten. *Milchwissenschaft, 28*, 620–628.

Robin, O. & Paquin, P. (1991). Evaluation of the particle size of fat globules in a milk model emulsion by photon correlation spectroscopy. *Journal of Dairy Science, 74*, 2440–2447.

Rowland, S. J. (1937). The creaming power of milk. The relationship between the denaturation of albumin and globulin and the reduction in creaming power. *Journal of Dairy Research, 8*, 195–202.

Rüegg, M. & Blanc, B. (1981). The fat globule size distribution in human milk. *Biochimica et Biophysica Acta, 666*, 7–14.

Sadiq, F. A., Li, Y., Liu, T. J., Flint, S., Zhang, G. H., Yuan, L., Pei, Z. P. & He, G. Q. (2016). The heat resistance and spoilage potential of aerobic mesophilic and thermophilic spore forming bacteria isolated from Chinese milk powders. *International Journal of Food Microbiology, 238*, 193–201.

Samuelson, E., Bengstsson, G., Nilsson, S. & Mattsson, N. (1954). Gräddsättning. *Svenska Mejeritidningen, 64*, 163–217.

Schmidmeier, C., O'Gorman, C., Dprapala, K. P., Waldron, D. S. & O'Mahony, J.A. (2019) Elucidation of factors responsible for formation of white flecks in reconstituted fat filled milk powders. *Colloids and Surfaces A – Physicochemical and Engineering Aspects, 575*, 245–255.

Schmidt, D. G., Buchheim, W. & Koops, J. (1971). An electron-microscopic study of fat-protein complexes in evaporated milk, using the freeze-etching technique. *Netherlands Milk and Dairy Journal, 25*, 200–216.

Sharma, R. & Dalgleish, D. G. (1994). Effects of heat treatments on the incorporation of milk serum proteins into the fat globule membrane of homogenised milk. *Journal of Dairy Research, 61*, 375–384.

Sharma, R. & Singh, H. (1998). Adsorption behaviour of commercial milk protein and milk powder products in low-fat emulsions. *Milchwissenschaft, 53*, 373–377.

Sharma, R. & Singh, H. (1999). Heat stability of recombined milk systems as influenced by the composition of fat globule surface layers. *Milchwissenschaft, 54*, 193–196.

Sharma, R., Singh, H. & Taylor, M. W. (1996a). Recombined milk: Factors affecting the protein coverage and composition of fat globule surface layers. *Australian Journal of Dairy Technology, 51*, 12–16.

Sharma, R., Singh, H. & Taylor, M. W. (1996b). Composition and structure of fat globule surface layers in recombined milk. *Journal of Food Science, 61*, 28–32.

Shi, Y., Smith, C. M. & Hartel, R. W. (2001). Compositional effects on milk fat crystallization. *Journal of Dairy Science, 84*, 2392–2401.

Shimizu, M., Yamauchi, K. & Kanno, C. (1980). Effect of proteolytic digestion of milk fat globule membrane proteins on stability of the globules. *Milchwissenschaft, 35*, 9–12.

Singh, H., Sharma, R., Taylor, M. W. & Creamer, L. K. (1996). Heat-induced aggregation and dissociation of protein and fat particles in recombined milk. *Netherlands Milk and Dairy Journal, 50*, 149–166.

Söderberg, I., Hernqvist, L. & Buchheim, W. (1989). Milk fat crystallization in natural milk fat globules. *Milchwissenschaft, 44*, 403–406.

Speroni, A. & Bertoni, G. (1984). L'affioramento del grasso del latte: Nuove proposte per la valutazione e l'interpretazione del fenomeno. *Science Tecn Latt-Cas, 35*, 97–108.

Stadhouders, J. & Hup, G. (1970). Complexity and specificity of euglobulin in relation to inhibition of bacteria and to cream rising. *Netherlands Milk and Dairy Journal, 24*, 79–95.

Stone, M. J. (1952a). The effect of temperature on the development of broken cream. *Journal of Dairy Research, 19*, 302–310.

Stone, M. J. (1952b). The action of lecithinase of *Bacillus cereus* on the globule membrane of milk fat. *Journal of Dairy Research, 19*, 311–315.

Stone, M. J. & Rowlands, A. (1952). 'Broken' or 'bitty' cream in raw and pasteurized milk. *Journal of Dairy Research, 19*, 51–62.

Strawbridge, K. B., Ray, E., Hallett, F. R., Tosh, S. M. & Dalgleish, D. G. (1995). Measurement of particle size distributions in milk homogenized by a microfluidizer: Estimation of populations of particles with radii less than 100 nm. *Journal of Colloid and Interface Science, 171*, 392–398.

Streuper, A. & Van Hooijdonk, A. C. M. (1986). Heat treatment of whipping cream. II. Effect on cream plug formation. *Milchwissenschaft, 41*, 547–552.

Su, J. & Everett, D. W. (2003). Adsorption of *K*-casein onto native milk fat globule, latex particle, and emulsion surfaces. *Food Hydrocolloids, 17*, 529–537.

Sugimoto, T., Mori, T., Mano, J., Mutoh, T. A., Shiinoki, Y. & Matsumura, Y. (2001). Effect of fat crystallization on the behavior of proteins and lipids at oil droplet surfaces. *Journal of the American Oil Chemists' Society, 78*, 183–188.

Sweetsur, A. W. M. & Muir, D. D. (1983). Effect of homogenization on the heat stability of milk. *Journal of Dairy Research, 50*, 291–300.

Tadros, T. F. & Vincent, B. (1983). Emulsion stability. In P. Becher (Ed.), *Encyclopedia of emulsion technology* (Vol. 1, pp. 129–268). New York: Marcel Dekker.

Tarassuk, N. P. & Koops, J. (1960). Inhibition of oxidized flavour in homogenised milk as related to the concentration of copper and phospholipids per unit of fat globule surface. *Journal of Dairy Science, 43*, 93–94.

Tatar, V., Mootse, H., Sats, A., Mahla, T., Kaart, T. & Poikalainen, V. (2015). Evaluation of size distribution of fat globules and fat and protein content in Estonian goat milk. *Agronomy Research, 13*, 1112–1119.

Ten Grotenhuis, E., Tuinier, R. & de Kruif, C. G. (2003). Phase stability of concentrated dairy products. *Journal of Dairy Science, 86*, 764–769.

Te Whaiti, I. E. & Fryer, T. F. (1975). Factors that determine the gelling of cream. *New Zealand Journal of Dairy Science and Technology, 10*, 2–7.

Thiebaud, M., Dumay, E., Picart, L., Guiraud, J. P. & Cheftel, J. C. (2003). High-pressure homogenization of raw bovine milk. Effects on fat globule size distribution and microbial inactivation. *International Dairy Journal, 13*, 427–439.

Timmen, H. & Patton, S. (1988). Milk fat globules: Fatty acid composition, size and in vivo regulation of fat liquidity. *Lipids, 23*, 685–689.

Tobin, J., Heffernan, S. P., Mulvihill, D. M., Huppertz, T. & Kelly, A. L. (2015). Applications of high-pressure homogenization and microfluidization for milk and dairy products. In N. Datta & P. M. Tomasula (Eds.), *Emerging dairy processing technologies: opportunities for the dairy industry* (pp. 93–114). John Wiley & Sons Ltd.

Tran, M., Roberts, R., Felix, T. L. & Harte, F. M. (2018). Effect of high-pressure-jet processing on the viscosity and foaming properties of pasteurized whole milk. *Journal of Dairy Science, 101*, 3887–3899.

Troy, H. C. & Sharp, P. F. (1928). Physical factors influencing the formation and fat content of gravity cream. *Journal of Dairy Science, 11*, 189–226.

Truong, T., Palmer, M., Bansal, N. & Bhandori, B. (2016). *Effect of milk fat globule size on the physical functionality of dairy products* (Springer Briefs in Food, Health and Nutrition). Cham: Springer.

Tuinier, R. & De Kruif, C. G. (2002). Stability of casein micelles in milk. *Journal of Chemical Physics, 117*, 1290–1295.

Uniacke-Lowe, T. (2011). *Studies on equine milk and comparative studies on equine and bovine milk systems*. PhD thesis. University College Cork, Cork, Ireland.

Uniacke-Lowe, T. & Fox, P. F. (2011). Milk: Equid milk. In J. W. Fuquay, P. F. Fox & P. L. H. McSweeney (Eds.), *Encyclopedia of dairy sciences* (Vol. 3, 2nd ed., pp. 518–529). San Diego: Academic.

Van Boekel, M. A. J. S. & Walstra, P. (1981). Stability of oil-in-water emulsions with crystals in the disperse phase. *Colloids and Surfaces, 3*, 109–118.

Van Boekel, M. A. J. S. & Walstra, P. (1989). *Physical changes in the fat globules in unhomogenized and homogenized milk* (Bulletin 238) (pp. 13–16). Brussels: International Dairy Federaration.

Van Boekel, M. A. J. S. & Folkerts, T. (1991). Effect of heat treatment on the stability of milk fat globules. *Milchwissenschaft, 46*, 758–765.

Van Boekel, M. A. J. S. & Walstra, P. (1995). Effect of heat treatment on chemical and physical changes to milk-fat globules. In P. F. Fox (Ed.), *Heat-induced changes in Milk* (2nd ed., pp. 51–65). Brussels: International Dairy Federation.

Van Vliet, T. & Walstra, P. (1980). Relationship between viscosity and fat content of milk and cream. *Journal of Texture Studies, 11*, 65–68.

Verwey, E. J. W. & Overbeek, J. T. G. (1948). *Theory of the stability of lyphobic colloids: The interaction of sol particles having an electric double layer*. Amsterdam: Elsevier.

Vignolles, M. L., Jeantet, R., Lopez, C. & Schuck, P. (2007). Free fat, surface fat and dairy powders: Interactions between process and product. A review. *Le Lait, 87*, 187–236.

Villamiel, M., Van Hamersveld, E. H. & De Jong, P. (1999). Review: Effect of ultrasound processing on the quality of dairy products. *Milchwissenschaft, 54*, 69–73.

Vithanage, N. R., Dissnayake, M., Bolge, G., Palombo, E. A., Yaeger, T. R. & Datta, N. (2016). Biodiversity of culturable psychrotrophic microbiota in raw milk attributable to refrigeration conditions, seasonality and their spoilage potential. *International Dairy Journal, 57*, 80–90.

Wade, T. & Beattie, J. K. (1997). Electroacoustic determination of size and zeta potential of fat globules in milk and cream emulsions. *Colloids and Surfaces B, 10*, 73–85.

Wade, T., Beattie, J., Rowlands, W. N. & Augustin, M. A. (1996). Electroacoustic determination of size and zeta potential of casein micelles in skim milk. *Journal of Dairy Research, 63*, 387–404.

Wahba, A., El-Hagarawy, I. S. & Zeidan, I. A. (1977). The effect of addition of globulin protein fractions on the creaming of buffaloes' and cow's milk. *Egyptian Journal of Dairy Science, 5*, 43–46.

Walstra, P. (1967). On the crystallization habit in milk fat globules. *Netherlands Milk and Dairy Journal, 21*, 166–191.

Walstra, P. (1969a). Studies on milk fat dispersion. II. The globule-size distribution of cow's milk. *Netherlands Milk and Dairy Journal, 23*, 99–110.

Walstra, P. (1969b). Studies on milk fat dispersion. V. The mean distance between fat globules. *Netherlands Milk and Dairy Journal, 23*, 245–249.

Walstra, P. (1975). Effect of homogenization on the fat globule size distribution in milk. *Netherlands Milk and Dairy Journal, 29*, 297–294.

Walstra, P. (1980). Effect of homogenization on milk plasma. *Netherlands Milk and Dairy Journal, 34*, 181–190.

Walstra, P. (1995). Physical chemistry of milk fat globules. In P. F. Fox (Ed.), *Advanced dairy chemistry, Vol. 2: Lipids* (pp. 131–178). London: Chapman and Hall.

Walstra, P. (1996). Emulsion stability. In P. Becher (Ed.), *Encyclopedia of emulsion technology* (Vol. 4, pp. 1–64). New York: Marcel Dekker Inc.

Walstra, P. (2003). *Physical chemistry of foods*. New York: Marcel Dekker, Inc.

Walstra, P. & Borggreve, G. J. (1966). Note on refractive index of individual milk fat globules. *Netherlands Milk and Dairy Journal, 20*, 140–144.

Walstra, P. & Jenness, R. (1984). *Dairy chemistry and physics*. New York: Wiley.

Walstra, P. & Oortwijn, H. (1975). Effect of globule size and concentration on creaming in pasteurized milk. *Netherlands Milk and Dairy Journal, 29*, 263–278.

Walstra, P. & Oortwijn, H. (1982). The membranes of recombined fat globules. 3. Mode of formation. *Netherlands Milk and Dairy Journal, 36*, 103–113.
Walstra, P., Oortwijn, H. & De Graaf, J. J. (1969). Studies on milk fat dispersion. I. Methods for determining globule-size distribution. *Netherlands Milk and Dairy Journal, 23*, 12–36.
Walstra P., Wouters J. T. M. & Geurts T. J. (2006). *Dairy science and technology*. CRC Press, Boca Raton, FL, USA, pp. 497–512.
Walstra, P., Van Vliet, T. & Kloek, W. (1995). Crystallization and rheological properties of milk fat. In P. F. Fox (Ed.), *Advanced dairy chemistry volume 2: Lipids* (pp. 179–211). London: Chapman and Hall.
Walstra, P., Geurts, T. J., Noomen, A., Jellema, A. & Van Boekel, M. A. J. S. (1999). *Dairy technology, principles of milk properties*. New York: Marcel Dekker, Inc.
Wayne, J. E. & Shoemaker, C. F. (1988). Rheological characterization of commercially processed fluid milks. *Journal of Texture Studies, 19*, 143–152.
Whitnah, C. H., Rutz, W. D. & Fryer, H. C. (1956). Some physicochemical properties of milk. III. Effects of homogenization pressures on the viscosity of whole milk. *Journal of Dairy Science, 39*, 1500–1505.
Wiking, L., Björck, L. & Nielsen, J. H. (2003). Influence of feed composition on stability of fat globules during pumping of raw milk. *International Dairy Journal, 13*, 797–803.
Wiking, L., Stagsted, J., Björck, L. & Nielsen, J. H. (2004). Milk fat globule size is affected by fat production in dairy cows. *International Dairy Journal, 14*, 909–913.
Willenbacher, N. & Georgieva, K. (2013). Rheology of disperse systems. In U. Brockel, W. Meier & G. Wagner (Eds.), *Product design and engineering: Formulation of gels and pastes*. Weinheim: Wiley-VCH Verlag GmbH & Co. KGaA.
Wu, H., Hulbert, G. J. & Mount, J. R. (2001). Effects of ultrasound on milk homogenization and fermentation with yoghurt starter. *Innovative Food Science and Emerging Technologies, 1*, 211–218.
Ye, A., Singh, H., Taylor, M. W. & Anema, S. (2002). Characterisation of protein components of natural and heat-treated fat globule membranes. *International Dairy Journal, 12*, 393–402.
Ye, A., Singh, H., Oldfield, D. J. & Anema, S. (2004). Kinetics of heat-induced association of β-lactoglobulin and α-lactalbumin with milk fat globule membrane in whole milk. *International Dairy Journal, 14*, 389–398.
Zou, X.-Q., Guo, Z., Huang, J.-H., Jin, Q.-Z., Cheong, L.-Z., Wang, X.-G. & Xu, X.-B. (2012). Human milk fat globules from different stages of lactation: A lipid composition analysis and microstructure characterization. *Journal of Agricultural and Food Chemistry, 60*, 7158–7167.

Composition, Fractionation, Techno-Functional Properties and Applications of Milk Fat Globule Membrane–Derived Material

6

D. Oliveira and J. A. O'Mahony

6.1 Introduction

The primary sources of polar lipids in the human diet include milk, soybeans, fish and eggs (Küllenberg *et al.* 2012). Milk fat occurs as globules with a non-polar lipid core, composed primarily of triglycerides and surrounded by the milk fat globule membrane (MFGM), which contains both phospholipids (PLs) and glycoproteins, collectively accounting for over 90% (in dry weight) of the membrane. Due to their origin, structure and original function in stabilising the fat globules in whole milk, MFGM-derived materials are efficient natural emulsifiers (Singh 2006) and are preferentially enriched in aqueous dairy phases, such as skimmed milk, buttermilk (BM) and butter serum (Rombaut *et al.* 2006a). However, the emulsifying properties of these MFGM-derived materials are strongly dependent on their content and profile of polar lipids and proteins, as well as the possible interactions between these components. The proteins of the MFGM represent 1–4% of total protein, and the MFGM PLs represent less than 1% of the total lipids in milk (Corredig *et al.* 2003). Due to its complex structure and component interactions, the MFGM and its derivatives, namely the specific membrane proteins and PLs, remain some of the least well understood components of milk. Despite the growing interest in the nutritional and functional properties of dairy PLs, the utilisation of MFGM to formulate PL-enriched ingredients remains poorly exploited commercially, and the study of MFGM-derived material sources, particularly other than those from BM, is still at an early stage of research. In addition, the starting material and the type of processing applied distinctly influence the composition of MFGM-derived fractions (Phan *et al.* 2014). Moreover, the lack of standardised isolation procedures and the use of different analytical approaches in such studies result in highly variable literature values for the composition of the MFGM-derived materials. This variability makes the comparison and discussion of sometimes

Marie Skłodowska Curie-Career FIT fellow funded by the European Union's Horizon 2020 research and innovation programme under the Marie Skłodowska-Curie grant agreement No. 713654.

D. Oliveira (✉)
School of Food and Nutritional Sciences, University College Cork, Cork, Ireland

Dairy Processing Technology Centre, University College Cork, Cork, Ireland

Universidade Católica Portuguesa, CBQF – Centro de Biotecnologia e Química Fina – Laboratório Associado, Escola Superior de Biotecnologia, Porto, Portugal
e-mail: doliveira@porto.ucp.pt

J. A. O'Mahony
School of Food and Nutritional Sciences, University College Cork, Cork, Ireland

Dairy Processing Technology Centre, University College Cork, Cork, Ireland

P. L. H. McSweeney et al. (eds.), *Advanced Dairy Chemistry, Volume 2*,
https://doi.org/10.1007/978-3-030-48686-0_6

contradictory results challenging, as emphasised by Holzmüller and Kulozik (2016).

Taking into consideration the state-of-the-art in the field of MFGM and MFGM-derived materials, this chapter aims to compile, and to assess critically, the available information on MFGM composition, particularly PLs and specific membrane proteins. In addition, the influence of the processing steps, the techno-functional properties and potential applications of MFGM-derived material are discussed.

6.2 Composition of the Milk Fat Globule Membrane

6.2.1 Milk Fat Globules and Milk Fat Globule Membrane

In milk, fat is present in the form of native milk fat globules (MFGs) and surrounded by a membrane referred to as the milk fat globule membrane (MFGM) (Figure 6.1). The MFG is a complex matrix composed of tri- (TGs), di- (DGs) and mono-glycerides (MGs), in addition to sterols and sterol esters, and serves as an energy store. MFGM lipids contain about 60% triglycerides and 0.3–2.3% sterols, 90% of which is cholesterol (Keenan and Mather 2006; Rombaut and Dewettinck 2006; Dewettinck *et al.* 2008). Bioactive sterols, such as lanosterol, lathosterol, desmosterol, stigmasterol and β-sitosterol, have also been identified (Fauquant *et al.* 2007). Other constituents are hexoses, hexosamines and sialic acids, which are associated exclusively with proteins and glycosphingolipid (Keenan and Mather 2006). MFGM is a surface-active membrane surrounding each of the MFGs, allowing them to remain colloidally dispersed in milk, mainly due to the polar lipid bilayer. It contains both PLs and glycoproteins, in addition to enzymes and other trace components (Keenan and Mather 2006; Mezouari *et al.* 2009; Ortega-Requena and Rebouillat 2015), distributed heterogeneously in different quantities within the MFGM (Dewettinck *et al.* 2008; Jiménez-Flores and Brisson 2008; Lopez *et al.* 2008). Because of its cellular membrane origin, the MFGM is the richest source of PLs, glycolipids, gangliosides and glycoproteins in milk. The general composition of milk lipids and their distribution between the MFG and the MFGM is shown in Table 6.1. The properties of the MFGM components, that is, specific membrane proteins and PLs, have been investigated and a number of associated health benefits have been reported. These biological activities associated with MFGM have triggered consideration of its potential as a nutraceutical in food applications (Spitsberg 2005). However, this potential has yet to be fulfilled, and some structural considerations of the MFGM-derived materials with regard to the suggested health benefits need to be further exploited (Jiménez-Flores and Brisson 2008). The utilisation of pure MFGM, extracted from raw cream and/or milk, as a functional ingredient, is currently deemed to be uneconomic (Danthine *et al.* 2000). Thus, the challenge is to develop techniques that allow isolation or enrichment of MFGM from sources other than raw milk, while maintaining its biological and functional properties.

6.2.2 Structure of the Milk Fat Globule Membrane

A tri-layered structure consisting of polar lipids and proteins is widely accepted as an appropriate description of the MFGM surrounding secreted fat globules in the lumen (Holzmüller and Kulozik 2016). The innermost membrane is a monolayer, containing PLs and proteins, in direct contact with the intracellular lipid core. In this monolayer, the hydrophobic tails of the polar lipids are in contact with the triglyceride-rich core. Secondly, there is an outer bilayer that originates from the secretory cell apical plasma membrane and where the outermost hydrophilic head groups of the polar lipids are in contact with the aqueous phase of milk. This second bilayer membrane contains glycosylated and non-glycosylated proteins, glycerophospholipids and sphingolipids, enzymes, cholesterol and other minor components. Between the inner membrane and this bilayer, there is electron-dense coat material, rich

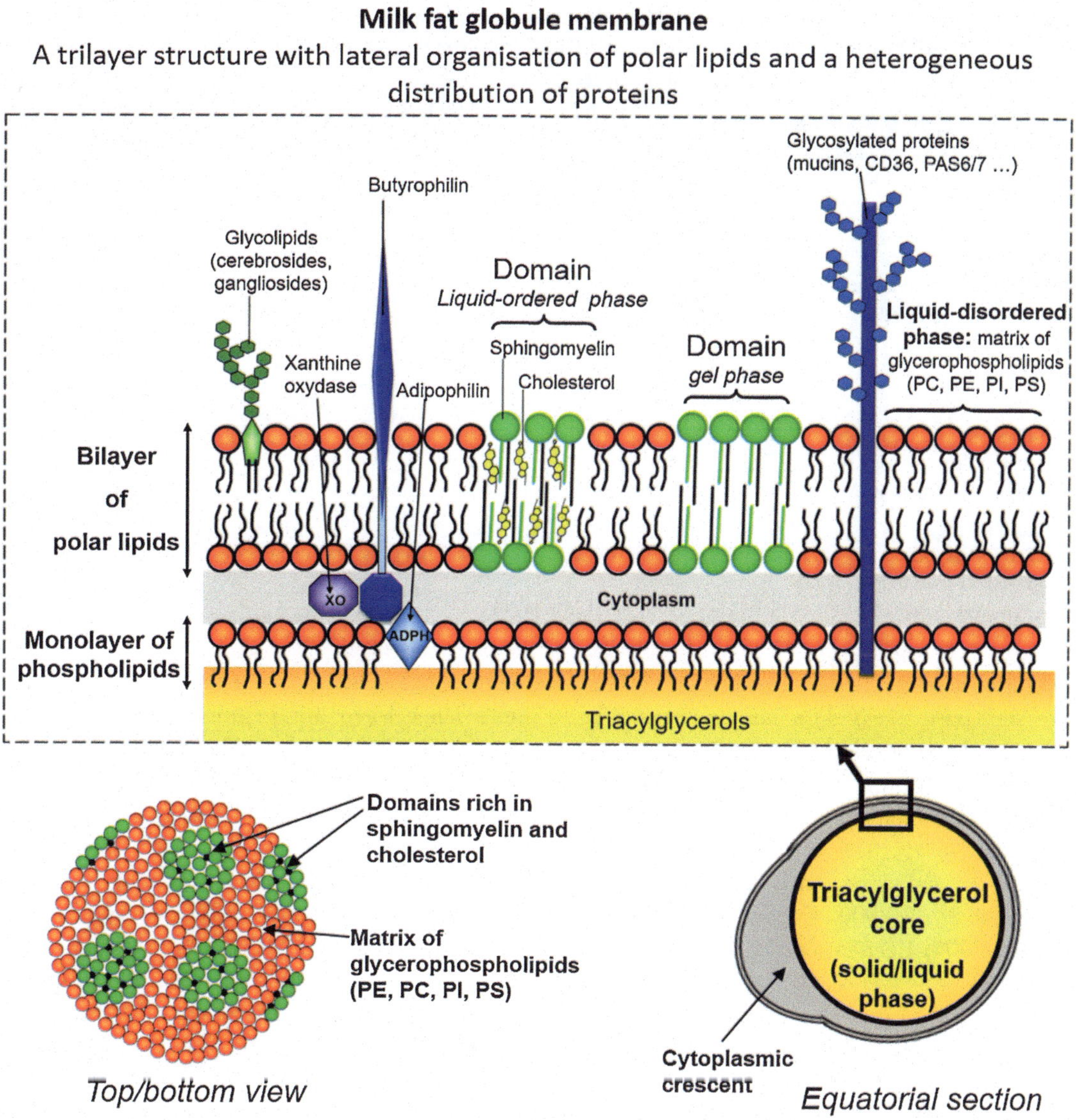

Figure 6.1. Superstructure of milk fat globules with a schematic representation of the milk fat globule membrane (MFGM). *PC* phosphatidylcholine, *PE* phosphatidylethanolamine, *PS* phosphatidylserine, *PI* phosphatidylinositol, *CD 36* cluster of differentiation 36, *PAS 6/7* lactadherin or periodic acid Schiff 6/7. Adapted from Lopez (2011), with permission.

in proteins (Rombaut *et al.* 2006a; Lopez *et al.* 2008; Vanderghem *et al.* 2010). A schematic representation of the MFGM is shown in Figure 6.1.

The emulsifying properties of these MFGM materials are strongly dependent on the content and type of polar lipids and proteins, as well as the possible interactions between these components, which can affect the distribution and composition of the PLs in the final matrix (Contarini and Povolo 2013) and consequently have a significant impact on their nutritional and technological functionalities. The MFGM structure could be considered to some extent, as a fingerprint of milk fat biosynthesis in the mammary epithelial cells (see Chapter 4). Extensive knowledge of the intracellular origin, composition and

Table 6.1. General composition of milk lipids and their distribution between the milk fat globule (MFG) and the milk fat globule membrane (MFGM)

Lipid class	Total fat (g·kg^{-1})	Fraction content in		
		MFG (%)	MFGM (%)	Skim (%)
Neutral glycerides				
Triacylglycerol	38.3–39.3	100[a]	n.p.	n.p.
Diacylglycerol	0.11–0.90	~90	~10	n.p.
Mono-acylglycerol	0.01–0.17	Trace	Trace	Trace
Free fatty acids	0.04–0.18	60	~10	30
Phospholipids (including sphingomyelin, SGM)	0.08–0.44	n.p.	65	35
Cerebrosides	0.4	n.p.	70	30
Gangliosides	0.004	n.p.	~70	~30
Sterols	n.p.	80	10	10
Cholesterol	0.12–0.18	n.p.	n.p.	n.p.
Cholesteryl ester	≤0.008	n.p.	n.p.	n.p.
Carotenoids + Vitamin A	0.0008	~95	~5	Trace

n.p. not provided.
[a]TAG reported to be found in the MFGM is likely to be residues from the core because MFGM separation procedures are imperfect (Michalski 2007).
Adapted from Jiménez-Flores and Brisson (2008) and Michalski (2007).

structure of MFG, MFGM and MFGM-derived material, particularly PLs and membrane proteins, has been developed and accumulated over the years. Details on the origin, nature, composition, structure, nutritional, processing and technological properties of the MFGM and MFGM-derived components have been reviewed (Danthine *et al.* 2000; Keenan 2001; Heid and Keenan 2005; Keenan and Mather 2006; Singh 2006; Dewettinck *et al.* 2008; Lopez *et al.* 2008; El-Loly 2011; Lopez *et al.* 2017c; Singh and Gallier 2017), including in human and other non-bovine species (Cilla *et al.* 2016; Yao *et al.* 2016a; Ali *et al.* 2018; Luo *et al.* 2018). Moreover, knowledge of milk fat synthesis and the MFG secretion process is essential to develop a better understanding of the structure of the MFGM.

6.2.3 Phospholipids of the Milk Fat Globule Membrane

When the MFGM is disrupted, for example, by homogenisation of milk, MFGM-derived material, that is, PLs and membrane proteins, are released into the serum phase, and thus, aqueous phases like skimmed milk, BM and butter serum are preferentially enriched with these components (Rombaut *et al.* 2006a). Not only do these materials act as natural emulsifying agents, preventing the flocculation and coalescence of the MFG, they also protect the milk fat against enzymatic action. From an emulsion science perspective, the MFGs are not a simple oil-in-water emulsion, as the MFGM surrounding the globules is not a simple monolayer of surface-active material. Instead, the MFGM has a unique structure and a unique composition that reflect the fat globule biosynthesis process in the mammary secretory cells (Singh and Gallier 2017). Both technological and biological properties of the MFGM-derived material have led to an interest in concentrating and isolating this material from dairy sources. However, data from the literature have shown major differences in the functionality of MFGM-derived materials, with these differences attributed to various factors, such as type and composition of the raw material used for isolation of MFGM (e.g. cream, BM and whey) (Fuller *et al.* 2013), the use of pre-treatments, especially heat treatments (Corredig and Dalgleish 1997; Lee and Sherbon 2002), and differences in the choice of concentration and isolation procedures used (mostly membrane filtration) (Astaire *et al.* 2003; Roesch *et al.* 2004; Rombaut *et al.* 2006b; Morin *et al.* 2007a). This variability makes it difficult to compare reports on the composition of the isolated MFGM-derived material. Therefore, more detailed information on the partitioning of PLs and proteins of the MFGM and their potential in development of functional ingredients is needed.

Milk contains approximately 0.01–0.04% (w/w) of PLs (Vanderghem *et al.* 2010) distributed in an approximate 60:40 ratio between the MFGM and skim milk membrane fraction (Mather 2011). These PLs are divided into two main groups: glycerol-phospholipids (GPs) and sphingolipids (SPs). GPs are derived from glycerol with a polar head group and two fatty acids esterified at the *sn*-1 and *sn*-2 positions of the glycerol backbone. They include principally phosphatidylcholine (PC, 35%), phosphatidylethanolamine (PE, 30%), phosphatidylserine (PS, 3%) and phosphatidylinositol (PI, 5%) (Vanderghem *et al.* 2010). SPs are derived from sphingosine, and sphingomyelin (SGM, 22%) is the dominant species and is composed of a hydrophilic phosphorylcholine head group and a fatty acid linked to the amide nitrogen of the sphingoid long-chain base (Duan 2011). PC, PE and SGM constitute 80–90% of the total PLs (Rombaut *et al.* 2004). Other SPs present in lower amounts in the MFGM are glucosylceramide, lactosylceramide and gangliosides. Galactosylceramide, ganglioside and SGM have been identified in mammalian tissue but not in plant tissue (Duan 2011). The distribution of PLs is asymmetrical throughout the MFGM; PE, PS and PI are concentrated in the inner surface, whereas PC, SGM and glycosphingolipid, such as cerebrosides and gangliosides, are preferentially located in the bilayer membrane (Lopez *et al.* 2008). The presence of SGM in foods has been reported to have antibacterial effects *in vitro* (Sprong *et al.* 2001), and it plays important roles in infant neural development (Contarini and Povolo 2013), inflammatory processes (Duan and Nilsson 2009), immunity and gut health (Nilsson 2016). Foods rich in sphingolipids are eggs, soybeans, milk and dairy products such as cream and cheese. SGM in eggs is present almost exclusively in the yolk and represents 2–3% of the polar lipid fraction, while in bovine and human milks, SGM constitutes 25% and 35% of the total polar lipid, respectively (Küllenberg *et al.* 2012; Contarini and Povolo 2013). However, this sphingolipid is absent in plant-derived products such as soybeans (Liu and Ma 2011). An overview of the PL profile of different PL-rich products is provided in Table 6.2.

For infants, the only source of sphingolipids is milk (~160 μmol/kg; Duan 2011). As SGM is more abundant in human milk than in bovine milk, formula-fed infants are likely to have lower SGM intakes (Cilla *et al.* 2016). However, the SGM content of an infant formula is dependent on the composition of the raw materials and processing conditions used in their manufacture (Moloney *et al.* 2018). Claumarchirant *et al.* (2016) reported that infant formulae manufactured with a whey-based ingredient enriched in MFGM had higher SGM levels than those produced without it, and these MFGM-enriched formulae were reported to have SGM levels more closely resembling human milk. Although there

Table 6.2. Total phospholipid (PL) composition (%) from different sources

Phospholipid	Soy	Egg	Milk	Lacprodan® PL-20[a]
Phosphatidylcholine (PC)	34	75	27	5.4
Phosphatidylethanolamine (PE)	21	15	25	5
Phosphatidylinositol (PI)	18	0.4	8	1.6
Sphingomyelin (SGM)	0	1.5	24	4.8
Phosphatidylserine (PS)	0.5	0	12	2.4
Phosphatidic acid (PA)	9	0	0	n.p.
Others	17.5	8.1	4	0.8

n.p. not provided.
[a]% of individual phospholipid from a 20% total phospholipid-enriched product available from Arla Food Ingredients, DK.
Adapted from Burling *et al.* (2009).

is no essential nutritional requirement for PLs and sphingolipids, recent research has provided evidence suggesting a relationship between dietary consumption of these lipid components and enhanced health (Küllenberg *et al.* 2012), more specifically in terms of gut health (Duan 2011), cardiovascular diseases (Michalski 2007) and several other bioactivities (Burling *et al.* 2009). Thus, more studies on the characterisation of the composition of MFGM constituents, which can be used to produce dairy products with improved nutritional properties, are needed.

Proteomic and lipidomic analyses of the MFGM have been used extensively; however, as outlined earlier, the composition of PLs is variable and strongly dependent on the method and conditions of isolation (Mezouari *et al.* 2009; Gallier *et al.* 2010a). Therefore, it is important to compare the efficiency of different methods of extraction of PLs and determine the PL profile of dairy products, taking into account the precise extraction approaches and conditions used. In addition, the variability of the PL profile between different mammalian species and within bovine milk (Tables 6.3 and 6.4) reveals that not only genetics but also animal diet influence the composition of MFGM-derived PLs (Lopez *et al.* 2008).

Table 6.3. Concentration (% of total PLs) of the main phospholipids (PLs) in milk fat of different mammalian species

	PE	PI	PS	PC	SGM
Species	Values are expressed as percentage of total PLs[a]				
Buffalo	24.5	19.7	6.6	24.3	24.9
Goat	31.7	6.3	8.3	28.5	25.2
Ewe	34.4	4.4	5.2	28.6	27.4
Human[b]	21.3	16.4	n.p.	19	43.3
Donkey	60.2	2.4	11.2	17.3	8.8
Mare	24.3	8.5	10.6	27.8	28.9
Human	21.7	4.5	9.6	29	35.2
Camel	34.3	4.9	10.5	22.1	28.1

PE phosphatidylethanolamine, *PI* phosphatidylinositol, *PS* phosphatidylserine, *PC* phosphatidylcholine, *SGM* sphingomyelin, *n.p.* not provided.
Adapted from Contarini and Povolo (2013).
[a]The total amount of polar lipid was calculated considering an average fat content of cow milk of 3.5%.
[b]The percentage of PI includes also PS.

Table 6.4. Polar lipid content and phospholipid (PL) composition of liquid bovine milk

Polar lipid	PE	PI	PS	PC	SGM
g/100 g fat	% of total PL				
0.69	38.6	n.p.	n.p.	32.2	29.2
0.36	32.3	9.3	10.5	27.3	20.5
n.p.	26.9	13.7	4.1	27.5	27.7
0.96	33.2	5.2	9.3	27.4	25.1
0.7	46.4	5.3	7.4	21.1	19.8
n.p.	32.6	7.6	5.3	33.2	21.3
n.p.	36.4	7.6	6.5	32.1	17.3
0.25–0.30	26.8	13.6	16.1	22	21.6
0.48[a,b]	28.5	14.1	n.p.	32.7	23
0.36	38.5	6.5	7.7	25.9	21.4
n.p.	26.4	3.4	2	42.8	25.5
0.69	36.9	6.1	6.3	27	23.7
n.p.	72.3	1.4	11.5	8	7.9
n.p.	33.8	3.9	10.6	30.5	21.2
0.65–0.89[a]	34.2	7.7	8.6	45.5	4.1

n.p. not provided.
Adapted from Contarini and Povolo (2013).
[a]The total amount of polar lipid was calculated considering an average fat content of cow milk of 3.5%.
[b]The percentage of PI includes also PS.

6.2.4 Proteins of the Milk Fat Globule Membrane

The proteins of the MFGM represent only 1–4% of total milk protein, and depending on the milk source and how it is processed, 25–70% of the MFGM proteins may vary in molecular weight (MW) from 15,000 to 240,000 Da (Dewettinck *et al.* 2008). Although factors such as diet, breed, health and stage of lactation of cows can affect the protein concentration (Keenan and Mather 2006), the variations are probably due to a combination of membrane modifications during secretion or lactation and methodological differences in the sample preparation prior to analysis. Removal of skim milk-derived proteins, for example, by washing cream, destabilisation of fat globules and concentration of released membrane fragments, may affect the apparent compositional and structural properties of the MFGM (Holzmüller and Kulozik 2016). Additionally, inconsistencies regarding the structural composition of the MFGM found in the literature may be related to the application of different isolation methods for MFGM material, a challenge that Eigel *et al.* (1984) tried to overcome by proposing a standardised MFGM isolation procedure and to define the MFGM-derived proteins. According to these authors, several criteria must be met to characterise a protein as an MFGM-derived protein. Firstly, MFGM-derived proteins must be associated with the cream fraction after three washing steps. Secondly, the proteins must sediment with the membrane components after ultracentrifugation or salting out of washed cream BM. Thirdly, MFGM proteins should be extractable by sodium dodecyl sulphate-polyacrylamide gel electrophoresis (SDS–PAGE) and visualised afterwards. However, despite the meaningful attempt to define MFGM proteins and to standardise the isolation process, the approach proposed by Eigel *et al.* (1984) has rarely been applied. To date, more than 100 MFGM-associated proteins have been reported, and the MFGM proteins that the majority of studies have focused on are mucin 1 (MUC 1) and 15 (MUC 15), periodic acid Schiff 3 (PAS 3), cluster of differentiation 36 (CD 36), lactadherin or periodic acid Schiff 6/7 (PAS 6/7), butyrophilin (BTN), the redox enzyme, xanthine oxidase/dehydrogenase (XO/XDH), adipophilin (ADPH) and the fatty acid-binding protein (FABP), the latter three are unglycosylated (El-Loly 2011; Holzmüller and Kulozik 2016).

Some of the MFGM proteins are an integral part of the membrane, for example, XDH/XO, and located in the dense proteinaceous layer of the MFGM, whereas ADPH is located principally in the inner face of the polar lipid bilayer and FABP is located in the monolayer close to the lipid core (Vanderghem *et al.* 2011). MUC1, MUC15, BTN and CD36 are mostly transmembrane proteins, peripherally or loosely attached within the trilaminar membrane including BTN and glycosylated proteins, respectively, whereas PAS 6/7 is not anchored into the membrane but is loosely adsorbed at the surface (Dewettinck *et al.* 2008; Vanderghem *et al.* 2011). Mather (2000) provided an extensive overview on the major proteins of the MFGM with amino acid sequences and a proposed nomenclature of the proteins.

Since polar lipids and proteins are closely associated within the MFGM, they may have the tendency to co-partition during dairy processing; therefore, dairy products rich in polar lipids may be expected to be enriched in MFGM-derived proteins (Dewettinck *et al.* 2008). However, not all MFGM proteins fully meet the criteria set out by Eigel *et al.* (1984). Thus, depending on the isolation approach and conditions used, the composition of MFGM-derived material differs greatly. It is therefore questionable whether or not the procedure proposed by Eigel *et al.* (1984) would be appropriate and applicable as a standard method, since the outcome would not represent the total amount of MFGM-derived proteins of native fat globules in milk (Holzmüller and Kulozik 2016).

6.3 Fractionation of the Milk Fat Globule Membrane

6.3.1 Laboratory Scale

With some exceptions, the first step in any analysis is the isolation/extraction of the target compounds and the most commonly used method for the extraction of lipids is solvent extraction. The solubility of lipid in organic solvents is controlled by the hydrophobic hydrocarbon chains of the fatty acid or other aliphatic moieties and any polar functional groups, such as phosphate or sugar residues, which are hydrophilic. Triacylglycerol (TGs) and cholesteryl esters (CEs) are neutral lipids that do not have a polar group and are, thus, readily soluble in non-polar solvents (e.g. hexane, toluene and cyclohexane) and also in moderately polar solvents (e.g. diethyl ether or chloroform), whereas they have relatively low solubility in polar solvents such as methanol. More polar lipids such as PLs and SPs are only slightly soluble in hydrocarbon-based solvents, but have good solubility in more polar solvents, such as methanol, ethanol and chloroform. The pH is also important in the extraction of some lipid subclasses, either because of their acid-base characteristics or because of degradation in specific pH environments. For example, acidic lipids such as phosphatidic acid and PS are extracted more efficiently under acidic conditions. Careful optimisation of the extraction conditions is also essential for more polar lipids such as polyphosphoinositides and gangliosides (Sethi and Brietzke 2017). Several solvents have been used for lipid extraction, but currently, the most commonly used is that proposed by Folch *et al.* (1957), which uses 20 volumes of 2:1 chloroform/methanol (CM). According to Iverson *et al.* (2001), Folch's method achieved better isolation yields than with other similar methods, for example, that of Bligh and Dyer (1959). One of the challenges with using chloroform is its toxicity, and therefore, the tendency is to replace it using, for example, methyl *tert*-butyl ether (MTBE), which has shown good results (Matyash *et al.* 2008). More recent approaches have been focused on the utilisation of pressurised liquid extraction (PLE) for the isolation of milk fat (Castro-Gómez *et al.* 2014). These authors did not find differences between the composition of TG and the total fatty acid profile using either the PLE or Folch's method.

Conventionally, thin-layer chromatography (TLC) was used to separate different classes of lipids, such as steroids, triglycerides, free fatty acids and PLs, from crude lipid mixtures. However, TLC has limited resolution and sensitivity (Zhang *et al.* 2014). Alternatively, solid-phase extraction (SPE) columns are increasingly being used due to the availability of a wide variety of commercial products and high reproducibility (Pimentel *et al.* 2016).

6.3.2 Industrial Scale

Regarding food-grade, industrial-scale separation, isolation and fractionation techniques, the recovery of MFGM-derived material without impairing their composition and nutritional value is still a challenge for food chemists and the dairy industry. A survey of the literature shows that a number of separation approaches have been evaluated using various samples with different starting compositions and processing histories. However, inconsistencies regarding the structural composition of the MFGM found in the literature may be related to the application of numerous isolation methods for MFGM material (Holzmüller and Kulozik 2016); therefore, it is difficult to make general statements on the efficiency of the different separation techniques. The structure, composition and nature of the interactions between the membrane components are still not fully understood, and the MFGM presents a complex mixture of lipids, proteins, lipoproteins, glycolipids and enzymes with different polarities and affinities. In this context, the challenges encountered in isolating the MFGM and its polar and protein fractions in their intact and pure forms are to be expected.

Traditionally, milk is separated into cream and skim milk by mechanical processing based on differences in colloidal particle velocity under centrifugal fields. The isolation method is generally performed through a four-step process involving (1) separation of the fat globules, (2) washing of cream, (3) release of MFGM from washed cream and (4) collection of MFGM material. The effectiveness of the process is reduced significantly if fat globules are damaged before MFGM isolation, thus producing exposed MFGM fragments to which caseins bind and are difficult to wash out entirely (Jukkola and Rojas 2017). Despite its usefulness, centrifugal separation induces high shear stress that can remove proteins and phospholipids from the MFGM surface and cause coalescence of fat globules (Zheng *et al.* 2014). The amount of MFGM material that is lost during cream washing is often neglected, and washing native MFG may selectively remove MFGM components from the interface and affect enzyme activity (Le *et al.* 2009). In particular, some proteins do not sediment by ultracentrifugation, and up to 20% of the total MFGM proteins remain in the supernatant after centrifugation at 100,000 g (Kanno and Kim 1990; Mather 2000, 2011). Mather (2000) characterised eight major MFGM proteins: mucin 1 (MUC 1), periodic acid Schiff 3 (PAS 3), cluster of differentiation 36 (CD 36), butyrophilin (BTN), lactadherin or periodic acid Schiff 6/7 (PAS 6/7), the redox enzyme xanthine oxidase/dehydrogenase (XO/XDH), adipophilin (ADPH) and the fatty acid-binding protein (FABP), after ultracentrifugation of washed cream BM. Analysis of protein profile using SDS–PAGE has shown that MUC 1 and

PAS 6/7 remain mostly in the supernatant, whereas CD 36 and BTN are found mainly in the pellet. XO/XDH was found in both fractions in similar amounts after washing of cream. Also, Zheng *et al.* (2014) showed that structural changes to the MFGM occurred on washing of cream. The authors stated that depletion of PLs and the increased extent of protein-stained surfaces (as determined using microscopic analysis) were indicative of damage and loss of integrity of the bilayer. Hence, in order to increase the purity of MFGM, while maintaining a high recovery of MFGM proteins and phospholipids, a large volume of wash solution is advisable rather than an increased number of washings (Le *et al.* 2009).

In addition to centrifugal isolation, MFGM material can be isolated from industrial dairy by-products such as BM, butter serum and whey, using membrane filtration, supercritical fluid extraction (SFE) or precipitation (Corredig *et al.* 2003; Damodaran 2010, 2011; Liu and Zhong 2014), thus causing less damage to shear-sensitive components. Moreover, industrial by-products are normally subjected to several treatments during processing, leading to differences in composition between associated MFGM materials and those from unheated raw milk. Microfiltration (MF) of raw milk prior to butter-making was suggested to enhance the isolation of MFGM (protein separation) from BM (Jukkola *et al.* 2016). More specifically, MFGM fragments are enriched in the serum phase, free of casein, which improves the isolation of MFGM and results in increased purity. Hence, membrane filtration processing can be considered as a suitable approach to achieve isolation of MFGM-derived components. However, under practical conditions, it is nearly impossible to obtain pure MFGM material surrounding native MFGs in raw milk without any losses. Rebouillat and Ortega-Requena (2015) suggested that further research should focus on the loss of the native structure of fat globules during MFGM isolation and on the change in MFGM functionality when milk is processed. Despite these technological difficulties, it should be emphasised that the MFGM fragments could be concentrated to some extent using membrane separation technologies. Fuller *et al.* (2013) extracted MFG-enriched in polar lipids from regular and whey BM powders by MF and SFE. Jukkola *et al.* (2016) separated MFGs from raw whole milk by MF using three diafiltration cycles with ultrafiltration permeate of skim milk, saline and water. MF seems to be a promising method for obtaining an MFGM-rich fraction, and several attempts have been made to optimise such processes using different pre-treatments to reduce the presence of caseins, which are generally concentrated together with the MFGM during filtration.

Damodaran (1995) patented a method to precipitate lipid from whey by adding sufficient chitosan to form a chitosan–lipid complex. The whey proteins, not complexing with chitosan, may then be recovered by conventional means such as ultrafiltration and spray drying, while the lipids from the chitosan–lipid complex could be extracted by either filtration or centrifugation. More recently, Damodaran (2015) patented another method to selectively separate MFGM fragments and MFG from whey. The method included the addition of a whey-soluble zinc salt to the whey, and adjusting the pH to less than 6.0; the concentration of zinc salt added to the whey was sufficient to cause MFGM fragments and MFGs contained in the whey to precipitate. Although the chitosan-based process is relatively simple and produces highly functional whey protein concentrate (WPC) and isolate (WPI), whey processors are unable to use this technology in the United States because chitosan is not approved as a 'GRAS' (generally regarded as safe) substance in the United States (although chitosan has GRAS status in Europe and Japan). Furthermore, recovery of MFGM from the chitosan–MFGM complex is not effective under mild conditions. Thus, there is still a need to develop a simple process to remove MFGM fragments from cheese whey, thereby enabling cost-effective and energy-efficient production of high-protein content whey-based ingredients such as whey protein concentrates (WPC) and isolates (WPI). From the point of view of processing of WPC and WPI ingredients, the need is firstly to reduce fouling of membrane filtration equipment by lipid components of MFGM, and secondly to remove, or reduce significantly the concentration of lipid material of MFGM origin in the final

WPC/WPI product, thereby limiting off-flavour development and discoloration of the final powder product (Damodaran, 2015). Therefore, the material characteristics of MFGM should be considered carefully when selecting the method and the technological parameters to be used for their recovery, and all isolation techniques utilised should be adjusted to the characteristics of the starting material and to the target fraction to be purified (Mezouari *et al.* 2009).

6.3.3 Sources of Milk Fat Globule Membrane–Derived Material

While MFGM material is present in virtually all dairy products containing milk fat, it is naturally enriched in BM (Hintze *et al.* 2011) as a result of the mechanical emulsion destabilisation process involved in butter manufacture, which disrupts the MFGM, releasing free fat as the globules coalesce. Although seen as a by-product, currently BM is the main source of MFGM-derived PLs and membrane proteins and is used to produce MFGM-enriched ingredients, thereby increasing the value of BM. The number of scientific publications on the chemistry, fractionation and functionality of BM has quadrupled over the past 20 years (PubMed), and the vast majority of these studies have focussed on either fractionating or concentrating various components of MFGM.

The separation of MFGM material from other dairy constituents, namely proteins, can be achieved by MF, which has been used successfully for the separation and fractionation of MFG (Goudedranche *et al.* 2000). However, the presence of skim milk solids, especially casein micelles in BM, restricts the concentration of MFGM, as the MFGM particles and casein micelles are comparable in size. A possible solution to separating the MFGM-derived material from the casein micelles, when using BM as feed material, is to selectively dissociate casein micelles, allowing casein proteins to permeate the MF membrane, together with the whey proteins, lactose and minerals, thereby allowing concentration of the MFGM material in the MF retentate stream. For example, Corredig *et al.* (2003) used sodium citrate to disperse the casein micelles and reduce the retention of caseins upon MF of BM, aiming for a PL-enriched retentate. After concentration and two diafiltration steps, while the relative concentration of casein was reduced compared with the other proteins present in the retentate, casein still represented 30% of the total protein. Morin *et al.* (2007a) produced BM with a lower casein content using MF of cream washed with skim milk ultrafiltrate, washing the cream prior to churning yielded BM with 74% less protein than regular BM. However, the percentage of PL was slightly lower for washed cream BM as compared with normal cream BM (3.71 vs. 4.10%). Jukkola *et al.* (2016) used MF for the separation of native MFG from whole milk, which allowed ~90% of milk proteins to be removed from the cream prior to buttermaking; when butter was made using this novel stream, the resulting BM was naturally enriched in MFGM components and was referred to as 'ideal buttermilk'.

In addition to BM (Sachdeva and Buchheim 1997; Astaire *et al.* 2003; Morin *et al.* 2007a), other by-products have been studied as sources of MFGM material, namely whey BM (WBM) (Morin *et al.* 2006) or butter serum (Rombaut *et al.* 2006b). However, the low volumes of whey cream produced and the higher susceptibility to oxidation of whey cream limited the commercial potential of this approach. Research on the recovery and purification of dairy-derived polar lipids, mostly using BM or whey-derived streams as raw material, is very active, and this promises to be an interesting area for ingredient development over the next 5–10 years.

6.4 Effects of Milk Processing Unit Operations on Milk Fat Globule Membrane Constituents

Destabilisation of MFGM can occur under the influence of a number of factors, including physiological, physical/mechanical and environmental. Besides environmental factors, including the presence of bacteria in milk (e.g. pathogens in mastitic milk), it is well known that processing conditions markedly affect the composition and stability of the MFGM. So far, the effects of various unit operations used in the processing of MFGM (e.g. agitation, homogenisation, heat treatment, concentration, drying and freezing) have been extensively investigated (Sharma and Dalgleish 1994; Kim and Jimenez-Flores 1995; Corredig and Dalgleish 1996, 1998b; Lee and Sherbon 2002; Ye *et al.* 2002; Evers 2004; Ye *et al.* 2004; Ye and Singh 2007; Murthy *et al.* 2016; Lopez *et al.* 2017b). Their main consequences are the loss of membrane components, adsorption of milk plasma components and enhanced chemical or enzymatic reactions, which in turn may affect the stability and functionality of the fat globule (Jukkola and Rojas 2017).

The preparation of emulsions containing MFGM-coated lipid droplets is of primary interest to mimic the interfacial properties of native MFG and to supply bioactive components for health benefits to consumers. Furthermore, emulsion droplets are subjected to changes in pH during digestion and in the manufacture of various dairy products, such as cheeses, yoghurts or creams. Hence, knowing how the properties of MFGM-coated lipid droplets vary as a function of pH is of considerable interest and relevance. Lopez *et al.* (2017b) have shown that the MFGM-rich ingredient used and described by Gassi *et al.* (2016) has excellent emulsifying capacity and that the properties of the emulsions are influenced by pH. Microstructural investigations revealed the presence of MFGM fragments attached at the surface of the lipid droplets, as well as polar lipid and glycoproteins. The work by Lopez *et al.* (2017b) demonstrated that the MFGM components are highly sensitive to pH and that the reactivity and stability of the MFGM-rich ingredient are influenced by both the lipid and protein components. Changes in pH affected the microstructural and rheological properties of the emulsions, including changes in the zeta potential of the MFGM-rich ingredient. Such changes can also affect the biological functionality, for example, the mechanisms of lipid digestion, by limiting the accessibility of the digestive gastric lipase to the TG core of the lipid droplets. Such detailed knowledge about the interfacial properties and microstructure of MFGM-coated lipid droplets as a function of pH will undoubtedly contribute to the development of food emulsions prepared with MFGM-rich ingredients, for example, infant milk formulae, dairy emulsions and spreads (Lopez *et al.* 2017b). However, despite the extensive efforts dedicated to the isolation of MFGM, deeper enquiries are required to better understand, predict and control damage and structural alteration of the MFGM during processing. Only very limited information is available about the effect of pH on MFGs and their membrane (i.e. MFGM). Moreover, the distinctions between MFG and MFGM damage, details on the composition of MFGM fragments released upon disruption, and the extent of their influence on the biological and techno-functional properties of the ingredients into which they are incorporated are still poorly understood. The factors that affect the structure and composition of MFGM, their effects on the MFGM and respective references are summarised in Table 6.5. An overview of the published data available on total and individual PLs from different dairy products, and their variability, even for a given product type, is given in Table 6.6. The following sections provide an overview of how the MFGM can be influenced when subjected to the more common milk processing unit operations, including mechanical, thermal and drying treatments.

6.4.1 Homogenisation and Churning

Homogenisation involves forcing milk between valve faces under pressure, resulting in a dramatic reduction in fat globule size that greatly decreases the separation rate of fat globules and the formation of a cream layer (Walstra 2003). Homogenisation has no impact on the composition of milk, but significantly affects its structure and the bioactivity of certain milk components (Michalski *et al.* 2002). Indeed, the rupture of fat globules caused by homogenisation creates a new interface that cannot be covered entirely by the

Table 6.5. Chemical, enzymatic, physiological and physical/mechanical factors that affect the composition and structure of the milk fat globule membrane

Factor	Effect	References
Chemical/enzymatic		
Processing origin	Loss of membrane components, adsorption of milk plasma components	Erwin and Randolph (1975), Evers (2004), McPherson and Kitchen (1983)
Bacterial origin	Production of lipases, phospholipases, proteinases, etc. may change MFGM	
Physiological		
Diet, breed, stage of lactation	Amount of membrane material, fat globule size distribution and relative distribution of membrane components	Lopez *et al.* (2008), Mondy and Keenan (1993), Murgiano *et al.* (2009), Reinhardt and Lippolis (2008), Ye *et al.* (2002)
Physical/mechanical (handling and processing)		
Air bubbles Ageing Cooling	Loss of membrane material Effects depend on temperature Loss of copper, loss of phospholipids, absorption of proteins	Walstra and Jenness (1984) Anderson *et al.* (1975), Baumrucker and Keenan (1973), Dickow *et al.* (2011), Patton *et al.* (1980)
Pumping, stirring, agitation	Depends on the degree of air incorporated High shear forces are required to change the MFGM in the absence of air Adsorption of serum protein, rupture of membrane	Mulder and Walstra (1974), Stannard (1975) Anderson *et al.* (1975), Cano-Ruiz and Richter (1997), Lee and Sherbon (2002), Mondy and Keenan (1993), Ye *et al.* (2004)
Homogenisation	Formation of PLs or membrane vesicles Redistribution of protein and lipids in globules	Michalski *et al.* (2002) Keenan *et al.* (1983)
Heating	Adsorption of copper, whey proteins and caseins Aggregation of BTN and XO, loss of PAS 6/7 and phospholipids	Corredig and Dalgleish (1996), Garcia-Amezquita *et al.* (2009), Kim and Jimenez-Flores (1995), Lee and Sherbon (2002), Sharma and Dalgleish (1994), Ye *et al.* (2002, 2004)
Spray drying	Reduces globule size and increases surface area for adsorption of serum proteins	Landstrom *et al.* (2000), Ye and Singh (2007)

MFGM milk fat globule membrane, *BTN* butyrophilin, *XO* xanthine oxidase, *PAS 6/7* lactadherin or periodic acid Schiff 6/7, *PLs* phospholipids.
Adapted from Elías-Argote (2011).

MFGM so that other surface-active components adsorb and contribute to the formation of a new membrane around the fat droplets. Cano-Ruiz and Richter (1997) estimated that only 10% of the total MFG after homogenisation is covered by the original MFGM material and stated that the newly formed fat globules exhibit different physical and chemical properties. Casein micelles are reported to be the major protein fraction adsorbed at the newly created interface, and they spread onto the newly created interface during homogenisation, even if part of the native MFGM remains associated with the fat droplets (Michalski and Januel 2006). The schematic representation put forward by Michalski (2007) for the organisation of native MFG, the MFGM and

Table 6.6. Total and individual phospholipid concentrations in different dairy products

Product	Total phospholipid content		Individual phospholipid distribution (g/100 g PL)				
	mg/100 g product	g/100 fat	PE	PC	PS	PI	SGM
Raw milk							
Raw ewes' milk	29.8	0.390	26.1–40.0	26.4–27.2	4.96–10.7	4.16–6.40	22.6–29.7
Raw goats' milk	27.6	0.710	19.9–41.4	27.2–31.9	3.2–14.0	4.00–9.37	16.1–29.2
Raw cows' milk	29.4–40.0	0.700–0.980	23.4–46.7	19.1–33.2	2.00–9.07	3.98–8.97	17.8–29.2
Liquid milk							
Skimmed milk	20.0–81.9	11.1–19.1					
Pasteurised semi-skimmed	18.8	1.30	35.0	20.2	8.90	7.90	17.0
UHT full-fat	21.2	0.600	34.0	20.5	9.10	7.90	19.5
UHT semi-skimmed	14.2	0.900	33.0	22.0	7.90	4.80	22.9
UHT skimmed	12.8	10.7	38.2	19.6	9.90	5.50	16.7
Sterilised semi-skimmed	16.0	1.00	34.3	24.2	7.70	5.10	17.4
Infant formula							
Infant formula (0–6 mo)	228–304		23.2–29.5	25.9–35.6	5.51–10.1	10.2–19.5	13.1–26.9
Infant formula (0–3 mo)	607		23.6	41.7	4.28	9.88	20.6
Others							
Unsweetened condensed milk	75.1	1.00					
Buttermilk and buttermilk derivatives							
Buttermilk	1.12–160	4.49–33.1	33.5–39.5	27.71–35.5	10.3–22.8	2.40–7.20	16.7–18.3
Buttermilk (skimmed/serum)	110	20.00	27.2	29.8	7.20	10.8	24.9
Buttermilk (acid)	160	33.05					
Buttermilk (reconstituted)	130	21.66					
Buttermilk (powder)			19.8	33.9	20.6	4.93	19.9
Buttermilk quark	310	29.06					
Butter serum (whole)	97.0	40.0					
Butter serum (skimmed)	93.0	49.0					
Fermented buttermilk – sweet	91.8	21.8					
Fermented buttermilk – sour	115.5	23.1					
Goat buttermilk		0.19	35.2	24.8	9.90	9.80	20.3
Cream, butter and butter products							
Cream	139–190	0.350–0.860	42.7	14.6	7.20	6.80	28.6
Butter	70.5–230	0.090–0.270	31.0	24.7	15.3	11.9	17.1
Butter serum	660–1250	14.8–48.4					
Butter oil	10.0	0.01	20.3	50.7			25.9
Goat butter serum		1.010	27.1	26.2	8.20	11.7	26.8
Yogurt and kefir							
Yoghurt, skimmed	17.9	5.50	31.1	19.9	7.90	6.30	24.9
Kefir semi-skimmed	34.0	2.30					

(continued)

Table 6.6. (continued)

Product	Total phospholipid content		Individual phospholipid distribution (g/100 g PL)				
	mg/100 g product	g/100 fat	PE	PC	PS	PI	SGM
Cheese whey							
Whey (*Cheddar*)	18.0	5.32					
Whey (Emmental)	22.0	45.2					
Mozzarella (whey)	19.1	6.20	40.6–41.1	19.0	9.30	3.70–4.60	15.7–16.4
Cheddar (whey)	17.6	5.30					
Cheeses							
Fresh cheese	310	29.1	39.9	21.7	8.20	6.50	14.1
Ricotta	279	2.70					
Quark – skimmed	32.4	24.70					
Quark – cream	58.1	0.90					
Cottage cheese	55.8–376	1.30–5.30					
Sphingolipid quark mix	480	10.60					
Fresh cheese	149	0.600					
Mozzarella (buffalo)	115	0.500					
Gouda light 8 weeks	93.9	0.400					
Gouda 8 weeks	151	0.500					
Gouda 36 mo	147	0.400					
Cheddar	154	0.500	38.0	20.3	8.50	7.70	16.3
Camembert	123	0.500					
Emmental	110	0.400					
Parmigiano 24 mo	111	0.400					

Non-provided values left blank; all products are from bovine milk if not stated otherwise.
mo months, *PL* phospholipid, *PE* phosphatidylethanolamine, *PC* phosphatidylcholine, *PS* phosphatidylserine, *PI* phosphatidylinositol, *SGM* sphingomyelin.
Adapted from Pimentel *et al.* (2016).

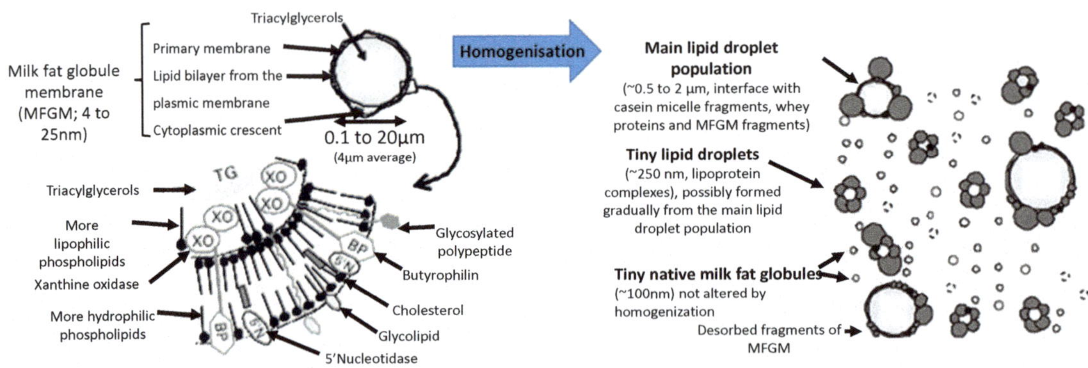

Figure 6.2. (**a**) Ultrastructure of the native milk fat globules and their membrane. (**b**) Proposed general organisation of the lipid particles after homogenisation. Adapted from Michalski (2007).

the newly created fat droplets during homogenisation of milk is shown in Figure 6.2. According to Lee and Sherbon (2002), in addition to the binding of caseins and whey proteins to MFGM, heat treatment and homogenisation cause the loss of original membrane components, such as triglycerides, PLs and proteins, from the MFGM. However, in contrast with heat treatment, homogenisation significantly reduces the size of fat globules, increasing their specific surface area (Lee and Sherbon 2002). The same authors showed that when raw milk was homogenised, significant amounts of casein proteins were associated with the membrane, but associations between whey proteins and MFGM components were detected only when milk was heated either before or after homogenisation. As a result of homogenisation, the protein content of the MFGM was increased about 3.5-fold, with adsorbed caseins being responsible for this increase. Bolling *et al.* (2005) have shown that natural cream homogenised before pasteurisation contained significantly ($P < 0.05$) more MFGM material, including a higher protein load associated with the MFGM, than natural cream or BM/aqueous phase formulations homogenised after pasteurisation.

The churning of cream, that is, during buttermaking, creates high shear at the surface of the fat globules, and loosely bound PLs on the outer side of the MFGM may be released into the skim milk, leading to a cream fraction rich in firmly bound PLs, mainly PE, and less concentrated in SGM (Morin *et al.* 2008). Residual MFGM material is usually found in skim milk, and since SGM and PC are located in the outer layers, they are more likely to migrate into the milk serum upon centrifugation (Anderson and Brooker 1975). Interaction of MFGM particles with skim milk proteins could modify the hydrophobicity of MFGM particles and, therefore, affect their distribution between the aqueous phase (BM) and the lipid phase (butter). However, the skim milks contained significantly less PE and significantly more SGM (Morin *et al.* 2008). Gallier *et al.* (2010a) confirmed these observations and reported that SGM (17.7 mol % vs 21.8 mol %) and PC (31.1 mol % vs 36.6 mol %) were less abundant in cream than in raw milk, respectively, and PE was three times less abundant in BM and more concentrated in cream. Since PE is more hydrophobic than PC and SGM, it may partition into the butter phase during butter manufacture and preparation of MFGM. Haddadian *et al.* (2018) reported that reducing the temperature and pH during churning increased the levels of PLs in the MFGM fractions.

6.4.2 Thermal Processing

The physical state of lipids, lipid–lipid and lipid–protein interactions and the organisation of the membrane (e.g. the presence of lipid domains and the associated functions) all depend on thermodynamic parameters (Veatch *et al.* 2008; Alessandrini and Facci 2014). The MFGM is submitted to a wide range of temperatures including its secretion at the physiological temperature (37 °C), thermal treatments, such as pasteurisation at 65–75 °C and storage at low temperatures (4–7 °C). The application of these thermal processes can cause changes to the structure of MFGM, which in turn can affect its technological properties, biological properties and physical stability (Murthy *et al.* 2016). However, the literature on this topic is inconsistent: according to Houlihan *et al.* (1992), heating alone has little or no effect on polar lipids, whereas Christie *et al.* (1987) did not detect PLs in ultra-high temperature (UHT)-treated milk powders, attributing such observations to the effect of heat during processing, in particular oxidation of polyunsaturated fatty acids (PUFA). Cano-Ruiz and Richter (1997) reported that the composition of the proteins forming the MFGM in homogenised milk was not affected by homogenisation pressure or fat concentration but were influenced by the heat treatment applied before homogenisation. Heat treatment of milk was reported as the main cause of changes in the MFGM by promoting interactions between plasma proteins and native MFGM components (Houlihan *et al.* 1992; Kim and Jimenez-Flores 1995). Rombaut and Dewettinck (2006) showed that the polar lipid content of various dairy streams, including BM, skim milk, whole milk and whey, varied considerably, suggesting that different processing strategies and technologies encouraged partitioning of polar lipids into particular streams and phases. For Morin *et al.* (2007b), pasteurisation is undoubtedly a critical step that changes the composition of MFGM, especially in terms of total lipid content. According to the same authors, BM produced from pasteurised cream contained a lower PL concentration, especially PE, PS and PI. The authors hypothesised that during churning of cream, the inner membrane of the fat globules is exposed to milk serum, inducing possible interactions as well as dislocation of components of the MFGM. Gassi *et al.* (2008) found no variation in total PL content of creams that underwent different heat treatments, while the PL content increased in their respective BMs by a factor of seven. However, Gallier *et al.* (2010a) stated that pasteurisation may be responsible for the different PL profiles in raw and processed milk.

According to Morin *et al.* (2008), pasteurisation of milk before or after skimming induced whey protein denaturation and increased the fraction of insoluble protein at pH 4.6 from 80.4% to approximately 83.6%. This result suggests that the solubility of MFGM-derived protein at pH 4.6 is strongly affected by pasteurisation. The partitioning of proteins in the MFGM, protein–lipid interactions and bioactivity of MFGM proteins are likely to depend on the physical state and packing of the polar lipid as well as on their mechanical properties, which in turn depend strongly on temperature (Alessandrini *et al.* 2011; Alessandrini and Facci 2014). Altogether, heat treatment can significantly alter the composition and functionality of MFGM. However, the exact changes that occur in the MFGM upon heating in either a milk or cream environment are not fully understood and warrant further investigation. This has led to the consideration of techniques such as pulsed electric field (PEF) treatment to replace traditional thermal pasteurisation, which may be effective in preserving MFGM bio-functionality (Xu *et al.* 2015).

Cooling and cold storage of milk may have a pronounced effect on fat globule stability and composition of the MFGM. Upon cooling, milk releases certain proteins and PLs (especially PC) from the surface of MFGM, thereby reducing its stability (Gallier *et al.* 2010b). According to Et-Thakafy *et al.* (2017), both the cooling kinetics and storage duration at a low temperature (4–20 °C) affect the shape, size and number of lipid domains, through the mechanisms of nucleation and growth.

The membranes surrounding the fat globules are affected by cold storage (temperature and temperature variations), which may have important technological consequences and affect the nutritional and biological functions of the MFGM surrounding fat globules, for example, in human milk donations stored in milk banks (Murthy *et al.* 2016). In addition, at temperatures less than 40 °C, the stability is further reduced by fat crystals that cause local structural changes to the membrane, which can be pierced upon deformation, thus leading to aggregation and partial coalescence of the fat globule (Jukkola and Rojas 2017). Haddadian *et al.* (2018) concluded that temperature and pH can influence the PL composition, most likely due to how the treatments impact on the different PLs that are located in different regions within the MFGM trilayer structure. Moreover, MFGM is involved in many interfacial mechanisms, such as adsorption of enzymes, bacteria, viruses, location of proteins and physical stability of milk fat globules. Changes in the morphology and physical properties of the lipid domains induced by temperature and time dynamics could affect the technological properties of milk fat globules (e.g. butter manufacture) and their nutritional functions, particularly the mechanisms of lipid digestion (Et-Thakafy *et al.* 2017).

According to Gallier *et al.* (2013a), milk processing and the ultrastructure of the milk fat globules appear to affect lipolysis of the milk fat globules. The changes in composition and structure of the surface of the MFG that occur during thermal treatment or homogenisation have an impact on interfacial processes such as interactions with the food matrix and the process of lipolysis. As lipolysis is an interfacial process, the ultrastructure of the MFG is likely to play a key role in the digestion and absorption of milk fat (Gallier *et al.* 2013b). Knowledge of the interfacial properties of MFGM polar lipids as a function of temperature may be helpful to their application in the stabilisation of emulsions (e.g. in infant milk formulae) or the formulation of liposomes for encapsulation (Murthy *et al.* 2016). Moreover, the incorporation of PLs from the MFGM in dairy products requires an in-depth understanding of the functional properties of PLs and the chemical and physical modifications of PLs during digestion. It is critical to understand which factors play a role in their stability and bio-efficacy (e.g. bioactivity, bioavailability and bioaccessibility) as delivery systems (Arranz and Corredig 2017).

6.4.3 Drying

Freeze drying and spray drying are conventional methods for converting milk into powder in the dairy industry. Freeze drying, besides being time consuming and expensive, is a specialist drying process, which is very rarely applied to milk to conserve its flavour, physical properties and bioactivities (Wang *et al.* 2013). On the other hand, spray drying, as a much faster drying process, is widely used in the dairy industry to convert concentrated milk into powders.

The influence of the drying method on the surface properties and morphological characteristics of MFG has been reported (Ye and Singh 2007; Yao *et al.* 2016b). For example, Yao *et al.* (2016b) measured and compared the microstructure and lipid composition of MFG when freeze dried or spray dried. The study revealed that the size of the fat globules increased as a result of coalescence and aggregation of the MFG during the drying processes and showed evidence of microstructural changes of the MFG undergoing freeze drying or spray drying. However, the present authors found only limited references addressing the PL composition and changes in the microstructure of freeze-dried or spray-dried MFGM material (Morin *et al.* 2007b; Zhu and Damodaran 2011; Ali *et al.* 2018). According to Morin *et al.* (2007b), spray drying of BM had a significant effect on PL content and composition. After spray drying, the PL content decreased by 38.2 and 40.6%, respectively, in BM from raw or pasteurised cream when compared with initial BMs. In addition, a reduction in PL content was also observed in MFGM isolates following spray drying. Also, Zhu and Damodaran (2011) observed morphological differences between freeze-dried and spray-dried MFGM, using light microscopy. The freeze-dried MFGM appeared as irregular flaky translucent sheets with sharp edges, whereas the spray-dried MFGM had more spherical particles. The authors concluded that the drying method can affect the state of the PLs in the MFGM fractions. In addition, the content of PS, an anionic phospholipid, found exclusively in the internal layer of the MFGM, decreased on freeze drying in comparison to the raw milk.

6.5 Techno-Functionality and Applications of Milk Fat Globule Membrane–Derived Material

Besides their nutritional properties, polar milk lipids have very interesting techno-functional properties in applications involving emulsifying and whipping (Dewettinck *et al.* 2008; Vanderghem *et al.* 2010). Polar lipids have been used in the food industry for a number of applications, such as baking improvers to aid fat dispersion, as anti-staling agents, as additives to chocolate to reduce viscosity and prevent crystallisation, as wetting enhancers for instant products and as stabilisers for margarine to prevent spattering and browning (Van Nieuwenhuyzen 1976, 1981; Szuhaj 1983; Rombaut *et al.* 2004). Non-food applications include drug delivery carriers, fat liquoring for leather softening and as raw materials for the production of ceramides and liposomes (Van Nieuwenhuyzen 1981; Kisel *et al.* 2001; Guo *et al.* 2005). Despite their modest concentration in the MFGM, PLs are critical in stabilising milk fat globules against coalescence, due to their polar and non-polar structure, making this MFGM material an efficient and natural emulsifier (Ortega-Requena and Rebouillat 2015). PC and PE tend to stabilise water-in-oil emulsions due to their low hydrophile–lipophile balance value, while SGM, PI and PS tend to form oil-in-water emulsions (Fedotova and Lencki 2008, 2010).

Besides the structural and compositional studies on the MFGM, work related to the biological and physicochemical properties has also been reported. These studies focussed mainly on the role of various components of native MFGM towards surface and techno-functional properties (Kanno *et al.* 1991; Corredig and Dalgleish

1998a; Levin *et al.* 2016b), gastric and intestinal digestion (Shimizu *et al.* 1979; Gallier *et al.* 2013a, b; Singh and Gallier 2017; Singh 2019), milk processing effects (Morin *et al.* 2007b; Ferreiro *et al.* 2016; Jukkola and Rojas 2017; Haddadian *et al.* 2018) and isolation methods (Mather *et al.* 1980; Jukkola *et al.* 2016; Pimentel *et al.* 2016). In addition, several patents (e.g. Gorewit 2002a, b) and scientific papers have been published regarding the isolation and concentration of MFGM components (Spitsberg and Gorewit 2002; Morin *et al.* 2006; Malik *et al.* 2015; Jukkola *et al.* 2016; Pimentel *et al.* 2016) and the production of milk polar lipid-rich ingredients (Costa *et al.* 2010; Phan *et al.* 2013; Gassi *et al.* 2016).

Normally, the level of MFGM material required as a technological aid, for example, emulsifier, is lower than that needed to support a health effect; therefore, on adding MFGM material to foods to confer various health benefits, for example, cholesterol-lowering and anti-inflammatory, it is expected that the addition level would be in excess of that required to confer a technological benefit (Dewettinck *et al.* 2008; El-Loly 2011; Cilla *et al.* 2016). Therefore, ingredients enriched in MFGM-derived components have potential for use as novel food ingredients with both technological and biological functionalities in formulated food systems. The potential sources of PL and MFGM-derived material-enriched ingredients are shown schematically in Figure 6.3. Good sources of MFGM-derived PLs are low-fat products, such as skimmed milk (11.1–19 g PL/100 g fat), BM (up to 33 g PL/100 g fat) and butter serum (14.8–48.4 g PL/100 g fat) (Pimentel *et al.* 2016). Thus, cream has a polar lipid content (expressed as a proportion of the total lipid) lower than that of skimmed milk, just as butter and cheese have a lower polar lipid content than BM and whey (Contarini and Povolo 2013). Therefore, by- and co-products of dairy processing, for example, BM, butter serum and whey BM, are increasingly been seen as attractive sources of milk PLs and MFGM-derived proteins. BM is to date the most studied and well-documented starting material for the production of PL-enriched products (Sachdeva and Buchheim 1997; Corredig *et al.* 2003; Shukla *et al.* 2004; Sodini *et al.*

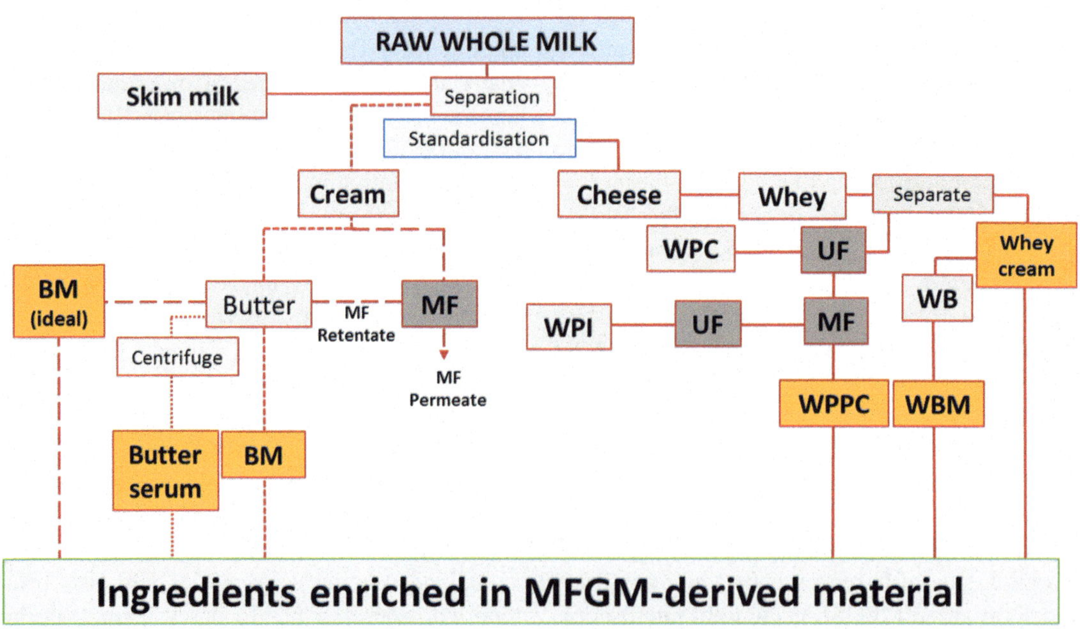

Figure 6.3. Schematic representation of the potential sources of phospholipid-enriched ingredients from raw, whole milk and its derivatives. *BM* buttermilk, *WBM* whey buttermilk, *WPPC* whey protein phospholipid concentrate, *WB* whey butter, *UF* ultrafiltration, *MF* microfiltration.

2006; Morin *et al.* 2007a; Vanderghem *et al.* 2010; Svanborg *et al.* 2015; Lopez *et al.* 2017a; Haddadian *et al.* 2018). Shukla *et al.* (2004) developed a functional dairy beverage based on BM and fruit juices and pulps that contained similar levels of protein to those found in milk and acceptable organoleptic characteristics. More recently, Konkoli *et al.* (2006) proposed a formulation for a nutritious and functional dairy beverage which includes BM (as MFGM material source) as an ingredient. Le *et al.* (2011) used MFGM material (isolated from BM) in yogurt and concluded that increasing both the polar lipids and proteins by the addition of MFGM material not only provides beneficial nutritional properties but also contributes to the technological properties of the product, such as improved water-holding capacity and increased adhesiveness of yoghurt gels. However, alternative sources of these MFGM-derived materials should be exploited, especially from low-value side-streams, which are currently under-utilised and present an opportunity for value-added ingredient formulation. Whey protein–phospholipid concentrate (WPPC), generated alongside WPC and WPI, is the most evident example of this. WPPC is a co-product of WPI, produced by the microfiltration of high-protein-content WPC, which separates the majority of the undenatured whey proteins from the fat, phospholipids, lactose and denatured whey proteins. Although displaying extensive variability, in 2015, the American Dairy Products Institute prepared a standard for WPPC composition: minimum of 50% protein (dry basis), minimum of 12% fat, maximum of 8% ash and maximum of 6% moisture (Levin *et al.* 2016a). Previous studies have demonstrated that this whey-based side-stream is a potential source of highly functional PLs (Zhu and Damodaran 2013), and a small number of food applications studies have been completed using this ingredient (Bund and Hartel 2013; Levin *et al.* 2016a, b). Nevertheless, knowledge of the origin, composition and properties of the MFGM-derived material is crucial in supporting development of new products with improved nutritional or functional properties, and the composition of MFGM materials must be carefully standardised before use in food products.

6.5.1 Milk and Whey Phospholipid-Enriched Products

There are a small number of milk-based PL-enriched ingredients available commercially. Lacprodan® PL-20 (Arla Food Ingredients, Viby, Denmark) is a milk protein concentrate (MPC) enriched with PLs and gangliosides. Some of the benefits of this ingredient, as claimed by the manufacturer, include stability to oxidation, milky taste and good emulsifying properties, in addition to nutritional benefits as a source of choline, PS and other biologically important lipids. The PL concentrates, PC 500™ (25% PLs), PC600™ (75% PLs) and PC700™ (60% PLs) are produced by Fonterra Co-operative Group Limited, New Zealand, and are all cream-derived ingredients rich in PL, with up to 5000 times the concentrations naturally present in milk. SureStart™ Lipid 100, another product made by Fonterra Co-operative Group Limited, is an MFGM complex lipid material, a by-product of the manufacture of anhydrous milk fat from cream.

Phan *et al.* (2013) have demonstrated the emulsifying properties of MFGM fragments concentrated from reconstituted BM using MF and their potential in the development of new products. To a much lesser extent, butter serum (Gassi *et al.* 2016; Lambert *et al.* 2016; Lopez *et al.* 2017a; Bourlieu *et al.* 2018), whey BM (Morin *et al.* 2006; Sodini *et al.* 2006; Costa *et al.* 2010; Konrad *et al.* 2013) and whey cream (Fuller *et al.* 2013) have been studied as starting materials for enrichment of MFGM components. Of these possible by-products, BM is the most thoroughly studied and is often used as the starting material for the production of PL- and protein-enriched products. However, the casein micelles in BM, which represent a major proportion of the dry matter, can present challenges as their diameter is comparable to that of MFGM fragments present in BM. Whey BM, the aqueous fraction obtained on churning of whey cream, which is separated from fresh liquid whey, and acid BM whey, the aqueous fraction obtained by acidification of sweet-cream BM, were considered suitable for the isolation of MFGM material by filtration due to the absence of casein micelles (Morin *et al.* 2006; Rombaut *et al.* 2007). Sodini *et al.* (2006) reported that whey BM had the best emulsifying properties and the lowest foaming capacity when compared to sweet and sour BMs, possibly due to a higher ratio of PL to protein in whey BM than the other two.

Up to 2015, whey powder (WP) remained the dominant (70%) whey product in terms of both volume and value, which together with WPC and WPI represented a global market value of ~$4.9 billion. The general trend in the whey ingredients market is strong growth for high-protein products, with WPC80 and WPI having the highest growth rates between 2011 and 2015 (Affertsholt and Pederson 2017). Considering this, thousands of tonnes of WPPC are generated annually as a co-product of WPI manufacture. Like other co-products, WPPC has a variable composition and, while it is the by-product of MF-based defatting of whey, it can be manufactured using different process parameters, and thus, commercial WPPC products, as described below, vary considerably in terms of chemical composition and consequently functionality (Burrington *et al.* 2014).

There are a number of whey-based PL-enriched products available commercially, for example, PRO-Cream (Prinova®, Carol Stream, Illinois, USA), Salibra® 700 (Glanbia Nutritionals, Kilkenny, Ireland) and Lacprodan® MFGM-10 (Arla Food Ingredients, Viby, Denmark). Lacprodan® MFGM-10 is produced from a whey protein fraction with a high concentration of bioactive proteins and lipids, which, according to the producer, has unique protein and lipid profile and includes bioactive compounds, such as lactoferrin, immunoglobulin G (IgG), sialic acid, PLs and gangliosides. Salibra® 700 is a value-added WPC ingredient with more than 20% bioactive whey-derived components and can be used in food systems such as yogurt, ice cream, low-fat products and nutritional beverages for various technological purposes, including emulsification and thickening. Blends of WPPC and delactosed permeate (DLP), a lactose by-product, have been used in several food formulations to replace other dairy products, emulsifiers, salt and eggs in food applications, such as ice cream, soups and confectionery products (Bund and Hartel 2013; Levin *et al.* 2016a). There is a trend in the development of functional foods towards including MFGM-rich ingredients as a source of PLs and sphingolipid, namely SGM, as well as proteins with beneficial properties. In addition, MFGM-derived material is also a source of antioxidants, such as vitamin E (Jensen and Nielsen 1996) and vitamin B2 (Kuchta *et al.* 2012).

In recent years, significant research efforts have been made to supplement infant formulae with new bioactive ingredients with health benefits for the infant (De Almagro Garcia *et al.* 2017). MFGM-enriched ingredients represent one such type of ingredient and provide a number of bioactive compounds, namely MFGM-derived PLs and proteins (Cilla *et al.* 2016; Claumarchirant *et al.* 2016; Moloney *et al.* 2018; Ortega-Anaya and Jiménez-Flores 2018). Therefore, if MFGM material could be enriched/isolated from currently under-utilised dairy by-products, such as WPPC, in an industrially feasible and economical way, it has considerable potential for use as a stand-alone functional food or as a functional food ingredient for inclusion in several food products, including infant formulae and geriatric foods.

6.6 Conclusions

Because of its cellular membrane origin, the MFGM is the richest source of PLs, glycolipids, gangliosides and glycoproteins in milk. Despite the growing interest in the nutritional and functional properties of dairy PLs, the utilisation of MFGM in development of PL-enriched ingredients is still poorly exploited commercially. Therefore, more studies are required for an exhaustive identification and quantification of these bioactive components, namely PLs and membrane-specific proteins, as the functional and nutritional relevance of the MFGM-derived material is closely related to its specific chemical composition, which is, in turn, controlled by numerous factors including the starting material and the type of processing applied. Destabilisation of MFGM can occur under the influence of a number of factors including physiological, physical/mechanical and environmental, which altogether affect its nutritional and techno-functional properties.

Currently, BM is the main source of MFGM-derived PLs and membrane proteins and is used to produce MFGM-enriched ingredients; however, recent studies have highlighted the potential of whey-based PL-enriched by-products as an alternative source of these compounds. The utilisation of these side-streams as starting materials to recover functional MFGM-derived components and then apply them in the production of new products and ingredients may bring great benefit and can potentially be used as a standalone functional food or as a functional food ingredient for application in several food matrices, including nutritional products such as infant formulae and geriatric foods and functional ingredients for use in applications such as bakery and dairy products.

References

Affertsholt, T. & Pederson, D. (2017). *The ingredient market. Whey and lactose: Market fueled by high protein. The world of food ingredients*. Source: Innovation market insights, 26–28. http://ewpa.euromilk.org/fileadmin/user_upload/Public_Documents/EWPA_Press_Room/Whey_and_Lactose_-_Market_Fueled_by_High_Protein.pdf. Accessed 1 Oct 2019.

Alessandrini, A. & Facci, P. (2014). Phase transitions in supported lipid bilayers studied by AFM. *Soft Matter, 10*(37), 7145–7164.

Alessandrini, A., Seeger, H. M., Di Cerbo, A., Caramaschi, T. & Facci, P. (2011). What do we really measure in AFM punch-through experiments on supported lipid bilayers? *Soft Matter, 7*(15), 7054–7064.

Ali, A. H., Wei, W., Abed, S. M., Korma, S. A., Mousa, A. H., Hassan, H. M., Qingzhe, J. & Wang, X. (2018). Impact of technological processes on buffalo and bovine milk fat crystallization behavior and milk fat globule membrane phospholipids profile. *LWT Food Science Technology, 90*, 424–432.

Anderson, M. & Brooker, B. E. (1975). Loss of material during the isolation of milk fat globule membrane. *Journal of Dairy Science, 58*(10), 1442–1448.

Arranz, E. & Corredig, M. (2017). Invited review: Milk phospholipid vesicles, their colloidal properties, and potential as delivery vehicles for bioactive molecules. *Journal of Dairy Science, 100*(6), 4213–4222.

Astaire, J., Ward, R., German, J. & Jiménez-Flores, R. (2003). Concentration of polar MFGM lipids from buttermilk by microfiltration and supercritical fluid extraction. *Journal of Dairy Science, 86*(7), 2297–2307.

Baumrucker, C. R. & Keenan, T. W. (1973). Membranes of mammary gland. VII. Stability of milk fat globule membrane in secreted milk. *Journal of Dairy Science, 56*, 1092–1094.

Bligh, E. G. & Dyer, W. J. (1959). A rapid method of total lipid extraction and purification. *Canadian Journal of Biochemistry and Physiology, 37*(1), 911–917.

Bolling, J. C., Duncan, S. E., Eigel, W. N. & Waterman, K. M. (2005). Processing effects on physicochemical properties of creams formulated with modified milk fat. *Journal of Dairy Science, 88*(4), 1342–1351.

Bourlieu, C., Cheillan, D., Blot, M., Daira, P., Trauchessec, M., Ruet, S., Gassi, J. Y., Beaucher, E., Robert, B., Leconte, N., Bouhallab, S., Gaucheron, F., Gésan-Guiziou, G. & Michalski, M. C. (2018). Polar lipid composition of bioactive dairy co-products buttermilk and butterserum: Emphasis on sphingolipid and ceramide isoforms. *Food Chemistry, 240*, 67–74.

Bund, R. & Hartel, R. (2013). Blends of delactosed permeate and pro-cream in ice cream: Effects on physical, textural and sensory attributes. *International Dairy Journal, 31*(2), 132–138.

Burling, H., Nilsson, Å. & Ohlsson, L. (2009). International news on fats, oils, and related materials. Milk phospholipids: A new ingredient for formulation of functional foods with bioactivity. *Inform – AOCS – International News on Fats, Oils, and Related Materials. AOCS, 20*, 494–496.

Burrington, K. J., Schoenfuss, T. & Patel, S. (2014). Coproducts of milk and whey processing. In S. Agarwal (Ed.), *Technical report: Dairy Research Institute*. Dairy Research Institute. https://www.thinkusadairy.org/resources-and-insights/resources-and-insights/application-and-technical-materials/technical-report-coproducts-of-milk-and-whey-processing. Accessed 1 Oct 2019.

Cano-Ruiz, M. E. & Richter, R. L. (1997). Effect of homogenisation pressure on the milk fat globule membrane proteins. *Journal of Dairy Science, 80*, 2732–2739.

Castro-Gómez, M. P., Rodriguez-Alcalá, L. M., Calvo, M. V., Romero, J., Mendiola, J. A., Ibañez, E. & Fontecha, J. (2014). Total milk fat extraction and quantification of polar and neutral lipids of cow, goat, and ewe milk by using a pressurized liquid system and chromatographic techniques. *Journal of Dairy Science, 97*(11), 6719–6728.

Christie, W. W., Noble, R. C. & Davies, G. (1987). Phospholipids in milk and dairy products. *International Journal of Dairy Technology, 40*(1), 10–12.

Cilla, A., Quintaes, K. D., Barbera, R. & Alegria, A. (2016). Phospholipids in human milk and infant formulas: Benefits and needs for correct infant nutrition. *Critical Reviews in Food Science, 56*, 1880–1892.

Claumarchirant, L., Cilla, A., Matencio, E., Sanchez-Siles, L. M., Castro-Gomez, P., Fontecha, J., Alegria, A. & Lagarda, M. J. (2016). Addition of milk fat globule membrane as an ingredient of infant formulas for resembling the polar lipids of human milk. *International Dairy Journal, 61*, 228–238.

Contarini, G. & Povolo, M. (2013). Phospholipids in milk fat: Composition, biological and technological significance, and analytical strategies. *International Journal of Molecular Sciences, 14*(2), 2808–2831.

Corredig, M. & Dalgleish, D. G. (1996). Effect of different heat treatments on the strong binding interactions between whey proteins and milk fat globules in whole milk. *Journal of Dairy Research, 63*, 441–449.

Corredig, M. & Dalgleish, D. G. (1997). Isolates from industrial buttermilk: Emulsifying properties of materials derived from the milk fat globule membrane. *Journal of Agricultural Food Chemistry, 45*, 4595–4600.

Corredig, M. & Dalgleish, D. G. (1998a). Characterization of the interface of an oil-in-water emulsion stabilized by milk fat globule membrane material. *Journal of Dairy Research, 65*, 465–477.

Corredig, M. & Dalgleish, D. G. (1998b). Effect of heating of cream on the properties of milk fat globule membrane isolates. *Journal of Agricultural Food Chemistry, 46*, 2533–2540.

Corredig, M., Roesch, R. R. & Dalgleish, D. G. (2003). Production of a novel ingredient from buttermilk. *Journal of Dairy Science, 86*, 2744–2750.

Costa, M. R., Elias-Argote, X. E., Jiménez-Flores, R. & Gigante, M. L. (2010). Use of ultrafiltration and supercritical fluid extraction to obtain a whey buttermilk powder enriched in milk fat globule membrane phospholipids. *International Dairy Journal, 20*, 598–602.

Damodaran, S. (1995). *Removing lipids from cheese whey using chitosan*. US patent application US5436014A. 1995-07-25.

Damodaran, S. (2010). Zinc-induced precipitation of milk fat globule membranes: A simple method for the preparation of fat-free whey protein isolate. *Journal of Agricultural Food Chemistry, 58*, 11052–11057.

Damodaran, S. (2011). Straightforward process for removal of milk fat globule membranes and production of fat-free whey protein concentrate from cheese whey. *Journal of Agricultural Food Chemistry, 59*, 10271–10276.

Damodaran, S. (2015). *Method to separate lipids from cheese whey and fat-free whey protein product formed thereby*. European patent EP2346342A1. 2015-09-09.

Danthine, S., Blecker, C., Paquot, M., Innocente, N. & Deroanne, C. (2000). Progress in milk fat globule membrane research: A review. *Dairy Science & Technology 80*(2), 209–222.

De Almagro Garcia, M. C., Moreno Munoz, J. A., Jimenez Lopez, J. & Rodriguez-Palmero Seuma, M. (2017). New ingredients in infant formula. Health and functional benefits. *Nutricion Hospitalaria, 34*, 8–12.

Dewettinck, K., Rombaut, R., Thienpont, N., Le, T. T., Messens, K. & Van Camp, J. (2008). Nutritional and technological aspects of milk fat globule membrane material. *International Dairy Journal, 18*, 436–457.

Dickow, J. A., Larsen, L. B., Hammersh, J. M. & Wiking, L. (2011). Cooling causes changes in the distribution of lipoprotein lipase and milk fat globule membrane proteins between the skim milk and cream phase. *Journal of Dairy Science, 94*, 646–656.

Duan, R.-D. (2011). Physiological functions and clinical implications of sphingolipids in the gut. *Journal of Digestive Diseases, 12*, 60–70.

Duan, R. D. & Nilsson, A. (2009). Metabolism of sphingolipids in the gut and its relation to inflammation and cancer development. *Progress in Lipid Research, 48*, 62–72.

Eigel, W. N., Butler, J. E., Ernstrom, C. A., Farrell, H. M., Harwalkar, V. R., Jenness, R. & Whitney, R. M. (1984). Nomenclature of proteins of cow's milk: Fifth revision. *Journal of Dairy Science, 67*, 1599–1631.

Elías-Argote, Xiomara E. (2011). Effects of Milk Processing on the Milk Fat Globule Membrane Constituents. Thesis, California State Polytechnic University, San Luis Obispo, CA. https://doi.org/10.15368/theses.2011.156. Available at: https://digitalcommons.calpoly.edu/theses/616/

El-Loly, M. (2011). Composition, properties and nutritional aspects of milk fat globule membrane – A review. *Polish Journal Food Nutrition Science, 61*, 7–32.

Erwin, R. E. & Randolph, H. E. (1975). Influence of mastitis on properties of milk XI. Fat globule membrane. *Journal of Dairy Science, 58*, 9–12.

Et-Thakafy, O., Guyomarc'h, F. & Lopez, C. (2017). Lipid domains in the milk fat globule membrane: Dynamics investigated *in situ* in milk in relation to temperature and time. *Food Chemistry, 220*, 352–361.

Evers, J. M. (2004). The milkfat globule membrane – Methodologies for measuring milkfat globule (membrane) damage. *International Dairy Journal, 14*, 747–760.

Fauquant, C., Briard-Bion, V., Leconte, N., Guichardant, M. & Michalski, M.-C. (2007). Membrane phospholipids and sterols in microfiltered milk fat globules. *European Journal of Lipid Science Technology, 109*, 1167–1173.

Fedotova, Y. & Lencki, R. W. (2008). The effect of phospholipids on milkfat crystallization behavior. *Journal of the American Oil Chemists' Society, 85*, 205–212.

Fedotova, Y. & Lencki, R. W. (2010). The effect of phospholipids on butter physical and sensory properties. *Journal of the American Oil Chemists' Society, 87*, 75–82.

Ferreiro, T., Martínez, S., Gayoso, L. & Rodríguez-Otero, J. L. (2016). Evolution of phospholipid contents during the production of quark cheese from buttermilk. *Journal of Dairy Science, 99*, 4154–4159.

Folch, J., Lees, M. & Sloane Stanley, G. H. (1957). A simple method for the isolation and purification of total lipides from animal tissues. *Journal of Biological Chemistry, 226*, 497–509.

Fuller, K. L., Kuhlenschmidt, T. B., Kuhlenschmidt, M. S., Jimenez-Flores, R. & Donovan, S. M. (2013). Milk fat globule membrane isolated from buttermilk or whey cream and their lipid components inhibit infectivity of rotavirus *in vitro*. *Journal of Dairy Science, 96*, 3488–3497.

Gallier, S., Gragson, D., Cabral, C., Jimenez-Flores, R. & Everett, D. W. (2010a). Composition and fatty acid distribution of bovine milk phospholipids from processed milk products. *Journal of Agricultural Food Chemistry, 58*, 10503–10511.

Gallier, S., Gragson, D., Jiménez-Flores, R. & Everett, D. (2010b). Using confocal laser scanning microscopy to probe the milk fat globule membrane and associated proteins. *Journal of Agricultural Food Chemistry, 58*, 4250–4257.

Gallier, S., Cui, J., Olson, T. D., Rutherfurd, S. M., Ye, A., Moughan, P. J. & Singh, H. (2013a). *In vivo* digestion of bovine milk fat globules: Effect of processing and interfacial structural changes. I. Gastric digestion. *Food Chemistry, 141*, 3273–3281.

Gallier, S., Zhu, X. Q., Rutherfurd, S. M., Ye, A., Moughan, P. J. & Singh, H. (2013b). *In vivo* digestion of bovine milk fat globules: Effect of processing and interfacial structural changes. II. Upper digestive tract digestion. *Food Chemistry, 141*, 3215–3223.

Garcia-Amezquita, L. E., Primo-Mora, A. R., Barbosa-C·Novas, G. V. & Sepulveda, D. R. (2009). Effect of nonthermal technologies on the native size distribution of fat globules in bovine cheese-making milk. Innovative *Food Science & Emerging Technologies, 10*, 491–494.

Gassi, J.-Y., Famelart, M.-H. & Lopez, C. (2008). Heat treatment of cream affects the physicochemical properties of sweet buttermilk. *Dairy Science & Technology, 88*, 369–385.

Gassi, J. Y., Blot, M., Beaucher, E., Robert, B., Leconte, N., Camier, B., Rousseau, F., Bourlieu, C., Jardin, J., Briard-Bion, V., Lambert, S., Gésan-Guiziou, G., Lopez, C. & Gaucheron, F. (2016). Preparation and characterisation of a milk polar lipids enriched ingredient from fresh industrial liquid butter serum: Combination of physico-chemical modifications and technological treatments. *International Dairy Journal, 52*, 26–34.

Gorewit, R. C. (2002a). *Bulk preparation of milk fat globule membranes*. US patent application US6391362B1. 2002-05-21.

Gorewit, R. C. (2002b). *Preparation of milk fat globule membranes in tablet or capsule form*. US patent application US6444247B1. 2002-09-03.

Goudedranche, H., Fauquant, J. & Maubois, J. L. (2000). Fractionation of globular milk fat by membrane microfiltration. *Dairy Science & Technology, 80*(1), 93–98.

Guo, Z., Vikbjerg, A. F. & Xu, X. (2005). Enzymatic modification of phospholipids for functional applications and human nutrition. *Biotechnology Advances, 23*(3), 203–259.

Haddadian, Z., Eyres, G. T., Bremer, P. & Everett, D. W. (2018). Polar lipid composition of the milk fat globule membrane in buttermilk made using various cream churning conditions or isolated from commercial samples. *International Dairy Journal, 81*, 138–142.

Heid, H. W. & Keenan, T. W. (2005). Intracellular origin and secretion of milk fat globules. *European Journal of Cell Biology, 84*, 245–258.

Hintze, K. J., Snow, D., Burtenshaw, I. & Ward, R. E. (2011). Nutraceutical properties of milk fat globular membrane. In M. Elnashar (Ed.), *Biotechnology of biopolymers* (pp. 321–342). London: IntechOpen Publishing.

Holzmüller, W. & Kulozik, U. (2016). Technical difficulties and future challenges in isolating membrane material from milk fat globules in industrial settings – A critical review. *International Dairy Journal, 61*, 51–66.

Houlihan, A. V., Goddard, P. A., Kitchen, B. J. & Masters, C. J. (1992). Changes in structure of the bovine milk fat globule membrane on heating whole milk. *Journal of Dairy Research, 59*, 321–329.

Iverson, S. J., Lang, S. L. & Cooper, M. H. (2001). Comparison of the Bligh and Dyer and Folch methods for total lipid determination in a broad range of marine tissue. *Lipids, 36*, 1283–1287.

Jensen, S. K. & Nielsen, K. N. (1996). Tocopherols, retinol, beta-carotene and fatty acids in fat globule membrane and fat globule core in cows' milk. *Journal of Dairy Research, 63*, 565–574.

Jiménez-Flores, R. & Brisson, G. (2008). The milk fat globule membrane as an ingredient: Why, how, when? *Dairy Science & Technology, 88*(1), 5–18.

Jukkola, A. & Rojas, O. J. (2017). Milk fat globules and associated membranes: Colloidal properties and processing effects. *Advances in Colloid and Interface Science, 245*, 92–101.

Jukkola, A., Partanen, R., Rojas, O. J. & Heino, A. (2016). Separation of milk fat globules via microfiltration: Effect of diafiltration media and opportunities for stream valorization. *Journal of Dairy Science, 99*, 8644–8654.

Kanno, C. & Kim, D. H. (1990). A simple procedure for the preparation of bovine milk fat globule membrane and a comparison of its composition, enzymatic activities, and electrophoretic properties with those prepared by other methods. *Agricultural and Biological Chemistry, 54*(11), 2845–2854.

Kanno, C., Shimomura, Y. & Takano, E. (1991). Physicochemical properties of milk fat emulsions stabilized with bovine milk fat globule membrane. *Journal of Food Science, 56*, 1219–1223.

Keenan, T. W. (2001). Milk lipid globules and their surrounding membrane: A brief history and perspectives for future research. *Journal of Mammary Gland Biology Neoplasia, 6*(3), 365–371.

Keenan, T. W. & Mather, I. H. (2006). Intracellular origin of milk fat globules and the nature of the milk fat globule membrane. In P. F. Fox & P. L. H. McSweeney (Eds.), *Advanced dairy chemistry*, 3rd edn. (Lipids) (Vol. 2, pp. 137–171). Berlin: Springer.

Kim, H. H. & Jimenez-Flores, R. (1995). Heat-induced interactions between the proteins of milk fat globule membrane and skim milk. *Journal of Dairy Science, 78*, 24–35.

Keenan, T. W., Moon, T-W. & Dylewski, D. P. (1983). Lipid globules retain globule membrane material after homogenization. *Journal of Dairy Science, 66*, 196–203.

Kisel, M. A., Kulik, L. N., Tsybovsky, I. S., Vlasov, A. P., Vorob'yov, M. S., Kholodova, E. A. & Zabarovskaya, Z. V. (2001). Liposomes with phosphatidylethanol as a carrier for oral delivery of insulin: Studies in the rat. *International Journal of Pharmaceutics, 216*(1–2), 105–114.

Konkoli, A., Baker, M. A., Norquist, P. R., Schoenfus, T. C., Springer, T. J. & Zacho, M. (2006). *Dairy beverage and method of preparation US patent application US Patent application*. 2006/0240149A1.

Konrad, G., Kleinschmidt, T. & Lorenz, C. (2013). Ultrafiltration of whey buttermilk to obtain a phospholipid concentrate. *International Dairy Journal, 30*, 39–44.

Kuchta, A. M., Kelly, P. M., Stanton, C. & Devery, R. A. (2012). Milk fat globule membrane – A source of polar lipids for colon health? A review. *International Journal of Dairy Technology, 65*, 315–333.

Küllenberg, D., Taylor, L. A., Schneider, M. & Massing, U. (2012). Health effects of dietary phospholipids. *Lipids in Health Disease, 11*(1), 3.

Lambert, S., Leconte, N., Blot, M., Rousseau, F., Robert, B., Camier, B., Gassi, J.-Y., Cauty, C., Lopez, C. & Gésan-Guiziou, G. (2016). The lipid content and microstructure of industrial whole buttermilk and butter serum affect the efficiency of skimming. *Food Research International, 83*, 121–130.

Landstrom, K., Alsins, J. & Bergenstâhl, B. (2000). Competitive protein adsorption between bovine serum albumin and [beta]-lactoglobulin during spray-drying. *Food Hydrocolloids, 14*, 75–82.

Le, T. T., Van Camp, J., Rombaut, R., van Leeckwyck, F. & Dewettinck, K. (2009). Effect of washing conditions on the recovery of milk fat globule membrane proteins during the isolation of milk fat globule membrane from milk. *Journal of Dairy Science, 92*, 3592–3603.

Le, T. T., van Camp, J., Pascual, P. A. L., Meesen, G., Thienpont, N., Messens, K. & Dewettinck, K. (2011). Physical properties and microstructure of yoghurt enriched with milk fat globule membrane material. *International Dairy Journal, 21*(10), 798–805.

Lee, S. J. & Sherbon, J. W. (2002). Chemical changes in bovine milk fat globule membrane caused by heat treatment and homogenisation of whole milk. *Journal of Dairy Research, 69*, 555–567.

Levin, M. A., Burrington, K. J. & Hartel, R. W. (2016a). Whey protein phospholipid concentrate and delactosed permeate: Applications in caramel, ice cream, and cake. *Journal of Dairy Science, 99*, 6948–6960.

Levin, M. A., Burrington, K. J. & Hartel, R. W. (2016b). Composition and functionality of whey protein phospholipid concentrate and delactosed permeate. *Journal of Dairy Science, 99*, 6937–6947.

Liu, D. & Ma, F. (2011). Soybean phospholipids. In D. Krezhova (Ed.), *Recent trends for enhancing the diversity and quality of soybean products*. ISBN: 978-953-307-533-4, InTech, Available from: http://www.intechopen.com/books/recent-trends-for-enhancing-the-diversity-and-quality-of-soybeanproducts/soybean-phospholipids.

Liu, G. & Zhong, Q. (2014). Removal of milk fat globules from whey protein concentrate 34% to prepare clear and heat-stable protein dispersions. *Journal of Dairy Science, 97*, 6097–6106.

Lopez, C. (2011). Milk fat globules enveloped by their biological membrane: Unique colloidal assemblies with a specific composition and structure. *Current Opinion in Colloid & Interface Science, 16*, 391–404. https://doi.org/10.1016/j.cocis.2011.05.007.

Lopez, C., Briard-Bion, V., Menard, O., Rousseau, F., Pradel, P. & Besle, J. M. (2008). Phospholipid, sphingolipid, and fatty acid compositions of the milk fat globule membrane are modified by diet. *Journal of Agricultural Food Chemistry, 56*, 5226–5236.

Lopez, C., Blot, M., Briard-Bion, V., Cirie, C. & Graulet, B. (2017a). Butter serums and buttermilks as sources of bioactive lipids from the milk fat globule membrane: Differences in their lipid composition and potentialities of cow diet to increase *n*-3 PUFA. *Food Research International, 100*, 864–872.

Lopez, C., Cauty, C., Rousseau, F., Blot, M., Margolis, A. & Famelart, M.-H. (2017b). Lipid droplets coated with milk fat globule membrane fragments: Microstructure and functional properties as a function of pH. *Food Research International, 91*, 26–37.

Lopez, C., Cauty, C., Rousseau, F., Blot, M., Margolis, A. & Famelart, M.-H. (2017c). Lipid droplets coated with milk fat globule membrane fragments: Microstructure and functional properties as a function of pH. *Food Research International, 91*, 26–37.

Luo, J., Huang, Z., Liu, H., Zhang, Y. & Ren, F. (2018). Yak milk fat globules from the Qinghai-Tibetan Plateau: Membrane lipid composition and morphological properties. *Food Chemistry, 245*, 731–737.

Malik, P., Danthine, S., A., P. and Blecker, C. (2015). Physical-chemical properties of milk fat globule membrane at different stages of isolation. Scientific Bulletin Series F Biotechnologies, XIX, pp.154–159.

Mather, I. H. (2000). A review and proposed nomenclature for major proteins of the milk-fat globule membrane. *Journal of Dairy Science, 83*, 203–247.

Mather, I. H. (2011). Milk Lipids: milk fat globule membrane. In J. Fuquay, P. F. Fox & P. L. H. McSweeney (Eds.), *Encyclopedia of dairy sciences* (2nd ed., pp. 680–690). Amsterdam: Elsevier.

Mather, I. H., Tamplin, C. B. & Irving, M. G. (1980). Separation of the proteins of bovine milk-fat-globule membrane by electrofocusing with retention of enzymatic and immunological activity. *European Journal of Biochemistry, 110*, 327–336.

Matyash, V., Liebisch, G., Kurzchalia, T. V., Shevchenko, A. & Schwudke, D. (2008). Lipid extraction by methyl-*tert*-butyl ether for high-throughput lipidomics. *Journal of Lipid Research, 49*, 1137–1146.

McPherson, A. V. & Kitchen, B. J. (1983). Reviews of the Progress of Dairy Science: The bovine milk fat globule membrane - its formation, composition, structure and behaviour in milk and dairy products. *Journal of Dairy Research, 50*, 107–133.

Mezouari, S., Pouliot, Y. & Britten, M. (2009). Lipids from the milk fat globule membrane as a health ingredient: Composition, properties and technological aspects. In M. Corredig (Ed.), *Dairy-derived ingredients: Food and nutraceutical uses* (pp. 344–367). Amsterdam: Elsevier.

Michalski, M. C. (2007). On the supposed influence of milk homogenisation on the risk of CVD, diabetes and allergy. *British Journal of Nutrition, 97*, 598–610.

Michalski, M. & Januel, C. (2006). Does homogenisation affect the human health properties of cow's milk? *Trends in Food Science & Technology, 17*, 423–437.

Michalski, M.-C., Michel, F. O. & Geneste, C. (2002). Appearance of submicronic particles in the milk fat globule size distribution upon mechanical treatments. *Lait, 82*, 193–208.

Moloney, C., Walshe, E., Phelan, M., Giuffrida, F., Badoud, F., Bertschy, E. & O'Regan, J. (2018). Sphingomyelin content of dairy protein ingredients and infant formula powders, and identification of bovine sphingomyelin species. *International Dairy Journal, 78*, 138–144.

Mondy, B. L. & Keenan, T. W. (1993). Butyrophilin and xanthine oxidase occur in constant molar proportions in milk lipid globule membrane but vary in amount with breed and stage of lactation. *Protoplasma, 177*, 32–36.

Morin, P., Pouliot, Y. & Jimenez-Flores, R. (2006). A comparative study of the fractionation of regular buttermilk and whey buttermilk by microfiltration. *Journal of Food Engineering, 77*, 521–528.

Morin, P., Britten, M., Jiménez-Flores, R. & Pouliot, Y. (2007a). Microfiltration of buttermilk and washed cream buttermilk for concentration of milk fat globule membrane components. *Journal of Dairy Science, 90*(5), 2132–2140.

Morin, P., Jimenez-Flores, R. & Pouliot, Y. (2007b). Effect of processing on the composition and microstructure of buttermilk and its milk fat globule membranes. *International Dairy Journal, 17*, 1179–1187.

Morin, P., Pouliot, Y. & Britten, M. (2008). Effect of buttermilk made from creams with different heat treatment histories on properties of rennet gels and model cheeses. *Journal of Dairy Science, 91*, 871–882.

Mulder, H. & Walstra, P. (1974). The milk fat globule. Emulsion science as applied to milk products and comparable foods. Wageningen, Netherlands, Centre for Agricultural Publishing and Documentation, Farnham Royal, Commonwealth Agricultural Bureaux, UK, Technical Communication, Commonwealth Bureau of Dairy Science and Technology, 1974, No. 4, pp. 296. https://www.cabdirect.org/cabdirect/abstract/19740410345.

Murgiano, L., Timperio, A. M., Zolla, L., Bongiorni, S., Valentini, A. & Pariset, L. (2009). Comparison of Milk Fat Globule Membrane (MFGM) Proteins of Chianina and Holstein Cattle Breed Milk Samples Through Proteomics Methods. *Nutrients, 1*, 302–315.

Murthy, A. V. R., Guyomarc'h, F. & Lopez, C. (2016). The temperature-dependent physical state of polar lipids and their miscibility impact the topography and mechanical properties of bilayer models of the milk fat globule membrane. *Biochimica et Biophysica Acta, 1858*, 2181–2190.

Nilsson, A. (2016). Role of sphingolipids in infant gut health and immunity. *Journal of Pediatrics, 173*, S53–S59.

Ortega-Anaya, J. & Jiménez-Flores, R. (2018). Symposium review: The relevance of bovine milk phospholipids in human nutrition – Evidence of the effect on infant gut and brain development. *Journal of Dairy Science, 102*(3), 2738–2748.

Ortega-Requena, S. & Rebouillat, S. (2015). Bigger data open innovation: Potential applications of value-added products from milk and sustainable valorization of by-products from the dairy industry. *Green Chemistry, 17*, 5100–5113.

Patton, S., Long, C. & Sooka, T. (1980). Effect of storing milk on cholesterol and phospholipid of skim milk. *Journal of Dairy Science, 63*, 697–700.

Phan, T. T. Q., Asaduzzaman, M., Le, T., Fredrick, E., Van der Meeren, P. & Dewettinck, K. (2013). Composition and emulsifying properties of a milk fat globule membrane enriched material. *International Dairy Journal, 29*, 99–106.

Phan, T. T. Q., Le, T. T., Van der Meeren, P. & Dewettinck, K. (2014). Comparison of emulsifying properties of milk fat globule membrane materials isolated from different dairy by-products. *Journal of Dairy Science, 97*(8), 4799–4810.

Pimentel, L., Gomes, A., Pintado, M. & Rodriguez-Alcala, L. M. (2016). Isolation and analysis of phospholipids in dairy foods. *Journal of Analytical Methods in Chemistry, 2016*, 9827369.

Rebouillat, S. & Ortega-Requena, S. (2015). Potential applications of milk fractions and valorization of dairy by-products: A review of the state-of-the-art available data, outlining the innovation potential from a bigger data standpoint. *Journal of Biomaterials and Nanobiotechnology, 6*, 176–203.

Reinhardt, T. A. & Lippolis, J. D. (2008). Developmental changes in the milk fat globule membrane proteome during the transition from colostrum to milk. *Journal of Dairy Science, 91*, 2307–2318.

Roesch, R. R., Rincon, A. & Corredig, M. (2004). Emulsifying properties of fractions prepared from commercial buttermilk by microfiltration. *Journal of Dairy Science, 87*, 4080–4087.

Rombaut, R. & Dewettinck, K. (2006). Properties, analysis and purification of milk polar lipids. *International Dairy Journal, 16*, 1362–1373.

Rombaut, R., Vanhoutte, B., Dewettinck, K. & Van Der Meeren, P. (2004). Phospholipids. In L. Nollet (Ed.), *Food analysis* (pp. 349–382). New York: Marcel Dekker.

Rombaut, R., Camp, J. V. & Dewettinck, K. (2006a). Phospho- and sphingolipid distribution during processing of milk, butter and whey. *International Journal of Food Science Technology, 41*, 435–443.

Rombaut, R., Dejonckheere, V. & Dewettinck, K. (2006b). Microfiltration of butter serum upon casein micelle destabilization. *Journal of Dairy Science, 89*, 1915–1925.

Rombaut, R., Dejonckheere, V. & Dewettinck, K. (2007). Filtration of milk fat globule membrane fragments from acid buttermilk cheese whey. *Journal of Dairy Science, 90*, 1662–1673.

Sachdeva, S. & Buchheim, W. (1997). Recovery of phospholipids from buttermilk using membrane processing. *Kieler Milchwirtschaftliche Forschungsberichte, 49*, 47–68.

Sethi, S. & Brietzke, E. (2017). Recent advances in lipidomics: Analytical and clinical perspectives. *Prostaglandins & Other Lipid Mediators, 128–129*, 8–16.

Sharma, S. K. & Dalgleish, D. G. (1994). Effect of heat treatments on the incorporation of milk serum proteins into the fat globule membrane of homogenized milk. *Journal of Dairy Research, 61*, 375–384.

Shimizu, M., Yamauchi, K. & Kanno, C. (1979). Proteolytic digestion of milk-fat globule-membrane proteins. *Milchwissenschaft, 34*, 666–668.

Shukla, F. C., Sharma, A. & Singh, B. (2004). Studies on the preparation of fruit beverages using whey and buttermilk. *Journal of Food Science Technology, 41*, 102–105.

Singh, H. (2006). The milk fat globule membrane – A biophysical system for food applications. *Current Opinion in Colloid & Interface Science, 11*, 154–163.

Singh, H. (2019). Symposium review: Fat globules in milk and their structural modifications during gastrointestinal digestion. *Journal of Dairy Science, 102*(3), 2749–2759.

Singh, H. & Gallier, S. (2017). Nature's complex emulsion: The fat globules of milk. *Food Hydrocolloids, 68*, 81–89.

Sodini, I., Morin, P., Olabi, A. & Jimenez-Flores, R. (2006). Compositional and functional properties of buttermilk: A comparison between sweet, sour, and whey buttermilk. *Journal of Dairy Science, 89*, 525–536.

Spitsberg, V. L. (2005). Bovine milk fat globule membrane as a potential nutraceutical. *Journal of Dairy Science, 88*, 2289–2294.

Spitsberg, V. L. & Gorewit, R. C. (2002). Isolation, purification and characterization of fatty-acid-binding protein from milk fat globule membrane: Effect of bovine growth hormone treatment. *Pakistan Journal of Nutrition, 1*, 43–48.

Sprong, R. C., Hulstein, M. F. & Van der Meer, R. (2001). Bactericidal activities of milk lipids. *Antimicrobial Agents and Chemotherapy, 45*(4), 1298–1301.

Stannard, D. J. (1975). The use of marker enzymes to assay the churning of milk. *Journal of Dairy Research, 42*, 241–246.

Svanborg, S., Johansen, A. G., Abrahamsen, R. K. & Skeie, S. B. (2015). The composition and functional properties of whey protein concentrates produced from buttermilk are comparable with those of whey protein concentrates produced from skimmed milk. *Journal of Dairy Science, 98*, 5829–5840.

Szuhaj, B. F. (1983). Lecithin production and utilization. *Journal of the American Oil Chemists' Society, 60*(2), 306–309.

Van Nieuwenhuyzen, W. (1976). Lecithin production and properties. *Journal of the American Oil Chemists' Society, 53*, 425–427.

Van Nieuwenhuyzen, W. (1981). The industrial uses of special lecithins: A review. *Journal of the American Oil Chemists' Society, 58*, 886–888.

Vanderghem, C., Bodson, P., Danthine, S., Paquot, M., Deroannes, C. & Blecker, C. (2010). Milk fat globule membrane and buttermilk : From composition to valorization. *Biotechnologie, Agronomie, Société et Environnement, 14*(3), 485–500.

Vanderghem, C., Francis, F., Danthine, S., Deroanne, C., Paquot, M., De Pauw, E. & Blecker, C. (2011). Study on the susceptibility of the bovine milk fat globule membrane proteins to enzymatic hydrolysis and organization of some of the proteins. *International Dairy Journal, 21*, 312–318.

Veatch, S. L., Cicuta, P., Sengupta, P., Honerkamp-Smith, A., Holowka, D. & Baird, B. (2008). Critical fluctuations in plasma membrane vesicles. *ACS Chemical Biology, 3*, 287–293.

Walstra, P. & Jenness, R. (1984). Dairy Chemistry and Physics. New York: John Wiley and Sons.

Walstra, P. (2003). *Physical chemistry of foods*. New York: Marcel Dekker.

Wang, H., Li, Y., Cheng, Y., Yin, L. & Li, L. (2013). Effect of the maillard reaction on angiotensin I-converting enzyme (ACE)-inhibitory activity of douchi during fermentation. *Food Bioprocess Technology, 6*, 297–301.

Xu, S., Walkling-Ribeiro, M., Griffiths, M. W. & Corredig, M. (2015). Pulsed electric field processing preserves the antiproliferative activity of the milk fat globule membrane on colon carcinoma cells. *Journal of Dairy Science, 98*, 2867–2874.

Yao, Y., Zhao, G., Xiang, J., Zou, X., Jin, Q. & Wang, X. (2016a). Lipid composition and structural characteristics of bovine, caprine and human milk fat globules. *International Dairy Journal, 56*, 64–73.

Yao, Y., Zhao, G., Yan, Y., Chen, C., Sun, C., Zou, X., Jin, Q. & Wang, X. (2016b). Effects of freeze drying and spray drying on the microstructure and composition of milk fat globules. *RSC Advances, 6*, 2520–2529.

Ye, A. A. S. G. & Singh, H. (2007). Behaviour of homogenized fat globules during the spray drying of whole milk. *International Dairy Journal, 17*, 374–382.

Ye, A., Singh, H., Taylor, M. W. & Anema, S. (2002). Characterization of protein components of natural and heat-treated milk fat globule membranes. *International Dairy Journal, 12*, 393–402.

Ye, A., Anema, S. G. & Singh, H. (2004). High-pressure-induced interactions between milk fat globule membrane proteins and skim milk proteins in whole milk. *Journal of Dairy Science, 87*, 4013–4022.

Zhang, L., Li, P., Sun, X., Wang, X., Xu, B., Wang, X., Ma, F., Zhang, Q. & Ding, X. (2014). Classification and adulteration detection of vegetable oils based on fatty acid profiles. *Journal of Agricultural Food Chemistry, 62*, 8745–8751.

Zheng, H., Jiménez-Flores, R. & Everett, D. W. (2014). Lateral lipid organization of the bovine milk fat globule membrane is revealed by washing processes. *Journal of Dairy Science, 97*, 5964–5974.

Zhu, D. & Damodaran, S. (2011). Composition, thermotropic properties, and oxidative stability of freeze-dried and spray-dried milk fat globule membrane isolated from cheese whey. *Journal of Agricultural Food Chemistry, 59*, 8931–8938.

Zhu, D. & Damodaran, S. (2013). Dairy lecithin from cheese whey fat globule membrane: Its extraction, composition, oxidative stability and emulsifying properties. *Journal of the American Oil Chemists' Society, 90*, 217–224.

Milk Fat: Chemical and Physical Modification

7

S. A. Hogan and T. F. O'Callaghan

7.1 Introduction

Milk fat exists naturally as an oil-in-water emulsion, is comprised of a wide range of lipid components and is a valuable commodity ingredient. Traditionally, milk fat has been used as an ingredient for a huge variety of food applications. However, recent advances in technologies for fat separation and modification have opened new avenues for its uses in nutritional formulations. The compositional make up of milk fat is impacted by a variety of factors, including farm management practices, breeding, diet, stage of lactation, parity and animal health (Alothman *et al.* 2019). The composition of milk fat impacts significantly on its nutritional and functional characteristics and affects important attributes of many food products, including textural, melting and sensory properties. Given the importance of milk fat as a food ingredient, significant commercial interest exists in enhancing its functional and quality attributes. This chapter provides an overview of the physical and chemical mechanisms available for the modification and fractionation of milk fat and its components.

S. A. Hogan (✉)
Food Chemistry and Technology Department, Teagasc Food Research Centre, Moorepark, Fermoy, Co. Cork, Ireland
e-mail: sean.a.hogan@teagasc.ie

T. F. O'Callaghan
School of Food and Nutritional Sciences, University College Cork, Cork, Ireland

7.2 Physical Modification of Milk Fat

7.2.1 Crystallisation of Milk Fat

Crystallisation of milk fat is a complex process involving solidification of fatty acids (FAs) at different rates and temperatures, due to the broad FA profile and positional distribution of milk triacylglycerols (TAGs). Milk fat has a very wide melting range (−40 to +40 °C) and highly variable levels of plasticity due to large variations in the solid–liquid fat ratio (FA composition, temperature and cooling rate dependent). Both FA chain length and degree of unsaturation have significant effects on the rheological behaviour of milk fat. Crystallisation is an exothermic, supramolecular process in which molecules moving in random fluid motion are converted to fixed molecular assemblies. It does not occur spontaneously due to the varying thermodynamic stability of different TAGs. Crystallisation results from nucleation (in supercooled or supersaturated systems) followed by crystal growth and rearrangement.

The early stages of conversion from a liquid to a solid phase involve the formation of nuclei and crystal growth by annexation of molecules on

P. L. H. McSweeney et al. (eds.), *Advanced Dairy Chemistry, Volume 2*,
https://doi.org/10.1007/978-3-030-48686-0_7

existing crystal faces. In practical terms, nucleation and growth occur simultaneously (Gibon 2006). A process of molecular ordering leads to stacking of lamellae (crystalline plates) and finally to crystal sub-cell packing. The overall process of diffusion of liquid to solid phases and organisation of molecules to enable crystal growth involves a number of kinetic factors over and above the thermodynamic ones that eventually determine polymorphic occurrence, crystal composition and purity (Himawan *et al.* 2006). Kinetic factors affecting crystallisation rate include clustering of molecules, molecular adsorption, solvation/de-solvation and conformational rearrangements (Wright and Marangoni 2006). In practical terms, crystallisation during processing is affected by seasonality (FA compositional change), temperature (and its rate of change), scale of operation, agitation and storage time (Wright and Marangoni 2006). A temperature-induced increase in the viscosity of milk fat reduces the rate of diffusion of TAG molecules to crystal surfaces and lowers latent heat mass transfer. Rapid cooling of milk fat melts (the liquefied product of melting) results in high numbers of small nuclei with the opposite effect prevailing during slow cooling. The rate of crystal growth is inversely proportional to temperature and melt viscosity (Lawler and Dimick 1998).

Crystal growth in milk fat is made complex by the wide range of FAs present (Table 7.1). Mulder (1953) proposed the theory of mixed crystal formation in which more than one FA species is contained in milk fat crystals. As such, the melting curve of milk fat cannot be derived from the sum of its component TAGs. Mixed or compound crystals are characterised by a melting range resulting from the different melting and crystallisation behaviours of FAs attached to the same TAG molecule (Fredrick *et al.* 2011). The fast cooling rates associated with high-volume industrial milk fat processing result in the formation of mixed crystals and unstable crystal formation as insufficient time is available for pure crystals to form. Fast cooling of TAGs favours the formation of small crystals and results in a 'firm' crystal network. In contrast, slow cooling tends to produce larger crystals with fewer contact points and hence a 'softer' network in dairy products like butter (Wiking *et al.* 2009).

The native form of milk fat, i.e., globular form, contained within the milk fat globule membrane has different TAG structures or stacking compared to the anhydrous form, i.e., bulk fat in non-emulsified form (Lopez *et al.* 2002, 2007). Globular fat provides more contact points between crystals and also pushes the crystals closer, leading to crystal aggregation. This causes development of a stronger, less brittle crystal network compared to similar TAG composition in anhydrous form.

The crystalline polymorphic structures formed by cooling milk fat at different rates lead to the formation of three main polymorphic types: α, β′ and β forms. Another polymorph, γ, is a short-lived form with a lower melting temperature and stability than the α type. Such polymorphism underlies the complex melting phenomena of milk fat. Of these, α to γ transitions are reversible, while α to β′ to β transitions are irreversible (Lopez *et al.* 2007). The α form is the least stable of the three main polymorphs and changes to β or β′ on heating. The β form is most thermodynamically stable and forms large, plate-like crystals, with poor connectivity and can confer undesirable textural properties such as brittleness and flakiness to butter and other dairy products. The β′ form is metastable and is desirable for the 'proper' texture of butter. Milk fat is considered to be a β′ fat as the majority of crystals remain in this form even after prolonged storage (Timms 1979). The melting temperature of α, β' and β polymorphs in milk fat are 20–22, 30 and 35 °C, respectively (Van Beresteyn 1972; Ten Grotenhuis *et al.* 1999).

The effects of heating and cooling rates on milk fat crystalline and polymorphic structures in anhydrous milk fat (AMF), butterfat and cream have been reported extensively (Schaap *et al.* 1975; Ten Grotenhuis *et al.* 1999; Lopez *et al.* 2001, 2002, 2005, 2007; Campos *et al.* 2003; Fredrick 2012; Van Aaken and Visser 2000). Van Aaken and Visser (2000) reported that the firm-

Table 7.1. Triacylglycerol composition (mass%) of milk fat and its primary olein and stearin fractions obtained by dry fractionation at 21 °C. Minor fatty acids (branched, *trans* and odd-numbered not shown)

Triacylglycerol[a]	Milk fat	Olein fraction	Stearin fraction	Triacylglycerol[a]	Milk fat	Olein fraction	Stearin fraction
BCM	0.19	0.26	0.15	CyPS	0.56	0.45	0.51
BCO	0.20	0.26	0.18	LaPdO/LaPdP/LMO	0.17	0.48	0.28
BCP	0.42	0.48	0.24	MPL	1.26	1.51	0.92
BMPo/BLaO	0.48	0.76	0.41	LaOO/MoPO	1.08	1.30	0.70
BLaP/BMM	1.15	1.54	0.85	MMO/LaPO	3.13	3.63	2.49
BPoO	0.46	0.78	0.22	COS/MPPo	1.35	1.18	1.31
BMPd	0.39	0.97	0.43	MMP/LaPP/CPS	3.21	1.91	5.19
BPPo	0.63	0.44	1.35	MPdO/MPdP	1.04	0.90	0.64
BMO/CaLaP	2.03	2.97	2.62	MOO/PPoO	3.08	3.18	2.00
BMP	3.19	4.63	0.44	MSL/PPL	1.44	1.28	0.61
BOO/CaMO	1.67	2.62	1.11	MPO/PPPo	5.24	5.10	4.60
BPO/CaMP	4.40	6.41	3.26	LaOS	1.47	0.88	1.69
BPP/BMS	5.39	7.33	3.77	MMS/MPP/LaPS	3.84	1.62	7.14
CaPdP	0.40	0.79	0.32	MMP/PdPP/OOO	1.63	1.44	1.53
CaOO	0.83	1.40	0.59	POO	4.14	4.04	2.91
CaPO/CLaO	2.62	3.78	1.90	PoSO	1.87	1.88	1.25
CaPP/BOS/CaMS	2.93	3.94	2.54	PPO/MSO	5.83	5.38	6.06
BPS	2.40	3.10	1.81	PPoS	1.45	0.60	2.74
CyOO	1.06	1.01	0.56	PPP/MPS	3.06	0.99	7.04
CMO	1.32	2.39	0.52	PdOS/PdPS	0.58	0.37	0.77
CyOP/CMP	2.27	1.09	1.80	SOO	1.24	1.35	0.78
CyPP	1.22	1.13	1.08	SSL	0.74	0.69	0.80
CaPS	1.18	0.14	0.83	PSO	3.45	3.18	3.72
LaOL	0.45	0.46	0.31	PoSS	0.89	0.30	1.71
COO/MMoO	0.63	0.59	0.41	PPS/MSS	2.32	0.45	5.47
CPO/LaMO	2.38	2.58	1.55	SSO	0.72	0.65	0.67
MMoP/CyOS	0.86	0.91	0.54	PSS	1.14	0.18	2.73
LaMP/CPP/CMS	2.68	2.26	3.40	SSS	0.23	0.07	0.53
Total					100.00	100.00	100.00

Reproduced with permission from Lopez *et al.* (2006).

[a]Abbreviations—*B* butyric acid (C4:0), *Ca* caproïc acid (C6:0), *Cy* caprylic acid (C8:0), *C* capric acid (C10:0), *La* lauric acid (C12:0), *M* myristic acid (C14:0), *Pd* pentadecanoic acid (C15:0), *P* palmitic acid (C16:0), *Po* palmitoleic acid (C16:1), *S* stearic acid (C18:0), *O* oleic acid (C18:1).

ness of milk fat is related to the kinetics of α-crystal conversion to β′ crystals in high- and medium-melting TAG fractions.

7.2.2 Fractionation of Milk Fat

Fractionation of milk fat is a separation process by which separate groups of TAGs can be obtained with distinct physical, chemical and functional characteristics. Physical modification of milk fat by physical means or blending results in products with altered FA compositions but without change to the underlying molecular structure, that is, the position of FAs within TAG molecules (Kaylegian 1999). The chemical composition of milk fat varies greatly with the season, breed of cow, stage of lactation and diet, which impact on the functional properties of the fat. Modification of fat composition to improve functional and nutritional characteristics is of considerable interest. Approaches to the modification of milk fat composition have included dietary manipulation, hydrogenation, enzymatic inter-esterification and others (Michalski *et al.* 2020). These have not always been commercially successful due to cost considerations and the public desire for natural, minimally processed foods (Lopez *et al.* 2006).

The differential scanning calorimetry (DSC) thermogram, shown in Figure 7.1, describes the relationships between heat flow and polymorphic transformations, summarised by Maneesha *et al.* (2020) from a number of different studies (Lopez *et al.* 2002, 2005, 2007; Ten Grotenhuis *et al.* 1999). The melting profile of milk fat, determined by DSC, typically shows three main endothermic peaks – the low melting fraction (LMF) below 10 °C, the medium melting fraction (MMF) from 10 to 19 °C and the high-melting fraction (HMF) above 20 °C (Kaylegian and Lindsay 1995). The LMF accounts for approximately 55% (w/w) of total milk fat and contains TAGs with one long-chain saturated FA (SFA) and two short-chain SFAs or *cis* unsaturated FAs (UFAs); the MMF accounts for about 35% (w/w) of milk fat and contains TAGs with two long-chain SFAs and one short-chain SFA or *cis* unsaturated FA; the HMF contains predominantly long-chain SFAs and accounts for approximately 10% (w/w) of total milk fat (Timms 1979; Marangoni and Lencki 1998; Buldo 2012).

Lopez *et al.* (2006) characterised the TAG profiles of stearin (TAGs in solid form) and olein (TAGs in liquid form) fractions obtained by dry fractionation of AMF in order to demonstrate the structural and crystalline distributions of milk FAs as a function of temperature. All milk TAGs were present in both stearin and olein fractions but in different proportions. The stearin fraction was enriched with long-chain SFA TAGs containing primarily: a) three long-chain SFAs, b) one or two medium-chain SFAs and a long-chain SFA and c) one long-chain monounsaturated FA (MUFA) and two long-chain SFAs. The long-chain SFAs of the stearin fraction were mainly myristic (C14:0), palmitic (C16:0) and stearic (C18:0) acids with melting points of 54, 63 and 70 °C, respectively.

The olein fraction consists predominantly of TAGs containing unsaturated and short-chain FAs, that is, a) two long-chain MUFAs and b) one MUFA and two medium-chain FAs. The olein fraction was further enriched with c) two long-chain SFAs or d) one long-chain SFA and one MUFA and e) two long-chain MUFAs. The lower

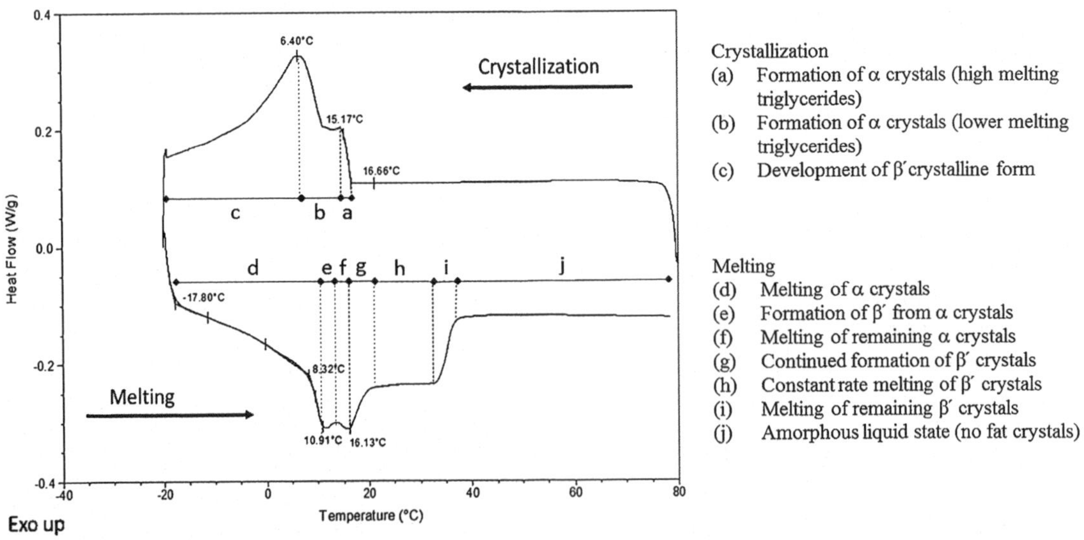

Figure 7.1. Representative DSC thermogram showing crystallisation and melting behaviour of anhydrous milk fat (AMF). Crystallisation: AMF equilibrated at 80 °C for 30 min followed by cooling to −20 °C at 5 °C min^{-1}. Melting: −20 to 80 °C at 5 °C min^{-1}. (Adapted from Michalski *et al.* 2020).

melting range of the olein fraction is due mainly to oleic (C18:1), linoleic (C18:2) and linolenic (C18:3) acids (melting points 16, −5 and – 12 °C, respectively) and the short-chain FAs such as butyric (C4:0), caproic (C6:0) and caprylic (C8:0) acids (melting points – 8, −4 and 19 °C, respectively) (Walstra 1987).

In general, the melting point of fat decreases with decreasing chain length and increasing degree of unsaturation. The melting range of milk fat can be lowered by increasing the concentration of short-chain FAs and long-chain unsaturated FAs (UFA) and by decreasing the concentration of long-chain SFAs (Kaylegian and Lindsay 1995). The melting point of butter oil is raised by increasing the amount of myristic and palmitic acids, whereas it is lowered by increasing oleic and linoleic acids (Ortiz-Gonzalez *et al.* 2007). Relatively little information is available in the literature on the correlation between FA profile and the crystallisation and melting behaviour of milk fat.

In summary, the stearin fraction tends to be enriched in long-chain SFAs (mainly myristic, palmitic and stearic acids) whereas the olein fraction is enriched in TAGs with at least one short-chain FA (butyric acid) and/or one UFA (oleic acid) (Lopez and Ollivon 2009). The olein fraction is higher in polyunsaturated FAs (PUFA), which are more prone to oxidation, and short-chain (butyric) FAs. Several types of crystals are observed, the structural properties of which are dependent on their TAG composition. Differences in the TAG composition of AMF, stearin and olein fractions are also reflected in their thermal properties, as determined by DSC (Lopez and Ollivon 2009).

In a second study on dry-fractionated milk fat, Lopez and Ollivon (2009) further characterised the thermal and polymorphic properties of TAGs by application of x-ray diffraction (XRD) and DSC techniques. In this way, temperature-dependent, solid–liquid phase diagrams can be generated to show the relative proportion of crystalline phases within the solid phase during heating (Figure 7.2.). Melting curves, determined by DSC, showed that the HMF present in milk fat and in the stearin fraction is absent from the olein fraction. The solid fat content (SFC) of milk fat and its fractions was calculated using DSC enthalpy data and reported to be 84, 86 and 88.8% for olein, milk fat and stearin fractions, respectively, at 4 °C. SFC at 20 °C was estimated to be 40, 58 and 66.4%, respectively. Such information is very useful for determining the functionality (spreadability of butter, shear-induced disruption of high-fat products during processing, destabilisation of emulsions through partial coalescence, mouth feel, etc.) of milk fat as a function of temperature. Although the information is restricted to the heating and cooling rates of the study in question, as thermal history must be considered in the determination of milk fat crystallisation behaviour, it provides a very useful addition to the literature.

Polymorphism of milk fat TAGs is monotropic, that is, inter-conversion of crystal types occurs only in the direction of more stable forms with higher melting points. Lopez and Ollivon (2009) also showed that groups of milk fat TAGs crystallise and melt separately and that TAGs of the stearin fraction separate naturally from those of the olein fraction during cooling and subsequent heating. Heating of milk fat is a complex process with melting, reorganisation and crystallisation occurring in tandem. The authors noted that the structures formed during cooling did not correspond to the structures that melted on subsequent heating. During heating, TAG molecules do not have time to reach equilibrium and undergo intermediate solid state organisational and polymorphic changes before final melting.

The melting behaviour of milk fat is likely to have nutritional and digestive implications as the extent of hydrolysis and absorption of solid TAG molecules remains open to question. Ayala-Bribiesca *et al.* (2017) showed that *Cheddar*-type cheese, prepared using fractions of AMF stearin, was less prone to disintegration during *in vitro* digestion. Analysis of digestion products also showed that calcium interacted with long-chain FAs to produce soaps, thereby reducing the bioavailability of both FAs and calcium. Provision of OPO-type TAGs (oleic, palmitic, oleic FAs) at *sn*-1, *sn*-2 and *sn*-3 positions,

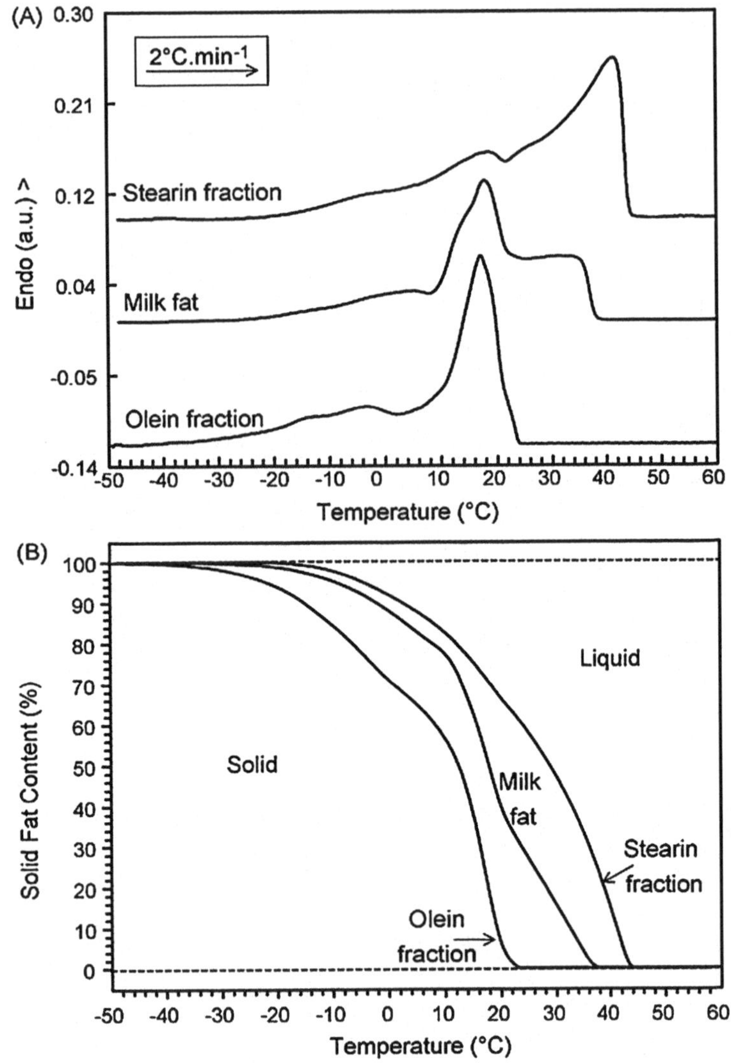

Figure 7.2. Thermal properties of milk fat, stearin and olein fractions following cooling from 17 °C to −50 °C at 1 °C min^{-1} followed by heating from −50 to 60 °C at 2 °C min^{-1}. (**a**) Differential scanning calorimetry (DSC) melting curves. (**b**) Solid fat content vs temperature profiles calculated by direct integration of the area of melting curves. (Reproduced with permission from Lopez and Ollivon 2009).

respectively, in infant formula in recent years represents an effort to minimise this problem in bottle-fed babies and attempts to mimic the high level of palmitic acid at the *sn*-2 position of human milk fat TAG molecules (Hogan and Kelly 2018).

Truong *et al.* (2015) examined nano-emulsions containing olein and stearin fractions in order to see how confinement in oil droplets might impact on crystalline structures. Previous work had shown that the structural behaviour of milk fat crystals in the disperse phase (native milk fat globules and emulsions) is highly complex (Michalski *et al.* 2004; Lopez *et al.* 2007; Fredrick *et al.* 2011).

From a practical perspective, protruding fat crystals in such nano- and micro-emulsion systems can have an important implication in modulating the microstructure of fat-based dairy products, with impacts on phase separation/partial crystallisation (Truong *et al.* 2014).

A sound technological understanding of the complexities in the thermal properties of milk fat is necessary to better understand its behaviour during processing and for the development of high-quality milk fat-containing food products.

7.2.3 Technologies and Approaches for Milk Fat Fractionation and Processing

A number of fractionation techniques are used to separate milk fat into a number of functional lipid fractions based on fat melting and crystallisation properties. A number of review articles and book chapters are available on the subject (Deffense 1993; Kaylegian and Lindsay 1995; Augustin and Versteeg 2006; Sichien *et al.* 2009; Illingworth *et al.* 2009). More recent technologies developed to fractionate milk fat include supercritical fluid extraction, ultrasound and high-pressure processing.

7.2.3.1 Dry Fractionation

Dry fractionation involves the controlled crystallisation of melted oil by a specific cooling programme, followed by separation of the solid from the liquid phase (Gibon 2006). It involves separation of molten fat into stearin and olein fractions by exploiting the crystallisation behaviour of TAGs. It does not use organic solvents, detergents or processing aids and preserves the desirable sensory properties of milk fat (albeit partitioned between liquid and solid products). Its simplicity and cost-effectiveness make it the most widely used commercial fractionation process. Approximately 1000 tonnes per day of milk fat was processed in this way, mostly in Europe, in 2009 (Sichien *et al.* 2009). The global milk fat fractionation industry is set to surpass €14bn by 2024 (Ahuja and Rawat 2018), with dry fractionation dominating the market with a value of approximately US$ 5.4bn in 2018 (Fior Market Reports 2018).

Dry fractionation involves production of suitably sized crystals by controlled cooling and agitation of molten fat, followed by separation of crystals from the mother liquor. The chemical and physical characteristics of fat fractions are dependent on milk fat FA profile, equipment design, fractionation process, separation technique and process parameters; including cooling and agitation rate, residence time in the crystalliser and fractionation temperature (Maneesha *et al.* 2020).

The two most commonly used commercial techniques for dry fractionation are the Tirtiaux (Deffense 1993; Tirtiaux 1976) and DeSmet (Kellens 1994) processes, both developed in Belgium. They involve initial heating of milk fat to 60–80 °C to ensure removal of all crystals, which may act as possible nuclei for (uncontrolled) crystallisation. The methods differ in how the milk fat melt is cooled. The Tirtiaux method involves a patented cooling process based on a constant temperature difference between the cooling liquid and the melt, carried out by gentle agitation at very low cooling rates in order to enable the formation of large, uniformly shaped β' crystals (optimum size between 200 and 300 μm), which are easier to separate than smaller crystals. This cooling step involves initial formation of crystal nuclei in buffering tanks, from where the oil is pumped to crystallisers with constant monitoring of temperature and agitation speed. Production of the desired crystal size distribution means a long crystallisation period of 16–20 h (Sichien *et al.* 2009). Following initial crystal formation the aim of the process is to grow crystals rather than create new ones. In contrast with the Tirtiaux method, the DeSmet process follows the principle of controlled, fast cooling with precise, stepwise lowering of temperature. The crystal sizes obtained with the DeSmet process are smaller and more heterogeneous than the Tirtiaux process. The production of smaller crystals requires higher efficiency filtration and centrifugation steps compared to the Tirtiaux process, although this is counteracted by much faster crystallisation times (6–8 h).

Single-stage fractionation processes typically produce two fractions and involve cooling of melted milk fat with crystallisation typically at 29 °C in order to obtain stearin and olein fractions. Separation of these fractions is usually done by filtration using vacuum filtration or membrane press filtration. The efficiency of the filtration step is crucial to overall fractionation efficiency. Membrane press filters achieve better separation than vacuum filters as the application of pressure (up to 3 MPa) to the stearin cake removes a greater proportion of the entrained olein fraction. The membrane plates are made with flexible membranes that allow the stearin to be squeezed to maximise the yield of the olein fraction. This is particularly important for multi-stage fractionation processes. Two types of filter press are available: (a) low-pressure presses that use compressed air, and (b) high-pressure presses that rely on hydraulic pressure (Illingworth *et al.* 2009).

In multi-stage crystallisation processes, fractions from the first separation are reheated and the process applied again. A typical three-stage temperature programme might start at 29 °C before the olein fraction is cooled and re-crystallised at 19 °C and subsequently at 8–10 °C. Fractionation step temperatures can be altered in order to accommodate milk fats with different initial hardness characteristics. Alteration of supercooling temperature and/or seeding may be required to initiate crystallisation in metastable solutions and improve fractionation efficiency. Dry fractionation processes are usually carried out under N_2 to prevent oxidation and loss of product quality.

7.2.3.2 Wet Fractionation

These procedures use organic solvents in the fractionation process. Although such methods confer certain advantages with respect to rapid crystallisation times, safety issues, toxicological and environmental concerns mean that such processes are not used commercially for milk fat fractionation.

7.2.3.3 Supercritical Fluid Extraction

In recent years, supercritical carbon dioxide (SC-CO_2) has received considerable attention for extraction and fractionation of lipid-containing food and plant materials. The technique is considered a sustainable technology as it does not use organic solvents. Instead, SC-CO_2 exploits the altered solvent properties of fluids in their critical state (i.e. above critical temperature and pressure), which exhibit properties intermediate between those of a liquid and a gas. Due to excellent solvent power, low viscosity, absence of surface tension and high diffusion rates, supercritical fluids are very efficient in separating lipid fractions. SC-CO_2 fractionation is based on molecular weight and dielectric constants of TAGs rather than their thermal properties. CO_2 (supercritical above 31 °C and 74 bar) is inexpensive, non-toxic and non-flammable, unlike organic solvents (Hartel and Kaylegian 2001; Rozzi and Singh 2002). Under normal operating conditions, the solvent used is a gas and can be easily removed from the fat fractions.

SC-CO_2 fractions obtained are more markedly different in FA and TAG composition than melt crystallisation (Sichien *et al.* 2009). On the other hand, the fractions obtained by SC-CO_2 fractionation are less distinct in terms of melting characteristics, thereby restricting the use of supercritical extraction for the production of fractions for specific food applications (Sichien *et al.* 2009). The use of SC-CO_2 in dairy processing has recently been reviewed by Amaral *et al.* (2017), and focuses predominantly on the efficacy of this technique as a low-heat treatment alternative for microbial inactivation.

Chen *et al.* (1992) fractionated milk fat into low-, intermediate- and high-melting fractions using different extraction pressures. Bhaskar *et al.* (1993) reported that fractions containing longer chain length FAs contained both long-chain UFAs and SFAs, and that the melting behaviour of such fractions reflects the combined thermal effects of both these higher molecular weight TAGs.

Arul *et al.* (1994) reported that increasing the density (pressure) of supercritical CO_2 increased the solubility of milk fat TAGs, but that the supercritical fluid became less selective for the extraction of short-, medium- or long-chain TAGs. Such differences were attributed to TAG-dependent solute–solvent interactions and vapour pressures.

Spano *et al.* (2004) used single-step SC-CO_2 fractionation to separate sheep milk fat into two fractions with distinctively different properties. Fatouh *et al.* (2007) and Büyükbeşe *et al.* (2013, 2018) fractionated buffalo butter oil and AMF, respectively, into a number of fractions using SC-CO_2. Overall, a decrease in the proportion of short-chain (C4:0 to C8:0), medium-chain (C10:0- C14:0) FAs was observed with increasing pressure. The relative proportions of long-chain FAs (C16:0 to C18:2) and total UFAs increased with greater $SC\text{-}CO_2$ pressure. Significant changes occurred to the FA profile of the fractions and led to distinct differences to their thermal profiles and solid fat content.

Varying the process conditions makes it possible to separate multiple fractions by this process. The solvent properties and fractionation potential of $SC\text{-}CO_2$ can also be altered by using co-solvents such as ethanol, which alter the solvent properties of the fluid and allow extraction of more polar lipids. Barry *et al.* (2017) used this approach to maximise the extraction of phospholipids in a novel buttermilk powder. $SC\text{-}CO_2$ has also been used to enhance the inter-esterification behaviour of lipases for development of structured lipids from AMF for use in infant milk formulae (Santos *et al.* 2017). Supercritical fluid fractionation of AMF and other dairy products remains a laboratory-scale/research technique, primarily due to high initial capital investment costs.

7.2.3.4 Short-Path Distillation

Short-path distillation involves the volatilisation of molecules into a vacuum. The controlling factor is the rate at which the molecules escape from the heated surface of the distilling liquid and are received by the cooled condenser surface. It exploits variation in the volatility of milk fat TAGs and has a high degree of molecular weight separation potential. The olein fraction (distillate) is enriched in short- and medium-chain FAs and depleted in long-chain FAs, while an opposite selectivity is observed in the stearin (retentate) fraction (Campos *et al.* 2003). Arul *et al.* (1994) reported that, in contrast with other techniques, a gradual increase in the concentration of unsaturated long-chain FAs in the stearin fraction is observed in short-path distillation. Issues relating to thermal degradation (lipid oxidation, loss of flavour) of milk fat during short-path distillation mean its use as a fractionation technique for milk fat has remained limited. The technique also allows removal and isolation of cholesterol, vitamins, sterols, mono-acylglycerols (MAGs), di-acylglycerols (DAGs), free fatty acids (FFAs) and flavour compounds (Jimenez-Flores 1997; Arul *et al.* 1994; Craven and Lencki 2007).

Short-path distillation has also been used to remove volatile products from lipase-catalysed reactions for the production of 'humanised' milk fat products (Li *et al.* 2010; Sørensen *et al.*, (2010); Kontkanen *et al.* 2011). Ketenoglu and Tekin (2015) recently reviewed various applications of molecular distillation techniques in food products.

7.2.3.5 Ultrasound Processing

Ultrasound is a novel technology that has received increased research attention in recent years. Applications include homogenisation, emulsification and crystallisation, inactivation of bacteria, de-foaming, de-fouling, denaturation and functional improvements of proteins and lactose crystallisation. The typical range of ultrasound frequency used in dairy processing is 20 kHz–10 MHz. Forces playing a role in ultrasound processing include cavitation, heating, dynamic agitation, shear stress and turbulence.

A number of studies have been conducted on rapid sono-crystallisation of fats (Patrick *et al.* 2004; Martini *et al.* 2008a, b; Suzuki *et al.* 2010; Hartel 2013; Chandrapala and Leong 2015; Gregersen *et al.* 2019). A number of reviews are also available on ultrasound processing in food production, two of which concentrate on its use in dairy products (Ashokkumar and Mason 2007; Chandrapala and Leong 2015).

Although concerns exist that oxidation of milk fat can be initiated at certain processing frequencies (due to cavitation-induced heat generation) and energy intensities by ultrasonication (Chouliara *et al.* 2010). Juliano *et al.* (2013) established that free radical formation is not related directly to frequency but rather to the combined effects of frequency, power and energy levels. Gregersen *et al.* (2019) also reported that lipid oxidation was not initiated by high-energy ultrasound treatment of AMF/rapeseed oil blends. Low-frequency conditions caused much lower hydroxyl radical formation in milk (Riener *et al.* 2009; Chouliara *et al.* 2010).

7.2.4 Factors Affecting Fractionation

The quality of the AMF raw material is critical to the efficiency of fractionation. Traces of milk serum (predominantly water and milk proteins), phospholipids and dissolved air can interfere with nucleation and crystal growth. As such, pre-fractionation processing of AMF such as separation, polishing (removal, with water, of proteins and soaps) and vacuum drying (dehydration) is important.

7.2.4.1 Cooling Rate

Fast cooling of TAGs favours the formation of small crystals. Slow cooling, in contrast, tends to produce larger crystals with fewer contact points between them (Wiking *et al.* 2009). Slower cooling of fats tends to provide a softer structure than faster cooling in dairy products, such as butter, and a more clearly defined crystal form (fewer mixed or compound crystals). Woodrow and DeMan (1968) and Lopez *et al.* (2005) reported that slow cooling promotes the β' crystalline form and that fast cooling leads to α crystallisation (Ten Grotenhuis *et al.* 1999). Once β′ crystals are formed, TAGs at the surface of α crystals dissolve and re-crystallise onto β′ crystals. The rate of this transformation is much slower at lower temperatures (Sato and Kuroda 1987; Ten Grotenhuis *et al.* 1999). Eventually, all the α crystals are converted to the β′ form. At temperatures between α and β′ crystal melting points, crystallisation occurs directly to the β′ form. After tempering for a sufficient period of time, milk fat invariably contains crystals in the more stable β′ form.

The ability of a fat to become supercooled (i.e. remain liquid at temperatures below its theoretical crystallisation temperature) means that fractionation temperatures must be reduced and agitation used to initiate nucleation and crystal growth (Illingworth *et al.* 2009). In the liquid state, order can persist at temperatures as much as 40 °C above the melting point of a fat (Hernqvist 1984). Once nucleation has begun it is important that the cooling rate and agitator speed are controlled in order to optimise nuclei numbers and that small crystals formed are not broken further to smaller crystals (Kloek *et al.* 2005). During the final stages of crystal growth agitation speed should keep crystals in suspension without breaking crystals, as smaller crystals prevent efficient separation. Small crystals produced in this way may also act as nucleation sites and lead to secondary nucleation (Buldo 2012). Seed material may be added to counteract supercooling particularly in multi-stage fractionation processes where high-melting crystals may already have been removed and so are unavailable to act as nucleation sites. Increasing mechanical agitation tends to promote higher numbers of small crystals (Kaufmann *et al.* 2012) and can accelerate the transition from α to β′ without affecting the onset time of the α form (Mazzanti *et al.* 2009).

7.2.4.2 Fractionation Temperature

Vanhoutte *et al.* (2002) reported that higher fractionation temperatures result in lower stearin yield and a higher melting point olein fraction. Fractionation temperature also affects stearin cake density and filtration efficiency. Lower fractionation temperatures can also result in lower levels of oil entrainment in the stearin cake and more compact clumps of smaller crystals (Sichien *et al.* 2009). Bonomi *et al.* (2012) reported that the crystallisation temperature has a greater effect than the cooling rate on thermal fractionation of milk fat.

7.2.5 Chemical Composition of Milk Fat Fractions

Illingworth *et al.* (2009) provided a useful summary of the main FA transitions during a representative three-stage milk fat fractionation process at 22 °C, 14 °C and 7 °C (Table 7.2). Milk fat fractionated at 22 °C yields stearin (H) and olein (S) fractions (where H and S refer to Hard (solid) and Soft (liquid), respectively). The S fraction is then re-crystallised at 14 °C to yield second stearin (SH) and second olein (SS) fractions. The SS

Table 7.2. Fatty acid distribution in milk fat and its fractions following three-stage fractionation at 22, 14 and 7 °C

	Saturated fatty acids				Unsaturated fatty acids		
Fraction	Short chain	Medium chain	Long chain	Total	Mono	Poly	Total
AMF	8.8	5.9	47.3	62.0	29.1	3.5	32.6
Stearin (H)[a]	4.3	4.6	60.0	68.9	21.3	2.8	24.1
Olein (S)[a]	8.1	5.5	47.2	60.8	29.9	3.6	33.5
2nd stearin fraction (SH)	8.0	5.0	52.8	65.8	25.7	3.2	28.9
2nd olein fraction (SS)	8.4	5.8	42.8	57.0	33.8	4.0	37.8
3rd stearin fraction (SSH)	7.9	5.1	49.4	62.4	28.4	4.1	32.5
3rd olein fraction (SSS)	8.7	6.2	37.8	52.7	36.3	4.3	40.6

Redrawn from Illingworth *et al.* (2009).
[a]H and S refer to 'Hard' (Stearin) and 'Soft' (Olein) fractions, respectively.

fraction is further re-crystallised to yield a third stearin (SSH) and third olein (SSS) fractions. The H stearin is enriched with long-chain FAs (predominantly palmitic and stearic acids) and is depleted in short- and medium-chain FAs and long-chain UFAs (predominantly oleic acid). The olein fractions from each stage, S, SS and SSS, show a gradual depletion in palmitic and stearic acids and become enriched with oleic acid. Overall, H fractions are enriched in SFAs, particularly the long-chain forms, which are associated with high melting temperatures. Medium-chain and short-chain FAs tend to be evenly distributed between fractions with the exception of short-chain SFAs, which are reduced in the H fraction. UFAs tend to be associated with S fractions. The nature of milk fat TAG molecules means that complete separation of FAs is not possible and all fractions will contain at least some of each milk FA. The efficiency of the separation process is critical as S lipids are entrained to some extent in H fractions and cannot be fully eliminated by standard (dry) fractionation techniques. Combining proper crystal development with a highly efficient separation process is state of the art (Gibon 2006).

Although the thermal properties of milk fat fractions are quite distinct, the FA composition of the various fractions does not differ hugely from the parent fat and is no greater than natural variation due to seasonality, breed, diet, etc. An advantage of this technology is that it allows such variations to be eliminated through recombination of fractionated products.

7.2.6 Applications of Fractionated Milk Products

Fractionation of milk fat is typically carried out using three to five temperature steps. These include very low (<10 °C), low (10–25 °C), middle (25–32 °C), high (32–50 °C) and very high (>50 °C) fractions. Commercial products such as 'Hard and Soft Stearin', 'Olein', 'Super Olein' and 'Top Olein' provide for a very wide range of applications in food products (Chandan 2011).

Milk fat and its fractions are used in a variety of dairy, bakery and confectionary products. The solid fat index (SFI) of butter is very similar to several other shortenings and can be used interchangeably with them (Huang *et al.* 2019).

Hard stearin (high melting point) fractions may be used to harden butter or spreads in warmer markets, in laminated pastries such as puff pastry, Danish pastry and croissants. In terms of functionality the fat phase must remain solid, but not be brittle, to enable the formation of separate layers and prevent these layers from sticking together (Munro and Illingworth 1986; Deffense 1993; Sichien *et al.* 2009). The hard fraction can be used to replace cocoa butter in chocolate (Bystrom and Hartel 1994; Ransom-Painter *et al.* 1997; Schmelzer and Hartel 2001) and to inhibit fat bloom (crystalline polymorphic change or phase separation) in chocolate (Grall and Hartel 1992; Hartel 1996; Pajin and Jovanovic 2005; Sichien *et al.* 2009). It is also used as a component of hardstock for margarine production in place of hydrogenated vegetable fats (Sichien *et al.* 2009). Milk fat is one of only a few fats that are compatible with cocoa butter, with the stearin fraction even more compatible than whole milk fat (Sichien *et al.* 2009). Stearin fractions are also used as a flavour and texture agent in milk chocolate and to improve whipping properties of cream.

The mid-range fractions however tend to have limited applications, apart from premium priced gourmet bakery products.

Liquid olein fractions are used in cakes, shortbread and cookies. They provide a desirable buttery flavour and are used in cold spreadable butter and dairy spreads to improve spreadability at refrigeration temperatures. Schäffer *et al.* (2000) reported that addition of up to 25% (w/w) of a milk fat fraction with a melting point of 10 °C significantly improved the spreadability of butter at low temperatures. A number of studies have established that the incorporation of 15–80% olein fraction in cream before butter manufacture improves the spreadability of butter (Deffense 1993; Burgess 2001; Hartel and Kaylegian 2001; Kaylegian 1999). The olein fraction can also be used to inhibit bloom in baked goods. Developments in fractionation technology, processes and applications have received relatively little research interest in recent years.

7.3 Effects of Minor Lipid Components on Milk Fat Crystallisation

The presence of minor lipid components can affect crystallisation processes of milk fat (Talbot *et al.* 2012). These may be naturally present in milk or added to enhance functionality such as surface gloss, rheological properties and polymorphic stability. Naturally occurring lipid components, other than TAGs, tend to delay crystallisation by adsorption to crystals (Vanhoutte *et al.* 2002; Wiking *et al.* 2009), without affecting polymorphism (Wright *et al.* 2000), and include MAGs, DAGs and phospholipids. The degree of saturation of MAG and DAG FAs impacts significantly on the crystallisation of bulk fat (Patel and Dewettinck 2015).

Minor lipid components added to milk fat include sorbitan esters, sucrose esters, propylene glycol esters, tweens, waxes and phytosterols (Smith *et al.* 2011). Sucrose esters co-crystallise with TAGs and can delay nucleation and inhibit crystal growth (Cerdeira *et al.* 2003). Martini *et al.* (2008a, b) showed that addition of sunflower oil waxes to AMF affected crystallisation onset temperature and organoleptic quality. Fedotova and Lencki (2008) reported that addition of phospholipids (in MFGM form) to butter increased the size of spherulites and that the level of phospholipid must be controlled in order to prevent coalescence of the aqueous phase. It is thought that addition of minor lipid components does not affect thermodynamic properties such as thermal behaviour and liquid–solid fat ratios (Wright *et al.* 2000; Fedotova and Lencki 2008). Patel and Dewettinck (2015) proposed that emulsifiers, when added at saturation levels, tend to crystallise before the bulk fat TAGs, and serve as nuclei for crystallisation. When added at concentrations below their solubility limit emulsifiers act as de-clustering components during molecular packing that can slow down crystallisation (Sato *et al.* 2013; Shimamura *et al.* 2013; Saitou *et al.* 2014). Zychowski *et al.* (2016) showed that phytosterol included into the milk fat phase of emulsions delayed nucleation and can be integrated into crystalline TAG units without affecting polymorphic form.

7.4 Removal of Cholesterol

Cholesterol is present in milk at a level from ca. 0.2 to 0.5% (w/w) (Jensen 2002; Fong *et al.* 2007; Martini *et al.* 2016). The long-perceived negative influence of dietary cholesterol on human health has undergone something of a revision in the scientific community in recent years (Lordan *et al.* 2018; Tsoupras *et al.* 2018; Maneesha *et al.* 2020). Drouin-Chartier *et al.* (2016) showed that literature evidence does not support the public perception that low-fat diets are advantageous, compared to regular- or high-fat diets, in terms of a wide spectrum of cardiometabolic disease risk factors. Gallier *et al.*

(2015) reported that breast-fed babies are exposed to higher levels of cholesterol than bottle-fed infants and that a reduced risk of cardiovascular disease in adults breast-fed as babies may be associated with early exposure to plasma cholesterol (Rueda 2014). Nevertheless, significant research effort has examined ways in which cholesterol can be removed from dairy products. These include extraction, distillation, adsorption and enzymatic modification techniques, or combinations thereof (Illingworth *et al.* 2009).

A considerable amount of research effort has focused on the use of β-cyclodextrin (GRAS status since 1998), which forms a complex with cholesterol (by simple mixing) and which can be removed subsequently by centrifugation and does not impact on the flavour of milk fat. Alonso *et al.* (2009) demonstrated the potential of β-cyclodextrin on a commercial scale (15,000 L) to remove cholesterol in pasteurised, whole milk. β-Cyclodextrin has also been shown to complex with other milk components. Ha *et al.* (2009a) found that losses of milk nutrients (amino acids, water soluble vitamins, short-chain FFAs and lactose) were negligible and that only trace amounts of residual β-cyclodextrin were found in milk following processing. Maskooki *et al.* (2013) reported losses of approximately 6.5, 10 and 2.5% (w/w) protein, fat and SNF, respectively, and that these were dependent on mixing temperature, β-cyclodextrin concentration and milk type (raw or homogenised). Dias *et al.* (2010) reported that co-precipitation was far more effective (up to 90% reduction) compared to kneading and mixing (< 30% removal) processes as a method for removing cholesterol from butter. Alonso *et al.* (2019) used β-cyclodextrin in a scale-up process for the manufacture of butter with a 90% decrease in cholesterol. Bhatia *et al.* (2019) used β-cyclodextrin to remove 90% of the cholesterol from ghee, albeit with a 75% reduction in vitamin D and a slight decrease in PUFAs. Tahir *et al.* (2013) demonstrated that β-cyclodextrin can be re-used up to 10 times without loss of efficiency (72% cholesterol removal). The cost of β-cyclodextrin, however, remains an obstacle to commercial use.

Supercritical CO_2 has also been examined as a means of cholesterol removal from dairy products. In early studies, Huber *et al.* (1996) and Bradley (1989) used supercritical CO_2 to remove up to 97 and 90% in AMF and milk, respectively. In recent studies, Chitra *et al.* (2015) and Dey *et al.* (2016) used supercritical CO_2 to remove 56 or 46% of cholesterol from whole milk powders. Ghosh *et al.* (2018) reported a 39% removal of cholesterol from cream powder with a 10.6% reduction in total fat. The higher volatility of cholesterol compared to milk fat TAGs means it can also be removed by steam distillation processes such as vacuum steam distillation and short-path distillation (Lanzani *et al.* 1994; Hammond 2006). A disadvantage of such processes is the loss of desired milk fat flavour compounds such as ketones and lactones, which can be trapped and re-incorporated, but such processing steps add extra cost.

Enzymes have also been used to remove cholesterol from milk fat. Serajzadeh and Alemzadeh (2008) used immobilised *L. acidophilus* to remove 86% of the cholesterol from milk. Serajzadeh and Alemzadeh (2010) subsequently used a cholesterol oxidase from *Nocardia erythropolis* to reduce cholesterol by 70%. The properties of cholesterol-reduced butter, ice-cream and cheese have also been reported (Ha *et al.* 2009b; Seon *et al.* 2009; Kim *et al.* 2006).

7.5 Chemical and Enzymatic Modification of Dairy Lipids

Lipid modification methods commonly used in the food industry include fractionation, hydrogenation and inter-esterification. As well as the aforementioned physical mechanisms for the modification of lipids, chemical processes for the modification of milk fat are widely used in an effort to achieve desirable functional, rheological or nutritional properties. Chemical inter-esterification modifies the physical properties of lipids by altering the distribution of FAs on the TAG glycerol backbone without changing the overall FA profile. In such circumstances, FAs can be exchanged both within (intra-esterification) and between (inter-esterification) TAGs. The resulting products can then be referred to as structured lipids or structured TAGs (Soumanou *et al.* 2013). In vegetable lipids, SFAs are found predominantly at the external *sn*-1 and *sn*-3 positions, with the UFAs at the *sn*-2 position, while in animal fats the *sn*-2 position of the TAG contains a high proportion of SFAs. While such molecular structures determine the physical properties of the fat, from a nutritional perspective this distribution can also influence its absorption, metabolism and passage into the tissues with implications for health (Berry 2009).

Hydrogenation is a chemical reaction used to increase the hardness of an oil by reducing the degree of unsaturation. It improves the oxidative stability and functionality of fats and alters the melting characteristics of vegetable oils for use in margarines and cooking fats. Hydrogenation is typically carried out at a high temperature in the presence of a metal catalyst and involves the addition of hydrogen atoms to double bonds in the FA chain (Balcão and Malcata 1998). As a consequence of this process, some *trans* FAs can be produced, which are associated with a number of undesirable nutritional characteristics (Kaylegian *et al.* 1993).

Inter-esterification is a chemically or enzymatically catalysed positional rearrangement of FAs on the glycerol backbone of TAGs (Kaylegian *et al.* 1993). Two principal forms of inter-esterification are typically used: chemical inter-esterification which can be random or directed, and enzymatic inter-esterification (Marangoni and Rousseau 1995). In the case of chemical inter-esterification, a catalyst such as metallic sodium or sodium methylate is used (Paula *et al.* 2015), whereas enzymatic modifications rely on random, regio-specific or FA-specific lipases. Other forms of modification include acidolysis and alcoholysis. In acidolysis, the reaction takes place between an acylglycerol and an FFA, whereas in alcoholysis an acyl group is exchanged between an acylglycerol and an alcohol (Kontkanen *et al.* 2011).

Comparatively, chemical inter-esterification requires more processing steps, a higher temperature and therefore more energy than enzymatic mechanisms; furthermore, as a result of the chemical process the product can become discoloured requiring further clean-up and purification; the enzymatic process is simpler and can take place in a fixed-bed reactor under mild conditions (Paula *et al.* 2015). Kaylegian *et al.* (1993) reported that when inter-esterification is performed under uncontrolled temperature conditions, this results in randomised arrangement of the FAs on the glyceride molecule, resulting in a harder milk fat, whereas under temperature-controlled conditions inter-esterification at temperatures less than 35 °C results in a softer milk fat.

Microbial lipases are widely used in different industrial applications, due to a broad substrate specificity and excellent regio- and stereo-selectivity under mild reaction conditions (Kontkanen *et al.* 2011). Compared to chemical inter-esterification, lipase-catalysed reactions allow greater control of end products. In either circumstance, the release of FFAs is of concern due to resultant negative sensory traits. Microbial lipases are typically extracellular enzymes and exhibit maximum activity in the temperature range of 30–40 °C, but lower and higher ranges have been observed (Gupta *et al.* 2004). Microbial-derived lipases can be classified into two specific groups differing in specificity: non-specific lipases are those which act independently of FA position, while 1,3-specific lipases catalyse reactions only at the outer positions (i.e. *sn*-1 and *sn*-3) of the glycerol backbone (Balcão and Malcata 1998). Without altering the FA residues at the *sn*-2 position. In contrast alkali catalysts re-arrange the FA residues at all positions (Posorske *et al.* 1988).

Lipase-induced modifications of milk fat have been used in the past to improve the spreadability of butter-based spreads (Marangoni and Rousseau 1998), preparation of human milk fat substitutes (Christensen and Hølmer 1993) and to enhance the nutritional properties of milk fat through modification of the ratio of saturated and unsaturated FAs. Marangoni and Rousseau (1998) demonstrated that the spreadability of butter can be improved significantly through the use of chemical or enzymatic inter-esterification.

While chemical inter-esterification reduced the hardness of butter by 43%, enzymatic inter-esterification reduced hardness by 9%. In both instances losses in butter-like flavour and the development of off-flavours were observed. In a series of studies, Rousseau *et al.* (1996a, b, c) demonstrated the impact of blending and chemical inter-esterification of butterfat on its melting behaviour, microstructure and rheology. Nunes *et al.* (2011) examined the effects of blending milk fat and soybean oil in a variety of proportions which were inter-esterified using an *Aspergillus niger* lipase. The results demonstrated that the highest inter-esterification yield was found at a ratio of 67% milk fat and 33% soybean oil, which resulted in a more spreadable product at refrigeration temperatures, with a 32% reduction in butter hardness. Paula *et al.* (2010) examined the properties of a milk fat and soybean blend enzymatically inter-esterified with an *sn*-1,3 specific lipase from *Rhizopus oryzae*. The resulting inter-esterified products had a lower consistency than the starting material producing a cold-spreadable milk fat. Balcão and Malcata (1997) examined the ability of lipase-catalysed acidolysis of butterfat with oleic acid to increase the UFA content and to reduce the levels of medium- to long-chain SFAs. Balcão *et al.* (1998) demonstrated alterations to the TAG structure of butterfat induced by the action of an immobilised (specific) lipase; and concluded that lipase-catalysed inter-esterification (accompanied to some extent by net hydrolysis) of butterfat is a technically feasible alternative to sodium methoxide-catalysed inter-esterification of butterfat for the purpose of producing butterfat with (somewhat) improved nutritional properties. Shin *et al.* (2010) produced a structured fat rich in α-linolenic acid (C18:3) using a blend of anhydrous butterfat, palm stearin and flax seed oil which were inter-esterified using immobilised lipases. When short-path distillation was subsequently used, the contents of tocopherols, cholesterol and phytosterol were reduced. Paula *et al.* (2015), examined the inter-esterification of milk fat with soybean oil in an effort to produce a product with increased PUFAs and more desirable physical properties. In this instance, a *Candida antarctica* lipase and *sn*-1,3-regioselective *Rhizopus oryzae* lipase were examined. The resultant inter-esterified blends had a reduced consistency (yield value) compared to that of the original product, and produced a more spreadable product than butter.

In recent years, efforts have included the modification of milk fat to more closely resemble human milk fat, both in FA profile and in the FA distribution of TAGs. Sproston and Akoh (2016) used immobilised lipases to produce a structured lipid with higher amounts of palmitic acid at the *sn*-2 position, enriched in the long-chain PUFAs docosahexaenoic acid (DHA, C22:6) and arachidonic acid (ARA, C20:4). In this study, three mechanisms for the production of structured fat were examined, including physical blending, enzymatic acidolysis and enzymatic inter-esterification. In this instance, both physical blending and inter-esterification were deemed suitable in terms of adequate reaction yields and incorporation of arachidonic and palmitic acids at the *sn*-2 position. Enzymatically modified butterfat has also been examined for its suitability as a human milk fat substitute; Sørensen *et al.* (2010) used a combination of enzymatic acidolysis, fractionation and subsequent deodorisation of a butterfat blended with soybean oil and rapeseed oil to produce a human milk fat substitute with similar FA structure and profile to that of human milk. Hogan and Kelly (2018) reported that an AMF-based Infant Milk Formula (IMF) powder, with balanced FA composition and ca. 33% of palmitic acid at the *sn*-2 position can be produced using a single-step inter-esterification process. Removal of enzymatic by-products (FFAs, MAGs and DAGs) by short-path distillation accounted for a 7% loss in total lipids and yielded a powder with excellent sensory and functional properties. Recent years have seen a significant increase in the number of IMF products containing milk fat and MFGM material, both of which are thought to provide nutritional benefit to babies (Timby *et al.* 2017; Huërou-Luron *et al.* 2018; Furse and Koulman 2019).

7.6 Conclusion

World milk production is estimated to reach 1000 million tonnes by 2025 (IDF 2015), making available approximately 35 million tonnes of milk fat. Such increases in output mean that alternative markets, products and processes are required to use surplus milk fat. The current outlook for the consumption of milk fat appears encouraging, after historic lows observed in the second half of the last century. The re-appraisal of the health benefits of milk fat and appreciation of its unique flavours should continue to open up new markets, although such developments need to be considered in the light of public opinion on sustainability, animal welfare and farming best practice.

Chemical and physical modification of milk fat allows improved control over this complex material and improves the range of applications for milk fat-based food products. Developments in research and technology of milk fat processing have remained relatively static in recent decades. High milk fat prices mean processors have not had to deviate from bulk production of cream and butter. In comparison, far greater effort has been targeted at the exploitation of other milk constituents, specifically proteins. It would appear that significant possibilities remain for innovation and diversification in milk fat processing. Fractionation technologies combined with enzymatic modification of milk fat suggest that a huge array of products and ingredients are possible, differing in functional and nutritional properties. Improved knowledge of the biological roles and digestive fates of lipid species should enhance such potential. Recent technological developments, such as ultrasound and supercritical fluid extraction, may ultimately contribute to production efficiency, impact positively on carbon footprint and improve the functional properties of milk fat-containing products. Improved understanding of the impacts of temperature, shear and pressure on the material characteristics and stability of milk fat as a function of composition is also required.

References

Ahuja, K. & Rawat, A. (2018). *Milk fat fractions market size by product* (Report ID: GMI2486) (p. 300). Delaware.

Alonso, L., Cuesta, P. & Gilliland, S. G. (2009). Analytical method for quantification of β-cyclodextrin in milk, cream and butter by LC. *Chromatographia, 69*, 1089–1092.

Alonso, L., Calvo, M. V. & Fontecha, J. (2019). A scale-up process for the manufacture of reduced-cholesterol butter using beta-cyclodextrin. *Journal of Food Process Engineering, 42*, 1–7.

Alothman, M., Hogan, S. A., Hennessy, D., Dillon, P., Kilcawley, K. N., O'Donovan, M., Tobin, J., Fenelon, M. A. & O'Callaghan, T. F. (2019). The "grass-fed" milk story: Understanding the impact of pasture feeding on the composition and quality of bovine milk. *Food, 8*, 350.

Amaral, G. V., et al. (2017). Dairy processing using supercritical carbon dioxide technology: Theoretical fundamentals, quality and safety aspects. *Trends in Food Science and Technology, 64*, 94–101.

Arul, J., Tardif, R., Boudreau, A., McGinnis, D. S. & Lencki, R. W. (1994). Solubility of milk fat triglycerides in supercritical carbon dioxide. *Food Research International, 21*, 459–467.

Ashokkumar, M. & Mason, T. J. (2007). *Sonochemistry. Kirk-Othmer encyclopedia of chemical technology* (pp. 1–34). New York: Wiley.

Augustin, M. A. & Versteeg, C. (2006). Milk fat: Physical, chemical and enzymatic modification. In P. F. Fox & P. L. H. McSweeney (Eds.), *Advanced dairy chemistry lipids* (Vol. 2, 3rd ed., pp. 293–332). New York: Springer.

Ayala-Bribiesca, E., Turgeon, S. L. & Britten, M. (2017). Effect of calcium on fatty acid bioaccessibility during in vitro digestion of Cheddar-type cheeses prepared with different milk fat fractions. *Journal of Dairy Science, 100*, 2454–2470.

Balcão, V. M. & Malcata, F. X. (1997). Lipase-catalyzed modification of butterfat via acidolysis with oleic acid. *Journal of Molecular Catalysis B: Enzymatic, 3*, 161–169.

Balcão, V. M. & Malcata, F. X. (1998). Lipase catalyzed modification of milkfat. *Biotechnology Advances, 16*, 309–341.

Balcão, V. M., Kemppinen, A., Malcata, F. X. & Kalo, P. J. (1998). Modification of butterfat by selective hydrolysis and interesterification by lipase: Process and product characterization. *Journal of the American Oil Chemists' Society, 75*, 1347–1358.

Barry, K. M., Dinan, T. G. & Kelly, P. M. (2017). Pilot scale production of a phospholipid-enriched dairy ingredient by means of an optimised integrated process employing enzymatic hydrolysis, ultrafiltration and super-critical fluid extraction. *Innovative Food Science and Emerging Technologies, 41*, 301–306.

Berry, S. E. E. (2009). Triacylglycerol structure and interesterification of palmitic and stearic acid-rich fats: An overview and implications for cardiovascular disease. *Nutrition Research Reviews, 22*, 3–17.

Bhaskar, A. R., Rizvi, S. S. H. & Sherbon, J. W. (1993). Anhydrous milk fat fractionation with continuous countercurrent supercritical carbon dioxide. *Journal of Food Science, 58*, 748–752.

Bhatia, P., Sharma, V., Arora, S. & Rao, P. S. (2019). Effect of cholesterol removal on compositional and the physicochemical characteristics of anhydrous cow milk fat (cow ghee). *International Journal of Food Properties, 22*, 1–8.

Bonomi, E. C., Luccas, V. & Kieckbusch, T. G. (2012). Characterization of the stearin obtained by thermal fractionation of anhydrous milk fat. *Process Engineering, 42*, 91–923.

Bradley, R. L. (1989). Removal of cholesterol from milk fat using supercritical carbon dioxide. *Journal of Dairy Science, 72*, 2834–2840.

Buldo, P. (2012). *Crystallization of fat in and outside milk fat globules – Effect of processing and storage conditions*. Ph.D. thesis, Aarhus University, Aarhus.

Burgess, K. J. (2001). Milk fats as ingredients. *International Journal of Dairy Technology, 54*, 56–60.

Büyükbeşe, D., Emre, E. E. & Kaya, A. (2013). Supercritical carbon dioxide fractionation of anhydrous milk fat. *Journal of the American Oil Chemists' Society, 91*, 169–177.

Buyukbese, D., Rousseau, D. & Kaya, A. (2018). Composition and shear crystallization of milkfat fractions extracted with supercritical carbon dioxide. *International Journal of Food Properties, 20*, S3015–S3026.

Bystrom, C. & Hartel, R. W. (1994). Evaluation of milk fat fractionation and modification techniques for creating cocoa butter replacers. *Lebensmittel-Wissenschaft und Technologie, 27*, 142–150.

Campos, R., Litwinenko, J. & Marangoni, A. (2003). Fractionation of milk fat by short-path distillation. *Journal of Dairy Science, 86*, 735–745.

Cerdeira, M., Martini, S., Hartel, R. W. & Herrera, M. L. (2003). Effect of sucrose ester addition on nucleation and growth behavior of milk fat– sunflower oil blends. *Journal of Agricultural and Food Chemistry, 51*, 6550–6557.

Chandan, R. C. (2011). Dairy ingredients for food processing: An overview. In R. C. Chandan & A. Kilara (Eds.), *Dairy ingredients for food processing* (pp. 3–33). Chicago: Wiley.

Chandrapala, J. & Leong, T. (2015). Ultrasonic processing for dairy applications: Recent advances. *Food Engineering Reviews, 7*, 143–158.

Chen, H., Schwartz, S. J. & Spanos, G. A. (1992). Fractionation of butter oil by supercritical carbon dioxide. *Journal of Dairy Science, 75*, 2659–2669.

Chitra, J., Deb, S. & Mishra, H. N. (2015). Selective fractionation of cholesterol from whole milk powder: Optimisation of supercritical process conditions. *International Journal of Food Science and Technology, 50*, 2467–2474.

Chouliara, E., Georgogianni, K. G., Kanellopoulou, N. & Kontominas, M. G. (2010). Effect of ultrasonication on microbiological, chemical and sensory properties of raw, thermized and pasteurized milk. *International Dairy Journal, 20*, 307–313.

Christensen, T. & Hølmer, G. (1993). Lipase catalyzed acyl-exchange reactions of butter oil. Synthesis of a human milk fat substitute for infant formulas. *Milchwissenschaft, 48*, 543–547.

Craven, R. & Lencki, R. (2007). Rapid analysis of acylglycerols in low molecular weight milk fat fractions. *Lipids, 42*, 473–482.

Deffense, E. (1993). Milk fat fractionation today: A review. *Journal of the American Oil Chemists' Society, 70*, 1193–1201.

Dey, P. I., Jayakumar, C. & Niwas Mishra, H. (2016). Optimization of process parameters for supercritical fluid extraction of cholesterol from whole milk powder using ethanol as co-solvent. *Journal of the Science of Food and Agriculture, 96*, 4885–4895.

Dias, H. M. A. M., Berbicz, F., Pedrochi, F., Baesso, M. L. & Matioli, G. (2010). Butter cholesterol removal using different complexation methods with beta-cyclodextrin, and the contribution of photoacoustic spectroscopy to the evaluation of the complex. *Food Research International, 43*, 1104–1110.

Drouin-Chartier, J. P., Cote, J. A., Labonte, M. E., Brassard, D., Tessier-Grenier, M., Desroches, S., Couture, P. & Lamarche, B. (2016). Comprehensive review of the impact of dairy foods and dairy fat on cardiometabolic risk. *Advances in Nutrition, 7*, 1041–1051.

Fatouh, A. E., Mahrana, G. A., El-Ghandoura, M. A. & Singh, R. K. (2007). Fractionation of buffalo butter oil by supercritical carbon dioxide. *LWT – Food Science and Technology, 40*, 1687–1693.

Fedotova, Y. & Lencki, R. W. (2008). The effect of phospholipids on milkfat crystallization behavior. *Journal of the American Oil Chemists' Society, 85*, 205–212.

Fior Market Report. (2018). *Milk fat fractions by technology (short-path distillation, supercritical fluid extraction, solvent fractionation, dry fractionation), product form, application, region global industry analysis, market size, share, growth, trends, and forecast 2019 to 2026* (*Report No. 385983*). Pune.

Fong, B. Y., Norris, C. S. & MacGibbon, A. K. H. (2007). Protein and lipid composition of bovine milk-fat-globule membrane. *International Dairy Journal, 17*, 275–288.

Fredrick, E., Van de Walle, D., Walstra, P., Zijtveld, J. H., Fischer, S., Van der Meeren, P. & Dewettinck, K. (2011). Isothermal crystallization behaviour of milk fat in bulk and emulsified state. *International Dairy Journal, 21*, 685–695.

Fredrick, E. (2012). Fat crystallization and partial coalescence in dairy creams: Role of monoacylglycerols. PhD Thesis, Ghent University, Ghent.

Furse, S. & Koulman, A. (2019). The lipid and glyceride profiles of infant formula differ by manufacturer, region and date sold. *Nutrition, 11*, 1122–1135.

Gallier, S., Vocking, K., Post, J. A., Van De Heijning, B., Acton, D., Van Der Beek, E. M. & Van Baalen, T. (2015). A novel infant milk formula concept: Mimicking the human milk fat globule structure. *Colloids and Surfaces B: Biointerfaces, 136*, 329–339.

Ghosh, M., Srivastava, S., Chitra, J. & Mishra, H. N. (2018). Advent of clean and green technology for preparation of low-cholesterol dairy cream powder: Supercritical fluid extraction process. *Food Quality Safety, 2*, 205–211.

Gibon, V. (2006). Fractionation of lipids for use in foods. In F. D. Gunstone (Ed.), *Modifying lipids for use in food* (pp. 201–233). Abington: Woodhead Publishing Ltd.

Grall, D. & Hartel, R. (1992). Kinetics of butterfat crystallization. *Journal of the American Oil Chemists' Society, 69*, 741–747.

Gregersen, S. B., Frydenberg, R. P., Hammershøj, M., Dalsgaard, T. K., Andersen, U. & Wiking, L. (2019). Application of high intensity ultrasound to accelerate crystallization of anhydrous milk fat and rapeseed oil blends. *European Journal of Lipid Science and Technology, 121*, 1–9.

Gupta, R., Gupta, N. & Rathi, P. (2004). Bacterial lipases: An overview of production, purification and biochemical properties. *Applied Microbiology and Biotechnology, 64*, 763–781.

Ha, H. J., Ahn, J., Min, S. G. & Kwak, H. S. (2009a). Properties of cholesterol-reduced ice cream made with cross-linked β-cyclodextrin. *International Journal of Dairy Technology, 62*, 452–457.

Ha, H. J., Jeon, S. S., Chang, Y. H. & Kwak, H. S. (2009b). Entrapment of nutrients during cholesterol removal from milk by crosslinked β –cyclodextrin. *Korean Journal for Food Science of AnimalResearch, 29*, 566–572.

Hammond, E. (2006). Filled and artificial dairy products and altered milk fats. In F. D. Gunstone (Ed.), *Modifying lipids for use in food* (pp. 462–487). Abington: Woodhead Publishing Ltd.

Hartel, R. W. (1996). Applications of milk-fat fractions in confectionery products. *Journal of the American Oil Chemists' Society 73*(8), 945–953.

Hartel, R. W. (2013). Advances in food crystallization. *Annual Review of Food Science and Technology, 4*, 277–292.

Hartel, R. & Kaylegian, K. (2001). Advances in milk fat fractionation: Technology and applications. In N. Garti & K. Sato (Eds.), *Crystallization processes in fats and lipid systems* (pp. 395–455). New York: Marcel Dekker.

Hernqvist, L. (1984). On the structure of triglycerides in the liquid state and fat crystallization. *Fette Seifen Anstrichtm, 86*, 297–300.

Himawan, C., Starov, V. M. & Stapley, A. G. F. (2006). Thermodynamic and kinetic aspects of fat crystallization. *Advances in Colloid and Interface Science, 122*, 3–33.

Hogan, S. A. & Kelly, P. (2018). *Structural modification of milk fat for use in infant formula*. 16th Euro Fed Lipid Congress, Belfast, September 16–19th, 2018.

Huang, Z., Stipkovits, L., Zheng, H., Serventi, L. & Brennan, C. S. (2019). Bovine milk fats and their replacers in baked goods: A review. *Food, 8*, 383–403.

Huber, W., Molero, A., Pereyra, C. & Martinez de la Ossa, E. (1996). Dynamic supercritical CO_2 extraction for removal of cholesterol from anhydrous milk fat. *International Journal of Food Science and Technology, 31*, 143–151.

Huërou-Luron, I. L., Lemaire, M. & Blat, S. (2018). Health benefits of dairy lipids and MFGM in infant formula. *Oils Crop Lipids, 25*, D306.

IDF. (2015). *The world dairy situation (2015)* (Bulletin no. 481/2015). Brussels: International Dairy Federation (INPA), Belgium. Wiley Online Library.

Illingworth, D., Patil, G. R. & Tamime, A. Y. (2009). Anhydrous milk fat manufacture and fractionation. In A. Y. Tamime (Ed.), *Dairy fats and related products* (pp. 108–166). Oxford: Wiley-Blackwell.

Jensen, R. G. (2002). The composition of bovine milk lipids: January 1995 to December 2000. *Journal of Dairy Science, 85*, 295–350.

Jiménez-Flores, R. (1997). Trends in research for alternate uses of milk fat. *Journal of Dairy Science 80*(10), 2644–2650.

Juliano, P., Temmel, S., Rout, M., Swiergon, P., Mawson, R. & Knoerzer, K. (2013). Creaming enhancement in a liter scale ultrasonic reactor at selected transducer configurations and frequencies. *Ultrasonics Sonochemistry, 20*, 52–62.

Kaufmann, N., Andersen, U. & Wiking, L. (2012). The effect of cooling rate and rapeseed oil addition on the melting behaviour, texture and microstructure of anhydrous milk fat. *International Dairy Journal, 25*, 73–79.

Kaylegian, K. (1999). The production of specialty milk fat ingredients. *Journal of Dairy Science, 82*, 1433–1439.

Kaylegian, K. E. & Lindsay, R. C. (1995). *Handbook of milkfat fractionation technology and application*. Illinois: AOCS Press.

Kaylegian, K. E., Hartel, R. W. & Lindsay, R. C. (1993). Applications of modified milk fat in food products. *Journal of Dairy Science, 76*, 1782–1796.

Kellens, M. (1994). Developments in fat fractionation technology. *Society of Chemical Industry, 42*, 1–11.

Ketenoglu, O. & Tekin, A. (2015). Applications of molecular distillation technique in food products. *Italian Journal of Food Science, 27*, 277–281.

Kim, J. J., Jung, T. H., Ahn, J. & Kwak, H. S. (2006). Properties of cholesterol-reduced butter made with β-cyclodextrin and added evening primrose oil and phytosterols. *Journal of Dairy Science, 89*, 4503–4510.

Kloek, W., van Vliet, T. & Walstra, P. (2005). Mechanical properties of fat dispersions prepared in a mechanical crystallizer. *Journal of Texture Studies, 36*, 544–568.

Kontkanen, H., Rokka, S., Kempinnen, A., Miettinen, H., Hellström, J., Kruus, K., Marnila, P., Alatossava, T. & Korhonen, H. (2011). Enzymatic and physical modification of milk fat: A review. *International Dairy Journal, 21*, 3–13.

Lanzani, A., Bondioli, P., Mariani, C., Folegatti, L., Venturini, S., Fedeli, E. & Barreteau, P. (1994). A new short-path distillation system applied to the reduction of cholesterol in butter and lard. *Journal of the American Oil Chemists' Society, 71*, 609–614.

Lawler, P. J. & Dimick, P. S. (1998). Crystallization and polymorphism of fats. In C. C. Akoh & D. B. Min (Eds.), *Food lipids; chemistry, nutrition and biotechnology* (pp. 229–250). New York: Marcel Dekker Inc.

Li, Y., Mu, H., Andersen, J., Xu, X., Meyer, O. & Ørngreen, A. (2010). New human milk fat substitutes from butterfat to improve fat absorption. *Food Research International, 43*, 739–744.

Lopez, C. & Ollivon, M. (2009). Triglycerides obtained by dry fractionation of milk fat: 2. Thermal properties and polymorphic evolutions on heating. *Chemistry and Physics of Lipids, 159*, 1–12.

Lopez, C., Lavigne, F., Lesieur, P., Bourgaux, C. & Ollivon, M. (2001). Thermal and structural behavior of milk fat. 1. Unstable species of anhydrous milk fat. *Journal of Dairy Science, 84*, 756–766.

Lopez, C., Bourgaux, C., Lesieur, P., Bernadou, S., Keller, G. & Ollivon, M. (2002). Thermal and structural behavior of milk fat. 3. Influence of cooling rate and droplet size on cream crystallization. *Journal of Colloid and Interface Science, 254*, 64–78.

Lopez, C., Lesieur, P., Bourgaux, C. & Ollivon, M. (2005). Thermal and structural behavior of anhydrous milk fat. 3. Influence of cooling rate. *Journal of Dairy Science, 88*, 511–526.

Lopez, C., Bourgaux, C., Lesieur, P., Riaublanc, A. & Ollivon, M. (2006). Milk fat and primary fractions obtained by dry fractionation 1. Chemical composition and crystallisation properties. *Chemistry and Physics of Lipids, 144*, 17–33.

Lopez, C., Bourgaux, C., Lesieur, P. & Ollivon, M. (2007). Coupling of time-resolved synchrotron X-ray diffraction and DSC to elucidate the crystallisation properties and polymorphism of triglycerides in milk fat globules. *Le Lait, 87*, 459–480.

Lordan, R., Tsoupras, A., Mitra, B. & Zabetakis, I. (2018). Dairy fats and cardiovascular disease: Do we really need to be concerned? *Food, 7*, 29–63.

Marangoni, A. G. M. & Lencki, R. W. (1998). Ternary phase behavior of milk fat fractions. *Journal of Agricultural and Food Chemistry, 46*, 3879–3884.

Marangoni, A. G. & Rousseau, D. (1995). Engineering triacylglycerols: The role of interesterification. *Trends in Food Science and Technology, 6*, 329–335.

Marangoni, A. & Rousseau, D. (1998). Chemical and enzymatic modification of butterfat and butterfat-canola oil blends. *Food Research International, 31*, 595–599.

Martini, S., Carelli, A. A. & Lee, J. (2008a). Effect of the addition of waxes on the crystallization behavior of anhydrous milk fat. *Journal of the American Oil Chemists' Society, 85*, 1097–1104.

Martini, S., Suzuki, A. & Hartel, R. (2008b). Effect of high intensity ultrasound on crystallization behavior of anhydrous milk fat. *Journal of the American Oil Chemists' Society, 85*, 621–628.

Martini, M., Salari, F. & Altomonte, I. (2016). The macrostructure of milk lipids: The fat globules. *Critical Reviews in Food Science and Nutrition, 56*, 1209–1221.

Maskooki, A. M., Beheshti, S. H., Valibeigi, S. & Feizi, J. (2013). Effect of cholesterol removal processing using beta -cyclodextrin on main components of milk. *International Journal of Food Science, 215305*, 1–6.

Mazzanti, G., Marangoni, A. G. & Idziak, S. H. J. (2009). Synchrotron study on crystallisation kinetics of milk fat under shear flow. *Food Research International, 42*, 682–694.

Michalski, M. C., Ollivon, M., Briard, V., Leconte, N. & Lopez, C. (2004). Native fat globules of different sizes selected from raw milk: Thermal and structural behavior. *Chemistry and Physics of Lipids, 132*, 247–261.

Michalski, M. C., O'Callaghan, T. F., Kelly, P. & Hogan, S. A. (2020). Milk fat: opportunities, challenges and innovation, *Critical Reviews in Food Science and Nutrition*. https://doi.org/10.1080/10408398.2020.1778631.

Mulder, H. (1953). The consistency of butter. In S. Blair (Ed.), *Foodstuffs: Their plasticity, fluidity and consistency* (pp. 91–123). Amsterdam: North-Holland Publishing Company.

Munro, D. & Illingworth, D. (1986). Milkfat-based food ingredients: Present and potential products. *Food Technology Australia, 38*, 335–337.

Nunes, G. F. M., de Paula, A. V., de Castro, H. F. & dos Santos, J. C. (2011). Compositional and textural properties of milkfat–soybean oil blends following enzymatic interesterification. *Food Chemistry, 125*, 133–138.

Ortiz-Gonzalez, G., Jimenez-Flores, R., Bremmer, D. R., Clark, J. H., DePeters, E. J., Schmidt, S. J. & Drackley, J. K. (2007). Functional properties of butter oil made from bovine milk with experimentally altered fat composition. *Journal of Dairy Science, 90*, 5018–5031.

Pajin, B. & Jovanovic, O. (2005). Influence of high-melting milk fat fraction on quality and fat bloom stability of chocolate. *European Food Research and Technology, 220*, 389–395.

Patel, A. R. & Dewettinck, K. (2015). Current update on the influence of minor lipid components, shear and presence of interfaces on fat crystallization. *Current Opinion in Food Science, 3*, 65–70.

Patrick, M., Blindt, R. & Janssen, J. (2004). The effect of ultrasonic intensity on the crystal structure of palm oil. *Ultrasonics Sonochemistry, 11*, 251–255.

Paula, A. V., Nunes, G. F. M., Freitas, L., de Castro, H. F. & Santos, J. C. (2010). Interesterification of milkfat and soybean oil blends catalyzed by immobilized *Rhizopus oryzae* lipase. *Journal of Molecular Catalysis B: Enzymatic, 65*, 117–121.

Paula, A. V., Nunes, G. F. M., Osório, N. M., Santos, J. C., de Castro, H. F. & Ferreira-Dias, S. (2015). Continuous enzymatic interesterification of milkfat with soybean oil produces a highly spreadable product rich in polyunsaturated fatty acids. *European Journal of Lipid Science and Technology, 117*, 608–619.

Posorske, L. H., LeFebvre, G. K., Miller, C. A., Hansen, T. T. & Glenvig, B. L. (1988). Process considerations of continuous fat modification with an immobilized lipase. *Journal of the American Oil Chemists' Society, 65*, 922–926.

Ransom-Painter, K. L., Williams, S. D. & Hartel, R. W. (1997). Incorporation of milk fat and milk fat fractions into compound coatings made from palm kernel oil. *Journal of Dairy Science, 80*, 2237–2248.

Riener, J., Noci, F., Cronin, D. A., Morgan, D. J. & Lyng, J. G. (2009). Characterisation of volatile compounds generated in milk by high intensity ultrasound. *International Dairy Journal, 19*, 269–272.

Rousseau, D., Hill, A. R. & Marangoni, A. G. (1996a). Restructuring butterfat through blending and chemical interesterification. 3. Rheology. *Journal of the American Oil Chemists' Society, 73*, 983–989.

Rousseau, D., Hill, A. R. & Marangoni, A. G. (1996b). Restructuring butterfat through blending and chemical interesterification. 2. Microstructure and polymorphism. *Journal of the American Oil Chemists' Society, 73*, 973–981.

Rousseau, D., Forestière, K., Hill, A. R. & Marangoni, A. G. (1996c). Restructing butterfat through blending and chemical interesterification. 1. Melting behavior and triacylglycerol modifications. *Journal of the American Oil Chemists' Society, 73*, 963–972.

Rozzi, N. & Singh, R. (2002). Supercritical fluids and the food industry. *Comprehensive Reviews in Food Science and Food Safety, 1*, 33–44.

Rueda, R. (2014). The role of complex lipids in attaining metabolic health. *Current Cardiovascular Risk Reports, 8*, 1–8.

Saitou, K., Homma, R., Katsuragi, Y. & Sato, K. (2014). Retardation of crystallization of diacylglycerol oils using polyglycerol fatty acid esters. *Journal of the American Oil Chemists' Society, 91*, 711–719.

Santos, S., Kelly, P. M. & Hogan, S. A. (2017). *Potential of anhydrous milk fat as human milk fat substitute*. The Society of Dairy Technology conference, University College, Cork, April10–12th.

Sato, K. & Kuroda, T. (1987). Kinetics of melt crystallization and transformation of tripalmitin polymorphs. *Journal of the American Oil Chemists' Society, 64*, 124–127.

Sato, K., Bayés-Garciá, L., Calvet, T., Cuevas-Diarte, M. A. & Ueno, S. (2013). External factors affecting polymorphic crystallization of lipids. *European Journal of Lipid Science and Technology, 115*, 1224–1238.

Schaap, J., Badings, H., Schmidt, D. & Frede, E. (1975). Differences in butterfat crystals, crystallized from acetone and from the melt. *Netherlands Milk and Dairy Journal, 29*, 242–252.

Schäffer, B., Szakály, S., Lõrinczy, D. & Belágyi, J. (2000). Structure of butter – IV: Effect of modification of cream ripening and fatty acid composition on the consistency of butter. *Milchwissenschaft, 55*, 132–135.

Schmelzer, J. M. & Hartel, R. W. (2001). Interactions of milk fat and milk fat fractions with confectionery fats. *Journal of Dairy Science, 84*, 332–344.

Seon, K. H., Ahn, J. & Kwak, H. S. (2009). The accelerated ripening of cholesterol-reduced Cheddar cheese by crosslinked β-cyclodextrin. *Journal of Dairy Science, 92*, 49–57.

Serajzadeh, S. & Alemzadeh, I. (2008). Milk cholesterol reduction using immobilized *Lactobacillus acidophilus* ATCC1643 in sodium-alginate. *International Journal of Food Engineering, 4*(12), 1–8.

Serajzadeh, S. & Alemzadeh, I. (2010). Low-cholesterol milk production using cholesterol oxidase from *Nocardia erythropolis*. *Australian Journal of Dairy Technology, 65*, 10–14.

Shimamura, K., Ueno, S., Miyamoto, Y. & Sato, K. (2013). Effects of polyglycerine fatty acid esters having different fatty acid moieties on crystallization of palm stearin. *Crystal Growth & Design, 13*, 4746–4754.

Shin, J. A., Akoh, C. C. & Lee, K. T. (2010). Physicochemical properties of lipase-catalyzed interesterified fat containing α-linolenic acid. *Journal of the American Oil Chemists' Society, 87*, 647–657.

Sichien, M., Thienpont, N., Fredrick, E., Trung Le, T., Van Camp, J. & Dewettinck, K. (2009). Processing means for milk fat fractionation and production of functional compounds. In M. Corredig (Ed.), *Dairy derived ingredients* (pp. 68–102). Cambridge: Woodhead Publishing.

Smith, K. W., Bhaggan, K., Talbot, G. & van Malssen, K. F. (2011). Crystallization of fats: Influence of minor components and additives. *Journal of the American Oil Chemists' Society, 88*, 1085–1101.

Sørensen, A.-D. M., Xu, X., Zhang, L., Kristensen, J. B. & Jacobsen, C. (2010). Human milk fat substitute from butterfat: Production by enzymatic interesterification and evaluation of oxidative stability. *Journal of the American Oil Chemists' Society, 87*, 185–194.

Soumanou, M. M., Pérignon, M. & Villeneuve, P. (2013). Lipase-catalyzed interesterification reactions for human milk fat substitutes production: A review. *European Journal of Lipid Science and Technology, 115*, 270–285.

Spano, V., Salis, A., Mele, S., Madau, P. & Mosduzzi, M. (2004). Note: Fractionation of sheep milk fat via supercritical carbon dioxide. *Food Science and Technology International, 10*, 421–425.

Sproston, M. J. & Akoh, C. C. (2016). Enzymatic modification of anhydrous milkfat with n-3 and n-6 fatty acids for potential use in infant formula: Comparison of methods. *Journal of the American Oil Chemists' Society, 93*, 251–265.

Suzuki, A., Lee, J., Padilla, S. & Martini, S. (2010). Altering functional properties of fats using power ultrasound. *Journal of Food Science, 75*, E208–E214.

Tahir, M. N., Kwon, C., Jeong, D., Cho, E., Paik, S. R. & Jung, S. (2013). Cholesterol reduction from milk using beta-cyclodextrin immobilized on glass. *Journal of Dairy Science, 96*, 4191–4196.

Ten Grotenhuis, E., van Aken, G. A., van Malssen, K. F. & Schenk, H. (1999). Polymorphism of milk fat studied by differential scanning calorimetry and real-time X-ray powder diffraction. *Journal of the American Oil Chemists' Society, 76*, 1031–1039.

Timby, N., Domellof, M., Lonnerdal, B. & Hernell, O. (2017). Supplementation of infant formula with bovine milk fat globule membranes. *Advances in Nutrition, 8*, 351–355.

Timms, R. E. (1979). The physical properties of milk fat with beef tallow and beef tallow fractions. *Australian Journal of Dairy Technology, 34*, 60–65.

Tirtiaux, F. (1976). Le fractionnement industriel des orps gras par cristallisation dirigée -, procede Tirtiaux. *Oléagineux, 31*, 279–285.

Talbot, G., Smith, K. & Bhaggan, K. (2012). Influence of minor components on fat crystallization. *Lipid Technology, 24*, 83–85.

Truong, T., Bansal, N., Sharma, R., Palmer, M. & Bhandari, B. (2014). Effects of emulsion droplet sizes on the crystallisation of milk fat. *Food Chemistry, 145*, 725–735.

Truong, T., Morgan, G. P., Bansal, N., Palmer, M. & Bhandari, B. (2015). Crystal structures and morphologies of fractionated milk fat in nanoemulsions. *Food Chemistry, 171*, 157–167.

Tsoupras, A., Lordan, R. & Zabetakis, I. (2018). Inflammation, not cholesterol, is a cause of chronic disease. *Nutrition, 10*, 604–642.

Van Aaken, G. A. & Visser, K. A. (2000). Firmness and crystallization of milk fat in relation to processing conditions. *Journal of Dairy Science, 83*, 1919–1932.

Van Beresteyn, E. C. H. (1972). Polymorphism in milkfat in relation to the solid/liquid ratio. *Netherlands Milk and Dairy Journal, 26*, 117–130.

Vanhoutte, B., Foubert, I., Duplacie, F., Huyghebaert, A. & Dewettinck, K. (2002). Effect of phospholipids on isothermal crystallisation and fractionation of milk fat. *European Journal of Lipid Science and Technology, 104*, 738–744.

Walstra, P. (1987). Fat crystallization. In J. M. W. Blanshard & P. Lillford (Eds.), *Food structure and behavior* (Food and technology: A series of monographs) (pp. 67–85). Academic.

Wiking, L., Fredrick, E. & Dewettinck, K. (2009). Phospholipids delay crystallisation of milk fat. *Milchwissenschaft, 4*, 376–379.

Woodrow, I. L. & DeMan, J. M. (1968). Polymorphism in milk fat shown by X-ray diffraction and infrared spectroscopy. *Journal of Dairy Science, 51*, 996–100.

Wright, A. J. & Marangoni, A. G. (2006). Crystallisation and rheological properties of milk fat. In P. F. Fox & P. L. H. McSweeney (Eds.), *Advanced dairy chemistry lipids* (Vol. 2, 3rd ed., pp. 245–292). New York: Springer.

Wright, A., Hartel, R., Narine, S. & Marangoni, A. (2000). The effect of minor components on milk fat crystallization. *Journal of the American Oil Chemists' Society, 77*, 463–475.

Zychowski, L. M., Logan, A., Augustin, M. A., Kelly, A. L., Zabara, A., O'Mahony, J. A., Conn, C. E. & Auty, M. A. (2016). Effect of phytosterols on the crystallization behavior of oil-in-water milk fat emulsions. *Journal of Agricultural and Food Chemistry, 64*, 6546–6554.

Crystallization and Rheological Properties of Milk Fat

8

K. D. Mattice, A. J. Wright, and A. G. Marangoni

8.1 Introduction

Fat is found in milk and cream as an oil-in-water emulsion, contained and stabilized within globules surrounded by a phospholipid and glycolipid membrane. The production of butter results from a phase inversion from an oil-in-water emulsion to a water-in-oil emulsion caused by destabilization and aggregation of the milk fat globules. Isolating milk fat by removal of water, proteins and other minor components from cream or butter forms a solid fat crystal network known as anhydrous milk fat (AMF). The crystallization behaviour and rheological properties of milk fat are extremely important to the processing and texture of many dairy and dairy-based foods. For example, the crystal network structure of butter depends on the composition and crystallization behaviour of the milk fat present. In turn, these properties determine the end use applications, spreadability, mouthfeel, appearance and even taste. Milk fat is also a common ingredient in food systems, contributing to mouthfeel, functionality, lubrication and flavour. This chapter will review our current understanding of milk fat crystallization, structure and mechanical properties, primarily concerning milk fat in the form of AMF or butter. Manipulation strategies for altering the properties of milk fat, including improved butter consistency, will also be discussed. In general, the crystallization of fat crystal networks is quite complex and can be discussed in great detail. While this chapter provides just an overview of crystallization, with details specific to milk fat, the general crystallization process has been reviewed extensively elsewhere (Marangoni and Wesdorp 2013a; O'Sullivan *et al.* 2014; Sangwal and Sato 2018).

K. D. Mattice · A. J. Wright · A. G. Marangoni (✉)
Department of Food Science, University of Guelph, Guelph, ON, Canada
e-mail: amarango@uogeulph.ca

8.2 Crystallization of Milk Fat

Milk fat is semi-solid in nature due to the presence of a large proportion of high melting triacylglycerols (TAGs). These TAGs form crystalline structures at room temperature, resulting in a network that confines the lower melting TAGs (in liquid state) within. The principal determinant of the consistency of semi-solid fats is the ratio of solid to liquid fat at a given temperature. Therefore, the quantity of TAGs capable of crystallization is critical to the texture of butter. Milk fat is composed of hundreds of unique and varied TAG species, resulting in complicated crystallization, melting and rheological behaviours (Jensen *et al.* 1991; Jensen 2002; Lindmark Månsson 2008). Fatty acid composition and the positional distribution are therefore determinant factors in the physical properties present as they dictate the melting temperature.

P. L. H. McSweeney et al. (eds.), *Advanced Dairy Chemistry, Volume 2*,
https://doi.org/10.1007/978-3-030-48686-0_8

Fat crystallization is an exothermic process, and is defined by the change from a liquid to a solid state. This involves nucleation, crystal growth and crystal rearrangements. Crystallization is influenced by both kinetic and thermodynamic factors (Sato 2001), and the kinetics of milk fat crystallization have been reviewed extensively (Grall and Hartel 1992; Walstra *et al.* 1995; Herrera *et al.* 1999; Fredrick *et al.* 2011). Kinetic parameters involved in milk fat crystallization include the initial clustering of liquid molecules into solid nuclei, molecular adsorption, diffusion, solvation/desolvation and conformational rearrangements. The surface-melt interfacial free energy and the chemical potential are the thermodynamic parameters that influence crystallization. The temperature at which a fat is crystallized is a major determinant of the reaction kinetics and resultant structure since it defines the supercooling and supersaturation of the solution, the driving forces for nucleation and crystal growth.

8.2.1 Nucleation of Milk Fat

When a fat is cooled to a temperature below its melting point, the molecules are 'super cooled'. Supercooling is considered equivalent to supersaturation, and is the thermodynamic driving force for crystallization to occur. Supercooling fats makes changing from liquid to solid state energetically favourable (Rousset 2002; Marangoni and Wesdorp 2013b). In this non-equilibrium state, molecules begin to aggregate into tiny clusters (i.e. embryos) which continuously form and dissolve until some critical size is reached. At this point, the cluster is referred to as a nucleus (Garside 1987; Rousset 2002; Himawan *et al.* 2006; Marangoni and Wesdorp 2013b). The critical radius, at which nuclei are stable, depends on temperature; for example, at low temperatures smaller clusters are stable because of decreases in the solubility of TAGs and increases in the free energy change (Marangoni and Wesdorp 2013b).

Three types of nucleation are generally discussed for fats: primary homogeneous, primary heterogeneous and secondary. Homogeneous nucleation occurs in pure solutions in the absence of foreign materials or interfaces. In milk fat, this type of nucleation is very rare, and more often, nucleation is heterogeneous in nature. Heterogeneous nucleation is catalysed by the presence of foreign particles or interfaces. It requires a much lower level of supercooling as foreign matter or impurities in the system can act as a site on which fat can crystallize, making it more energetically favourable to crystallize directly on that surface rather than form nuclei (Garside 1987; Boistelle 1988; Coupland 2002; Rousset 2002) — as such, the rate of crystallization is increased. Secondary nucleation is very important in milk fat crystallization (Walstra 1998); during secondary nucleation, nuclei form upon contact with existing crystals or crystal fragments (Marangoni and Wesdorp 2013b) and this is promoted by agitation, which breaks apart existing crystals, increasing their surface area. Differences observed between crystallization of milk fat in the bulk and emulsified states can be explained by differences in nucleation. In bulk fats, only a small number of nuclei are needed to induce crystallization. However, when the same fat is emulsified as cream, crystallization occurs in small droplets and each fat droplet must contain a nucleus or impurity in order to crystallize, the probability of which is low (Fredrick *et al.* 2011). Consequently, emulsified fat requires a greater supercooling (i.e. a lower temperature) in order to nucleate.

8.2.2 Growth of Milk Fat Crystals

Once nucleation has begun, the crystal size depends on whether the dominating force is crystal growth or further nucleation. Compared to nucleation, the free energy required for crystal growth is relatively low (Sato 2001). Therefore, larger crystals form when TAGs predominantly attach to existing crystals or nuclei, which is favoured when slow cooling rates are used. The rate of crystal growth is determined by the degree of supersaturation, the rate of molecular diffusion to the crystal surface and the time required for TAG molecules to fit into the growing crystal lattice (Marangoni and Wesdorp 2013b). However, in a multicomponent fat, the supersaturation for each TAG is small (Walstra 1998). This, combined with the fact that there is competition between similar molecules for the same sites in a crystal lattice, means that milk fat crystallization is especially slow (Skoda and Van den Tempel 1967; Knoester *et al.* 1968; Grall and Hartel 1992).

The Avrami model is commonly used to fit isothermal fat crystallization data, as the variables in this model are used to identify and quantify the nature of crystal growth in a system (Sharples 1996; Marangoni 1998; Wright *et al.* 2000). The equation is in the following form:

$$\frac{\mathrm{SFC}(t)}{\mathrm{SFC}_{\max}} = 1 - e^{-k(t)^n} \quad (8.1)$$

where k is the Avrami constant that represents the crystallization rate constant, and n is the Avrami exponent, or index of crystallization. SFC(t) corresponds to the solid fat content (%) at a particular time and $\mathrm{SFC}_{\max}$ corresponds to the maximum solid fat content (%) achieved at a specific temperature. The Avrami exponent n pertains to the geometry of crystal growth, or the crystal growth mechanism. Specific values are expected based on the type of nucleation (instantaneous or sporadic) and growth (rods, disks or spheres) that exist in the system (Marangoni and Wesdorp 2013b). The Avrami constant, k, can be considered a crystallization rate constant, and takes into account both nucleation and crystal growth; larger values indicate faster crystallization rates and vice versa. Using this equation, Campos *et al.* (2002) determined that slowly cooled milk fat demonstrates sporadic nucleation and a crystal growth with greater dimensionality (greater n value). In contrast, fast cooled milk fat exhibited more instantaneous nucleation with growth of lower dimensionality (lower n value). A faster crystallization rate was also observed (greater k value) when fast cooled. Though integer values for n were not obtained in this study, it has been suggested that fractional values indicate the simultaneous development of two, or more, types of crystals within the system (Marangoni and Wesdorp 2013b). Mulder proposed the theory of mixed crystal formation to explain the complex crystallization behaviour of milk fat (Mulder 1953). Mixed crystals have a lower density and a lower enthalpy of fusion than pure crystals of the same polymorph. Mixed crystals tend to form in natural fats, including the solid phase of milk fat due to the complex mixtures of TAGs (Marangoni and Lencki 1998; Lopez *et al.* 2002; Truong *et al.* 2015).

Milk fat is composed primarily of TAGs, while minor lipids, including partial acylglycerols, free fatty acids, cholesterol and phospholipids, are also present in small quantities (Wright *et al.* 2000; Wright and Marangoni, 2003). Crystallization proceeds at a more rapid rate when the minor components are removed from milk fat. Specifically, at low degrees of supercooling (20 °C and above) the minor components delay crystallization either at the nucleation stage or during early crystal growth. Although crystallization is delayed, polymorphism is unaffected by the minor components. Additionally, the microstructure was similar between 5 and 25 °C in the absence or presence of the minor components. Therefore, the effect of the minor components is kinetic, rather than thermodynamic. The samples also reached the same SFC values, leading to similar rheological properties between the fats; neither the storage modulus nor the yield force of milk fat was affected by removal of the minor components (Wright and Marangoni 2003).

8.2.3 Crystallization and Melting

Because milk fat contains over 400 different TAG species, each with its own melting temperature, it demonstrates a wide melting range, from −40 to 40 °C (Jensen *et al.* 1991; Lindmark Månsson 2008; Tomaszewska-Gras 2013). This results in a wide range of plasticity where both solid and liquid fats are present. Milk fat composition is often discussed in terms of the three different fractions of TAGs, which are chemically and physically distinct (Marangoni and Lencki 1998; Fredrick *et al.* 2011; Wright *et al.* 2011). These main TAG fractions are distinguished by their melting behaviour: the low-melting fraction (LMF), middle-melting fraction (MMF) and high-melting fraction (HMF), corresponding to the three endothermic peaks observed in milk fat by differential scanning calorimetry (DSC), seen in Figure 8.1. The LMF is liquid at room temperature, due to its substantial content of long-chain unsaturated and short-chain saturated fatty acids. Conversely, HMF melts at temperatures greater than room temperature, resulting from many long-chain saturated fatty acids and a much lower content of long-chain unsaturated and short-chain saturated fatty acids. Accordingly, the MMF is characterized by an intermediate melting temperature.

8.2.4 Polytypism and Polymorphism

When TAGs crystallize, different solid-state structures are possible. These solid-state structures arise due to both different geometric packing arrangements of the fatty acyl chains within the TAG crystal, as well as the longitudinal stacking arrangement of TAG bilayers, referred to as lamellae. Polytypism refers to the different stacking geometries of TAGs within crystalline lamellae (Sato 2018a). The lamellar layer thickness depends on both the length of the TAG molecules and the angle of tilt between the chain axis and the basal lamellar plane. Polytypism is indicated with a 2L or 3L designation for a bilayer or trilayer TAG arrangement, respectively (Lawler and Dimick 2008; Sato 2018a). This designation refers to the fact that the length of the stacked TAGs is either two times or three times the length of the fatty acids present in the TAG molecule (2L or 3L), for bilayers and trilayers, respectively (Figure 8.2a). The bilayer arrangement of fatty acid chains is the most common packing structure for natural fats, including milk fat. However, Lopez *et al.* (2001) found evidence of the coexistence of both the bilayer and trilayer lamellar structures (Lopez *et al.* 2001, 2007). Truong *et al.* (2015) found that emulsified stearin fractions (highly saturated) tended to organize into 2L structures, but emulsified olein fractions (highly unsaturated) formed a mixture of 2L and 3L crystal structures (Truong *et al.* 2015). It has also been described that 3L structures are formed by long-chain monounsaturated and/or mixed TAGs, while 2L structures are formed mainly by long-chain trisaturated TAGs (Lopez *et al.* 2008).

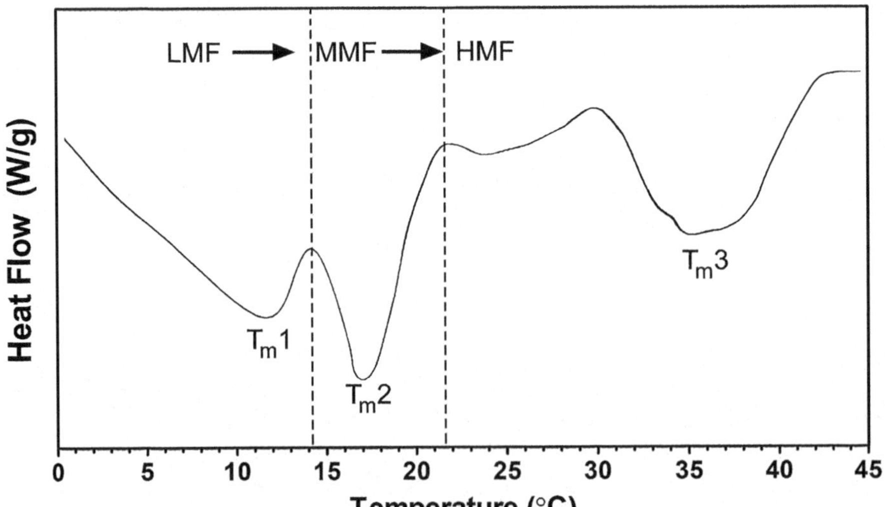

Figure 8.1. Typical differential scanning calorimetry melting endotherm of milk fat with corresponding melting temperatures (T_m) labelled. LMF, low melting fraction; MMF, medium melting fraction; HMF, high melting fraction.

Figure 8.2. Visualization of (**a**) bilayer and trilayer arrangements; (**b**) subcell structures observed in fats, corresponding to the different polymorphic crystals.

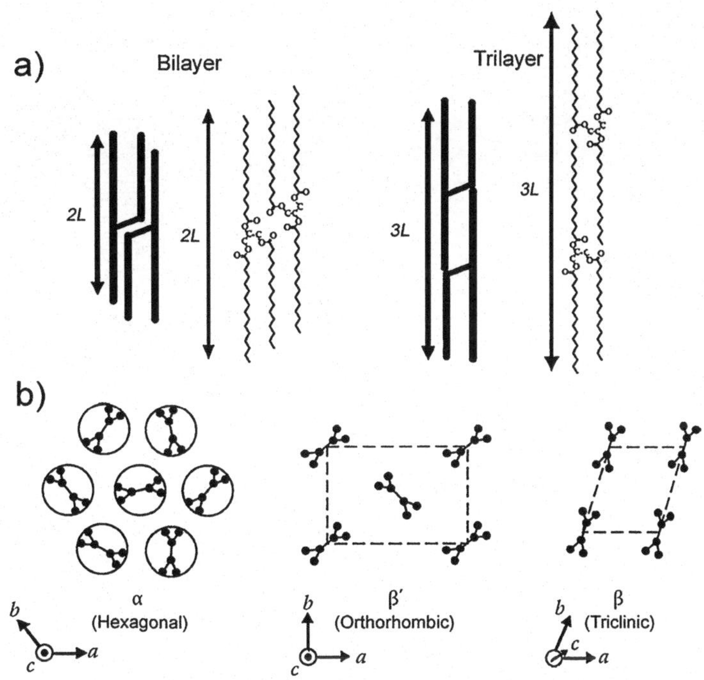

Polymorphism, on the other hand, arises due to geometric packing arrangement of the long hydrocarbon chains within the constituent fatty acids. In this field, the packing arrangement is characterized by the subcell concept, a remnant of early long-chain hydrocarbon crystallography (Chapman 1961; Marangoni and Wesdorp 2013a; Sato 2018b). The three major subcell packing arrangements in fats, which define a polymorphic form are α (hexagonal), β′ (orthorhombic) and β (triclinic), listed in increasing order of melting point, density and stability (Figure 8.2b). The β polymorph is most thermodynamically stable. The polymorphic form(s) present will directly influence the melting point of a fat and have also been correlated to macroscopic rheological properties, therefore making the polymorphism of fat relevant in terms of functionality and potential applications (Marangoni and Wesdorp 2013a). The presence of different polymorphic forms can therefore explain the multiple melting phenomena observed in fats.

The crystallization behaviour of milk fat is anything but simple. Nucleation of milk fat typically occurs in the metastable α-form as these crystals require a lower activation energy for nucleation (Lopez *et al.* 2005; Fredrick *et al.* 2011; Sangwal and Sato 2018). This unstable form rapidly converts to the β′ form (Cisneros *et al.* 2006). Overall, many different compound crystalline subcell types have been identified in milk fat: sub-α, α, β′1, β′2, β1 and β2, with the proportions largely depending on the applied cooling rate (Lopez *et al.* 2002, 2007; Truong *et al.* 2015). Milk fat is still considered to be a β′ tending fat, as crystals have a tendency to transform to the β′ form, and remain in this conformation even after prolonged storage (Timms 1979; ten Grotenhuis *et al.* 1999; Rønholt *et al.* 2014).

Investigation of milk fat crystallization during cooling using synchrotron X-rays has helped define each stage of crystal formation, including the initial formation of α crystals and the subsequent transformation to β′ crystals (Lopez *et al.* 2007; Mazzanti *et al.* 2009). Figure 8.3 illustrates this polymorphic conversion as a function of time. Additionally, the coupling of DSC with synchrotron X-ray analysis has uncovered the relationship between crystal structure and temperature (Ollivon *et al.* 2006). Upon heating, Lopez *et al.* (2007) showed that α polymorphs are the first to melt, followed by β′, reaching the

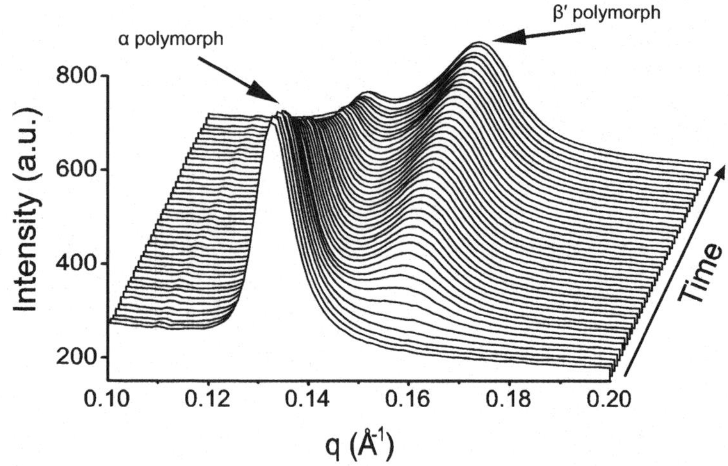

Figure 8.3. The evolution of polymorphism from α to β′ in time as measured using X-ray diffraction, plotted as the diffractive intensity vs the scattering vector length (*q*) over time.

liquid state at ~40 °C. Endothermic peaks recorded by DSC were associated with the successive melting of the different crystalline structures. However, this work demonstrated that structural rearrangement will occur upon heating. Milk fat TAGs experienced polymorphic transformation, beginning at ~12 °C, such that TAGs present in the 3L α and β′ forms were rearranging into 2L β′ structures instead of melting (Lopez *et al.* 2007). This was observed by X-ray diffraction, with the associated peaks increasing in height until melting commenced at ~20 °C. The authors also established that milk fat heated to only intermediate temperatures, and cooled, contained greater proportions of these 2L β′ crystals, demonstrating their stability. Crystals in this 2L β′ form are associated with the greatest melting temperatures; therefore, this behaviour reveals the potential for manipulation of structural and thermal properties via tempering.

8.2.5 Microstructure of Milk Fat Crystal Networks

The microstructure of fat crystal networks is best understood by considering the levels of structure that develop during crystallization. Fat crystal networks develop from initial nucleation sites which grow into crystals (there may be further nucleation during growth). Crystal growth proceeds rapidly, where the conditions present determine the number and size of the crystals formed (Figure 8.4a). Growing crystals become primary particles, or microstructural elements (collection of primary crystallites or single crystals), of generally uniform size (<5 μm). These microstructural elements then aggregate into clusters, or microstructures (>100 μm). A representation of how the microstructures pack in a regular, homogenous, space-filling manner can be seen in Figure 8.4b. Microstructures constitute the largest structural building block of the fat crystal network, and the liquid phase of the network is interspersed between the microstructural elements and microstructures.

The fractal dimension of a network is a measure of the order in the spatial distribution of the solid mass in the network as well as the degree of fill of such space. Determining the fractal dimension is therefore a quantitative method of characterizing the uniformity of the mass distribution in a crystal network. In the method of box-counting fractal dimension, for example, a value that is close to 2 indicates a homogenous mass distribution. Conversely, a value that is less than 2 indicates a network with heterogenous mass distribution. Determining differences in the fractal dimension can be beneficial to understanding the structure of fat crystal networks as this can be used

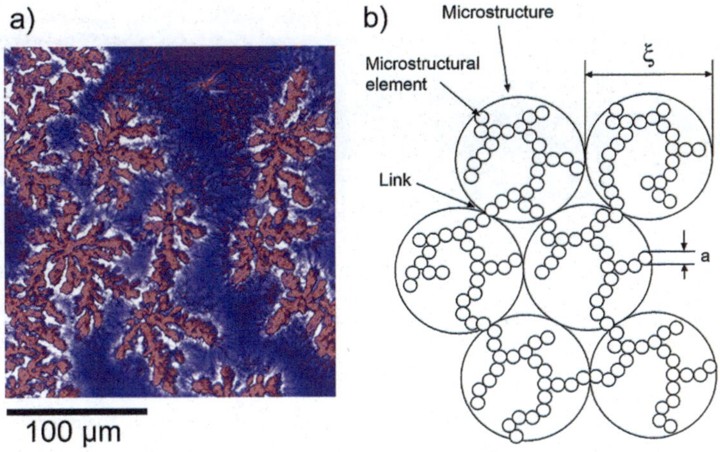

Figure 8.4. (**a**) Microstructure of fat crystal networks: crystals imaged using polarized light microscopy and phase contrast microscopy; (**b**) schematic model of a fat crystal network showing microstructural elements (small circles) arranged into microstructures (large circles) with liquid oil interspersed. In the figure: *a* corresponds to the length of one microstructural element and *ξ* corresponds to the length of one microstructure.

to describe microstructural changes and the rate of nucleation (Marangoni and Mcgauley 2003).

8.2.6 Nanoscale Structure of Milk Fat

The majority of structural characterization in milk fat has been focused on solid-state structure in terms of polymorphism, as well as some microstructural work using microscopy. However, characterization of the crystalline structure at the nanoscale is also important. As mentioned previously (Section 8.2.3), milk fat crystals self-assemble into lamellar structures. Cryogenic transmission electron microscopy (cryo-TEM) has allowed for visualization of how these lamellar structures stack homoepitaxially, forming crystalline nanoplatelets (CNPs), the primary crystal in a fat crystal network (Figure 8.5a) (Acevedo and Marangoni 2010). These CNPs are typically composed of the HMF fraction, but mixed crystals are also possible. Characterization of the morphology and structure of CNPs, as well as the structure of aggregated CNPs, is significant to fully understanding milk fat functionality. An in-depth investigation into the nanoscale structural features of milk fat TAGs by Ramel, Peyronel and Marangoni (2016) was carried out using ultra-small angle X-ray scattering (USAXS) and cryo-TEM. This work revealed smooth CNPs, of approximately 130 nm, that formed 1.14 μm aggregates (Peyronel *et al.* 2014; Ramel *et al.* 2016). The findings also suggested that further aggregation occurs at larger scales, resulting in large aggregates with diffuse surfaces (Pink *et al.* 2015). Cryo-TEM results suggested that nanoplatelet size in statically cooled milk fat averaged 600–900 nm in length and 200–300 nm in width (Figure 8.5b). However, evaluation of CNPs in other TAG systems has demonstrated that crystal size distribution is greatly affected by shear and cooling rate during crystallization (Maleky *et al.* 2011; Acevedo *et al.* 2012). Ramel *et al.* (2016) also established that milk fat CNPs are composed of approximately six to seven stacked lamellae, based on a reported lamellar size of ~4.2 nm for β′ polymorphs and an obtained 27 nm domain value (Woodrow and DeMan 1968; Truong *et al.* 2015). Similarly, a height of ~5 nm was reported in the determination of lamellae thickness for isolated HMF crystals, determined using atomic force microscopy (Sebben *et al.* 2019).

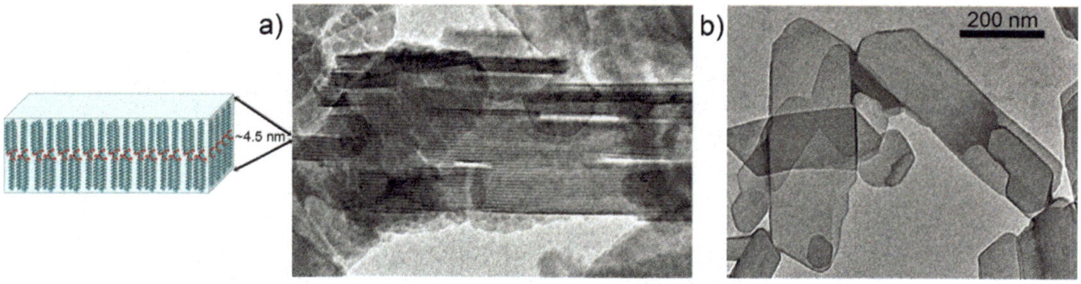

Figure 8.5. Nanoscale structure of milk fat networks: (**a**) epitaxially stacked triacylglycerol lamellae forming crystalline nanoplatelets; (**b**) micrograph of crystalline nanoplatelets obtained by cryo-transmission electron microscopy.

8.3 Rheology of Milk Fat Crystal Networks

8.3.1 Methods to Determine the Rheological Properties of Milk Fat

Methods that characterize the rheological properties of milk fat can probe for many different physical properties. These methods can also be used to correlate qualitative observations with quantitative values for comprehensive comparison. The most common properties of interest are hardness and spreadability, where values obtained are used as indicators of sensory attributes and for quality control purposes. Rheological measurements are also performed to characterize viscoelasticity by comparing the values of the storage modulus (G'), which describes the elastic or solid properties, to the loss modulus (G''), which describes the viscous or liquid properties. These values are obtained over a predetermined range of increasing shear strain, frequency, temperature or time.

8.3.1.1 Small Deformation Rheological Testing of Milk Fat

Butter and milk fat exhibit viscoelastic behaviour at small stresses. To probe this behaviour, a very small deformation force is applied to a sample and the relationships between stress, strain and time are monitored (Shama and Sherman 1970; Shukla and Rizvi 1995). Here, the strain applied is low enough that the sample remains intact (i.e. non-destructive). Small deformation testing is performed in the linear viscoelastic region (LVR) where there exists a linear relationship between stress (τ) and shear strain (γ). Small deformation tests are also very sensitive and provide valuable information on the structure of a sample. Static compression tests can be used to determine Young's modulus (E), which describes the relationship between stress and strain upon uniaxial deformation for a sample. Oscillatory rheometry involves positioning samples between two plates, where one plate is stationary and the other oscillates at specified frequency and amplitude, determining G' and G'' values simultaneously. This method offers the benefits of significant control and precision at low strain values, and measures the dependence of rheological properties on different variables, such as temperature, time, amplitude and frequency. Time sweeps are used to monitor structure formation or deformation as a function of time. Temperature sweeps are commonly used to examine the rheological events associated with melting or crystallization. Amplitude sweeps elucidate differences in material functionality, where the responses of G' and G'' are observed with the application of increasing shear strain (γ). Similarly, frequency sweeps measure the response with increasing angular frequency (ω), providing information about the time dependence of the changes in the structure of the material.

8.3.1.2 Large Deformation Rheological Testing

Large deformation rheological testing is used to analyse the rheological behaviour of materials outside their linear viscoelastic region. With greater amounts of strain applied, this type of analysis provides information about how materials break down; for example, applying large deformations to butter samples is particularly useful as butter is subjected to non-linear shear rates during processing and spreading by consumers. At stresses/strains outside of the LVR (the onset of non-linear rheological behaviour), it is possible to record the critical strain level, or yield stress. This point corresponds to a yield strain (γ_y) and yield stress (σ_y) value and is the point at which the material begins to break down. This point can be defined in a number of ways, though common definitions include the point where G' and G'' crossover ($\tan \delta = 1$), or as the point where a 5% decrease in G' (or $G'/G'' = 0.95$) occurs (Macias-Rodriguez 2018). From a structural perspective, the yield stress has been described as the point at which the shear force applied moves interacting microstructural elements apart to the extent that interactions no longer exist.

Different types of rheometric tests, including sectility, extrusion, compression and oscillatory shear, can be used to apply large deformations. Sectility, or the ability to cut the material, is determined by forcing a stretched steel wire through a sample, with the load force required to cut the sample recorded (Dixon and Williams 1977; Rohm 1992). Extrusion is said to mimic the action of spreading by forcing the fat through a small hole. The force required to sustain the motion is measured, which comprises both the force necessary to push the sample through and the force necessary to overcome the friction along the walls of the extruder (Prentice 1972). Compression testing involves a uniform stress being applied to the top and bottom of a sample placed between two flat plates. The analysis of fats using steady shear compression is prone to shortcomings due to wall slip and edge fracture (Macias-Rodriguez 2018). It has also been noted that compression tests may not provide an accurate determination of the elastic properties at low yield strains due to limitations in method precision at these levels (Kloek *et al.* 2005). Considering the fact that the LVR of milk fat or butter exists entirely in a range of low shear strains, this method is not recommended. Non-linear oscillatory shear tests are now considered superior due to the great ease and control inherent in obtaining data (Hyun *et al.* 2011). Results from large amplitude oscillatory shear tests provide useful information on how the material dissipates energy, which can be used to describe fat functionality (Macias-Rodriguez, 2018).

8.3.1.3 Penetrometry-Based Testing

Penetrometry has been the most common method used to evaluate butter texture and consistency, as it is simple and economical to use. Performed using a penetrometer or texture profile analyser, either the depth to which a penetrating body (e.g. a cone, needle or sphere) falls during a specified length of time, or the rate at which the body falls is measured (Kamel and DeMan 1975; DeMan and Beers 1987; Lee and Martini 2018). A standard method for determining firmness of plastic fats by penetrometry is AOCS Official Method Cc16-60 (2017). Penetrometry results can be translated into 'spreadability' or 'hardness' values. Hardness values are often taken as the force required to penetrate samples with a cone-shaped probe to a specified depth. There are also specialized probes to specifically measure spreadability, where perfectly fitting male and female fixtures are pressed together and the force to displace the sample, or to sustain the motion, is measured (Texture Technologies Corp. 2019). Efforts have also been made to correlate instrumental results with testing by sensory panels, including by Rousseau and Marangoni (1998a), who found good agreement between sensory spreadability and the depth of cone penetration, using AOCS Method Cc16-60 (Rohm and Ulberth 1989; Rousseau and Marangoni 1998a).

Many studies have worked to relate obtained hardness values to the yield stress of fats (Haighton 1959; Hayakawa and DeMan 1982; Lefebvre 1983; Narine and Marangoni 1999b). The work is commonly summarized as determining a 'yield value' (C) according to the following equation:

$$C = \frac{K w}{d^{1.6}} \tag{8.2}$$

where w is the weight of the cone, d is the penetration depth and K is a factor depending on the cone angle. The constant n varies with sample structure, and a value of 1.6 is often used for plastic fats (Haighton 1959; DeMan and Beers 1987; Wright *et al.* 2001).

8.3.2 Rheology of Milk Fat

The rheology of creams and milks has been found to correspond with that of emulsions, where Newtonian or non-Newtonian behaviour is observed, depending on composition, temperature and processing conditions (McCarthy 2011). Here, milk fat's contribution to rheology is in terms of concentration, fat globule size and temperature, or whether it exists in the molten or crystallized state. In both AMF and butter, the underlying fat crystal assembly of the different melting fractions gives these materials their structural integrity, and determines their rheological and textural properties. The structure of butter is complicated by the presence of water droplets as they tend to weaken the structure. Therefore, much of the rheological characterization of milk fat is performed using AMF.

Several rheological characteristics of milk fat and butter have practical significance. Spreadability and hardness are two very important characteristics of plastic fats that have a large impact on consumer acceptability and perceived sensory properties. To be spreadable, butter should possess an SFC between 20% and 40%, typically corresponding to temperatures between 11 °C and 20 °C, corresponding to an apparent yield value of approximately 30–60 kPa (Rohm and Raaber 1991). Outside this temperature range, milk fat becomes too soft or too hard for many applications. For milk fat, the ratio of high melting TAGs to medium and lower melting TAGs is a significant determining factor for firmness. Therefore, differences in TAG composition and fatty acid distribution can be critical to such understanding. Consequently, the SFC and microstructure are affected, resulting in a direct relationship between composition, crystallization and rheological properties (Rousseau and Marangoni 1998a; Singh *et al.* 2004; Hurtaud and Peyraud 2007).

The importance of temperature on milk fat rheology can be easily understood when thinking of the difference in properties at room temperature (greater viscous component, spreadable) and at refrigeration temperature (greater elastic component, easily fractured). This is caused by the different ratios of solid to liquid TAGs at each of these temperatures. Changes in the ratio of solid to liquid TAGs, and therefore the SFC, are often referred to for the explanation of physical properties. However, SFC is not the sole factor controlling the functionality of a fat (Marangoni and Rousseau 1998). Fat polymorphism can also be used as an indicator for fat functionality, and optimum plasticity has been correlated to the presence of β′ crystals (Macias-Rodriguez and Marangoni 2016a). Conversion to β polymorphs can occur but these crystals are often larger and associated with hard and brittle textures (Baldwin *et al.* 1972; DeMan *et al.* 1991). The desire for β′ crystals has become standard for the optimum functionality of plastic fats.

Analysis of rheology of milk fat must consider also time as a factor due to setting. Setting refers to the continued increase in the firmness over time for recently manufactured butter. Increases in firmness and SFC sometimes occur for months because of continued crystallization and crystal aggregation (de Man and Wood 1959; van Aken and Visser 2000). The extent of setting depends on several variables, including composition, storage temperature, storage time, initial hardness and manufacturing conditions (Prentice 1972; Shukla and Rizvi 1995).

The elastic modulus of a fat crystal network has been shown to directly relate to hardness (Narine and Marangoni 1999c). It is expected that $G' > G''$ for butter and milk fat samples at room temperature and lower, given their solid state, with each moduli typically in the order of G', $G'' \approx$ 104–106 Pa for solid fats (Macias-Rodriguez 2018). However, greater temperatures cause the partial or complete melting of TAGs, which results in a shift to $G'' \approx G'$ or $G'' > G'$. Frequency sweeps have shown that solid fat networks, including butter, have little to no frequency dependence, characteristic of viscoelastic solids of this nature (Wright *et al.* 2001; Buldo and Wiking 2012; Macias-Rodriguez and Marangoni 2016b).

Milk fat is viscoelastic in nature. With small stresses, potential energy is stored because the network of the sample is distorted. This type of deformation is reversible and elastic. Greater stresses cause crystals to rearrange or fracture, leading to irreversible changes in structure. These changes may not be permanent because recrystallization does occur gradually. However, the sample will likely never return to its original configuration. The use of oscillatory rheometry easily determines the onset of sample deformation through tests at increasing shear strains. The critical strain for butter occurs at low strain levels of 0.001% to 0.1% (Rohm and Weidinger 1993; Shukla and Rizvi 1995; Wright *et al.* 2001; Macias-Rodriguez 2018). An extensive review of the literature did not find results from any non-linear oscillatory shear experiments (above these critical strain levels) on milk fat to date, although TAG shortenings of different compositions have been analysed. Macias-Rodriguez, Peyronel and Marangoni (2017) established the importance of crystal aggregate structure for fat functionality by relating findings to non-linear viscoelastic properties (Macias-Rodriguez *et al.* 2017). The authors found notable differences between the non-linear viscoelasticity of laminating and cake shortenings, primarily noting enhanced non-linear viscous dissipation in laminating shortenings compared to non-linear elasticity observed for cake shortenings. This was correlated to the three-hierarchy homogenous network structure made up of layered crystal aggregates in laminating shortenings. Only a two-hierarchy heterogeneous network comprised of distorted spherical-like crystal aggregates was observed in cake shortening. Noting the similarities in nanoscale structure between characterizations of laminating shortening by Macias-Rodriguez *et al.* (2017) and of milk fat by Ramel *et al.* (2016), it is reasonable to predict that the non-linear viscoelastic properties of milk fat could be similar to the results reported for laminating shortenings. This fact is supported by butter's known functionality and success as a laminating fat in many baked goods.

Work softening occurs in plastic fats (e.g. butter) and is dependent on both the amount of force or deformation applied and also on the testing time (DeMan and Beers 1987). When a force is applied (i.e. when the sample is compressed), there is an initial elastic response which can be represented by a Hookean spring (DeMan and Beers 1987). Forces that exceed the yield stress cause structural breakdown, and the material becomes much softer. When strain continues to increase, in a time-dependent fashion, following the initial deformation, bonds between the crystals in the fat network break and reform, though the original structure and consistency is never regained (Shama and Sherman 1970). Thixotropic hardening then occurs with network restructuring during rest (Macias-Rodriguez and Marangoni 2016b).

8.3.3 Modelling Milk Fat Rheology

Solidified milk fat and butter display non-Newtonian behaviour, acting as a plastic material with a yield value (DeMan and Beers 1987). As previously described, milk fat is a viscoelastic material through its wide melting range and possesses both solid and liquid-like characteristics. Several models to describe the complex rheological behaviour of milk fat have been proposed and Figure 8.6 shows the corresponding stress–strain curves for the models discussed.

The simplest model assumes ideal elastic behaviour (Figure 8.6a). At a stress below the yield stress, the sample behaves perfectly elastically. In this region, a modulus of elasticity can be determined. For milk fat, the continuous network of fat crystals contributes solid or elastic properties to the material in this region. At the yield stress, the sample flows and it continues to flow until the stress is lowered again to below the yield stress value. Therefore, both the elastic modulus and yield stress describe the behaviour of a plastic material and can be determined easily by compression testing (Narine and Marangoni 1999b).

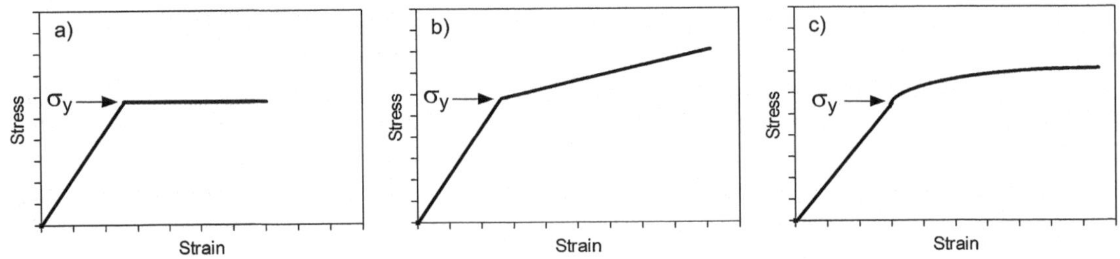

Figure 8.6. Stress–strain curves showing yield stress (σ_y) for (**a**) elastic–plastic behaviour, (**b**) the Elliot and Ganz (1971) model and (**c**) the viscous Maxwell–Bingham model of Diener and Heldman (1968).

Given the complexity of milk fat and the wide range of melting temperatures of the TAGs present, we would expect that the liquid oil portion of milk fat plays a significant role in the rheology of milk fats. Consideration of the viscous elements, in addition to the elastic elements, should lead to a more accurate description of milk fat rheology. Therefore, butter, and other unctuous materials, may be qualitatively described by a modified Bingham body (Elliott and Ganz 1971; Elliott and Green 1972) which consists of viscous, plastic and elastic elements in series. The stress–strain behaviour for the model proposed by Elliot and Ganz (1971) is presented in Figure 8.6b.

Diener and Heldman (1968) proposed a more complex model to describe how butter behaves when a low level of strain is applied. The model consists of plastic and viscous elements in parallel, coupled in series with a viscous element in parallel with a combination of a viscous and an elastic element (Figure 8.6c) (Diener and Heldman 1968). Diener and Heldman also attributed the elasticity in butter to the fat globule membrane and the viscosity to the flow of the surrounding liquid fat.

8.3.4 Relating Structure to Rheology

To model the mechanical properties of a network in terms of its structure, a good definition of the levels of structure that exist within that network is essential and a logical starting point. For fat crystal networks, the hierarchical organization of structure (detailed in Section 8.2.5) is often taken. Based on this description of a fat crystal network, it makes sense that its macroscopic properties should depend significantly on the nature of the microstructures since this level of structure is closest to the macroscopic world (Narine and Marangoni 1999a, c).

A scaling theory was proposed to relate the Young's modulus (E) of a fat to the spatial distribution of mass within a fat crystal network, and the volume fraction of solid fat (Φ = SFC/100) present (Marangoni and Rousseau 1996; Marangoni 2000; Narine and Marangoni 2001). This is represented as

$$E = \lambda \Phi^{\mu} \tag{8.3}$$

where λ is a pre-exponential parameter and, in three-dimensional space:

$$\mu = \frac{1}{3 - D} \tag{8.4}$$

where D corresponds to the mass fractal dimension for the spatial distribution of mass within the network and is not related to the roughness of crystallite interfaces or the spatial distribution of mass within crystallites. The Young's modulus can be related to the shear elastic modulus (G') by assuming a Poisson's ratio of 0.5, and thus, $E = 3$ G. This theory was based on the assumption that when the network is stressed, the links between the microstructures are more likely to be stressed than the microstructures themselves or the structures within them. This is reminiscent of the old adage 'the strength of a chain lies in its weakest link', as the weakest link here is the links between the microstructures. Thus, this theory is called the weak-link theory. A schematic of a fat network under extension when the weak-link theory is applicable is provided in Figure 8.7.

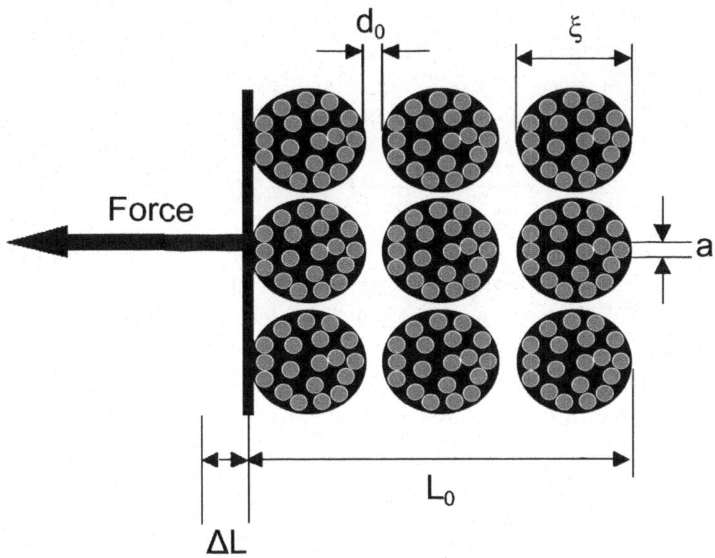

Figure 8.7. Schematic of fat crystal network under extension when the weak-link theory is applicable, where a is the length of one microstructural element, ξ is the length of one microstructure, d_0 is the initial distance between microstructures, L_0 is the length of the material and ΔL is the change in length after force is applied.

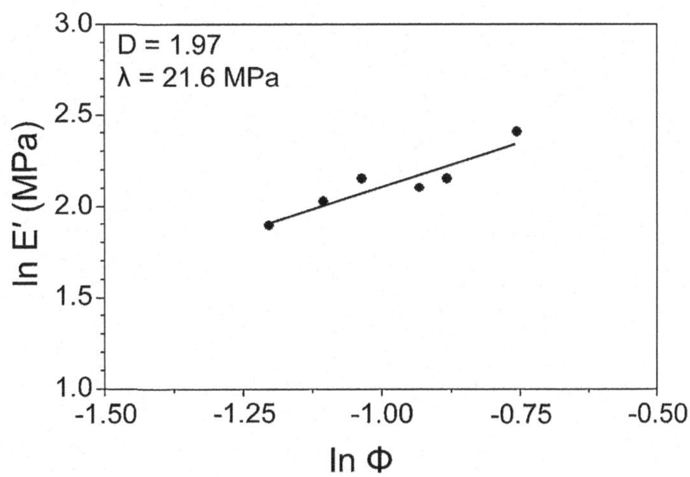

Figure 8.8. Plot of the relationship between natural logarithm of the dynamic compressional modulus (E') as a function of the natural logarithm of the volume fraction (Φ) for milk fat at 5 °C.

The fractal dimension of a microstructural network can be determined rheologically by diluting a fat with an oil that does not appreciably dissolve the fat under the test conditions (preferably at a low temperature and crystallized rapidly to prevent fractionation). The exact range of dilutions required will depend on the SFC range of interest, since different SFC ranges will yield different structures with different fractal dimensions (Awad *et al.* 2004). Typically, dilutions in the range 70–100% (w/w) milk fat are used. By measuring the storage modulus by small-deformation dynamic rheological techniques under shear (G') or compression (E'), and the SFC by pulsed NMR, it is possible to plot:

$$\log E' = \log \lambda + \mu \log \tag{8.5}$$

From the slope and y-intercept of such a plot, it is possible to determine D and λ, as shown in Figure 8.8.

High fractal dimensions are associated with more ordered distributions and higher degrees of fill. Therefore, the fractal dimension of a crystal network is an important parameter in terms of its relation to mechanical strength. However, the values of the pre-exponential term, λ (and the solid fat content), are equally important (Narine and Marangoni 1999a; Marangoni 2000; Marangoni and Rogers 2003). For spherical microstructures:

$$\lambda \sim \frac{6\delta}{a\varepsilon^*} = \frac{A}{2\pi a\varepsilon^* d_0^2} \quad (8.6)$$

where δ is the crystal-melt interfacial tension, A is Hamacker's constant, a is the diameter of a microstructural element, ε^* is the macroscopic strain at the limit of linearity and d_0 is the average equilibrium distance between microstructures. This model identifies key network parameters that are important in determination of the value of λ. The final equation provides impetus for the development of investigations of relationships between TAG composition and polymorphism and the various parameters of the model in Equation (8.6).

In the work of Marangoni and Rogers (2003), an expression for the yield stress (σ^*) of a fat was also derived, assuming that $\sigma^* = E{\cdot}\varepsilon^*$, namely,

$$\sigma^* = \frac{6\delta}{a}^{\frac{1}{3-D}} \quad (8.7)$$

For a plastic fat, the yield stress was defined as the stress at the limit of linearity in a small deformation rheological test. Therefore, this expression shows that the force required to break down a fat crystal network depends on many structural parameters. In this work, agreement between theory and experiment was found to be good (Marangoni and Rogers, 2003).

From the discussion above, values of δ, a and D can be manipulated by changing the processing conditions or chemical composition. By defining the network characteristics responsible for mechanical strength of a fat product, the model provides an array of indicators, which can be monitored during the development of tailored fat crystal networks or as indicators for quality control purposes. Moreover, the compression storage modulus (E') of milk fat was found to be directly proportional to the yield force (Figure 8.9), in agreement with the model above. It is therefore possible to map the effects of structural changes on small deformation rheological behaviour, to large deformation rheological behaviour, and therefore, possibly, texture (Rousseau and Marangoni 1998a).

8.4 Modifying the Crystallization and Rheological Properties of Milk Fat

Milk fat and butter can be tailored to have desired properties and functionalities. The narrow temperature range for optimal functionality often drives efforts to modify milk fat rheology. Treatments are frequently aimed at improving cold spreadability without compromising room temperature stability. There is also some desire for milk fat that remains firm at higher temperatures, particularly for the application of laminating in bakery products. To modify the textural and rheological properties of butter, both composition and processing conditions can be manipulated.

8.4.1 Manipulating Composition

Several factors influence the chemical composition of milk fat, including TAG and fatty acid composition. The crystallization properties, firmness and consistency of butter, therefore, depend on these variables. Milk fat composition can depend on the breed of cow (Beaulieu and Palmquist 1995; Maurice-van Eijndhoven *et al.* 2011; Soyeurt *et al.* 2011), stage of lactation (Kelsey *et al.* 2003; Stoop *et al.* 2009), season (Lock and Garnsworthy 2003; Heck *et al.* 2009; Larsen *et al.* 2014) and climate (Mcdowell *et al.* 1975; Collomb *et al.* 2002). Efforts to manipulate milk fat composition are also commonly made by modifications to diet (Ashes *et al.* 1997; Lock and Bauman 2004; Kalač and Samková 2010; Liu *et al.* 2016). The enrichment of, and incorporation of, healthful lipids into milk fat is also attractive from a nutritional standpoint, leading to attempts at improving the fatty acid profile of cow's milk via diet (Mansbridge and Blake 1997; Singh *et al.* 2004; Caroprese *et al.* 2010; Zachut *et al.* 2010; Pappritz *et al.* 2011). A recent study

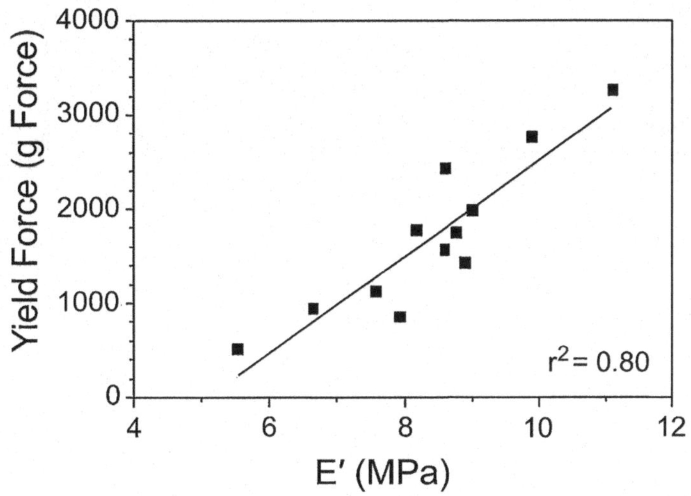

Figure 8.9. Relationship between yield stress value (in g force) and compressive storage modulus (E', in MPa) for milk fat.

looked at the incorporation of omega-3 fatty acid rich diets on the properties of butter produced from the resultant milk (Vanbergue *et al.* 2018). This work found that butter produced from this modified milk was less firm than controls, as determined by both instrumental and sensory testing.

Modifying the level of unsaturation in milk fat can be achieved through the basic method of blending (Rousseau *et al.* 1996b, c; De *et al.* 2007; Ramel and Marangoni 2016); for example, when milk fat was blended with canola oil, changes in melting point and hardness index were observed (Rousseau *et al.* 1996b, c). The combination of AMF and lower melting TAGs from high oleic sunflower oil also resulted in dairy spreads with decreased solid fat content and, subsequently, hardness (Lázaro *et al.* 2019).

Milk fat fractionation has been studied by several groups (Arul *et al.* 1988; Marangoni and Lencki 1998; van Aken *et al.* 1999; Illingworth 2002; Campos *et al.* 2003; Vanhoutte *et al.* 2003). The most common and simplest method of fractionation is dry fractionation. Here, partial crystallization of milk fat is achieved by slowly cooling to specific temperatures. This leads to the separation into two distinct fractions: the high melting stearin fraction and the lower melting, liquid olein fraction. Increasing the proportions of the stearin fraction provides structural integrity, while the olein fraction can serve to reduce hardness (Figure 8.10) (Kaylegian and Lindsay 1992; Queirós *et al.* 2016). Distinct properties can be achieved by fractionating milk fat and then recombining the fractions in various proportions (Kaylegian and Lindsay 1992; Illingworth 2002; De *et al.* 2007). This has been shown to modify the crystallization and microstructural properties of milk fat, such that target microstructures can be achieved through the blending of specific fractions (Ramel and Marangoni 2016). From a functional standpoint, fractionation is an excellent method for improving the physical properties of butter or for tailoring milk fat to specific applications (Shukla *et al.* 1994; Kaylegian 1999; Pal *et al.* 2001; Ramel and Marangoni 2016; Berti *et al.* 2018).

Interesterification involves the exchange of fatty acids within and between TAGs, which can be achieved using either chemical or enzymatic (i.e. with lipases) means. Interesterification of milk fat and butter can be used to produce products with desired functionalities and health benefits (Kalo *et al.* 1986; Balcao and Malcata 1998; Marangoni and Rousseau 1998; Rousseau and Marangoni 1998b; Rodrigues and Gioielli 2003; De *et al.* 2007; Shin *et al.* 2009). Interesterification of butter or milk fat with liquid oils has two main benefits: the increase of unsaturated fatty acids in the composition and the resultant softening of the material. For example, milk fat enzymatically interesterified with either canola oil and soybean oil resulted in products with enhanced spreadability and consistency (Rousseau *et al.*

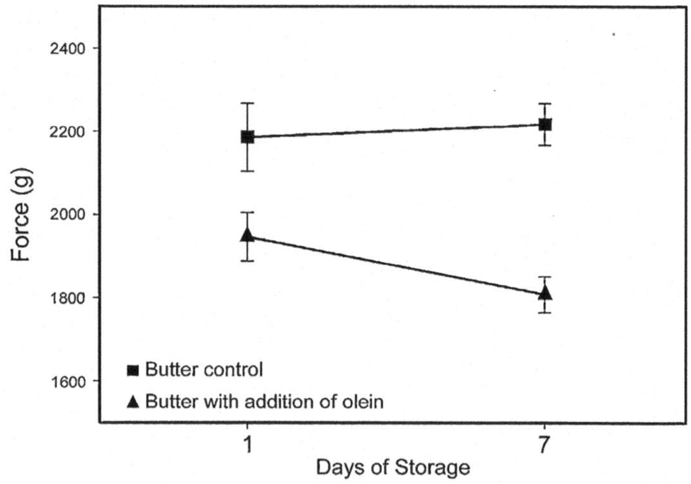

Figure 8.10. Firmness of control butter and butter with addition of 50% olein fraction, after 1 and 7 days of refrigerated storage. (Adapted from Queirós *et al.* 2016).

1996; Nunes *et al.* 2011). However, one disadvantage to interesterification is the reported reduction in butter flavour (Rousseau and Marangoni 1998a).

Recent work has described how dissolving CO_2 into milk fat can modify crystallization and texture (Truong *et al.* 2017a). CO_2 is soluble in milk fat, particularly in the LMF, as compared to the HMF (Truong *et al.* 2017b). It was determined that the presence of solubilized CO_2 in AMF causes rapid nucleation and crystallization upon cooling, evidenced by the reduction of crystal size and the existence and stability of α polymorphs. As a result, milk fat at room temperature increased in hardness with increasing CO_2 concentration.

8.4.2 Manipulations During Processing

Altering processing conditions can be an easy means of manipulating crystallization and, ultimately, the rheology of milk fat. The external factors that influence the crystallization of milk fat include temperature, cooling rate, batch size, churning parameters and storage conditions (Grall and Hartel 1992; Herrera and Hartel 2000a, b, c; van Aken and Visser 2000). Overall, parameters that are able to alter the kinetics of fat crystallization possess the greatest ability to influence physical properties.

8.4.2.1 Cooling Rate

Cooling rate influences milk fat in many ways, making it difficult to isolate its effect on any one parameter. Rapid cooling influences the number of nuclei formed during crystallization, the size of fat crystals and, in turn, the texture of butter. During slow cooling, extensive crystal growth occurs. During rapid cooling, crystallization proceeds more quickly and nucleation events predominate over crystal growth processes. Cooling rate also affects the ratio of solid to liquid fat present because of mixed crystal formation (Mulder 1953). High cooling rates force TAGs to rapidly arrange, often forming crystals that tend to be less stable (α polymorph). The formation of many small crystals also results in a high crystal surface area. The formation of many small crystals increases milk fat hardness, as the greater surface area allows for a greater proportion of liquid fat to be absorbed and immobilized. In turn, the amount of liquid fat that is available to form the continuous phase in butter is reduced, resulting in a firmer consistency. In contrast, slower cooling rates allow TAGs plenty of time to arrange and form larger, stable crystals (β′ or β polymorphs). A spherulitic crystal microstructure is observed in milk fat cooled from the melt to 5 °C at a rate of 0.1 °C/min (Figure 8.11a). In contrast, the same milk fat cooled quickly from the melt to 5 °C at a rate of 1 °C/min, will exhibit a granular microstructure (Figure 8.11b) (Campos *et al.* 2002; Lopez *et al.* 2005). Additionally, fast

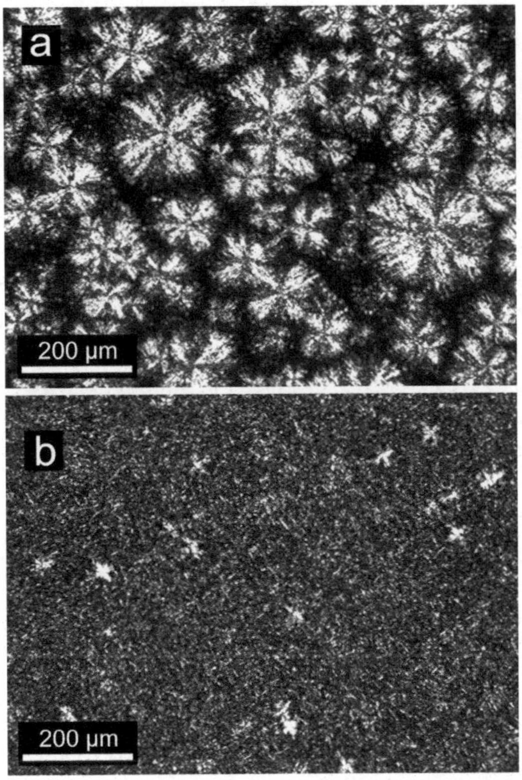

Figure 8.11. Effect of cooling rate on the microstructure of milk fat cooled to 5 °C (**a**) slowly (0.1 °C/min) and (**b**) quickly (1.0 °C/min).

cooling causes these microstructural features to be distributed in a less orderly fashion.

The fractal dimension of milk fat crystal networks decreased from 2.5 to 2.0 when the cooling rate was increased (Wright *et al.* 2001). Concurrently, the particle-related constant, λ, increases. These results demonstrate how a faster cooling rate leads to a decrease in the order in the spatial distribution of mass within the microstructural network. This would result in a lower value of D, and a decrease in the average particle diameter, which would result in a higher value of λ, as predicted by our model (Section 8.3.3) (Marangoni 2000). These microstructural changes were correlated with a much higher yield force value, with the rapidly cooled milk fat yielding a firmer fat (Campos *et al.* 2002).

8.4.2.2 Churning and Shear

The physical properties are also influenced by the application of shear. The production process of butter from cream inherently involves shear, as churning is responsible for the destabilizing and agglomeration of milk fat globules, causing a phase inversion from an oil-in-water (cream) to a water-in-oil emulsion. The thermal treatment, known as ripening or aging, is also critical to butter production as this governs the extent and rate of milk fat globule partial coalescence (Buldo *et al.* 2013; Lee and Martini 2018). The kinetics of crystallization are influenced by the shear involved in churning as, fundamentally, shear causes an increase in heat and mass transfer. As a result, the frequency of molecular collisions increases, causing crystals to break or to form many smaller crystal entities. This typically leads to crystal networks with increased strength.

At the nanoscale, high shear rates can decrease crystal nanoplatelet size (Tran and Rousseau 2016). It has also been suggested that shear can change the rate of nucleation and the alignment of nuclei formed, inducing the formation of specific crystal orientations in the direction of the shear field (Mazzanti *et al.* 2003; Maleky and Marangoni 2008; Acevedo *et al.* 2012). While the application of shear provides some advantages, there is an upper limit as extensive shear forces can compromise the structure, inhibiting growth of larger crystals and aggregation. Below this limit, crystal aggregation is promoted, making optimization of the shear process necessary based on the desired properties of the system. Shear was observed to cause a faster conversion between α and β' polymorphs in milk fat crystals (Mazzanti *et al.* 2009). During milk fat fractionation, shear was found to influence both the composition and the structure of the filtered crystals (Breitschuh and Windhab 1998).

8.4.2.3 High-Intensity Ultrasound

High-intensity ultrasound, which uses sound waves at high frequencies (20–100 kHz) to transmit stress waves from one molecule to another, has also been used to modify the functionality of solid fats. This process can be used to induce crystallization, producing many smaller crystals (Martini *et al.* 2008). Resulting from the alteration in crystal size, milk fat treated by high intensity ultrasound can demonstrate greater hardness at higher temperatures (24–28 °C) (Suzuki *et al.*

2010). Though a study by Frydenberg *et al.* (2013) found that while many smaller crystals were formed, high-intensity ultrasound resulted in AMF with decreased hardness. The authors attributed this result to a greater degree of supercooling during crystallization compared to previous studies. Therefore, the use of high-intensity ultrasound could provide a method for tailoring milk fat functionality to specific applications.

8.5 Crystallization and Rheological Properties of Milk Fat Within Food Systems

Milk fat is an important ingredient in many dairy products, including whipped cream, ice cream, butter, baked goods and cheese. Each of these foods has specific requirements of milk fat, and, in many cases, the properties that achieve these requirements have been studied extensively. The presence of non-fat ingredients and severe processing conditions (e.g. high temperatures and shear) make it reasonable to question whether crystallization and rheological properties of milk fat change when milk fat is crystallized within food systems. Determination of milk fat behaviour within a food matrix is therefore significant, as this better reflects the properties existing at the time of consumption. It also leads to a better understanding of the interactive role of fats within food systems, establishing relationships between milk fat crystallization, microstructure and mechanical properties with the properties of finished food products.

8.5.1 Thermal Behaviour

The characteristic melting of milk fat is attributed to many desired sensory qualities of butter and other dairy products. Therefore, it can be beneficial to confirm this behaviour remains when incorporated in a complex food system. The presence of endothermic peaks during melting has confirmed that milk fat is partially crystallized within cheeses at specified storage temperatures. Similar to bulk milk fat, melting properties within cheese products have been found to depend on the cooling rate. Faster cooling rates have been shown to lower the crystallization temperatures of each of the milk fat fractions (Gliguem *et al.* 2009). This result can be attributed to rapid crystallization, as this has been shown to lead to the formation of mixed crystals, associated with peak broadening in melting endotherms (Lopez *et al.* 2002). Similar results were obtained by Ramel and Marangoni (2017a, b) for processed cheese and cream cheese products. Milk fat is known to contain distinct melting fractions, corresponding to LMF, MMF and HMF fractions. A direct comparison found that while AMF displayed three melting peaks, processed cheese products displayed only two melting peaks (Ramel and Marangoni 2017a, b). This results from different fractions melting concurrently. An investigation by Gliguem *et al.* (2011) identified the importance of thermal history on the melting properties of milk fat contained in processed cheese products (Gliguem *et al.* 2011). In this work, samples heated from −10 °C to 50 °C showed two separate endothermic peaks. The same samples were also heated to 60 °C to erase the thermal history prior to analysis, and then demonstrated at least four endothermic peaks. This study displayed that processing parameters can still have a large impact on the thermal behaviour of milk fat contained within food systems.

Outside of cheese systems, measurement of the melting behaviour of butter baked into croissants determined that crystallization within the baked matrix results in heterogeneous crystallization, based on peak width (Mattice and Marangoni 2017). The high temperatures involved in baking inherently erase all crystallization and thermal history. Upon cooling, analysis of the crystallization kinetics revealed that the presence of gelatinized wheat starch in these baked systems causes an increase in crystallization rate (Mattice and Marangoni 2018a). This result is similar to those reported for cheese, where the presence of matrix components causes changes in the crystallization kinetics, leading to the formation of mixed crystals.

8.5.2 Polytypism and Polymorphism

As discussed previously (Section 8.2.4), fat crystal polymorphism has practical consequences for product quality and functionality. Comparison of milk fat in bulk with that contained within a food matrix also allows for the identification of interactions or matrix influence on polymorphic form and stability. In addition to thermal analysis, the identification of different polymorphic forms within cheese products has led to the conclusion that milk fat, at least partially, crystallizes within cheese systems. One study reported that only α and β′ crystals were present in processed cheese products (Gliguem *et al.* 2009, 2011). Though, Ramel and Marangoni (2017a, b) found that some β crystals were present within processed cheese and cream cheese matrices (Figure 8.12a) (Ramel and Marangoni 2017a, b). Lopez *et al.* (2008) also identified the presence of β crystals in Emmental cheese. Further investigation indicated that the formation of β polymorphs is primarily a result of physical restraints imposed by the protein matrix, forcing the rearrangement into the most stable crystal polymorph (Ramel and Marangoni 2017b). Both 2L and 3L lamellar structures have been identified in cheese products, with the more compact 2L structures related to the more stable β′ and β crystals (Lopez *et al.* 2008; Gliguem *et al.* 2009; Ramel and Marangoni 2017a).

Mattice and Marangoni (2017) determined that butter contained in and baked in croissants crystallized into β′ crystals; however, the components of the matrix caused polymorphic instability and significant conversion to β crystals occurred within as little as 1 week of storage (Mattice and Marangoni 2017). Evidence of matrix influence was shown when butter baked under the same conditions and stored did not display this conversion (Figure 8.12b). The results of the study suggested that the instability was caused by wheat starch gelatinization and subsequent retrogradation inherent in baking.

8.5.3 Solid Fat Content

Given the importance of SFC on the physical and rheological properties of milkfat and butter, the SFC would also contribute greatly to the physical and rheological properties of milk fat-containing products. Many additional factors must be con-

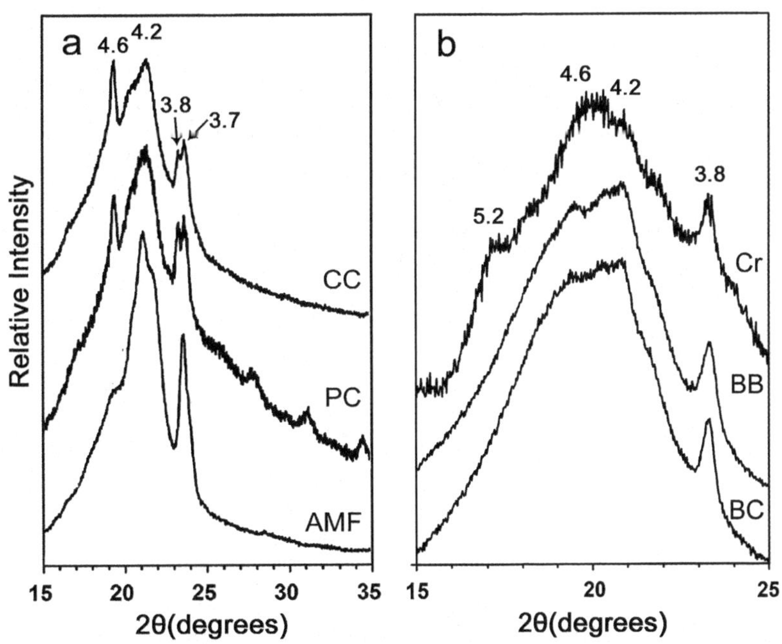

Figure 8.12. Polymorphism determined by wide-angle X-ray diffraction of (**a**) cheese products (*CC* cream cheese, *PC* processed cheese loaf, *AMF* anhydrous milk fat) and (**b**) butter and croissants (*Cr* croissant, *BB* baked butter, *BC* butter control). Peaks are labelled corresponding to their *d*-spacings. Characteristic peaks of β′ polymorphs: 4.2, 3.8 and 3.7 Å and characteristic peak of β polymorphs: 4.6 Å.

sidered when measuring the SFC of fat contained within a food matrix, as the direct method of SFC measurement by nuclear magnetic resonance spectroscopy (AOCS Official Method Cd 16b-93) will find the amount of total solids in the sample, fat or otherwise (Mattice and Marangoni 2018b). Therefore, many studies simply discuss the relative magnitudes of SFC, referring to rheological changes at different temperatures.

Solid milk fat becomes quite rigid at refrigeration temperatures, and this has been deemed significant for cheese texture, where a higher SFC increases the hardness or firmness of cheeses. In contrast, low SFC, with the majority of fat in the liquid state, results in a decrease or, at least, no increase in hardness (Brighenti *et al.* 2008; Rogers *et al.* 2010; Yang *et al.* 2011). This was also demonstrated with the replacement of AMF with canola oil in model cheese systems, where it was found that the solid nature of milk fat at low temperatures provides superior matrix reinforcement compared with liquid oil at the same temperature (Ramel and Marangoni 2018a, b). One study determined that Emmental cheese has an SFC of approximately 55–56% when stored at 4 °C after ripening (Lopez *et al.* 2006). In this investigation, the SFC was determined by comparing the enthalpy of melting of the fat contained in the cheese, compared to that of AMF extracted from the cheese. Their obtained results indicated that the enthalpy of melting (in J/g of fat) was almost two times greater for extracted AMF than for milk fat contained within the cheese. In non-cheese systems, the SFC of butter contained within croissants was also determined to be significantly lower than that measured in bulk (Mattice and Marangoni 2017). It was hypothesized that an interaction between matrix components causes some of the TAGs normally solid in the range of 10–40 °C to remain in the liquid state.

8.5.4 Rheology

The analysis of physical and rheological properties is required to discover the true relationship between milk fat and the physical properties of the food. In cheeses, it has been shown that the reinforcement of the protein matrix by solid milk fat globules governs many of the physical properties of cheeses, including the softening of cheese as milk fat melts with increasing temperatures (Gliguem *et al.* 2009; Ramel and Marangoni 2018a, b). This was demonstrated by Ramel and Marangoni (2018) in model cheese systems, where the replacement of milk fat with liquid canola oil caused the loss of any structural reinforcement, therefore causing cheese containing canola oil to demonstrate lower G' and G'' values (Figure 8.13). Gliguem *et al.* (2009) were able to correlate the softening within certain temperature ranges with the melting of specific crystal forms (i.e. α or β′). The authors also determined that processed cheese cooled at slower cooling rates exhibits greater G' and G'' values, compared to those cooled quickly, though a precise explanation for the behaviour was not

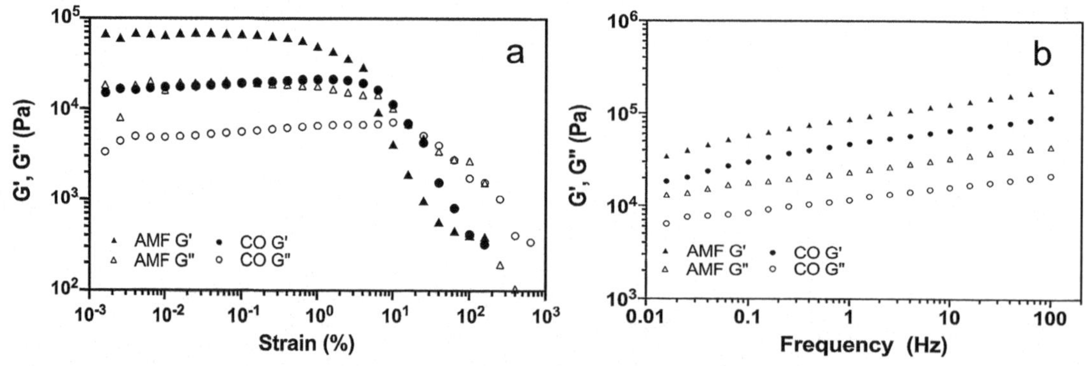

Figure 8.13. Rheological characterizations of imitation cheese prepared with anhydrous milk fat (AMF) or canola oil (CO): (**a**) amplitude sweep at a constant frequency (1 Hz) and temperature (7 °C); (**b**) frequency sweep at a constant strain (0.1%) and temperature (7 °C).

offered. In another study, Gliguem *et al.* (2011) found that, similar to bulk milk fat, the rheological properties of processed cheeses depend on the thermal history. This study demonstrated the importance of the processing parameters used, as processed cheese samples were found to have approximately 25% higher G' values, compared to the same samples that had been heated and re-cooled. Ramel and Marangoni (2017a, b) were able to connect the yield strain of processed cheese samples to their microstructure using oscillatory rheometry. The authors found that indications of a weaker structure, namely the presence of larger fat globules of varying size, corresponded to a lower yield strain value (Ramel and Marangoni 2017a, b).

8.6 Conclusions

The crystallization behaviour of milk fat is complex, owing, in large part, to its complicated composition. By manipulating composition and crystallization conditions, milk fat and dairy products with unique structures and mechanical properties can be designed. Understanding the relationships between composition, crystallization, structure, rheology and texture is a powerful tool in this regard. The characterization of milk fat structure at many scales (e.g. microstructure, nanoscale structure, molecular packing), in addition to extensive analysis of physical properties and rheology, has allowed for comprehensive understanding of the interrelationships between microstructure, functionality and product quality. Analysis of the role milk fat plays within different food systems reveals detailed relationships between fat and matrix components, as well as what must be accounted for should milk fat be reduced, removed, manipulated or replaced in a food product.

Bibliography

Acevedo, N. C. & Marangoni, A. G. (2010). Characterization of the nanoscale in triacylglycerol crystal networks. *Crystal Growth & Design, 10*, 3327–3333.

Acevedo, N. C., Block, J. M. & Marangoni, A. G. (2012). Critical laminar shear-temperature effects on the nano- and mesoscale structure of a model fat and its relationship to oil binding and rheological properties. *Faraday Discussions, 158*, 171–194.

AOCS. (2017a). *Official methods and recommended practices of the American Oil Chemists' Society* (7th ed.). Urbana: AOCS Press. Method Cc16-60.

AOCS. (2017b). *Official methods and recommended practices of the American Oil Chemists' Society* (7th ed.). Urbana: AOCS Press. Method Cd 16b-93.

Arul, J., Boudreau, A., Makhlouf, J., Tardif, R. & Bellavia, T. (1988). Fractionation of anhydrous milk fat by short-path distillation. *Journal of the American Oil Chemists' Society, 65*, 1642–1646.

Ashes, J. R., Gulati, S. K. & Scott, T. W. (1997). Potential to Alter the content and composition of milk fat through nutrition. *Journal of Dairy Science, 80*, 2204–2212.

Awad, T.S., Rogers, M.A. & Marangoni, A.G. (2004). Scaling behavior of the elastic modulus in colloidal networks of fat crystals. *Journal of Physical Chemistry B, 108*, 171–179.

Balcao, V. M. & Malcata, F. X. (1998). Lipase catalyzed modification of milkfat. *Biotechnology Advances, 16*, 309–341.

Baldwin, R. R., Baldry, R. P. & Johansen, R. G. (1972). Fat systems for bakery products. *Journal of the American Oil Chemists' Society, 49*, 473–477.

Beaulieu, A. & Palmquist, D. (1995). Differential effects of high fat diets on fatty acid composition in milk of Jersey and Holstein cows. *Journal of Dairy Science, 78*, 1336–1344.

Berti, J., Grosso, N. R., Fernandez, H., Pramparo, C. & Gayol, M. F. (2018). Sensory quality of milk fat with low cholesterol content fractioned by molecular distillation. *Journal of the Science of Food and Agriculture, 98*, 3478–3484.

Boistelle, R. (1988). Fundamentals of nucleation and crystal growth. In N. Garti & K. Sato (Eds.), *Crystallization and polymorphism of fats and fatty acids* (pp. 189–226). New York: Marcel Dekker.

Breitschuh, B. & Windhab, E. J. (1998). Parameters influencing cocrystallization and polymorphism in milk fat. *Journal of the American Oil Chemists' Society, 75*, 897–904.

Brighenti, M., Govindasamy-Lucey, S., Lim, K., Nelson, K. & Lucey, J. A. (2008). Characterization of the rheological, textural, and sensory properties of samples of commercial US cream cheese with different fat contents. *Journal of Dairy Science, 91*, 4501–4517.

Buldo, P. & Wiking, L. (2012). The role of mixing temperature on microstructure and rheological properties of butter blends. *Journal of the American Oil Chemists' Society, 89*, 787–795.

Buldo, P., Kirkensgaard, J. J. K. & Wiking, L. (2013). Crystallization mechanisms in cream during ripening and initial butter churning. *Journal of Dairy Science, 96*, 6782–6791.

Campos, R., Narine, S. S. & Marangoni, A. G. (2002). Effect of cooling rate on the structure and mechanical properties of milk fat and lard. *Food Research International, 35*, 971–981.
Campos, R. J., Litwinenko, J. W. & Marangoni, A. G. (2003). Fractionation of milk fat by short-path distillation. *Journal of Dairy Science, 86*, 735–745.
Caroprese, M., Marzano, A., Marino, R., Gliatta, G., Muscio, A. & Sevi, A. (2010). Flaxseed supplementation improves fatty acid profile of cow milk. *Journal of Dairy Science, 93*, 2580–2588.
Chapman, D. (1961). The polymorphism of glycerides. *Chemical Reviews, 62*, 433–456.
Cisneros, A., Mazzanti, G., Campos, R. & Marangoni, A. G. (2006). Polymorphic transformation in mixtures of high- and low-melting fractions of milk fat. *Journal of Agricultural and Food Chemistry, 54*, 6030–6033.
Collomb, M., Ueli, B., Sieber, R., Jeangros, B. & Bosset, J. (2002). Composition of fatty acids in cow's milk fat produced in the lowlands, mountains and highlands of Switzerland using high-resolution gas chromatography. *International Dairy Journal, 12*, 649–659.
Coupland, J. N. (2002). Crystallization in emulsions. *Current Opinion in Colloid & Interface Science, 7*, 445–450.
de Man, J. M. & Wood, F. (1959). Hardness of butter II. Influence of setting. *Journal of Dairy Science, 42*, 56–61.
De, B. K., Hakimji, M., Patel, A., Sharma, D., Desai, H. & Kumar, T. (2007). Plastic fats and margarines through fractionation, blending and interesterification of milk fat. *European Journal of Lipid Science and Technology, 109*, 32–37.
DeMan, J. M. & Beers, A. M. (1987). Fat crystal networks: Structure and rheological properties. *Journal of Texture Studies, 18*, 303–318.
DeMan, L., DeMan, J. M. & Blackman, B. (1991). Physical and textural characteristics of some North American shortenings. *Journal of the American Oil Chemists' Society, 68*, 63–69.
Diener, R. G. & Heldman, D. R. (1968). Methods of determining rheological properties of butter. *Transactions of ASAE, 11*, 444–447.
Dixon, B. & Williams, T. (1977). Measurement of butter firmness by sectility testing. *Australian Journal of Dairy Technology, 32*, 177–179.
Elliott, J. & Ganz, A. (1971). Modification of food characteristics with cellulose hydrocolloids. I: Rheological characterization of an organoleptic property (unctuousness). *Journal of Texture Studies, 2*, 220–229.
Elliott, J. & Green, C. (1972). Modification of food characteristics with cellulose hydrocolloids. II: The modified bingham body – A useful rheological model. *Journal of Texture Studies, 3*, 194–205.
Fredrick, E., Van de Walle, D., Walstra, P., Zijtveld, J. H., Fischer, S., Van Der Meeren, P. & Dewettinck, K. (2011). Isothermal crystallization behaviour of milk fat in bulk and emulsified state. *International Dairy Journal, 21*, 685–695.
Frydenberg R.P., Hammershøj M., Andersen U. & Wiking L. (2013). Ultrasonication affects crystallization mechanisms and kinetics of anhydrous milk fat. Crystal growth & design, 13, 5375–5382.
Garside, J. (1987). General principles of crystallization. In J. M. W. Blanshard & P. Lillford (Eds.), *Food structure and behavior; food and technology: A series of monographs*. Cambridge, MA: Academic.
Gliguem, H., Ghorbel, D., Lopez, C., Michon, C., Ollivon, M. & Lesieur, P. (2009). Crystallization and polymorphism of triacylglycerols contribute to the rheological properties of processed cheese. *Journal of Agriculture and Food Research, 57*, 3195–3203.
Gliguem, H., Lopez, C., Michon, C., Lesieur, P. & Ollivon, M. (2011). The viscoelastic properties of processed cheeses depend on their thermal history and fat polymorphism. *Journal of Agricultural and Food Chemistry, 59*, 3125–3134.
Grall, D. S. & Hartel, R. W. (1992). Kinetics of butterfat crystallization. *Journal of the American Oil Chemists' Society, 69*, 741–747.
Haighton, A. (1959). The measurement of the hardness of margarine and fats with cone penetrometers. *Journal of the American Oil Chemists' Society, 36*, 345–348.
Hayakawa, M. & DeMan, J. (1982). Interpretation of cone penetrometer consistency measurements of fats. *Journal of Texture Studies, 13*, 201–210.
Heck, J. M. L., van Valenberg, H. J. F., Dijkstra, J. & van Hooijdonk, A. C. M. (2009). Seasonal variation in the Dutch bovine raw milk composition. *Journal of Dairy Science, 92*, 4745–4755.
Herrera, M. L. & Hartel, R. W. (2000a). Effect of processing conditions on physical properties of a milk fat model system: Microstructure. *Journal of the American Oil Chemists' Society, 77*, 1197–1205.
Herrera, M. L. & Hartel, R. W. (2000b). Effect of processing conditions on physical properties of a milk fat model system: Rheology. *Journal of the American Oil Chemists' Society, 77*, 1189–1196.
Herrera, M. L. & Hartel, R. W. (2000c). Effect of processing conditions on crystallization kinetics of a milk fat model system. *Journal of the American Oil Chemists' Society, 77*, 1177–1188.
Herrera, M. L., de Leon Gatti, M. & Hartel, R. W. (1999). A kinetic analysis of crystallization of a milk fat model system. *Food Research International, 32*, 289–298.
Himawan, C., Starov, V. M. & Stapley, A. G. F. (2006). Thermodynamic and kinetic aspects of fat crystallization. *Advances in Colloid and Interface Science, 122*, 3–33.
Hurtaud, C. & Peyraud, J. L. (2007). Effects of feeding Camelina (seeds or meal) on milk fatty acid composition and butter spreadability. *Journal of Dairy Science, 90*, 5134–5145.
Hyun, K., Wilhelm, M., Klein, C. O., Soo Cho, K., Gun Nam, J., Hyun Ahn, K., Jong Lee, S., Ewoldt, R. H. & McKinley, G. H. (2011). A review of nonlinear oscillatory shear tests: Analysis and application of

large amplitude oscillatory shear (LAOS). *Progress in Polymer Science, 36*, 1697–1753.
Illingworth, D. (2002). Fractionation of fats. In A. G. Marangoni & S. S. Narine (Eds.), *Physical properties of lipids* (pp. 411–447). New York: Marcel Dekker.
Jensen, R. G. (2002). The composition of bovine milk lipids: January 1995 to December 2000. *Journal of Dairy Science, 85*, 295–350.
Jensen, R. G., Ferris, A. M. & Lammi-Keefe, C. J. (1991). The composition of milk fat. *Journal of Dairy Science, 74*, 3228–3243.
Kalač, P. & Samková, E. (2010). The effects of feeding various forages on fatty acid composition of bovine milk fat: A review. *Czech Journal of Animal Science, 55*, 521–537.
Kalo, P., Vaara, K. & Antila, M. (1986). Changes in triglyceride composition and melting properties of butter fat solid fraction/rapeseed oil mixtures induced by lipase catalysed inter-esterification. *Fette, Seifen, Anstrichmittel, 88*, 362–365.
Kamel, B. S. & DeMan, J. (1975). Measurement of fat consistency by the penetration method. *Canadian Institute of Food Science and Technology, 8*, 117–121.
Kaylegian, K. E. (1999). The production of specialty milk fat ingredients. *Journal of Dairy Science, 82*, 1433–1439.
Kaylegian, K. E. & Lindsay, R. C. (1992). Performance of selected milk fat fractions in cold-spreadable butter. *Journal of Dairy Science, 75*, 3307–3317.
Kelsey, J. A., Corl, B. A., Collier, R. J. & Bauman, D. E. (2003). The effect of breed, parity, and stage of lactation on conjugated linoleic acid (CLA) in milk fat from dairy cows. *Journal of Dairy Science, 86*, 2588–2597.
Kloek, W., Van Vliet, T. & Walstra, P. (2005). Large deformation behavior of fat crystal networks. *Journal of Texture Studies, 36*, 516–543.
Knoester, M., De Bruyne, P. & Van den Tempel, M. (1968). Crystallization of triglycerides at low supercooling. *Journal of Crystal Growth, 3*, 776–780.
Larsen, M. K., Andersen, K. K., Kaufmann, N. & Wiking, L. (2014). Seasonal variation in the composition and melting behavior of milk fat. *Journal of Dairy Science, 97*, 4703–4712.
Lawler, P. J. & Dimick, P. S. (2008). Crystallization and polymorphism of fats. In C. Akoh & D. Min (Eds.), *Food lipids: Chemistry, nutrition and biotechnology* (3rd ed., pp. 245–263). New York: Marcel Dekker.
Lázaro, R., Viriato, S., Queirós, M. D. S., Isabel, M., Neves, L., Paula, A., Ribeiro, B. & Gigante, M. L. (2019). Improvement in the functionality of spreads based on milk fat by the addition of low melting triacylglycerols. *Food Research International, 120*, 432–440.
Lee, J. & Martini, S. (2018). Effect of cream aging temperature and agitation on butter properties. *Journal of Dairy Science, 101*, 7724–7735.
Lefebvre, J. (1983). Finished product formulation. *Journal of the American Oil Chemists' Society, 60*, 295–300.
Lindmark Månsson, H. (2008). Fatty acids in bovine milk fat. *Food & Nutrition Research, 52*, 1821.
Liu, S., Zhang, R., Kang, R., Meng, J. & Ao, C. (2016). Milk fatty acids profiles and milk production from dairy cows fed different forage quality diets. *Animal Nutrition, 2*, 329–333.
Lock, A. L. & Bauman, D. E. (2004). Modifying milk fat composition of dairy cows to enhance fatty acids beneficial to human health. *Lipids, 39*, 1197–1206.
Lock, A. L. & Garnsworthy, P. C. (2003). Seasonal variation in milk conjugated linoleic acid and D 9-desaturase activity in dairy cows. *Livestock Production Science, 79*, 47–59.
Lopez, C., Lavigne, F., Lesieur, P., Bourgaux, C. & Ollivon, M. (2001). Thermal and structural behavior of milk fat: 1. Unstable species of anhydrous milk fat. *Journal of Dairy Science, 84*, 756–766.
Lopez, C., Bourgaux, C., Lesieur, P., Bernadou, S. & Ollivon, M. (2002). Thermal and structural behavior of milk fat 3. Influence of cooling rate and droplet size on cream crystallization. *Journal of Colloid and Interface Science, 78*, 64–78.
Lopez, C., Lesieur, P., Bourgaux, C. & Ollivon, M. (2005). Thermal and structural behavior of anhydrous milk fat. 3. Influence of cooling rate. *Journal of Dairy Science, 88*, 511–526.
Lopez, C., Briard-Bion, V., Camier, B. & Gassi, J.-Y. (2006). Milk fat thermal properties and solid fat content in Emmental cheese: A differential scanning calorimetry study. *Journal of Dairy Science, 89*, 2894–2910.
Lopez, C., Bourgaux, C., Lesieur, P. & Ollivon, M. (2007). Coupling of time-resolved synchrotron X-ray diffraction and DSC to elucidate the crystallisation properties and polymorphism of triglycerides in milk fat globules. *Le Lait, 87*, 459–480.
Lopez, C., Briard-Bion, V., Beaucher, E. & Ollivon, M. (2008). Multiscale characterization of the organization of triglycerides and phospholipids in Emmental cheese: From the microscopic to the molecular level. *Journal of Agricultural and Food Chemistry, 56*, 2406–2414.
Macias-Rodriguez, B. A. (2018). Nonlinear rheology of fats using large amplitude oscillatory shear tests. In A. G. Marangoni (Ed.), *Structure-function analysis of edible fats* (2nd ed., pp. 169–196). London: AOCS Press.
Macias-Rodriguez, B. A. & Marangoni, A. G. (2016a). Physicochemical and rheological characterization of roll-in shortenings. *Journal of the American Oil Chemists' Society, 93*, 575–585.
Macias-Rodriguez, B. A. & Marangoni, A. G. (2016b). Rheological characterization of triglyceride shortenings. *Rheologica Acta, 55*, 767–779.
Macias-rodriguez, B. A., Peyronel, F. & Marangoni, A. G. (2017). The role of nonlinear viscoelasticity on the functionality of laminating shortenings. *Journal of Food Engineering, 212*, 87–96.
Maleky, F. & Marangoni, A. G. (2008). Process development for continuous crystallization of fat under laminar shear. *Journal of Food Engineering, 89*, 399–407.

Maleky, F., Smith, A. K. & Marangoni, A. G. (2011). Laminar shear effects on crystalline alignments and nanostructure of a triacylglycerol crystal network. *Crystal Growth & Design, 11*, 2335–2345.

Mansbridge, R. & Blake, J. (1997). Nutritional factors affecting the fatty acid composition of bovine milk. *British Journal of Nutrition, 78*, S37–S47.

Marangoni, A. G. (1998). On the use and misuse of the Avrami equation in characterization of the kinetics of fat crystallization. *Journal of the American Oil Chemists' Society, 75*, 1465–1467.

Marangoni, A. G. (2000). Elasticity of high-volume-fraction fractal aggregate networks: A thermodynamic approach. *Physical Review B, 62*, 13951–13955.

Marangoni, A. G. & Lencki, R. W. (1998). Ternary phase behavior of milk fat fractions. *Journal of Agricultural and Food Chemistry, 46*, 3879–3884.

Marangoni, A. G. & Mcgauley, S. E. (2003). Relationship between crystallization behavior and structure in cocoa butter. *Crystal Growth & Design, 3*, 95–108.

Marangoni, A. G. & Rogers, M. A. (2003). Structural basis for the yield stress in plastic disperse systems. *Applied Physics Letters, 82*, 3239–3241.

Marangoni, A. G. & Rousseau, D. (1996). Is plastic fat rheology governed by the fractal nature of the fat crystal network? *Journal of the American Oil Chemists' Society, 73*, 991–994.

Marangoni, A. G. & Rousseau, D. (1998). Chemical and enzymatic modification of butterfat and butterfat-canola oil blends. *Food Research International, 31*, 595–599.

Marangoni, A. G. & Wesdorp, L. H. (2013a). Nucleation and crystalline growth kinetics. In A. G. Marangoni & L. H. Wesdorp (Eds.), *Structure and properties of fat crystal networks* (2nd ed., pp. 27–100). Boca Raton: CRC Press.

Marangoni, A. G. & Wesdorp, L. H. (2013b). Crystallography and polymorphism. In A. G. Marangoni & L. H. Wesdorp (Eds.), *Structure and properties of fat crystal networks* (2nd ed., pp. 1–26). Boca Raton: CRC Press.

Martini, S., Suzuki, A. & Hartel, R. (2008). Effect of high intensity ultrasound on crystallization behavior of anhydrous milk fat. *Journal of the American Oil Chemists' Society, 85*, 621–628.

Mattice, K. D. & Marangoni, A. G. (2017). Matrix effects on the crystallization behaviour of butter and roll-in shortening in laminated bakery products. *Food Research International, 96*, 54–63.

Mattice, K. D. & Marangoni, A. G. (2018a). Gelatinized wheat starch influences crystallization behaviour and structure of roll-in shortenings in laminated bakery products. *Food Chemistry, 243*, 396–402.

Mattice, K. D. & Marangoni, A. G. (2018b). Fat crystallization and structure in bakery, meat, and cheese systems. In A. G. Marangoni (Ed.), *Structure-function analysis of edible fats* (2nd ed., pp. 287–311). London: AOCS Press.

Maurice-van Eijndhoven, M. H. T., Hiemstra, S. J. & Calus, M. P. L. (2011). Short communication: Milk fat composition of 4 cattle breeds in the Netherlands. *Journal of Dairy Science, 94*, 1021–1025.

Mazzanti, G., Guthrie, S. E., Sirota, E. B., Marangoni, A. G. & Idziak, S. H. J. (2003). Orientation and phase transitions of fat crystals under shear. *Crystal Growth & Design, 3*, 721–725.

Mazzanti, G., Marangoni, A. G. & Idziak, S. H. J. (2009). Synchrotron study on crystallization kinetics of milk fat under shear flow. *Food Research International, 42*, 682–694.

McCarthy, O. J. (2011). Rheology of liquid and semi-solid milk products. In J. W. Fuquay (Ed.), *Encyclopedia of dairy sciences* (2nd ed., pp. 520–531). San Diego: Academic.

Mcdowell, R. E., Hooven, N. W. & Camoens, J. K. (1975). Effect of climate on performance of Holsteins in first lactation. *Journal of Dairy Science, 59*, 965–971.

Mulder, H. (1953). The consistency of butter. In S. Blair (Ed.), *Foodstuffs: Their plasticity, fluidity and consistency*. Amsterdam: North Holland Publishing Company.

Narine, S. & Marangoni, A. G. (1999a). Mechanical and structural model of fractal networks of fat crystals at low deformations. *Physical Review E, 60*, 6991–7000.

Narine, S. S. & Marangoni, A. G. (1999b). Relating structure of fat crystal networks to mechanical properties: A review. *Food Research International, 32*, 227–248.

Narine, S. S. & Marangoni, A. G. (1999c). Fractal nature of fat crystal networks. *Physical Review E, 59*, 1908–1920.

Narine, S. S. & Marangoni, A. G. (2001). Elastic modulus as an indicator of macroscopic hardness of fat crystal networks. *LWT – Food Science and Technology, 34*, 33–40.

Nunes, G. F. M., Veloso de Paula, A., Ferreira de Castro, H. & César dos Santos, J. (2011). Compositional and textural properties of milkfat – Soybean oil blends following enzymatic interesterification. *Food Chemistry, 125*, 133–138.

O'Sullivan, C., Acevedo, N., Peyronel, F. & Marangoni, A. G. (2014). Fat nanostructure. In A. G. Marangoni & D. Pink (Eds.), *Edible nanostructures: A bottom-up approach* (pp. 6–40). Cambridge: Royal Society of Chemistry.

Ollivon, M., Keller, G., Bourgaux, C., Kalnin, D., Villeneuve, P. & Lesieur, P. (2006). DSC and high resolution X-ray diffraction coupling. *Journal of Thermal Analysis and Calorimetry, 85*, 219–224.

Pal, P. K., Bhattacharyya, D. K. & Ghosh, S. (2001). Modifications of butter stearin by blending and interesterification for better utilization in edible fat products. *Journal of the American Oil Chemists' Society, 78*, 31–36.

Pappritz, J., Meyer, U., Kramer, R., Weber, E., Rehage, J., Flachowsky, G., Dänicke, S., Pappritz, J., Meyer, U., Kramer, R., Weber, E., Rehage, J., Flachowsky, G. & Dänicke, S. (2011). Effects of long-term supplementation of dairy cow diets with rumen-protected conjugated linoleic acids (CLA) on performance, metabolic

parameters and fatty acid profile in milk fat. *Archives of Animal Nutrition, 65*, 89–107.
Peyronel, F., Ilavsky, J., Pink, D. A. & Marangoni, A. G. (2014). Quantification of the physical structure of fats in 20 minutes: Implications for formulation. *Lipid Technology, 26*, 223–226.
Pink, D. A., Peyronel, F., Quinn, B., Singh, P. & Marangoni, A. G. (2015). Condensation versus diffusion. A spatial- scale-independent theory of aggregate structures in edible oils: Applications to model systems and commercial shortenings studied via rheology and USAXS. *Journal of Physics D: Applied Physics, 48*, 384003.
Prentice, J. H. (1972). Rheology and texture of dairy products. *Journal of Texture Studies, 3*, 415–458.
Queirós, M. S., Grimaldi, R. & Gigante, M. L. (2016). Addition of olein from milk fat positively affects the firmness of butter. *Food Research International, 84*, 69–75.
Ramel, P. R. & Marangoni, A. G. (2016). Engineering the microstructure of milk fat by blending binary and ternary mixtures of its fractions. *RSC Advances, 6*, 41189–41194.
Ramel, P. R. & Marangoni, A. G. (2017a). Characterization of the polymorphism of milk fat within processed cheese products. *Food Structure, 12*, 15–25.
Ramel, P. R. & Marangoni, A. G. (2017b). Insights into the mechanism of the formation of the most stable crystal polymorph of milk fat in model protein matrices. *Journal of Dairy Science, 100*, 6930–6937.
Ramel, P. R. & Marangoni, A. G. (2018a). Engineering the rheological and thermomechanical properties of model imitation cheese using particle fillers. *Journal of Food Engineering, 235*, 9–15.
Ramel, P. R. & Marangoni, A. G. (2018b). Processed cheese as a polymer matrix composite: A particle toolkit for the replacement of milk fat with canola oil in processed cheese. *Food Research International, 107*, 110–118.
Ramel, P. R. R., Peyronel, F. & Marangoni, A. G. (2016). Characterization of the nanoscale structure of milk fat. *Food Chemistry, 203*, 224–230.
Rodrigues, J. N. & Gioielli, L. A. (2003). Chemical interesterification of milkfat and milkfat-corn oil blends. *Food Research International, 36*, 149–159.
Rogers, N. R., Mcmahon, D. J., Daubert, C. R., Berry, T. K. & Foegeding, E. A. (2010). Rheological properties and microstructure of Cheddar cheese made with different fat contents. *Journal of Dairy Science, 93*, 4565–4576.
Rohm, H. (1992). Influence of sectility speed on the relationship between sensory and instrumental evaluation of edible fat firmness. *Zeitschrift für Leb. und Forsch., 194*, 240–243.
Rohm, H. & Raaber, S. (1991). Hedonic spreadability optima of selected edible fats. *Journal of Sensory Studies, 6*, 81–88.
Rohm, H. & Ulberth, F. (1989). Use of magnitude estimation in sensory texture analysis of butter. *Journal of Texture Studies, 20*, 409–418.
Rohm, H. & Weidinger, K.-H. (1993). Rheological behaviour of butter at small deformations. *Journal of Texture Studies, 24*, 157–172.
Rønholt, S., Kirkensgaard, J. J. K., Mortensen, K. & Knudsen, J. C. (2014). Effect of cream cooling rate and water content on butter microstructure during four weeks of storage. *Food Hydrocolloids, 34*, 169–176.
Rousseau, D. & Marangoni, A. G. (1998a). The effects of interesterifcation on physical and sensory attributes of butterfat and butterfat-canola oil spreads. *Food Research International, 31*, 381–388.
Rousseau, D. & Marangoni, A. G. (1998b). Tailoring the textural attributes of butter fat/canola oil blends via Rhizopus arrhizus lipase-catalyzed interesterification. 2. Modifications of physical properties. *Journal of Agricultural and Food Chemistry, 46*, 2375–2381.
Rousseau, D., Foresti, K., Hill, A. R. & Marangoni, A. G. (1996). Restructuring butterfat through blending and chemical interesterification. 1. Melting behaviour and triacylglycerol modifications. *Journal of the American Oil Chemists' Society, 73*, 963–972.
Rousseau, D., Hill, A. R. & Marangoni, A. G. (1996b). Restructuring butterfat through blending and chemical interesterification. 2. Microstructure and polymorphism. *Journal of the American Oil Chemists' Society, 73*, 973–981.
Rousseau, D., Hill, A. R. & Marangoni, A. G. (1996c). Restructuring butterfat through blending and chemical interesterification. 3. Rheology. *Journal of the American Oil Chemists' Society, 73*, 983–989.
Rousset, P. (2002). Modeling crystallization kinetics of triacylglycerols. In A. G. Marangoni & S. S. Narine (Eds.), *Physical properties of lipids* (pp. 9–44). New York: Marcel Dekker.
Sangwal, K. & Sato, K. (2018). Nucleation and crystallization kinetics of fats. In A. G. Marangoni (Ed.), *Structure-function analysis of edible fats* (2nd ed., pp. 21–72). San Diego: Elsevier.
Sato, K. (2001). Crystallization behaviour of fats and lipids – A review. *Chemical Engineering Science, 56*, 2255–2265.
Sato, K. (2018a). Polymorphism of lipid crystals. In K. Sato (Ed.), *Crystallization of lipids: Fundamentals and applications in food, cosmetics, and pharmaceuticals* (pp. 17–60). Hoboken: Wiley-Blackwell.
Sato, K. (2018b). Introduction: Relationships of structures, properties, and functionality. In K. Sato (Ed.), *Crystallization of lipids: Fundamentals and applications in food, cosmetics, and pharmaceuticals* (pp. 1–15). Hoboken: Wiley-Blackwell.
Sebben, D. A., Gao, N., Gillies, G., Beattie, D. A. & Krasowska, M. (2019). Fractionation and characterisation of hard milk fat crystals using atomic force microscopy. *Food Chemistry, 279*, 98–104.
Shama, F. & Sherman, P. (1970). The influence of work softening on the viscoelastic properties of butter and margarine. *Journal of Texture Studies, 1*, 196–205.
Sharples, A. (1996). Overall kinetics of crystallization. In A. Sharples (Ed.), *Introduction to polymer crystallization* (pp. 44–59). London: Edward Arnold.

Shin, J.-A., Akoh, C. & Lee, K.-T. (2009). Production and physicochemical properties of functional-butterfat through enzymatic interesterification in a continuous reactor. *Journal of Agricultural and Food Chemistry, 57*, 888–900.

Shukla, A. & Rizvi, S. (1995). Viscoelastic properties of butter. *Journal of Food Science, 60*, 902–905.

Shukla, A., Bhaskar, A. R., Rizvi, S. S. H. & Mulvaney, S. J. (1994). Physicochemical and rheological properties of butter made from supercritically fractionated milk fat. *Journal of Dairy Science, 77*, 45–54.

Singh, A. P., Avramis, C. A., Kramer, J. K. G. & Marangoni, A. G. (2004). Algal meal supplementation of the cows' diet alters the physical properties of milk fat. *Journal of Dairy Research, 71*, 66–73.

Skoda, W. & Van den Tempel, M. (1967). Growth kinetics of triglyceride crystals. *Journal of Crystal Growth, 1*, 207–217.

Soyeurt, H., Dehareng, F., Gengler, N., Mcparland, S., Wall, E., Berry, D. P. & Coffey, M. (2011). Mid-infrared prediction of bovine milk fatty acids across multiple breeds, production systems, and countries. *Journal of Dairy Science, 94*, 1657–1667.

Stoop, W. M., Bovenhuis, H., Heck, J. M. L. & van Arendonk, J. A. M. (2009). Effect of lactation stage and energy status on milk fat composition of Holstein-Friesian cows. *Journal of Dairy Science, 92*, 1469–1478.

Suzuki, A., Lee, J., Padilla, S. & Martini, S. (2010). Altering functional properties of fats using power ultrasound. *Journal of Food Science, 75*, E208–E214.

ten Grotenhuis, E., van Aken, G. A., van Malssen, K. F. & Schenk, H. (1999). Polymorphism of milk fat studied by differential scanning calorimetry and real-time X-ray powder diffraction. *Journal of the American Oil Chemists' Society, 76*, 1031–1039.

Texture Technologies Corp. (2019). *Spreadability*. Accessed Feb 2019. http://texturetechnologies.com/application-studies/spreadability

Timms, R. E. (1979). The physical properties of blends of milk fat with beef tallow and beef tallow fractions. *Australian Journal of Dairy Technology, 34*, 60.

Tomaszewska-Gras, J. (2013). Melting and crystallization DSC profiles of milk fat depending on selected factors. *Journal of Thermal Analysis and Calorimetry, 113*, 199–208.

Tran, T. & Rousseau, D. (2016). Influence of shear on fat crystallization. *Food Research International, 81*, 157–162.

Truong, T., Morgan, G. P., Bansal, N., Palmer, M. & Bhandari, B. (2015). Crystal structures and morphologies of fractionated milk fat in nanoemulsions. *Food Chemistry, 171*, 157–167.

Truong, T., Palmer, M., Bansal, N. & Bhandari, B. (2017a). Investigation of solubility of carbon dioxide in anhydrous milk fat by lab-scale manometric method. *Food Chemistry, 237*, 667–676.

Truong, T., Palmer, M., Bansal, N. & Bhandari, B. (2017b). Effect of solubilised carbon dioxide at low partial pressure on crystallisation behaviour, microstructure and texture of anhydrous milk fat. *Food Research International, 95*, 82–90.

van Aken, G. A. & Visser, K. A. (2000). Firmness and crystallization of Milk fat in relation to processing conditions. *Journal of Dairy Science, 83*, 1919–1932.

van Aken, G. A., ten Grotenhuis, E., van Langevelde, A. J. & Schenk, H. (1999). Composition and crystallization of milk fat fractions. *Journal of the American Oil Chemists' Society, 76*, 1323–1331.

Vanbergue, E., Hurtaud, C., Peyraud, J., Beuvier, E., Duboz, G. & Buchin, S. (2018). Effects of n-3 fatty acid sources on butter and hard cooked cheese; technological properties and sensory quality. *International Dairy Journal, 82*, 35–44.

Vanhoutte, B., Dewettinck, K., Vanlerberghe, B. & Huyghebaert, A. (2003). Monitoring milk fat fractionation: Filtration properties and crystallization kinetics. *Journal of the American Oil Chemists' Society, 80*, 213–218.

Walstra, P. (1998). Secondary nucleation in triglyceride crystallization. *Progress in Colloid and Polymer Science, 108*, 4–8.

Walstra, P., Van Vliet, T. & Kloek, W. (1995). Crystallization and rheological properties of milk fat. In P. F. Fox (Ed.), *Advanced dairy chemistry. Volume 2: Lipids* (pp. 179–211). London: Chapman and Hall.

Woodrow, I. L. & DeMan, J. (1968). Polymorphism in milk fat shown by X-ray diffraction and infrared spectroscopy. *Journal of Dairy Science, 51*, 996–1000.

Wright, A. J. & Marangoni, A. G. (2003). The effect of minor components on milk fat microstructure and mechanical properties. *Journal of Food Science, 68*, 182–186.

Wright, A. J., Hartel, R. W., Narine, S. S. & Marangoni, A. G. (2000). The effect of minor components on milk fat crystallization. *Journal of the American Oil Chemists' Society, 77*, 463–475.

Wright, A. J., Scanlon, M., Hartel, R. W. & Marangoni, A. G. (2001). Rheological properties of milkfat and butter. *Journal of Food Science, 66*, 1056–1071.

Wright, A. J., Marangoni, A. G. & Hartel, R. W. (2011). Milk lipids|rheological properties and their modification. In J. W. Fuquay (Ed.), *Encyclopedia of dairy sciences* (2nd ed., pp. 704–710). San Diego: Academic.

Yang, X., Rogers, N. R., Kendricks Berry, T. & Foegeding, E. A. (2011). Modeling the rheological properties of Cheddar cheese with different fat contents at various temperatures. *Journal of Texture Studies, 42*, 331–348.

Zachut, M., Arieli, A., Lehrer, H., Livshitz, L., Yakoby, S. & Moallem, U. (2010). Effects of increased supplementation of n-3 fatty acids to transition dairy cows on performance and fatty acid profile in plasma, adipose tissue, and milk fat. *Journal of Dairy Science, 93*, 5877–5889.

Role of Milk Fat in Dairy Products 9

D. S. Waldron, W. Hoffmann, W. Buchheim,
D. J. McMahon, H. Douglas Goff, S. V. Crowley,
C. Moloney, J. O'Regan, F. Giuffrida,
I. Celigueta Torres, and P. Siong

9.1 Role of Milk Fat in Butter

D. S. Waldron
School of Food and Nutritional Sciences
University College Cork
Cork, Ireland
e-mail: d.waldron@ucc.ie

The popularity of butter is on the rise again with consumption in EU28 up to 9% since 2013 (CLAL 2019). In the 1970s, the association between saturated fats and cardiovascular disease triggered a move away from butter to vegetable oil alternatives. In recent years, customers have allowed fat, and butter, back into their diets.

Conventional dietary guidelines, which recommend a very low intake of fats, have not accounted for the diverse range of food sources, processing or specific individual complex nutrients and structures in dairy products. Following a review of butter consumption and cardiometabolic outcomes, Pimpin *et al.* (2016) found that there was no association with cardiovascular disease in a balanced diet.

Butter is defined as a "product with a milk-fat content of not less than 80% but less than 90%, a maximum water content of 16% and a maximum dry non-fat milk material content of 2%" (EU 2013). This simple definition belies a complex microstructure of a continuous crystal network of partially coalesced fat globules, dispersed in a liquid oil phase and interrupted by intact fat globules, water and air droplets (Juriaanse and Heertje 1988). The result is an intricate interaction of texture, melting characteristics, colour, aroma and flavour.

Butter holds an almost unique place on our dining table. It possesses a delicate sweet creamy flavour, with a fat content that can act as an effective solvent to accept the flavour of other ingredients and aid their dispersion. Butter acts as a

D. S. Waldron (✉) · S. V. Crowley
School of Food and Nutritional Sciences,
University College Cork, Cork, Ireland
e-mail: d.waldron@ucc.ie

W. Hoffmann
Department of Safety and Quality of Milk and Fish Products, Max Rubner-Institut, Kiel, Germany

W. Buchheim
Kiel, Germany
e-mail: wo-hoffmann@kabelmail.de

D. J. McMahon
Department of Nutrition, Dietetics and Food Sciences, Utah State University, Logan, UT, Germany

H. Douglas Goff
Department of Food Science, University of Guelph, Guelph, ON Canada

C. Moloney · J. O'Regan
Nestlé Development Centre Nutrition, Limerick, Ireland

F. Giuffrida
Nestlé Research Centre, Lausanne, Germany

I. Celigueta Torres · P. Siong
Nestlé Product Technology Centre, Confectionery, Nestlé, York, UK

P. L. H. McSweeney et al. (eds.), *Advanced Dairy Chemistry, Volume 2*,
https://doi.org/10.1007/978-3-030-48686-0_9

spreadable moisture barrier in a sandwich, but with ideal melt-in-the-mouth characteristics. As a laminate between dough layers, butter traps steam to yield the *millefeuille* of croissants and puff pastry (Mattice and Marangoni 2017). As an emulsion base for *roux*, or sautéing vegetables in a *beurre* blanc, *noisette* or *noir*, the heat sensitivity of milk solids yields a gastronomic playground as Maillard browning reactions are exploited (van Boekel 1998; Kato 2003; McGee 2004).

No essential difference exists to facilitate this wide range of applications. The multiple natures are determined by the three-dimensional solid fat crystal network, surrounded by the liquid oil phase and other milk solids (DeMan and Beers 1987).

9.1.1 Milk Fat and the Milk Fat Globule Membrane (MFGM)

Fat in milk is organized as spherical lipid droplets ranging in size from 0.1 to 10 μm, with a mean diameter of around 4 μm. The core of the droplet is comprised mainly of triglycerides (TAGs) (98%) and smaller quantities of diglycerides, monoglycerides and cholesterol esters and is stabilized by an outer biological membrane, the milk fat globule membrane (MFGM), composed of polar lipids, cholesterol and proteins. The MFGM is asymmetric in composition, having an inner layer of neutral lipids, phosphatidylethanolamine, phosphatidylserine and phosphatidylinositol and an outer layer of polar lipids, phospholipids (phosphatidylcholine and sphingomyelin), glycolipids (cerebrosides and gangliosides) and sphingolipid domains (Deeth 1997; Lopez *et al.* 2011). The phospholipids and glycolipids, together with the protein components, are important for the emulsification role of the MFGM (Thomé and Eriksson 1973).

The MFGM is sensitive to physiological, chemical, biological (enzymatic) and physical (mechanical) changes. Typical handling processes such as ageing, agitation, separation, foaming and temperature treatment can easily cause damage and loss of MFGM material (see review by Evers 2004). The MFGM cannot be regarded as intact at the beginning of buttermaking. The separation of milk to skimmed milk and cream causes some damage, resulting in loss of surface proteins and destabilizing the MFGM (Holzmüllera *et al.* 2016).

9.1.2 Buttermaking: Whipping of Cream

At the beginning of buttermaking, the cream is whipped. Air is incorporated into this oil-in-water emulsion (35–40% fat) by a shear action forming and breaking bubbles, which are initially stabilized by the adsorption of soluble proteins, especially β-casein and whey proteins, to the air/water interface (Brooker 1986). A secondary process involves the collision of fat globules inducing aggregation and partial coalescence at the interface to create a continuous enveloping matrix to stabilize the foam (Anderson *et al.* 2005; Goff 1997). The aqueous phase becomes trapped between the fat globules by capillary action. This process can only occur if there has been sufficient cooling of the cream to achieve a suitable solid to liquid fat ratio. Otherwise, the whipping process would cause a release of excess butter oil, which would spread across the interface leading to eventual bubble collapse (Prins 1986; Brooker 1993; Holtrum 2004).

9.1.3 Cooling and Crystallization

Cooling and crystallization of fat within the MFGM occurs differently than in bulk oil environments (Herrera and Hartel 2000; Fredrick *et al.* 2011; Pérez-Martínez *et al.* 2012). Within the MFGM is an environment thought to be free from impurities, which may otherwise trigger early seed crystal formation, meaning that the crystallization behaviour is affected by the triacylglycerol (TAG) composition alone (Lopez *et al.* 2001). This can generally be split into three fractions: a low melting fraction (LMF) melting in the range of −25 °C to 10 °C, a medium melting fraction (MMF) from 10 °C to 19 °C and a high melting fraction (HMF) above 20 °C (Deffense 1993). As the cream is chilled, supercooling occurs within each range, and nucleation of seed fat crystals begins. Slow cooling rates favour the formation of a small number of larger crystals with a wide size distribution (Lopez *et al.* 2001; Rønholt *et al.* 2012). Before the crystals aggregate, they have time to become organized into a layered structure creating an internal crystalline shell that gives the globule stability. These large crystals have the potential to grow during ageing and can later pierce the MFGM when subjected to shear forces during churning (Rønholt *et al.* 2014a). Fast cooling rates, on the other hand, are shown to induce rapid nucleation of a large number of small uniform crystals, which aggregate more quickly, forming bulges in the outer surface of the MFGM (Moens *et al.* 2019). These bulges are more susceptible to rupture on collision resulting in greater partial coalescence with other globules.

Fat crystallizes in one of three forms: α, β′ or β, in increasing order of stability (Chapman 1962). Initial crystallization in the unstable α-form occurs in cream <20 °C and transitions to the desirable β′-form during ageing (Van Aken and Visser 2000; Fredrick *et al.* 2011). After cooling, cream is held at cool temperatures (typically 10–12.8 °C) to continue the crystallization of the fat for a period of 12–15 h. Lopez *et al.* (2002), using X-ray diffraction, noted that cooling cream yielded a triple layer α-phase, which transforms in the following hours into double and triple layer β′-forms. The temperature and composition of the TAGs within the globule determine the amount of fat that converts to a crystalline state or solid fat content (SFC).

9.1.4 Tempering of Milk Fat

The balance of unsaturated (UFA) and saturated fatty acids (SFA) in milk throughout the year affects the SFC for buttermaking (Phelan *et al.* 1982; Cullinane *et al.* 1984b; Couvreur *et al.* 2006). These variations are mainly attributed to changing feeding systems (grass vs mixed ration) during a season. Without modification of the buttermaking process, changes to this balance would result hard and brittle butter in the winter and soft and oily butter in the summer. To counteract these variations in preparation for churning, the pasteurized cream can be tempered through a thermal programme to manipulate size and distribution of the solid and liquid phases. "Hard" winter cream requires cooling to 8 °C to induce as few crystals as possible, before heating to 20 °C to melt the bulk oil phase leaving only the hard-fat crystals (Szakaly and Schaffer 1988; Bylund 1995). This cream is cooled back to 16 °C where any crystallizing fat will adhere to the existing crystals. This yields a higher volume of free liquid oil to soften the resultant butter. In summer, the Alnarp "6-12-6" (cold-warm-cold) method or variations can be used to crystallize a greater proportion of the total fat content. Rapid cooling leads to many small crystals and a larger crystal surface area. As a result, more liquid fat will be adsorbed to the crystal surfaces leaving less liquid fat available to form the continuous oil phase during churning and working, and a firmer butter results. Care should be taken to ensure that sufficient liquid oil remains to act as bridges between the solid fat crystals to create butter grains (Rønholt *et al.* 2014c; Lee and Martini 2018).

9.1.5 Churning and Working the Butter Grains

As churning continues, the MFGM in contact with the air interface ruptures, and the core liquid fat spreads over the surface (Evers 2004). The continued high shear forces of churning increase the air volume making the fat lamellae between bubbles progressively thinner. The fat globules concentrated in the serum phase form larger aggregates, leading to distortion and penetration of the MFGM by solid fat crystals, coalescence and subsequent leakage of internal liquid oils (Schmidt and van Hooydonk 1980; Brooker 1986). The stability of the crystals may be influenced by the ageing temperature, or starting temperature, which helps them to withstand the effect of shear and churning time (Rønholt *et al.* 2012). The spread of liquid fat collapses the foam and triggers an inversion to a water-in-oil emulsion and the formation of butter grains, with the consequent release of the bulk water volume as buttermilk containing MFGM fragments.

The butter grain mass is worked gently to disperse entrapped water (and salt, if added) in the continuous phase and serves to release additional fat crystals and liquid oil from the remaining fat globules. The deformation of the worked butter causes stress in the moisture droplets, disrupting larger droplets and resulting in discrete water droplets of 5–10 μm (Walstra *et al.* 2006). Water droplets are prevented from recoalescing by the continuous fat lamella and are too small to support microbial growth.

9.1.6 Changes During Storage

After production, butter is stored to allow it to set but is often held for prolonged periods in either a refrigerated or frozen state for commercial reasons. During the initial period, more of the liquid fat becomes integrated into the solid fat matrix, and the SFC increases. Within the first day of storage, Rønholt *et al.* (2014b) found that all α-crystals had transformed to β′-form. However, no further increase was noted in the elastic modulus (G′) or hardness. Similar findings were observed by Méndez-Cid *et al.* (2017) for butter stored for 9 months at 4 °C and 12 °C.

Milk possesses a range of antioxidant components, including the sulphur-containing amino acid cysteine, carotenoids, vitamins, metals (selenium and zinc) and antioxidant enzyme systems, such as superoxide dismutase, catalase and glutathione peroxidase (Khan *et al.* 2019). β-Carotene comprises 90% of the carotenoids in cows' milk (Ollilainen *et al.* 1989, Hulshof *et al.* 2006). It is acquired in the diet and stored in the fat globule core (Jensen and Nielsen 1996), shielded by the intact MFGM, but contributes to the rich colour of butter and cheese when exposed. It is a strong quenching antioxidant, binding light and acting as a scavenger of free radicals, such as singlet and triplet oxygen (Foote 1976; Min and Boff 2002). β-Carotene is partially metabolized by the cow to form retinol (vitamin A_1) but is entirely converted in buffalo, goat, sheep milk resulting in white cheese and butter products from these animals.

Méndez-Cid *et al.* (2017) recorded fat oxidation in samples held at 4 °C and 12 °C, with and without salt, rising from 0.39 to 0.90 and 1.36 meq O_2/kg fat for unsalted butter, respectively, and from 0.38 to 1.72 and 3.75 meq O_2/kg fat, respectively, for salted butter. Salt, which is added to butter as a preservative and for flavour, has been identified as a pro-oxidant in food systems with the chloride ion being the active agent possibly through the inhibition of antioxidant enzymes and increasing catalytic activity of metals (Osinchak *et al.* 1992; Cui *et al.* 2018). Salted butter stored at 5 °C and -20 °C over 18 months showed a similar oxidative pattern (Krause *et al.* 2008), with a stale off-flavours developing in the refrigerated samples within 6–9 months. Off-flavours were not evident in the frozen samples until 12–18 months.

Rancid off-flavours in butter are caused by free fatty acids (FFA) released from TAGs by the action of lipases. Low levels of short-chain FAA may be perceived as pleasant, but when storage conditions permit the release of high concentrations of stronger off-flavours result causing a downgrading of the butter (McNeill *et al.* 1986). Indigenous milk lipoprotein lipase is associated with the casein micelle (Deeth 2006) and therefore absent from the lipid fraction. In addition, this enzyme is heat labile and almost totally inactivated by pasteurization. However, heat-resistant psychrotrophic bacterial lipases can remain active (Woo and Lindsay 1984). Psychrophilic bacteria, whose growth is favoured by extended storage at low temperature, can produce heat-stable lipases. Their growth is estimated to be low due to the dispersed and isolated nature of minute water droplets, despite the relatively high moisture content and water activity (a_w) (Voysey *et al.* 2009). Production variation can result in a coarse dispersion of moisture and, combined with surface contamination, may allow the post-production growth of *Pseudomonas* spp. and moulds causing hydrolytic rancidity and surface taint (Jay 2000; Suryavanshi and Ghosh 2010).

9.1.7 Rheological Properties

Texture is one of the main factors that determines the acceptance of butter as a product. It influences butter's appearance, spreadability, mouthfeel, taste and its suitability for use in a wide range of applications. The composition and structure of the milk fat are mainly responsible for the rheological properties of butter.

The spreadability and melting characteristics of milk fat and butter are complex. Milk fat contains thousands of TAGs with different fatty acid combinations, each with their own melting point within the range of -40 to 40 °C. Within the butter, this balance of aggregated solid fat surrounded by liquid fat gives it a viscous nature. On testing, butter displays non-Newtonian behaviour and behaves as a plastic with yield values above which the butter will flow/spread (DeMan and Beers 1987). Generally, a butter with a SFC of between 20 and 40% is expected to be spreadable. It should be noted that a minimum of 9% SFC is required to prevent movement and coalescing of moisture within butter (Rousseau *et al.* 2003). For use as a "roll-in" laminate layer for Danish pastry and croissants, a SFC of 10–40% over a temperature range of 33.3–10.0 °C is desired to allow the butter to spread evenly without tearing the dough (Baldwin *et al.* 1971). Rolling of the dough to create 15 layers is performed in a three-step process with intervals of cooling at 2–4 °C to retard softening. SFC content also acts to soften bread and protect baked products against staling; saturated TAGs complex with amylopectin to prevent starch recrystallization and moisture migration (Smith and Johansson 2003).

The temperature at which butter reaches critical SFC levels is, of course, dependent on its milk fat composition, which is largely determined by the feeding system and diet of the cow, combined with age, breed, species, genotype, pregnancy and stage of lactation (Chilliard *et al.* 2007). The type of feed system is dictated by the geographic and cultural factors such as climate, land availability, feedstock and needs of the animal. In Ireland and New Zealand, the alternation between winter-indoor silage/concentrate and summer-outdoor fresh grass grazing (Cullinane *et al.* 1984a, Auldist *et al.* 1998) has a significant effect on the fatty acid compositions. Unsaturated fatty acid (UFA) levels peak in summer (April–June in the northern hemisphere) mainly due to increases in oleic (18:1), linoleic (18:0), γ-linoleic (18:3) and palmitoleic (C16:1) acids and a decrease in palmitic acid (C16:0) (Cullinane *et al.* 1984a). The resultant summer butter was found to be softer (137 kPa at 4 °C, 37 kPa at 15 °C) compared with the firmer winter (December–February) butter (412 kPa at 4 °C, 137 kPa at 15 °C) (Cullinane *et al.* 1984b). Butter is deemed to have acceptable spreadability in the range of 30–60 kPa (Rohm and Raaber 1991). Couvreur *et al.* (2006) and O'Callaghan *et al.* (2016) confirmed the benefit of feeding proportions of fresh pasture in place of maize silage with a similar correlation of a decrease in the C:16:0/C18:1 ratio with increasing grass supplementation, improved butter texture and a decrease in off-odours. The butter from pasture-fed cows also gave significantly lower atherogenic (AI)[1] and thrombogenic index (TI)[2] scores showing an added health benefit.

9.1.8 Flavour and Aroma

Butter should have a mild, sweet, clean and pleasant flavour and delicate aroma making it well suited as a food ingredient (Bradley and Smukowski 2009). The profile of butter includes a fine balance of short-chain fatty acids, such as butanoic acid (buttery, cheesy, sweaty), γ- and δ-lactones (sweet, peach, coconut), indole and skatole (mothball/faecal), phenols and hydrocarbons with a phytoene skeleton (Kato 2003, Mallia *et al.* 2008). Milk fat can act as an excellent solvent for flavours and as a carrier in cooking in a multitude of ways. Controlled fermentation with bacteria introduces a lactic acid taste and aromas of diacetyl, butanoic acid and δ-decalactone to create the distinctive lactic or cultured cream.

On heating butter and flour (white *roux*) to 120 °C, Kato (2003) detected an overlay of a slightly roasted aroma of aldehydes, carboxylic acids and lactones. At 160–180 °C (brown *roux*), flavour develops further and is strongly influenced by the Amadori rearrangement associated with Maillard browning of fats, sugars and proteins: furans, cyclic ketoenols (sweet, fruity, floral) and nitrogen-containing pyrazines (nutty, roasted) (Shigematsu *et al.* 1977, Maga and Katz 2009).

---

[1] *Atherogenic index: the sum of the proportion of lauric and palmitic and four times the myristic levels, divided by the total concentration of saturated fatty acids*

[2] *Thrombogenic index: an index of the tendency to form blood clots in blood vessels*

Unfortunately, butter is also susceptible to acquiring off-flavours and aromas. Consumption of highly aromatic weeds or flowers, or exposure to strong environmental odours in air, up to 30 min to 24 h prior to milking can be detected in the final butter due to direct transfer via the rumen, breakdown products or changes in the biochemistry of the cow (Urbach 1990, Viallon *et al.* 2000). During the processing of the milk, care is taken to prevent exposure to cleaning chemicals, foul water or any other compound that could taint the flavour. Some consideration should be given when selecting a milk supply for delicate products; high-fat milk products or powders made from milk from pasture are less susceptible to oxidation than from dry feed, while milk with higher levels of methyl ketones and free fatty acids may benefit butter shortening in baking or certain cheese types (Urbach 1990, 1997).

Once produced, it is essential to minimize the exposure of butter to light and oxygen (Koyuncu and Tuncturk 2017). Dairy products are sensitive to light due to the photosensitizer riboflavin (vitamin B_2), which under the influence of light produces methional (musty, potato) in a reaction with methionine to create a sunlight (burnt, oxidized) aroma (Patton 1954). Small amounts of riboflavin are retained in the aqueous phase of butter: 0.34 mg/kg of a total of 60–63.4 mg/kg in milk (USDA 2018). Foil laminates offer the greatest protection and are far superior to parchment or transparent options (Emmons *et al.* 1986). While this is the standard form in Ireland, novel consumer-driven options include parchment only or butter sculptures in clear packaging. Lozano *et al.* (2007) confirmed foil laminates as the best method of preventing styrene and aroma migration into butter during long-term storage. Buttermaking, both artisanal and industrial, has the ability to produce consistent and highly functional products to meet consumer needs and desires. Knowledge of the important role of milk fat composition and behaviour is essential in understanding the complexities of its correct use.

9.2 Creams

W. Hoffmann (✉)
Department of Safety and Quality of Milk and Fish Products
Max Rubner-Institut
Kiel, Germany

W. Buchheim
Kiel, Germany
e-mail: wo-hoffmann@kabelmail.de

9.2.1 Introduction

Cream is a fluid milk product comparatively rich in fat, in the form of a fat-in-skimmed milk emulsion, obtained by physical separation from milk (Codex Alimentarius Commission 2018). This simple definition does not reflect that the word "cream" has for a long time been considered a premium product or a value-enhancing ingredient in milk products and other foods. The special "creaminess" results from the fine dispersion of the fat globules in the hydrophilic phase and depends strongly on the fat content. In separated cream, the diameter of fat globules ranges between ca. 1 and 8 μm. During further processing to the different cream products, this typical oil-in-water (o/w) emulsion is modified or even converted into another physical state. Modification can be achieved by homogenization, which markedly reduces the average fat globule size and improves creaminess. On the other hand, mechanical treatment of chilled cream causes destabilization (i.e. coalescence of the fat globules). This treatment and the concurrent entrapment of air are essential for whipping cream into a stable foam.

The fat content of cream products varies from about 10 to 50%. Products with a low, internationally not-yet standardized, fat content are "coffee cream" (≥10% fat, Germany), "half-and-half cream" (≥10.5% fat, USA), "half cream" (≥12% fat, UK) or "light cream" (≥12% fat,

France). Traditional whipping cream has 30 to 40% fat, whereas double cream contains about 50% fat. Creams of high fat content are also essential ingredients in dairy or non-dairy products such as some fresh cheese varieties or cream liqueurs. Cream powders refer to a group of milk powder products containing 40–75% fat. These products are used as ingredients for bakery products, chocolate, ice cream or sweet desserts.

Butter is manufactured from cream (30–80% fat) by phase inversion. Reviews on cream, cream processing and cream products have been published by Towler (1994), Early (1998), Kessler (2002), Smiddy *et al.* (2009) and Hoffmann (2015a, b). Two older IDF Bulletins (IDF 1992, 1996) dealt with pasteurized and Ultra High Temperature (UHT) creams.

In summary, the significance of milk fat in the different cream products is based on fat content, fat distribution, the physical state of the fat and, last but not least, the chemical, physical and sensory properties of the non-fat ingredients. In the following, interactions between these factors are described for the most important cream products.

9.2.2 Coffee Cream

In many countries, coffee cream is still a popular long-life product. However, foamed milk in coffee specialities such as "cappuccino" or "café latte" offered in coffee bars or brewed in automatic coffee machines at work or at home increasingly competes with coffee cream. In this section, "coffee cream" does not mean a national statutory term but simply an appropriate description of the functional properties. Such creams contain usually 10 or 12% fat, less frequently 15 or 18%. Traditionally, coffee creams are sterilized in bottles or cans. During the last 30 years, continuous flow sterilization in a UHT plant, followed by aseptic packaging, has replaced the former process to a large extent. The products need good stability both during storage and in hot coffee beverages. A shelf-life of several months at ambient temperature requires particularly low creaming and sedimentation in the package, which is facilitated by a lower fat content (10 or 12%) and optimized processing conditions, mainly heat treatment and homogenization. Scientific bases for the manufacture of such products were published by Buchheim *et al.* (1986), Abrahamsson *et al.* (1988) and Geyer and Kessler (1989). Summaries were given by Hoffmann (2004, 2015a, b).

The different creams may contain stabilizing salts, which can be added as an aqueous solution after standardization and preheating (high-temperature pasteurization at 90–95 °C). They raise the pH and/or complex Ca^{2+}, resulting in reduced aggregation of casein micelles during sterilization and in hot coffee beverages. Sodium phosphates have a reduced buffering capacity and increased ion exchange ability with an increasing degree of condensation (chain length). Trisodium citrate has both buffering and sequestering properties and is used also. Whereas phosphates and citrates are essential additives in traditionally sterilized cream, high-quality flow-sterilized creams (containing 10 or 12% fat) may be produced without additives or hydrocolloids.

Homogenization of cream results in the formation of a secondary fat globule membrane, consisting predominantly of micellar casein and denatured whey protein (Walstra *et al.* 1999). To obtain desirable product properties, the formation of larger, thermally induced protein aggregates and, particularly, fat/protein complexes must be avoided (Buchheim *et al.* 1986). The number and dimensions of the particles are influenced more by temperature than by heating time during flow sterilization. In general, such adverse structures are reduced by flow sterilization at ≤130 °C rather than at UHT temperatures (≥135 °C). Frequently, a second two-stage homogenization step is performed after heating in order to disrupt heat-induced fat/protein aggregates. Sensory effects of a lower heating temperature and a necessary prolonged heating time (≥1 min) are a more pronounced cooked flavour and a more brownish colour of the resulting cream. However, these effects are of minor relevance in the coffee beverage.

Physical properties of flow-sterilized cream can be controlled by homogenizing conditions. Usually, one homogenizer is integrated upstream (i.e. before flow sterilization) and one downstream. Each of the two homogenizers often

Figure 9.1. Electron micrograph of flow-sterilized coffee cream; f: homogenized fat globules.

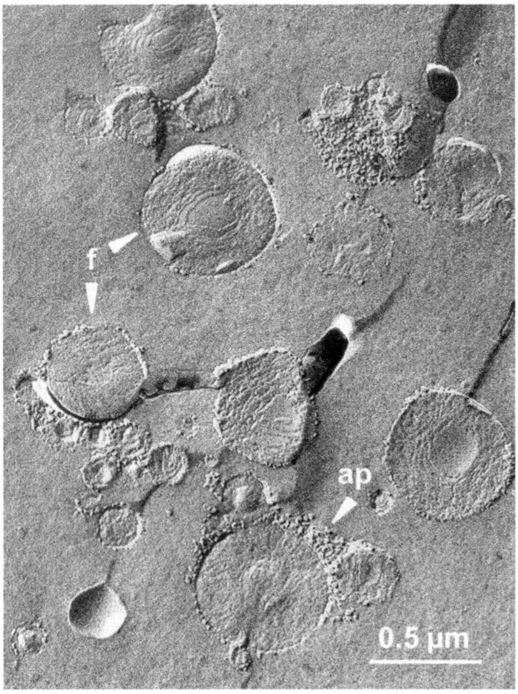

Figure 9.2. Electron micrograph of floccules in a coffee cream after feathering in a hot coffee solution (enlarged compared to Figure 9.1); f, fat globules; ap, aggregated protein.

operates at a total pressure (single- or double-stage) of about 20 MPa at 70 °C. A good coffee cream should have a narrow fat globule size distribution with a volume-mean diameter preferably between 0.4 and 0.6 µm and a very low degree of aggregation (Figure 9.1). This results in a product with low viscosity, high whitening power, slow creaming and high "coffee stability", i.e. resistance against feathering in hot coffee beverages.

The coffee stability is particularly important for the quality of the product. It is affected by the coffee brand and concentration, while high temperature (≥70 °C), low water hardness, low pH (about 5.0) and a high concentration of sulphates accelerate protein coagulation and, hence, fat/protein aggregation. Therefore, it must be ensured that this feathering remains invisible to the naked eye (Buchheim *et al.* 1986; Hoffmann *et al.* 1996) (Figure 9.2). The probability of feathering increases with the fat content of the cream. Coffee cream in small deep-drawn polystyrene (PS) cups lose 10–15% of their weight during their shelf-life of about 4 months which can result in floccules of condensed cream droplets.

9.2.3 Whipping Cream

Whereas the processing of long-life coffee cream is characterized by high-pressure homogenization and severe heat treatment, traditionally pasteurized whipping cream is produced carefully with less thermal input (ca. 85 °C for 10 s) and without homogenization (Figure 9.3). However, the demand for a longer shelf-life has led to a subsequent high-temperature pasteurization (≥110 °C) or even UHT heating with additional low-pressure homogenization. The higher thermal load results in more cooked flavour. The aim of UHT treatment is to produce sterile cream with a shelf-life of up to 3 months without refrigerated storage. Usually, indirect heating at ≥135 °C for a few seconds is applied in order to limit thermally induced physical, chemical and sensory changes. The homogenization effect

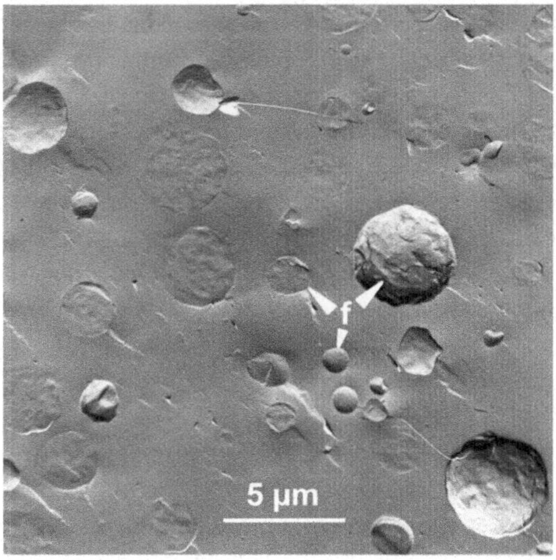

Figure 9.3. Electron micrograph of fluid whipping cream; f: fat globules.

Figure 9.4. Electron micrograph of whipped cream; a, air cell; f, fat globules; I, interfacial layer.

must be moderate in order to retain acceptable whipping properties. Therefore, a downstream two-stage homogenization of cream at a total pressure of not more than 4 MPa is used frequently. Milk fat globule size of unhomogenized and slightly homogenized cream has an influence of whipping properties (Eden *et al.* 2016).

To improve the whipping properties of cream with a longer shelf-life, both stabilizers and emulsifiers may be added, if legally permitted. In particular, emulsifiers such as monoacylglycerols reduce the extended whipping time of homogenized cream (Anderson and Brooker 1988; Smiddy *et al.* 2009). The addition of stabilizers, most commonly hydrocolloids such as carrageenans (Kovácová *et al.* 2010), also in combination with milk constituents (whey proteins and high-melting fat fractions; see Precht *et al.* 1988) can slow down creaming during the shelf-life of cream and improve the stiffness and stability of whipped cream. Hydrocolloids exert their positive effect by increasing the viscosity of the milk serum phase (Smiddy *et al.* 2009).

A sufficient cooling of cream containing ≥30% fat is indispensable before whipping since the transformation of the original oil-in-water (o/w) emulsion into a stable foam requires that part of the fat is solid. The initial stage of whipping involves stabilization of the trapped air bubbles by a temporary interfacial film, largely of soluble whey proteins and non-micellar β-casein. During the second stage of whipping, the bubbles are reduced in size, and the overrun remains nearly constant. On mechanical treatment, fat globules increasingly lose at least segments of their natural membrane, thereby exposing strongly hydrophobic surface areas of pure fat. Subsequently, these partly destabilized fat globules adsorb at the air/serum interface of the air bubbles (Figure 9.4). The leakage of liquid fat from mechanically stressed and deformed fat globules supports globule agglomeration and partial coalescence. These agglomerates also interact with the air bubbles and may form bridges between them. The whole process of foam formation results in a partly coalesced fat globule network, which stabilizes the air cells, traps the serum phase and forms the characteristic stiff texture. These highly dynamic and concurrent processes also apply on the whole to UHT whipping cream subjected to low-pressure homogenization. Details of the interactions and processes during whipping are described by Anderson *et al.* (2005), Anderson and Brooker (1988), Buchheim (1991), Buchheim and Dejmek (1997), Smith *et al.* (2000), van Aken (2001) and Goff and Vega (2007). Han *et al.* (2018) presented confocal scanning laser micrographs; the whole whipping process could be depicted and distinguished visually by different colours for fat and protein.

The whipping properties of creams are assessed by whipping time, increase in volume (expressed as overrun), foam firmness and subsequent serum leakage. Comparative studies require standardized temperature and procedure. Most test whipping devices are modifications of that described originally in 1937 by Mohr and Baur (see Hoffmann 2015b). Rheological properties of whipped cream can also be determined using a controlled-stress rheometer with parallel-plate geometry at 10 °C. The low temperature causes the fat to congeal on the plate, and a torque sweep is performed to establish the linear visco-elastic region. The subsequent frequency sweep uses the torque value from the middle of the linear region. The elastic modulus G', the viscous modulus G" and the resulting tan δ as a function of frequency are determined (Smith *et al.* 2000; Jacubczyk and Niranjan 2006).

Whipping of a typical cream increases the volume by 80–125% by inclusion of ambient air. UHT-treated creams can also be aerated by means of suitable propellants (e.g. N_2O), resulting in a volume increase in the range of about 300–600%. Such convenience products contain emulsifiers, stabilizers and usually also sugar and flavourings. These products are aseptically filled into sterilized aluminium or tin-plate cans. An obvious advantage of aerosol whipped cream is the speed and ease of foam production in controllable portions. The amount of gas dissolved in the cream must be sufficiently high to obtain satisfactory foaming properties, but is not allowed to exceed 1.5–2.0 MPa in the cans, particularly for household use, to avoid the risk of explosion. Compared with regular whipped cream, more fat globules adsorb at the air interfaces, and, simultaneously, agglomeration of fat globules is reduced substantially which results in an impaired network formation between the air bubbles. Due to the different foaming process, aerosol creams develop only soft foams with low stability (Buchheim 1991; Wijnen 1997; Smiddy *et al.* 2009).

Whipping cream ranks among premium food products and is consumed for its pure flavour. High-quality raw milk and separated cream must be handled carefully to minimize damage to the natural fat globule membrane. Excessive agitation and pumping should be avoided, and the flow velocity should not exceed the critical shear rate (Kessler 2002). Incorporated air bubbles increase the risk of damaged fat globules or can act as centres for fat globule aggregation and subsequent coalescence. During crystallization, fat globules are most sensitive to mechanical treatment. As a result of the partial or complete loss of the protective membrane, both indigenous and bacterial lipases catalyse the hydrolysis of exposed fat to fatty acids, imparting rancid taints (Kosinski 1996). When raw cream is homogenized without being subjected immediately to high-temperature pasteurization, indigenous milk lipoprotein lipase penetrates the secondary membrane of fat globules and hydrolyses triglycerides to free fatty acids within a few minutes, resulting in intense rancidity (Walstra *et al.* 1999).

Active extracellular bacterial lipases and proteinases of *Pseudomonas* spp. and most other Gram-negative psychrotrophs may be present, even in UHT cream, if refrigerated raw milk was stored for a prolonged period. They can contribute to rancid and tallowy flavours and also to bitty cream or serious physical changes such as gelation (Castberg 1992; Driessen and van den Berg; 1992; Houlihan 1992; Kosinski 1996).

Flavour defects in cream may occur not only during manufacture but also during transport or storage until the best-before date. UHT whipping cream with its long shelf-life at ambient temperature is particularly susceptible to off-flavours. Hence, adequate packaging materials must be chosen. Protection against oxygen and/or light is most important as they may induce oxidation of unsaturated fatty acids, leading to flavour deterioration. Stress history relating to light exposure during processing, packaging or retail display of cream increases the tendency of oxidation also during subsequent dark storage. Riboflavin acts as a photosensitizer and may initiate generation of radicals or other reactive species resulting in formation of lipid and protein oxidation products (Westermann *et al.* 2009). Paper cartons with a coating of polyethylene and an aluminium foil laminated to the inner carton layer are often used. Appropriate filling conditions should also be selected to minimize the oxygen content of the package and the cream. However, a certain level of residual oxygen may be beneficial as the UHT process exposes free sulphydryl groups and releases hydrogen sulphide from β-lactoglobulin, thus creating the typical cooked flavour. During storage, oxidation of these groups occurs, and most of the cooked flavour disappears. A balanced antioxidative/oxidative action of sulphur groups and oxygen will probably help to ensure cream products of good taste and odour (Eyer *et al.* 1996).

An important factor for physical stability of cream is the temperature of the cream during transport and storage. Even a brief warming to ≥30 °C supports creaming during subsequent storage at 20 °C and may lead to a distinct thickening after cooling before whipping (Hoffmann 1999).

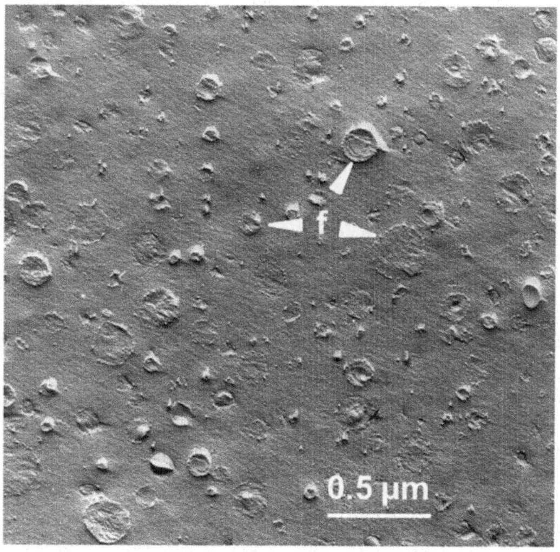

Figure 9.5. Electron micrograph of cream liqueur; f: homogenized fat globules.

9.2.4 Cream Liqueurs

Cream liqueurs combine the flavour of alcoholic drinks with the texture of cream in products. Cream liqueurs have a shelf-life of several years at ambient temperature. During that period, the liqueur must be resistant to both microbiological and physical changes. The microbiological safety is guaranteed by a sufficient concentration of alcohol (≥14%) together with a high sugar content (about 19%). Avoiding serious phase separation is the more demanding challenge. This can be achieved by optimal composition and processing. The addition of sodium caseinate (ca. 3%), trisodium citrate (ca. 0.2%) and possibly low-molecular-weight emulsifiers like monoacylglycerols (ca. 0.1%) stabilizes the o/w emulsion of the added cream (e.g. 16% of 48% fat cream) in the liqueur. This composition appears to represent many commercially available products. However, each manufacturer decides on the preferred components, and concentrations thereof, which enable a product of desired organoleptic properties (Smiddy *et al.* 2009; Hoffmann 2015a).

In the final liqueur product, more than 98% of the fat globules should have a diameter of <0.8 µm, resulting in enhanced viscosity, creaminess and whitening power (Banks and Muir 1988). The typical volume-mean diameter of about 0.2 µm is, by far, the smallest of all dairy products (Buchheim and Dejmek 1997) (Figure 9.5). This is achieved by multiple passes through a standard radial diffuser homogenizer or by a single pass through a high-pressure homogenizer (HPH) at 50–150 MPa. HPH may allow the use of lower pre-emulsion feed temperatures than conventional homogenization processes and has potential as a means of improving the shelf-life stability of cream liqueurs. However, too severe homogenization results in overprocessing of the products (Heffernan *et al.* 2009 and 2011).

Compared to unhomogenized cream, the total fat surface area increases by a factor of about 20 (up to ca. 40 m^2/g fat) in cream liqueurs. Therefore, sodium caseinate is dissolved first in hot water before adding the cream, sugar, citrate and a complementary emulsifier (if necessary). No other protein than sodium caseinate is able to provide the required long-term emulsion stability (Lynch and Mulvihill 1997). α_{S1}-Casein seems to be the fraction of sodium caseinate that is most soluble in ethanol (O'Kennedy *et al.* 2001; Mezdour *et al.* 2008). Trisodium citrate, a useful stabilizer for several dairy products, such as evaporated milk or sterilized coffee cream, complexes the Ca^{2+} and concurrently increases the pH. In cream liqueurs, trisodium citrate prevents the interaction between sodium caseinate and available calcium. Otherwise, gelation and syneresis during storage would occur. If a cream liqueur with a substantially higher alcohol content than 14% is produced (e.g. 19%), addition of further alcohol after homogenization of cream (and other ingredients) is required in order to produce a stable emulsion. The manufacture of cream liqueurs ends with filling into brown glass bottles to prevent light-induced off-flavour. Very occasionally, during long-term storage, the formation of a non-redispersible cream or fat plug in the neck of the bottle may occur (Dickinson *et al.* 1989). The fatty solid-like cohesive structure of this plug points to unfavourable ambient temperatures, possibly accompanied by excessive mechanical agitation. The formation of neck-plug may be similar in origin to the thickening of whipping cream after warming for a short period (≥30 °C) and subsequent cooling.

9.2.5 Cultured Cream

Cultured or sour(ed) creams find various applications as valuable ingredients. They are used in cake mixes, as dip bases for snacks and vegetables and to complement sauces and dressings. Cultured creams are manufactured in many countries, and their fat content generally ranges from 10 to more than 40%. According to Codex Alimentarius Standards (Codex Commission 2018), fermented creams are soured by the action of suitable microorganisms, normally mesophilic lactic acid bacteria, to a pH of about 4.5. Cultures commonly used include *Lactococcus (Lc.) lactis* subsp. *lactis*, *Lc. lactis* subsp. *cremoris*, citrate-positive strains of *Lc. lactis* and *Leuconostoc mesenteroides* subsp. *cremoris*, with *Lactobacillus acidophilus* also being increasingly used (Smiddy *et al.* 2009). Acidified creams are obtained by the action of food-grade acids and/or acidity regulators, often by a combination of lactic and citric acid, rarely glucono-δ-lactone. Lactic acid has bacteriostatic effects, while citric acid can be fermented by many organisms (Born 2013). Lactic acid-producing bacteria may also be added (e.g. acidified sour cream with ≥18% milk fat in the USA; Code of Federal Regulations Title 21, Part 131). The production of cultured cream is largely equivalent to that of other fermented milk products. It starts with the standardization of the fat content and may include enrichment of non-fat milk dry matter and hydrocolloids, if legally permitted. These ingredients improve texture and prevent syneresis of the final product. Adequate processing conditions and a higher fat content reduce the need for supplementation. The homogenization pressure required for cream decreases with increasing fat content. Homogenization after high-temperature pasteurization results in better consistency compared to upstream treatment. The fat globules participate directly in the following fermentation process and are integrated in the developed network (Buchheim and Dejmek 1997) (Figure 9.6). Normally, the use of mesophilic lactic acid bacteria results in long fermentation times (14–24 h). Typical cultured cream products should be uniform (without creaming), creamy and viscous with a slightly acidic, mild “cheesy” or “buttery” flavour. Advantages of direct acidification are elimination of culture-handling problems, improved production efficiency and extended shelf-life. Such creams have a similar appearance and texture as those of cultured sour creams, but the latter have a superior flavour (Smiddy *et al.* 2009). Cultured creams may also develop a nearly plastic consistency by modified composition and/or appropriate production and may then be used as low-fat spreads (o/w type).

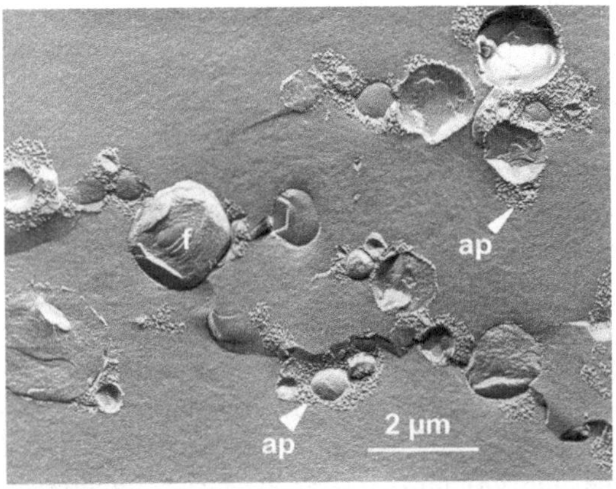

Figure 9.6. Electron micrograph of cultured cream (10% fat); f, fat globules; ap, aggregated protein.

9.2.6 Recombined Cream

These products are obtained by recombining milk products with or without the addition of potable water resulting in end product characteristics similar to those of natural cream (Codex Alimentarius Commission 2018). Over the last two decades, recombined creams have received increasing interest because of their obvious advantages in industrial production compared to fresh cream (van Lent *et al.* 2008). Unlike natural cream, the raw materials can easily be stored and transported to regions where fresh milk is not readily available and/or where suitable storage facilities are scarce. Moreover, the composition and manufacturing conditions of recombined creams can be modified for product development goals with different functional properties. Different dairy products have been used in studies over the years to emulsify anhydrous milk fat into recombined cream. Using cream residue powder instead of skim milk powder as protein source, such stabilized recombined creams mimicked fresh cream best as regards both emulsion and whipping properties (van Lent *et al.* 2008). An addition of different monoacylglycerols with fatty acids differing in chain length and/or saturation degree may influence the microstructural arrangement of the milk fat inside the fat globules and hence the whipping properties of the recombined cream (Frederick *et al.* 2013a, b). Milk fat globule membrane material from different sources has also the potential to improve whipping properties (Phan *et al.* 2014).

9.3 Role of Milk Fat in Dairy Products: Cheese

D. J. McMahon
Department of Nutrition, Dietetics and Food Sciences
Utah State University
Logan, UT, USA

Milk fat in cheese contributes to its flavour, texture, colour and functional behaviour. Its importance becomes apparent in any attempt at making a lower fat version of a cheese. Fat has multiple effects in cheese, including indirect effects on metabolic functioning of the microbiota during cheese ageing. This section will focus on the role of fat in *Cheddar*-like cheeses and *pasta filata* cheeses as these have been best studied.

9.3.1 Milk Fat and Cheese Flavour

Cheeses can be divided into two categories: those in which free fatty acids play a dominant role in their flavour profile and those in which excessive production of free fatty acids is considered a flavour defect. Fatty acids are released from milk fat by enzyme-catalysed lipolysis (1) through addition of lipases usually via rennet paste or pregastric esterases or (2) from production of lipases by cheese microbiota during ripening. In such cases, the contribution of fat to the flavour profile of cheese depends on fatty acid composition of the milk and activity and specificity of the lipase enzymes. The shorter-chain fatty acids such as butyric, capric, caproic and caprylic are most volatile and impact flavour the most. Their relative content in cow, sheep and goat's milk imparts slightly different flavour profiles to cheeses made from those milks. This field of lipolysis in cheese has been reviewed by Collins *et al.* (2003), and the remainder of this section will discuss the role of fat on cheeses made without exogenous lipases.

A common problem when fat content of cheese is lowered is that flavour changes and consumer liking decreases (Childs and Drake 2009). There is an obvious decrease in dairy or milk fat flavour (Drake *et al.* 2010), yet development of adverse flavours is not directly related to fat content. Carunchia Whetstine *et al.* (2006) found that removal of fat after ageing did not change cheese flavour profile or levels of volatile and aroma-active compounds. Flavour differences when lowering fat content of cheese are caused by altered biochemistry and a different array or balance of volatile compounds produced during ageing (Milo and Reineccius

1997) and differences in cheese texture (Drake *et al.* 2010).

Low-fat cheese contains the same key odorants as full-fat cheese but at different relative concentrations (Drake *et al.* 2010). Low-fat cheeses lack characteristic milk fat, sulphur and brothy flavours and instead are characterized by rosy and burnt flavours (reminiscent of burnt sugar). Rosy flavour was attributed to an increase in phenylethanal. The burnt flavour was attributed to an increase in furanones (especially homofuraneol) and 1-octen-3-one. Low-fat cheeses also had lower levels of methanethiol, a crucial precursor of compounds contributing to aged cheese flavour.

Bitterness observed in low-fat cheeses may be a cheese matrix effect rather than difference in levels of bitter compounds (Drake *et al.* 2010) as fat can alter sensory perception of bitter hydrophobic peptides. A slower rate of their release from the cheese mass during chewing would make them less noticeable in a full-fat cheese than in a low-fat cheese. Similarly, higher levels of fat in cheese present a more polar matrix which may influence detection thresholds of key odorants and influence perceived flavour.

The largest difference resulting in altered biochemistry during cheese ageing of lower-fat cheeses is that instead of being at 4.5 to 5.0% salt concentration, a low-fat cheese with the same overall 1.8% salt level may only have 3.5% salt in moisture. This alters survival and growth of bacteria as well as their metabolic activity (McMahon 2010; McMahon *et al.* 2014). If the salt concentration of a low-fat cheese is increased to a level comparable to that of a full-fat cheese, then apart from being overly salty and lacking in milk fat flavour, its other flavour attributes are similar to that of full-fat cheese.

9.3.2 Milk Fat and Cheese Colour

Removing fat from cheese imparts a translucent appearance and an increased intensity and atypical colour when annatto is added (Sipahioglu *et al.* 1999; Wadhwani *et al.* 2012). This has been attributed to fewer light-scattering centres (Pastorino *et al.* 2002) especially when fat is lowered by more than 50%. Other changes to cheese composition such as lowering calcium/protein ratio to soften the cheese and hold more moisture also contribute to this loss of opacity.

When desired flavour notes (such as buttery, nutty and cheese and milky attributes) are missing from a low-fat cheese, consumer attention is drawn to the cheese colour. In evaluations of low-fat cheeses made with the same flavour profile but different levels of annatto and titanium oxide to provide whiteness, the translucency of low-fat cheese was detrimental to consumer liking (Wadhwani *et al.* 2012). Even though trained panelists tasting cheese under red light perceived no differences in flavour, consumers rated the level of sharpness based upon colour and opacity of the cheese. Cheeses that were more opaque and whiter were considered to be mild to medium in flavour, while cheeses with the most annatto added were considered to have more flavour.

9.3.3 Milk Fat and Cheese Texture

Considerations on how fat influences cheese texture must take into account both the temperature and how texture is defined. In the cold, solid fat particles exert a reinforcing effect on the protein matrix within the cheese and increase its hardness when measured using small strain tests. However, when the cheese is being chewed and subject to large-scale deformations, the fat particles provide weak points within the protein network, allowing it to be fractured into micro-scale particles and formed into a smooth mass. Raising the temperature softens the fat particles in cheese, and they have less reinforcing effect (Rogers *et al.* 2010) and help in formation of a smooth mass when chewed such that individual particles are less easily detected (Rogers *et al.* 2009).

Lower-fat cheeses are reported to be perceived as being waxier, more fracturable and chewy, hard and springy, less sticky and cohesive, less meltable and less smooth than full-fat cheese (Johnson *et al.* 2009). If cheese is looked at as a material consisting of a hydrated protein matrix with interspersed fat particles, less fat results in less interruptions in the protein matrix and less interference of long-range interactions between those proteins leading to a more elastic texture. When the elastic nature of cheese is expressed as hand springiness/rate of recovery (a 30% nonfracture deformations), there are clear distinctions between full-fat, 50% reduced fat, and low-fat (90% reduced fat) *Cheddar* cheeses (Rogers *et al.* 2009). More springiness is maintained during storage in low-fat cheese than full-fat cheese. Also, as they age, full-fat cheeses can be broken down into smaller particles during chewing, becoming more cohesive and adhesive than low-fat cheese. This results in a greater smoothness of mouth coating, while low-fat cheese requires more chewing before it can be swallowed.

Sensory textural properties include mixing of the cheese with saliva during oral processing being broken down during mastication as it is chewed and lubricated by saliva. To achieve similar bolus formation, a low-fat cheese should have similar breakdown patterns and saliva interactions (Rogers *et al.* 2010). This includes warming of the cheese to body temperature as it is chewed and the consequent partially melting of the fat in cheese.

When cheeses were made with the same protein/moisture ratio and same calcium/protein ratio but differing fat contents (3% to 33%), Rogers *et al.* (2010) found no differences in rheological properties when measured at 25 °C. At 10 °C when the fat is solid, cheese rigidity increases with fat content. Cheese structure can be viewed as a continuous protein gel network disrupted with interspersed fat globules. This supports a filled gel model for fat in hard and semi-hard cheeses in which storage modulus is a combination of gel network elasticity and phase volume, filler particle elasticity and phase volume and interactions (or lack of) between the filler particle and the gel network (Sala *et al.* 2009; Rogers *et al.* 2010). Stiffer filler particles (cold fat globules) produce firmer materials by reinforcing the gel network. Filler particles with similar elasticity to the gel network (such as warm fat globules) will not show any differences as the filler particle volume increases. This filled gel model for cheese was confirmed by Barden *et al.* (2015) who made cheese in which fat globules were replaced with Sephadex beads. While the Sephadex beads had greater reinforcing effect and greater energy recovery than milk fat, there were no differences based on filler type on critical strain point (the level of strain where damage or long-term relaxations take place in the network). The only influence on critical strain was filler volume with full-fat cheeses having lower critical strain. Filler particles provide sites for stress concentration and initiate fracture, thereby lowering the strain required for initial fracture (Barden *et al.* 2015).

A low-fat cheese would have similar rheological properties to a full-fat cheese if (1) a filler particle with similar properties to fat is added at a phase volume equal to the fat phase and (2) the casein gel phase has the same chemistry as in the full-fat version. Filler particles that are rigid and do not melt will result in mechanical properties of cheese that are similar to milk fat at cold temperatures but different at room temperature. The ideal filler particle would melt or soften at body temperature, just as occurs in a full-fat cheese.

9.3.4 Milk Fat and Cheese Manufacture

During manufacture of rennet-coagulated cheeses, casein micelles aggregate and form a network structure that is driven by hydrophobic (entropic-driven) interactions. The shrinkage of the curd network structure continues throughout cheesemaking causing whey to be expelled as chains of casein micelles merge with one another forming thicker protein strands in which individual casein micelles can no longer be identified (Oberg *et al.* 1993; Merrill *et al.* 1996). Within curd particles, the fat globules act as an inert filler because of their hydrophilic fat globule membrane that is non-attractive to associations taking place between the hydrophobic proteins.

A consequence of fat globules blocking the protein strand coalescence is a small volume of serum (whey) is retained as well (McMahon and Oberg 2017). When fat content is lowered, there are fewer of these serum reservoirs. Although a low-fat cheese may have higher total moisture content, when measured in proportion to the protein (or on the basis of fat-free substance), moisture content is lowered compared to a full-fat cheese, and the make procedure has to be adjusted to counteract this effect (Merrill *et al.* 1994). Also, since some of the cream is separated, a low-fat cheese contains smaller fat globules and fewer clumps of fat globules (Rogers *et al.* 2010). Hence, there are fewer interrupting points in the protein matrix, and the ability to retain larger pools of serum within the protein is reduced.

A continuation of the influence of fat content on moisture content of cheese is observed when the cheese curd is cooked and stretched during the manufacture of *pasta filata* cheeses (Oberg *et al.* 1993; McMahon and Oberg 2011, 2017). As cheese curd is heated and reaches a temperature of 50 to 55 °C, the proteins in the cheese curds become plasticized and coalesce into larger strands that are oriented in the direction of stretching. This results in a redistribution of the water and fat during stretching and moulding with the larger strands (or fibres) of protein being separated by channels containing water, water-soluble cheese components, bacteria and fat globules (Figure 9.7) (McMahon *et al.* 1999).

The coalescence of the protein matrix strands continues until the fat globules are sufficiently packed to physically inhibit fusion of the protein matrix. Thus, closely packed clusters of fat globules and bacteria are formed and oriented into channels between the parallel-aligned network of protein fibres. The size and number of these fat-serum channels are dependent on fat content rather than moisture content; with total moisture content of the cheese being a combination of water holding capacity of the protein matrix plus water that is retained in the fat-serum channels (McMahon and Oberg 2017). With fewer fat-serum channels, a lower-fat *Mozzarella* cheese has less water that can be expressible by centrifugation (McMahon *et al.* 1999).

During storage of *Mozzarella* cheese, moisture in the fat-serum channels becomes absorbed into the protein matrix as it expands with new protein matrix material completely filling the spaces between the fat globules (Figure 9.8) (Oberg *et al.* 1993; McMahon *et al.* 1999; McMahon and Oberg 2017). So not only does fat act to interrupt fusion of the protein matrix in *Mozzarella* cheese, it also provides a space to retain moisture in addition to that contained within the protein matrix.

Another impact of having multiple fat-serum channels in *Mozzarella* cheese is that it introduces parallel weak points that allow fractures to easily propagate as the cheese is pulled apart. This is the basis for making string cheese. If fat content is lowered, there are fewer fat-serum channels, more protein-protein interactions and very little stringiness (Mulvaney *et al.* 1997). Adding xanthan gum to hot low-fat *Mozzarella* cheese before it is extruded mimics the fat-serum channels by preventing protein strand coalescence and allowing the cheese to be pulled apart into strings (Oberg *et al.* 2015).

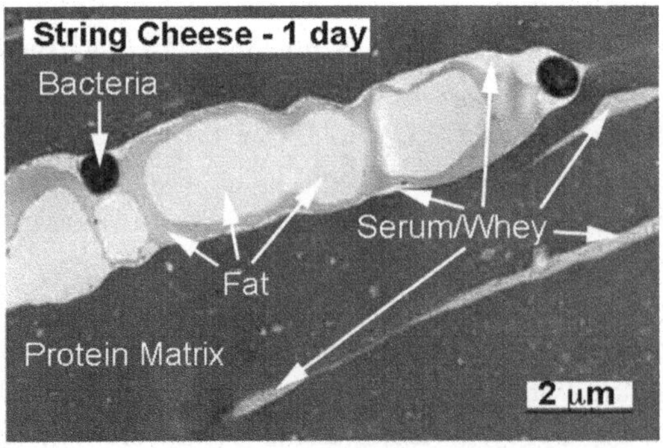

Figure 9.7. Transmission electron micrograph of 1-day-old string cheese (Courtesy of Dr. Almut Vollmer, Western Dairy Center, Utah State University, Logan, UT, USA).

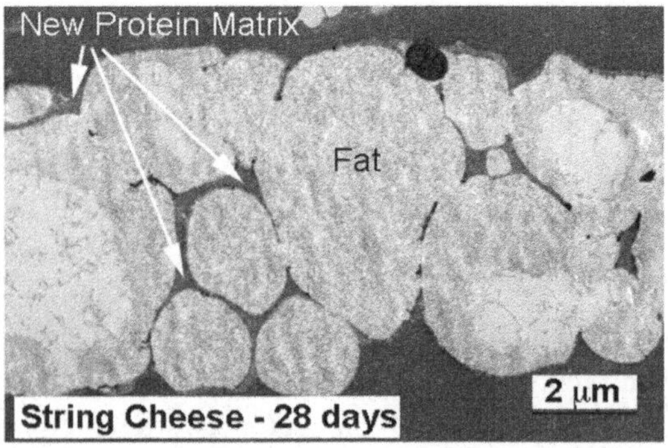

Figure 9.8. Transmission electron micrograph of string cheese after 28-day cold storage (Courtesy of Dr. Almut Vollmer, Western Dairy Center, Utah State University, Logan, UT, USA).

9.3.5 Milk Fat and Cheese Melting

Melting of cheese is a combination of softening of fat and the balance between protein/protein and protein/water interactions (McMahon *et al.* 1999). Fat plays an initial role by melting as the cheese is heated to 40 °C. The liquified fat does not affect further softening and flow as the cheese is heated to 60 to 70 °C (Lucey *et al.* 2003). Fat is important in baking applications, especially when *Mozzarella* is baked on a pizza in a hot-air convection oven. Low-fat cheese lacks melting and stretching characteristics when baked on a pizza because fat contributes to pliability and flowability (Lucey *et al.* 2003; Johnson *et al.* 2009).

Extensive lowering of fat content increases the amount of scorching during convection baking at 250 °C to 300 °C. During baking, cheese exposed on top of a pizza can lose up to 50% of its moisture (Rudan and Barbano 1998). This can be mediated by spraying the cheese shreds with a thin layer of oil (Rudan and Barbano 1998) or by making the cheese so that most of the fat is easily expressed as the cheese shreds soften (Wadhwani *et al.* 2011).

Normally there is ready release of fat from within the cheese, and this free oil lubricates the cheese surface and prevents excessive dehydration of the cheese shreds, so they can flow together to form a mass of melted cheese covering the pizza (Rudan and Barbano 1998; Wadhwani *et al.* 2011). If there is not enough oil covering the shreds, there is too much moisture loss before melting and the surface chars and blisters, preventing flow and leaving dry hard cheese shreds to remain on the pizza surface.

During curd manufacture, the *pasta filata* process, and during storage, the fat globules in the cheese can experience physical stress causing breakage of the milk fat globule membrane (Oberg *et al.* 1993; Lopez, *et al.* 2006; McMahon *et al.* 2009) allowing for easy release of liquified fat during heating. In contrast, fat droplets in low-fat cheeses tend to be smaller and more finely dispersed and trapped within the protein matrix, with fewer fat-serum channels. Hence, there is not enough expressed oil to prevent excessive dehydration, skin formation and charring of the cheese (Wadhwani *et al.* 2011).

9.4 Role of Milk Fat in Ice Cream

H. Douglas Goff
Department of Food Science
University of Guelph
Guelph, ON, Canada

9.4.1 Overview of Ice Cream Ingredients and Manufacture

Fat and fat structure development in ice cream and related frozen dairy desserts are critical to optimal structure and physical properties, stability, flavour and texture. A brief review of the functionality of fat in ice cream follows, with citations to the most recent and most pertinent references. Readers are referred to Clarke (2012), Goff (2002, 2016) and Goff and Hartel (2013) for more detailed information.

The term "ice cream" in its broadest sense includes most whipped dairy products that are manufactured by freezing and are consumed in the frozen state, including ice cream that contains either dairy or non-dairy fats; premium, higher-fat versions; "light", lower-fat versions; ice milk; and frozen yogurt. Ice cream mix formulations specify the content of fat, milk solids-not-fat, sweeteners, stabilizers, emulsifiers and water that are desired (Figure 9.9). Dairy and other ingredients used to supply these components are chosen on the basis of availability, cost, legislation and desired quality. Common ingredients include cream, butter or vegetable fats, as the main sources of fat; condensed skim or whole milk, skim or whole milk powder, milk protein concentrates and/or whey powder or whey protein products, as the sources of concentrated milk solids-not-fat (Goff 2016); sucrose and corn starch hydrolysates, or sugar alcohols, as the sweeteners; polysaccharides, such locust bean gum, guar gum, carboxymethyl cellulose and/or carrageenan, as the stabilizers; egg yolk or mono- and diglycer-

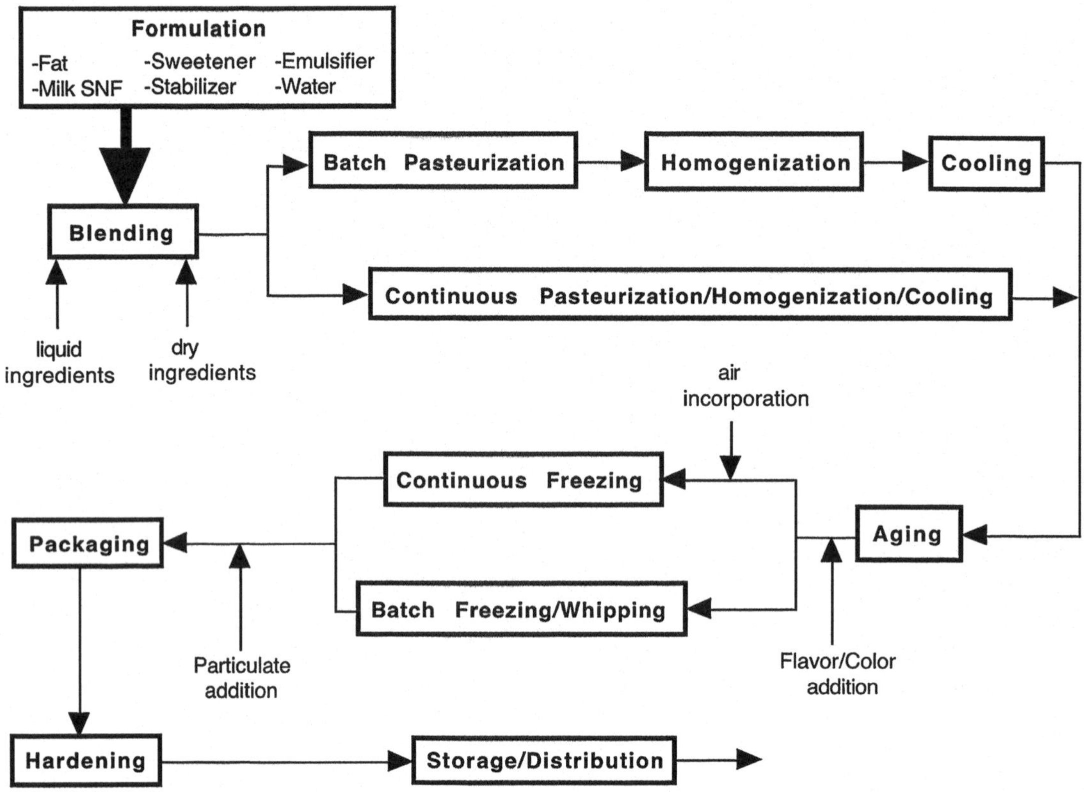

Figure 9.9. Flow diagram for the production of ice cream.

ides, as the emulsifiers; and milk or water, as the main sources of water in the formulation to balance the components (Goff and Hartel 2013). Usually one mix is used for the production of a variety of flavours.

The manufacturing process for most of these products is similar and involves the following steps (Figure 9.9): preparation of a liquid mix by blending of ingredients, pasteurization (65 °C for 30 min or 80 °C for 25 s), homogenization, cooling to 4 °C and ageing of the cold, liquid mix for 4–24 h; adding flavouring; concomitantly whipping and freezing this mix dynamically under high shear to a soft, semi-frozen slurry with an air phase volume of 30–55% (overrun, defined as air/mix ratio, of 40 to 120%) and a draw temperature of about −5 °C; incorporation of discrete flavouring ingredients to this partially frozen mix; packaging the product; and further quiescent freezing (hardening) of the product in blast air to below −25 °C (Goff and Hartel 2013). Homogenization is responsible for the formation of the fat emulsion by forcing the hot mix through a small orifice under a pressure of 14 to 18 MPa, perhaps with a second stage of 3–4 MPa. Ageing allows for hydration of milk proteins and stabilizers (some increase in viscosity occurs during the ageing period), crystallization of the fat globules and a membrane rearrangement due to competitive displacement of adsorbed proteins by small-molecule surfactants. The concomitant aeration and freezing process involves numerous physical changes, including the action of proteins and surfactants in forming and stabilizing the foam phase; partial coalescence of the fat emulsion, causing both adsorption of fat at the air interface and the formation of fat globule clusters that stabilize the lamellae between air bubbles; and freeze concentration of the premix by the removal of water from solution in the form of ice.

9.4.2 Sources of Fat in Ice Cream

The fat component of frozen dairy dessert mixes increases the richness of flavour, is a good carrier and synergist for added flavour compounds, produces a characteristic smooth texture by lubricating the palate, helps to give structure through the process of partial coalescence and foam stabilization and aids in producing desirable melting properties (Goff and Hartel 2013). The fat content can be used as an indicator of the perceived quality and/or value of ice cream. Ice cream must have a minimum fat content of 10% in many legal jurisdictions. Premium ice creams generally have a fat content of 14 to 18%. However, many high-quality ice cream products are also in the market with <10% fat (carrying such descriptors as reduced-fat, light, low-fat and non-fat or fat-free, as appropriate for the legal jurisdiction), where both structure and flavour considerations have been satisfied by other means, e.g. high in protein (Daw and Hartel 2015) or through processing (Tekin *et al.* 2017).

The use of milk fat as a fat source for ice cream formulations is widespread in North America, Australia, New Zealand and much of Europe. The triglycerides in milk fat have a wide melting range (+40° to −40 °C). The crystallization pattern of milk fat is also very complex, due in part to the large variation in fatty acids and a large number of different triglycerides present. Consequently, there is always a combination of liquid and crystalline fat at refrigeration and sub-zero temperatures, which is critical for structure formation, as will be discussed subsequently. The volatile, short-chain fatty acids also contribute to the unique flavour of milk fat. The best source of milk fat in ice cream for high-quality flavour is fresh cream. Other sources of milk fat include sweet (unsalted) butter, anhydrous milk fat (butter oil), frozen cream or condensed milk blends.

Vegetable fats are used extensively as fat sources in ice cream in the UK, parts of Europe, the Far East and Latin America and to a small but increasing extent North America. Five factors of great interest in the selection of fat sources are (1) the crystal structure of the fat; (2) the rate at which the fat crystallizes during dynamic temperature conditions; (3) the temperature dependence of the melting profile of the fat, especially at chilled and freezer temperatures; (4) the content of high melting point triglycerides (which can cause a waxy, greasy mouthfeel); and (5) the flavour and purity of the oil. It is important that the fat droplet contains an intermediate ratio of liquid/solid fat at the time of freezing. It is difficult to quantify this ratio as it is dependent on a number of composition and manufacturing factors; however, 50–67% crystalline fat at 4–5 °C is a good working rule (Sung and Goff 2010; Mendez-Velasco and Goff 2011). Palm kernel, coconut or palm fats, or fractions thereof, or blends of these fats and other oils are often used in ice cream manufacture, selected to take into account physical characteristics, flavour, availability, stability during storage and cost. Mendez-Velasco and Goff (2011) and Zulim Botega *et al.* (2013a, b) explored options for enhancing the unsaturated fat content, including through oil gelation.

9.4.3 Contribution of Fat to the Structure of Ice Cream

The texture of ice cream is one of its most important quality attributes. Texture is the sensory manifestation of structure; thus, establishment and maintenance of optimal ice cream structure are critical to maximal textural quality. The colloidal structure of ice cream begins with the mix as a simple emulsion, with a discrete phase of partially crystalline fat globules surrounded by an interfacial layer comprised of proteins and surfactants (Figure 9.10a). The continuous serum phase consists of unadsorbed casein micelles in suspension in a solution of sugars, unadsorbed whey proteins, salts and high-molecular-weight polysaccharides. During the dynamic freezing stage of manufacture, the mix emulsion is foamed, creating a dispersed phase of air bubbles (Figure 9.10b), and is partially frozen, forming another dispersed phase of ice crystals. Air bubbles and ice crystals usually range in size from 20 to 50 μm and are surrounded by a temperature-dependent unfrozen continuous matrix of sugars,

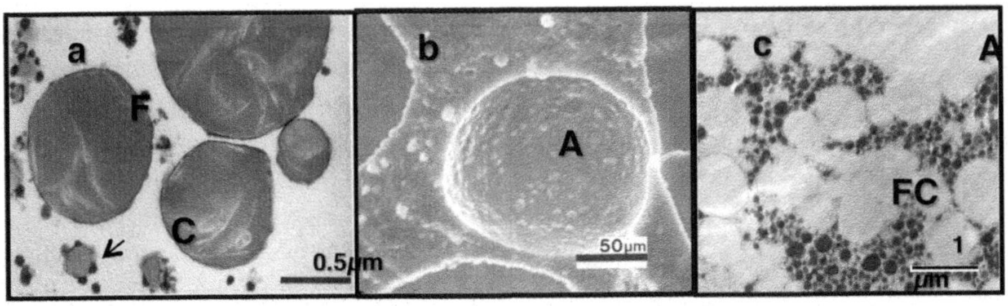

Figure 9.10. The structure of ice cream mix and ice cream. (**a**) Fat globules (F) in mix with crystalline fat within the globule and adsorbed casein micelles (C), as viewed by thin-section transmission electron microscopy. (**b**) Close-up of an air bubble (A) with adsorbed fat, as viewed by low-temperature scanning electron microscopy. (**c**) Air bubble (A) with adsorbed fat cluster (FC) that extends into the unfrozen phase, as viewed by thin-section transmission electron microscopy with freeze substitution and low-temperature embedding.

proteins, salts, polysaccharides and water (Goff 2002; Xinyi *et al.* 2010; Clarke 2012; Goff and Hartel 2013). In addition, the partially crystalline fat phase in the mix at refrigeration temperatures undergoes partial coalescence during the concomitant whipping and freezing process, resulting in a network of agglomerated fat (Mendez-Velasco and Goff 2012a), which adsorbs to the air bubbles and extends into the unfrozen phase, producing a fat network structure throughout the product (Figure 9.10c).

The development of structure and texture in ice cream is sequential, basically following the manufacturing steps. To describe the role of fat in the structure thoroughly, it is necessary to begin with the formation of the emulsion at the time of homogenization and the role of the ingredients present at the time of homogenization, with particular reference to the fat, proteins and emulsifiers. After preheating or pasteurization, the mix is at a temperature sufficient to have melted all the fat present, and the fat is passed through one or two homogenizing valves. The creation of a large population of small, discrete droplets is a prerequisite for the development of structure during dynamic freezing, utilizing these droplets. Thus, homogenization conditions can have a large impact on ice cream structure (Koxholt *et al.* 2001; Ruger *et al.* 2002; Hayes *et al.* 2003; Biasutti *et al.* 2013). Immediately following homogenization, the newly formed fat globules are practically devoid of membranous material and readily adsorb amphiphilic molecules from solution, including casein micelles, non-micellar β-casein, whey proteins, phospholipids, lipoprotein molecules, components of the original milk fat globule membrane and any added surfactants. These species all compete for space at the fat surface. The membrane formed during homogenization continues to develop during the ageing step, and rearrangement occurs until the lowest possible energy state is reached. The transit time through a homogenization valve is in the order of 10–5 to 10–6 s. Protein adsorption or unfolding at the newly formed interface may take minutes or even hours to complete. It is clear, therefore, that the membrane immediately formed upon homogenization is a function of the microenvironment at the time of its creation and that the recombined membrane of the fat globule in the aged mix is not fully developed until well into the ageing process.

Small molecular weight surfactants are not needed in an ice cream mix to stabilize the fat emulsion, due to an excess of protein and other amphiphilic molecules in solution. If a mix is homogenized without added surfactants, both the whey proteins and the caseins will form this new fat globule membrane, with the caseins contributing much more than the whey proteins to the bulk of the adsorbed protein (Zhang and Goff 2004). However, if added surfactants, such as monoglycerides or sorbitan esters or phospholipids, are present, they have the ability to reduce the interfacial tension between the fat and the water phases to a lower value than do the proteins. Thus they become preferentially adsorbed to the surface of the fat, and the mixed membrane of surfactant and protein gives rise to the appropriate membrane for subsequent partial coalescence of the fat globules (Davies *et al.* 2000, 2001; Sourdet *et al.* 2002, 2003; Mendez-Velasco and Goff 2012b; Warren and Hartel 2018). As the interfacial tension is lowered and proteins are eliminated from the surface of the fat, reducing the surface excess (quantity of adsorbed material, mg m^{-2}), the actual membrane becomes weaker to subsequent destabilization due to this reduction of steric stabilization, although the emulsion is thermodynamically favoured due to the lowering of the interfacial tension and the net free energy of the system. Fat globules with reduced steric stabilization also adsorb to air interfaces, enhancing foam stability (Goff *et al.* 1999; Zhang and Goff 2004).

Crystallization of fat also occurs during ageing, creating a highly intricate structure of needle-like crystals within the globule. The triglycerides with high melting points crystallize first and continue to be surrounded by liquid oil of those with lower melting points. Crystallization of emulsified milk fat at refrigeration temperature reaches equilibrium within 1.5 h (Adleman and Hartel 2002; Relkin *et al.* 2003). A partially crystalline fat droplet is necessary for optimal fat structure formation to occur during freezing (Davies *et al.* 2000, 2001). This has been attributed to the protrusion of crystals into the aqueous phase, which causes a surface distortion of the globule. These protrusions can pierce the film between two globules upon close approach. As the crystals are preferentially wetted by the lipid phase, clumping is inevitable.

The next stage of structure development occurs during the concomitant whipping and freezing step. Air either is incorporated through a lengthy whipping process (batch freezers) or is injected under pressure (continuous freezers). The air bubbles are formed through a combination of comminution and interfacial adsorption (Sofjan and Hartel 2004; Xinyi *et al.* 2010). If the fat globules are sufficiently unstable to shear, as a result of reduction in membrane surface excess and steric stabilization due to added small-molecule surfactants, the aeration and ice crystallization processes cause the emulsion to undergo partial coalescence or fat destabilization, during which clusters of the fat globules form and build an internal fat structure or network in the frozen product. Bolliger *et al.* (2000a) showed a direct relationship between protein content (mg m^{-2}), resulting from displacement by emulsifiers, and partial coalescence. The incorporation of air alone, or shearing action alone, independent of freezing, is not sufficient to cause the high degree of fat destabilization that occurs when ice crystallization and air incorporation occur simultaneously (Kokubo *et al.* 1996, 1998).

Fat destabilization results in the beneficial properties of dryness (shape retention) upon extrusion during the manufacturing stages (which facilitates packaging and novelty moulding, for example); a smooth, creamy texture in the frozen dessert; and resistance to melt-down or good stand-up properties (necessary for soft-serve operations) (Bolliger *et al.* 2000a; Daw and Hartel 2015; Warren and Hartel 2018). The clusters of fat globules formed during the process of partial coalescence are responsible for adsorbing to and stabilizing the air cells (Goff *et al.* 1999; Barfod 2001; Zhang and Goff 2004, 2005) and creating a semi-continuous network or matrix of fat throughout the product that crosses the lamellae between the air cells (Koxholt *et al.* 2001; Muse and Hartel 2004). Hence, a finer distribution of air bubbles, resulting in thinner lamellae, also helps to produce optimal shape retention during extrusion and melting (Bolliger *et al.* 2000b). Optimal formation of fat structure and air bubble size may also help to slow down ice recrystallization (Barfod 2001; Soukoulis and Fisk 2016). If an ice cream mix is subjected to excessive shearing action or contains too much emulsifier, the formation of objectionable butter particles can occur as the emulsion is churned beyond the optimum level.

9.4.4 Contribution of Fat to Ice Cream Texture and Flavour

Fat contributes greatly to the flavour and texture of ice cream. Several recent papers have discussed flavour and textural aspects of various fat sources, non-fat sources to provide fat-like properties and the effect of fat on sensory properties and flavour perception of ice cream (Roland *et al.* 1999; Hyvonen *et al.* 2003; Amador *et al.* 2017; Rolon *et al.* 2017; Tekin *et al.* 2017).

In addition to positive aspects, numerous flavour and textural defects may be associated with the fat phase of ice cream. Such flavour defects are usually related to either autoxidation of the fat resulting in oxidized flavours (cardboardy, painty, metallic) or, especially in the case of milk fat, lipolysis of free fatty acids from triglycerides by the action of lipases (known in the dairy industry as rancidity). A significant content of free butyric acid gives rise to very undesirable, rancid flavours. These defects tend to be present in the raw ingredients used in ice cream manufacture, rather than promoted by the ice cream manufacturing process itself. However, processing problems can also occur during ice cream mix manufacture, e.g. rapid agitation/foaming of raw milk or cream, that can give rise to these fat flavour defects (Goff and Hartel 2013).

The fat phase can also account for textural defects associated with the fat content (too high or too low) or degree of partial coalescence. Fat contributes smoothness to the finished product. Low-fat mixes must therefore compensate for this lack of inherent smoothness by altering the ratio of other components, particularly the protein, polysaccharide stabilizer and emulsifier components. On the other hand, mixes high in fat, such as the premium products, typically have a heavier, dense texture, related to both high fat and lower-than-normal air contents. Partial coalescence of the fat emulsion modifies the textural perception, giving ice cream a creamier texture, in addition to its role in structure. If too much destabilization has occurred, the ice cream will taste greasy, and a defect known as "does not melt" may occur (Goff and Hartel 2013). This results from a network of fat that gives sufficient structure to the product to hold its shape without collapse in the absence of the ice phase, after warming to a temperature sufficient to melt the ice.

9.5 Role of Milk Fat in Dairy Powders

S. V. Crowley
School of Food and Nutritional Sciences
University College Cork
Cork, Ireland

9.5.1 Introduction

The fat content of liquid milk fluctuates due to physiological and environmental factors (e.g. lactation, seasonality), but milk can be considered to contain an average of ~3.5% fat (Fox and McSweeney 1998). Liquid milk is often processed into various types of powder, which are less perishable and more efficient to transport, in addition to being highly flexible to use. Lipid levels can vary greatly depending on the type of dairy powder and the procedures involved in their processing. This sub-chapter covers the role of milk fat in dairy powders, with a particular focus on those with functional properties affected markedly by surface fat. Much of the discussion will thus relate to commodity-type dairy powders like whole milk powder (WMP) and cream powder (CP), in addition to milk-derived ingredients such as whey protein concentrate (WPC) and milk protein concentrate (MPC).

WMP contains ~26% fat, making this its second most abundant constituent after lactose and arguably the most critical from a functionality perspective (Buma 1971a,b,c,d). Although CP has over double the fat content (55–70%) of WMP, both have similar levels of fat at particle surfaces (>98% of surface composition), which has a marked negative influence on their functional properties (Kim *et al.* 2005). The comparatively trace level of fat (≤1%) in skim milk powder (SMP) is not known to cause significant technical challenges, which is also the case for other low-fat dairy powders (e.g. caseins, caseinates). High protein (i.e. ≥80%) varieties of WPC and MPC can contain as much as 2–5% fat, which although low on a mass basis can lead to a host of undesirable changes when overrepresented at particle surfaces (Gaiani *et al.* 2009; Kim *et al.* 2005). Thus, the influence of milk fat on important functional properties of dairy powders, including oxidative stability, flowability and wettability, is now generally agreed to be due to the fact that it is spatially localized at particle surfaces. What follows is a discussion of dairy powders that have been found to be affected by this issue. Powders in which milk fat has been replaced by vegetable oils to create "fat-filled milk powders" (Kelly *et al.* 2014; Vignolles *et al.* 2007) will not be discussed.

9.5.2 Commodity Dairy Powders

Liquid milk that has been standardized to a specific fat content can be processed into a milk powder, typically (though not always) by evaporation followed by spray drying. Additional processing steps are necessary in the production of WMP compared to SMP, namely, homogenization and lecithination. For SMP, agglomeration is sufficient in most cases to ensure instant-like properties; however, WMP typically requires a combination of agglomeration and lecithination (Skanderby *et al.* 2009). Due to the high fat content of WMP, it has a greater tendency to float, as fat is hydrophobic and has a lower density than water. Such issues may be further exacerbated in less common milk powders with higher fat contents, such as CP. Lecithin, a surface-active agent, is mixed with butteroil at $<0.5\%$ (w/w) prior to dosing of the WMP during lecithination (Skanderby *et al.*, 2009).

In a series of papers (Buma 1971a, b, c, d), T. J. Buma investigated the factors responsible for the levels of free fat in spray-dried WMP and the relationship between free fat and powder functionality. The work led Buma to develop a physical model of WMP particles that distinguished four types of fat, namely, *surface fat*, *outer layer fat*, *capillary fat* and *dissolution fat* (Buma 1971d). While all forms are extractable using a suitable apolar solvent, the first two, and particularly *surface fat*, were considered to have the most important consequences for powder functionality. Increasing total fat to 20% reduced the flowability of the milk powders, due to a corresponding increase in surface fat, though further increases had a limited effect (Buma 1971b); Buma thus concluded that the coverage of a critical surface area is sufficient to cause maximal cohesion between particles, which has been supported by later studies (Fitzpatrick *et al.* 2004; Kim *et al.* 2005). Particle structuration during atomization was considered by Buma to have the greatest impact on fat distribution in milk powders (Buma 1971c). Buma demonstrated that a high fat content and even a high free fat level are not necessary conditions for a powder to exhibit defects (e.g. poor flowability, solubility); on the other hand, if a powder had a high quantity of fat pooled at its particle surfaces, then such negative properties may be observed. As the major solid component, amorphous lactose plays an important role in encapsulating fat; however, storage conditions need to be carefully controlled and monitored to prevent lactose crystallization, which can lead to the release of fat (Vega and Roos 2006).

In the manufacture of milk chocolate, interactions between milk fat and cocoa butter are desirable. To promote these interactions, it is preferable that WMP has a high free fat content, which helps promote the development of optimal hardness or "snap" (Liang and Hartel 2004). Conventional spray drying results in a very low free fat content, while roller drying promotes high free fat, yielding powders with some desirable chocolate-making properties (Keogh and Twomey 2002); however, as demonstrated by Liang and Hartel (2004), roller-dried WMP has other characteristics (e.g. high particle density) that can promote undesirable phenomena in chocolate, such as bloom. Procedures to prepare WMP with high levels of free fat using spray drying are also available (Skanderby *et al.* 2009).

The localization of fat at particle surfaces commences during spray drying, although surface fat may accumulate further during storage. A high free fat level is associated with the formation of caked material in cyclones. With the exception of homogenization, processing steps prior to drying seem to have a limited impact on free fat. Homogenized concentrates have been shown to yield WMPs with lower free fat levels (Buma 1971d). In single droplet drying experiments, Foerster *et al.* (2016) suggested that disruption of the lipid-water interface during atomization is primarily responsible for the over-representation of fat on the surface of spray-dried milk powder particles; however, the authors could not prevent these effects by adjusting the atomization conditions. In a later study, the authors reported a successful reduction in surface fat levels through the incorporation of a hydrocolloid into the milk feed before drying (Foerster *et al.* 2017).

9.5.3 Milk- and Whey-Derived Ingredients

Surface fat has recently been found to play a significant role in the functional properties of protein ingredients. For the purpose of this sub-chapter, emphasis will be placed on ingredients made using membrane filtration, which is known to concentrate protein and fat simultaneously. "MPC" and "WPC" will be used broadly to refer to any protein-based ingredients made by ultrafiltration of skim milk or whey, respectively; thus, no distinction is made between "concentrates" and "isolates" for this purpose.

The feed material for membrane processes used to make MPC and WPC is low in fat (pretreated skim milk and whey, respectively); however, the membranes used to concentrate protein are not permeable to residual fat, which is also concentrated as a result. For example, as the protein content of MPC ingredients increases, so does the fat content (Crowley *et al.* 2014). Despite this, the concentration of protein far exceeds that of fat, such that an 85% protein MPC will contain ~2% fat. In protein ingredients such as MPC and WPC, such a small quantity of fat can be overrepresented at particle surfaces, which can influence powder characteristics. Kim *et al.* (2005) found that particle surfaces in a 6% fat WPC consisted of 53% fat, compared to only 18% surface fat in a 1% fat SMP. This overrepresentation of fat (relative to bulk composition) is common in dairy protein ingredients.

Some evidence suggests that fat migrates towards the surface during storage, particularly at elevated temperatures. Gaiani *et al.* (2009) found that the presence of fat at the surface of micellar casein powder particles (which can be considered similar to an MPC) increased during storage at both 20 °C and 50 °C, although more rapidly at the elevated temperature. The increase in surface fat observed corresponded to a decrease in powder wettability. While milk fat is completely liquid at 50 °C, it is only partially so at 20 °C, suggesting that lipids are likely to be more easily "mobilized" at high temperatures. In laboratory studies, it is now common to store MPC and MCC at refrigeration temperature to prevent such effects.

9.5.4 Impact of Fat on Powder Functionality

Even when fat is present at low levels in dairy powders, it can have significant effects on product functionality. The impact of fat is most pronounced in lipid oxidation, powder flowability and rehydration.

Although the fatty acids in milk fat are predominately saturated, the presence of polyunsaturated fatty acids renders products such as WMP susceptible to oxidation (Mahmoodani *et al.* 2018). Oxygen molecules react with free radicals formed during this chemical reaction, which is catalysed by UV light and the presence of metals. Taking these factors into account, it is clear that packaging and storage conditions are important in efforts to prevent oxidation. Thus, flushing packages with nitrogen, the use of opaque materials and limiting the exposure to metals are often successful.

Several studies have shown that the lipids have an important role in determining the flowability of dairy powders (Fitzpatrick *et al.* 2004; Kim *et al.* 2005). In general, the higher the level of fat, the poorer the flow behaviour; thus, WMP is more cohesive than SMP, but WMP is slightly less cohesive than cream powder (Kim *et al.* 2005). As mentioned already, however, it is not total fat *per se* that causes these issues but, rather, that fat which is present at particle surfaces. The resultant cohesive interactions were found to increase as temperature was elevated, presumably as the lipids become more liquid-like (Vignolles *et al.* 2007).

It has long been known that high-fat dairy powders can be difficult to solubilize effectively. This issue pertains to the early (wetting and sinking) not the later (dispersion and dissolution) stages of the rehydration process. This problem can be related to the hydrophobicity of fat and its low density; hence, powder particles coated in fat have a tendency to penetrate liquid surfaces very slowly. This is why most WMPs, and many WPC powders, are now lecithinated. The mechanism by which lecithin aids the rehydration of powders can be understood as follows: lecithin, an amphiphilic molecule, reduces the surface tension of the liquid in which the powder is to be dispersed, thereby increasing the rate at which the powder penetrates the surface. However, this form of explanation is challenged by a recent study (Mitchell *et al.* 2019), in which it was demonstrated that a reduction in the surface tension of water using a surfactant negatively affected the wetting/sinking of WMP; the authors argue that their finding is in accordance with the Washburn equation, in which a slower capillary penetration rate is predicted for systems with decreased surface tension. Nevertheless, the application of lecithin to improve the rehydration of WMPs and WPCs has achieved practical success, though further studies are required to validate the mechanisms responsible.

9.5.5 Conclusion

Lipids are a major constituent of WMP on a mass basis. In such high-fat dairy powders, lipids are a critical consideration with respect to functionality. However, it is now well documented that only a small proportion of this fat is necessary to cause product defects. Thus, issues associated with WMP, including slow wetting and poor flowability, also arise in MPCs and WPCs with far lower fat contents. In both cases, the spatial distribution of fat seems to be the key, with the presence of fat at particle surfaces being a critical factor.

Dairy materials are dehydrated to extend their life and facilitate their transit. Surface fat can accelerate product deterioration, particularly when powders are exposed to high ambient temperatures during transport and storage. Investigations of the processing factors that enable the creation of particle structures that inhcf future research.

9.6 The Role of Lipids in Infant Milk Formula

C. Moloney and J. O'Regan
Nestlé Development Centre Nutrition
Limerick, Ireland

F. Giuffrida
Nestlé Research Centre
Lausanne, Switzerland

9.6.1 Infant Formula: A Brief History

Human milk is the ideal source of nutrition for infants, providing for all of the dietary requirements of the neonate. When a mother cannot or chooses not to breastfeed, infant formula is a suitable replacement for milk. For millennia, young children have been fed with animal milks – either alone or in combination with other sources of nutrition, such as honey – as an alternative to breastmilk. However, the first formulations to resemble modern infant formula emerged in the nineteenth century, which were made possible due to the advances in analytical methodology that allowed food chemists to determine the compositions of different milks (Stevens *et al.* 2009). In 1865, Justus von Liebig patented a formulation consisting of bovine milk, wheat and malt flours and potassium bicarbonate, and in 1867, Henri Nestlé produced an infant cereal that built upon von Liebig's concept (Stevens *et al.* 2009). Since the pioneering nutritional developments of von Liebig and Nestlé, infant milk formula has advanced exponentially, but the research inspiration and focus have always remained the same: to make a formulation as close as possible to that of human milk.

9.6.2 Human Milk Lipids and Their Influence on Infant Formula Design

Human milk is composed of water, lactose and oligosaccharides, lipids, proteins and various micronutrients. The vast majority of lipids are dispersed throughout milk as part of spherical fat globules consisting of a neutral lipid core surrounded by the milk fat globule membrane (MFGM), which includes polar lipids and cholesterol as part of its composition and prevents the fat globules from coalescing (Garcia and Innis 2013). Carotenoids and other fat-soluble nutrients are also typically dissolved within these fat globules. Human and bovine milk contain similar levels of total lipids (3–4%), which, in both cases, is composed of >97.5% as triacylglycerols (TAGs). However, their lipid profiles have key differences that make bovine milk fat unsuitable as a sole source of lipids in infant formula – particularly the deficit of unsaturated fatty acids in bovine milk – compared to human milk. Until the 1970s, bovine milk was often used as a lipid source in infant formula, before being largely replaced by vegetable oils due to their high levels of unsaturated fatty acids, such as linoleic acid (Innis 2011). Progressive efforts to replicate the lipid profile of human milk have been behind the evolution of the lipid profile of modern infant formula, which will be discussed below for the different classes of lipids.

Table 9.1. Comparison of lipid profiles of human and bovine milk)

Component	Human milk	Bovine milk
Total lipid, g/L	38	37
Lipid class, % w/w of total lipids		
Monoacylglycerols	Trace	0.03
Diacylglycerols	0.70	0.3
Triacylglycerols	98.2	98.3
Free fatty acids	0.40	0.1
Cholesterol	0.25	0.3
Polar lipids	0.56	0.55
Fatty acids, %w/w of total lipids		
$C_{4:0}$ (butyric acid)	0.2	3.9
$C_{6:0}$ (caproic acid)	0.2	2.5
$C_{8:0}$ (caprylic acid)	0.5	1.5
$C_{10:0}$ (capric acid)	1.0	3.2
$C_{12:0}$ (lauric acid)	4.4	3.6
$C_{14:0}$ (myristic acid)	6.3	11.1
$C_{16:0}$ (palmitic acid)	22.0	27.9
$C_{16:1}$ (palmitoleic acid)	3.7	1.5
$C_{18:0}$ (stearic acid)	8.1	12.2
$C_{18:1}$ (oleic acid)	34.0	21.1
$C_{18:2}$ (linoleic acid)	10.9	1.4
$C_{18:3}$ (linolenic acid)	0.3	1.0
$C_{20:4}$ (arachidonic acid)	0.5	–
$C_{22:6}$ (docosahexaenoic acid)	0.32	–

Adapted from MacGibbon and Taylor (2006), Tayor and MacGibbon (2011), Garcia *et al.* (2012).

9.6.2.1 Neutral Milk Lipids

Fatty Acid Composition and the Use of Vegetable Oils

The primary role of lipids in milk is to provide energy – human milk provides 50–60% of the calorific requirements of the neonate (Jensen 1999) – and the vast majority of this energy is provided by neutral lipids, mostly in the form of TAG. These TAGs are composed of a diverse array of fatty acids; human milk contains approximately 200 different fatty acids, compared to ~400 in bovine milk, though only ~15 of these fatty acids are present at more than trace levels (Lindmark Månsson 2008; Hageman *et al.* 2019), and their typical profiles are compared in Table 9.1. Although the composition of human milk is well understood, the exact lipid profile can be influenced by factors such as genetics and diet, with some regional differences observed; for example, the milk of Chinese mothers can vary geographically even within the same country (Giuffrida *et al.* 2016). In general, as well as a relatively low level of small- and medium-chain fatty acids (i.e. shorter than 14 carbons), human milk is characterized by high levels of saturated palmitic acid (16:0), monounsaturated oleic acid (18:1 n-9) and polyunsaturated linoleic acid (18:2 n-6) and α-linolenic acid (18:3 n-3). These long-chain polyunsaturated fatty acids (LC-PUFAs) are important nutritional components of milk, serving as precursors for the biosynthesis of other n-6 and n-3 LC-PUFAs, including arachidonic acid (ARA, 20:4 n-6) and docosahexaenoic acid (DHA, 22:6 n-3), which are key in the brain development and function of infants (Hadley *et al.* 2016; Lauritzen *et al.* 2016). It had previously been thought that infants could synthesize adequate amounts of ARA and DHA from their respective precursors, linoleic acid and α-linolenic acid. However, it has been shown that human milk also contains ARA and DHA primarily deriving from the diet of the mother; and there is evidence that infants may benefit from formulae that contain ARA and DHA, in addition to their precursors (Carver 2003).

In order to mimic the human milk profile as closely as possible, the lipid fraction of infant formula is typically composed of a blend of different vegetable fats and oils; the typical sources are described in Table 9.2. An infant formula devised and labelled SMA, "simulated milk adapted", is thought to have been the first such infant nutritional product to contain a fat blend derived from animal and vegetable fats in order to better approximate the human milk profile (Gerstenberger and Ruh 1919). In order to mimic the fatty acid profile of human milk, the vegetable oils used tend to be rich in palmitic, oleic, linoleic and α-linolenic acids. Although currently no single oil can provide a fatty acid composition identical to that of human milk, when combined in the correct proportions, a good distribution of the fatty acid profile can be achieved. Most vegetable oils, with the exception of palm oil, do not contain high levels of palmitic acid; therefore, palm oil is typically required to achieve palmitic acid levels similar to those of human milk (Hageman *et al.* 2019). However, re-emerging milk fat sources such as anhydrous milk fat (AMF) represent a potential alternative source of palmitic acid, which may in time reduce or eliminate this dependency on palm oil.

Due to the importance of PUFAs in the infant diet, infant formula is often fortified with ARA and DHA, typically at ratio ranging from 1:1 to 2:1. Fish oil is probably the most available source of DHA and is suitable to manufacture infant formulae if the level of eicosapentaenoic acid (EPA, 20:5 n-3) is low. Indeed, EPA is a polyunsaturated fatty acid that is present only in traces (<0.3%; Yuhas *et al.* 2006) in human milk and is known to be antagonistic to the functions of ARA (Carlson *et al.* 1992). The technology to isolate and refine significant quantities of ARA and DHA from microalgae and fungal oils, egg yolk derived-lipids and marine oils has been developed and commercialized, and these form the basis of most ARA and DHA ingredients used to fortify infant formula.

There is no consensus on how much ARA and DHA should be supplied to the infant diet from formula; the Food and Agriculture Organization recommend that ARA should supply 0.2–0.3% of

Table 9.2. Typical fatty acid profiles of vegetable oils commonly used in infant formula manufacture

	%w/w of total fatty acids											
Lipid source	$C_{4:0}$	$C_{6:0}$	$C_{8:0}$	$C_{10:0}$	$C_{12:0}$	$C_{14:0}$	$C_{16:0}$	$C_{16:1}$	$C_{18:0}$	$C_{18:1}$	$C_{18:2}$	$C_{18:3}$
Canola oil	–	–	–	–	–	–	5	–	2	58	20	8
Canola oil, high oleic	–	–	–	–	–	–	4	–	2	70	15	3
Coconut oil	–	–	8	6	47	18	9	–	3	7	2	–
Palm oil	–	–	–	–	–	1	43	–	5	39	10	–
Palm olein	–	–	–	–	–	1	39	–	4	42	11	–
Palm oil, *sn-2* enriched	–	–	–	–	–	1	42[a]	–	3	44	7	–
Safflower oil	–	–	–	–	–	–	6	–	2	16	72	–
Soybean oil	–	–	–	–	–	–	11	–	4	22	53	6
Sunflower oil	–	–	–	–	–	–	6	–	3	27	60	–
Sunflower oil, high oleic	–	–	–	–	–	–	4	–	3	79	11	–

[a]~45% of the $C_{16:0}$ is found in the *sn-2* position.

total energy, with DHA comprising 0.10–0.18% of energy, and that, within these ranges, the precise ARA:DHA ratio is unimportant (FAO 2010). More recently, the European Food Safety Authority concluded that though there have been some demonstrated benefits to supplementing infant formulae with DHA, there is no practical requirement to add ARA, even in the presence of DHA (European Food Safety Authority 2014). The latest Codex standards dictate that if ARA and DHA are added to infant formula, then the ARA:DHA ratio should be at least 1 (Codex Alimentarius 2016). As further research is completed in this area, the recommended intakes of these key nutrients are likely to continue to evolve.

Fatty Acid Positional Distribution

Although the fatty acid composition of TAG varies widely, human milk TAGs often include palmitic acid at the central position (*sn-2*) on the glycerol backbone; 50–70% of palmitic acid chains are esterified in this position (Bar-Yoseph *et al.* 2013; Garcia and Innis 2013). These are often combined with unsaturated oleic acid chains at the *sn-1* and *sn-3* positions, and this configuration of oleic acid moieties (O) esterified either side of a palmitic acid chain (P) yields a characteristic OPO TAG structure. Palmitic acid is of particular importance in human milk – comprising approximately 10% of the energy supply of the infant by itself (Innis 2016) – and its effective absorption during digestion is crucial. When palmitic acid is found at the *sn-2* positon, it is less susceptible to being cleaved from the glycerol by gastric lipases – only ~22% of *sn-2* fatty acids are hydrolysed (Karupaiah and Sundram 2007). In this case, it remains on the glycerol backbone, and the resulting monoacylglycerol is well-absorbed by the infant; this preferential absorption of palmitic acid may improve intestinal comfort, calcium absorption and bone health (Bar-Yoseph *et al.* 2013; Garcia and Innis 2013).

A significant disadvantage of using vegetable oils in infant formula is that the TAGs present tend to contain palmitic acid at the *sn-1* and *sn-3* positions. When long-chain fatty acids such as palmitic acid are released from these positions, they are not absorbed, as well as chains of shorter fatty acids, and instead tend to bind calcium to form poorly absorbed calcium soaps. These soaps can lead to harder stools and constipation, as well as reduced bone mineralization (Nelson *et al.* 1996; Yao *et al.* 2014). To remedy this, ingredients have been developed with increased levels of palmitic acid esterified at the *sn-2* position – one such ingredient is Betapol®, produced by Lipid Nutrition BV, Wormerveer, Netherlands. These ingredients are typically produced by enzymatic interesterification of fractionated palm oil in the presence of an oleic acid source, such as soybean oil or high oleic sunflower oil, resulting in an oil containing high levels of OPO (Zou *et al.* 2016). These OPO-enriched oils, sometimes referred to as *sn-2* palmitate or *sn-2* fat, have allowed the commercialization of infant formula containing OPO levels similar to those of human milk; such formulations have shown clinical benefits in infants, more closely resembling the outcomes observed in breastfed infants (Yao *et al.* 2014; Béghin *et al.* 2018).

The Recent Re-emergence of Milk Fat

With the increasing prevalence of vegetable oils in infant formula throughout the twentieth century, the use of bovine milk fat fell largely out of favour; however, in recent years, fractions rich in milk fat have begun to find increased usage in infant formula as more evidence emerges of the potential benefits of bovine milk fat in the infant diet. Formulations containing a combination of vegetable and bovine milk lipids have been shown to be tolerated as well as those containing vegetable oils alone, with no impact on growth (Gianni *et al.* 2018), and in animal models, bovine milk lipids have been shown to retard beneficially the digestion of β-lactoglobulin while increasing the abundance of peptides from β-casein digestion (Le Huërou-Luron *et al.* 2018). Three main types of milk fat addition to infant formula can be considered: (i) through the direct addition of whole milk or cream; (ii) as complex milk lipids, which include polar lipids (PLs) and other MFGM components; and (iii) as anhydrous milk fat, which is composed mainly of TAG and cholesterol and depleted in PLs. AMF is produced from either cream or butter that is concentrated, homogenized and separated to remove residual proteins and as much water as possible to result in a material that is almost entirely composed of lipids – by definition, a minimum of 99.8% fat and a maximum of 0.1% water (Mortensen 2011) – and the fatty acid profile of AMF is similar to that of bovine milk. Bovine milk contains a relatively high level of palmitic acid (~30% of the total fatty acids), approximately 40–45% of which is found at the *sn-2* position (Tzompa-Sosa *et al.* 2014; Lindmark Månsson 2008); therefore, the increasing trend of AMF addition to infant formula also represents a means of naturally increasing the amount of *sn-2* palmitate present in the diet of the infant.

9.6.2.2 Polar Lipids

The PL fraction comprises 0.2–2.0% (typically ~0.5–0.8%) of human milk lipids and consists predominantly of phospholipids and sphingolipids found in the MFGM (Garcia *et al.* 2012; Jensen 1999). Although phospholipids and sphingolipids are structurally distinct, they are sometimes collectively referred to as phospholipids in the literature. True phospholipids are structurally composed of a glycerol backbone combined with fatty acids, a phosphoric acid group and a hydroxy compound (Contarini and Povolo 2013). The most commonly occurring hydroxy compounds in phospholipids are ethanolamine, choline, serine and inositol, and, thus, the primary milk phospholipids are phosphatidylethanol amine (PE), phosphatidylcholine (PC), phosphatidylinositol (PI) and phosphatidylserine (PS).

Sphingolipids consist of a long-chain, sphingoid base, usually combined with a fatty acid to form a ceramide, along with a phosphoric acid head, though sugars or alcohols may in some cases be attached. Sphingomyelin (SM) is the most common type of sphingolipid found in milk and consists of a sphingosine base bonded to a fatty acid chain of varying length (most commonly 16:0) and a phosphocholine head group (Byrdwell and Perry 2007; Contarini and Povolo 2013). Glycosphingolipids are a glycosylated family of sphingolipids and may be acidic – for example, gangliosides, which are sialylated sphingolipids – or neutral, such as cerebrosides, which are ceramides with single sugar moieties attached (Jensen 1996; Liu *et al.* 2018). The dominant gangliosides are mono-sialylated (GM_3) or di-sialylated (GD_3), with the most common cerebrosides being galactocerebrosides and glucocerebrosides (Giuffrida *et al.* 2014; Jensen 1996).

Despite their relatively low abundance, PLs are critical in the early development of infants as they play important roles in processes such as brain myelination (Tanaka *et al.* 2013) and lipid and cholesterol digestion, absorption and transport (Nilsson 2016) and are important in lipid membrane integrity (van Meer *et al.* 2008). Gangliosides may have immune and anti-infection properties by promoting the growth of *Bifidobacterium* spp. and binding to pathogenic targets (Rueda 2007).

Owing to the benefits associated with PLs, their levels in infant formula have become a major recent research focus. Conventional infant formula generally has two main sources of PLs: (i) dairy-derived protein ingredients (e.g. skim milk powder and whey protein concentrate), which naturally contain MFGM fragments released during various dairy processes such as separation and homogenization, and (ii) soybean lecithin, which is sometimes used as an emulsifier in infant formula. However, while soybean lecithin is a source of phospholipids, it does not contain SM, which is a particularly abundant and important PL in human milk. This means that effective enhancement of the PL profile of infant formula must come through the enrichment of MFGM components rather than the addition of soybean lecithin or other non-milk-derived PL sources. Several such MFGM-enriched "complex lipid" ingredients are already commercially available for inclusion in infant formula, including the SureStart™ range (Fonterra, Palmerston North, New Zealand).

Although the PL content of bovine milk is similar to that of human milk, the relative proportions of the five major PLs differ, with SM more abundant in human milk (Table 9.3). The PL content of conventional infant formula is typically lower than that of human or bovine milk, as ingredients low in milk fat are typically used in formulating (e.g. whey protein concentrates and isolates). The emerging trend towards the addition

Table 9.3. Comparison of the relative levels of the five major polar lipids in human and bovine milk

Polar lipid	Human milk mg/L	Bovine milk mg/L
Phospholipids		
Phosphatidylcholine	24.5	28.7
Phosphatidylethanolamine	18.3	31.8
Phosphatidylserine	8.1	10.0
Phosphatidylinositol	3.8	3.7
Sphingolipids		
Sphingomyelin	29.7	20.0

Adapted from Garcia *et al.* (2012).

of AMF in infant formula will not impact upon PL content, as AMF contains neutral lipids almost exclusively. However, the emergence of MFGM-enriched dairy ingredients has allowed for the investigation of the efficacy of MFGM supplementation of infant formula as a means of bringing increased benefit to the infant. A limited number of relatively small-scale clinical trials have suggested that supplementation of infant formula with MFGM or SM may lead to better outcomes compared to conventional formulations in terms of normal growth and cognitive function and development (Tanaka *et al.* 2013; Gurnida *et al.* 2012; Timby *et al.* 2014a) as well as reduced incidence of infection (Timby *et al.* 2015) and cardiovascular risk markers (Timby *et al.* 2014b). While some benefits can be linked to individual MFGM components – e.g. increasing SM content leading to better cognitive development (Tanaka *et al.* 2013) – many of the benefits associated with MFGM are likely as a result of the combination of factors present as part of its composition.

These innovations are made possible by technological advances that allow commercial-scale production of safe, high-quality ingredients at a production volume required by infant formula manufacturers. MFGM can be isolated from cream by various means, including centrifugation, churning or using mild detergents (Mather 2011); however, rather than relatively pure MFGM isolates, MFGM-enriched whey or lipid fractions are more commonly used in infant formula, the manufacturing processes of which are often proprietary. Some milk protein fractions may also be selectively enriched in MFGM components due to the fractionation processes used to enrich the desired protein components. One such ingredient, α-lactalbumin-enriched whey protein concentrate, has been shown to provide SM levels in infant formula similar to human milk levels, without the need for the addition of a specific MFGM-enriched ingredient (Moloney *et al.* 2018). This is made possible by a novel production process that combines specific pH and temperature to precipitate α-lactalbumin, which can then be separated by membrane filtration, which also serves largely to retain lipids and MFGM components. A combination of MFGM-enriched fractions and novel dual-purpose ingredients, such as the aforementioned α-lactalbumin-enriched whey protein concentrate, is likely to be employed in the near future to meet the desire to enhance PL levels in infant formula.

9.6.2.3 Other Lipid Components: Carotenoids and Cholesterol

Cholesterol is mostly found as part of the MFGM and is a vital component of lipid membranes, as well as a precursor of bile acids and steroid hormones (Ohlsson 2010). It is the most abundant sterol in milk, and a systematic review established that breastfeeding was associated with lower total blood cholesterol concentrations later in life compared with formula feeding suggesting that the level of cholesterol in human milk may contribute to homeostasis of the sterol in adults who were breastfed as infants (Pfrieger 2003). Bovine milk (300 mg/L) can contain more than double the concentration of cholesterol found in human milk (90–150 mg/L), and infant formula typically contains low levels (<5 mg/L) due to the lack of animal fat (Hageman *et al.* 2019). While no known efforts have been made to increase the cholesterol concentration of infant formula to levels similar to those of human milk, the addition of MFGM components is expected to lead to a concomitant increase in cholesterol, making this likely an area of increased focus for research in the near future.

Carotenoids are not synthesized in the human body but are usually present in human milk as a result of a mother's diet that includes carotenoid-rich food sources. These compounds – including lutein, α- and β-carotene, zeaxanthin and β-cryptoxanthin – are involved in promoting vitamin A function, cognitive performance and eye development (Eggersdorfer and Wyss 2018). Though there is no regulatory requirement to do so, infant formula is sometimes fortified with lutein and β-carotene, the two most abundant carotenoids in human milk (Lipkie *et al.* 2015). Failure to fortify infant formula with key carotenoids may lead to much lower plasma carotenoid levels in the infant, sometimes below detectable levels (Zielińska *et al.* 2017). Feeding with infant formula supplemented with carotenoids has been shown to increase plasma β-carotene, lutein and lycopene concentrations in infants, compared to infants fed with unfortified formula; higher levels of fortification corresponded with higher plasma levels of these nutrients, which may fall within the range of infants fed with human milk through such supplementation (Mackey *et al.* 2012). Though β-cryptoxanthin and zeaxanthin are sometimes detected in infant formula, these components are not typically fortified, and their presence is due to their natural occurrence in some of the ingredients used in infant formula manufacture.

9.6.3 Future Perspectives

As research continues into the benefits of LC-PUFAs, it is possible that the minimum recommended levels of ARA and DHA will be increased in order to maximize any associated benefit. This will likely pose a technical issue to infant formula manufacturers as these lipids are easily oxidized and, depending on the source, can have a "fishy" flavour note. These two factors can combine to cause undesirable sensory characteristics, and increasing LC-PUFA content will exacerbate this issue. Possible solutions to this problem will lie in the use of encapsulated lipids or alternative processing technologies to segregate and protect LC-PUFA until they reach the gastrointestinal tract, thereby maintaining consumer acceptance while enhancing nutritional benefits.

Palm oil is an important component of infant formula and particularly as a substrate for the production of OPO-enriched oils. Concern around the environmental impact of palm oil production is likely to continue, driving manufacturers towards more sustainable sourcing and production practices. This will also encourage ingredient manufacturers to accelerate the development of technologies to generate OPO-enriched oils from alternative substrates, including vegetable and animal fat sources.

Differences in the structure and physical attributes of the fat globules represent a major difference between human milk and infant formula; fat droplets in human milk are larger than in infant formula (Baumgartner *et al.* 2017), and this may influence the lipolysis rate and particle, in turn affecting postprandial responses (Armand *et al.* 1999; Michalski *et al.* 2006; Baumgartner *et al.* 2017) which is believed to impact metabolic health later in life (Oosting *et al.* 2012; Baars *et al.* 2016). Mimicking the physical structure of human milk will require further research to fill this gap and deliver optimal nutrition to infants who cannot be breastfed. However, achieving this may introduce technological complications, as larger fat globules will have a greater tendency to undergo creaming and coalescence in accordance with Stokes' law (Wilbey 2011). Considering the strict regulations governing infant formula composition, the use of traditional stabilizers may not be feasible, which may potentially require the development of novel techniques and technologies to successfully stabilize these larger, more human-like fat globules.

9.7 Milk Fat and Chocolate

I. Celigueta Torres and P. Siong
Nestlé Product Technology Centre
Confectionery, Nestlé
York, UK

9.7.1 Introduction

Chocolate is a complex suspension of cocoa solids, sugar crystals and milk powder, accounting for 70% of the total mass, dispersed in a fat con-

Table 9.4. Typical ingredient breakdown composition for milk, dark and white chocolate recipes

Chocolate type	Cocoa liquor or mass (%)	Sugar (%)	Cocoa butter (%)	Vegetable fats (%)	Non-fat milk solids[a] (%)	Milk fat (%)	Lecithin (%)	Flavour (%)	Total fat content (%)
Milk chocolate	8–12	34–58	18–25	0–5	12–18	3.5–6.5	0.3–0.5	0–0.5	26–38
White chocolate	0	37–50	22–35	0–5	18–24	4–8	0.2–0.5	0–0.5	29–40
Dark chocolate	45–80	20–55	0–5	0–5	0	0–5	0–0.5	0–0.5	29–49

[a]Dry solids obtained from dehydrated milk, i.e. milk powder, cream, buttermilk powder or milk fat.

tinuous matrix primarily consisting of cocoa butter, emulsifiers and often other fat fractions, e.g. milk fat, cocoa butter equivalents (CBEs).

Cocoa solid is the major ingredient in chocolate and originates from the beans of cocoa tree (*Theobroma cacao* L.) found originally in Central and South America but now cultivated in other geographical regions, such as Africa and Asia. There are four main cultivars of this tree used to produce cocoa beans for chocolate: Forastero, Trinitario, Criollo and Nacional, which differ in their composition and contribute to the flavour and quality of the final product (Fowler and Coutel 2017).

Initially, cocoa was consumed as a beverage, as early as 600 BC in Mesoamerican civilizations. This consumption continued through the centuries, when Aztecs of Mexico drank a cocoa-spiced beverage that was believed to provide them with strength and to act as an aphrodisiac (Grivetti *et al.* 2000). Such a drink was only reserved for high society ranks, such as the emperor, great warriors and priests (Lippi 2013). Cacao was first brought to Spain by Hernán Cortés in the sixteenth century, and its consumption spread afterwards throughout Europe. The use of chocolate in the semisolid emulsion form known today is believed to date back to 1875, when Daniel Peter of Vevey (Switzerland) created an economical way of removing water from milk (Beckett 2017) to obtain a new texture that is nowadays familiar to consumers.

9.7.1.1 Milk Fat and Legislation of Chocolate

Chocolate is categorized into three types: dark, milk and white, depending on the content of cocoa solids, milk fat and cocoa butter (Afoakwa *et al.* 2007). There is a fourth type of chocolate mass, called chocolate equivalent, which is an inexpensive replacement for chocolate, generally made with cocoa powder and vegetable fat, e.g. palm kernel oil, instead of cocoa liquor and cocoa butter. The differences in recipe of these four types of masses result in a variety of textures and sensory perception of the final chocolate, to accommodate consumer preference. Typical composition of different chocolate types is shown in Table 9.4.

The Codex Alimentarius Commission establishes the standards for chocolate and cocoa products to facilitate trade across countries. Today, the World Trade Organization (WTO) and the European Union (EU) participate in Codex work, and they may choose to accept Codex standards in full or with concrete deviations (Wood 2017). As a result, the legal minimum of milk solids and milk fat differs between countries, and this needs to be closely monitored by companies when exporting their chocolate products to other countries so as to abide to local regulations. In many countries, besides cocoa butter, milk fat is the only type of fat that is permitted in chocolate.

As an example, Codex standards (Codex 2003) require a minimum milk solid content of 12–14% and 2.5–3.5% of milk fat in milk chocolate, whereas the European Union indicates a minimum of 14% dry milk solids and at least 3.5% milk fat in the recipe (EU 2000). On the contrary, the USA requires a minimum content of milk solids of 12% and 3.39% milk fat (FDA 2018). Similarly, white chocolate needs to have a minimum of 14% milk solids and 2.5% milk fat

Table 9.5. Approximate composition of ingredients containing milk fat that are used in chocolate

Milk ingredient	Protein (%) as total N × 6.38 in d.m.[a]	Fat (%)	Lactose (%)	Moisture (%)	Minerals (%)	Other components [b](%)
Anhydrous milk fat	0.0	99.8	0.0	0.1	0.0	0.0
Anhydrous butter oil	0.0	99.5	0.0	0.3	0.0	0.0
Whole milk powder	24.7	26.3	40.0	3.2	5.8	0.0
Skim milk powder	34.0	0.5	54.0	3.5	8.0	0.0
High-fat cream powder	15.0	55.0	24.6	1.9	3.5	0.0
Crumb	7.0	10.0	12.0	1.0	0.0	70.0
Buttermilk powder	30.8	8.0	50.4	3.8	7.0	0.0

[a]d.m. = dry matter.
[b]Sucrose, cocoa butter, etc.

according to Codex standards. In the EU, legislation for milk content in white chocolate is as for milk chocolate, whereas other components such as defatted cocoa solids or cocoa butter requirements will differ. Finally, the different regulation bodies for dark chocolate mainly focus on cocoa solid content, and neither the Codex standards nor the EU establishes a minimum of milk fat to be present in the recipe. In general, in those countries where regulations permit, up to 5% of the cocoa butter can be replaced with cocoa butter equivalents (CBEs), which are compatible with cocoa butter. Based on European chocolate legislation, CBEs that are made from palm, shea, mango, kokum, Illipe and sal are permitted to be used in chocolate (EU 2000). White chocolate does not contain any non-fat cocoa solids, i.e. from cocoa liquor or powder, and only milk ingredients, cocoa butter and/or vegetable fats are present.

9.7.1.2 Milk Fat Ingredients in Chocolate

Milk fat influences the flavour, texture and quality of chocolate, contributing to a smooth texture and glossy appearance. Traditionally, it can be added to chocolate using whole milk powder (WMP, 26% fat), through incorporation of anhydrous milk fat (AMF, >99.8% fat) or as milk crumb (a cooked mixture of liquid milk, sugar and cocoa liquor) (Minifie 1989). Some manufacturers also use high-fat cream powders (42–75% fat) for the production of special high-fat chocolates (Bolenz *et al.* 2003). Buttermilk powder (8% fat) is another dairy ingredient that contains milk fat and that may be used in chocolate. Skimmed milk powder with less than 1% fat is always used in combination with AMF. The reason for choosing one milk fat-containing ingredient or another depends not only on the nutritional, texture and sensory requirements of the chocolate but also on the cost of the final product. The composition of typical dairy ingredients used for milk fat addition in chocolate is shown in Table 9.5.

Traditionally, milk fat is used in the confectionery industry as AMF or anhydrous butteroil (Hartel 1996) due to the low moisture that is required for chocolate manufacture. AMF is obtained from fresh cream or butter that is not expected to sell within the time to maintain acceptable quality and was developed as a method to store milk fat (Sichien *et al.* 2009; Early 2012). The process involves the removal of water, proteins and other minor components to obtain a product with a minimum fat content of 99.8%, of which 98.8% are triglycerides (TAGs) and 0.1% water (Codex 1999). This allows milk fat to be used over a longer shelf-life (up to 12 months) without a loss of quality. Besides AMF, anhydrous butter oil has the same minimum 99.8% fat content, but it is manufactured from cream and butter of different ages (Rønholt *et al.* 2013). AMF is the preferred option for confectionery applications, due to its higher quality and less susceptibility to oxidation.

Crumb is an ingredient widely used in the manufacture of milk chocolate in the UK, Australia and USA and provides a characteristic caramel flavour to chocolate (Stewart and Timms 2002; Beckett 2003). The chocolate crumb process was developed in the UK in the 1930s to enhance the shelf-life of milk used for chocolate manufacture. Despite the numerous methods to manufacture crumb, the process starts by mixing and heating liquid milk and sugar to remove moisture and guarantee a low water activity in the final crumb. Then, cocoa liquor is added, and the mixture is dried at a high temperature in the range from 75 to 105 °C to reach a final moisture content of 0.8–1.5%.

Although many variants exist, i.e. full milk crumb, white crumb, etc., the main ingredients of crumb are liquid milk, sugar and cocoa liquor, which are heated at high temperatures and dried to a moisture content of about 1%. This intermediate raw material can then be added to cocoa butter and other ingredients to create the final milk chocolate (Edwards 1984). The milk fat content of crumb can vary from 8 to 14%, which, in combination with its cocoa butter content, results in better flow properties than when ingredients with lower free fat contents are used to manufacture chocolate (Skytte and Kaylegian 2017).

9.7.1.3 Free Versus Bound Milk Fat

The form that fat adopts in each of these milk ingredients, namely, whether free or bound, is of paramount importance to the processing and sensory properties of the chocolate. Fat that can be extracted by organic fat solvents under standardized conditions is defined a "free fat. This type of fat is typically the most preferred because it is available to be incorporated to the fat continuous phase of chocolate, making it easier for the mass to flow and later crystallize.

With regard to WMP, the structure of its particles varies depending on its manufacture, i.e. roller- or spray-dried. Particles from roller-dried WMP are more compact, with sharp edges, and typically have >90% free fat content due to the shearing and scraping action of the knives on the drum roll as the film dries (Liang and Hartel 2004; Dewettinck *et al.* 1996). Roller-dried WMP is preferred by chocolate manufacturers, due to its high free fat content, which means that less fat is needed to be added during processing of chocolate (refining and conching) to coat particles and improve its flow. In addition, the higher heat load given to powders in the drum rolls during roller drying is believed to provide an improved flavour and taste to milk chocolate. In contrast, spray-dried WMP contains <10% free fat (Aguilar and Ziegler 1994), the rest of the fat being entrapped in vacuoles and a lactose-protein matrix. This makes spray-dried WMP less efficient for chocolate manufacture, requiring up to 2.5% extra cocoa butter to be added during processing to provide the same flow properties compared to roller-dried WMP (Verhey 1986). Although the use of roller-dried WMP may be preferred from a processing point of view, its high free fat content renders it more prone to oxidation, resulting in a shorter shelf-life.

Dairy manufacturers use different techniques to increase the free fat content of high free fat (HFF) milk powders or in specialized spray-dried milk powders. Pre-treating the milk powder with high shear and elevated temperatures (75 °C) has shown to help crystallize lactose just above its glass transition temperature and induce immobilized fat in the powder particle to be part of the continuous phase of the chocolate emulsion (Franke and Heinzelmann 2008; Koc *et al.* 2003; Twomey and Keogh 1998). Another way to increase free fat is to use high homogenization pressure before evaporation (Skytte and Kaylegian 2017), or controlling the protein and solid fat content of the raw milk (Twomey *et al.* 2000). Despite all the pretreatment methods that are reported in literature, the most common way manufacturers ensure milk fat is 100% free in the chocolate recipe is by addition of AMF in combination with skimmed milk powder (SMP), since SMP has very low fat content (0.5% fat). This mixture of dairy ingredients provides manufacturers an economical alternative to roller-dried WMP and HFF. However, the anhydrous milk fat is still susceptible to oxidative rancidity, leading to off-flavour development.

9.7.1.4 Manufacture of Chocolate

Traditionally, chocolate is manufactured following a set of common steps consisting of mixing, refining, conching and tempering, followed by moulding, demoulding, packaging and storage (Figure 9.11). During the initial mixing, a suspension is formed between the dry ingredients, i.e. sugar and milk powder, and the liquid ingredients, i.e. cocoa liquor, cocoa butter, milk fat and surfactants. Milk is always added to chocolate in a dehydrated form, either as a powder ingredient or as crumb, since moisture is detrimental for the quality and processing of chocolate, as it reduces flow properties and modifies the mouthfeel of chocolate. The final chocolate has a very low moisture content (less than 1.5%) that ensures microbiological stability.

The fat content at the mixing step is key to achieve the desired particle size and throughput during refining. However, the optimal fat content at the refining step is recipe-dependant. As is shown in Figure 9.12, recipes with spray-dried WMP (left) generally need a higher total fat content at the refining step, since it is expected that the free fat content will not be as high as when a combination of AMF and SMP is used (right).

The particle size of the sugar and milk particles present in the agglomerated mass is then reduced to less than 30 μm by grinding. Final particle size influences the rheological and sensory properties of the chocolate (Afoakwa *et al.* 2007). The preferred grinding method for high throughputs of chocolate is roll refining, where chocolate is forced to pass a series of gaps that decrease in width (2- and 5-roll refiners) (Ziegler and Hogg 2017). Minor chocolate manufacturers often use ball mill refining, where agglomerated chocolate is forced towards grinding balls and the mass is recirculated until the desired particle size is achieved (Alamprese *et al.* 2007). The stress applied to the chocolate mass during refining leads to the breakdown of agglomerates and particles, creating smaller particles with varied morphology and distribution within the fat continuous phase. The creation of such particles is also depending on the chocolate recipe, mostly the ratio of crystalline (sugar) to amorphous (milk) material and the amount of free fat available to coat the increased surface area of particles. As previously mentioned, availability of milk fat is an important parameter at the refining step, where the less free fat available will result in drier chocolate masses and that can result in problems achieving the desired particle size. The new particles created during refining will ultimately affect the flow properties of chocolate.

After refining, the chocolate mass is conched, a step with two equally important aims: first, to develop flavour and, second, to convert the crumbly paste, flake or powder that is obtained during refining to a flowable liquid that can be poured into a mould (Beckett *et al.* 2017). The desired viscosity of chocolate is obtained through three conching phases, dry, pasty and liquid phases, although not all phases occur in all types of recipes. During the dry phase, the moisture content of the chocolate mass needs to be reduced from the initial 1.6% in milk chocolates to <1%, through mixing and heating with temperatures as high as 80 ° C. The moisture is released slowly while trying to avoid the formation of agglomerates. During this evaporation, unwanted flavour compounds are also eliminated (Beckett *et al.* 2017). In the following pasty phase, the viscosity of the chocolate mass starts to fall due to the moisture removal and solid particles becoming now coated with fat and surface-active ingredients, such as lecithin, with the aim to modify viscosity. As a result, a pasty mass is obtained, and flavour will continue to be developed through heating. Finally, the last additions of fat and emulsifier are done in the liquid phase, in order to mix the added ingredients in a short time. The final viscosity of the chocolate will depend on both the shearing intensity in the conche and the final additions of fat and emulsifier (Beckett *et al.* 2017). The desired outcome is a product that has as low viscosity as possible. Nevertheless, this usually means higher amount of fat to cover the surface of the particles, resulting in more expensive chocolate.

After conching, the chocolate is ready to be tempered, a thermal treatment within the range of 27–29 °C (depending on the recipe) to ensure the formation of βV fat crystals with correct size. During tempering, these stable crystals will be

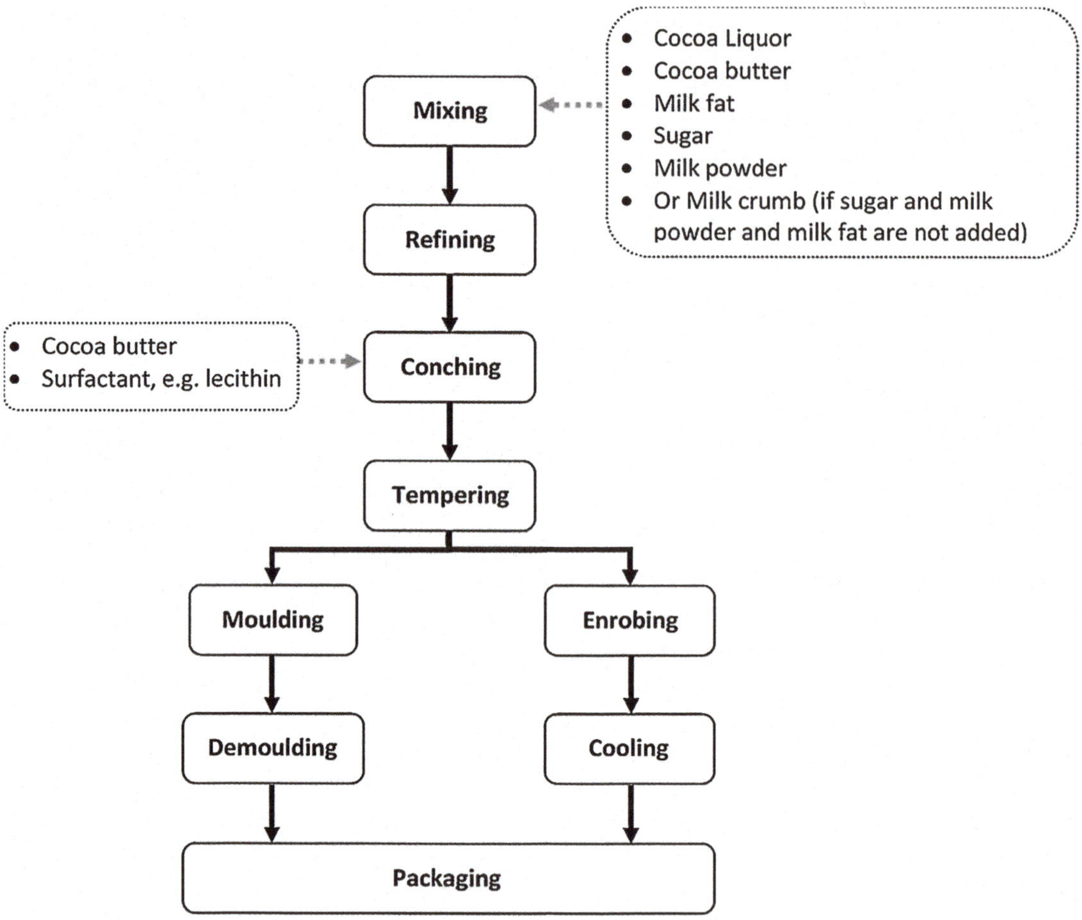

Figure 9.11. Chocolate processing scheme for chocolate manufacture.

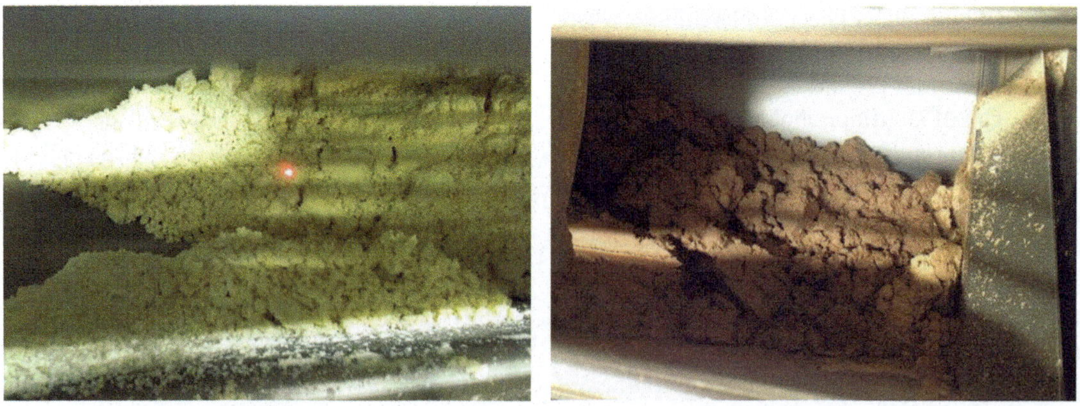

Figure 9.12. On the left, a white chocolate recipe made using spray-dried WMP (29.3% total fat at refining). On the right, a milk chocolate recipe made using AMF and SMP (24.5% total fat at refining).

subsequently used as seeds to obtain a homogenous fat crystal network during the cooling stage (Windhab 2017). At this time, the tempered chocolate mass can be used in moulding or enrobing (coating of products).

9.7.2 The Role of Milk Fat in Chocolate

Milk fat is used in a wide variety of confectionery products, such as milk, dark and white chocolate, as well as in fillings, e.g. caramel, toffee and cream. Besides cocoa butter, milk fat is the only fat that is permitted to be used globally in chocolate, as compared to other vegetable fats that cannot be added with the resulting product being called "chocolate". Not only it imparts a distinctive buttery flavour to chocolate applications but also reduces hardness, affecting therefore consumer satisfaction. Moreover, addition of milk fat into chocolate reduces the incidence of fat bloom, a defect of chocolate that results in a white or grey appearance and crumbly texture (Marangoni. 2002). Despite these advantages, milk fat has a high cost compared to other fats and oils, and its use needs to be always assessed for rheological suitability in the chocolate product (Hartel 1996).

In general, milk fat is compatible with cocoa butter, allowing a maximum addition of 30% of milk fat, calculated on the basis of total fat, without notable effect on the cocoa butter polymorphism (Metin and Hartel 1996). However, a eutectic effect in solid state is observed when mixing milk fat into chocolate (Marangoni. 2002). The mixture of TAGs from both milk fat and cocoa butter decreases the melting point of the fat mixture below the melting point of either components. Another factor changing the melting properties of milk fat-cocoa butter blend is the shifts in polymorphism. Changes in the crystal structure and stability of TAGs from both fats will cause crystallization as separate entities rather than as a mixture. All these factors will ultimately have a detrimental effect on the rheological behaviour of the chocolate mass, as well as on the hardness of the chocolate. Therefore, the total amount of milk fat that is often added into chocolate will depend on the chocolate recipe and tempering conditions (i.e. time and temperature).

9.7.2.1 Effect of Milk Fat Composition Variation in Chocolate

Composition of milk is very complex, and it is affected by herd (genetic variation), stage of lactation, seasonal variation and feeding regime of cows (Palmquist *et al.* 1993). Apart from the changes in the composition of protein, lactose and minerals of the milk, characteristics of milk fat are also greatly affected. Consequently, not only the fat content will vary but also the fatty acid and triglyceride composition, impacting, therefore, the solid fat content of the milk, which is particularly affected by the ratio of oleic acid to palmitic acid. The proportion of short-chain fatty acids in milk fat is low at the beginning of the cow's lactation, and it gradually increases during lactation (Palmquist *et al.* 1993). Since short-chain fatty acids play an important role in butter-like flavour, the use of milk fat obtained during early stage of lactation may also have a detrimental impact on chocolate flavour. However, the dietary regime of cows has been shown to have a greater influence on solid fat content of milk fat than stage of lactation (Twomey *et al.* 2000; Rowney and Christian 1996). Dairy cow diets are often composed of fresh forages (grass), conserved forages (silage, hay) and concentrates (plant seeds), in different ratios depending on season and geographical area (Elgersma *et al.* 2006). The proportion of fresh grass in the feed seems to be linearly correlated with a decreasing amount of milk fat content. In addition, if the proportion of grass in the forage is higher than 30% of the total cow's feed, the amount of unsaturated fatty acids increases (Couvreur *et al.* 2006). This level of unsaturated fatty acids can cause flavour defects, such as oxidation that ultimately may influence the flavour delivery in chocolate. Nowadays, it is possible to standardize the feeding regimes of cows to avoid seasonal variation in the fatty acid composition.

9.7.2.2 Use of Milk Fat Fractions in Chocolate

Over the years, milk fat fractionation has offered opportunities for creative use of this important dairy ingredient. From the complex composition of milk fat, and its broad melting range from −40 to +40 °C (Sichien *et al.* 2009), three main triglyceride groups can be generated via fractionation: low (LMFs), medium (MMFs) and high (HMFs) melting fractions. Two extra milk fractions are added by some authors, which include a very high melting fraction (VHMF), melting above 45 °C, and a very low melting fractions (VLMF), melting below 10 °C (Kaylegian and Lindsay 1995). Table 9.6 summarizes the potential benefit of using milk fat fractions in confectionery applications. HMFs can be used in milk and dark chocolate because they improve texture, hardness, gloss, snap and mouthfeel (Early 2012). Other potential applications for MMFs and LMFs could be confectionery fillings, e.g. caramels or toffees, due to their colour, flavour and softness.

Though several studies have been carried out exploring the structural benefits of using milk fat fractions in chocolate (Barna *et al.* 1992; Hartel 1996; Schmelzer and Hartel 2001; Kaylegian *et al.* 1993), added cost is still a primary concern. Therefore, besides the more traditional contribution of milk fat for bloom retardation, texture modification and flavour, the use of milk fat fractions in chocolate has been rather limited.

The most common industrial way to fractionate milk fat is dry crystallization, separating TAGs based on differences in their melting points (Kaylegian 1999). This is the fractionation method preferred by chocolate manufacturers, due to the growing consumer trend towards natural and clean label as solvent fractionation would not yield clean label fractions and there is a very limited choice of solvents suitable for extraction. As it can be seen in Figure 9.13, the three milk fat fractions melt at different temperatures: LMF melts below 10 °C, MMF between 10 and 21 °C and HMF above 21 °C (Vanhoutte *et al.* 2002). Timms (1980) similarly studied three fractions from untampered native milk fat using differential scanning calorimetry (DSC). His results showed that HMF melts at >50 °C and MMF in the range of 35 to 40 °C and LMF is melting at <15 °C. These melting ranges differed quite considerably from those obtained by Vanhoutte *et al.* (2002), highlighting the variability in melting behaviour that one can expect from milk fat fractions.

With regard to fractionation, Timms (1980) calculated the yield of HMF to be around 5%, 25% the yield of MMF and 70% the yield of LMF. On the other hand, Marangoni and Lencki (1998) reported 12% of HMF, 33% MMF and 55% LMF using solvent fractionation. These fractions are chemically distinct; generally long-chain saturated fatty acids are found in HMF, while TAGs of MMF consist of two long-chain saturated fatty acids and one short-chain or *cis*-unsaturated fatty acid, and one long-chain saturated fatty acid and two short-chain or *cis*-unsaturated fatty acids are found in TAGs of the LMF (Marangoni and Lencki 1998; Timms 1980).

As described before, these fractions are known to improve performance of milk fat in chocolate and confectionery applications, due to their different physical and chemical properties, melting point and solid fat content (SFC) (Bystrom and

Table 9.6. Potential benefits of using milk fat fractions in confectionery applications

Milk fat fractions[a]	Benefit	Application
HMF	Improved texture	Milk chocolate, dark chocolate, confectionery fillings, coatings
	Improved firmness	
	Antibloom	
	Gloss	
	Snap	
	Flavour and mouthfeel	
MMF	Softness	Confectionery fillings
	Flavour	
	Colour enhancement	
LMF	Softness	Chocolate used for enrobing, confectionery fillings
	Flavour	
	Colour enhancement	

[a]HMF = high melting fraction, MMF = middle melting fraction, LMF = low melting fraction.

Hartel 1994; Dimick *et al.* 1996). From these differences, the three fractions have very diverse melting characteristic as featured by the solid fat content profiles in Figure 9.14. Equally the fractions exhibit unique crystallization performance producing unique polymorphic forms. The use of these milk fat fractions, individually or as a blend, may allow for changing the rheological characteristics of the final chocolate. However, their suitability needs to be carefully evaluated to ensure the final product has the desired quality attributes, without incurring into a considerable cost increase.

9.7.3 Contribution of Milk Fat to Microstructure and Texture of Chocolate

Functional attributes that are commonly associated with milk fat include flavour and physical properties, such as structure formation, hardness, spreadability, layering, shortening and lubricity. However, these attributes are very dependent on the type of food system in which milk fat is used. For example, butter is used for its shortening properties in cookies to yield a tender crumb but is used in pastries for its layering properties, which promotes the characteristic flakiness of croissants and puff pastry. In chocolate, the inclusion of milk fat is done to modify the processing and subsequently impart the desired texture to the finished products, i.e. mouthfeel and melting behaviour.

9.7.3.1 Factors Affecting Crystallization of Milk Fat in Cocoa Butter

Traditionally, the addition of milk fat to chocolate to impart a creamy texture and desired milk flavour has notable structural implications. This is attributed to the fact that the composition of milk fat is among the most complex of fats and oils, comprising no less than 100 fatty acid types and 28 triglyceride species (Lopez *et al.* 2006; Shi *et al.* 2001; Van Aken and Visser 2000). This undoubtedly accounts for the highly variable crystallization behaviour of milk fat. The solidification of chocolate with milk fat added is dependent on the co-crystallization of the cocoa butter-milk fat phases, which ultimately affects the appearance and physical properties of chocolate (Koyano *et al.* 1990). In general, any crystallization of fat involves two stages: nucleation followed by crystal growth. Without considering tempering, which is necessary to induce the formation of the desired polymorphic form in cocoa butter, the crystallization of milk fat within a blend with cocoa butter remains sensitive to factors like temperature (undercooling), agitation, cooling rate and minor lipids. The impact of minor lipids is driven by concentration, which can fluctuate drastically, and this is discussed below. With regard to the temperature used, very low temperatures are normally avoided as they promote the formation of unstable polymorphs. Simultaneously, the rapid cooling of the fat after melting generates small regular crystals. On the contrary, slow cooling to warmer temperatures results in fewer but larger crystals. Therefore, using high agitation for recipes containing milk fat can help in breaking large crystals formed into smaller ones (a form of secondary nucleation). Low agitation rate conversely will result in slow crystal growth rates as both heat and mass transfer become less efficient.

9.7.3.2 Microstructure in Relation to Processing Conditions

Milk fat is partially solid at temperatures between approximately 5 and 25 °C, and its consistency is due to the presence of a network of fat crystals in liquid fat (Precht 1988; De Man and Wood 1959). In this network, the fat crystals are connected by solid connections, i.e. primary bonds. During crystallization, these bonds are formed by sintering, either when growing crystals come into contact with each other or via flocculation of small crystal nuclei between two fat crystals (Johansson and Bergenståhl 1995). Rupture of primary bonds by mechanical treatment leads to a softening of the fat (Heertje 1993). However, the firmness will quickly increase again due to a reorganization of the fat crystals into a relatively weak network held together by van der Waals

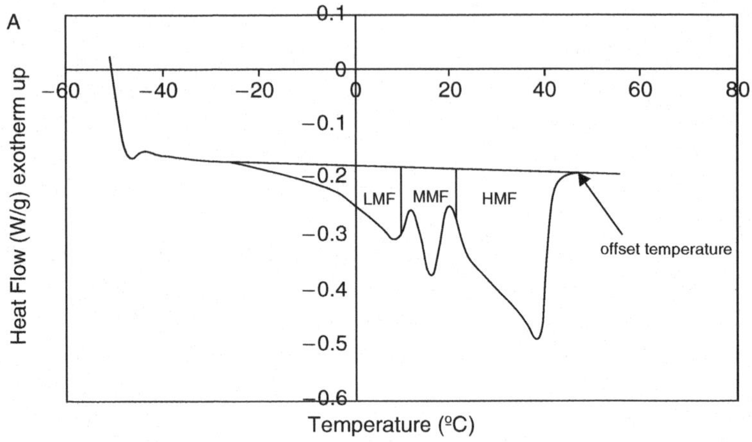

Figure 9.13. Differential scanning calorimetry (DSC) melting profile of milk fat showing different fractional components. Scanning rate: 5 °C/min (from Vanhoutte *et al.* 2002).

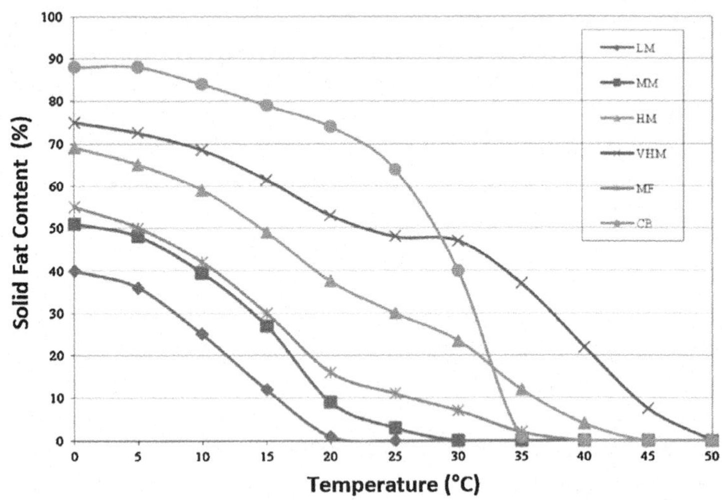

Figure 9.14. Solid fat content curves for a range of milk fat fractions. CB, cocoa butter; MF, milk fat; VHMF, very high melting fraction; HMF, high melting fraction, MMF, middle melting fraction; LMF, low melting fraction (from Metin and Hartel 2012).

forces, i.e. secondary bonds. This step is followed by the slow ongoing recrystallization processes, which ultimately leads to the formation of new primary bonds (Pedersen 1991; Van Aken and Visser 2000). Compared with most other fats, milk fat sets into its final firmness relatively slowly, mainly due to the very large number of triglyceride components with very large differences in fatty acid composition. This would lead to structural incompatibilities, which obstruct the incorporation of TAGs in the growing crystals. As milk fat may constitute a maximum of 20% in the total fat content in chocolate, the more crucial consideration in terms of structure will be its eutectic effect upon mixing with cocoa butter.

9.7.3.3 Eutectic Effect in Milk Fat–Cocoa Butter Blend

The occurrence of a eutectic effect when adding milk fat (or some of its fractions) is caused by the incompatibility of TAGs of milk fat and cocoa butter, the key fat components in chocolate. Hence, the use of non-fractionated milk fat in chocolate is limited to around 5% (Beckett 2008), before undesirable softening of the product is observed. This behaviour is primarily attributed to the LMF and MMF of milk fat, which interact with the TAGs of cocoa butter (Timms and Parekh 1980). The LMF of milk fat can act as a solvent dissolving the cocoa butter depending on the storage temperature of chocolate. On the other hand, the MMF forms a eutectic mix with the cocoa butter TAGs, combining with other low melting polymorphic cocoa butter crystal to form a semi-solid fraction. While a severe eutectic effect exhibited by HMF was previously reported by Hartel (1996) (Figure 9.15), Marangoni (2002) attributed this occurrence to residual MMF. Likewise, no eutectic effect with cocoa butter was previously observed for HMF by Kaylegian *et al.* (1993).

Hydrogenated milk fats are also good bloom inhibitors, although they still have been reported to soften the chocolate at ambient temperatures (Campbell *et al.* 1969; Timms and Parekh 1980). Hydrogenated milk fat exhibits lower solubility effects caused by the LMF. However, this advantage is compromised by the corresponding increase in the eutectic effects of the MMF that contributes to softening (Timms and Parekh 1980). With growing concern of hydrogenated fat ingredients by consumers, the industry is gradually moving away from this process.

9.7.3.4 Minor Components in Milk Fat

The TAGs and minor lipid components in milk fat such as diglycerides (DAGs), monoglycerides (MAGs), cholesterol, individual fatty acid components and phospholipids exhibit a range of characteristics that can be used to provide unique functional properties for applications with added milk fat. Wright *et al.* (2000a, b) and Herrera *et al.* (1999) verified that the minor lipids in milk fat slowed the nucleation period at temperatures above 25 °C. However, differences in the composition of the milk fat samples used have led to contradictory results between these studies. Mazzanti *et al.* (2004) found that the minor lipids present in milk fat can slow the early stages of crystallization, reducing the rate of crystal growth and generating unstable β′ form.

To understand the impact of minor lipids on the crystallization behaviour and physical properties of chocolate recipes, a study was conducted by Tietz and Hartel (2000) using a blend of cocoa butter and milk fat (10%). Evaluations were made with (a) complete removal, (b) normal level (2.5% w/w) and (c) double level (5.0% w/w) of minor lipids in the milk fat. Results showed that removing the minor lipids from milk fat leads to delayed nucleation. In addition, irregularly shaped primary and secondary crystals were observed trapping a significant amount of liquid fat. When the blend was then used in a chocolate recipe, it led to rapid bloom formation. In contrast, normal levels of minor lipids in milk fat resulted in quick onset of nucleation, generating spherical and uniform crystals that led to denser packing of the fat crystal network. Similar to the effects seen with the removal of minor lipids, doubling the level of minor lipids slowed nucleation, reduced the crystallization rate and led to a rapid development of bloom in chocolate. From this study, it is suggested that minor lipids may be acting as catalytic sites of nucleation at low levels but, when present in higher concentrations, may interfere with crystallization.

Based on the latest research studies, there is a consensus that the presence of milk fat minor lipids in chocolate at natural or slightly higher concentrations would give rise to uniform crystals and help reduce fat bloom. On the contrary, the exact impact of minor lipids on the crystallization and polymorphic habit remains unclear, as there is limited data on the model system containing milk fat, minor lipids and cocoa butter (Metin and Hartel 2005).

9.7.4 Contribution of Milk Fat to Flavour of Chocolate

Chocolate has a complex flavour profile composed of numerous volatile and partly odour-active compounds. Although these compounds are not unique to chocolate, their concentration and combination in different ratios result in the main flavour notes associated with chocolate: fruity, spicy, floral, cocoa, acidity, bitterness, astringency, woody and also some off-flavours (Engeseth *et al.* 2018). Flavour in chocolate is the outcome of the combination of individual raw materials, i.e. cocoa, milk ingredients, sugar, etc., and the processing steps involved in the manufacture of chocolate. Likewise, each of these ingredients has been subjected to specific processing

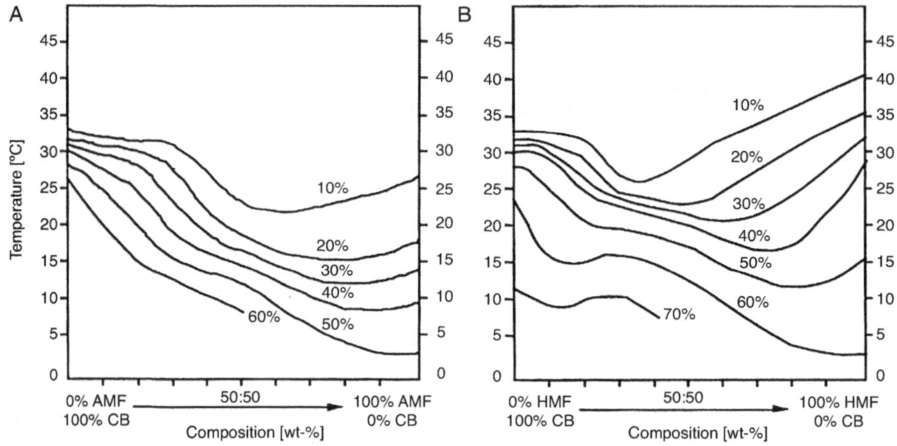

Figure 9.15. Isosolid phase diagrams of mixtures of cocoa butter (CB) with (A) anhydrous milk fat (AMF) and (B) high melting fraction (HMF) of milk fat (from Hartel 1996).

steps, e.g. roasting of cocoa beans, which are likely to affect the final flavour of chocolate. To develop a desirable flavour in chocolate, understanding of the key aroma and taste-active compounds in the finished product is important, as well as those in the raw materials. For a long time, industry and researchers considered conching of chocolate as the critical step where flavour development occurs. Nowadays, it is generally recognized that during conching, the flavour components in the chocolate mass are redistributed (Ziegleder 2017), and this can help to modulate the final flavour of the conched mass. As a result, a deep knowledge on the composition of sensorially active components in raw materials in combination with changes in conching conditions can help manufacturers to achieve the desired flavour in chocolate.

The most important raw materials for flavour development in chocolate are cocoa beans and milk ingredients (milk powder and milk fat). Fermentation, drying and roasting of cocoa beans are considered as some of the most important processing steps for chocolate flavour (Ziegleder 2017). Likewise, the flavour of milk fat ingredients is largely dependent on the cream used for its preparation. Key odorants of milk fat ingredients that contribute to butter-like flavour are diacetyl, δ-decalactone (coconut, peach) and butanoic acid (sweaty) (Schieberle *et al.* 1993). Their concentration in the final chocolate may promote differences in the overall flavour of chocolate and could act as an indicator of buttery flavour note.

In the case of milk chocolate, flavour is due to a balance between the primary flavour compounds originating from the cocoa mass and the volatiles found in milk ingredients. The flavour of milk ingredients used in chocolate, i.e. milk powders, AMF, or anhydrous butteroil, depends not only on the composition of such ingredients but also on how they have been manufactured. The most important aroma compounds derived from milk fat origin in milk chocolate have been identified as γ- and particularly δ-lactones (sweet, coconut) (Schlutt *et al.* 2007; Schieberle *et al.* 1993). In general, chocolates made with larger amounts of free fat are better perceived in terms of flavour by consumers than those with bound fat, e.g. using spray-dried WMP (Bolenz *et al.* 2003). In addition, free fatty acids present in milk ingredients, as well as the lactones (milky, sweet, coconut), seem to be fundamental contributors to flavour. The components formed from β-hydroxy fatty acids released from the TAGs when milk fat is heated (Skytte and Kaylegian 2017) are lactones and are responsible for milky, buttery or creamy odour (Shiratsuchi *et al.* 1994b). Moreover, aldehydes, aromatic hydrocarbons and some heterocyclic compounds are likely to participate indirectly in milk flavour (Shiratsuchi *et al.* 1994b).

The use of milk crumb in the manufacture of milk chocolate provides a characteristic caramel flavour to chocolate, widely used in the UK, in Australia and in the USA (Stewart and Timms 2002; Beckett 2003). This specific flavour is difficult to develop using other processing conditions. The process relies on the Maillard reaction occurring between proteins found in milk and cocoa and reducing sugars found in milk (lactose) in a high moisture environment. The crumb heating process will result in key flavour compounds such as furfural (sweet, woody, baked bread), maltol (sweet caramel, toffee), lactones and methyl ketones (Ziegleder 2017). The final crumb flavour will depend on the moisture content, the temperature and the time spent at each stage. Since most of the flavour of the chocolate is developed during the crumb process, conching in this type of chocolate is not needed for flavour purposes, and only a liquefaction is required to achieve the correct viscosity of the chocolate mass. In addition to the role of cocoa liquor for flavour formation in crumb processing, it is worth mentioning that the high polyphenol content in cocoa liquor acts as an antioxidant to prevent milk fat from becoming rancid and contributing a sour or cheesy flavour in the final chocolate (Beckett 2003). For milk chocolates where crumb is not used as an ingredient, it is not possible to develop the same caramel and flavour notes during conching as those obtained during crumb processing. This is due to the low moisture content present in the chocolate mass during conching. In milk chocolate, milk notes are usually represented by δ-lactones, 2,3-butanedione, 1-octen-3-one and 2,4-decadienal (Schnermann and Schieberle 1997).

With regard to novel milk fat ingredients, there is extensive research on the use of milk fat fractions to enhance butter flavour in milk chocolate (Hartel 1996). LMF tends to increase milk flavour, whereas higher melting ones seem to improve bloom stability (Full *et al.* 1996). On the contrary, chemical or enzymatic modification performed on milk fat to improve its rheological behaviour, e.g. interesterification, often leads to a decreased flavour perception in the products in which it is incorporated (Weihe 1961; Rousseau and Marangoni 1998).

9.7.4.1 Off-Flavours from Milk Fat Ingredients

Care must be taken on production and storage of dairy ingredients used in chocolate, since off-flavours generated from cow's diet, environment, heat treatment, bacterial spoilage, oxidation of lipids, etc. (Shiratsuchi *et al.* 1994a) could be transferred to the final chocolate, affecting its sensory perception. This was observed in stale milk chocolates, where an increase in short-chain free fatty acids and volatile lipid oxidation products, 3,5-octadien-2-ones, was found. These are formed by residual enzymatic activity of lipases and lipoxidases, despite the low water activity of chocolate (Ziegleder 2017).

Factors such as herd, stage of lactation and dietary feed affect the composition of milk fat, and this will finally influence its flavour and stability. Dietary fat fed to cows was shown to affect the flavour of milk fat, as well as its oxidative stability (Palmquist *et al.* 1993). High-fat diets decrease the content of components exhibiting coconut flavour (δ-octalactone and δ-decalactone), peach flavour (δ-dodecalactone) and blue cheese (methyl ketones), whereas these components increase with low-fat diets, e.g. alfalfa (Urbach 1990). Off-flavours originating from animal feeding can cause aroma defects, e.g. fermented silage, musty silage and alfalfa (Mallia *et al.* 2008). In addition, oxidative flavours can occur easily, depending on cows' feed. Thus, milk from cows fed pasture is less susceptible to oxidation than milk from cows on dry stored feed. Likewise, milk fat with more than 20% of linoleic acid (C18:2), originating from low-fat feed, results in oxidized off-flavours (Edmondson *et al.* 1974).

Among the several off-flavour compounds that can occur in chocolate, those described as cardboardy or fishy are related to oxidation of milk fat (Skytte and Kaylegian 2017). The main factor needed to start the oxidative chemical reaction is the availability of oxygen.

Nonetheless, other variables such as storage temperature, light, α-tocopherol level and metallic contamination, especially copper, will also speed the rate of the reaction (Keogh and Higgins 1986). In general, lipid oxidation can be promoted in two ways: firstly, when double bonds of unsaturated fatty acids uptake oxygen yielding hydroperoxides that are then further converted into off-flavour components such as aldehydes and ketones and, secondly, by the action of lipolytic enzymes on TAGs that can produce DAGs, MAGs and particularly free fatty acids (FFA), which are more prone to oxidation. Some of the shorter FFA may contribute to off-flavours, e.g. butanoic acid (sweaty), in products such as AMF (Keogh and Higgins 1986; Schieberle *et al.* 1993). Lipolysis can also be beneficial in milk chocolate, and it is characteristic of American-type chocolates, such as Hershey's, where lipolysed milk fat is used to obtain a chocolate with cheesy, sour or tangy or butyric flavour (Hayes *et al.* 2016; Martin Jr. 1988).

Oxidation in AMF is measured by the peroxide value, where intermediate compounds related to oxidative status of the fat are detected (Keogh and Higgins 1986). Peroxide values below 0.2 meq O_2/kg fat guarantee a high-quality AMF product with no risk of oxidation, although the maximum allowed value is 0.3 meq O_2/kg (Codex 1999). Considering that the use of antioxidants in AMF is not permitted by Codex standards (Codex 1999), mechanisms that are used to reduce the rate of oxidation and contribute to longer shelf-life include:

- Removal of oxygen by flushing AMF with nitrogen
- Using oxygen barrier packaging
- Storage at chilled temperatures, i.e. 4 °C
- Avoiding the presence of oxidation catalysts (Skytte and Kaylegian 2017)

Therefore, only high-quality premium AMF is generally used in confectionery in order to avoid changes in colour, flavour, aroma or nutritive value of the product.

9.7.5 Bloom-Retarding Effect of Milk Fat in Chocolate

A common quality defect that is observed in chocolate industry is development of bloom. Several factors have to date been identified that are known to contribute to occurrence of bloom in chocolate. These factors are (a) improper processing conditions, (b) use of incompatible ingredients and (c) excessive heat exposure. Any of these factors can itself lead to bloom, as well as their interaction can result in more complex situations that generate the following bloom types: (a) heat damage bloom, (b) migration bloom and (c) storage bloom. It is to rectify the latter two forms of bloom that milk fat has been widely used by chocolatiers and manufacturers.

9.7.5.1 Cause and Mechanism of Migration Bloom?

A good overview on the interplay of the various factors that contribute to fat bloom was systematically discussed by Ziegler (2009). Three mechanisms are discussed relative to fat bloom: polymorphic transformations, liquid-mediated recrystallization (thermally induced) and oil migration. The contribution of each mechanism to bloom in a given recipe may vary, and as reported, the occurrence of any one of these phenomena does not necessarily lead to bloom formation.

The mechanism that drives the appearance of migration bloom first stems from the presence of significantly increased semi-liquid fat fraction in chocolate recipes. This increase can be brought by incorporation of nuts, soft indulgent filled chocolate or contaminants containing incompatible fats, such as lauric-based fats. As the semi-liquid fraction tends to melt readily at around 20 °C (typical ambient temperature in temperate climates), when recipes with these components are made and exposed to temperature cycling where the maximum temperature does not exceed 25 °C, liquefaction of this semi-liquid fraction will occur. As a consequence, the mobile liquid fat will gradually move through the matrix of the chocolate.

Several hypothesized mechanisms have been put forward and examined through the years to explain the kinetics of oil migration in filled and solid chocolates, but the exact mechanisms have not been completely elucidated yet. The hypothesis of Fickian diffusion has been widely proposed as the key driver of oil migration (Galdámez *et al.* 2009; Lee *et al.* 2010; McCarthy and McCarthy 2008; De Clercq *et al.* 2014; Dahlenborg *et al.* 2015b; Maleky *et al.* 2012; Ghosh *et al.* 2002; Guiheneuf *et al.* 1997). On the contrary, there is evidence suggesting the capillary flow model (Aguilera *et al.* 2004; Choi *et al.* 2005), correlating with observations made on porosity in chocolate (Rousseau 2006; Smith and Dahlman 2005). Other groups suggest a combination of both hypotheses (Rousseau and Smith 2008; Deka *et al.* 2006; Reinke *et al.* 2015), and most recent studies have even pointed at a third hypothesis known as pressure-driven convective flow (Dahlenborg *et al.* 2011, 2015a; Dahlenborg 2014; Altimiras *et al.* 2007). Increasingly, it has been accepted that most likely an interaction of all these different hypotheses provides the best fit to experimental data.

In confectionery, the use of nuts and indulgent low melting fats is unavoidable, as these ingredients impart unique sensory delight critical to the product. Some work has been invested to understand the optimized usage level of these ingredients acceptable for product quality (Rothkopf and Danzl 2015). Comparatively more work has been done on understanding the nature of migration with the aim of developing solutions in immobilizing liquid fat or oil within recipes. Further challenges come with the expanding footprint of chocolate into tropical markets, and this is gradually shifting the focus to addressing migration bloom at tropical ambient temperatures (>25 °C). Nevertheless, formulating shelf-stable filled shelf-stable filled chocolate products under tropical conditions will unquestionably restrict the type of fat-rich ingredients, not to mention the stability of the chocolate itself. It has been observed that exposing praline and other filled chocolates to temperatures between 25 and 30 °C can confer unusual bloom resistance (Walter and Cornillon 2001; Juul 2010). On the contrary, increasing the incubation temperature beyond 30 °C will see the likelihood of heat damage bloom depending on the duration of exposure.

As molten fat moves to the surfaces of the product, dissolution of cocoa butter crystal within the chocolate matrix can happen. Particularly vulnerable are the small crystals that exhibit higher solubility in liquid fat/oil. Following the drop in surrounding temperatures, the dissolved cocoa butter can come out of solution growing onto larger crystal (acting as seed) as explained by Ostwald ripening process (Ziegler *et al.* 2004). As this liquid-mediated recrystallization occurs uncontrollably on the chocolate surface, these growing enlarged crystals will appear as light powdery dusting on the chocolate surface as illustrated in Figure 9.16. Adding to this random crystal growths would be minute crystals that were carried up to the surface with the molten fat before depositing there contributing to more nucleating centres.

Though Ziegleder (1996) has shown a clear correlation between the onset of fat bloom resulting from oil migration, additional factors seem to influence the recrystallization that subsequently lead to the appearance of bloom. The key driver influencing recrystallization lies in the composition of the liquid fat phase that is usually a mixture of filling fats, cocoa butter and nut oil. An important physical property which is affected by the composition of the migrated liquid fat phase is the solubility of the solid fat components therein. In the presence of high amount of nut oil or low melting eutectic blend, fat bloom is usually being suppressed since it is difficult for the cocoa butter to crystallize (Ziegler *et al.* 2004). Hence this observation has led to the proposal of saturation of filling fat with cocoa butter to reduce the dissolution of the cocoa butter by the migrated liquid fat phase. It is noted that though blooming has been retarded, the softening of the chocolate undoubtedly indicates that bulk transfer of the soft filling fat or nut nevertheless has occurred.

Conversely, limited presence of nut oil in the recipe tends to promote polymorphic transformation. Smith *et al.* (2007) demonstrated that hazelnut oil promotes the transition from βV to βVI in cocoa butter. Even small amounts of hazelnut oil (1%) can accelerate the polymorphic transition. Transformation from low polymorphic form to higher form crystal is similarly observed by Stewart (2017) where β' to βV cocoa butter crystal occurs in the presence of hazelnut oil. Similarly, the solution will be further enhanced if this is coupled with a mean to block cocoa butter recrystallization or polymorphic transformation.

9.7.5.2 Function of Milk Fat in Retarding Migration Bloom and Storage Bloom

The addition of milk fat to chocolate formulations to improve bloom resistance is now a common industrial practice The addition of 2 to 3% AMF has been recommended to combat migration bloom in dark chocolate particularly in the recipes with nut inclusion (Minifie 1989). The mechanism through which milk fat can retard bloom formation has been studied using two different approaches, (a) by slowing down the rate of cocoa butter recrystallization (Timms 2003) and (b) by retarding the form V to VI transition, as milk fat TAGs are primarily β-stable (Lohman and Hartel 1994).

The ability of milk fat to prevent the recrystallization of cocoa butter from solution derives primarily from the eutectic effect when mixed with cocoa butter. However, excessive addition of milk fat to the recipe would in turn lead to increased liquid fat phase when exposed to elevated temperature. For example, a 10% milk fat addition to milk chocolate leads to a higher migration rate compared to chocolates containing no milk fat (Choi *et al.* 2005).

The inhibitory effect of milk fat on cocoa butter crystallization would also affect the processability of chocolate recipe requiring the latter to be tempered at different conditions. A rule of thumb is to temper milk chocolate at slightly lower temperatures than dark chocolate to offer more undercooling to allow more cocoa butter crystallization (Metin and Hartel 2012). Generally it is well accepted that a reduction of 1 °C in the cooling or crystallization zone will be adequate (Manning and Dimick 1985).

9.7.5.3 Impact of Milk Fat in Processing Conditions

By tempering chocolate containing milk fat, a more densely packed crystal network can be achieved into which the added milk fat can be incorporated. This network can serve to immobilize or delay any liquid fat phase comprising of oil or any low melting fat migrating through the chocolate matrix (Ghosh *et al.* 2002; Svanberg *et al.* 2011; Dibildox-Alvarado *et al.* 2004). A good temper can also ensure a substantial amount of the cocoa butter is crystallized in the stable βV form. This will significantly improve the structural integrity of the chocolate making liquefaction of the fat network less likely to occur when the chocolate is exposed to warmer conditions.

Figure 9.16. Pictures and scanning electron micrographs of the surface of dark chocolate pralines: unbloomed sample stored at 20 °C (left) and bloomed sample stored at 23 °C for 3 months (right) (from Delbaere *et al.* 2016).

However, optimization of tempering conditions to counter migration should never be overlooked in formulation of any chocolate involving the use of milk fat, as this is a critical step in compensating the risk of decreased temperability of the chocolate caused by the eutectic effect of mixing milk fat and cocoa butter (Reddy *et al.* 1996; Liang and Hartel 2004; Hartel 1996). However, inconsistent results have been found when assessing the impact of milk fat on bloom development in chocolate (Bricknell and Hartel 1998; Lohman and Hartel 1994) due to the variability in the tempered status of chocolate used by different authors. This points to the fact that the processing conditions of the chocolate are an essential factor in developing the full antibloom potential of any AMF or milk fat fractions used.

Between the two key processes of tempering and cooling that directly impact on creating the desired structure in chocolate, considerable focus has been placed on tempering either by conventional tempering or through seeding (Barna *et al.* 1992; Reddy *et al.* 1996; Svanberg *et al.* 2011).

Barna *et al.* (1992) found that tempering procedures had to be altered to produce good chocolate and that these alterations were based more on the replacement level of milk fat than on the type of fraction used. As the milk fat content was increased, the tempering temperatures had to be decreased to overcome the inhibition of cocoa butter crystallization caused by the milk fat. However, as is widely known, rapid cooling at low temperature will cause the formation of metastable form. This can lead to lower quantity of βV crystal formation since the bulk of the cocoa butter can be locked in the less stable form which tend to nucleate more rapidly. Under this circumstance, even with the reheating of the chocolate mass to remove the less stable crystal, there will be inadequate βV crystals to achieve the critical level for seeding the chocolate. In other words, there is always a limit below which the crystallization temperature within the temperer cannot be reduced in order to generate adequate βV crystals. Hence, Barna *et al.* (1992) reported tempering difficulties while working on recipes with 20% replacement of cocoa butter by HMF. For other fractions and AMF, 30% replacement level was observed by Barna *et al.* (1992) as the limit. However, other findings set the limit at 20% (Sabariah *et al.* 1998) to 25% of AMF (Brown 2009), at which the crystallization efficiency of a more complex fat blend fat is considered such as involving the use of cocoa butter equivalents (CBEs).

As discussed earlier, the need to lower the crystallization temperature when using conventional tempering to process recipes with high milk fat can severely limit the formation of stable βV crystals. However, this limitation can be circumvented by using seeding technology. The advantage of this technology is the generation of the seed is independent on the crystallization efficiency of the fat system. With availability of sufficient seed, a stable fat crystal network can then be generated during cooling, providing a binding matrix for any liquid fat, as illustrated in Figure 9.17 (Svanberg *et al.* 2011).

With adequate seeds generated in the tempering process, cooling is applied to use these crystals to crystallize the bulk liquid cocoa butter. During cooling, the TAGs from the molten phase will be deposited onto the seeds. These growing crystals eventually join to form an interconnecting network (Afoakwa *et al.* 2009). Depending on the temperature and cooling rate, crystallization of the matrix can vary widely. This can give rise to different polymorphic forms and random crystal size (Kamphuis 2009). For example, when cooling is done at low temperatures, this tends to favour the formation of metastable crystal (βIV), resulting in a less compact network. Consequently, any massive nucleation of βIV will naturally compete for the bulk liquid cocoa butter, limiting the latter from growing onto the βV seeds.

In order to reap the benefits of milk fat in bloom retarding while lessening the adverse effect on the crystallization of cocoa butter, researchers are increasingly looking into milk fat fractions (Kaylegian *et al.* 1993; Lohman and Hartel 1994; Dimick *et al.* 1996) to bridge the gap as previously discussed. The use of HMF and MMF with modified tempering condition is reported to give better resistance to bloom than AMF as compared to LMF (Reddy *et al.* 1996);

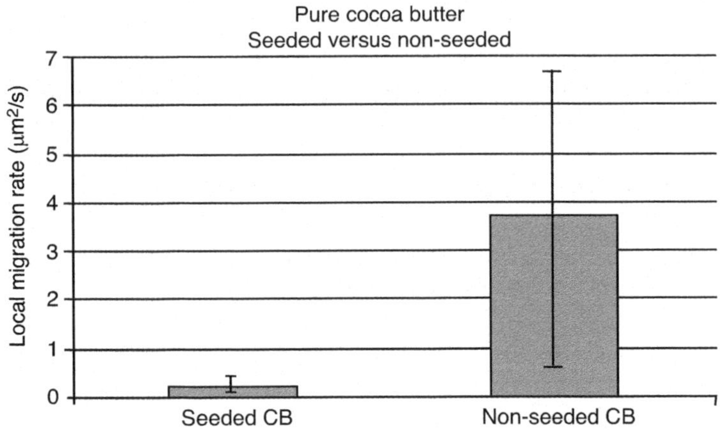

Figure 9.17. Calculated local diffusion rate of boron-dipyrromethene (BODIPY) dye ($\mu m^2/s$) in seeded versus non-seeded cocoa butter immediately after cooling. Error bars represent standard deviation of fluorescence recovery after photobleaching (FRAP) measurements of three sample replicates (from Svanberg *et al.* 2011).

the eutectic effect of HMF and MMF with cocoa butter (Hartel 1996; Bystrom and Hartel 1994) remains a key challenge in using these milk fat fractions. Studies by other teams however show no eutectic effect for HMF (Kaylegian *et al.* 1993; Marangoni and Lencki 1998) showing little or no reduction in hardness of the chocolate while conferring observable bloom resistance (Lohman and Hartel 1994).

References

References for Section 9.1

Anderson, M., Brooker, B. E. & Needs, E. C. (2005). The role of proteins in the stabilization/destabilization of dairy foams. In *Food emulsions and foams*. Cambridge: Woodhead Publishing Ltd. 100–109.

Auldist, M. J., Walsh, B. J. & Thomson, N. A. (1998). Seasonal and lactational influences on bovine milk composition in New Zealand. *Journal of Dairy Research, 65*, 4001–4411.

Baldwin, R. R., Baldry, R. P. & Johansen, R. G. (1971). Fat systems for bakery products. *Journal of the American Oil Chemists' Society, 49*(8), 473–477.

Bradley, R. L. & Smukowski, M. (2009). Chapter 6: Butter. In *The sensory evaluation of dairy products* (2nd ed.). Eds. Clark, S., Costello, M., Drake, M-A. & Bodyfelt, F. Washington, DC: Springer. 135–165.

Brooker, B. E. (1986). The development of structure in whipped cream. *Food Structure, 5*(2), 277–285.

Brooker, B. E. (1993). The stabilisation of air in foods containing fat – a review. *Food Structure, 12*, 115–122.

Bylund, G. (1995). *Dairy processing handbook*. Lund: Tetra Pak Processing Systems AB.

Chapman, D. (1962). The polymorphism of glycerides. *Chemical Reviews, 62*, 433–456.

Chilliard, Y., Glasser, F., Ferlay, A., Bernard, L., Rouel, J. & Doreau, M. (2007). Diet, rumen biohydration and nutritional quality of cow and goat milk fat. *European Journal of Lipid Science and Technology, 109*, 828–855.

CLAL (2019). Butter: Per capita consumption. CLAL S.R.L. Available from https://www.clal.it/en/?section=tabs_consumi_procapite

Couvreur, S., Hurtaud, C., Lopez, C., Delaby, L. & Peyraud, J.-L. (2006). The linear relationship between the proportion of fresh grass in the cow diet, milk fatty acid composition, and butter properties. *Journal of Dairy Science, 89*, 1956–1969.

Cui, L., Fan, J., Sun, Y., Zhu, Z. & Yi, J. (2018). The pro-oxidant effect of salts on the lipid oxidation of lecithin-stabilized oil-in-water emulsions. *Food Chemistry, 252*, 28–32.

Cullinane, N., Aherne, S., Connolly, J. F. & Phelan, J. A. (1984a). Seasonal variation on the triglyceride and fatty acid composition of Irish butter. *Irish Journal of Agricultural and Food Research, 8*(1), 1–12.

Cullinane, N., Condon, D., Eason, D., Phelan, J. A. & Connolly, J. F. (1984b). Influence of season and processing parameters on the physical properties of Irish butter. *Irish Journal of Agricultural and Food Research, 8*(1), 13–25.

Deeth, H. C. (1997). The role of phospholipids in the stability of milk fat globules. *Australian Journal of Dairy Technology, 42*, 44–46.

Deeth, H. C. (2006). Lipoprotein lipase and lipolysis in milk. *International Dairy Journal, 16*, 555–562.

Deffense, E. (1993). Milk fat fractionation today: A review. *Journal of the American Oil Chemists' Society, 70*(12), 1193–1201.

DeMan, J. M. & Beers, A. M. (1987). Fat crystal networks: Structure and rheological properties. *Journal of Texture Studies, 18*(4), 303–318.

Emmons, D. B., Froehlich, D. A., Paquette, G. J., Butler, G., Beckett, D. C., Modler, H. W., Brackenridge, P.

& Daniels, G. (1986). Light transmission characteristics of wrapping materials and oxidation of butter by fluorescent light. *Journal of Dairy Science, 69*(9), 2248–2267.

EU. (2013). Council Regulation (EU) No 1308/2013 of 17 December 2013 Establishing a Common Organisation of the Markets in Agricultural Products and Repealing Council Regulations (Eec) No 922/72, (Eec) No 234/79, (Ec) No 1037/2001 and (Ec) No 1234/2007. *OJ, L 347*, 671–854.

Evers, J. M. (2004). The milkfat globular membrane - compositional and structural changes post secretion by the mammary secretory cell. *International Dairy Journal, 14*, 661–674.

Foote, C. S. (1976). Photosensitized oxidation and singlet oxygen: Consequences in biological systems. In W. A. Pryor (Ed.), *Free radicals in biology*. New York: Academic.

Fredrick, E., Van de Walle, D., Walstra, P., Zijtveld, J. H., Fischer, S., Van der Meeren, P. & Dewettinck, K. (2011). Isothermal crystallization behaviour of milk fat in bulk and emulsified state. *International Dairy Journal, 21*, 685–695.

Goff, H. D. (1997). Instability and partial coalescence in whippable dairy emulsions. *Journal of Dairy Science, 80*(10), 2620–2630.

Herrera, M. L. & Hartel, R. W. (2000). Effect of processing conditions of physical properties of a milk fat model system: Rheology. *Journal of the American Oil Chemists' Society, 77*(11), 1189–1196.

Holtrum, N. E. (2004). *Emulsion droplet spreading at air/water interfaces: Mechanisms and relevance to the whipping of cream.* PhD thesis, Wageningen University.

Holzmüllera, W., Müllera, M., Himberta, D. & Kulozik, U. (2016). Impact of cream washing on fat globules and milk fat globule membrane proteins. *International Dairy Journal, 59*, 52–61.

Hulshof, P. J. M., van Roekel-Jansen, T., van de Bovenkamp, P. & West, C. E. (2006). Variation in retinol and carotenoid content of milk and milk products in the Netherlands. *Journal of Food Composition and Analysis, 19*, 67–75.

Jay, J. M. (2000). *Modern food microbiology*. New York: Chapman and Hall.

Jensen, S. K. & Nielsen, K. N. (1996). Tocopherols, retinol, β-carotene and fatty acids in fat globule membrane and fat globule core in cows' milk. *Journal of Dairy Research, 63*, 1–10.

Juriaanse, A. C. & Heertje, I. (1988). Microstructure of shortenings, margarine and butter – a review. *Food Structure, 7*(2), 181–188.

Kato, Y. (2003). Chemical and sensory changes in flavor of roux prepared from wheat flour and butter by heating to various temperatures. *Food Science and Technology Research, 9*(3), 264–270.

Khan, I. T., Bule, M., Ullah, R., Nadeem, M., Asif, S. & Niaz, K. (2019). The antioxidant components of milk and their role in processing, ripening, and storage: Functional food. *Veterinary World, 12*, 12–33.

Koyuncu, M. & Tuncturk, Y. (2017). Effect of packaging method and light exposure on oxidation and lipolysis in butter. *Oxidation Communications, 40*(2), 785–798.

Krause, A. J., Miracle, R. E., Sanders, T. H., Dean, L. L. & Drake, M. A. (2008). The effect of refrigerated and frozen storage on butter flavor and texture. *Journal of Dairy Science, 91*(2), 455–465.

Lee, J. & Martini, S. (2018). Effect of cream aging temperature and agitation on butter properties. *Journal of Dairy Science, 101*(9), 7724–7735.

Lopez, C., Lavigne, F., Lesieur, P., Bourgaux, P. & Ollivon, M. (2001). Thermal and structural behavior of milk fat. 1. Unstable species of anhydrous milk fat. *Journal of Dairy Science, 84*(4), 756–766.

Lopez, C., Bourgaux, C., Leisieur, P. & Ollivon, M. (2002). Crystalline structures formed in cream and anhydrous milk fat at 4°C. *Lait, 82*, 317–335.

Lopez, C., Briard-Bion, V., Ménard, O., Beaucher, E., Rousseau, F., Fauquant, J., Lecontea, N. & Robert, B. (2011). Fat globules selected from whole milk according to their size: Different compositions and structure of the biomembrane, revealing sphingomyelin-rich domains. *Food Chemistry, 125*, 355–368.

Lozano, P. R., Miracle, E. R., Krause, A. J., Drake, M. & Cadwallader, K. R. (2007). Effect of cold storage and packaging material on the major aroma components of sweet cream butter. *Journal of Agricultural and Food Chemistry, 55*(19), 7840–7846.

Maga, J. A. & Katz, I. (2009). Pyrazines in foods: An update. *Critical Reviews in Food Science and Nutrition, 16*(1), 1–48.

Mallia, S., Escher, F. & Schlichtherle-Cerny, H. (2008). Aroma-active compounds of butter: A review. *European Food Research and Technology, 226*, 315–325.

Mattice, K. D. & Marangoni, A. G. (2017). Matrix effects on the crystallization behaviour of butter and roll-in shortening in laminated bakery products. *Food Research International, 96*, 54–63.

McGee, H. (2004). *On food and cooking – The science and Lore of the kitchen*. New York: Scribner.

McNeill, G. P., O'Donoghue, A. & Connolly, J. F. (1986). Quantification and identification of flavour components leading to lipolytic rancidity in stored butter. *Irish Journal of Agricultural and Food Research, 10*(1), 1–10.

Méndez-Cid, F. J., Centeno, J. A., Martínez, S. & Carballo, J. (2017). Changes in the chemical and physical characteristics of cow's milk butter during storage: Effects of temperature and addition of salt. *Journal of Food Composition and Analysis, 63*, 121–132.

Min, D. B. & Boff, J. M. (2002). Chemistry and reaction of singlet oxygen in foods. *Comprehensive Reviews in Food Science and Food Safety, 1*, 58–72.

Moens, K., Tzompa-Sosa, D. A., Van de Walle, D., Van der Meeren, P. & Dewettinck, K. (2019). Influence

of cooling rate on partial coalescence in natural dairy cream. *Food Research International, 120*, 819–828.
O'Callaghan, T. F., Faulkner, H., McAuliffe, S., O'Sullivan, M. G., Hennessy, D., Dillon, P., Kilcawley, K. N., Stanton, C. & Ross, R. P. (2016). Quality characteristics, chemical composition, and sensory properties of butter from cows on pasture versus indoor feeding systems. *Journal of Dairy Science, 99*(12), 9441–9460.
Ollilainen, V., Heininen, M., Linkola, E., Varo, P. & Koivistoinen, P. (1989). Carotenoids and retinoids in Finnish foods: Dairy products and eggs. *Journal of Dairy Science, 72*(9), 2257–2265.
Osinchak, J. E., Hultin, H. O., Zajicek, O. T., Kelleher, S. D. & Huang, C.-H. (1992). Effect of NaCl on catalysis of lipid oxidation by the soluble fraction of fish muscle. *Free Radical Biology and Medicine, 12*, 35–41.
Patton, S. (1954). The mechanism of sunlight flavor formation in milk with special reference to methionine and riboflavin. *Journal of Dairy Science, 37*(4), 446–452.
Pérez-Martínez, J. D., Reyes-Hernández, J., Dibildox-Alvarado, E. & Toro-Vazquez, J. F. (2012). Physical properties of cocoa butter/vegetable oil blends crystallized in a scraped surface heat exchanger. *Journal of the American Oil Chemists' Society, 89*(2), 199–209.
Phelan, J. A., O'Keeffe, A. M., Keogh, M. K. & Kelly, P. M. (1982). Studies of milk composition and its relationship to some processing criteria: 1. Seasonal changes in the composition of Irish milk. *Irish Journal of Agricultural and Food Research, 6*(1), 1–11.
Pimpin, L., Wu, J. H. Y., Haskelberg, H., Del Gobbo, L. & Mozaffarian, D. (2016). Is butter back? A systematic review and meta-analysis of butter consumption and risk of cardiovascular disease, diabetes, and total mortality. *PLoS One, 11*(6), e0158118.
Prins, A. (1986). Some aspects of aerated milk products. *Netherlands Milk and Dairy Journal, 40*, 203–215.
Rohm, H. & Raaber, S. (1991). Hedonic spreadability optima of selected edible fats. *Journal of Sensory Studies, 6*(2), 81–88.
Rønholt, S., Kirkensgaard, J. J. K., Pedersen, T. B., Mortensen, K. & Knudsen, J. C. (2012). Polymorphism, microstructure and rheology of butter. Effects of cream heat treatment. *Food Chemistry, 135*, 1730–1739.
Rønholt, S., Buldo, P., Mortensen, K., Andersen, U., Knudsen, J. C. & Wiking, L. (2014a). The effect of butter grains on physical properties of butter-like emulsions. *Journal of Dairy Science, 97*(4), 1929–1938.
Rønholt, S., Kirkensgaard, J. J. K., Mortensen, K. & Knudsen, J. C. (2014b). Effect of cream cooling rate and water content on butter microstructure during four weeks of storage. *Food Hydrocolloids, 34*, 169–176.
Rønholt, S., Madsen, A. S., Kirkensgaard, J. J. K., Mortensen, K. & Knudsen, J. C. (2014c). Effect of churning temperature on water content, rheology, microstructure and stability of butter during four weeks of storage. *Food Structure, 2*(1–2), 14–26.
Rousseau, D., Zilnik, L., Khan, R. & Hodge, S. (2003). Dispersed phase destabilization in table spreads. *Journal of the American Oil Chemists' Society, 80*(10), 957–961.
Schmidt, D. G. & van Hooydonk, A. C. M. (1980). A scanning electron microscopal investigation of the whipping of cream. *Scanning Electron Microscopy, 3*, 653–658.
Shigematsu, H., Shibata, S., Kurata, T., Kato, H. & Fujimaki, M. (1977). Thermal degradation products of several Amadori compounds. *Agricultural and Biological Chemistry, 41*(12), 2377–2385.
Smith, P. R. & Johansson, J. (2003). Influences of the proportion of solid fat in a shortening on loaf volume and staling of bread. *Journal of Food Processing and Preservation, 28*, 359–367.
Suryavanshi, M. V. & Ghosh, J. S. (2010). Spoilage of white unsalted butter by psychrophilic lipolysis of *Pseudomonas aeruginosa* NCIM 2036. *British Journal of Dairy Sciences, 1*(1), 26–29.
Szakaly, S. & Schaffer, B. (1988). Structure of butter. II. Influence of technological modification of the liquid/solid fat ratio on butter consistency. *Milchwissenschaft, 43*, 561–564.
Thomé, K. E. & Eriksson, G. (1973). The foaming properties of cream. 1. Substances in milk increasing the whippability of creams. *Milchwissenschaft, 28*, 554–558.
Urbach, G. (1990). Effect of feed on flavor in dairy foods. *Journal of Dairy Science, 73*(12), 3639–3650.
Urbach, G. (1997). The flavour of milk and dairy products: II. Cheese: Contribution of volatile compounds. *International Journal of Dairy Technology, 50*(3), 79–89.
USDA. (2018). *National nutrient database for standard reference. NDB1145. Butter, without salt*. Washington, DC: Agricultural Research Service.
Van Aken, G. A. & Visser, K. A. (2000). Firmness and crystallization of milk fat in relation to processing conditions. *Journal of Dairy Science, 83*, 1919–1932.
van Boekel, M. A. J. S. (1998). Effect of heating on Maillard reactions in milk. *Food Chemistry, 62*(4), 403–414.
Viallon, C., Martin, B., Verdier-Metz, I., Pradel, P., Garel, J.-P., Coulon, J.-B. & Berdagué, J.-L. (2000). Transfer of monoterpenes and sesquiterpenes from forages into milk fat. *Lait, 80*(6), 635–641.
Voysey, P. A., Anslow, P. A., Bridgwater, K. J., Lavender, B. & Watson, L. (2009). The effects of butter characteristics on the growth of *Listeria monocytogenes*. *International Journal of Dairy Technology, 62*(3), 326–330.
Walstra, P., Wouters, J. T. M. & Geurts, T. J. (2006). *Dairy science and technology*. Boca Raton: CRC Press.
Woo, A. H. & Lindsay, R. C. (1984). Characterization of lipase activity in cold-stored butter. *Journal of Dairy Science, 67*(6), 1194–1198.

References for Section 9.2

Abrahamsson, P., Frennborn, P., Dejmek, P. & Buchheim, W. (1988). Effects of homogenization and heating conditions on physico-chemical properties of coffee cream. *Milchwissenschaft, 43*, 762–765.

Anderson, M. & Brooker, B. E. (1988). Dairy foams. In E. Dickinson & G. Stainsby (Eds.), *Advances in food emulsions and foams* (pp. 221–256). London: Elsevier Applied Science.

Anderson, M., Brooker, B. E. & Needs, E. C. (1987). The role of proteins in the stabilization/destabilization of dairy foams. In E. Dickinson (Ed.), *Food emulsions and foams* (pp. 100–109). London: Royal Society of Chemistry.

Banks, W. & Muir, D. D. (1988). Stability of alcohol-containing emulsions. In E. Dickinson & G. Stainsby (Eds.), *Advances in food emulsions and foams* (pp. 257–283). London: Elsevier Applied Science.

Born, B. (2013). Cultured/sour cream. In R. C. Chandan & A. Kilara (Eds.), *Manufacturing yogurt and fermented milks* (2nd ed., pp. 381–391). Hoboken: Wiley.

Buchheim, W. (1991). Mikrostruktur von aufschlagbaren Emulsionen (Microstructure of whippable emulsions). *Kieler Milchwirtschaftl. Forsch.ber, 43*, 247–272.

Buchheim, W. & Dejmek, P. (1997). Milk and dairy-type emulsions. In S. E. Friberg & K. Larsson (Eds.), *Food emulsions* (3rd ed., pp. 235–278). New York: Marcel Dekker.

Buchheim, W., Falk, G. & Hinz, A. (1986). Ultrastructural aspects and physico-chemical properties of UHT-treated coffee cream. *Food Microstructure, 5*, 181–192.

Castberg, H. B. (1992). *Lipase activity* (Bulletin 271) (pp. 18–20). Brussels: International Dairy Federation.

Codex Alimentarius Commission. (2018). *Standard for creams and prepared creams CXS 288–1976,* last amended in 2018. www.codexalimentarius.org

Contarini, G. & Povolo, M. (2013). Phospholipids in Milk Fat: Composition, Biological and Technological Significance, and Analytical Strategies. *International Journal of Molecular Sciences, 14*(2), 2808–2831.

Dickinson, E., Narhan, S. K. & Stainsby, G. (1989). Stability of alcohol-containing emulsions in relation to neck-plug formation in commercial cream liqueurs. *Food Hydrocolloids, 3*, 85–100.

Driessen, F. M. & van den Berg, M. G. (1992). *Microbiological aspects of pasteurized cream* (Bulletin 271) (pp. 4–10). Brussels: International Dairy Federation.

Early, R. (1998). Liquid milk and cream. In R. Early (Ed.), *The technology of dairy products* (2nd ed., pp. 1–49). London: Blackie Academic & Professional.

Eden, J., Dejmek, P., Löfgren, R., Paulsson, M. & Glantz, M. (2016). Native milk fat globule size and its influence on whipping properties. *International Dairy Journal, 61*, 176–181.

Eyer, H. K., Rattray, W. & Gallmann, P. U. (1996). *The packaging of UHT-treated cream* (Bulletin 315) (pp. 23–24). Brussels: International Dairy Federation.

Frederick, E., Heyman, B., Moens, K., Fischer, S., Verwijlen, T., Moldenaers, P., van der Meeren, P. & Kewettinck, K. (2013a). Monoacylglycerols in dairy recombined cream: 2. The effect on partial coalescence and whipping properties. *Food Research International, 51*, 936–945.

Frederick, E., Moens, K., Heyman, B., Fischer, S., van der Meeren, P. & Kewettinck, K. (2013b). Monoacylglycerols in dairy recombined cream: 1. The effect of milk fat crystallization. *Food Research International, 51*, 892–898.

Geyer, S. & Kessler, H. G. (1989). Influence of individual milk constituents on coffee cream feathering in hot coffee. *Milchwissenschaft, 44*, 284–288.

Goff, H. D. & Vega, C. (2007). Structure-emerging of ice-cream and foam-based foods. In D. J. McClements (Ed.), *Understanding and controlling the microstructure of complex foods* (pp. 557–574). Cambridge: Woodhead Publishing.

Gurnida, D. A, Rowan, A.M. Idjradinata, P., Muchtadi, D. and Sekarwana, N. (2012). *Early Human Development, 88*(8), 595–601.

Han, J., Zhou, X., Cao, J., Sun, B., Li, Y. & Zhang, L. (2018). Microstructural evolution of whipped cream in whipping process observed by confocal laser scanning microscopy. *International Journal of Food Properties, 21*, 593–605.

Heffernan, S. P., Kelly, A. L. & Mulvihill, D. M. (2009). High-pressure homogenised cream liqueurs. Emulsification and stabilization efficiency. *Journal of Food Engineering, 95*, 525–531.

Heffernan, S. P., Kelly, A. L., Mulvihill, D. M., Lambrich, U. & Schuchmann, H. P. (2011). Efficiency of a range of homogenisation technologies in the emulsification and stabilization of cream liqueurs. *Innovative Food Science and Emerging Technologies, 12*, 628–634.

Hoffmann, W. (1999). Lagerstabilität von H-Schlagsahne (Storage stability of UHT whipping cream). *Kieler Milchwirtschaftl. Forsch.ber, 51*, 125–136.

Hoffmann, W. (2001). Technology and quality of coffee cream. *Eur. Dairy Magazine, 16*(5), 41–42.

Hoffmann, W. (2015a). *Cream: Manufacture. Reference module in food sciences* (pp. 1–8). Elsevier. https://doi.org/10.1016/B978-0-08-10056-5.00701-0.

Hoffmann, W. (2015b). *Cream: Products. Reference module in food sciences* (pp. 1–6). Elsevier. https://doi.org/10.1016/B978-0-08-10056-5.00702-2.

Hoffmann, W., Moltzen, B. & Buchheim, W. (1996). Photometric measurement of coffee cream stability in hot coffee solutions. *Milchwissenschaft, 51*, 191–194.

Houlihan, A. V. (1992). *Enzymatic activity other than lipase* (Bulletin 271) (pp. 21–25). Brussels: International Dairy Federation.

IDF (1992). *Monograph on the pasteurization of cream* (Bulletin 271). Brussels: International Dairy Federation.

IDF (1996). *UHT-Cream* (Bulletin 315). Brussels: International Dairy Federation.

Jacubzyk, E. & Niranjan, K. (2006). Transient development of whipped cream properties. *Journal of Food Engineering, 77*, 79–83.

Kessler, H. G. (2002). *Food and bio process engineering – Dairy technology* (pp. 385–424). München: A. Kessler.

Kosinski, E. (1996). *Raw material quality* (Bulletin 315) (pp. 12–16). Brussels: International Dairy Federation.

Kovácovà, R., Stetina, J. & Curda, L. (2010). Influence of processing and κ-carrageenan on properties of whipping cream. *Journal of Food Engineering, 99*, 471–478.

Lynch, A. G. & Mulvihill, D. M. (1997). Effect of sodium caseinate on the stability of cream liqueurs. *International Journal of Dairy Technology, 50*, 1–7.

MacGibbon, A. K. H. & Taylor, M. W. (2006). Composition and Structure of Bovine Milk Lipids. In: Fox P. F., McSweeney P. L. H. (eds) *Advanced dairy chemistry volume 2 lipids* (3rd ed., pp. 1–42). Springer, Boston, MA.

Mezdour, S., Boyaval, P. & Korolzcuk, J. (2008). Solubility of α_{S1}-, β- and κ-casein in water-ethanol solutions. *Dairy Science Technology, 88*, 313–325.

O'Kennedy, B. T., Cribbin, M. & Kelly, P. M. (2001). Stability of sodium caseinate to ethanol. *Milchwissenschaft, 56*, 680–684.

Phan, T. T. Q., Moens, K., Le, T. T., van der Meeren, P. & Kewettinck, K. (2014). Potential of milk fat globule membrane enriched materials to improve the whipping properties of recombined cream. *International Dairy Journal, 39*, 16–23.

Precht, A., Peters, K.-H. & Petersen, J. (1988). Improvement of storage stability and foaming properties of cream by addition of carrageenan and milk constituents. *Food Hydrocolloids, 2*, 491–506.

Smiddy, M. A., Kelly, A. L. & Huppertz, T. (2009). Cream and related products. In A. Y. Tamime (Ed.), *Dairy fats and related products* (pp. 61–85). Oxford: Wiley Blackwell.

Smith, A. K., Goff, H. D. & Kakuda, Y. (2000). Microstructure and rheological properties of whipped cream as affected by heat treatment and addition of stabilizer. *International Dairy Journal, 10*, 295–301.

Towler, C. (1994). Developments in cream separation and processing. In R. K. Robinson (Ed.), *Modern dairy technology* (Vol. 1, 2nd ed., pp. 61–106). London: Chapman & Hall.

Taylor, M. W. & MacGibbon, A. K. H. (2011). Milk Lipids: General Characteristics. In: Fuquay, J. W., Fox, P. F., McSweeney, P. L. H. (eds) *Encyclopedia of dairy sciences* (2nd ed., pp. 649–654). Academic Press. San Diego, CA.

van Aken, G. A. (2001). Aeration of emulsions by whipping. *Colloids and Surfaces A: Physicochemical and Engineering Aspects, 190*, 333–354.

van Lent, K., Le, C. T., Vanlerberghe, B. & van der Meeren, P. (2008). Effect of formulation on the emulsion and whipping properties of recombined dairy cream. *International Dairy Journal, 18*, 1003–1010.

Walstra, P., Geurts, T. J., Noomen, A., Jellema, A. & van Boekel, M. A. J. S. (1999). *Dairy technology*. New York: Marcel Dekker.

Westermann, S., Møller, J. K. S. & Skibsted, L. H. (2009). Effect of light exposure on radical formation in dairy cream during subsequent dark storage. *Milchwissenschaft, 64*, 51–55.

Wijnen, M. E. (1997). *Instant foam physics: Formation and stability of aerosol whipped creams*. PhD thesis, Wageningen Agricultural University, Wageningen.

References for Section 9.3

Barden, L., Osborne, J., McMahon, D. J. & Foegeding, E. A. (2015). Investigating the filled gel model in Cheddar cheese through use of Sephadex beads. *Journal of Dairy Science, 98*, 1502–1516.

Carunchia Whetstine, M. E., Drake, M. A., Nelson, B. K. & Barbano, D. (2006). Flavor profiles of full fat, reduced fat and cheese fat made from aged Cheddar with the fat removed using a novel process. *Journal of Dairy Science, 89*, 505–517.

Childs, J. L. & Drake, M. A. (2009). Consumer perception of fat reduction in cheese. *Journal of Sensory Studies, 24*, 02–921.

Collins, Y. F., McSweeney, P. L. H. & Wilkinson, M. G. (2003). Lipolysis and free fatty acid catabolism in cheese: A review of current knowledge. *International Dairy Journal, 13*, 841–866.

Drake, M. A., Miracle, R. E. & McMahon, D. J. (2010). Impact of fat reduction on flavor and flavor chemistry of Cheddar cheeses. *Journal of Dairy Science, 93*, 5069–5081.

Johnson, M. E., Kapoor, R., McMahon, D. J., McCoy, D. R. & Narasimmon, R. G. (2009). Reduction of sodium and fat levels in natural and processed cheeses: Scientific and technological aspects. *Comprehensive Reviews in Food Science and Food Safety, 8*(3), 252–268.

Lopez, C., Briard-Bion, V., Camier, B. & Gassi, J. Y. (2006). Milk fat thermal properties and solid fat content in Emmental cheese: A differential scanning calorimetry study. *Journal of Dairy Science, 89*, 2894–2910.

Lucey, J. A., Johnson, M. E. & Horne, D. S. (2003). Perspectives on the basis of the rheology and texture properties of cheese. *Journal of Dairy Science, 86*, 2725–2743.

McMahon, D. J., Motawee, M. M. & McManus, W. R. (2009). Influence of brine concentration and temperature on composition, microstructure, and yield of feta cheese. *Journal of Dairy Science, 92* (9), 4169–4179.

McMahon, D. J. (2010). Issues with lower fat and lower salt cheeses. *Australian Journal of Dairy Technology, 65*, 200–205.

McMahon, D. J. & Oberg, C. J. (2011). Cheese: *Pasta-filata* cheeses: Low-moisture part-skim Mozzarella

(Pizza cheese). In J. W. Fuquay, P. F. Fox & P. L. H. McSweeney (Eds.), *Encyclopedia of dairy sciences* (Vol. 1, 2nd ed., pp. 737–744). San Diego: Academic.

McMahon, D. J. & Oberg, C. J. (2017). *Pasta-filata* cheese. In Cheese Technology and Major Cheese Groups, P. F. Fox, P. L. H. McSweeney, P. Cotter & D. W. Everett (Eds.), *Cheese: Chemistry physics and microbiology* (Vol. 2, 4th ed., pp. 1041–1068). London: Elsevier Publishers.

McMahon, D. J., Fife, R. L. & Oberg, C. J. (1999). Water partitioning in Mozzarella cheese and its relationship to cheese meltability. *Journal of Dairy Science, 82*, 1361–1369.

McMahon, D. J., Oberg, C. J., Drake, M. A., Farkye, N., Moyes, L. V., Arnold, M. R., Ganesan, B., Steele, J. & Broadbent, J. R. (2014). Effect of sodium, potassium, magnesium, and calcium salt cations on pH, proteolysis, organic acids, and microbial populations during storage of full fat Cheddar cheese. *Journal of Dairy Science, 97*, 4780–4798.

Merrill, R. K., Oberg, C. J. & McMahon, D. J. (1994). A method for manufacturing reduced fat Mozzarella cheese. *Journal of Dairy Science, 77*, 1783–1789.

Merrill, R. K., Oberg, C. J., McManus, W. R., Kalab, M. & McMahon, D. J. (1996). Microstructure and physical properties of a reduced fat Mozzarella cheese made using *Lactobacillus casei* adjunct culture. *LWT – Food Science and Technology, 29*, 721–728.

Milo, C. & Reineccius, G. A. (1997). Identification and quantification of potent odorants in regular fat and low fat mild Cheddar cheese. *Journal of Agricultural and Food Chemistry, 45*, 3590–3594.

Mulvaney, S., Rong, S., Barbano, D. M. & Yun, J. J. (1997). Systems analysis of the plastization and extrusion processing of Mozzarella cheese. *Journal of Dairy Science, 80*, 3030–3039.

Oberg, C. J., McManus, W. R. & McMahon, D. J. (1993). Microstructure of Mozzarella cheese during manufacture. *Journal of Food Structure, 12*, 251–258.

Oberg, E. N., Oberg, C. J., Motawee, M. M., Martini, S. & McMahon, D. J. (2015). Increasing stringiness of low-fat Mozzarella string cheese using polysaccharides. *Journal of Dairy Science, 98*, 243–4254.

Pastorino, A. J., Dave, R. I., Oberg, C. J. & McMahon, D. J. (2002). Temperature effect on structure-opacity relationships of nonfat Mozzarella cheese. *Journal of Dairy Science, 85*, 2106–2113.

Rogers, N. R., Drake, M. A., Daubert, C. R., McMahon, D. J., Bletch, T. K. & Foegeding, E. A. (2009). The effect of aging on low-fat, reduced-fat, and full-fat Cheddar cheese texture. *Journal of Dairy Science, 92*, 4756–4772.

Rogers, N. R., McMahon, D. J., Daubert, C. R., Berry, T. K. & Foegeding, E. A. (2010). Rheological properties and microstructure of Cheddar cheese made with different fat contents. *Journal of Dairy Science, 93*, 4565–4576.

Rudan, M. A. & Barbano, D. M. (1998). A model of Mozzarella cheese melting and browning during pizza baking. *Journal of Dairy Science, 81*, 2312–2319.

Sipahioglu, O., Alvarez, V. B. & Solano-Lopez, C. (1999). Structure, physico-chemical and sensory properties of Feta cheese made with tapioca starch and lecithin as fat mimetics. *International Dairy Journal, 9*, 783–789.

Wadhwani, R., McManus, W. R. & McMahon, D. J. (2011). Improvement in melting and baking properties of low fat Mozzarella cheese. *Journal of Dairy Science, 94*, 1713–1723.

Wadhwani, R., McManus, W. R. & McMahon, D. J. (2012). Color of low-fat cheese influences flavor perception and consumer liking. *Journal of Dairy Science, 95*, 2336–2346.

References for Section 9.4

Adleman, R. & Hartel, R. W. (2002). Lipid crystallization and its effect on the physical structure of ice cream. In N. Garti & K. Sato (Eds.), *Crystallization processes in fats and lipid systems* (pp. 381–427). New York: Marcel Dekker.

Amador, J., Hartel, R. W. & Rankin, S. (2017). The effects of fat structures and ice cream mix viscosity on physical and sensory properties of ice cream. *Journal of Food Science, 82*, 1851–1860.

Barfod, N. M. (2001). The emulsifier effect. *Dairy Industries International, 66*, 32–33.

Biasutti, M., Venir, E., Marino, M., Maifreni, M. & Innocente, N. (2013). Effects of high pressure homogenization of ice cream mix on the physical and structural properties of ice cream. *International Dairy Journal, 32*, 40–45.

Bolliger, S., Goff, H. D. & Tharp, B. W. (2000a). Correlation between colloidal properties of ice cream mix and ice cream. *International Dairy Journal, 10*, 303–309.

Bolliger, S., Kornbrust, B., Goff, H. D., Tharp, B. W. & Windhab, E. J. (2000b). Influence of emulsifiers on ice cream produced by conventional freezing and low temperature extrusion processing. *International Dairy Journal, 10*, 497–504.

Clarke, C. (2012). *The science of ice cream* (2nd ed.). Cambridge: Royal Society of Chemistry.

Davies, E., Dickinson, E. & Bee, R. D. (2000). Shear stability of sodium caseinate emulsions containing monoglyceride and triglyceride crystals. *Food Hydrocolloids, 14*, 145–153.

Davies, E., Dickinson, E. & Bee, R. D. (2001). Orthokinetic destabilization of emulsions by saturated and unsaturated monoglycerides. *International Dairy Journal, 11*, 827–836.

Daw, E. & Hartel, R. W. (2015). Fat destabilization and melt-down of ice creams with increased protein content. *International Dairy Journal, 43*, 33–41.

Goff, H. D. (2002). Formation and stabilization of structure in ice cream and related products. *Current Opinion in Colloid & Interface Science, 7*, 432–437.

Goff, H. D. (2016). Milk proteins in ice cream. In P. L. H. McSweeney & J. A. O'Mahony (Eds.), *Advanced dairy chemistry – 1B. Proteins. Applied aspects* (4th ed., pp. 329–345). New York: Springer.

Goff, H. D. & Hartel, R. W. (2013). *Ice Cream* (7th ed.). New York: Springer.

Goff, H. D., Verespej, E. & Smith, A. K. (1999). A study of fat and air structures in ice cream. *International Dairy Journal, 9*, 817–829.

Hayes, M. G., Lefrancois, A. C., Waldron, D. S., Goff, H. D. & Kelly, A. L. (2003). Influence of high pressure homogenisation on some characteristics of ice cream. *Milchwissenschaft, 58*, 519–523.

Hyvonen, L., Linna, M., Tuorila, H. & Dijksterhuis, G. (2003). Perception of melting and flavor release of ice cream containing different types and contents of fat. *Journal of Dairy Science, 86*, 1130–1138.

Kokubo, S., Sakurai, K., Hakamata, K., Tomita, M. & Yoshida, S. (1996). The effect of manufacturing conditions on the de-emulsification of fat globules in ice cream. *Milchwissenschaft, 51*, 262–265.

Kokubo, S., Sakurai, K., Iwaki, S., Tomita, M. & Yoshida, S. (1998). Agglomeration of fat globules during the freezing process of ice cream manufacturing. *Milchwissenschaft, 53*, 206–209.

Koxholt, M. M. R., Eisenmann, B. & Hinrichs, J. (2001). Effect of the fat globule sizes on the meltdown of ice cream. *Journal of Dairy Science, 84*, 31–37.

Mendez-Velasco, C. & Goff, H. D. (2011). Enhancement of fat colloidal interactions for the preparation of ice creams high in unsaturated fat. *International Dairy Journal, 21*, 540–547.

Mendez-Velasco, C. & Goff, H. D. (2012a). Fat structure in ice cream: A study on the types of fat interactions. *Food Hydrocolloids, 29*, 152–159.

Mendez-Velasco, C. & Goff, H. D. (2012b). Fat structures as affected by unsaturated or saturated monoglyceride and their effect on ice cream structure, texture and stability. *International Dairy Journal, 24*, 33–39.

Muse, M. R. & Hartel, R. W. (2004). Ice cream structural elements that affect melting rate and hardness. *Journal of Dairy Science, 87*, 1–10.

Relkin, P., Sourdet, S. & Fosseux, P.-Y. (2003). Fat crystallization in complex food emulsions: Effects of adsorbed milk proteins and of a whipping process. *Journal of Thermal Analysis and Calorimetry, 71*, 187–195.

Roland, A. M., Phillips, L. G. & Boor, K. J. (1999). Effects of fat content on the sensory properties, melting, colour, and hardness of ice cream. *Journal of Dairy Science, 82*, 32–38.

Rolon, M. L., Bakke, A. J., Coupland, J. N., Hayes, J. E. & Roberts, R. F. (2017). Effect of fat content on the physical properties and consumer acceptability of vanilla ice cream. *Journal of Dairy Science, 100*, 5217–5227.

Ruger, P. R., Baer, R. J. & Kasperson, K. M. (2002). Effect of double homogenization and whey protein concentrate on the texture of ice cream. *Journal of Dairy Science, 85*, 1684–1692.

Sofjan, R. P. & Hartel, R. W. (2004). Effects of overrun on structural and physical properties of ice cream. *International Dairy Journal, 14*, 255–262.

Soukoulis, C. & Fisk, I. (2016). Innovative ingredients and emerging technologies for controlling ice recrystallization, texture, and structure stability in frozen dairy desserts: A review. *Critical Reviews in Food Science and Nutrition, 56*, 2543–2559.

Sourdet, S., Relkin, P., Fosseux, P.-Y. & Aubry, V. (2002). Composition of fat protein layer in complex food emulsions at various weight ratios of casein-to-whey proteins. *Lait, 80*, 567–580.

Sourdet, S., Relkin, P. & Cesar, B. (2003). Effects of milk protein type and pre-heating on physical stability of whipped and frozen emulsions. *Colloids and Surfaces, B: Biointerfaces, 31*, 55–64.

Sung, K. K. & Goff, H. D. (2010). Effect of solid fat content on structure in ice creams containing palm kernel oil and high oleic sunflower oil. *Journal of Food Science, 75*, C274–C279.

Tekin, E., Sahin, S. & Sumnu, G. (2017). Physicochemical, rheological and sensory properties of low-fat ice cream designed by double emulsions. *European Journal of Lipid Science and Technology, 119*(1–9), 1600505.

Warren, M. M. & Hartel, R. W. (2018). Effects of emulsifier, overrun and dasher speed on ice cream microstructure and melting properties. *Journal of Food Science, 83*, 639–647.

Xinyi, E., Pei, Z. J. & Schmidt, K. A. (2010). Ice cream: Foam formation and stabilization – A review. *Food Review International, 26*, 122–137.

Zhang, Z. & Goff, H. D. (2004). Protein distribution at air interfaces in dairy foams and ice cream as affected by casein dissociation and emulsifiers. *International Dairy Journal, 14*, 647–657.

Zhang, Z. & Goff, H. D. (2005). On fat destabilization and composition of the air interface in ice cream containing saturated and unsaturated monoglyceride. *International Dairy Journal, 15*, 495–500.

Zulim Botega, D. C., Marangoni, A. G., Smith, A. K. & Goff, H. D. (2013a). Development of formulations and processes to incorporate wax oleogels in ice cream. *Journal of Food Science, 78*, C1845–C1851.

Zulim Botega, D. C., Marangoni, A. G., Smith, A. K. & Goff, H. D. (2013b). The potential application of rice bran wax oleogel to replace solid fat and enhance unsaturated fat content in ice cream. *Journal of Food Science, 78*, C1334–C1339.

References for Section 9.5

Buma, T. J. (1971a). Free fat in spray-dried whole milk. 4. Significance of free fat for other properties of practical importance. *Netherlands Milk Dairy Journal, 25*, 88–106.

Buma, T. J. (1971b). Free fat in spray-dried whole milk. 5. Cohesion. Determination, influence of particle size, moisture content and free-fat content. *Netherlands Milk Dairy Journal, 25*, 107–122.

Buma, T. J. (1971c). Free fat in spray-dried whole milk. 8. The relation between free-fat content and particle porosity of spray-dried whole milk. *Netherlands Milk Dairy Journal, 25*, 123–104.

Buma, T. J. (1971d). Free fat in spray-dried whole milk. 10. A final report with a physical model for free fat in

spray-dried milk. *Netherlands Milk Dairy Journal, 25*, 159–174.
Crowley, S. V., Gazi, I., Kelly, A. L., Huppertz, T. & O'Mahony, J. A. (2014). Influence of protein concentration on the physical characteristics and flow properties of milk protein concentrate powders. *Journal of Food Engineering, 135*, 31–38.
Fitzpatrick, J. J., Iqbal, T., Delaney, C., Twomey, T. & Keogh. (2004). Effect of powder properties and storage conditions on the flowability of milk powders with different fat contents. *Journal of Food Engineering, 64*, 435–444.
Foerster, M., Gengenbach, T., Woo, M. W. & Selomulya, C. (2016). The impact of atomization on the surface composition of spray-dried milk droplets. *Colloids and Surfaces. B, Biointerfaces, 140*, 460–471.
Foerster, M., Liu, C., Gengenbach, T., Woo, M. W. & Selomulya, C. (2017). Reduction of surface fat formation on spray-dried milk powders through emulsion stabilization with λ-carrageenan. *Food Hydrocolloids, 70*, 163–180.
Fox, P. F. & McSweeney, P. L. H. (1998). *Dairy chemistry and biochemistry*. London: Blackie Academic & Professional.
Gaiani, C., Schuck, P., Scher, J., Ehrhardt, J. J., Arab-Tehrany, E., Jacquot, M. & Banon, S. (2009). Native phosphocaseinate powder during storage: Lipids released onto the surface. *Journal of Food Engineering, 94*, 130–134.
Kelly, G. M., O'Mahony, J. A., Kelly, A. L. & O'Callaghan, D. J. (2014). Physical characteristics of spray-dried dairy powders containing different vegetable oils. *Journal of Food Engineering, 122*, 122–129.
Keogh, K. & Twomey, M. (2002). Effect of milk composition on spray-dried high-fat milk powders and their use in chocolate. *Lait, 82*, 531–539.
Kim, E. H.-J., Chen, X. D. & Pearce, D. (2005). Effect of surface composition on the flowability of industrial spray-dried dairy powders. *Colloids and Surfaces. B, Biointerfaces, 46*, 182–187.
Liang, B. & Hartel, R. W. (2004). Effects of milk powders in milk chocolate. *Journal of Dairy Science, 87*, 20–31.
Mahmoodani, F., Perera, C. O., Abernethy, G., Fedrizzi, B. & Chen, H. (2018). Lipid oxidation and vitamin D3 degradation in simulated whole milk powder as influenced by processing and storage. *Food Chemistry, 261*, 149–156.
Mitchell, W. R., Forny, L., Althaus, T., Niederreiter, G., Palzer, S., Hounslow, M. J. & Salman, A. D. (2019). Surface tension-driven effects in the reconstitution of food powders. *Chemical Engineering Research and Design, 146*, 464–469.
Skanderby, M., Westergaard, V., Partridge, A. & Muir, D. D. (2009). Dried milk products. In A. Y. Tamine (Ed.), *Dairy powders and concentrated products* (pp. 180–234). West Sussex: Wiley-Blackwell.
Vega, C. & Roos, Y. H. (2006). Invited review: Spray-dried dairy and dairy-like emulsions – Compositional considerations. *Journal of Dairy Science, 89*, 383–401.
Vignolles, M.-L., Jeantet, R., Lopez, C. & Schuck, P. (2007). Free fat, surface fat and dairy powders: Interactions between process and product. A review. *Lait, 87*, 187–236.

References for Section 9.6

Armand, M., Pasquier, B., Andre, M., Borel, P., Senft, M., Peyrot, J., Salducci, J., Portugal, H., Jaussan, V. & lairon, D. (1999). Digestion and absorption of 2 fat emulsions with different droplet sizes in the human digestive tract. *American Journal of Clinical Nutrition, 70*, 1096–1106.
Baars, A., Oosting, A., Engels, E., Kegler, D., Kodde, A., Schipper, L., Verkade, H. J. & van der Beek, E. M. (2016). Milk fat globule membrane coating of large lipid droplets in the diet of young mice prevents body fat accumulation in adulthood. *British Journal of Nutrition, 115*, 1930–1937.
Bar-Yoseph, F., Lifshitz, Y. & Cohen, T. (2013). Review of *sn*-2 palmitate oil implications for infant health. *Prostaglandins Leukot. Essent Fatty Acids, 89*, 139–143.
Baumgartner, S., van de Heijning, B. J. M., Acton, D. & Mensisk, R. P. (2017). Infant milk fat droplet size and coating effect postprandial responses in healthy adult men: A proof-of-concept study. *European Journal of Clinical Nutrition, 71*, 1108–1113.
Béghin, L., Marchandise, X., Lien, E., Bricout, M., Bernet, J.-P., Lienhardt, J.-F., Jeannerot, F., Menet, V., Requillart, J.-C., Marx, J., De Groot, N., Jaeger, J., Steenhout, P. & Turck, D. (2018). Growth, stool consistency and bone mineral content in healthy term infants fed *sn*-2-palmitate-enriched starter infant formula: A randomized, double-blind, multicentre clinical trial. *Clinical Nutrition, 38*, 1023–1030. https://doi.org/10.1016/j.clnu.2018.05.015.
Byrdwell, W. C. & Perry, R. H. (2007). Liquid chromatography with dual parallel mass spectrometry and ^{31}P nuclear magnetic resonance spectroscopy for analysis of sphingomyelin and dihydrosphingomyelin II. Bovine milk sphingolipids. *Journal of Chromatography A, 1146*, 64–185.
Carlson, S. E., Cooke, R. J., Werkman, S. H. & Tolley, E. A. (1992). First year growth of preterm infants fed standard compared to marine oil n-3 supplemented formula. *Lipids, 27*, 901–907.
Carver, J. D. (2003). Advances in nutritional modifications of infant formulas. *American Journal of Clinical Nutrition, 77*, 1550S–1554S.
Codex Alimentarius. (2016). Standard for infant formula and formulas for special medical purposes intended for infants. *Codex Stan*, 72–1981. Revision: 2007.
Eggersdorfer, M. & Wyss, A. (2018). Carotenoids in human nutrition and health. *Archives of Biochemistry and Biophysics, 652*, 18–26.
European Food Safety Authority (2014). Scientific opinion on the essential composition of infant and follow-on formulae. *EFSA Journal, 12*, 3760.

FAO (2010). Fats and fatty acids in human nutrition – Report of an expert consultation. *FAO Food and Nutrition Paper, 91*, 63–76.

Garcia, C. & Innis, S. (2013). Structure of the human milk fat globule. *Lipid Technology, 25*, 223–226.

Garcia, C., Lutz, N., Confort-Gouny, S., Cozzone, P., Armand, M. & Bernard, M. (2012). Phospholipid fingerprints of milk from different mammalians determined by 31P NMR: Towards specific interest in human health. *Food Chemistry, 135*, 1777–1783.

Gerstenberger, H. J. & Ruh, H. O. (1919). Studies in the adaptation of an artificial food to human milk. II: A report of three years clinical experience with the feeding of SMA. *American Journal of Diseases of Children, 171*, 1–9.

Gianni, M. L., Roggero, P., Baudry, C., Fressange-Mazda, C., le Ruyet, P. & Mosca, F. (2018). No effect of adding dairy lipids or long chain polyunsaturated fatty acids on formula tolerance and growth in full tern infants: A randomized controlled trial. *BMC Pediatrics, 18*, 10.

Giuffrida, F., Masserey-Elmelegy, I., Thakkar, S. K., Marmet, C. & Destaillats, F. (2014). Longitudinal evolution of the concentration of gangliosides GM_3 and GD_3 in human milk. *Lipids, 49*, 997–1004.

Giuffrida, F., Cruz-Hernandez, C., Bertschy, E., Fontannaz, P., Masserey-Elmelegy, I., Tavazzi, I., Marmet, C., Sanchez-Bridge, B., Thakkar, S. K., De Castro, C. A., Vynes-Pares, G., Zhang, Y. & Wang, P. (2016). Temporal changes of human breast milk lipids of Chinese mothers. *Nutrients, 8*, 715.

Hadley, K. B., Ryan, A. S., Forsyth, S., Gautier, S. & Salem, N., Jr. (2016). The essentiality of arachidonic acid in infant development. *Nutrients, 8*, 216.

Hageman, J. H. J., Danielsen, M., Nieuwenhuizen, A. G., Feitsma, A. L. & Dalsgaard, T. K. (2019). Comparison of bovine milk fat and vegetable fat for infant formulae: Implications for infant health. *International Dairy Journal, 92*, 37–49.

Innis, S. M. (2011). Dietary triacylglycerol structure and its role in infant nutrition. *Advances in Nutrition, 2*, 275–283.

Innis, S. M. (2016). Palmitic acid in early human development. Critical Reviews of Food Science and Nutrition, 56 (12), 1952–1959.

Jensen, R. (1996). The lipids in human milk. *Progress in Lipid Research, 35*, 53–92.

Jensen, R. G. (1999). Lipids in human milk. *Lipids, 34*, 1243–1271.

Karupaiah, T. & Sundram, K. (2007). Effects of sterospecific positioning of fatty acids in triacylglycerol structures in native and randomized fats: A review of their nutritional implications. *Nutrition, Metabolism, 4*, 16.

Lauritzen, L., Bambilla, P., Mazzocchi, A., Harsløf, L. B. S., Ciappolino, V. & Agostoni, C. (2016). DHA effects in brain development and function. *Nutrients, 8*, 6.

Le Huërou-Luron, I., Bouzerzour, K., Ferret-Bernard, S., Ménard, O., Le Normand, L., Perrier, C., Le Bourgot, C., Jardin, J., Bourlieu, C., Carton, T., Le Ruyet, P., Cuinet, I., Bonhomme, C. & Dupont, D. (2018). A mixture of milk and vegetable lipids in infant formula changes gut digestion, mucosal immunity and microbiota composition in neonatal pigs. *European Journal of Nutrition, 57*, 463–476.

Lindmark Månsson, H. (2008). Fatty acids in bovine milk fat. *Food & Nutrition Research, 52*. https://doi.org/10.3402/fnr.v52i0.1821

Lipkie, T. E., Morrow, A. L., Jouni, Z. E., McMahon, R. J. & Ferruzzi, M. G. (2015). Longitudinal survey of carotenoids in human milk from urban cohorts in China, Mexico and the USA. *PLoS One, 10*, e0127729.

Liu, Z., Rochfort, S. & Cocks, B. (2018). Milk lipidomics: What we know and what we don't. *Progress in Lipid Research, 71*, 70–85.

Mackey, A. D., Albrecht, D., Oliver, J., Williams, T., Long, A. C. & Price, P. T. (2012). Plasma carotenoid concentrations of infants are increased by feeding a milk-based infant formula supplemented with carotenoids. *Journal of the Science of Food and Agriculture, 93*, 1945–1952.

Mather, I. H. (2011). Milk fat globule membrane. In J. W. Fuquay, P. F. Fox & P. L. H. McSweeney (Eds.), *Encyclopedia of dairy sciences* (Vol. 3, 2nd ed., pp. 680–690). Oxford: Academic.

Michalski, M. C., Soares, A. F., Lopez, C., Leconte, N., Briard, V. & Geloen, A. (2006). The supramolecular structure of milk fat influences plasma triacylglycerols and fatty acid profile in the rat. *European Journal of Nutrition, 45*, 215–224.

Moloney, C., Walshe, E., Phelan, M., Giuffrida, F., Badoud, F., Bertschy, E. & O'Regan, J. (2018). Sphingomyelin content of dairy protein ingredients and infant formula powders, and identification of bovine sphingomyelin species. *International Dairy Journal, 78*, 138–144.

Mortensen, B. K. (2011). Anhydrous milk fat/butter oil and ghee. In J. W. Fuquay, P. F. Fox & P. L. H. McSweeney (Eds.), *Encyclopedia of dairy sciences* (Vol. 1, 2nd ed., pp. 515–521). Oxford: Academic.

Nelson, S. E., Rogers, R. R., Frantz, J. A. & Ziegler, E. E. (1996). Palm olein in infant formula: Absorption of fat and minerals by normal infants. *American Journal of Clinical Nutrition, 64*, 291–296.

Nilsson, A. (2016). Role of sphingolipids in infant gut health and immunity. *Journal of Pediatrics, 173S*, S53–S59.

Ohlsson, L. (2010). Dairy products and plasma cholesterol levels. *Food & Nutrition Research, 54*, 5124.

Oosting, A., Kegler, D., Wopereis, H. J., Teller, I. C., van de Heijning, B. J., Verkade, H. J. & van der Beek, E. M. (2012). Size and phospholipid coating of lipid droplets in the diet of young mice modify body fat accumulation in adulthood. *Pediatric Research, 72*, 362–369.

Pfrieger, F. W. (2003). Cholesterol homeostasis and function in neurons of the central nervous system. *Cellular and Molecular Life Sciences, 60*, 1158–1171.

Rueda, R. (2007). The role of dietary gangliosides on immunity and the prevention of infection. *British Journal of Nutrition, 98*, S68–S73.

Stevens, E. E., Patrick, T. E. & Pickler, R. (2009). A history of infant feeding. *Journal of Perinatal Education, 18*, 32–39.

Tanaka, K., Hosozawa, M., Kudo, N., Yoshikawa, N., Hisata, K., Shoji, H., Shinohara, H. & Shimizu, T. (2013). The pilot study: Sphingomyelin-fortified milk has a positive association with the neurobehavioral development of very low birth weight infants during infancy, randomised control trial. *Brain & Development Journal, 35*, 45–52.

Timby, N., Domellöf, E., Hernell, O., Lönnerdal, B. & Domellöf, M. (2014a). Neurodevelopment, nutrition, and growth until 12 mo of age in infants fed a low-energy, low-protein formula supplemented with bovine milk fat globule membranes: A randomized controlled trial. *American Journal of Clinical Nutrition, 99*, 860–868.

Timby, N., Lönnerdal, B., Hernell, O. & Domellöf, M. (2014b). Cardiovascular risk factors until 12 mo of age in infants fed a formula supplemented with bovine milk fat globule membrane. *Pediatric Research, 76*, 394–400.

Timby, N., Hernell, O., Vaarala, O., Merlin, M., Lönnerdal, B. & Domellöf, M. (2015). Infections in infants fed formula supplemented with bovine milk fat globule membranes. *Journal of Pediatric Gastroenterology and Nutrition, 60*, 384–389.

Tzompa-Sosa, D. A., van Aken, G. A., van Hooijdonk, A. C. M. & van Valenberg, H. J. F. (2014). Influence of C16:0 and long-chain saturated fatty acids on normal variation of bovine milk fat triacylglycerol structure. *Journal of Dairy Science, 97*, 4542–4551.

Van Meer, G., Voelker, D. R. & Feigenson, G. W. (2008). Membrane lipids: Where they are and how they behave. *Nature Reviews. Molecular Cell Biology, 9*, 112–124.

Wilbey, R. A. (2011). Homogenisation of milk – Principles and mechanisms of homogenisation, effects and assessment of efficiency: Valve homogenisers. In J. W. Fuquay, P. F. Fox & P. L. H. McSweeney (Eds.), *Encyclopedia of dairy sciences* (Vol. 2, 2nd ed., pp. 750–754). Oxford: Academic.

Yao, M., Lien, E. L., Capeding, M. R. Z., Fitzgerald, M., Ramanujam, K., Yuhas, R., Northington, R., Lebumfacil, J., Wang, L. & DeRusso, P. A. (2014). Effects of term infant formulas containing high *sn*-2 palmitate with and without oligofructose on stool composition, stool characteristics and bifidogenicity. *Journal of Pediatric Gastroenterology and Nutrition, 59*, 440–448.

Yuhas, R., Pramuk, K. & Lien, E. L. (2006). Human milk fatty acid composition from nine countries varies most in DHA. *Lipids, 41*, 851–858.

Zielińska, M. A., Wesołowska, A., Pawlus, B. & Hamułka, J. (2017). Health effects of carotenoids during pregnancy and lactation. *Nutrients, 9*, 838.

Zou, L., Pande, G. & Akoh, C. C. (2016). Infant formula fat analogs and human milk fat: New focus on infant developmental needs. *Annual Review of Food Science and Technology, 7*, 139–165.

References for Section 9.7

Afoakwa, E. O., Paterson, A. & Fowler, M. (2007). Factors influencing rheological and textural qualities in chocolate – A review. *Trends in Food Science and Technology, 18*, 290–298.

Afoakwa, E. O., Paterson, A., Fowler, M. & Vieira, J. (2009). Influence of tempering and fat crystallization behaviours on microstructural and melting properties in dark chocolate systems. *Food Research International, 42*, 200–209.

Aguilar, C. A. & Ziegler, G. R. (1994). Physical and microscopic characterization of dry whole milk with altered lactose content. 1. Effect of lactose concentration. *Journal of Dairy Science, 77*, 1189–1197.

Aguilera, J. M., Michel, M. & Mayor, G. (2004). Fat migration in chocolate: Diffusion or capillary flow in a particulate solid? – A hypothesis paper. *Journal of Food Science, 69*, R167–R174.

Alamprese, C., Datei, L. & Semeraro, Q. (2007). Optimization of processing parameters of a ball mill refiner for chocolate. *Journal of Food Engineering, 83*, 629–636.

Altimiras, P., Pyle, L. & Bouchon, P. (2007). Structure-fat migration relationships during storage of cocoa butter model bars: Bloom development and possible mechanisms. *Journal of Food Engineering, 80*, 600–610.

Barna, C. M., Hartel, R. W. & Metin, S. (1992). Incorporation of milk-fat fractions into milk chocolate. *Manufacturing Confectioner, 72*, 107–116.

Beckett, S. T. (2003). Is the taste of British milk chocolate different? *International Journal of Dairy Technology, 56*, 139–142.

Beckett, S. T. (2008). *The science of chocolate* (2nd ed.). London: Royal Society of Chemistry.

Beckett, S. T. (2017). Traditional chocolate making. In S. T. Beckett, M. S. Fowler & G. R. Ziegler (Eds.), *Industrial chocolate manufacture and use* (5th ed., pp. 1–8). New York: Wiley.

Beckett, S. T., Paggios, K. & Roberts, I. (2017). Conching. In S. T. Beckett, M. S. Fowler & G. R. Ziegler (Eds.), *Industrial chocolate manufacture and use* (5th ed., pp. 241–297). Oxford: Wiley.

Bolenz, S., Thiessenhusen, T. & Schäpe, R. (2003). Influence of milk components on properties and consumer acceptance of milk chocolate. *European Food Research and Technology, 216*, 28–33.

Bricknell, J. & Hartel, R. W. (1998). Relation of fat bloom in chocolate to polymorphic transition of cocoa butter. *Journal of the American Oil Chemists' Society, 75*, 1609–1615.

Brown, M. (2009). Biscuits and bakery products. In G. Talbot (Ed.), *Science and Technology of Enrobed and Filled Chocolate, confectionery and bakery products* (1st ed., pp. 152–162). Cambridge: Woodhead Publishing.

Bystrom, C. E. & Hartel, R. W. (1994). Evaluation of milk fat fractionation and modification techniques for creating cocoa butter replacers. *LWT – Food Science Technology, 27*, 142–150.

Campbell, L. B., Andersen, D. A. & Keeney, P. G. (1969). Hydrogenated milk fat as an inhibitor of the fat bloom defect in dark chocolate. *Journal of Dairy Science, 52*, 976–979.

Choi, Y. J., McCarthy, K. L. & McCarthy, M. J. (2005). Oil migration in a chocolate confectionery system evaluated by magnetic resonance imaging. *Journal of Food Science, 70*, E312–E317.

Codex (1999). *Codex standard for milkfat products. CODEX STAN A-2-1973, rev.1–1999 Codex Alimentarius* (pp. 280–1973). Rome: FAO/WHO.

Codex (2003). *Codex standard for chocolate and chocolate products. CODEX STAN 87–1981, Rev. 1–2003 Codex Alimentarius*. Rome: FAO/WHO.

Couvreur, S., Hurtaud, C., Lopez, C., Delaby, L. & Peyraud, J. L. (2006). The linear relationship between the proportion of fresh grass in the cow diet, milk fatty acid composition, and butter properties. *Journal of Dairy Science, 89*, 1956–1969.

Dahlenborg, H. (2014). *Fat bloom on chocolate confectionery systems – From core to surface*. PhD thesis, University of Lund, Sweden.

Dahlenborg, H., Millqvist-Fureby, A., Bergenståhl, B. & Kalnin, D. J. E. (2011). Investigation of chocolate surfaces using profilometry and low vacuum scanning electron microscopy. *Journal of the American Oil Chemists' Society, 88*, 773–783.

Dahlenborg, H., Millqvist-Fureby, A. & Bergenståhl, B. (2015a). Effect of particle size in chocolate shell on oil migration and fat bloom development. *Journal of Food Engineering, 146*, 172–181.

Dahlenborg, H., Millqvist-Fureby, A. & Bergenståhl, B. (2015b). Effect of shell microstructure on oil migration and fat bloom development in model pralines. *Food Structure, 5*, 51–65.

De Clercq, N., Depypere, F., Delbaere, C., Nopens, I., Bernaert, H. & Dewettinck, K. (2014). Influence of cocoa butter diacylglycerols on migration induced fat bloom in filled chocolates. *European Journal of Lipid Science and Technology, 116*, 1388–1399.

De Man, J. M. & Wood, F. W. (1959). Hardness of butter. II. Influence of setting. *Journal of Dairy Science, 42*, 56–61.

Deka, K., MacMillan, B., Ziegler, G. R., Marangoni, A. G., Newling, B. & Balcom, B. J. (2006). Spatial mapping of solid and liquid lipid in confectionery products using a 1D centric SPRITE MRI technique. *Food Research International, 39*, 365–371.

Delbaere, C., Van de Walle, D., Depypere, F., Gellynck, X. & Dewettinck, K. (2016). Relationship between chocolate microstructure, oil migration, and fat bloom in filled chocolates. *European Journal of Lipid Science and Technology, 118*, 1800–1826.

Dewettinck, K., De Moor, H. & Huyghebaert, A. (1996). The free fat content of dried milk products and flow properties of milk chocolate. *Milchwissenschaft, 51*, 25–28.

Dibildox-Alvarado, E., Rodrigues, J. N., Gioielli, L. A., Toro-Vazquez, J. F. & Marangoni, A. G. (2004). Effects of crystalline microstructure on oil migration in a semisolid fat matrix. *Crystal Growth & Design, 4*, 731–736.

Dimick, P. S., Ziegler, G. R., Full, N. A. & Reddy, S. Y. (1996). Formulation of milk chocolate using milkfat fractions. *Australian Journal of Dairy Technology, 51*, 123–126.

Early, R. (2012). Dairy products and milk-based food ingredients. In D. Baines & R. Seal (Eds.), *Natural food additives, ingredients and flavourings* (1st ed., pp. 417–445). Cambridge: Woodhead Publishing.

Edmondson, L. F., Yoncoskie, R. A., Rainey, N. H., Douglas, F. W. & Bitman, J. (1974). Feeding encapsulated oils to increase the polyunsaturation in milk and meat fat. *Journal of the American Oil Chemists' Society, 51*, 72–76.

Edwards, W. P. (1984). Uses for dairy ingredients in confectionery. *International Journal of Dairy Technology, 37*, 122–125.

Elgersma, A., Tamminga, S. & Ellen, G. (2006). Modifying milk composition through forage. *Animal Feed Science and Technology, 131*, 207–225.

Engeseth, N. J., Pangan, A. C. & Fernando, M. (2018). Current context on chocolate flavor development — A review. *Current Opinion in Food Science, 21*, 84–91.

EU (2000). *Directive 2000/36/EC of the European Parliament and of the Council of 23 June 2000 relating to cocoa and chocolate products intended for human consumption*. Brussels: European Union.

FDA (2018). *Code of Federal Regulations – Title 21 – food and drugs. Part 163*. Cacao: Products.

Fowler, M. & Coutel, F. (2017). Cocoa beans: From tree to factory. In S. T. Beckett, M. S. Fowler & G. R. Ziegler (Eds.), *Industrial chocolate manufacture and use* (5th ed., pp. 9–49). Oxford: Wiley.

Franke, K. & Heinzelmann, K. (2008). Structure improvement of milk powder for chocolate processing. *International Dairy Journal, 18*, 928–931.

Full, N. A., Reddy, S. Y., Dimick, P. S. & Ziegler, G. R. (1996). Physical and sensory properties of milk chocolate formulated with anhydrous milk fat fractions. *Journal of Food Science, 61*, 1068–1084.

Galdámez, J. R., Szlachetka, K., Duda, J. L. & Ziegler, G. R. (2009). Oil migration in chocolate: A case of non-Fickian diffusion. *Journal of Food Engineering, 92*, 261–268.

Ghosh, V., Ziegler, G. R. & Anantheswaran, R. C. (2002). Fat, moisture, and ethanol migration through chocolates and confectionary coatings. *Critical Reviews in Food Science and Nutrition, 42*, 583–626.

Grivetti, L. E., Dillinger, T. L., Barriga, P., Escárcega, S., Jimenez, M. & Lowe, D. S. (2000). Food of the gods: Cure for humanity? A cultural history of the medicinal and ritual use of chocolate. *Journal of Nutrition, 130*, 2057S–2072S.

Guiheneuf, T. M., Couzens, P. J., Wille, H. J. & Hall, L. D. (1997). Visualisation of liquid triacylglycerol migration in chocolate by magnetic resonance imaging. *Journal of the Science of Food and Agriculture, 73*, 265–273.

Hartel, R. W. (1996). Applications of milk-fat fractions in confectionery products. *Journal of the American Oil Chemists' Society, 73*, 945–953.

Hayes, J. E., Running, C. A. & Ziegler, G. R. (2016). Degree of free fatty acid saturation influences chocolate rejection in human assessors. *Chemical Senses, 42*, 161–166.

Heertje, I. (1993). Microstructural studies in fat research. *Food Structure, 12*, 77–94.

Herrera, M. L., de León Gatti, M. & Hartel, R. W. (1999). A kinetic analysis of crystallization of a milk fat model system. *Food Research International, 32*, 289–298.

Johansson, D. & Bergenståhl, B. (1995). Sintering of fat crystal networks in oil during post-crystallization processes. *Journal of the American Oil Chemists' Society, 72*, 911–920.

Juul, B. M. (2010). Beat migration bloom by optimizing your process. A new look at the tempering process to reduce migration bloom in chocolate products, *Manufacturing Confectioner*, 68–75.

Kamphuis, H. J. (2009). Production and quality standards of cocoa mass, cocoa butter and cocoa powder. In S. T. Beckett, M. S. Fowler & G. R. Ziegler (Eds.), *Industrial chocolate manufacture and use* (4th ed., pp. 121–141). Oxford: Wiley.

Kaylegian, K. E. (1999). The production of specialty milk fat ingredients. *Journal of Dairy Science, 82*, 1433–1439.

Kaylegian, K. E. & Lindsay, R. C. (1995). *Handbook of milkfat fractionation technology and applications* (1st ed.). Urbana: AOCS Press.

Kaylegian, K. E., Hartel, R. W. & Lindsay, R. C. (1993). Applications of modified milk fat in food products. *Journal of Dairy Science, 76*, 1782–1796.

Keogh, M. K. & Higgins, A. C. (1986). Anhydrous milk fat: 1. Oxidative stability aspects. *Iranian Journal of Food Science and Technology, 10*, 11–22.

Koc, A. B., Heinemann, P. H. & Ziegler, G. R. (2003). A process for increasing the free fat content of spray-dried whole milk powder. *Journal of Food Science, 68*, 210–216.

Koyano, T., Hachiya, I. & Sato, K. (1990). Fat polymorphism and crystal seeding effects on fat bloom stability of dark chocolate. *Food Structure, 9*, 231–240.

Lee, W. L., McCarthy, M. J. & McCarthy, K. L. (2010). Oil migration in 2-component confectionery systems. *Journal of Food Science, 75*, E83–E89.

Liang, B. & Hartel, R. W. (2004). Effects of milk powders in milk chocolate. *Journal of Dairy Science, 87*, 20–31.

Lippi, D. (2013). Chocolate in history: Food, medicine, medi-food. *Nutrients, 5*, 1573–1584.

Lohman, M. H. & Hartel, R. W. (1994). Effect of milk fat fractions on fat bloom in dark chocolate. *Journal of the American Oil Chemists' Society, 71*, 267–276.

Lopez, C., Bourgaux, C., Lesieur, P., Riaublanc, A. & Ollivon, M. (2006). Milk fat and primary fractions obtained by dry fractionation: 1. Chemical composition and crystallisation properties. *Chemistry and Physics of Lipids, 144*, 17–33.

Maleky, F., McCarthy, K. L., McCarthy, M. J. & Marangoni, A. G. (2012). Effect of cocoa butter structure on oil migration. *Journal of Food Science, 77*, E74–E79.

Mallia, S., Escher, F. & Schlichtherle-Cerny, H. (2008). Aroma-active compounds of butter: A review. *European Food Research and Technology, 226*, 315–325.

Manning, D. & Dimick, P. (1985). Crystal morphology of cocoa butter. *Food Microstructure, 4*, 249–265.

Marangoni, A. G. (2002). The functionality of milkfat fractions in confectionery and plastic fats. In J. Welti-Chanes & J. M. Aguilera (Eds.), *Engineering and food for the 21st century* (1st ed., pp. 125–140). Boca Raton: CRC Press.

Marangoni, A. G. & Lencki, R. W. (1998). Ternary phase behavior of milk fat fractions. *Journal of Agricultural and Food Chemistry, 46*, 3879–3884.

Martin, R. A., Jr. (1988). Chocolate. *Advances in Food Research, 31*, 211–342.

Mazzanti, G., Guthrie, S. E., Sirota, E. B., Marangoni, A. G. & Idziak, S. H. J. (2004). Effect of minor components and temperature profiles on polymorphism in milk fat. *Crystal Growth & Design, 4*, 1303–1309.

McCarthy, K. L. & McCarthy, M. J. (2008). Oil migration in chocolate-peanut butter paste confectionery as a function of chocolate formulation. *Journal of Food Science, 73*, 266–273.

Metin, S. & Hartel, R. W. (1996). Crystallization behavior of blends of cocoa butter and milk fat or milk fat fractions. *Journal of Thermal Analysis, 47*, 1527–1544.

Metin, S. & Hartel, R. (2005). Crystallization of fats and oils. In F. Shahidi (Ed.), *Bailey's industrial oil and fat products* (6th ed.). New York: Wiley.

Metin, S. & Hartel, R. W. (2012). 15 - Milk fat and cocoa butter. In N. Garti & N. R. Widlak (Eds.), *Cocoa butter and related compounds* (1st ed., pp. 365–392). Urbana: AOCS Press.

Minifie, B. W. (1989). *Chocolate, cocoa, and confectionery: Science and technology* (3rd ed.). New York: Van Nostrand Reinhold.

Palmquist, D. L., Denise Beaulieu, A. & Barbano, D. M. (1993). Feed and animal factors influencing milk fat composition. *Journal of Dairy Science, 76*, 1753–1771.

Pedersen, A. (1991). *Texturization of milk fat* (Bulletin 260) (pp. 10–12). Brussels: International Dairy Federation.

Precht, D. (1988). Fat crystal structure in cream and butter. In N. G. A. K. Saito (Ed.), *Crystallization and polymorphism of fats and fatty acids* (pp. 329–357). New York: Marcel Dekker.

Reddy, Y. S., Full, N., Dimick, P. S. & Ziegler, G. R. (1996). Tempering method for chocolate containing milk-fat fractions. *Journal of the American Oil Chemists' Society, 73*, 723–727.

Reinke, S. K., Roth, S. V., Santoro, G., Vieira, J., Heinrich, S. & Palzer, S. (2015). Tracking structural changes in lipid-based multicomponent food materials due to oil migration by microfocus small-angle X-ray scattering. *ACS Applied Materials Interface, 7*, 9929–9936.

Rønholt, S., Mortensen, K. & Knudsen, J. C. (2013). The effective factors on the structure of butter and other milk fat-based products. *Comprehensive Reviews in Food Science and Food Safety, 12*, 468–482.

Rothkopf, I. & Danzl, W. (2015). Changes in chocolate crystallization are influenced by type and amount of introduced filling lipids. *European Journal of Lipid Science and Technology, 117*, 1714–1721.

Rousseau, D. (2006). On the porous mesostructure of milk chocolate viewed with atomic force microscopy. *LWT – Food Science and Technology, 39*, 852–860.

Rousseau, D. & Marangoni, A. G. (1998). The effects of interesterification on physical and sensory attributes of butterfat and butterfat–canola oil spreads. *Food Research International, 31*, 381–388.

Rousseau, D. & Smith, P. (2008). Microstructure of fat bloom development in plain and filled chocolate confections. *Soft Matter, 4*, 1706–1712.

Rowney, M. & Christian, M. (1996). Effect of cow diet and stage of lactation on the composition of milkfat for cheese manufacture. *Australian Journal of Dairy Technology, 51*, 118–122.

Sabariah, S., Md Ali, A. R. & Chong, C. L. (1998). Physical properties of Malaysian cocoa butter as affected by addition of milkfat and cocoa butter equivalent. *International Journal of Food Sciences and Nutrition, 49*, 211–218.

Sala, G., Van Vliet, T., Stuart, M.C., Van de Velde, F. & Van Aken, G.A. (2009). Deformation and fracture of emulsion-filled gels: Effect of gelling agent concentration and oil droplet size. *Food Hydrocolloids, 23*(7), 1853–1863.

Schieberle, P., Gassenmeier, K., Guth, H., Sen, A. & Grosch, W. (1993). Character impact odour compounds of different kinds of butter. *LWT – Food Science and Technology, 26*, 347–356.

Schlutt, B., Moran, N., Schieberle, P. & Hofmann, T. (2007). Sensory-directed identification of creaminess-enhancing volatiles and semivolatiles in full-fat cream. *Journal of Agricultural and Food Chemistry, 55*, 9634–9645.

Schmelzer, J. M. & Hartel, R. W. (2001). Interactions of milk fat and milk fat fractions with confectionery fats. *Journal of Dairy Science, 84*, 332–344.

Schnermann, P. & Schieberle, P. (1997). Evaluation of key odorants in milk chocolate and cocoa mass by aroma extract dilution analyses. *Journal of Agricultural and Food Chemistry, 45*, 867–872.

Shi, Y., Smith, C. M. & Hartel, R. W. (2001). Compositional effects on milk fat crystallization. *Journal of Dairy Science, 84*, 2392–2401.

Shiratsuchi, H., Shimoda, M., Imayoshi, K., Noda, K. & Osajima, Y. (1994a). Off-flavor compounds in spray-dried skim milk powder. *Journal of Agricultural and Food Chemistry, 42*, 1323–1327.

Shiratsuchi, H., Shimoda, M., Imayoshi, K., Noda, K. & Osajima, Y. (1994b). Volatile flavor compounds in spray-dried skim milk powder. *Journal of Agricultural and Food Chemistry, 42*, 984–988.

Sichien, M., Thienpont, N., Fredrick, E., Trung Le, T., Van Camp, J. & Dewettinck, K. (2009). Processing means for milk fat fractionation and production of functional compounds. In M. Corredig (Ed.), *Dairy-derived ingredients: Food and nutraceutical uses* (1st ed., pp. 68–102). Cambridge: Woodhead Publishing.

Skytte, U. P. & Kaylegian, K. E. (2017). Ingredients from milk. In S. T. Beckett, M. S. Fowler, & G. R. Ziegler (Eds.), *Industrial chocolate manufacture and use* (5th ed., pp. 102–134). Oxford: Wiley.

Smith, P. R. & Dahlman, A. (2005). The use of atomic force microscopy to measure the formation and development of chocolate bloom in pralines. *Journal of the American Oil Chemists' Society, 82*, 165–168.

Smith, K., Cain, F. W. & Talbot, G. (2007). Effect of nut oil migration on polymorphic transformation in a model system. *Food Chemistry, 102*, 656–663.

Stewart, I. D. (2017). *The effect of low melting oils on the crystallisation of confectionery fats*. PhD thesis, Loughborough University, United Kingdom.

Stewart, I. M. & Timms, R. E. (2002). Fats for chocolate and sugar confectionery. In K. K. Rajah (Ed.), *Fats in food technology* (1st ed., pp. 159–189). Sheffield: Sheffield Academic Press.

Svanberg, L., Ahrné, L., Lorén, N. & Windhab, E. (2011). Effect of pre-crystallization process and solid particle addition on cocoa butter crystallization and resulting microstructure in chocolate model systems. *Procedia Food Science, 1*, 1910–1917.

Tietz, R. A. & Hartel, R. W. (2000). Effects of minor lipids on crystallization of milk fat-cocoa butter blends and bloom formation in chocolate. *Journal of the American Oil Chemists' Society, 77*, 763–771.

Timms, R. E. (1980). The phase behavior and polymorphism of milkfat, milkfat fractions and fully hardened milkfat. *Australian Journal of Dairy Technology*, 47–52.

Timms, R. E. (2003). *Confectionery fats handbook: Properties, production and applications* (1st ed.). Cambridge: The Oily Press, Woodhead Publishing.

Timms, R. E. & Parekh, J. V. (1980). The possibilities for using hydrogenated, fractionated or interesterified milk fat in chocolate. *LWT – Food Science and Technology, 13*, 177–181.

Twomey, M. & Keogh, K. (1998). Milk powder in chocolate. *Farm Food, 8*, 9–11.

Twomey, M., Keogh, M. K., O'Kennedy, B. T., Auty, M. & Mulvihill, D. M. (2000). Effect of milk composition on selected properties of spray-dried high-fat and skim-milk powders. *Irish Journal of Agricultural and Food Research, 39*, 79–94.

Urbach, G. (1990). Effect of feed on flavor in dairy foods. *Journal of Dairy Science, 73*, 3639–3650.

Van Aken, G. A. & Visser, K. A. (2000). Firmness and crystallization of milk fat in relation to processing conditions. *Journal of Dairy Science, 83*, 1919–1932.

Vanhoutte, B., Dewettinck, K., Vanlerberghe, B. & Huyghebaert, A. (2002). Monitoring milk fat fractionation: Effect of agitation, temperature, and residence time on physical properties. *Journal of the American Oil Chemists' Society, 79*, 1169–1176.

Verhey, J. (1986). Physical properties of dried milk in relation to chocolate manufacture. *Netherlands Milk and Dairy Journal, 40*, 261–268.

Walter, P. & Cornillon, P. (2001). Influence of thermal conditions and presence of additives on fat bloom in chocolate. *Journal of the American Oil Chemists' Society, 78*, 927–932.

Weihe, H. D. (1961). Interesterified butter oil. *Journal of Dairy Science, 44*, 944–947.

Windhab, E. J. (2017). Tempering. In S. T. Beckett, M. S. Fowler, & G. R. Ziegler (Eds.), *Industrial chocolate manufacture and use* (5th ed., pp. 314–355). Oxford: Wiley.

Wood, R. (2017). Legal aspects of chocolate manufacture. In S. T. Beckett, M. S. Fowler, & G. R. Ziegler (Eds.), *Industrial chocolate manufacture and use* (5th ed., pp. 675–694). Oxford: Wiley.

Wright, A. J., Hartel, R. W., Narine, S. S. & Marangoni, A. G. (2000a). The effect of minor components on milk fat crystallization. *Journal of the American Oil Chemists' Society, 77*, 463–475.

Wright, A. J., Narine, S. S. & Marangoni, A. G. (2000b). Comparison of experimental techniques used in lipid crystallization studies. *Journal of the American Oil Chemists' Society, 77*, 1239–1242.

Ziegleder, G. (2017). Flavour development in cocoa and chocolate. In S. T. Beckett, M. S. Fowler, & G. R. Ziegler (Eds.), *Industrial chocolate manufacture and use, 5th edition* (5th ed., pp. 185–215). Oxford: Wiley.

Ziegleder, G., Moser, C. & Geier-Greguska, J. (1996). Kinetik der fettmigration in schokoladenprodukten teil I: grundlagen und analytik. *European Journal of Lipid Science and Technology, 98*, 196–199.

Ziegler, G. (2009). Product design and shelf-life issues: Oil migration and fat bloom. In G. Talbot (Ed.), *Science and technology of enrobed and filled chocolate, confectionery and bakery products* (1st ed., pp. 185–210). Cambridge: Woodhead Publishing.

Ziegler, G. R. & Hogg, R. (2017). Particle size reduction. In S. T. Beckett, M. S. Fowler, & G. R. Ziegler (Eds.), *Industrial chocolate manufacture and use* (5th ed., pp. 216–240). New York: Wiley.

Ziegler, G., Shetty, A. & Anantheswaran, R. (2004). Nut oil migration through chocolate. *Manufacturing Confectioner, 84*(9), 118–126.

10 Nutritional Significance of Milk Lipids: From Bioactive Fatty Acids to Supramolecular Structures Impacting Metabolism

E. Guillocheau, M. Milard, V. Rioux, C. Bourlieu, and M.-C. Michalski

10.1 Introduction

Milk lipids generally refer to the cow's milk lipids that dominate the strongly increasing world milk production. Milk products are important foods in the Western diet and as such, they convey numerous nutrients and minerals. However, milk fat still suffers from a bad image in the lay public because it is also rich in saturated fatty acids, which are perceived as associated with deleterious health effects only. Strikingly, recent meta-analyses report lack of correlation between milk products consumption and the occurrence of diabetes on the one hand, and lack of correlation between milk fat consumption and cardiovascular diseases on the other hand; some analyses even report protective effects of milk fat consumption on cardiovascular health. To best outline the state of the art compared with previous editions, this chapter will therefore not provide a broad epidemiological review of the associations between milk fat consumption and several disease risks that can be found elsewhere. Instead, the chapter will first focus on the major recent scientific advances on the impact of specific molecular fatty acid species (among which specific odd and *trans* species) and supramolecular structures of milk fat on several aspects of lipid metabolism and metabolic disease risk. In this context, we will highlight possible mechanisms, including those related to the impact of lipids in the gut (Figure 10.1). Second, we present the recent advances in infant nutrition science related to knowledge gained on the importance of milk fat globule structure, breastmilk versus infant formula differential in composition and structure, on metabolic programming and the recent advances in this field.

Authors Etienne Guillocheau and Marine Milard contributed equally.

E. Guillocheau · V. Rioux
Laboratory of Biochemistry and Human Nutrition, Agrocampus Ouest, Rennes, France

M. Milard · M.-C. Michalski (✉)
Univ Lyon, CarMeN laboratory, INRAE, INSERM, INSA Lyon, CRNH-RA, Pierre-Bénite, France
e-mail: marie-caroline.michalski@inrae.fr

C. Bourlieu
INRAE, CIRAD, UMR1208 IATE, Montpellier, France

10.1.1 Milk Fatty Acids: From Specific Biochemical Functions to Physiological Effects and Nutritional Significance

The detailed fatty acid composition of ruminant milk lipids will not be presented in this chapter because the reader can find it detailed in earlier chapters of this book. Here, we will focus on some major composition characteristics that can impact the nutritional and health effects of milk and dairy products as highlighted by recent

P. L. H. McSweeney et al. (eds.), *Advanced Dairy Chemistry, Volume 2*,
https://doi.org/10.1007/978-3-030-48686-0_10

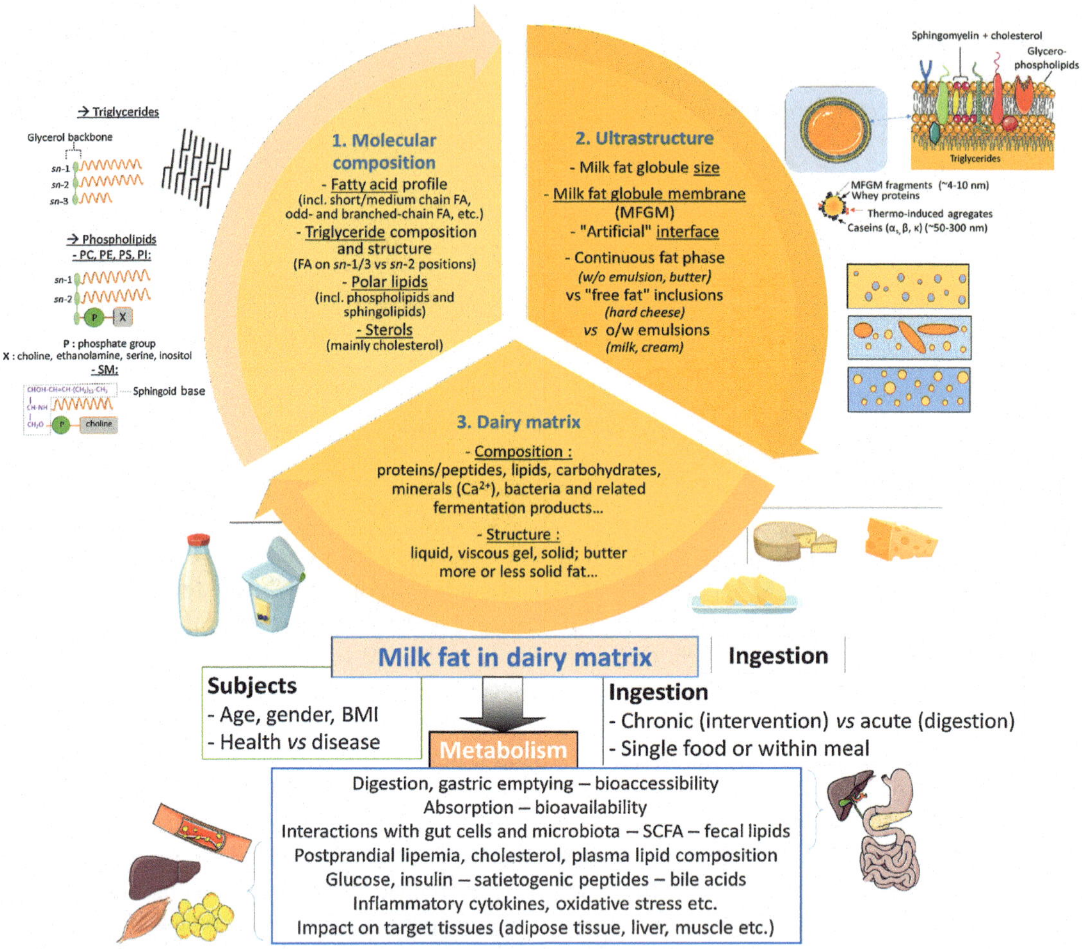

Figure 10.1. Overview of several molecular and supramolecular structures of lipids in milk fat and dairy matrices and possible impacts on metabolism. Polar lipids include phospholipids and sphingolipids. Buttermilk contains MFGM fragments. "Artificial" interface can be found in homogenized fat droplets (milk protein interface) or in formulated fat droplets (using emulsifiers and stabilizers, such as proteins, esters of partial glycerides, or sugars and gums). *BMI* body mass index, *MFGM* milk fat globule membrane, *o/w* oil-in-water, *P* phosphate group, *PC* phosphatidylcholine, *PE* phosphatidylethanolamine, *PS* phosphatidylserine, *PI* phosphatidylinositol, *SCFA* short-chain fatty acids, *SM* sphingomyelin, *X* specific head group of the phospholipid molecule, *w/o* water-in-oil. Schemes are not to scale (Adapted from Michalski (2009), Bourlieu and Michalski (2015), Vors *et al.* (2016) and Michalski *et al.* (2020). Pieces of artwork of dairy products were designed by macrovector/Freepik, those of body organs are derivatives of artworks kindly provided by the Servier Medical Art database licensed under a Creative Commons Attribution 3.0 Unported License.

scientific data. Ruminant milk contains only 3–5% fat, but its fatty acid composition is unique. The major milk fatty acids (palmitic and oleic acids) are universally found in natural dietary fats. The proportion of total saturated fatty acids (60–65% of fatty acids) is relatively high, albeit with the original presence of short-chain (C4:0) and medium-chain (C6:0, C8:0, C10:0) fatty acids that can reach up to 10–15% of fatty acids. The group of long-chain saturated fatty acids is exceptionally rich in myristic acid (C14:0; 9–12% of fatty acids). Early observational studies have suggested that diets rich in saturated fatty acids are positively associated with increased risks of cardiovascular diseases (CVD) (Keys *et al.* 1966, 1984). However, recent epidemiological studies have shown that there is no significant evidence for concluding that intake of dairy fat is associated with an increased risk of CVD (Drouin-Chartier *et al.* 2016a, b; Dehghan *et al.* 2018).

Some other milk fatty acids (*trans*, odd-chain and branched-chain) cannot be synthesized by humans. Even if they are minor components of dairy fat, they are therefore considered as biomarkers of dairy product consumption when measured in human plasma phospholipids or red blood cells (Pranger *et al.* 2018). They are also often used to study the correlations between dairy fat dietary intake level and risk factors for human health (Imamura *et al.* 2018). Among other nutrients specific to milk, these original fatty acids are integral to mechanisms explaining the inverse associations found between dairy product intakes, and lower risk of both type 2 diabetes (Elwood *et al.* 2010; Tong *et al.* 2011; Aune *et al.* 2013) and metabolic syndrome (Chen *et al.* 2015; Kim and Je 2016). Milk fat is characterized by its relatively low content of essential polyunsaturated fatty acids, linoleic acid (LA or C18:2 $n-6$; 2–3% of fatty acids) for the ω6 family and α-linolenic acid (ALA or C18:3 $n-3$, <1% of fatty acids) for the ω3 family. These values may be slightly increased by modifying the diet of lactating animals (Kliem and Shingfield 2016). Table 10.1 provides an attempt to synthesize the important and specific biological roles of selected individual milk fatty acids and the subsequent elucidated or putative physiological significance in terms of nutritional interest.

10.1.2 Short- and Medium-Chain Saturated Fatty Acids

Milk is characterized by the presence of saturated SCFAs (C2:0–C5:0) including notably butyric acid. Because it is a specific histone deacetylase inhibitor, butyric acid selectively induces growth arrest and apoptosis in a variety of cancer cells (Chen *et al.* 2006; Kolar *et al.* 2007). It seems, therefore, that the delivery of an adequate amount of butyrate to the appropriate site could protect against early tumorigenic events (Sengupta *et al.* 2006), leading to its described role against colorectal cancer (Li *et al.* 2017). It should be noted that the supply of butyric acid is not limited to milk fat, it is also highly produced during fermentation of fibers in the colon (Sun and Zhu 2018).

Medium-chain saturated fatty acids (MCFAs) also constitute an important and original feature of characteristic nutrients present in milk and dairy products (Jensen *et al.* 1990). This class of fatty acids includes caproic acid (hexanoic acid, C6:0), caprylic acid (octanoic acid, C8:0), and capric acid (decanoic acid, C10:0) (Bach and Babayan 1982). For instance, caprylic acid represents 1–2% of cow milk fatty acids (Jensen *et al.* 1990) and 3% in goat milk (Alonso *et al.* 1999). MCFAs are primarily esterified at the *sn*-3 position of triglycerides (triacylglycerols, TAGs) in cow milk (Jensen *et al.* 1990). MCFAs display distinct physicochemical and metabolic properties from those of long-chain saturated fatty acids (LCFAs $\geq$ 12 carbons), leading to specific physiological effects (Bach and Babayan 1982). First, a part of MCFAs coming from dietary medium-chain TAGs (MCTs) are quickly released after ingestion under the action of preduodenal lipases (Clark *et al.* 1969), allowing direct absorption by the stomach mucosa (Perret 1980; Lai and Ney 1998; Lemarié *et al.* 2016b). Second, small intestinal cells absorb MCFAs (and LCFAs) after the subsequent action of duodenal pancreatic lipases on both dietary remaining MCTs and long-chain TAGs (LCTs). However, unlike LCFAs that are re-esterified with 2-monoglycerides into TAGs and incorporated into chylomicrons entering the lymphatic system, MCFAs are directly transferred to the portal circulation and transported as free fatty acids (FFAs) with albumin to the liver (You *et al.* 2008). Third, hepatic MCFAs are rapidly subjected to mitochondrial β-oxidation (Ooyama *et al.* 2009), because they easily enter the mitochondria independently of the carnitine transport system, as opposed to LCFAs (Papamandjaris *et al.* 1998). These metabolic properties of MCFAs (rapid gastro-intestinal hydrolysis and absorption, specific transport through the portal vein and rapid beta-oxidation in the liver) lead to a high catabolism and low tissue storage, especially

Table 10.1. Specific biochemical functions of milk fatty acids and subsequent demonstrated or postulated nutritional significance

Milk fatty acid	Specific biochemical function or metabolic utilization	Physiological effects and nutritional significance
Butyric acid (C4:0)	Specific histone deacetylase inhibitor	Induction of growth arrest and apoptosis in a variety of cancer cells Role against colorectal cancer
Caproic acid (C6:0) Caprylic acid (C8:0) Capric acid (C10:0)	Quick release after ingestion under the action of preduodenal lipases Direct absorption by the stomach mucosa Direct transfer to the portal circulation Rapid mitochondrial β-oxidation	High catabolism and low tissue storage Diminished deposition of fat
Caprylic acid (C8:0)	Ghrelin octanoylation	Regulation of growth hormone secretion, appetite stimulation and food intake, gastric acid secretion, gastric motility, glucose homeostasis and adiposity
Myristic acid (C14:0)	Myristoylation of about 100 human proteins	Regulation of protein subcellular localization, protein–protein interaction or protein–membrane interaction Increase the activity of desaturases and contribute to the increase in long-chain polyunsaturated fatty acids
Palmitic acid (C16:0)	Palmitoylation (*S*-acylation) of hundreds of proteins	Regulation of protein subcellular localization, protein–protein interaction or protein–membrane interaction
Stearic acid C18:0	Active Δ9-desaturation to oleic acid (C18:1 *n* – 9)	Neutral effect on cholesterol metabolism
Odd-chain fatty acids Pentadecanoic acid (C15:0) Heptadecanoic acid (C17:0)	Synthesis of odd-numbered VLCFAs[a] Anaplerotic intermediates for the citric acid cycle Store excess propionic acid	Protection against the risk of type 2 diabetes and NAFLD[b]
Branched-chain fatty acids Phytanic acid	Natural ligand of PPAR[c] and RXR[d]	Increases glucose uptake by rat hepatocytes
Trans fatty acid *Trans*-vaccenic acid (*trans*-11 C18:1) *Trans*-palmitoleic acid (*trans*-9 C16:1)	Rapid Δ9-desaturated into rumenic acid Ligand of PPAR-α and PPAR-γ[c] Comes from dietary *trans*-vaccenic acid retroconversion	Increases the expression of FAS[e] and ACC[f] Decreases visceral adipose tissue Lower risk of type 2 diabetes Lower risk of metabolic syndrome

[a]*VLCFAs* very long-chain fatty acids, [b]*NAFLD* non-alcoholic fatty liver disease, [c]*PPAR* peroxisome proliferator activated receptor, [d]*RXR* retinoid X receptor, [e]*FAS* fatty acid synthase, [f]*ACC* acetyl-CoA carboxylase.

in adipose tissue (Bach *et al.* 1996; Lemarié *et al.* 2015). Dietary MCFAs have therefore been associated with beneficial physiological effects compared with LCFAs. Indeed, a diminished deposition of fat was reported in rats overfed with MCT diets, compared with LCT diets (Geliebter *et al.* 1983; Baba *et al.* 1987; Han *et al.* 2003). In overweight humans, intakes of equal-caloric diets rich in MCFAs were shown to decrease adiposity and increase energy expenditure compared to similar diets rich in LCFAs (Tsuji *et al.* 2001; St-Onge *et al.* 2008).

More recently, caprylic acid was shown to specifically acylate ghrelin (Kojima *et al.* 1999), a mechanism called ghrelin octanoylation. Ghrelin is a 28 amino-acid peptide expressed mainly in the gastrointestinal tract, especially in the stomach (Hosoda *et al.* 2000) and the only known peptide hormone with an orexigenic effect. Octanoylated ghrelin binds the growth hormone secretagogue receptor 1a (GHS-R1a) located in the pituitary gland and the hypothalamus (Howard *et al.* 1996; Wang *et al.* 2014). Octanoylated ghrelin is therefore involved in the regulation of many relevant biological processes, including the secretion of growth hormone (GH), the stimulation of appetite and food intake, the regulation of gastric acid secretion, gastric motility, glucose homeostasis and adiposity (Delporte 2013). Altogether, these data suggest that dietary caprylic acid, absorbed at the gastric level, may regulate octanoylated ghrelin production, circulating concentration and functions (Lemarié *et al.* 2018), including appetite regulation. This regulation may be crucial at specific physiological stages such as growth in early life.

10.1.3 Long-Chain Saturated Fatty Acids

Saturated fatty acids (SFAs) with a long chain represent about 50% of total milk fatty acids, with lauric (C12:0), myristic (C14:0), palmitic (C16:0), and stearic (C18:0) acids accounting for 3–4%, 10–12%, 25–35% and 10% of total milk fatty acids, respectively. In ruminant milk, these SFAs are esterified on specific positions of the TAGs, with lauric and myristic acids being mainly esterified at the *sn*-2 position whereas palmitic and stearic acid are equally esterified between the *sn*-1 and *sn*-2 positions. This stereospecific positioning of SFAs in TAGs may have an impact in infant nutrition (Innis 2011) since the *sn*-2 fatty acid is left as a monoglyceride during the cleavage by digestive lipases, which facilitates its intestinal absorption.

Early observational studies have suggested that diets high in saturated fatty acids cause an increase in plasma total and low-density lipoprotein (LDL)-cholesterol and are positively associated with increased risks of CVD (Keys *et al.* 1966, 1984), questioning the role of reducing intakes of total saturated fat and dairy fat for prevention. However, more recent studies have shown inverse association (Gillman *et al.* 1997; Mozaffarian *et al.* 2004) and a number of recent meta-analyses (Siri-Tarino *et al.* 2010; O'Sullivan *et al.* 2013; Chowdhury *et al.* 2014; de Souza *et al.* 2015) and prospective studies (Dehghan *et al.* 2017) also suggested that there is no significant evidence for concluding that saturated fat and/or dairy fat (de Oliveira Otto *et al.* 2012; Praagman *et al.* 2015; Drouin-Chartier *et al.* 2016a, b; Dehghan *et al.* 2018) intake is associated with an increased risk of CVD. One proposed explanation for these inconsistent findings is that the association between SFAs and CVD differs across types of saturated fatty acids, depending on their specific chain length (Praagman *et al.* 2016). For example, when considered individually, intakes of lauric (C12:0), and myristic (C14:0) acids were inversely associated with myocardial infarction (Praagman *et al.* 2019) and several years before it was shown that short- and medium-chain SFAs were not deleterious for cholesterolemia and CVD (Hu *et al.* 1999).

Beyond the controversy and debate regarding the role of some SFAs in CVD, it seems important to remember that SFAs can no longer be considered as a collective in terms of structure, metabolism, cellular functions, and physiological effects (Legrand and Rioux 2015). Moreover, human data reporting the balance between intake and *de novo* synthesis of SFAs are lacking. However, recent findings suggest that individual long-chain SFAs possess specific properties associated with important biological functions, like fatty acid acylation of proteins, which corresponds to the co- or post-translational covalent linkage of a SFA, activated in the form of acyl-CoA, to an amino acid residue of the substrate protein (Rioux 2016).

Indeed, the cellular fatty acids covalently bound to proteins are mainly SFAs. Palmitoylation (*S*-acylation) corresponds to the reversible attachment of palmitic acid (C16:0) to the side chain of a cysteine residue via a thioester bond (Blaskovic *et al.* 2014). N-terminal myristoylation refers to the covalent attachment of myristic acid (C14:0) by an amide bond to the N-terminal glycine of many eukaryotic and viral proteins (Johnson *et al.* 1994). As also described above, octanoylation (*O*-acylation) typically concerns the formation of an ester bond between octanoic acid (caprylic acid, C8:0) (Lemarié *et al.* 2016a) and the side chain of a serine residue of the stomach ghrelin peptide (Kojima *et al.* 1999). An increasing number of proteins (enzymes, hormones, receptors, oncogenes, tumor suppressors, proteins involved in signal transduction, eukaryotic and viral structural proteins) have been shown to undergo fatty acid acylation. The acyl moiety can mediate protein subcellular localization, protein–protein interaction or protein–membrane interaction. Therefore, through the covalent modification of proteins, these particular saturated fatty acids exhibit emerging specific and important roles in modulating protein functions.

For example, myristic acid was shown 15 years ago to trigger a specific and dose-dependent increasing effect on Δ6-desaturase activity in cultured rat hepatocytes (Jan *et al.* 2004). Myristoylation of one protein (NADH-cytochrome b5 reductase) of the whole Δ6-desaturase complex (Rioux *et al.* 2011) was proposed to explain how dietary myristic acid increased the overall conversion of α-linolenic acid to longer, highly unsaturated fatty acids, such as eicosapentaenoic (EPA), and docosahexaenoic (DHA) acids in animal studies (Rioux *et al.* 2005, 2008; Legrand *et al.* 2010; Ezanno *et al.* 2015).

10.1.4 Odd-Chain and Branched-Chain Saturated Fatty Acids

Odd-chain SFAs, such as pentadecanoic (C15:0) and heptadecanoic (C17:0) acids, are produced by rumen microbial metabolism and are therefore present in milk fat. Their plasma concentration in humans is now considered as a biomarker of dairy fat intake (Pfeuffer and Jaudszus 2016). In addition, a number of studies on cardiometabolic diseases have shown that plasma concentrations of these odd-chain SFAs are associated with lower such risks, although the mechanisms are not fully understood. Recent meta-analyses showed that C15:0 and C17:0 are protective against the risk of type 2 diabetes (de Souza *et al.* 2015; Imamura *et al.* 2018). Moreover, a recent cohort study (Yoo *et al.* 2017) found a negative correlation between serum levels of C15:0 and C17:0, and severity of NAFLD measured by the NAFLD activity scores and hepatocyte ballooning score. In the same study, mice fed the MCD diet (methionine and choline deficient), but supplemented with C15:0, showed reduced aspartate transaminase activity and hepatic infiltration of ceroiladen macrophages, compared to MCD-treated animals. This result suggests that C15:0 can serve as a promising biomarker for the diagnosis of NAFLD but may also be used for therapeutic treatment. Among the suggested mechanisms, C15:0 and C17:0 may be used for synthesis of odd-numbered VLCFAs, provide anaplerotic intermediates for the citric acid cycle or store excess propionic acid (Pfeuffer and Jaudszus 2016).

Among the branched-chain fatty acids in milk, phytanic acid (multi-branched with 4-methyl; 3,7,11,15-tetramethylhexadecanoic acid) appears now as an interesting bioactive compound (Wanders *et al.* 2011). Phytanic acid is a derivative of phytol, itself an important component of chlorophyll. Once released from the porphyrin nucleus of chlorophyll by rumen bacteria (the digestive system of humans is not able perform this cleavage), the phytol is converted to phytanic acid, which explains its particular presence in milk and ruminant products (Che *et al.* 2013). After absorption by humans, and due to its particular structure, phytanic acid cannot be β-oxidized in the mitochondria. Instead, it is α-oxidized in the peroxisomes into pristanic acid. The knowledge of phytanic acid was obtained after discovery of its accumulation in the cells and plasma of patients with Refsum disease (hereditary deficiency of the peroxisomal phytanoyl-CoA hydroxylase), causing lesions in the body, retina and the central nervous system. At non-toxic doses, phytanic acid and pristanic acid appeared to be natural ligands of the different isoforms of the PPAR and RXR (Zomer *et al.* 2000); for example, phytanic acid has been shown to increase glucose uptake by rat primary hepatocytes (Heim *et al.* 2002).

10.1.5 *Trans* Fatty Acids (and CLAs)

Dairy fat contains natural *trans* fatty acids such as *trans*-vaccenic acid (*trans*-11 C18:1, or *trans*-C18:1 *n*−7, TVA) but also conjugated linoleic acids (CLAs) like rumenic acid (*cis*-9,*trans*-11 C18:2, RMA) for which benefits towards inflammation, obesity, and type 2 diabetes are highly suspected (Field *et al.* 2009). This chapter will focus only on *trans*-vaccenic and *trans*-palmitoleic acids (*trans*-C16:1 *n*−7, or *trans* Δ9-C16:1, TPA) since CLAs are already presented in another chapter of this book (Chapter 3). Current evidence on the effect of total ruminant *trans* fatty acids (R-TFA) on coronary heart diseases (CHD) in humans is inconclusive. Most epidemiological studies have suggested inverse or no association between R-TFA intake and CHD (Gebauer *et al.* 2011; Baer 2012; Gayet-Boyer *et al.* 2014), but controversy still exists in relation to nutritional doses. Apart from the cardiovascular parameters, interesting results concerning type 2 diabetes and metabolic syndrome have been described for TVA and TPA.

10.1.5.1 *Trans*-Vaccenic Acid

TVA is by far the main *trans*-C18:1 fatty acid present in ruminant milk fat, accounting for between 1.5% and 4% of total fatty acids in dairy products (Precht *et al.* 2001; Kuhnt *et al.* 2011). In ruminant and non-ruminant mammals, TVA is well known to be Δ9-desaturated into the specific CLA rumenic acid (*cis* 9,*trans* 11 CLA) by the Stearoyl-CoA desaturase (Griinari *et al.* 2000; Santora *et al.* 2000; Turpeinen *et al.* 2002). TVA may additionally be Δ13-desaturated into the *trans*-11,*cis*-13-CLA isomer by Fatty Acid Desaturase 3 (FADS3), as recently demonstrated *in vitro* (Rioux *et al.* 2013) and *in vivo* (Garcia *et al.* 2017, 2018), but the physiological effects of this particular CLA isomer are not known. Several nutritional studies on TVA specific supplementation have been carried out, both on rodents and on humans. TVA supplementation led to interesting results; a decrease in circulating TAGs was observed, as well as a decrease in hepatic lipids compared to a positive control (Wang *et al.* 2008, 2009). In the liver, TVA supplementation decreased the expression of fatty acid synthase (FAS) and acetyl-CoA carboxylase (ACC), which could explain the decrease in NAFLD (Wang *et al.* 2009; Jacome-Sosa *et al.* 2014). In addition, TVA supplementation led to a decrease in size and weight of the visceral adipose tissue in obese JCR:LA-cp rats (Jacome-Sosa *et al.* 2014) but also in fa/fa Zucker rats (Mohankumar *et al.* 2013). This impact on visceral adipose tissue was probably due to the ability of TVA to act as a ligand for PPAR-α and PPAR-γ (Wang *et al.* 2012). Regarding insulin resistance, the results are less clear; a decrease was reported in the JCR:LA-cp rat model (Jacome-Sosa *et al.* 2014) and pancreatic function was improved (Wang *et al.* 2016), but no impact on the HOMA-IR was found either on the Zucker rat (Mohankumar *et al.* 2013) or on humans (Gebauer *et al.* 2015). The immune function has also been explored, and TVA supplementation had a favorable impact, especially on splenocytes (Blewett *et al.* 2009). On cardiovascular parameters, the results are still contradictory, as described above. Studies on the hamster, with different doses of TVA, did not allow the conclusion of a deleterious impact of TVA on HDL-cholesterol nor on LDL-cholesterol (Meijer *et al.* 2001; Tyburczy *et al.* 2009). In contrast, Gebauer *et al.* (2015) identified a TVA supplementation of 3% of total energy in humans for 24 days led to a significant increase in LDL-cholesterol and a significant increase in the total cholesterol/HDL cholesterol ratio.

10.1.5.2 *Trans*-Palmitoleic Acid

Although it represents only 0.04% of total fatty acids in bovine milk (Destaillats *et al.* 2000), *trans*-palmitoleic acid (*trans*-C16:1 *n*−7 or *trans* Δ9-C16:1, TPA) is described as the most abundant *trans*-C16:1 isomer in dairy products (Luna *et al.* 2009). It is therefore deemed as a biomarker of dairy fat intake (Sun *et al.* 2007; Micha *et al.* 2010; von Schacky *et al.* 2017). As recently demonstrated, it also comes from liver peroxisomal retroconversion of dietary *trans*-vaccenic acid (Jaudszus *et al.* 2014; Guillocheau *et al.* 2019), which itself represents 1–2% of total dairy fatty acids. TPA was previously identified as a biomarker of metabolic health in several epidemiological studies, both observational and prospective. At the observational level first, inverse associations were found between TPA content in adipose tissue and skinfold thickness (Smit *et al.* 2010). Consistent with these outcomes, circulating levels of TPA were inversely associated with body mass index (BMI), systolic blood pressure (Da Silva *et al.* 2014) and insulin-resistance among non-obese people (Da Silva *et al.* 2015). Kratz and colleagues consistently associated high circulating levels of TPA to lower hepatic TAGs, higher glucose tolerance and higher insulin sensitivity (Kratz *et al.* 2014). In strong agreement with these results, TPA was negatively correlated with the NAFLD score (Han *et al.* 2002). In addition, it was recently found in the PREDIMED study that high levels of circulating TPA are associated with lower systemic metabolic inflammation (Muralidharan *et al.* 2019).

Second, from a prospective point of view, high levels of plasma phospholipid TPA were prospectively associated with lower insulin resistance, lower presence of atherogenic dyslipidemia and lower incidence of type 2 diabetes in the Cardiovascular Health Study (CHS) cohort (Mozaffarian *et al.* 2010). In the Multi-Ethnic Study of Atherosclerosis (MESA) cohort, plasma phospholipid TPA was found to be inversely linked with triglyceridemia, fasting insulin, hypertension, and lower risk of type 2 diabetes again (Mozaffarian *et al.* 2013). In line with these outcomes, Yakoob and colleagues prospectively associated high levels of plasma phospholipid TPA with lower risk of type 2 diabetes in both Nurses' Health Study and Health Professionals Follow-Up Study cohorts (Yakoob *et al.* 2016). Finally, two meta-analyses consistently associated high levels of circulating TPA to lower risk of type 2 diabetes (de Souza *et al.* 2015; Imamura *et al.* 2018). Taken together, there is epidemiological evidence of probable beneficial physiological impacts of dietary TPA in humans. However, due to the lack of availability of pure TPA in high amounts, no data currently exists regarding nutritional studies and few hypotheses on possible mechanisms.

10.2 Milk Polar Lipids and the Milk Fat Globule Membrane

The health effects of dairy products containing fat can be explained by their qualitative and quantitative lipid composition. Many studies are interested in the effects of the fatty acid profile of milk fat, as detailed above. However, metabolic effects of milk fat could also be explained in part by the milk polar lipids naturally present in the milk fat globule membrane (MFGM) (Figure 10.2).

Milk polar lipids are bioactive molecules normally present in all biological membranes. Therefore, the MFGM, consisting of a specific lipid trilayer (an inner monolayer derived from the endoplasmic reticulum of the lactating cell, and an outer bilayer from the cell membrane) (Argov *et al.* 2008; Bourlieu and Michalski 2015), naturally contains polar lipids. The inner monolayer is mainly composed of PI and PS. The outer bilayer includes an inner leaf rich in PE, PC, sphingomyelin (SM) and PS, and an outer leaf rich in PC, SM, cholesterol and PS. As in mammalian biological membranes, cholesterol, and SM are co-localized in the MFGM in the so-called lipid rafts (Lopez 2010). Altogether, milk polar lipids present a phospholipid composition of the order of ~31% PE, ~29% PC, ~25% SM, ~7% PI, and ~ 6% PS (Michalski 2009; Contarini and Povolo 2013; Castro-Gomez *et al.* 2015). Of note, polar lipids (PL) generally contain more polyunsaturated fatty acids (PUFAs) and also longer chain FA (specifically C22:0–C23:0), and the individual FA profile of each class of milk PL is specific: SM is characterized by high amounts of long-chain FA (C22:0–C24:0), PC is dominated by C16 and C18:1 *cis*-9, PE by C18:1, whereas C18:0 and C18:1 *cis*-9 are major FA in PI and PS (Fauquant *et al.* 2007; Fong *et al.* 2007). Moreover, like for milk TAGs, the manipulation of cow's diet to enrich milk fat in PUFA consequently modulates the fatty acid (FA) profile of milk PL. Lopez *et al.* (2008) demonstrated that supplementation of a maize silage diet with extruded linseed significantly decreased SFAs in MFGM, while increasing unsaturated FA. This supplementation significantly increased monoun-

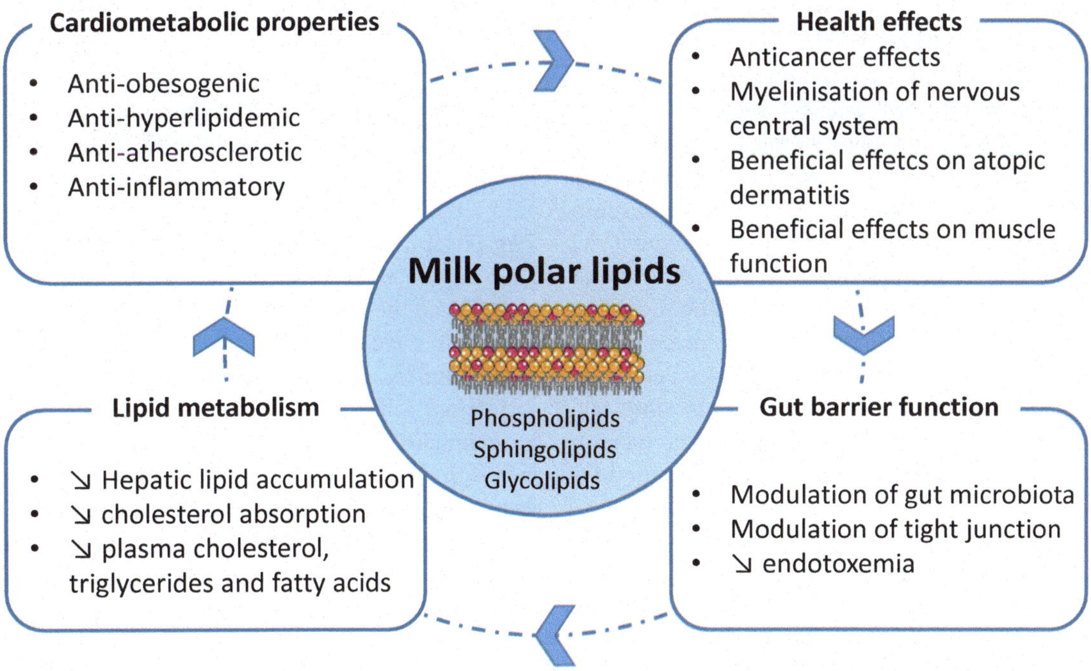

Figure 10.2. Potential beneficial effects of milk polar lipid supplementation against metabolic disorders and some other health impacts. Most effects were demonstrated up to date in preclinical models. See details and summary of clinical trials in the text.

saturated FA, including vaccenic acid and PUFAs. Importantly, the specificity of the MFGM is its SM content of ~25%, compared with vegetable polar lipids devoid of SM and egg polar lipids having <5% of egg SM. Milk SM is composed of a wide range of molecular species and milk polar lipids also contain ceramides (Bourlieu *et al.* 2018), gangliosides, and cerebrosides (of importance notably in infant nutrition, see Section 10.4). The observed metabolic effects of whole dairy products might thus be explained in part by their content of milk polar lipids.

10.2.1 Effects of Milk Polar Lipids on Lipid Metabolism

Several studies have examined the metabolic effects of milk polar lipids or MFGM (including ~25% polar lipids) on lipid metabolism. Wat *et al.* (2009) reported that in mice the addition of milk polar lipids (1.2% by weight in the diet via a MFGM-rich ingredient) in a high-fat diet (21% milk fat and 0.15% cholesterol), compared to a high-fat diet devoid of polar lipids, reduced liver weight, decreased plasma and hepatic lipids and decreased the expression of genes involved in hepatic synthesis of FAs after 8 weeks of diet (Wat *et al.* 2009). In addition, an intake of 1.25% by weight of milk polar lipids in a high-fat diet (21% milk fat and 0.15% cholesterol) during 8 weeks in rats contributed to significant reductions in plasma and hepatic TAG concentrations (Kamili *et al.* 2010). Moreover, rats fed with a low-fat diet (5% corn oil or milk fat) and rich in sucrose, supplemented with 0.6% milk polar lipids (again via MFGM) had an inhibition of lipid synthesis and an increase in hepatic lipid catabolism compared to control diet devoid of milk polar lipids (Zhou *et al.* 2012). A short-term study where mice were force-fed with an emulsion containing milk polar lipids demonstrated an earlier postprandial lipemic kinetics compared with emulsion containing soybean polar lipids (Lecomte *et al.* 2015). *In vivo* and *in vitro* studies suggested that this difference in postprandial kinetics is due to a stimulation of digestive lipolysis by milk polar lipids versus soybean polar lipids (Lecomte *et al.* 2015). However, the incorporation of 1.2% by weight of milk polar lipids or polar lipids from soy in a high-fat diet (20% lipid from palm oil) for 8 weeks in mice did not affect fasting TAG and cholesterol levels (Lecomte *et al.* 2016). Ohlsson *et al.* (2009) evaluated the effect in healthy humans of the daily consumption of a dairy drink including an ingredient rich in milk polar lipids (2.8 g/d via MFGM-rich buttermilk concentrate) for 4 weeks. The ingredient rich in milk polar lipids did not induce a significant effect on plasma lipids after 4 weeks compared to baseline, while the egg polar lipid (egg lecithin)-based control drink induced an increase in plasma lipids (Ohlsson *et al.* 2009). In addition, in comparison with egg lecithin, milk polar lipids tended to decrease the cholesterol content of chylomicrons during the first phases of digestion (Ohlsson *et al.* 2010a, b). In another clinical study, supplementation with 3 g/d of milk polar lipids within different dairy products for 10 days induced a significant decrease in total plasma cholesterol concentration compared to baseline; however, a concomitant decrease of HDL-cholesterol concentration was observed (which is not a favorable effect), and we must point out a major limitation of this trial that is the lack of a control group (Keller *et al.* 2013). A supplementation with a buttermilk-based drink for 4 weeks (including <200 mg milk PL/d) was reported to induce a 3% decrease in serum total cholesterol compared with a skim milk–based drink after 4 weeks (Conway *et al.* 2013). Using a dietary intervention with a non-dairy matrix (muffin), Rosqvist *et al.* (2015) demonstrated that using cream as lipid ingredient, that is, containing the MFGM including 190 mg/d of milk PL, did not alter plasma lipid profile after intervention compared with the butteroil control muffins devoid of MFGM (TAG only).

Regarding mechanisms involved, various *in vitro* and *in vivo* studies in mice have demonstrated that SM and its metabolites participate in the reduction of intestinal absorption of cholesterol (Nilsson and Duan 2006). *In vitro*, on Caco-2 cells, the supplementation of milk SM with egg phosphatidylcholine induced a significant reduction in cholesterol, correlated positively with the reduction of its absorption and esterification. The anti-cholesterolemic effects observed with total milk polar lipids were also demonstrated with milk SM (Noh and Koo 2004). In mice, a decrease in cholesterol uptake was observed during consumption of a chow diet containing 0.1% by weight of milk SM compared to egg phosphatidylcholine (Eckhardt *et al.* 2002). Noh *et al.* compared the efficacy of milk SM and egg SM on the inhibition of intestinal cholesterol absorption and other lipids. Rats received in the duodenum by lymphatic cannula a lipid emulsion (triolein, cholesterol, labelled cholesterol, α-tocopherol, and taurocholate) supplemented or not with 90 μmol of SM and an inhibition of cholesterol absorption and other lipids was observed with the two SM sources. In contrast, milk SM had a stronger inhibitory effect than that of egg (Noh and Koo 2004). Another study confirmed that egg SM supplementation contributes to the reduction of plasma cholesterol and TAGs in APOE*3 hyperlipidemic mice (Duivenvoorden *et al.* 2006).

Most recently, a new clinical trial demonstrated that a 4-week dietary intervention with 3–5 g of milk polar lipids per day (via isolipidic cream cheeses more or less enriched with a butterserum concentrate rich in MFGM) significantly reduced an array of lipid markers of cardiovascular risk in overweight postmenopausal women compared with control cream cheese devoid of polar lipids, including LDL-cholesterol (−8.7% with 5 g milk polar lipids/d, which is a clinically meaningful change), total:HDL cholesterol ratio and ApoB:ApoA1 ratio (Vors *et al.* 2019). This was associated with a lower concentration of TAG and cholesterol of chylomicrons in plasma after a high-sucrose and high-fat meal, compared with the control diet (cream cheese devoid of milk polar lipids, with milk TAGs only, which did not modify the profile of circulating cholesterol). The authors demonstrated that this could be explained by an ileal co-excretion of cholesterol with milk SM, a lower intestinal absorption of cholesterol, and a higher conversion of cholesterol to coprostanol (a non-absorbable compound) in feces (Vors *et al.* 2019). This opens the way for the development of new nutritional strategies based on dairy product enrichment with milk polar lipids to manage cardiometabolic risk factors in at-risk populations. This also prompts investigation of the possible role of the natural polar lipid content of dairy products (15–30 fold lower than in the above-described enriched cream cheeses) in maintaining blood cholesterol homeostasis in the general population.

10.2.2 Effects of Milk Polar Lipids on Gut Barrier Function

Obesity and type 2 diabetes have been associated with adipose tissue hypertrophy but also with low-grade inflammation, so-called metabolic inflammation (Shoelson *et al.* 2007). In this context, nutritional strategies are needed to improve the quality of ingested foods and the impact of milk fat in dairy has to be examined. In addition to effects observed on lipid metabolism and the reduction of the risk of cardiovascular or metabolic diseases, milk fat, milk polar lipids and SM have demonstrated beneficial effects in relation to metabolic inflammation. Considering mechanisms, dietary lipids are able to modulate the intestinal barrier and the gut microbiota. A causal link exists between intestinal microbiota and adipose tissue inflammation. Indeed, germ-free mice are protected from obesity when they are subjected to a hyperlipidemic diet, and have a lower adipose tissue inflammation (Caesar *et al.* 2015). Moreover, the low-grade inflammation observed in obesity is correlated with an increase in the presence of endotoxin (LPS, proinflammatory compounds from Gram negative bacteria) in the bloodstream. Indeed, Cani *et al.* (2007) have demonstrated that a multi-week hyperlipidic diet in mice increases endotoxemia, which is further negatively correlated with the number of *Bifidobacteria* in the gut. Modulation of the gut microbiota can thus reduce endotoxemia as has been demonstrated by the administration of antibiotics to obese or hyperlipidic mice, thereby leading to a decrease in low-grade inflammation (Cani *et al.* 2008). Additionally, the establishment of metabolic endotoxemia and low-grade inflammation appears to be correlated with changes in the intestinal barrier, including the gut microbiota. Indeed, an increase in intestinal permeability, favoring paracellular translocation of bacterial compounds, is observed following hyperlipidic diets in rodents (Cani *et al.* 2008; de La Serre *et al.* 2010) and was also recently demonstrated in obese humans (Genser *et al.* 2018). The effects of milk polar lipids on the intestinal barrier remain poorly studied to date even though a decrease in inflammation is observed in several *in vivo* and *in vitro* studies. Snow *et al.* (2011) have shown that a 5-week diet in mice (5% by weight of lipid and 0.6% by weight of milk polar lipids) induces a decrease in the inflammatory reaction and prevents the increase in gut permeability induced by the injection of LPS. Mice fed a diet consisting of 10% by weight of MFGM, followed by an injection of LPS, showed a decrease in gut permeability compared to an injected control group. Strikingly, after 48 h, only mice supplemented with milk polar lipids survived after the LPS injection (Snow *et al.* 2011). Another *in vivo* study in mice fed a semi-synthetic (fiber-poor and starch-rich) high fat diet showed that supplementation with milk polar lipids (1.2% by weight) did not induce hypertrophy of white adipose tissue compared to soy lecithin. However, milk polar lipids contributed to an increase in the number of mucus-secreting goblet cells in the colon (Lecomte *et al.* 2016).

Recently, several studies in mice examined the impact on the intestinal barrier and gut microbiota of milk polar lipids. Mice fed a chow-based (fiber-rich) high-fat diet (48% kcal) supplemented with milk polar lipids (1.6% w/w) showed a significantly lower weight gain, despite increased energy intake in comparison with control high-fat diet (Milard *et al.* 2019a, b). Furthermore, a supplementation with 1.1% (w/w) of milk polar lipids induced higher abundance of *Bifidobacterium* spp. and in particular *B. animalis*, which is a beneficial bacterium. This is consistent with the results of Norris *et al.* (2016) who showed lower weight gain, lower endotoxemia and higher content of Gram positive bacteria, in particular *Bifidobacterium* spp., after 4 weeks of semi-synthetic (fiber-poor, starch-rich) high-fat diet (45% of total energy intake) supplemented with 0.25% (w/w) of milk SM. A decrease in the number of *Bifidobacteria* spp. is correlated with metabolic disorders (Cani *et al.* 2007) and endotoxemia (Cani *et al.* 2008). The increase in *Bifidobacterium* spp. with milk polar lipid supplementation can thus be considered as a beneficial mechanism of action. With a chow-based high-fat diet supplemented with milk polar lipids, a positive correlation has even been observed between *Bifidobacterium* spp. and *Akkermansia muciniphila* (Milard *et al.* 2019a), the latter gut bacteria being described to present very favorable metabolic effects (Hansen *et al.* 2012; Schneeberger *et al.* 2015). Of note, the lower weight gain observed with milk polar lipids and milk SM incorporated in high-fat diets (45–48% of total energy intake) was not observed with 0.1% by weight of milk SM in a high-fat diet (60% kcal) during 10 weeks (Norris *et al.* 2017). Furthermore, another study did not demonstrate any impact on weight gain after 8 weeks of a semi-purified, high-fat diet (40% kcal) supplemented with 1.2% of milk polar lipids (Lecomte *et al.* 2016). The lower endotoxemia with 0.25% (w/w) of milk SM in a high-fat diet (45% of total energy intake) observed by Norris *et al.* suggests an impact on the intestinal barrier. However, in this study, intestinal permeability and tight junction expression (implicated in paracellular permeability) were not impacted by diet (Norris *et al.* 2016). An *in vitro* study was performed to evaluate the specific impact of milk SM on intestinal cells and for that, Caco-2/TC7 cells, as a model of the intestinal barrier, were incubated with mixed micelles with or without milk SM. Results showed that milk SM increases the gene expression of tight junction proteins such as Occludin and ZO-1 (of note, this study was not an *in vitro* model of altered gut barrier). However, mixed micelles supplemented with milk polar lipids did not lead to modulation of expression of tight junctions (Milard *et al.* 2019b). Synergistic or individual effects of the different milk polar lipid species on intestinal tight junctions and the gut barrier cannot be eliminated, but also a different bioaccessibility for milk SM by intestinal epithelial cells may prevail when pure or with other polar lipids. On the other hand, concerning the impact of dietary sources of SM on the intestine, 7-day-old pups fed with 0.5% milk SM or PC presented a better maturation of the intestine with the dairy SM, as indicated by the enzymatic activity of the intestinal mucosa (the lactase in the jejunum and the ileum), the number of mature cells, the location of vacuoles in the villi. These results probably reflect faster bowel development (Motouri *et al.* 2003). Such effects were also observed after MFGM administration in rats (Bhinder *et al.* 2016), deserving further exploration of the importance of breast milk polar lipids on infant intestinal development and metabolism (see Section 10.4).

10.2.3 Bioactivity of Sphingolipids and Their Potential Metabolic Effects

Trophic effects of lipid residues from milk polar lipids in the distal gut need to be considered. Studies have focused more precisely on SM since Nilsson (1969) established that 25% of SM was found in feces of rats in the undegraded form (10%), ceramide form (30–90%), and free sphingosine form (3–6%) (Nilsson 1969). More recently, human ileostomy content have shown that 20–25% of dietary SM was found intact at the end of the small intestine (Ohlsson *et al.* 2010a, b; Vors *et al.* 2019). The consumption of sphingolipids (including dairy sources in addition to other sources of sphingolipids) can be of fundamental importance in nutrition. Data from the literature reveal a potential impact of SM hydrolysis products, including ceramides and sphingosine and other sphingoid-based lipids. These components can notably have an effect on the inhibition of growth, differentiation and apoptosis (Vesper *et al.* 1999).

SM is naturally a major component of the biological membranes of mammals, including humans. Specific studies on SM metabolism began more extensively in the 1990s due to works demonstrating significant effects of SM hydrolysis products on cell proliferation and differentiation. SM and its hydrolysis products (e.g., ceramide, sphingosine, and sphingosine-1-phosphate) appear to have second messenger functions and participate in cell growth and contribute to the suppression of oncogenesis (Kolesnick 1991; Hannun and Linardic 1993). In contrast, ceramides from sphingolipid degradation are signals involved in the regulation of cell growth, induction of apoptosis and inflammation (Maceyka and Spiegel 2014). The degradation products of SM are active metabolites involving effects of mucosal growth and immune maturation (Nilsson 2016). In contrast, the abnormal accumulation of lipids, including endogenous ceramides, can lead to insulin resistance, inflammation, and cell death by lipotoxicity (Chaurasia and Summers 2015; Norris and Blesso 2017). In this context, milk SM displays a specific fatty acid distribution, with C16:0–C24:0 acyl chains, and major sphingoid bases ranging from d16:0 to d19:0. Milk SM has anti-inflammatory effects on macrophage *in vitro* (after LPS stimulation) and because these anti-inflammatory effects have also been demonstrated with sphingoid base (d18:1), this suggests that the observed effects are in part due to this sphingoid base (Norris *et al.* 2017b). Of note, egg SM, which also contains d18:1 sphingoid base, has similar effects to milk SM on tight junction expression after incubation on Caco-2/TC7 cells, a model of intestinal barrier (Milard *et al.* 2019b). A recent review by Norris *et al.* (2019) describes the protective properties of milk sphingomyelin against dysfunctional lipid metabolism, gut dysbiosis, and inflammation reported to date in cell or animal models and in humans (Norris *et al.* 2019).

10.2.4 Other Health Effects of Milk Polar Lipids

In addition to metabolic effects on lipid metabolism, intestinal development and inflammation, other effects of milk polar lipids have been demonstrated. Spitsberg (2005) mentioned a nutraceutical potential of MFGM on health, in particular on cholesterolemia and colon cancer (Spitsberg 2005). Regarding the beneficial effects on colon cancer, rats administered drinking milk have demonstrated a decreased number of intestinal tumors (Sachez Negrette *et al.* 2007). Milk polar lipids could explain these effects, as a rat model of colon cancer fed during 13 weeks with 2.5% by weight of MFGM in the diet presented decreased aberrant crypt foci and no increase in candidate colon cancer gene (Snow *et al.* 2010). Among polar lipids that could contribute to these effects, milk SM is again an interesting candidate effector. Studies have also demonstrated beneficial effects of milk SM regarding colon cancer. Indeed, 0.1% by weight of milk SM incorporated in a semi-purified diet (5% by weight of corn oil) decreases colonic aberrant foci and the incidence of colon tumors (Dillehay *et al.* 1994; Schmelz *et al.* 1996; Mazzei *et al.* 2011).

Additionally, cutaneous effects have been observed with dietary supplementation with milk SM in preclinical models. Indeed, a diet supplementation with milk SM (9 g via 120 g/kg of a dairy compound concentrated in SM) during 12 weeks protected the skin in mice (Haruta-Ono *et al.* 2012) and improved atopic dermatitis through a mechanism involving decreasing systemic and peripheral inflammation in NC/Nga mice (model spontaneously developing dermatitis) (Takeshima *et al.* 2014). Dermatological studies involving supplementation of mice diets with milk polar lipids have also shown that they participate in the improvement of the cutaneous barrier function, due to incorporation of ceramide (Oba *et al.* 2015). The authors also demonstrated that milk SM was involved in modulating the epidermis structure and could prevent imbalances in the skin barrier after irradiation with UV-B (Morifuji *et al.* 2015). Effects on muscular function and neurological development were also observed. In SAMP1 and ICR mice subjected to physical activity, a diet supplemented with 1% of MFGM (including 16.1% of milk polar lipids) improved the deficits of muscular function (Arakawa *et al.* 1988). Additionally, Sprague–Dawley rats receiving a lipid emulsion containing milk SM demonstrated that the latter contributed to the myelination of the central nervous system during development (Oshida *et al.* 2003).

10.3 Milk Fat Structure: From the Impact of Milk Fat Globule Structure on Lipemia to the Dairy Matrix Concept in Nutrition

During the hours following the consumption of a meal (so-called postprandial phase), dietary lipids are digested by gastric and pancreatic lipases in order to be ultimately absorbed by the small intestine. The amount of lipids circulating in the bloodstream (so-called lipemia) thereby acutely increases and later decreases. This is first due to the sequential appearance of TAG-rich lipoproteins (TGRL) produced by the intestine (chylomicrons carrying meal-derived TAG) and the liver (very low-density lipoproteins, VLDL). Second, TGRL are hydrolyzed by circulating lipases and their remnants are further cleared from the bloodstream. If TGRL are overproduced and their clearance is inefficient, postprandial lipemia remains too high for a long time and blood vessels are thereby exposed to an excessive amount of TGRL (Vors *et al.* 2014). Such postprandial hyperlipemia has recently been recognized as an independent risk factor in the development of CVD (Nordestgaard *et al.* 2007), something which is particularly notable in countries consuming an unbalanced Western-type diet. Therefore, the kinetics of postprandial lipemia in an important metabolic parameter, which encompasses not only the area under the curve of postprandial plasma TAG and/or TGRL, but also the time of peak appearance, the maximum concentration and the time taken to recover to baseline lipemia. Recent research interest has focused on elucidating how the kinetics of postprandial lipemia can be modulated by the physicochemical structure of lipids, and by the food matrix, thereby contributing to the prevention of metabolic risk (Keogh *et al.* 2011; Vors *et al.* 2013; Grundy *et al.* 2016; Thorning *et al.* 2016; Drouin-Chartier *et al.* 2017). This concept is particularly consistent regarding the different structural levels of milk lipids, from the basic structure of lipid molecules to the supramolecular organization of milk fat globules (Michalski 2009; Bourlieu and Michalski 2015).

10.3.1 Potential Impact of Milk Fatty Acids and Triglyceride Structure on Digestion and Postprandial Lipemia

Among the main FA in milk fat, the short- and medium-chain FA mostly escape the chylomicron-secretion pathway during the digestion and intestinal absorption phase. Instead, they are directly absorbed by the portal vein to be oriented towards β-oxidation in the liver (Small 1991; Mu and Porsgaard 2005) and thereby do not contribute significantly to postprandial lipemia. The CLA components of milk fat are naturally located at the external positions of milk TAG (*sn*-1, *sn*-3), which has been shown to improve their intestinal absorption and β-oxidation in rats (Chardigny *et al.* 2003) and may contribute to the low atherogenic effects of CLA (Valeille *et al.* 2006).

Regarding the structure of milk TAG, that is, the location of the different types of FA on the three positions of the TAG glycerol backbone, a high proportion of palmitic acid is located on the *sn*-2 position (Michalski 2009). This contributes to an overall good bioavailability of palmitic acid from milk, which can be favorable for specific populations such as neonates when they cannot be breastfed (see Section 10.4). More broadly, considering the very diverse composition of milk FA, and the wide diversity of milk TAG species and structures, this can induce kinetics of TAG digestion and absorption differing from those observed with pure structured TAG or with fats presenting more simple structures or only a few different FA (Mu and Porsgaard 2005).

Interestingly though, a small proportion of the long-chain saturated FA located at *sn*-1 and *sn*-3 positions of milk TAG typically remain unabsorbed by the intestine, due to the formation of calcium soaps excreted in feces, with less potential to contribute to postprandial lipemia (Lorenzen *et al.* 2007, 2014; Soerensen *et al.*

2014). In rats, a higher fecal loss of saturated FA (indicating a lower intestinal absorption of these FA) is observed after an experimental cheese rich in calcium compared to a regular-calcium experimental cheese (Ayala-Bribiesca *et al.* 2017). In humans, a postprandial study with 18 healthy volunteers showed that isocaloric meals with medium (350 mg) or high (about 800 mg) calcium content (incorporated in meal by yogurt and milk) induced a postprandial chylomicronemia that was about 20% lower than after a low calcium meal (about 70 mg) (Lorenzen *et al.* 2007, 2014; Soerensen *et al.* 2014). In a short-term (2 weeks) crossover dietary intervention trial in 15 healthy men, volunteers consumed a control diet with 500 mg of calcium per day, followed by two diets enriched with calcium (1700 mg per day) either by cheese, or by half-skimmed milk (Soerensen *et al.* 2014). Compared to the control low-calcium diet that induced a fecal lipid loss of about 4 g/d, both calcium-enriched diets induced a higher fecal lipid loss of about 5.5 g/d on average; interested readers will find further details of such studies in the review of Christensen *et al.* (2009). Further research will now be necessary to explore the impact of calcium from different dairy products, compared with other foods or from dietary supplements, on the postprandial fate, absorption and bioavailability of saturated FA, notably from milk lipids, and the consequences for the kinetics of postprandial lipemia in humans but also on gut microbiota.

Finally, the composition and structure of milk TAG results in the fact that up to 5% of milk TAG can still be crystalized at 37 °C (body temperature; Lopez 2011). It cannot be ruled out that this could impact milk fat digestion, absorption, and postprandial lipemia. Using pure tripalmitin in a static *in vitro* digestion model, Bonnaire *et al.* (2008) demonstrated that digestion kinetics were slower when tripalmitin was in the solid state compared with a liquid (supercooled) state at 37 °C. Using vegetable fats of simple composition, it has been shown in adult humans that stearic-rich fats induced a lower intestinal absorption due to their higher melting temperature (Berry and Sanders 2005; Berry *et al.* 2007); moreover, fat absorption also appeared to be correlated with solid fat content at 37 °C. In fact, palmitic and stearic acid have a high melting temperature, so as the TAG carrying them: this results in a partially solid/crystallized state of these lipids in the digestive tract. This can lower lipase activity and promote a lower postprandial lipemia compared with oils that are entirely liquid (Sanders *et al.* 2011). This concept was recently confirmed with milk fat. Rats fed an experimental cheese manufactured with a high-melting point milk fat fraction (about 42 °C) had a lower plasma triglyceridemia than rats fed the cheese with the lowest melting temperature (Ayala-Bribiesca *et al.* 2017). Moreover, there was more unabsorbed palmitic acid in the feces of rats fed the high-melting point fat-cheese, that is, the more solid cheese TAG were less absorbed by the rat intestine. Most recently, Danthine *et al.* (2019) demonstrated that even a small fraction of solid fat may impact digestion kinetics: incorporating 5.7% (w/w) of a mixture of tripalmitin, triolein, and tricaprylin (in proportion 10:7:3) in anhydrous milkfat resulted in the presence of partially crystallized fat at 37 °C (i.e., at body temperature), which delayed digestive lipolysis onset *in vitro* (Danthine *et al.* 2019). As a next step, clinical trials should be conducted in humans to further elucidate the importance of the physical state (i.e., liquid:solid) and crystallized structures of milk TAG on digestion, absorption and postprandial lipemia in humans, as well as fecal residues of milkfat and their impact on the gut.

10.3.2 Milk Fat Globule and Milkfat Emulsified Structure: Impact on Digestion and Postprandial Metabolism and Metabolic Fate of Ingested Fatty Acids

Milk is a natural emulsion, the physicochemical structure of which can be modified by food processing and formulation, which has raised questions about the metabolic impact of such process-induced modifications (Michalski and Januel 2006). Different interfacial composition of milkfat droplets (milk proteins vs MFGM) can notably modulate the digestion kinetics of differently processed milks (Bourlieu and Michalski 2015). Several studies have shown that during simulated duodenal digestion *in vitro*, the lipolysis of milkfat is higher in homogenized milk compared with native unhomogenized milk fat globules, notably due to increased surface area of fat droplets (Devle *et al.* 2014; Garcia *et al.* 2014; Bourlieu *et al.* 2015a, b; Tunicket al. 2016; Lamotheet al. 2017; Van Hekken *et al.* 2017) (see Section 10.4). *In vivo* in rats, the structure of fat, and its digestion in the stomach and in the small intestine, is different with cream from raw milk or from pasteurized milk or from pasteurized and homogenized milk (Gallier *et al.* 2013a, b). Homogenized cream slowed down gastric protein digestion and Gallier *et al.* (2013a) suggested that this was due to the entrapment of milk fat globules in the protein aggregate network possibly decreasing accessibility of gastric lipase to the milk TAG substrate. Conversely, protein digestion was complete in the small intestine regardless of cream processing. However, cream from homogenized milk resulted in an apparently greater lipolysis of milkfat, because in the lower half of rat small intestine, myristic, palmitic, oleic and linoleic acids concentrations were significantly lower (43–71% lower) when pasteurized cream was homogenized compared with unhomogenized (Gallier *et al.* 2013b). There was also an increased appearance of fat crystals in the lower half of the rat small intestine. Of note regarding potential mechanisms, homogenized fat droplets or native milk fat globules could indirectly impact lipid digestion due to their effect on milk gel structure in the stomach. Indeed, it has been shown in dairy gels similar to yogurts that homogenized fat droplets can act as structure promoters by interacting positively with the proteinaceous network, while native milk fat globules act as inert fillers or even structure breakers if they are large enough (e.g., >5 μm) (Michalski *et al.* 2002). It was also observed *in vitro* that the smallest native milk fat globules (~1 μm), collected by microfiltration of whole milk (Michalski *et al.* 2006a, b), were more efficiently hydrolyzed by digestive lipases than the largest native milk fat globules (~5 μm) (Garcia *et al.* 2014). The impact of the size of milk fat globules and of milk homogenization on fatty acid bioavailability and postprandial lipid metabolism are deserving of further investigation, particularly in humans.

Interest is also growing in the impact of emulsified versus non-emulsified milk fat structure on the intestinal absorption of milk FA and on postprandial lipid metabolism. In rats, a vegetable oil emulsion induces (i) increased intestinal lipid absorption (measured by the cumulative output of lymphatic lipids) compared with the same amount of non-emulsified bulk oil (Couedelo *et al.* 2011, 2015) and (ii) increased peak of postprandial lipemia (Laugerette *et al.* 2011). Still in rats but using milk fat gavaged in the form of different structures, the lymphatic absorption of milk FA is faster after gavage with homogenized cream than with butter (continuous lipid phase), and a higher cumulative FA absorption is observed over 8 h using homogenized cream (Fruekilde and Hoy 2004). Different kinetics of postprandial lipemia were also observed after rats were gavaged with native or homogenized milk fat globules, and with milkfat emulsion droplets covered with caseins or with soy phospholipids (Michalski *et al.* 2005, 2006a, b). A lower lipemia was observed (i) using homogenized versus native milk fat globules and (ii) using milk fat droplets covered with caseins versus soybean lecithin, which also induced different kinetics of milk FA β-oxidation (Michalski *et al.* 2005).

In nine healthy normal-weight and nine non-diabetic obese men, a proof-of-concept crossover clinical trial (LIPINFLOX) demonstrated that the emulsified structure of fat in the meal can modulate the kinetics of postprandial lipemia and also the metabolic fate of ingested FA, that is, β-oxidation versus storage (Vors *et al.* 2013). After ingestion of a test breakfast containing 40 g of milkfat either unemulsified (spread on bread) or emulsified (in skimmed milk), the characteristics of postprandial lipid metabolism were explored: lipemia and chylomicron properties, non-esterified FA (NEFA) release in plasma, β-oxidation. An earlier and sharper peak of chylomicron lipids was observed after emulsified versus non-emulsified fat. Moreover, chylomicron lipids were also cleared from the bloodstream earlier after emulsified versus non-emulsified fat. This phenomenon appeared more pronounced in obese men than in normal-weight men (Vors *et al.* 2013). This shows that the enhancement by emulsification of digestive lipolysis and intestinal lipid absorption can be observed after a realistic mixed meal in humans, in contrast to previous studies where emulsion digestion and absorption using pure emulsion boluses were sometimes fed via a nasogastric tube (Armand *et al.* 1999; Keogh *et al.* 2011). Nutritional intervention studies of several weeks should now be conducted to determine the metabolic impact of such modified postprandial lipemia profile according to milk fat structure at breakfast, considering that high prolonged lipemia, with a poor chylomicron clearance, is an independent cardiovascular risk factor (Nordestgaard *et al.* 2007; Langsted and Nordestgaard 2015). Moreover, in the LIPINFLOX trials, milkfat was enriched with homogeneous ^{13}C-TAG (stable isotope lipid tracers, tricaprylin, triolein and tripalmitin in proportions similar to the corresponding FA types in milkfat), enabling the performance of a $^{13}CO_2$ breath test to explore the final metabolic fate of ingested FAs: β-oxidation versus storage. Vors *et al.* (2013) thereby demonstrated that the β-oxidation of exogenous dietary FA was higher after the breakfast containing emulsified fat compared to non-emulsified spread fat. In obese men, exogenous FA β-oxidation was approximately 60% of ingested FA after emulsion compared with only 40% after spread fat. Fecal loss of tracers was otherwise similar regardless of the structure of the fat (i.e., emulsified or not). A plausible mechanism is the observed appearance of exogenous NEFA in plasma (so-called spillover) that was higher after emulsion than after spread milkfat in both obese and normal-weight volunteers. This NEFA spillover could be due to a faster intestinal absorption of emulsified lipids, resulting in an earlier vascular lipolysis of chylomicrons, which generate an excess of circulating NEFA of dietary origin. These readily available NEFA serve as a substrate for peripheral tissues, explaining the higher contribution of exogenous FA to the total lipid β-oxidation in volunteers after ingestion of emulsion compared with spread fat (Vors *et al.* 2013). These results led to the new concept of "fast *vs* slow fat" structures to understand their metabolic impact, which is particularly relevant for milk fat present in very different types of structures in milk and in the wide variety of dairy products prepared using different processes (Michalski 2009).

10.3.3 Impact of Milk Fat in Different Dairy Matrices

Several recent reviews summarize results on the impact of different dairy products on postprandial lipid metabolism and related metabolic disorders (Lamarche 2008; Nestel 2008; German *et al.* 2009; Labonte *et al.* 2013). In relation to the digestion step, as reviewed by Guo and collaborators (2017a, b), both the firmness and protein density of a dairy gel can affect lipid release kinetics during *in vitro* digestion (Guo *et al.* 2017a, b). Moreover, the lipids of milk fat globules in cheese are trapped within the matrix and this was shown to affect the kinetics of lipolysis according to cheese matrix properties, such as protein matrix density, fat globule size distribution, matrix porosity, cohesiveness and elasticity (Lamothe *et al.* 2012; Fang *et al.* 2016). Altogether, cheese fat content, proteolysis, hardness and chewiness were reported to positively affect the matrix disintegration in simulated *in vitro* digestion conditions, that is, *cheddar* > light *cheddar* > *mozzarella* (Fang *et al.* 2016). In rats, lymphatic absorption of lipids was reported to be both slower and of lower magnitude after gavage with butter < cream cheese < regular or sour cream (Fruekilde and Hoy 2004). Moreover, in type 2 diabetic human subjects, consuming a meal containing 30 g lipid as butter delays the increase of plasma TG compared with *mozzarella* or milk (Clemente *et al.* 2003). In a randomized cross-over clinical trial, 14 healthy young men were assigned during 3 weeks to a diet with 20% of total energy from dairy fat, as either whole milk, butter or hard cheese. On day 4 of each intervention period, no effect was observed on

FA quantity and profile of secreted chylomicrons in the postprandial phase. However, after 3 weeks, plasma LDL-cholesterol was lower after the cheese diet than after the butter diet (Tholstrup *et al.* 2004). Fermentation of the dairy matrix also impacts postprandial lipid metabolism: a slower gastric emptying, together with a higher peak of plasma TG returning faster to baseline, was observed with fermented milk than with liquid milk, which can be explained by the increased viscosity induced by fermentation (Sanggaardet al. 2004). The cheesemaking process also alters the structure of milk fat globules, notably in hard-pressed cheese in which coalescence and the formation of "free fat inclusions" are observed (Michalski *et al.* 2004, 2007; Lopez *et al.* 2015). This could contribute to modulation of the fate of cheese lipids during digestion and associated bioaccessibility and bioavailability, but this remains to be elucidated in considerably more detail as highlighted in a recent review by Thorning *et al.* (2017). In this respect, a recent clinical trial in healthy volunteers demonstrated that the consumption of soft cream cheese, *Cheddar* cheese and butter (control) as part of a test meal could modulate postprandial lipemia (Drouin-Chartier *et al.* 2017), albeit to a lesser extent than previously demonstrated with cream-like dairy emulsion versus butterfat (Vors *et al.* 2013). Postprandial TG were higher at 2 h and lower at 6 h after cream cheese meal versus *Cheddar* cheese meal, and results suggested that cream cheese induced lower and smaller chylomicron secretion than *Cheddar* cheese. The cheese matrix can thus modulate postprandial lipemia: whether this modifies further fatty acid bioavailability and metabolic impact remains to be mechanistically studied, notably through the setup of randomized controlled trials. Some mechanisms of the matrix effect can be due to interactions between the composition of proteins, carbohydrates and/or fibers and lipid digestion and absorption, as well as calcium as explained above.

Importantly, the recent epidemiological data support the dairy matrix concept in nutrition. De Oliveira Otto *et al.* (2012) showed in humans that SFA consumed as meat would be associated with increased cardiovascular risk factors whereas SFA from dairy would decrease the risk of CVD (de Oliveira Otto *et al.* 2012). Drouin-Chartier *et al.* (2016a, b) also described an inverse correlation between cheese consumption and risks of cardiovascular disease and stroke (Drouin-Chartier *et al.* 2016a, b). A recent postprandial study by this team also demonstrates that cream cheese in a meal induces faster kinetics of postprandial lipemia than hard cheese or butter, that is, with a faster clearance of circulating TAGs (Drouin-Chartier *et al.* 2017). Such associations can be explained by the fact that cheese products are generally rich in lipids, but also proteins, calcium and products of fermentation, such that fat globules are entrapped in a more or less rigid matrix; moreover, such products also contain MFGM and its constituent polar lipids. A recent report from an international expert committee concluded that the "dairy matrix effect" on cardiometabolic health should be further explored rather than considering cheese as a rich source of SFA only (Thorning *et al.* 2017). Specific lipid residues and other derived compounds can notably impact the gut microbiota, and also the intestinal mucosa. The same panel indicated in an analysis article that dietary guidelines should now focus more on real foods containing SFA than on SFA as a bulk nutrient category (Astrup *et al.* 2019). The role of several characteristics of the dairy matrix (according to the several types of dairy processes, matrix composition and structure, etc.) on the digestibility of dairy fat and potential health consequences has recently been reviewed (Michalski *et al.* 2020). Altogether, the complex structure of dairy foods, including their various lipids, and the mechanisms by which milk lipids are released from these matrices, are of utmost importance because they can directly impact different physiological functions (e.g., gastric emptying and microbial fermentation in the gut) and ultimately on cardiometabolic disease risks.

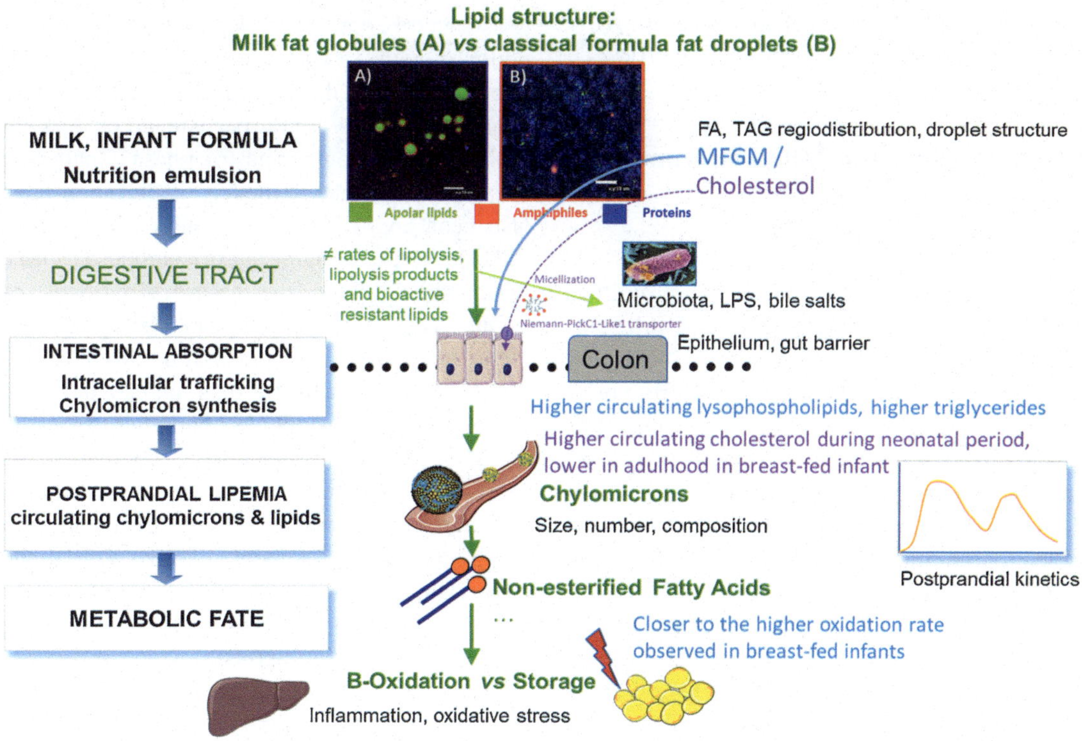

Figure 10.3. Differences in lipid structure between human milk and classical infant formula and the potential implications of these differences on digestion, metabolic fate and programming (Adapted from Bourlieu and Michalski (2015).

10.4 Significance of Milk Fat Structure in Infant Nutrition

Milk fat structure at several scales (fatty acyl chain, distribution on glycerol backbone, association of lipid molecules within the membrane and globular supramolecular assembly) finely tune the digestive behavior and biological fate of milk fat (Michalski 2009; Gallier *et al.* 2014; Bourlieu and Michalski 2015) (Figure 10.3). The specificity of milk fat architecture seems especially important during the neonatal period and infancy. Presenting a detailed picture of the structure–function relationships played by milk fat during infancy is beyond the scope of this chapter and can be found elsewhere (Michalski 2013; Bourlieu *et al.* 2015a, b, 2017); therefore we will focus mainly on key features of milk fat structure conveying specific nutrients that contribute to infant growth and immune system development. One of the most important features of human milk chemical composition is its high content in unsaturated fatty acids and in long-chain polyunsaturated fatty acids, especially DHA and arachidonic acid (ARA), which are centrally important in brain growth, cognitive skills development and motor and retinal functions during the neonatal period (Yehuda *et al.* 2005), such long-chain PUFAs are not present in bovine milk. Conversely, human and bovine milks share similarities, such as their high proportion of palmitic acid located at the *sn*-2 position in TAGs (whereas this FA is mainly located on the *sn-1* and *sn-3* positions in vegetable-based lipids) (Innis 2011), their content of medium-chain FA that is rapidly metabolized (Section 10.2) and can be considered as an important factor for fat absorption in conditions of limiting ability to digest fat. In this respect, a low amount of biliary salts is certainly the most important limiting factor in the neonatal period. Indeed during this period of life, pancreatic insufficiency is partially compensated for by carboxylesterase and pancreatic lipase related of type 2 activities (Lindquist and Hernell 2010; Bourlieu *et al.* 2014; Shani-Levi *et al.* 2017).

10.4.1 Recent Knowledge on Human Milk Composition and Structure

Because breastmilk is the gold standard for infant nutrition, data on this biological fluid is still being published. However, the viewpoint of such active research has recently been refocussed from basic chemical composition to more comprehensive characterization, including its evolution during lactation, macronutrient variability, stability during storage, leptin content and its effect on metabolic programming. Some examples of these very recent publications are summarized in Table 10.2.

Knowledge about human milk structure is mostly inferred from bovine milk, and there are still considerable gaps in our knowledge. Indeed, gaining a good understanding of human milk structure is essential as more and more elements have indicated that the structure of milk fat triggers its digestive behavior and impacts lipid metabolism programming (Figure 10.3). However, considerable differences in structure persist between infant formula products, which are mainly based on submicron-sized droplets (0.5 μm) of vegetable fats stabilized mainly by proteins. Human milk naturally contains milk fat globules with volume-weighted mean diameters in the range 3–5 μm, stabilized by the trilayered membrane based on polar lipids (SM, traces of ceramides and glycerophospholipids), cholesterol and proteins (Bourlieu *et al.* 2015a, b, 2017; Hageman *et al.* 2019).

10.4.2 Specific Behavior of Human Milk Fat Globules During Digestion

Some *in vitro* studies on human milk and infant formula have helped to shed light on their gastrointestinal digestive behavior and how the structure and organization of the oil globule interface modulate digestibility. Static *in vitro* digestion of human milk, in comparison to whey protein concentrate and infant formula, has underlined the resistance of some MFGM proteins (e.g., osteopontin, mucin) to gastric proteolysis. Such relative resistance to digestion is a key element that contributes to the gastric persistence of milk fat globules throughout the gastric phase (Chatterton *et al.* 2004). Dynamic models of digestion have also been applied to human milk. Zhang *et al.* (2014) using a dynamic gastric model (DGM), designed to mimic the gastric conditions of infants aged 9–12 months, established that the lower susceptibility to pepsin of the MGFM proteins was attributable by their glycosylation or the presence of disulphide bonds (Zhang *et al.* 2014). Fondaco *et al.* (2015) compared the behavior of different infant formula and human milk during dynamic gastro-intestinal digestion using the TIM-1 model (TNO, Zeist, the Netherlands) (Fondaco *et al.* 2015). Human milk and infant formula differed notably in terms of particle size distribution (main modes of 7 μm vs ~0.4 μm, respectively). Infant formula was characterized by a rapid initial release of FFA in the jejunal and ileal compartments, whereas a lag was observed for human milk. After this lag, when correcting the rate of release of FFA for human milk by its specific surface area, human milk presented the highest rate of lipolysis. Altogether these results suggested that, in addition to the specific surface of the emulsion, other key factors such as the content of endogenous lipase (e.g., bile salt stimulated lipase, BSSL) and the unique interfacial properties of MFGM in human milk contribute to elevated levels of lipolysis after an initial lag. These results corroborated previous data of Armand *et al.* indicating reduced lipolysis in the gastric aspirate of preterm infants after administration of infant formula, compared to human milk (Armand *et al.* 1996). De Oliveira and collaborators (2015, 2016) used an *in vitro* dynamic system (DIDGI®) to study the gastrointestinal digestion of pooled mature human milk, in either raw or pasteurized form, mimicked either in full term or preterm newborns. In the full term, pasteurization by inactivating BBSL resulted at any time of the digestion in lower lipolysis levels, limited gastric destabilization, accelerated β-casein proteolysis but did not modify FFA release (de Oliveira *et al.* 2015, 2016).

Table 10.2. Examples of recent studies investigating human milk components and their evolution during lactation

Samples of human milk	Hypothesis/objective	Main conclusions	References
150 mature human milk samples from 25 mothers – term and preterm infants – storage at either −20 °C or −80 °C over 4, 12, or 24 weeks	Long freezing storage of human milk affect its macronutrients	Fat and energy consistently higher in the −80 °C group compared to −20 °C, significant loss of fat (0.3 g/100 mL) and energy (2.3 kcal/100 mL) when human milk is stored at −20 °C over 24 weeks, the same trend is observed to a lesser extent for fat and protein at −80 °C.	Orbach *et al.* (2019)
Mature human milks	Describe macronutrient changes in breast milk over long lactation period (up to 48 months)	Macronutrients of human milk after 18 months lactation are modified (fat and protein increase while carbohydrate decreases). Macronutrients remain stable from 24 to 48 months.	Czosnykowska-Lukacka *et al.* (2018)
1119 human milk samples from 443 individual donors (retrospective observational study)	Describe nutrient variability in human milk to optimize fortification strategy and explain the basis (subject or temporal) of its variation	Fat is the most variable nutrient, responsible for 80% of the energy intake variability; subject effect variability is stronger than time effect; random pooling reduced variability, however most protein fortified human milk did not meet target daily delivery of proteins (3.5 g/kg) without exceeding volumes of 160 mL/kg/day.	John *et al.* (2019)
1917 preterm human milk samples from 225 mothers; divided in 2 groups: A (24–30 GA) $n = 969$, B (31–35 GA), $n = 948$.	Evaluate the concentration of macronutrients in pooled preterm human milk during the first 2 months of lactation in relation with gestational age (GA)	Human milk composition did not differ with GA; protein content was low and decreased from 1.5 to 1.6 g/100 mL (1 week) to 1.1 g/100 mL (after 3 weeks) and then remained stable. Fat and carbohydrate remained stable over the study.	Maly *et al.* (2019),
(–)	Review on leptin content and its effect on metabolic programming	In animal models (murine) supplementation with leptin over lactation protected against overweight and obesity when exposed to high fat diet; these beneficial effects were due to greater sensitivity to central and peripheral action of leptin. Some indirect evidence of benefit of leptin exposure during lactation in humans have been obtained, such as negative correlation between breastmilk leptin concentration and BMI – bodyweight gain of their infants later on.	Palou *et al.* (2018)
96 mature human milk samples from 16 mothers (from 4 weeks to 7 months postpartum)	The distribution of human milk bioactive factors (cytokines, chemokines and epidermal growth factor) in terms of phase distribution (aqueous vs lipid) evolve throughout lactation	Bioactive factors detected in each phase over a wide range of concentration during the study; presence of cytokine Flt-3 L, chemokine (MDC) in all phases and nearly all bioactive factors in the lipid phase.	Vass *et al.* (2019)

In vivo data comparing the gastric behavior of either raw or pasteurized human milk, or pasteurized and pasteurized-homogenized human milk, were also obtained in the preterm newborns. In a first trial, conducted over a 6-day period in which each infant was its own control (n = 12, GA = 30.0 ± 1.1 weeks, BW = 1.4 ± 0.3 kg, medium age = 27 ± 12 weeks), pasteurization did inactivate the endogenous lipase BSSL but did not impact gastric lipolysis level (12.6 ± 4.7% after 90 min of digestion) nor human milk gastric emptying (half-life of ~30 min). This rapid gastric emptying was in agreement with the average values reported for human milk in preterm infants (25–47 min) and out of line with the longest time reported for infant formula (72 min on average). In a second trial (n = 8, GA 29.5 ± 1.5 weeks, BW 1.73 ± 0.48 kg, medium age = 32 ± 21 weeks), it was shown that homogenization increased gastric lipolysis rate and slowed down gastric emptying (de Oliveira *et al.* 2017).

10.4.3 Towards a Role of Milk Fat Globule Structure on Metabolic Programming

Altogether, the above-described impacts of dairy emulsion structure on the upper gastro-intestinal digestion are in the forefront of most systemic effects (Figure 10.3). Indeed, the striking difference in terms of size and interfacial composition between classic infant formula and human milk was investigated in an animal model (mice) by Oosting *et al.* (2012) and has been shown to be involved in influencing body composition and programming of lipid metabolism (Oosting *et al.* 2012). When comparing a concept infant formula designed to be 'more breastmilk like' (Nuturis®; WO2013135739A1) to a standard formula administrated during the neonatal period (from 16 to 42 days) to mice, followed by a Western diet challenge, a lower fat accumulation, lower fasting plasma leptin, resistin, glucose and lipid (TG and total cholesterol) in the Nuturis® group than in the group fed conventional formula was observed at adulthood (Oosting *et al.* 2012). The mechanisms underlying these nutritional programming effects of Nuturis® remain at this stage unclear and very likely multifactorial. However, the authors hypothesize that emulsion droplet size and interfacial composition/structure could modify digestion kinetics. However, at this stage, the fact that kinetics of lipid digestion could further modulate plasma TG appearance (Michalski *et al.* 2006a, b) and β-oxidation rate (Michalski *et al.* 2005) was already established in animal models and later confirmed in adult humans (Vors *et al.* 2013). More recently, the volatile organic compounds in breath (VOC, n = 29 healthy adults) were found to be significantly modified 240 min after ingestion of the Nuturis® infant formula compared to classic infant formula (Smolinska *et al.* 2019). Such modification can be used as non-invasive indicator of modification in metabolism, but remains difficult to relate to a specific metabolic process. Indeed, VOC can be emitted during upper digestion or further by gut microbiota activity leading to VOC release into blood followed by lung excretion. To extend the understanding of the programming effect of large emulsion droplets in infant formula, Oosting *et al.* (2014) have shown, still in a mouse model, that the neonatal administration of Nuturis® infant formula induced, in adulthood, a reduced adipocyte size without affecting their number in epididymal white adipose tissue (Oosting *et al.* 2014). This was accompanied by the modulation of several transcription factors involved in metabolic regulation such as peroxisome proliferator-activated receptor gamma, CCAAT/enhancer-binding protein and retinoid X receptor. Also, in a mouse model, Nuturis® infant formula triggered improved performance in short-term memory tasks during adolescence and adulthood, while brain phospholipids remained unaffected (Schipper *et al.* 2016). Efforts to disentangle this programming effect of the physical structure of lipids were undertaken by Baars *et al.* (2016). These authors questioned whether the protective effect against obesity in adulthood was arising from the droplet size and/or from the membrane surrounding the droplets of the emulsions ingested during the neonatal period. Only the administration of large droplets with phospholipid coating, that is, the combination of the two factors in the Nuturis® formula, during the neonatal period reduced fat accumulation in adulthood. In addition, a direct effect of MFGM contained in the concept infant formula was also weakened by a study comparing the growth of infants fed with a standard infant formula versus a formula enriched with MFGM fragments (Timby *et al.* 2014a, b). The study did not reveal any differences in infant growth, weight gain and body fat at 12 months. Conversely, human milk lipids affected preadipocyte differentiation *in vitro*, in the absence of the standard adipogenic compounds, which was not observed with infant formula lipids (Fujisawa *et al.* 2013). Thus, among the structural aspects of human milk lipids, both lipid droplet size, as well as the MFGM coating, may contribute to its reported protective effect against obesity (Baars *et al.* 2016). Nuturis® infant formula has been tested in an interventional clinical trial in healthy term infants (NTR3683) for its tolerance, safety and impact on growth up to 17 months (Breij *et al.* 2019). Control infant formula (n = 108), concept infant formula (n = 115) were isocaloric and had similar protein content (1.3 g/100 mL), lipid content (3.4 g/100 mL) and mixture of prebiotics. A control breast-fed group (n = 88) was also included in the study. Equivalent growth, tolerance and number/severity of adverse events were observed in both formula-fed groups. Stool consistency was closer to breast-fed group in concept large droplet infant formula-fed group than in the control infant formula. No difference in plasma levels of vitamins A and E were detected between the two infant formula groups, suggesting similar bioavailability of liposoluble vitamins regardless of the emulsion droplet size, which was in line with previous data reported by Borel *et al.* (2001). Unfortunately, the impact of the concept infant formula on body composition, which was modified in preclinical models, was not investigated in the infant intervention study. This issue of the quality of growth and body composition is specifically challenging when considering preterm infants who tend to have more fat mass and less fat-free mass, which can predispose them to metabolic syndrome and cardiovascular problems later in life (Strydom *et al.* 2019). This difference in body composition was demonstrated very recently in a prospective cohort study, showing that formula-fed very low birth weight (VLBW) preterms developed higher amounts of adipose tissue and lower amounts of fat-free mass than the full-term infants. In comparison, breast-fed VLBW infants shared similar body composition with full-term infants (Mol *et al.* 2019), there again underlining that breast milk is the gold standard for neonatal nutrition, and that infant formula must mimic it at all scales, to try to approach as much as possible its main biological effects.

10.4.4 Metabolic Importance of Milk Triglycerides and MFGM for Infants

A straightforward strategy to produce more biomimetic infant formula has been to supplement classic infant formula with bovine milk fat fractions. Milk fat TAGs can be added in the form of anhydrous milk fat or in the form of cream (which also brings phospholipids and better preserves the globular structure of bovine fat). Addition of milk fat TAGs at levels of up to 50% of the total fat has been advocated in addition to other sources of fats, bringing polyunsaturated components with balanced ω6/ω3 ratio by various research groups (Bourlieu *et al.* 2015a, b; Delplanque *et al.* 2015, Hernell *et al.* 2016, Lonnerdal and Hernell 2016; Delplanque *et al.* 2019). Milk fat TAGs enriched with ALA would help with restoration of DHA in brain in animal models (Delplanque *et al.* 2011, 2013).

MFGM extracts can also be added directly to infant formula (Le Huerou-Luron *et al.* 2019). Indeed, from a technological perspective, MFGM fractions can be concentrated from by-products of the butter industry, such as buttermilk and butterserum (El-Loly 2011; Conway *et al.* 2014; Gassi *et al.* 2016; Bourlieu *et al.* 2018). These compounds can be added to infant formula as emulsifiers and stabilizers of emulsions, as an alternative to the use of non-dairy emulsifiers or milk proteins, to better approach human milk ultrastructure. In terms of clinical impact, six double-blinded randomized controlled clinical trials in infants and children have been conducted to assess the safety and efficacy of such infant formula supplementation with bovine MFGM and are exhaustively summarized in Hernell *et al.* (2016). Results are mostly positive and have demonstrated benefits in terms of cognition, behavior, gut health, and immunity (Zavaleta *et al.* 2011; Gurnida *et al.* 2012; Veereman-Wauters *et al.* 2012; Billeaud *et al.* 2014; Poppitt *et al.* 2014; Timby *et al.* 2014a, b, 2015; Ortega-Anaya and Jimenez-Flores 2019). Although the intervention strategy and the MFGM concentrates added are still too heterogeneous, these results provide evidence of the beneficial effects of individual components of MFGM. More precisely, Timby and co-workers have recently investigated in detail how MFGM may impact infant metabolism. To this aim, a cohort of infants (n = 160, breastfed from birth to 2 months, intervention period from 2 to 6 months), was assigned either to an MFGM-supplemented, low-energy, low-protein experimental formula (EF) or a standard formula (SF) and compared to a group of breast-fed infants (n = 80). During the intervention period, formula fed infants displayed higher growth velocity, which was coupled with higher plasma insulin and blood urea nitrogen levels. Conversely, breast-fed infants had higher levels of serum TAGs and cholesterol, higher LDL/HDL ratios and leptin/fat mass ratios. Lower plasma insulin levels reported in breast-fed infants reflected increased lipolytic rate in adipose tissue and increased FA oxidation in muscle and liver, compared with formula-fed infants. Infants that had been assigned the EF formula had distinct circulating lipidomic profiles, with higher serum cholesterol and TAG levels, thus approaching the breast-fed group in terms of lipid metabolism (Timby *et al.* 2014a, b). He *et al.* (2019) went further in the investigation of this cohort, submitting serum and plasma samples to comprehensive metabolomic analyses (He *et al.* 2019). The authors established that the different feeding induced a shift in metabolism during the intervention period, which can be considered as a feeding metabolic signature: breast-fed infants displayed higher fatty acid beta-oxidation rate and medium/long-chain acylcarnitines; formula-fed infants had higher rate of amino acid catabolism, odd-chain fatty acid synthesis through the C3-acylcarnitine precursor, and higher circulating urea. This shift disappeared with the introduction of complementary feeding. MFGM-supplemented infants improved the profile of circulating lysophospholipids, making it more similar to the one of breast-fed infants. A trend towards lower circulating amino acids and higher choline, betaine, and ketone bodies was also observed in the MFGM-supplemented infant group, again suggesting that MFGM helps in reducing the gap between formula-fed and breast-fed infants in terms of metabolism. The lipidomic impact of serum/plasma and erythrocyte membrane in infants fed formula supplemented with MFGM, a conventional formula or breastfed, was also studied by Grip *et al.* (2018). This study demonstrated that the effect of supplementation in MFGM on lipidome partially explains the cognitive and immunological benefits (Grip *et al.* 2018).

10.4.5 Towards a Role for Milk Sterols in Infant Blood Cholesterol Homeostasis

Among the numerous constituents of MFGM, cholesterol is a centrally important one. Sterols range from 10 to 25 mg/100 mL in human milk, with cholesterol accounting for approximately 90% of total sterol content, with levels generally declining as lactation progresses (Jensen 1999; Michalski 2013). Sterol content is not affected by changes in maternal diet (Jensen 1999). It is the precursor of steroid hormones and other related derivatives in humans. It was suspected very early that milk fat cholesterol ingested during the neonatal period, and thus before the total maturation of hepatobiliary system, contributed to cholesterol homeostasis later on in the adult (Reiser *et al.* 1979). Owen *et al.* (2002) indeed established that the high cholesterol content in human milk leads to transiently higher total serum cholesterol concentration in infancy (Owen *et al.*

2002). However, later on in adulthood, the grown-up breastfed infants have lower total serum cholesterol concentration in comparison with a formula-fed group. Vegetable-based formula-fed infants (i.e., which receive 3–5 times lower cholesterol content than breast-fed infants) have increased cholesterol synthesis rates in comparison to breast-fed infants which can be perceived as a kind of adaptation towards low dietary uptake. Timby *et al.* (2014b) showed, in a double-blinded randomized trial, that increasing cholesterol intake between 2 and 6 months of age via bovine MFGM addition in infant formula induced a higher total serum cholesterol concentration, similar to the one observed in breast-fed infants without modification of the LDL/HDL ratio (Timby *et al.* 2014b). According to the recent results of Vors *et al.* (2019) in adults, it cannot be ruled out that mechanisms such as increased fecal coprostanol, and activation of the transintestinal cholesterol export, may be involved (Vors *et al.* 2019).

Besides cholesterol, recent research has been dedicated to enhancing understanding of the impact of phytosterols included in vegetable-based fats of most infant formulas (such as campesterol, brassicasterol, stigmasterol, and sitosterol) on cholesterol absorption during the neonatal period. Babawale *et al.* (2018) set the hypothesis that these phytosterols may interfere with cholesterol absorption and contribute to the difference of cholesterol metabolic phenotypes observed between vegetable-based formula-fed infants and breast-fed infants (Babawale *et al.* 2018). The mechanisms of inhibition of cholesterol absorption suspected was competition for micelle formation and for transporter Niemann-PickC1-Like1 across the ileal brush border. These authors investigated, using piglet models, the cholesterol absorption induced by four isocaloric formula containing either high or low cholesterol or phytosterol contents. A 7-day old piglet model (n = 32, 8 piglets per group, differential feeding over 21 days) was used to test the inhibitory effects of phytosterols on cholesterol absorption during postnatal feeding. The authors validated their hypothesis as both the ileal cholesterol digestibility and hepatic cholesterol synthesis were increased in the piglets fed low phytosterol content. This metabolic modification was accompanied by reduced nuclear expression of sterol regulatory element binding transcription factor 2 (SREBP2) in this group in comparison with piglets fed high phytosterol content formula.

10.5 Conclusion and Future Prospects

Milk fat contains thousands of distinct fatty acids, including minor odd and *trans* species, whose importance in biochemical processes and potential metabolic effects have been underlined in recent scientific studies. At the supramolecular level, milk fat globules are unique bioassemblies, finely tuned at different scales, from structured TAGs up to sphingolipids from the MFGM to release progressively, during digestion, various lipids of nutritional interest. These fat globules are surrounded by the natural MFGM, whose structure and composition, notably including polar lipids rich in SM, can modulate gastrointestinal lipolysis, postprandial lipemia, and metabolic fate of dietary lipids. Recent human data also support a role for milk polar lipids of the MFGM in the regulation of blood cholesterol by mechanisms involving their fate in the gut. The specific composition and structure of milk lipids is also important in neonatal nutrition, as they were shown to contribute to programming of lipid metabolism in young mice. Recent research has highlighted that the presence of bioactive lipids, such as SM, could have a beneficial metabolic impact on lipid metabolism and contribute to intestinal and systemic health. Long-term effects of milk lipids, including their specific fatty acids, and also fat globule structure, and MFGM polar lipids in different dairy matrices, should now be studied on metabolic players such as low-grade inflammation and gut microbiota in humans. This opens the way to strategies of valuing the unique richness of milk fat through the complexity of its natural globules and could be of interest for target populations at risk, such as infant, elderly, and individuals with metabolic syndrome.

References

Alonso, L., Fontecha, J., Lozada, L., Fraga, M. J. & Juárez, M. (1999). Fatty acid composition of caprine milk: Major, branched-chain, and *trans* fatty acids. *Journal of Dairy Science, 82*, 878–884.

Arakawa, S., Ito, M. & Tejima, S. (1988). Promoter function of carrageenan on development of colonic tumors induced by 1,2-dimethylhydrazine in rats. *Journal of Nutritional Science and Vitaminology (Tokyo), 34*, 577–585.

Argov, N., Lemay, D. G. & German, J. B. (2008). Milk fat globule structure & function; nanosciece comes to milk production. *Trends in Food Science and Technology, 19*, 617–623.

Armand, M., Hamosh, M., Mehta, N. R., Angelus, P. A., Philpott, J. R., Henderson, T. R., Dwyer, N. K., Lairon, D. & Hamosh, P. (1996). Effect of human Milk or formula on gastric function and fat digestion in the premature infant. *Pediatric Research, 40*, 429–437.

Armand, M., Pasquier, B., Andre, M., Borel, P., Senft, M., Peyrot, J., Salducci, J., Portugal, H., Jaussan, V. & Lairon, D. (1999). Digestion and absorption of 2 fat emulsions with different droplet sizes in the human digestive tract. *American Journal of Clinical Nutrition, 70*, 1096–1106.

Astrup, A., Bertram, H. C., Bonjour, J. P., de Groot, L. C., de Oliveira Otto, M. C., Feeney, E. L., Garg, M. L., Givens, I., Kok, F. J., Krauss, R. M., Lamarche, B., Lecerf, J. M., Legrand, P., McKinley, M., Micha, R., Michalski, M. C., Mozaffarian, D. & Soedamah-Muthu, S. S. (2019). WHO draft guidelines on dietary saturated and trans fatty acids: Time for a new approach? *BMJ, 366*, l4137.

Aune, D., Norat, T., Romundstad, P. & Vatten, L. J. (2013). Dairy products and the risk of type 2 diabetes: A systematic review and dose-response meta-analysis of cohort studies. *American Journal of Clinical Nutrition, 98*, 1066–1083.

Ayala-Bribiesca, E., Turgeon, S. L. & Britten, M. (2017). Effect of calcium on fatty acid bioaccessibility during in vitro digestion of Cheddar-type cheeses prepared with different milk fat fractions. *Journal of Dairy Science, 100*, 2454–2470.

Baars, A., Oosting, A., Engels, E., Kegler, D., Kodde, A., Schipper, L., VERKADE, H. J. & van der Beek, E. M. (2016). Milk fat globule membrane coating of large lipid droplets in the diet of young mice prevents body fat accumulation in adulthood. *British Journal of Nutrition, 115*, 1930–1937.

Baba, N., Bracco, E. F. & Hashim, S. A. (1987). Role of brown adipose tissue in thermogenesis induced by overfeeding a diet containing medium chain triglyceride. *Lipids, 22*, 442–444.

Babawale, E. A., Jones, P. J. H., Mercer, K. E., Lin, H. X., Yeruva, L., Bar Yoseph, F. & Rutherfurd, S. M. (2018). Modulating sterol concentrations in infant formula influences cholesterol absorption and synthesis in the neonatal piglet. *Nutrients, 10*, 13.

Bach, A. C. & Babayan, V. K. (1982). Medium-chain triglycerides: An update. *American Journal of Clinical Nutrition, 36*, 950–962.

Bach, A. C., Ingenbleek, Y. & Frey, A. (1996). The usefulness of dietary medium-chain triglycerides in body weight control: Fact or fancy? *Journal of Lipid Research, 37*, 708–726.

Baer, D. J. (2012). What do we really know about the health effects of natural sources of *trans* fatty acids? *American Journal of Clinical Nutrition, 95*, 267–268.

Berry, S. E. & Sanders, T. A. (2005). Influence of triacylglycerol structure of stearic acid-rich fats on postprandial lipaemia. *Proceedings of the Nutrition Society, 64*, 205–212.

Berry, S. E., Miller, G. J. & Sanders, T. A. (2007). The solid fat content of stearic acid-rich fats determines their postprandial effects. *American Journal of Clinical Nutrition, 85*, 1486–1494.

Bhinder, G., Garcia, C., Gräf, F., Crowley, S., Bosman, E., Ryz, N., Berkmann, J., Dyer, R., Innis, S. & Vallance, B. (2016). The impact of formula fortified with Milk fat globule membrane on the neonatal intestine. *The FASEB Journal, 30*, lb244–lb244.

Billeaud, C., Puccio, G., Saliba, E., Guillois, B., Vaysse, C., Pecquet, S. & Steenhout, P. (2014). Safety and tolerance evaluation of Milk fat globule membrane-enriched infant formulas: A randomized controlled Multicenter non-inferiority trial in healthy term infants. *Clinical Medicine Insights Pediatrics, 8*, 51–60.

Blaskovic, S., Adibekian, A., Blanc, M. & van der Goot, G. F. (2014). Mechanistic effects of protein palmitoylation and the cellular consequences thereof. *Chemistry and Physics of Lipids, 180*, 44–52.

Blewett, H. J., Gerdung, C. A., Ruth, M. R., Proctor, S. D. & Field, C. J. (2009). Vaccenic acid favourably alters immune function in obese JCR:LA-cp rats. *British Journal of Nutrition, 102*, 526–536.

Bonnaire, L., Sandra, S., Helgason, T., Decker, E. A., Weiss, J. & McClements, D. J. (2008). Influence of lipid physical state on the in vitro digestibility of emulsified lipids. *Journal of Agricultural and Food Chemistry, 56*, 3791–3797.

Borel, P., Pasquier, B., Armand, M., Tyssandier, V., Grolier, P., Alexandre-Gouabau, M.-C., Andre, M., Senft, M., Peyrot, J., Jaussan, V., Lairon, D. & Azais-Braesco, V. (2001). Processing of vitamin a and E in the human gastrointestinal tract. *American Journal of Physiology. Gastrointestinal and Liver Physiology, 280*, G95–G103.

Bourlieu, C. & Michalski, M. C. (2015). Structure-function relationship of the milk fat globule. *Current Opinion in Clinical Nutrition and Metabolic Care, 18*, 118–127.

Bourlieu, C., Ménard, O., Bouzerzour, K., Mandalari, G., Macierzanka, A., Mackie, A. R. & Dupont, D. (2014). Specificity of infant digestive conditions: Some clues for developing relevant in vitro models. *Critical Reviews in Food Science and Nutrition, 54*, 1427–1457.

Bourlieu, C., Bouzerzour, K., Ferret-Bernard, S., Le Bourgot, C., Chever, S., Menard, O., Deglaire, A., Cuinet, I., Le Ruyet, P., Bonhomme, C., Dupont, D. & Le Huerou-Luron, I. (2015a). Infant formula interface and fat source impact on neonatal digestion and gut microbiota. *European Journal of Lipid Science and Technology, 117*, 1500–1512.

Bourlieu, C., Menard, O., De La Chevasnerie, A., Sams, L., Rousseau, F., Madec, M. N., Robert, B., Deglaire, A., Pezennec, S., Bouhallab, S., Carriere, F. & Dupont, D. (2015b). The structure of infant formulas impacts their lipolysis, proteolysis and disintegration during in vitro gastric digestion. *Food Chemistry, 182*, 224–235.

Bourlieu, C., Deglaire, A., De Oliveira, S. C., Ménard, O., Le Gouar, Y., Carrière, F. & Dupont, D. (2017). Towards infant formula biomimetic of human milk structure and digestive behaviour. *OCL, 24*, D206.

Bourlieu, C., Cheillan, D., Blot, M., Daira, P., Trauchessec, M., Ruet, S., Gassi, J. Y., Beaucher, E., Robert, B., Leconte, N., Bouhallab, S., Gaucheron, F., Gesan-Guiziou, G. & Michalski, M. C. (2018). Polar lipid composition of bioactive dairy co-products buttermilk and butterserum: Emphasis on sphingolipid and ceramide isoforms. *Food Chemistry, 240*, 67–74.

Breij, L. M., Abrahamse-Berkeveld, M., Vandenplas, Y., Jespers, S. N. J., de Mol, A. C., Khoo, P. C., Kalenga, M., Peeters, S., van Beek, R. H. T., Norbruis, O. F., Schoen, S., Acton, D., Hokken-Koelega, A. C. S. & Group, M. S. (2019). An infant formula with large, milk phospholipid–coated lipid droplets containing a mixture of dairy and vegetable lipids supports adequate growth and is well tolerated in healthy, term infants. *American Journal of Clinical Nutrition, 109*, 586–596.

Caesar, R., Tremaroli, V., Kovatcheva-Datchary, P., Cani, P. D. & Backhed, F. (2015). Crosstalk between gut microbiota and dietary lipids aggravates WAT inflammation through TLR Signaling. *Cell Metabolism, 22*, 658–668.

Cani, P. D., Neyrinck, A. M., Fava, F., Knauf, C., Burcelin, R. G., Tuohy, K. M., Gibson, G. R. & Delzenne, N. M. (2007). Selective increases of bifidobacteria in gut microflora improve high-fat-diet-induced diabetes in mice through a mechanism associated with endotoxaemia. *Diabetologia, 50*, 2374–2383.

Cani, P. D., Bibiloni, R., Knauf, C., Waget, A., Neyrinck, A. M., Delzenne, N. M. & Burcelin, R. (2008). Changes in gut microbiota control metabolic endotoxemia-induced inflammation in high-fat diet-induced obesity and diabetes in mice. *Diabetes, 57*, 1470–1481.

Castro-Gomez, P., Garcia-Serrano, A., Visioli, F. & Fontecha, J. (2015). Relevance of dietary glycerophospholipids and sphingolipids to human health. *Prostaglandins Leukot. Essent. Fatty Acids, 101*, 41–51.

Chardigny, J. M., Masson, E., Sergiel, J. P., Darbois, M., Loreau, O., Noel, J. P. & Sebedio, J. L. (2003). The position of rumenic acid on triacylglycerols alters its bioavailability in rats. *Journal of Nutrition, 133*, 4212–4214.

Chatterton, D. E. W., Rasmussen, J. T., Heegaard, C. W., Sorensen, E. S. & Petersen, T. E. (2004). In vitro digestion of novel milk protein ingredients for use in infant formulas: Research on biological functions. *Trends in Food Science and Technology, 15*, 373–383.

Chaurasia, B. & Summers, S. A. (2015). Ceramides – Lipotoxic inducers of metabolic disorders. *Trends in Endocrinology and Metabolism, 26*, 538–550.

Che, B. N., Kristensen, T., Nebel, C., Dalsgaard, T. K., Hellgren, L. I., Young, J. F. & Larsen, M. K. (2013). Content and distribution of phytanic acid diastereomers in organic milk as affected by feed composition. *Journal of Agricultural and Food Chemistry, 61*, 225–230.

Chen, J., Ghazawi, F. M., Bakkar, W. & Li, Q. (2006). Valproic acid and butyrate induce apoptosis in human cancer cells through inhibition of gene expression of Akt/protein kinase B. *Molecular Cancer, 5*, 71.

Chen, G. C., Szeto, I. M. Y., Chen, L. H., Han, S. F., Li, Y. J., van Hekezen, R. & Qin, L. Q. (2015). Dairy products consumption and metabolic syndrome in adults: Systematic review and meta-analysis of observational studies. *Scientific Reports, 5*, 14606.

Chowdhury, R., Warnakula, S., Kunutsor, S., Crowe, F., Ward, H. A., Johnson, L., Franco, O. H., Butterworth, A. S., Forouhi, N. G., Thompson, S. G., Khaw, K. T., Mozaffarian, D., Danesh, J. & Di Angelantonio, E. (2014). Association of dietary, circulating, and supplement fatty acids with coronary risk: A systematic review and meta-analysis. *Annals of Internal Medicine, 160*, 398–406.

Christensen, R., Lorenzen, J. K., Svith, C. R., Bartels, E. M., Melanson, E. L., Saris, W. H., Tremblay, A. & Astrup, A. (2009). Effect of calcium from dairy and dietary supplements on faecal fat excretion: A meta-analysis of randomized controlled trials. *Obesity Reviews, 10*, 475–486.

Clark, S. B., Brause, B. & Holt, P. R. (1969). Lipolysis and absorption of fat in the rat stomach. *Gastroenterology, 56*, 214–222.

Clemente, G., Mancini, M., Nazzaro, F., Lasorella, G., Rivieccio, A., Palumbo, A. M., Rivellese, A. A., Ferrara, L. & Giacco, R. (2003). Effects of different dairy products on postprandial lipemia. *Nutrition, Metabolism, and Cardiovascular Diseases, 13*, 377–383.

Contarini, G. & Povolo, M. (2013). Phospholipids in milk fat: Composition, biological and technological significance, and analytical strategies. *International Journal of Molecular Sciences, 14*, 2808–2831.

Conway, V., Couture, P., Richard, C., Gauthier, S. F., Pouliot, Y. & Lamarche, B. (2013). Impact of buttermilk consumption on plasma lipids and surrogate markers of cholesterol homeostasis in men and women. *Nutrition, Metabolism, and Cardiovascular Diseases, 23*, 1255–1262.

Conway, V., Gauthier, S. F. & Pouliot, Y. (2014). Buttermilk: Much more than a source of milk phospholipids. *Animal Frontiers, 4*, 44–51.

Couedelo, L., Boue-Vaysse, C., Fonseca, L., Montesinos, E., Djoukitch, S., Combe, N. & Cansell, M. (2011). Lymphatic absorption of alpha-linolenic acid in rats fed flaxseed oil-based emulsion. *British Journal of Nutrition, 105*, 1026–1035.

Couedelo, L., Amara, S., Lecomte, M., Meugnier, E., Monteil, J., Fonseca, L., Pineau, G., Cansell, M., Carriere, F., Michalski, M. C. & Vaysse, C. (2015). Impact of various emulsifiers on ALA bioavailability and chylomicron synthesis through changes in gastrointestinal lipolysis. *Food & Function, 6*, 1726–1735.

Czosnykowska-Lukacka, M., Krolak-Olejnik, B. & Orczyk-Pawilowicz, M. (2018). Breast milk macronutrient components in prolonged lactation. *Nutrients, 10*, E1893.

Da Silva, M. S., Julien, P., Couture, P., Lemieux, S., Vohl, M. C. & Rudkowska, I. (2014). Associations between dairy intake and metabolic risk parameters in a healthy French-Canadian population. *Applied Physiology, Nutrition, and Metabolism, 39*, 1323–1331.

Da Silva, M. S., Julien, P., Pérusse, L., Vohl, M. C. & Rudkowska, I. (2015). Natural rumen-derived *trans* fatty acids are associated with metabolic markers of cardiac health. *Lipids, 50*, 873–882.

Danthine, S., Vors, C., Agopian, D., Durand, A., Guyon, R., Carrière, F., Knibbe, C., Létisse, M. & Michalski, M. C. (2019). Homogeneous triacylglycerol tracers have an impact on the thermal and structural properties of dietary fat and its lipolysis rate under simulated physiological conditions. *Chemistry and Physics of Lipids, 225*, 104815.

de La Serre, C. B., Ellis, C. L., Lee, J., Hartman, A. L., Rutledge, J. C. & Raybould, H. E. (2010). Propensity to high-fat diet-induced obesity in rats is associated with changes in the gut microbiota and gut inflammation. *American Journal of Physiology. Gastrointestinal and Liver Physiology, 299*, G440–G448.

de Oliveira Otto, M. C., Mozaffarian, D., Kromhout, D., Bertoni, A. G., Sibley, C. T., Jacobs, D. R. & Nettleton, J. A. (2012). Dietary intake of saturated fat by food source and incident cardiovascular disease: The multiethnic study of atherosclerosis. *American Journal of Clinical Nutrition, 96*, 397–404.

de Oliveira, S. C., Deglaire, A., Ménard, O., Bellanger, A., Rousseau, F., Henry, G., Dirson, E., Carrière, F., Dupont, D. & Bourlieu, C. (2015). Holder pasteurization impacts the proteolysis, lipolysis and disintegration of human milk under in vitro dynamic term newborn digestion. *Food Research International, 88*, 263–275.

de Oliveira, S. C., Bourlieu, C., Ménard, O., Bellanger, A., Henry, G., Rousseau, F., Dirson, E., Carriere, F., Dupont, D. & Deglaire, A. (2016). Impact of pasteurization of human milk on preterm newborn in vitro digestion: Gastrointestinal disintegration, lipolysis and proteolysis. *Food Chemistry, 211*, 171–179.

de Oliveira, S. C., Bellanger, A., Ménard, O., Pladys, P., Le Gouar, Y., Henry, G., Dirson, E., Rousseau, F., Carrière, F., Dupont, D., Bourlieu, C. & Deglaire, A. (2017). Impact of homogenization of pasteurized human milk on gastric digestion in the preterm infant: A randomized controlled trial. *Clinical Nutrition ESPEN, 20*, 1–11.

de Souza, R. J., Mente, A., Maroleanu, A., Cozma, A. I., Ha, V., Kishibe, T., Uleryk, E., Budylowski, P., Schünemann, H., Beyene, J. & Anand, S. S. (2015). Intake of saturated and *trans* unsaturated fatty acids and risk of all cause mortality, cardiovascular disease, and type 2 diabetes: Systematic review and meta-analysis of observational studies. *BMJ, 351*, h3978.

Dehghan, M., Mente, A., Zhang, X., Swaminathan, S., Li, W., Mohan, V., Iqbal, R., Kumar, R., Wentzel-Viljoen, E., Rosengren, A., Amma, L. I., Avezum, A., Chifamba, J., Diaz, R., Khatib, R., Lear, S., Lopez-Jaramillo, P., Liu, X., Gupta, R., Mohammadifard, N., Gao, N., Oguz, A., Ramli, A. S., Seron, P., Sun, Y., Szuba, A., Tsolekile, L., Wielgosz, A., Yusuf, R., Hussein Yusufali, A., Teo, K. K., Rangarajan, S., Dagenais, G., Bangdiwala, S. I., Islam, S., Anand, S. S. & Yusuf, S. (2017). Associations of fats and carbohydrate intake with cardiovascular disease and mortality in 18 countries from five continents (PURE): A prospective cohort study. *Lancet, 390*, 2050–2062. On behalf of the Prospective Urban Rural Epidemiology (PURE) study investigators.

Dehghan, M., Mente, A., Rangarajan, S., Sheridan, P., Mohan, V., Iqbal, R., Gupta, R., Lear, S., Wentzel-Viljoen, E., Avezum, A., Lopez-Jaramillo, P., Mony, P., Varma, R. P., Kumar, R., Chifamba, J., Alhabib, K. F., Mohammadifard, N., Oguz, A., Lanas, F., Rozanska, D., Bostrom, K. B., Yusoff, K., Tsolkile, L. P., Dans, A., Yusufali, A., Orlandini, A., Poirier, P., Khatib, R., Hu, B., Wei, L., Yin, L., Deeraili, A., Yeates, K., Yusuf, R., Ismail, N., Mozaffarian, D., Teo, K., Anand, S. S. & Yusuf, S. (2018). Association of dairy intake with cardiovascular disease and mortality in 21 countries from five continents (PURE): A prospective cohort study. *Lancet, 392*, 2288–2297. On behalf of the Prospective Urban Rural Epidemiology (PURE) study investigators.

Delplanque, B., Du, Q., Leruyet, P., Agnani, G., Pages, N., Gripois, D., Ould-Hamouda, H., Carayon, P. & Martin, J. C. (2011). Brain docosahexaenoic acid (DHA) levels of young rats are related to alpha-linolenic acid (ALA) levels and fat matrix of the diet: Impact of dairy fat*. *OCL, 18*, 293–296.

Delplanque, B., Du, Q., Agnani, G., Le Ruyet, P. & Martin, J. C. (2013). A dairy fat matrix providing alpha-linolenic acid (ALA) is better than a vegetable fat mixture to increase brain DHA accretion in young rats. *Prostagl. Leukotr. Essent. Fatty Acids, 88*, 115–120.

Delplanque, B., Gibson, R., Koletzko, B., Lapillonne, A. & Strandvik, B. (2015). Lipid quality in infant nutrition: Current knowledge and future opportunities.

Journal of Pediatric Gastroenterology and Nutrition, 61, 8–17.

Delplanque, B., Du, Q., Martin, J. C. & Guesnet, P. (2019). Lipids for infant formulas. *Cahiers de Nutrition et de Dietetique, 54*, 44–51.

Delporte, C. (2013). Structure and physiological actions of ghrelin. *Scientifica, 2013*, 518909.

Destaillats, F., Wolff, R. L., Precht, D. & Molkentin, J. (2000). Study of individual *trans*- and *cis*-16:1 isomers in cow, goat, and ewe cheese fats by gas-liquid chromatography with emphasis on the *trans*-delta3 isomer. *Lipids, 35*, 1027–1032.

Devle, H., Ulleberg, E. K., Naess-Andresen, C. F., Rukke, E.-O., Vegarud, G. & Ekeberg, D. (2014). Reciprocal interacting effects of proteins and lipids during ex vivo digestion of bovine milk. *International Dairy Journal, 36*, 6–13.

Dillehay, D. L., Webb, S. K., Schmelz, E. M. & Merrill, A. H., Jr. (1994). Dietary sphingomyelin inhibits 1,2-dimethylhydrazine-induced colon cancer in CF1 mice. *Journal of Nutrition, 124*, 615–620.

Drouin-Chartier, J. P., Brassard, D., Tessier-Grenier, M., Cote, J. A., Labonte, M. E., Desroches, S., Couture, P. & Lamarche, B. (2016a). Systematic review of the association between dairy product consumption and risk of cardiovascular-related clinical Outcomes. *Advances in Nutrition, 7*, 1026–1040.

Drouin-Chartier, J. P., Côté, J. A., Labonté, M. E., Brassard, D., Tessier-Grenier, M., Desroches, S., Couture, P., Lamarche, B. & B. (2016b). Comprehensive review of the impact of dairy foods and dairy fat on cardiometabolic risk. *Advances in Nutrition, 7*, 1041–1051.

Drouin-Chartier, J. P., Tremblay, A. J., Maltais-Giguere, J., Charest, A., Guinot, L., Rioux, L. E., Labrie, S., Britten, M., Lamarche, B., Turgeon, S. L. & Couture, P. (2017). Differential impact of the cheese matrix on the postprandial lipid response: A randomized, cross-over, controlled trial. *American Journal of Clinical Nutrition, 106*, 1358–1365.

Duivenvoorden, I., Voshol, P. J., Rensen, P. C., van Duyvenvoorde, W., Romijn, J. A., Emeis, J. J., Havekes, L. M. & Nieuwenhuizen, W. F. (2006). Dietary sphingolipids lower plasma cholesterol and triacylglycerol and prevent liver steatosis in APOE*3Leiden mice. *American Journal of Clinical Nutrition, 84*, 312–321.

Eckhardt, E. R. M., Wang, D. Q. H., Donovan, J. M. & Carey, M. C. (2002). Dietary sphingomyelin suppresses intestinal cholesterol absorption by decreasing thermodynamic activity of cholesterol monomers. *Gastroenterology, 122*, 948–956.

El-Loly, M. (2011). Composition, properties and nutritional aspects of milk fat globule membrane – A review. *Polish Journal of Food and Nutrition Sciences, 61*, 7–32.

Elwood, P. C., Pickering, J. E., Givens, D. I. & Gallacher, J. E. (2010). The consumption of milk and dairy foods and the incidence of vascular disease and diabetes: An overview of the evidence. *Lipids, 45*, 925–939.

Ezanno, H., Beauchamp, E., Catheline, D., Legrand, P. & Rioux, V. (2015). Beneficial impact of a mix of dairy fat with rapeseed oil on n-6 and n-3 PUFA metabolism in the rat: A small enrichment in dietary alpha-linolenic acid greatly increases its conversion to DHA in the liver. *European Journal of Lipid Science and Technology, 117*, 281–290.

Fang, X., Rioux, L.-E., Labrie, S. & Turgeon, S. L. (2016). Commercial cheeses with different texture have different disintegration and protein/peptide release rates during simulated in vitro digestion. *International Dairy Journal, 56*, 169–178.

Fauquant, C., Briard-Bion, V., Leconte, N., Guichardant, M. & Michalski, M.-C. (2007). Membrane phospholipids and sterols in microfiltered milk fat globules. *European Journal of Lipid Science and Technology, 109*, 1167–1173.

Field, C. J., Blewett, H. H., Proctor, S. & Vine, D. (2009). Human health benefits of vaccenic acid. *Applied Physiology, Nutrition, and Metabolism, 34*, 979–991.

Fondaco, D., AlHasawi, F., Lan, Y., Ben-Elazar, S., Connolly, K. & Rogers, M. (2015). Biophysical aspects of lipid digestion in human breast Milk and Similac((TM)) infant formulas. *Food Biophysics, 10*, 282–291.

Fong, B. Y., Norris, C. S. & MacGibbon, A. K. H. (2007). Protein and lipid composition of bovine milk-fat-globule membrane. *International Dairy Journal, 17*, 275–288.

Fruekilde, M. B. & Hoy, C. E. (2004). Lymphatic fat absorption varies among rats administered dairy products differing in physico-chemical properties. *Journal of Nutrition, 134*, 1110–1113.

Fujisawa, Y., Yamaguchi, R., Nagata, E., Satake, E., Sano, S., Matsushita, R., Kitsuta, K., Nakashima, S., Nakanishi, T., Nakagawa, Y. & Ogata, T. (2013). The lipid fraction of human milk initiates adipocyte differentiation in 3T3-L1 cells. *Early Human Development, 89*, 713–719.

Gallier, S., Cui, J., Olson, T. D., Rutherfurd, S. M., Ye, A., Moughan, P. J. & Singh, H. (2013a). In vivo digestion of bovine milk fat globules: Effect of processing and interfacial structural changes. I. Gastric digestion. *Food Chemistry, 141*, 3273–3281.

Gallier, S., Zhu, X. Q., Rutherfurd, S. M., Ye, A., Moughan, P. J. & Singh, H. (2013b). In vivo digestion of bovine milk fat globules: Effect of processing and interfacial structural changes. II. Upper digestive tract digestion. *Food Chemistry, 141*, 3215–3223.

Gallier, S., Laubscher, A. & Jimenez-Flores, R. (2014). Chapter 4 – The Milk fat globule membrane: Structure, methodology for its study, and functionality. In M. Boland, M. Golding & H. Singh (Eds.), *Food structures, digestion and health* (pp. 107–142). San Diego: Academic Press.

Garcia, C., Antona, C., Robert, B., Lopez, C. & Armand, M. (2014). The size and interfacial composition of milk fat globules are key factors controlling triglycerides bioavailability in simulated human gastro-duodenal digestion. *Food Hydrocolloids, 35*, 494–504.

Garcia, C., Duby, C., Catheline, D., Toral, P. G., Bernard, L., Legrand, P. & Rioux, V. (2017). Synthesis of the suspected *trans*-11,*cis*-13 conjugated linoleic acid isomer in ruminant mammary tissue by FADS3-catalyzed Δ13-desaturation of vaccenic acid. *Journal of Dairy Science, 100*, 783–796.

Garcia, C., Guillocheau, E., Richard, L., Drouin, G., Catheline, D., Legrand, P. & Rioux, V. (2018). Conversion of dietary *trans*-vaccenic acid to *trans*11,*cis*13-conjugated linoleic acid in the rat lactating mammary gland by Fatty Acid Desaturase 3-catalyzed methyl-end Δ13-desaturation. *Biochemical and Biophysical Research Communications, 505*, 385–391.

Gassi, J. Y., Blot, M., Beaucher, E., Robert, B., Leconte, N., Camier, B., Rousseau, F., Bourlieu, C., Jardin, J., Briard-Bion, V., Lambert, S., Guesan-Guiziou, G., Lopez, C. & Gaucheron, F. (2016). Preparation and characterisation of a milk polar lipids enriched ingredient from fresh industrial liquid butter serum: Combination of physico-chemical modifications and technological treatments. *International Dairy Journal, 52*, 26–34.

Gayet-Boyer, C., Tenenhaus-Aziza, F., Prunet, C., Marmonier, C., Malpuech-Brugère, C., Lamarche, B. & Chardigny, J. M. (2014). Is there a linear relationship between the dose of ruminant *trans*-fatty acids and cardiovascular risk markers in healthy subjects: Results from a systematic review and meta-regression of randomised clinical trials. *British Journal of Nutrition, 112*, 1914–1922.

Gebauer, S. K., Chardigny, J. M., Jakobsen, M. U., Lamarche, B., Lock, A. L., Proctor, S. D. & Baer, D. J. (2011). Effects of ruminant *trans* fatty acids on cardiovascular disease and cancer: A comprehensive review of epidemiological, clinical, and mechanistic studies. *Advances in Nutrition, 2*, 332–354.

Gebauer, S. K., Destaillats, F., Dionisi, F., Krauss, R. M. & Baer, D. J. (2015). Vaccenic acid and *trans* fatty acid isomers from partially hydrogenated oil both adversely affect LDL cholesterol: A double-blind, randomized controlled trial. *American Journal of Clinical Nutrition, 102*, 1339–1346.

Geliebter, A., Torbay, N., Bracco, E. F., Hashim, S. A. & Van Itallie, T. B. (1983). Overfeeding with medium-chain triglyceride diet results in diminished deposition of fat. *American Journal of Clinical Nutrition, 37*, 1–4.

Genser, L., Aguanno, D., Soula, H. A., Dong, L., Trystram, L., Assmann, K., Salem, J. E., Vaillant, J. C., Oppert, J. M., Laugerette, F., Michalski, M. C., Wind, P., Rousset, M., Brot-Laroche, E., Leturque, A., Clement, K., Thenet, S. & Poitou, C. (2018). Increased jejunal permeability in human obesity is revealed by a lipid challenge and is linked to inflammation and type 2 diabetes. *Journal of Pathology, 246*, 217–230.

German, J. B., Gibson, R. A., Krauss, R. M., Nestel, P., Lamarche, B., van Staveren, W. A., Steijns, J. M., de Groot, L. C., Lock, A. L. & Destaillats, F. (2009). A reappraisal of the impact of dairy foods and milk fat on cardiovascular disease risk. *European Journal of Nutrition, 48*, 191–203.

Gillman, M. W., Cupples, L. A., Millen, B. E., Ellison, R. C. & Wolf, P. A. (1997). Inverse association of dietary fat with development of ischemic stroke in men. *JAMA, 278*, 2145–2150.

Griinari, J. M., Corl, B. A., Lacy, S. H., Chouinard, P. Y., Nurmela, K. V. & Bauman, D. E. (2000). Conjugated linoleic acid is synthesized endogenously in lactating dairy cows by Delta(9)-desaturase. *Journal of Nutrition, 130*, 2285–2291.

Grip, T., Dyrlund, T. S., Ahonen, L., Domellof, M., Hernell, O., Hyotylainen, T., Knip, M., Lonnerdal, B., Oresic, M. & Timby, N. (2018). Serum, plasma and erythrocyte membrane lipidomes in infants fed formula supplemented with bovine milk fat globule membranes. *Pediatric Research, 84*, 726–732.

Grundy, M. M.-L., Lapsley, K. & Ellis, P. R. (2016). A review of the impact of processing on nutrient bioaccessibility and digestion of almonds. *International Journal of Food Science and Technology, 51*, 1937–1946.

Guillocheau, E., Garcia, C., Drouin, G., Richard, L., Catheline, D., Legrand, P. & Rioux, V. (2019). Retroconversion of dietary *trans*-vaccenic (*trans*-C18:1 n-7) acid to *trans*-palmitoleic acid (*trans*-C16:1 n-7): Proof of concept and quantification in both cultured rat hepatocytes and pregnant rats. *Journal of Nutritional Biochemistry, 63*, 19–26.

Guo, Q., Bellissimo, N. & Rousseau, D. (2017a). The physical state of emulsified edible oil modulates its in vitro digestion. *Journal of Agricultural and Food Chemistry, 65*, 9120–9127.

Guo, Q., Ye, A., Bellissimo, N., Singh, H. & Rousseau, D. (2017b). Modulating fat digestion through food structure design. *Progress in Lipid Research, 68*, 109–118.

Gurnida, D. A., Rowan, A. M., Idjradinata, P., Muchtadi, D. & Sekarwana, N. (2012). Association of complex lipids containing gangliosides with cognitive development of 6-month-old infants. *Early Human Development, 88*, 595–601.

Hageman, J. H. J., Danielsen, M., Nieuwenhuizen, A. G., Feitsma, A. L. & Dalsgaard, T. K. (2019). Comparison of bovine milk fat and vegetable fat for infant formula: Implications for infant health. *International Dairy Journal, 92*, 37–49.

Han, J., Farmer, S. R., Kirkland, J. L., Corkey, B. E., Yoon, R., Pirtskhalava, T., Ido, Y. & Guo, W. (2002). Octanoate attenuates adipogenesis in 3T3-L1 preadipocytes. *Journal of Nutrition, 132*, 904–910.

Han, J., Hamilton, J. A., Kirkland, J. L., Corkey, B. E. & Guo, W. (2003). Medium-chain oil reduces fat mass and down-regulates expression of adipogenic genes in rats. *Obesity Research, 11*, 734–744.

Hannun, Y. A. & Linardic, C. M. (1993). Sphingolipid breakdown products: Anti-proliferative and tumor-suppressor lipids. *Biochimica et Biophysica Acta, 1154*, 223–236.

Hansen, C. H., Krych, L., Nielsen, D. S., Vogensen, F. K., Hansen, L. H., Sorensen, S. J., Buschard, K. & Hansen,

A. K. (2012). Early life treatment with vancomycin propagates Akkermansia muciniphila and reduces diabetes incidence in the NOD mouse. *Diabetologia, 55*, 2285–2294.

Haruta-Ono, Y., Setoguchi, S., Ueno, H. M., Higurashi, S., Ueda, N., Kato, K., Saito, T., Matsunaga, K. & Takata, J. (2012). Orally administered sphingomyelin in bovine milk is incorporated into skin sphingolipids and is involved in the water-holding capacity of hairless mice. *Journal of Dermatological Science, 68*, 56–62.

He, X., Parenti, M., Grip, T., Domellof, M., Lonnerdal, B., Hernell, O., Timby, N. & Slupsky, C. M. (2019). Metabolic phenotype of breast-fed infants, and infants fed standard formula or bovine MFGM supplemented formula: A randomized controlled trial. *Scientific Reports, 9*, 13.

Heim, M., Johnson, J., Boess, F., Bendik, I., Weber, P., Hinziker, W. & Flühmann, B. (2002). Phytanic acid, a natural peroxisome proliferator-activated receptor (PPAR) agonist, regulates glucose metabolism in rat primary hepatocytes. *The FASEB Journal, 16*, 718–720.

Hernell, O., Timby, N., Domellöf, M. & Lönnerdal, B. (2016). Clinical benefits of Milk fat globule membranes for infants and children. *Journal of Pediatrics, 173*, S60–S65.

Hosoda, H., Kojima, M., Matsuo, H. & Kangawa, K. (2000). Ghrelin and des-acyl ghrelin: Two major forms of rat ghrelin peptide in gastrointestinal tissue. *Biochemical and Biophysical Research Communications, 279*, 909–913.

Howard, A. D., Feighner, S. D., Cully, D. F., Arena, J. P., Liberator, P. A., Rosenblum, C. I., Hamelin, M., Hreniuk, D. L., Palyha, O. C., Anderson, J., Paress, P. S., Diaz, C., Chou, M., Liu, K. K., McKee, K. K., Pong, S. S., Chaung, L. Y., Elbrecht, A., Dashkevicz, M., Heavens, R., Rigby, M., Sirinathsinghji, D. J., Dean, D. C., Melillo, D. G., Patchett, A. A., Nargund, R., Griffin, P. R., DeMartino, J. A., Gupta, S. K., Schaeffer, J. M., Smith, R. G. & Van der Ploeg, L. H. (1996). A receptor in pituitary and hypothalamus that functions in growth hormone release. *Science, 273*, 974–977.

Hu, F. B., Stampfer, M. J., Manson, J. E., Ascherio, A., Colditz, G. A., Speizer, F. E., Hennekens, C. H. & Willett, W. C. (1999). Dietary saturated fats and their food sources in relation to the risk of coronary heart disease in women. *American Journal of Clinical Nutrition, 70*, 1001–1008.

Imamura, F., Fretts, A., Marklund, M., Ardisson Korat, A. V., Yang, W. S., Lankinen, M., Qureshi, W., Helmer, C., Chen, T. A., Wong, K., Bassett, J. K., Murphy, R., Tintle, N., Yu, C. I., Brouwer, I. A., Chien, K. L., Frazier-Wood, A. C., del Gobbo, L. C., Djoussé, L., Geleijnse, J. M., Giles, G. G., de Goede, J., Gudnason, V., Harris, W. S., Hodge, A., Hu, F., Koulman, A., Laakso, M., Lind, L., Lin, H. J., McKnight, B., Rajaobelina, K., Risérus, U., Robinson, J. G., Samieri, C., Siscovick, D. S., Soedamah-Muthu, S. S., Sotoodehnia, N., Sun, Q., Tsai, M. Y., Uusitupa, M., Wagenknecht, L. E., Wareham, N. J., Wu, J. H., Micha, R., Forouhi, N. G., Lemaitre, R. N., Mozaffarian, D., Acids, F. & Outcomes Research Consortium (FORCE). (2018). Fatty acid biomarkers of dairy fat consumption and incidence of type 2 diabetes: A pooled analysis of prospective cohort studies. *PLoS Medicine, 15*, e1002670.

Innis, S. M. (2011). Dietary triacylglycerol structure and its role in infant nutrition. *Advances in Nutrition, 2*, 275–283.

Jacome-Sosa, M. M., Borthwick, F., Mangat, R., Uwiera, R., Reaney, M. J., Shen, J., Quiroga, A. D., Jacobs, R. L., Lehner, R., Proctor, S. D. & Nelson, R. C. (2014). Diets enriched in *trans*-11 vaccenic acid alleviate ectopic lipid accumulation in a rat model of NAFLD and metabolic syndrome. *Journal of Nutritional Biochemistry, 25*, 692–701.

Jan, S., Guillou, H., D'Andrea, S., Daval, S., Bouriel, M., Rioux, V. & Legrand, P. (2004). Myristic acid increases Δ6-desaturase activity in cultured rat hepatocytes. *Reproduction, Nutrition, Development, 44*, 131–140.

Jaudszus, A., Kramer, R., Pfeuffer, M., Roth, A., Jahreis, G. & Kuhnt, K. (2014). *Trans* palmitoleic acid arises endogenously from dietary vaccenic acid. *American Journal of Clinical Nutrition, 99*, 431–435.

Jensen, R. G. (1999). Lipids in human milk. *Lipids, 34*, 1243–1271.

Jensen, R. G., Ferris, A. M., Lammi-Keefe, C. J. & Henderson, R. A. (1990). Lipids of bovine and human milks: A comparison. *Journal of Dairy Science, 73*, 223–240.

John, A., Sun, R., Maillart, L., Schaefer, A., Spence, E. H. & Perrin, M. T. (2019). Macronutrient variability in human milk from donors to a milk bank: Implications for feeding preterm infants. *PLoS One, 14*, 12.

Johnson, D. R., Bhatnagar, R. S., Knoll, L. J. & Gordon, J. I. (1994). Genetic and biochemical studies of protein N-myristoylation. *Annual Review of Biochemistry, 63*, 869–914.

Kamili, A., Wat, E., Chung, R. W., Tandy, S., Weir, J. M., Meikle, P. J. & Cohn, J. S. (2010). Hepatic accumulation of intestinal cholesterol is decreased and fecal cholesterol excretion is increased in mice fed a high-fat diet supplemented with milk phospholipids. *Nutrition & Metabolism (London), 7*, 90.

Keller, S., Malarski, A., Reuther, C., Kertscher, R., Kiehntopf, M. & Jahreis, G. (2013). Milk phospholipid and plant sterol-dependent modulation of plasma lipids in healthy volunteers. *European Journal of Nutrition, 52*, 1169–1179.

Keogh, J. B., Wooster, T. J., Golding, M., Day, L., Otto, B. & Clifton, P. M. (2011). Slowly and rapidly digested fat emulsions are equally satiating but their triglycerides are differentially absorbed and metabolized in humans. *Journal of Nutrition, 141*, 809–815.

Keys, A., Aravanis, C., Blackburn, H. W., Van Buchem, F. S., Buzina, R., Djordjevic, B. D., Dontas, A. S., Fidanza, F., Karvonen, M. J., Kimura, N., Lekos, D., Monti, M., Puddu, V. & Taylor, H. L. (1966).

Epidemiological studies related to coronary heart disease: Characteristics of men aged 40-59 in seven countries. *Acta Medica Scandinavica Supplementum, 460*, 1–392.

Keys, A., Menotti, A., Aravanis, C., Blackburn, H., Djordevic, B. S., Buzina, R., Dontas, A. S., Fidanza, F., Karvonen, M. J., Kimura, N., Mohacek, I., Nedeljkovic, S., Puddu, V., Punsar, S., Taylor, H. L., Conti, S., Kromhout, D. & Toshima, H. (1984). The seven countries study: 2,289 deaths in 15 years. *Preventive Medicine, 13*, 141–154.

Kim, Y. & Je, Y. (2016). Dairy consumption and risk of metabolic syndrome: A meta-analysis. *Diabetic Medicine, 33*, 428–440.

Kliem, K. E. & Shingfield, K. J. (2016). Manipulation of milk fatty acid composition in lactating cows: Opportunities and challenges. *European Journal of Lipid Science and Technology, 118*, 1661–1683.

Kojima, M., Hosoda, H., Date, Y., Nakazato, M., Matsuo, H. & Kangawa, K. (1999). Ghrelin is a growth-hormone-releasing acylated peptide from stomach. *Nature, 402*, 656–660.

Kolar, S. S. N., Barhoumi, R., Lupton, J. R. & Chapkin, R. S. (2007). Docosahexaenoic acid and butyrate synergistically induce colonocyte apoptosis by enhancing mitochondrial Ca^{2+} accumulation. *Cancer Research, 67*, 5561–5568.

Kolesnick, R. N. (1991). Sphingomyelin and derivatives as cellular signals. *Progress in Lipid Research, 30*, 1–38.

Kratz, M., Marcovina, S., Nelson, J. E., Yeh, M. M., Kowdley, K. V., Callahan, H. S., Song, X., Di, C. & Utzschneider, K. M. (2014). Dairy fat intake is associated with glucose tolerance, hepatic and systemic insulin sensitivity, and liver fat but not β-cell function in humans. *American Journal of Clinical Nutrition, 99*, 1385–1396.

Kuhnt, K., Baehr, M., Rohrer, C. & Jahreis, G. (2011). *Trans* fatty acid isomers and the *trans*-9/*trans*-11 index in fat containing foods. *European Journal of Lipid Science and Technology, 113*, 1281–1292.

Labonte, M. E., Couture, P., Richard, C., Desroches, S. & Lamarche, B. (2013). Impact of dairy products on biomarkers of inflammation: A systematic review of randomized controlled nutritional intervention studies in overweight and obese adults. *American Journal of Clinical Nutrition, 97*, 706–717.

Lai, H. C. & Ney, D. M. (1998). Gastric digestion modifies absorption of butterfat into lymph chylomicrons in rats. *Journal of Nutrition, 128*, 2403–2410.

Lamarche, B. (2008). Review of the effect of dairy products on non-lipid risk factors for cardiovascular disease. *Journal of the American College of Nutrition, 27*, 741s–746s.

Lamothe, S., Corbeil, M. M., Turgeon, S. L. & Britten, M. (2012). Influence of cheese matrix on lipid digestion in a simulated gastro-intestinal environment. *Food & Function, 3*, 724–731.

Lamothe, S., Remillard, N., Tremblay, J. & Britten, M. (2017). Influence of dairy matrices on nutrient release in a simulated gastrointestinal environment. *Food Research International, 92*, 138–146.

Langsted, A. & Nordestgaard, B. G. (2015). Nonfasting lipid profiles: The way of the future. *Clinical Chemistry, 61*, 1123–1125.

Laugerette, F., Vors, C., Geloen, A., Chauvin, M. A., Soulage, C., Lambert-Porcheron, S., Peretti, N., Alligier, M., Burcelin, R., Laville, M., Vidal, H. & Michalski, M. C. (2011). Emulsified lipids increase endotoxemia: Possible role in early postprandial low-grade inflammation. *Journal of Nutritional Biochemistry, 22*, 53–59.

Le Huerou-Luron, I., Lemaire, M. & Blat, S. (2019). Health benefits of dairy lipids and MFGM in infant formula. *Cahiers de Nutrition et de Dietetiqu, 54*, 52–60.

Lecomte, M., Bourlieu, C., Meugnier, E., Penhoat, A., Cheillan, D., Pineau, G., Loizon, E., Trauchessec, M., Claude, M., Menard, O., Geloen, A., Laugerette, F. & Michalski, M. C. (2015). Milk polar lipids affect in vitro digestive lipolysis and postprandial lipid metabolism in mice. *Journal of Nutrition, 145*, 1770–1777.

Lecomte, M., Couedelo, L., Meugnier, E., Plaisancie, P., Letisse, M., Benoit, B., Gabert, L., Penhoat, A., Durand, A., Pineau, G., Joffre, F., Geloen, A., Vaysse, C., Laugerette, F. & Michalski, M. C. (2016). Dietary emulsifiers from milk and soybean differently impact adiposity and inflammation in association with modulation of colonic goblet cells in high-fat fed mice. *Molecular Nutrition & Food Research, 60*, 609–620.

Legrand, P. & Rioux, V. (2015). Specific roles of saturated fatty acids: Beyond epidemiological data. *European Journal of Lipid Science and Technology, 117*, 1489–1499.

Legrand, P., Beauchamp, E., Catheline, D., Pédrono, F. & Rioux, V. (2010). Short chain saturated fatty acids decrease circulating cholesterol and increase tissue pufa content in the rat. *Lipids, 45*, 975–986.

Lemarié, F., Beauchamp, E., Dayot, S., Duby, C., Legrand, P. & Rioux, V. (2015). Dietary caprylic acid (C8:0) does not increase plasma acylated ghrelin but decreases plasma unacylated ghrelin in the rat. *PLoS One, 10*, e0133600.

Lemarié, F., Beauchamp, E., Legrand, P. & Rioux, V. (2016a). Revisiting the metabolism and physiological functions of caprylic acid (C8:0) with special focus on ghrelin octanoylation. *Biochimie, 120*, 40–48.

Lemarié, F., Cavalier, J. F., Garcia, C., Boissel, F., Point, V., Catheline, D., Legrand, P., Carrière, F. & Rioux, V. (2016b). Effect of preduodenal lipase inhibition in suckling rats on dietary octanoic acid (C8:0) gastric absorption and plasma octanoylated ghrelin concentration. *Biochimica et Biophysica Acta, 1861*, 1111–1120.

Lemarié, F., Beauchamp, E., Drouin, G., Legrand, P. & Rioux, V. (2018). Dietary caprylic acid and ghrelin O-acyltransferase activity to modulate octanoylated ghrelin functions: What is new in this nutritional field? Prostag. Leukot. Ess. *Fatty Acids, 135*, 121–127.

Li, Q., Ding, C., Meng, T., Lu, W., Liu, W., Hao, H. & Cao, L. (2017). Butyrate suppresses motility of colorectal cancer cells via deactivating Akt/ERK signaling in histone deacetylase dependent manner. *Journal of Pharmacological Sciences, 135*, 148–155.

Lindquist, S. & Hernell, O. (2010). Lipid digestion and absorption in early life: An update. *Current Opinion in Clinical Nutrition Metabolic Care, 13*, 314–320.

Lonnerdal, B. & Hernell, O. (2016). An opinion on staging of infant formula: A developmental perspective on infant feeding. *Journal of Pediatric Gastroenterology and Nutrition, 62*, 9–21.

Lopez, C. (2010). Lipid domains in the milk fat globule membrane: Specific role of sphingomyelin. *Lipid Technology, 22*, 175–178.

Lopez, C. (2011). Milk fat globules enveloped by their biological membrane: Unique colloidal assemblies with a specific composition and structure. *Current Opinion in Colloid and Interface Science, 16*, 391–404.

Lopez, C., Briard-Bion, V., Menard, O., Rousseau, F., Pradel, P. & Besle, J. M. (2008). Phospholipid, sphingolipid, and fatty acid compositions of the Milk fat globule membrane are modified by diet. *Journal of Agricultural and Food Chemistry, 56*, 5226–5236.

Lopez, C., Cauty, C. & Guyomarc'h, F. (2015). Organization of lipids in milks, infant milk formulas and various dairy products: Role of technological processes and potential impacts. *Dairy Science & Technology, 95*, 863–893.

Lorenzen, J. K., Nielsen, S., Holst, J. J., Tetens, I., Rehfeld, J. F. & Astrup, A. (2007). Effect of dairy calcium or supplementary calcium intake on postprandial fat metabolism, appetite, and subsequent energy intake. *American Journal of Clinical Nutrition, 85*, 678–687.

Lorenzen, J. K., Jensen, S. K. & Astrup, A. (2014). Milk minerals modify the effect of fat intake on serum lipid profile: Results from an animal and a human short-term study. *British Journal of Nutrition, 111*, 1412–1420.

Luna, P., Rodríguez-Pino, V. & de la Fuente, M. A. (2009). Occurrence of C16:1 isomers in milk fats from ewes fed with different dietary lipid supplements. *Food Chemistry, 117*, 248–253.

Maceyka, M. & Spiegel, S. (2014). Sphingolipid metabolites in inflammatory disease. *Nature, 510*, 58–67.

Maly, J., Burianova, I., Vitkova, V., Ticha, E., Navratilova, M., Cermakova, E. & Grp, P. M. S. (2019). Preterm human milk macronutrient concentration is independent of gestational age at birth. *Archives of Disease in Childhood. Fetal and Neonatal Edition, 104*, F50–F56.

Mazzei, J. C., Zhou, H., Brayfield, B. P., Hontecillas, R., Bassaganya-Riera, J. & Schmelz, E. M. (2011). Suppression of intestinal inflammation and inflammation-driven colon cancer in mice by dietary sphingomyelin: Importance of peroxisome proliferator-activated receptor gamma expression. *Journal of Nutritional Biochemistry, 22*, 1160–1171.

Meijer, G. W., van Tol, A., van Berkel, T. J. & Weststrate, J. A. (2001). Effect of dietary elaidic versus vaccenic acid on blood and liver lipids in the hamster. *Atherosclerosis, 157*, 31–40.

Micha, R., King, I. B., Lemaitre, R. N., Rimm, E. B., Sacks, F., Song, X., Siscovick, D. S. & Mozaffarian, D. (2010). Food sources of individual plasma phospholipid trans fatty acid isomers: The cardiovascular health study. *American Journal of Clinical Nutrition, 91*, 883–893.

Michalski, M. C. (2009). Specific molecular and colloidal structures of milk fat affecting lipolysis, absorption and postprandial lipemia. *European Journal of Lipid Science and Technology, 111*, 413–431.

Michalski, M. C. (2013). Lipids and milk fat globules properties in human milk. In S. Zibadi, R. R. Watson & V. R. Prudy (Eds.), *Handbook of dietary and nutritionnel aspects of breast milk* (pp. 310–329). Wagueningen: Wagueningen Academic Pusblishers.

Michalski, M. C. & Januel, C. (2006). Does homogenization affect the human health properties of cow's milk? *Trends in Food Science and Technology, 17*, 423–437.

Michalski, M. C., Cariou, R., Michel, F. & Garnier, C. (2002). Native vs. damaged milk fat globules: Membrane properties affect the viscoelasticity of milk gels. *Journal of Dairy Science, 85*, 2451–2461.

Michalski, M. C., Camier, B., Briard, V., Leconte, N., Gassi, J.-Y., Goudédranche, H., Michel, F. & Fauquant, J. (2004). The size of native milk fat globules affects physico-chemical and functional properties of Emmental cheese. *Lait, 84*, 343–358.

Michalski, M. C., Briard, V., Desage, M. & Geloen, A. (2005). The dispersion state of milk fat influences triglyceride metabolism in the rat-a $^{13}CO_2$ breath test study. *European Journal of Nutrition, 44*, 436–444.

Michalski, M. C., Leconte, N., Briard-Bion, V., Fauquant, J., Maubois, J. L. & Goudedranche, H. (2006a). Microfiltration of raw whole milk to select fractions with different fat globule size distributions: Process optimization and analysis. *Journal of Dairy Science, 89*, 3778–3790.

Michalski, M. C., Soares, A. F., Lopez, C., Leconte, N., Briard, V. & Geloen, A. (2006b). The supramolecular structure of milk fat influences plasma triacylglycerols and fatty acid profile in the rat. *European Journal of Nutrition, 45*, 215–224.

Michalski, M. C., Camier, B., Gassi, J. Y., Briard-Bion, V., Leconte, N., Famelart, M. H. & Lopez, C. (2007). Functionality of smaller vs control native milkfat globules in emmental cheeses manufactured with adapted technologies. *Food Research International, 40*, 191–202.

Michalski, M. C., Rioux, L. E. & Turgeon, S. (2020). Role of the matrix on the digestibility of dairy fat and health consequences. In T. Truong, C. Lopez, S. Prakash & B. Bhandari (Eds.), *Dairy fat products and functionality: Fundamental science and technology*. Springer Nature Switzerland AG 2020.

Milard, M., Laugerette, F., Durand, A., Buisson, C., Meugnier, E., Loizon, E., Louche-Pelissier, C., Sauvinet, V., Garnier, L., Viel, S., Bertrand, K., Joffre,

F., Cheillan, D., Humbert, L., Rainteau, D., Plaisancie, P., Bindels, L. B., Neyrinck, A. M., Delzenne, N. M. & Michalski, M. C. (2019a). Milk polar lipids in a high-fat diet can prevent body weight gain: Modulated abundance of gut bacteria in relation with fecal loss of specific fatty acids. *Molecular Nutrition & Food Research, 63*, e1801078.

Milard, M., Penhoat, A., Durand, A., Buisson, C., Loizon, E., Meugnier, E., Bertrand, K., Joffre, F., Cheillan, D., Garnier, L., Viel, S., Laugerette, F. & Michalski, M. C. (2019b). Acute effects of milk polar lipids on intestinal tight junction expression: Towards an impact of sphingomyelin through the regulation of IL-8 secretion? *Journal of Nutritional Biochemistry, 65*, 128–138.

Mohankumar, S. K., Hanke, D., Siemens, L., Cattini, A., Enns, J., Shen, J., Reaney, M., Zahradka, P. & Taylor, C. G. (2013). Dietary supplementation of *trans*-11-vaccenic acid reduces adipocyte size but neither aggravates nor attenuates obesity-mediated metabolic abnormalities in fa/fa Zucker rats. *British Journal of Nutrition, 109*, 1628–1636.

Mol, N., Zasada, M. & Kwinta, P. (2019). Does type of feeding affect body composition in very low birth weight infants? – a prospective cohort study. *Pediatrics and Neonatology, 60*, 135–140.

Morifuji, M., Oba, C., Ichikawa, S., Ito, K., Kawahata, K., Asami, Y., Ikegami, S., Itoh, H. & Sugawara, T. (2015). A novel mechanism for improvement of dry skin by dietary milk phospholipids: Effect on epidermal covalently bound ceramides and skin inflammation in hairless mice. *Journal of Dermatological Science, 78*, 224–231.

Motouri, M., Matsuyama, H., Yamamura, J., Tanaka, M., Aoe, S., Iwanaga, T. & Kawakami, H. (2003). Milk sphingomyelin accelerates enzymatic and morphological maturation of the intestine in artificially reared rats. *Journal of Pediatric Gastroenterology and Nutrition, 36*, 241–247.

Mozaffarian, D., Rimm, E. B. & Herrington, D. M. (2004). Dietary fats, carbohydrate, and progression of coronary atherosclerosis in postmenopausal women. *American Journal of Clinical Nutrition, 80*, 1175–1184.

Mozaffarian, D., Cao, H., King, I. B., Lemaitre, R. N., Song, X., Siscovick, D. S. & Hotamisligil, G. S. (2010). *Trans*-palmitoleic acid, metabolic risk factors, and new-onset diabetes in U.S. adults: A cohort study. *Annals of Internal Medicine, 153*, 790–799.

Mozaffarian, D., de Oliveira Otto, M. C., Lemaitre, R. N., Fretts, A. M., Hotamisligil, G., Tsai, M. Y., Siscovick, D. S. & Nettleton, J. A. (2013). *Trans*-palmitoleic acid, other dairy fat biomarkers, and incident diabetes: The multi-ethnic study of atherosclerosis (MESA). *American Journal of Clinical Nutrition, 97*, 854–861.

Mu, H. & Porsgaard, T. (2005). The metabolism of structured triacylglycerols. *Progress in Lipid Research, 44*, 430–448.

Muralidharan, J., Papandreou, C., Sala-Vila, A., Rosique-Esteban, N., Fitó, M., Estruch, R., Angel Martínez-González, M., Corella, D., Ros, E., Razquín, C., Castañer, O., Salas-Salvadó, J. & Bulló, M. (2019). Fatty acids composition of blood cell membranes and peripheral inflammation in the PREDIMED study: A cross-sectional analysis. *Nutrients, 11*, 576.

Nestel, P. J. (2008). Effects of dairy fats within different foods on plasma lipids. *Journal of the American College of Nutrition, 27*, 735s–740s.

Nilsson, Å. (1969). The presence of sphingomyelin- and ceramide-cleaving enzymes in the small intestinal tract. *Biochem. Biophys. Acta (BBA) – Lipids Lipid Metabolic, 176*, 339–347.

Nilsson, Å. (2016). Role of sphingolipids in infant gut health and immunity. *Journal of Pediatrics, 173*, S53–S59.

Nilsson, Å. & Duan, R. D. (2006). Absorption and lipoprotein transport of sphingomyelin. *Journal of Lipid Research, 47*, 154–171.

Noh, S. K. & Koo, S. I. (2004). Milk sphingomyelin is more effective than egg sphingomyelin in inhibiting intestinal absorption of cholesterol and fat in rats. *Journal of Nutrition, 134*, 2611–2616.

Nordestgaard, B. G., Benn, M., Schnohr, P. & Tybjaerg-Hansen, A. (2007). Nonfasting triglycerides and risk of myocardial infarction, ischemic heart disease, and death in men and women. *JAMA, 298*, 299–308.

Norris, G. H. & Blesso, C. N. (2017). Dietary and endogenous sphingolipid metabolism in chronic inflammation. *Nutrients, 9*, E1180.

Norris, G. H., Jiang, C., Ryan, J., Porter, C. M. & Blesso, C. N. (2016). Milk sphingomyelin improves lipid metabolism and alters gut microbiota in high fat diet-fed mice. *Journal of Nutritional Biochemistry, 30*, 93–101.

Norris, G. H., Porter, C. M., Jiang, C. & Blesso, C. (2017a). Dietary Milk sphingomyelin reduces systemic inflammation in diet-induced obese mice and inhibits LPS activity in macrophages. *Beverages, 3*, 37.

Norris, G. H., Porter, C. M., Jiang, C., Millar, C. L. & Blesso, C. N. (2017b). Dietary sphingomyelin attenuates hepatic steatosis and adipose tissue inflammation in high-fat-diet-induced obese mice. *Journal of Nutritional Biochemistry, 40*, 36–43.

Norris, G. H., Milard, M., Michalski, M. C. & Blesso, C. N. (2019). Protective properties of milk sphingomyelin against dysfunctional lipid metabolism, gut dysbiosis, and inflammation. *Journal of Nutritional Biochemistry, 73*, 108224.

O'Sullivan, T. A., Hafekost, K., Mitrou, F. & Lawrence, D. (2013). Food sources of saturated fat and the association with mortality: A meta-analysis. *American Journal of Public Health, 103*, e31–e42.

Oba, C., Morifuji, M., Ichikawa, S., Ito, K., Kawahata, K., Yamaji, T., Asami, Y., Itou, H. & Sugawara, T. (2015). Dietary Milk sphingomyelin prevents disruption of skin barrier function in hairless mice after UV-B irradiation. *PLoS One, 10*, e0136377.

Ohlsson, L., Burling, H. & Nilsson, A. (2009). Long term effects on human plasma lipoproteins of a formulation enriched in butter milk polar lipid. *Lipids in Health and Disease, 8*, 44.

Ohlsson, L., Burling, H., Duan, R. D. & Nilsson, A. (2010a). Effects of a sphingolipid-enriched dairy formulation on postprandial lipid concentrations. *European Journal of Clinical Nutrition, 64*, 1344–1349.

Ohlsson, L., Hertervig, E., Jonsson, B. A., Duan, R. D., Nyberg, L., Svernlov, R. & Nilsson, A. (2010b). Sphingolipids in human ileostomy content after meals containing milk sphingomyelin. *American Journal of Clinical Nutrition, 91*, 672–678.

Oosting, A., Kegler, D., Wopereis, H. J., Teller, I. C., van de Heijning, B. J. M., Verkade, H. J. & van der Beek, E. M. (2012). Size and phospholipid coating of lipid droplets in the diet of young mice modify body fat accumulation in adulthood. *Pediatric Research, 72*, 362–369.

Oosting, A., van Vlies, N., Kegler, D., Schipper, L., Abrahamse-Berkeveld, M., Ringler, S., Verkade, H. J. & Van Der Beek, E. M. (2014). Effect of dietary lipid structure in early postnatal life on mouse adipose tissue development and function in adulthood. *British Journal of Nutrition, 111*, 215–226.

Ooyama, K., Kojima, K., Aoyama, T. & Takeuchi, H. (2009). Decrease of food intake in rats after ingestion of medium-chain triacylglycerol. *Journal of Nutritional Science and Vitaminology, 55*, 423–427.

Orbach, R., Mandel, D., Mangel, L., Marom, R. & Lubetzky, R. (2019). The effect of deep freezing on human Milk macronutrients content. *Breastfeeding Medicine, 14*, 172–176.

Ortega-Anaya, J. & Jimenez-Flores, R. (2019). Symposium review: The relevance of bovine milk phospholipids in human nutrition-evidence of the effect on infant gut and brain development. *Journal of Dairy Science, 102*, 2738–2748.

Oshida, K., Shimizu, T., Takase, M., Tamura, Y., Shimizu, T. & Yamashiro, Y. (2003). Effects of dietary sphingomyelin on central nervous system myelination in developing rats. *Pediatric Research, 53*, 589–593.

Owen, C. G., Whincup, P. H., Odoki, K., Gilg, J. A. & Cook, D. G. (2002). Infant feeding and blood cholesterol: A study in adolescents and a systematic review. *Pediatrics, 110*, 597–608.

Palou, M., Pico, C. & Palou, A. (2018). Leptin as a breast milk component for the prevention of obesity. *Nutrition Reviews, 76*, 875–892.

Papamandjaris, A. A., MacDougall, D. E. & Jones, P. J. (1998). Medium chain fatty acid metabolism and energy expenditure: Obesity treatment implications. *Life Sciences, 62*, 1203–1215.

Perret, J. P. (1980). Gastric lipolysis of maternal milk triglycerides, and gastric absorption of medium chain fatty acids in the young rabbit. *Journal of Physiology, Paris, 76*, 159–166.

Pfeuffer, M. & Jaudszus, A. (2016). Pentadecanoic and heptadecanoic acids: Multifaceted odd-chain fatty acids. *Advances in Nutrition, 7*, 730–734.

Poppitt, S. D., McGregor, R. A., Wiessing, K. R., Goyal, V. K., Chitkara, A. J., Gupta, S., Palmano, K., Kuhn-Sherlock, B. & McConnell, M. A. (2014). Bovine complex Milk lipid containing gangliosides for prevention of rotavirus infection and diarrhoea in northern Indian infants. *Journal of Pediatric Gastroenterology and Nutrition, 59*, 167–171.

Praagman, J., Franco, O. H., Ikram, M. A., Soedamah-Muthu, S. S., Engberink, M. F., van Rooij, F. J. A., Hofman, A. & Geleijnse, J. M. (2015). Dairy products and the risk of stroke and coronary heart disease: The Rotterdam study. *European Journal of Nutrition, 54*, 981–990.

Praagman, J., Beulens, J. W., Alssema, M., Zock, P. L., Wanders, A. J., Sluijs, I. & van der Schouw, Y. T. (2016). The association between dietary saturated fatty acids and ischemic heart disease depends on the type and source of fatty acid in the European Prospective investigation into Cancer and nutrition-Netherlands cohort. *American Journal of Clinical Nutrition, 103*, 356–365.

Praagman, J., Vissers, L. E. T., Mulligan, A. A., Laursen, A. S. D., Beulens, J. W. J., van der Schouw, Y. T., Wareham, N. J., Hansen, C. P., Khaw, K. T., Jakobsen, M. U. & Sluijs, I. (2019). Consumption of individual saturated fatty acids and the risk of myocardial infarction in a UK and a Danish cohort. *International Journal of Cardiology, 279*, 18–26.

Pranger, I. G., Joustra, M. L., Corpeleijn, E., Muskiet, F. A. J., Kema, I. P., Oude Elferink, S. J. W. H., Singh-Povel, C. & Bakker, S. J. L. (2018). Fatty acids as biomarkers of total dairy and dairy fat intakes: A systematic review and meta-analysis. *Nutrition Reviews, 77*, 46–63.

Precht, D., Molkentin, J., Destaillats, F. & Wolff, R. L. (2001). Comparative studies on individual isomeric 18:1 acids in cow, goat, and ewe milk fats by low-temperature high-resolution capillary gas-liquid chromatography. *Lipids, 36*, 827–832.

Reiser, R., O'Brien, B. C., Henderson, G. R. & Moore, R. W. (1979). Studies on a possible function for cholesterol in milk. *Nutrition Reports International, 19*, 835–849.

Rioux, V. (2016). Fatty acid acylation of proteins: Specific roles for palmitic, myristic and caprylic acids. *OCL, 23*, D304.

Rioux, V., Catheline, D., Bouriel, M. & Legrand, P. (2005). Dietary myristic acid at physiologically relevant levels increases the tissue content of C20:5 n-3 and C20:3 n-6 in the rat. *Reproduction, Nutrition, Development, 45*, 599–612.

Rioux, V., Catheline, D., Beauchamp, E., Le Bloc'H, J., Pédrono, F. & Legrand, P. (2008). Substitution of dietary oleic acid for myristic acid increases the tissue storage of α-linolenic acid and the concentration of docosahexaenoic acid in the brain, red blood cells and plasma in the rat. *Animal, 2*, 636–644.

Rioux, V., Pédrono, F. & Legrand, P. (2011). Regulation of mammalian desaturases by myristic acid: N-terminal myristoylation and other modulations. *Biochimica et Biophysica Acta, 1811*, 1–8.

Rioux, V., Pedrono, F., Blanchard, H., Duby, C., Boulier-Monthean, N., Bernard, L., Beauchamp, E., Catheline, D. & Legrand, P. (2013). *Trans*-vaccenate is Δ13-

desaturated by FADS3 in rodents. *Journal of Lipid Research, 54*, 3438–3452.
Rosqvist, F., Smedman, A., Lindmark-Mansson, H., Paulsson, M., Petrus, P., Straniero, S., Rudling, M., Dahlman, I. & Riserus, U. (2015). Potential role of milk fat globule membrane in modulating plasma lipoproteins, gene expression, and cholesterol metabolism in humans: A randomized study. *American Journal of Clinical Nutrition, 102*, 20–30.
Sachez Negrette, M., Montenegro, M. A., Catuogno, M. S., Lertora, W. J. & Guanziroli, M. (2007). Decrease of intestinal tumors induced by 1,2-dimethylhydrazine in rats fed with cow milk and buffalo milk. *Biocell, 31*, 391–396.
Sanders, T. A., Filippou, A., Berry, S. E., Baumgartner, S. & Mensink, R. P. (2011). Palmitic acid in the sn-2 position of triacylglycerols acutely influences postprandial lipid metabolism. *American Journal of Clinical Nutrition, 94*, 1433–1441.
Sanggaard, K. M., Holst, J. J., Rehfeld, J. F., Sandstrom, B., Raben, A. & Tholstrup, T. (2004). Different effects of whole milk and a fermented milk with the same fat and lactose content on gastric emptying and postprandial lipaemia, but not on glycaemic response and appetite. *British Journal of Nutrition, 92*, 447–459.
Santora, J. E., Palmquist, D. L. & Roehrig, K. L. (2000). *Trans*-vaccenic acid is desaturated to conjugated linoleic acid in mice. *Journal of Nutrition, 130*, 208–215.
Schipper, L., van Dijk, G., Broersen, L. M., Loos, M., Bartke, N., Scheurink, A. J. & van der Beek, E. M. (2016). A postnatal diet containing phospholipids, processed to yield large, phospholipid-coated lipid droplets, affects specific cognitive Behaviors in healthy male mice. *Journal of Nutrition, 146*, 1155–1161.
Schmelz, E. M., Dillehay, D. L., Webb, S. K., Reiter, A., Adams, J. & Merrill, A. H., Jr. (1996). Sphingomyelin consumption suppresses aberrant colonic crypt foci and increases the proportion of adenomas versus adenocarcinomas in CF1 mice treated with 1,2-dimethylhydrazine: Implications for dietary sphingolipids and colon carcinogenesis. *Cancer Research, 56*, 4936–4941.
Schneeberger, M., Everard, A., Gomez-Valades, A. G., Matamoros, S., Ramirez, S., Delzenne, N. M., Gomis, R., Claret, M. & Cani, P. D. (2015). Akkermansia muciniphila inversely correlates with the onset of inflammation, altered adipose tissue metabolism and metabolic disorders during obesity in mice. *Scientific Reports, 5*, 16643.
Sengupta, S., Muir, J. G. & Gibson, P. R. (2006). Does butyrate protect from colorectal cancer? *Journal of Gastroenterology and Hepatology, 21*, 209–218.
Shani-Levi, C., Alvito, P., Andrés, A., Assunção, R., Barberá, R., Blanquet-Diot, S., Bourlieu, C., Brodkorb, A., Cilla, A., Deglaire, A., Denis, S., Dupont, D., Heredia, A., Karakaya, S., Giosafatto, C. V., Mariniello, L., Martins, C., Ménard, O., El, S. N., Vegarud, G. E., Ulleberg, E. K. & Lesmes, U. (2017). Extending in vitro digestion models to specific human populations: Perspectives, practical tools and bio-relevant information. *Trends in Food Science and Technology, 60*, 52–63.
Shoelson, S. E., Herrero, L. & Naaz, A. (2007). Obesity, inflammation, and insulin resistance. *Gastroenterology, 132*, 2169–2180.
Siri-Tarino, P. W., Sun, Q., Hu, F. B. & Krauss, R. M. (2010). Meta-analysis of prospective cohort studies evaluating the association of saturated fat with cardiovascular disease. *American Journal of Clinical Nutrition, 91*, 535–546.
Small, D. M. (1991). The effects of glyceride structure on absorption and metabolism. *Annual Review of Nutrition, 11*, 413–434.
Smit, L. A., Willett, W. C. & Campos, H. (2010). Trans-fatty acid isomers in adipose tissue have divergent associations with adiposity in humans. *Lipids, 45*, 693–700.
Smolinska, A., Baranska, A., Dallinga, J. W., Mensink, R. P., Baumgartner, S., van de Heijning, B. J. M. & van Schooten, F. J. (2019). Comparing patterns of volatile organic compounds exhaled in breath after consumption of two infant formulae with a different lipid structure: A randomized trial. *Scientific Reports, 9*, 554.
Snow, D. R., Jimenez-Flores, R., Ward, R. E., Cambell, J., Young, M. J., Nemere, I. & Hintze, K. J. (2010). Dietary milk fat globule membrane reduces the incidence of aberrant crypt foci in Fischer-344 rats. *Journal of Agricultural and Food Chemistry, 58*, 2157–2163.
Snow, D. R., Ward, R. E., Olsen, A., Jimenez-Flores, R. & Hintze, K. J. (2011). Membrane-rich milk fat diet provides protection against gastrointestinal leakiness in mice treated with lipopolysaccharide. *Journal of Dairy Science, 94*, 2201–2212.
Soerensen, K. V., Thorning, T. K., Astrup, A., Kristensen, M. & Lorenzen, J. K. (2014). Effect of dairy calcium from cheese and milk on fecal fat excretion, blood lipids, and appetite in young men. *American Journal of Clinical Nutrition, 99*, 984–991.
Spitsberg, V. L. (2005). Invited review: Bovine milk fat globule membrane as a potential nutraceutical. *Journal of Dairy Science, 88*, 2289–2294.
St-Onge, M. P., Bosarge, A., Goree, L. L. & Darnell, B. (2008). Medium chain triglyceride oil consumption as part of a weight loss diet does not lead to an adverse metabolic profile when compared to olive oil. *Journal of the American College of Nutrition, 27*, 547–552.
Strydom, K., Van Niekerk, E. & Dhansay, M. A. (2019). Factors affecting body composition in preterm infants: Assessment techniques and nutritional interventions. *Pediatrics and Neonatology, 60*, 121–128.
Sun, X. & Zhu, M. J. (2018). Butyrate inhibits indices of colorectal carcinogenesis via enhancing α-ketoglutarate-dependent DNA demethylation of mismatch repair genes. *Molecular Nutrition & Food Research, 62*, 1700932.
Sun, Q., Ma, J., Campos, H. & Hu, F. B. (2007). Plasma and erythrocyte biomarkers of dairy fat intake and risk of ischemic heart disease. *American Journal of Clinical Nutrition, 86*, 929–937.

Takeshima, K., Mori, Y., Sakai, F., Haruta-Ono, Y., Kadooka, Y., Yagasaki, K. & Miura, Y. (2014). Ameliorative effect of dietary milk-derived sphingomyelin concentrates on atopic-like dermatitis in NC/Nga mice. *Journal of Functional Foods, 8*, 53–61.

Tholstrup, T., Hoy, C. E., Andersen, L. N., Christensen, R. D. K. & Sandstrom, B. (2004). Does fat in milk, butter and and cholesterol differently? *Journal of the American College of Nutrition, 23*, 169–176.

Thorning, T. K., Raben, A., Bendsen, N. T., Jørgensen, H. H., Kiilerich, P., Ardö, Y., Lorenzen, J. K., Kristiansen, K. & Astrup, A. (2016). Importance of the fat content within the cheese-matrix for blood lipid profile, faecal fat excretion, and gut microbiome in growing pigs. *International Dairy Journal, 61*, 67–75.

Thorning, T. K., Bertram, H. C., Bonjour, J. P., de Groot, L., Dupont, D., Feeney, E., Ipsen, R., Lecerf, J. M., Mackie, A., McKinley, M. C., Michalski, M. C., Remond, D., Riserus, U., Soedamah-Muthu, S. S., Tholstrup, T., Weaver, C., Astrup, A. & Givens, I. (2017). Whole dairy matrix or single nutrients in assessment of health effects: Current evidence and knowledge gaps. *American Journal of Clinical Nutrition, 105*, 1033–1045.

Timby, N., Domellöf, E., Hernell, O., Lönnerdal, B. & Domellöf, M. (2014a). Neurodevelopment, nutrition, and growth until 12 mo of age in infants fed a low-energy, low-protein formula supplemented with bovine milk fat globule membranes: A randomized controlled trial. *American Journal of Clinical Nutrition, 99*, 860–868.

Timby, N., Lönnerdal, B., Hernell, O. & Domellöf, M. (2014b). Cardiovascular risk markers until 12 mo of age in infants fed a formula supplemented with bovine milk fat globule membranes. *Pediatric Research, 76*, 394–400.

Timby, N., Hernell, O., Vaarala, O., Melin, M., Lönnerdal, B. & Domellöf, M. (2015). Infections in infants fed formula supplemented with bovine Milk fat globule membranes. *Journal of Pediatric Gastroenterology and Nutrition, 60*, 384–389.

Tong, X., Dong, J. Y., Wu, Z. W., Li, W. & Qin, L. Q. (2011). Dairy consumption and risk of type 2 diabetes mellitus: A meta-analysis of cohort studies. *European Journal of Clinical Nutrition, 65*, 1027–1031.

Tsuji, H., Kasai, M., Takeuchi, H., Nakamura, M., Okazaki, M. & Kondo, K. (2001). Dietary medium-chain triacylglycerols suppress accumulation of body fat in a double-blind, controlled trial in healthy men and women. *Journal of Nutrition, 131*, 2853–2859.

Tunick, M. H., Ren, D. X., Van Hekken, D. L., Bonnaillie, L., Paul, M., Kwoczak, R. & Tomasula, P. M. (2016). Effect of heat and homogenization on in vitro digestion of milk. *Journal of Dairy Science, 99*, 4124–4139.

Turpeinen, A. M., Mutanen, M., Aro, A., Salminen, I., Basu, S., Palmquist, D. L. & Griinari, J. M. (2002). Bioconversion of vaccenic acid to conjugated linoleic acid in humans. *American Journal of Clinical Nutrition, 76*, 504–510.

Tyburczy, C., Major, C., Lock, A. L., Destaillats, F., Lawrence, P., Brenna, J. T., Salter, A. M. & Bauman, D. E. (2009). Individual *trans* octadecenoic acids and partially hydrogenated vegetable oil differentially affect hepatic lipid and lipoprotein metabolism in golden Syrian hamsters. *Journal of Nutrition, 139*, 257–263.

Valeille, K., Ferezou, J., Parquet, M., Amsler, G., Gripois, D., Quignard-Boulange, A. & Martin, J. C. (2006). The natural concentration of the conjugated linoleic acid, cis-9,trans-11, in milk fat has antiatherogenic effects in hyperlipidemic hamsters. *Journal of Nutrition, 136*, 1305–1310.

Van Hekken, D. L., Tunick, M. H., Ren, D. X. & Tomasula, P. M. (2017). Comparing the effect of homogenization and heat processing on the properties and in vitro digestion of milk from organic and conventional dairy herds. *Journal of Dairy Science, 100*, 6042–6052.

Vass, R. A., Kemeny, A., Dergez, T., Ertl, T., Reglodi, D., Jungling, A. & Tamas, A. (2019). Distribution of bioactive factors in human milk samples. *International Breastfeeding Journal, 14*, 10.

Veereman-Wauters, G., Staelens, S., Rombaut, R., Dewettinck, K., Deboutte, D., Brummer, R. J., Boone, M. & Le Ruyet, P. (2012). Milk fat globule membrane (INPULSE) enriched formula milk decreases febrile episodes and may improve behavioral regulation in young children. *Nutrition, 28*, 749–752.

Vesper, H., Schmelz, E. M., Nikolova-Karakashian, M. N., Dillehay, D. L., Lynch, D. V. & Merrill, A. H., Jr. (1999). Sphingolipids in food and the emerging importance of sphingolipids to nutrition. *Journal of Nutrition, 129*, 1239–1250.

von Schacky, C., Passow, A. & Kiefl, R. (2017). Trans-fatty acid levels in erythrocytes in Europe. *European Journal of Nutrition, 56*, 1719–1723.

Vors, C., Pineau, G., Gabert, L., Drai, J., Louche-Pélissier, C., Defoort, C., Lairon, D., Désage, M., Danthine, S., Lambert Porcheron, S., Vidal, H., Laville, M. & Michalski, M.-C. (2013). Modulating absorption and postprandial handling of dietary fatty acids by structuring fat in the meal: A randomized crossover clinical trial. *American Journal of Clinical Nutrition, 97*, 23–36.

Vors, C., Nazare, J. A., Michalski, M. C. & Laville, M. (2014). Intérêt de la phase postprandiale pour la santé de l'Homme. *Obésité, 9*, 31–41.

Vors, C., Lecomte, M. & Michalski, M. C. (2016). Impact of lipid emulsified structure on the metabolic fate of dietary fatty acids. *Cahiers de Nutrition et de Dietetique, 51*, 238–247.

Vors, C., Joumard-Cubizolles, L., Lecomte, M., Combe, E., Ouchchane, L., Drai, J., Raynal, K., Joffre, F., Meiller, L., Le Barz, M., Gaborit, P., Caille, A., Sothier, M., Domingues-Faria, C., Blot, A., Wauquier, A., Blond, E., Sauvinet, V., Gesan-Guiziou, G., Bodin, J. P., Moulin, P., Cheillan, D., Vidal, H., Morio, B., Cotte, E., Morel-Laporte, F., Laville, M.,

Bernalier-Donadille, A., Lambert-Porcheron, S., Malpuech-Brugere, C. & Michalski, M. C. (2020). Milk polar lipids reduce lipid cardiovascular risk factors in overweight postmenopausal women: Towards a gut sphingomyelin-cholesterol interplay. *Gut, 69*, 487–501.

Wanders, R. J. A., Komen, J. & Ferdinandusse, S. (2011). Phytanic acid metabolism in health and disease. *Biochimica et Biophysica Acta, 1811*, 498–507.

Wang, Y., Lu, J., Ruth, M. R., Goruk, S. D., Reaney, M. J., Glimm, D. R., Vine, D. F., Field, C. J. & Proctor, S. D. (2008). *Trans*-11 vaccenic acid dietary supplementation induces hypolipidemic effects in JCR:LA-cp rats. *Journal of Nutrition, 138*, 2117–2122.

Wang, Y., Jacome-Sosa, M. M., Ruth, M. R., Goruk, S. D., Reaney, M. J., Glimm, D. R., Wright, D. C., Vine, D. F., Field, C. J. & Proctor, S. D. (2009). *Trans*-11 vaccenic acid reduces hepatic lipogenesis and chylomicron secretion in JCR:LA-cp rats. *Journal of Nutrition, 139*, 2049–2054.

Wang, Y., Jacome-Sosa, M. M., Ruth, M. R., Lu, Y., Shen, J., Reaney, M. J., Scott, S. L., Dugan, M. E. R., Anderson, H. D., Field, C. J., Proctor, S. D. & Vine, D. F. (2012). The intestinal bioavailability of vaccenic acid and activation of peroxisome proliferator-activated receptor-α and -γ in a rodent model of dyslipidemia and the metabolic syndrome. *Molecular Nutrition & Food Research, 56*, 1234–1246.

Wang, Q., Liu, C., Uchida, A., Chuang, J. C., Walker, A., Liu, T., Osborne-Lawrence, S., Mason, B. L., Mosher, C., Berglund, E. D., Elmquist, J. K. & Zigman, J. M. (2014). Arcuate AgRP neurons mediate orexigenic and glucoregulatory actions of ghrelin. *Molecular Metabolism, 3*, 64–72.

Wang, X., Gupta, J., Kerslake, M., Rayat, G., Proctor, S. D. & Chan, C. B. (2016). *Trans*-11 vaccenic acid improves insulin secretion in models of type 2 diabetes *in vivo* and *in vitro*. *Molecular Nutrition & Food Research, 60*, 846–857.

Wat, E., Tandy, S., Kapera, E., Kamili, A., Chung, R. W., Brown, A., Rowney, M. & Cohn, J. S. (2009). Dietary phospholipid-rich dairy milk extract reduces hepatomegaly, hepatic steatosis and hyperlipidemia in mice fed a high-fat diet. *Atherosclerosis, 205*, 144–150.

Yakoob, M. Y., Shi, P., Willett, W. C., Rexrode, K. M., Campos, H., Orav, E. J., Hu, F. B. & Mozaffarian, D. (2016). Circulating biomarkers of dairy fat and risk of incident diabetes mellitus among men and women in the United States in two large prospective cohorts. *Circulation, 133*, 1645–1654.

Yehuda, S., Rabinovitz, S., Mostofsky, DI. (2005). Essential fatty acids and the brain: From infancy to aging, SPARK Workshop 2004, Neurobilogy of aging, 26: S98–S102, Bar Harbor, ME. PubMed 16226347. https://doi.org/10.1016/j.neurobiolaging.2005.09.013

Yoo, W., Gjuka, D., Stevenson, H. L., Song, X., Shen, H., Yoo, S. Y., Wang, J., Fallon, M., Ioannou, G. N., Harrison, S. A. & Beretta, L. (2017). Fatty acids in non-alcoholic steatohepatitis: Focus on pentadecanoic acid. *PLoS One, 12*, e0189965.

You, Y. Q., Ling, P. R., Qu, J. Z. & Bistrian, B. R. (2008). Effects of medium-chain triglycerides, long-chain triglycerides, or 2-monododecanoin on fatty acid composition in the portal vein, intestinal lymph, and systemic circulation in rats. *Journal of Parenteral and Enteral Nutrition, 32*, 169–175.

Zavaleta, N., Kvistgaard, A. S., Graverholt, G., Respicio, G., Guija, H., Valencia, N. & Lonnerdal, B. (2011). Efficacy of an MFGM-enriched complementary food in diarrhea, anemia, and micronutrient status in infants. *Journal of Pediatric Gastroenterology and Nutrition, 53*, 561–568.

Zhang, Q., Cundiff, J. K., Maria, S. D., McMahon, R. J., Wickham, M. S. J., Faulks, R. M. & van Tol, E. A. F. (2014). Differential digestion of human Milk proteins in a simulated stomach model. *Journal of Proteome Research, 13*, 1055–1064.

Zhou, A. L., Hintze, K. J., Jimenez-Flores, R. & Ward, R. E. (2012). Dietary fat composition influences tissue lipid profile and gene expression in Fischer-344 rats. *Lipids, 47*, 1119–1130.

Zomer, A. W., van Der Burg, B., Jansen, G. A., Wanders, R. J., Poll-The, B. T. & van Der Saag, P. T. (2000). Pristanic acid and phytanic acid: Naturally occurring ligands for the nuclear receptor peroxisome proliferator-activated receptor alpha. *Journal of Lipid Research, 41*, 1801–1807.

11 Stability and Spoilage of Lipids in Milk and Dairy Products

H. C. Deeth

11.1 Introduction

The stability and spoilage of lipids in milk and dairy products relate to the susceptibility of the lipids to two types of degradation: hydrolytic or lipolytic and oxidative. Each type of degradation results in flavoursome products which, in excess, cause hydrolytic rancidity and oxidative rancidity, respectively. Although sometimes mistakenly considered to be similar or the same, presumably because of the use of the common term "rancidity", they are distinctly different in both mechanism and effect. Both lipolysis and oxidation have been, and still are, significant issues for the dairy industry.

Although the basics of both phenomena remain constant, the focus of each has shifted over time. Lipolysis has become more of an issue since the widespread introduction of mechanical handling of milk. In oxidation, the focus used to be on copper-induced oxidation, but this has now greatly diminished because stainless steel has replaced copper-containing "dairy" metals; light-induced oxidation is now more prevalent because of the retail display of milk packaged in clear plastic containers in cabinets with fluorescent lighting. Another development has been in animal genetics which has recently led to researchers finding genetic associations with the susceptibility of milk to both lipolysis and oxidation.

A major difference between the mechanisms of lipolysis and oxidation is that lipolysis is almost entirely caused by enzymes (lipases) while oxidation is almost entirely non-enzymic, although enzymes have been implicated in some cases of oxidation. Consequently, there appears to be little overlap of the susceptibilities of milk and dairy products to lipolysis and oxidation or of ways of combatting the degradations. One point in common, however, is that feeding and breeding can influence the susceptibilities of raw milk to both phenomena. A further connection between lipolysis and oxidation is that the free fatty acids (FFAs) produced by lipolysis are more susceptible to oxidation than esterified fatty acids (Frega *et al.* 1999).

11.2 Lipolysis

11.2.1 Background

Lipolysis is the hydrolysis of lipids and, in milk and dairy products, is catalysed by enzymes, lipases. The focus here is on lipolysis of triglycerides (TGs), the major lipid type in milk fat, but lipolysis of phospholipids can occur through the action of phospholipases. The products of lipolysis of TGs are FFAs and the partial glycerides,

H. C. Deeth (✉)
School of Agriculture and Food Sciences, University of Queensland, Brisbane, QLD, Australia
e-mail: h.deeth@uq.edu.au

P. L. H. McSweeney et al. (eds.), *Advanced Dairy Chemistry, Volume 2*,
https://doi.org/10.1007/978-3-030-48686-0_11

di- and monoglycerides. Their significance in the dairy industry is that the FFAs, particularly the short-chain acids, butyric (C4), caprylic (C6) and caproic (C8), are very flavoursome and, in excess, impart an unpleasant flavour to milk and milk products. The flavour is variously described as rancid, unclean, astringent, bitter and butyric. The partial glycerides do not alter the flavour but have surface-active properties which reduce the foaming capacity of milk, an unwelcome effect when the milk is used for making cappuccino or latte coffee.

An understanding of lipolysis in milk and dairy products requires an understanding of the lipases involved. Two main types of lipase are responsible: the lipases which occur naturally in the milk as it leaves the animal and bacterial lipases produced by bacteria which contaminate and grow in the milk. Normal mature milk of cows and goats contains only one lipase, a lipoprotein lipase (LPL), but the milk of humans contains both LPL and a bile salt-stimulated lipase. In general, LPL is the only lipase which causes lipolysis in raw cows' or goats' milk, while bacterial lipases are responsible for any lipolysis which occurs in heat-treated milks and dairy products made from heated milk.

Lipases and lipolysis in milk and milk products have been reviewed by Deeth and FitzGerald (1983, 1995, 2006), Weihrauch (1988), Olivecrona *et al.* (2003) and Deeth (2006, 2011, 2015).

11.2.2 Raw Milk

11.2.2.1 Lipoprotein Lipase

Since the 1970s, it has been accepted that the only lipase in normal cows' and goats' milk is LPL (Olivecrona *et al.* 1975, 2003). Before that, several reports suggested the presence of multiple enzymes. These reports were reviewed by Schwartz and Parks (1974). Small amounts of other lipases appear to be present in abnormal milk, colostrum (Driessen 1976) and mastitic milk (Azzara and Dimick 1985).

LPL originates in the mammary gland where it is involved in the lipolysis of TGs of circulating lipoproteins to produce components necessary for synthesis of milk TGs in the mammary gland. The reason for LPL being in milk is unclear as it does not take part in digestion of milk by the newborn because it is inactivated by the acidic conditions of the stomach. It therefore appears to "leak" into the milk during secretion from the mammary gland.

As the name implies, LPL is activated by lipoproteins. This can be demonstrated by adding blood serum containing lipoproteins to raw milk. This causes the lipase to act on the TGs and produce FFAs and partial glycerides. Without these lipoprotein co-factors, the fat globule membrane acts as a barrier to prevent such lipolysis. This is fortunate as all raw milk contains sufficient LPL to produce a level of FFAs which would render the milk undrinkable in a very short time, if the conditions were optimal.

The optimum pH for LPL activity against long-chain TGs in the presence of co-factors is ~8.0, but in the absence of co-factors, the optimum pH is 8.8–9.0. However, the LPL activity against unactivated substrates is much lower than in the presence of co-factors. LPL is also active against short-chain TGs such as tributyrin, but in this case activating co-factors are not required.

The FFAs produced during lipolysis tend to inhibit the lipolysis, and the rate of reaction slows down with time. This product inhibition is presumably due to the shielding of the substrate from the LPL by the FFAs. Because of this, a fatty acid acceptor such as bovine serum albumin is normally included in LPL assays.

Most of the LPL in cows' milk is normally associated with the casein fraction although small amounts are in soluble form in the milk serum and some is associated with the milk fat globule membrane (MFGM). This differs from the situation in goats' milk where about half of the LPL is associated with the MFGM (Chilliard *et al.* 1984). LPL is mostly attached to casein by electrostatic bonding through negative charges interacting with positive charges on caseins such as κ-casein and can be dissociated with salt or the highly negatively charged heparin.

LPL is a rather unstable enzyme, being inactivated by acid, UV light, oxidizing agents and heat. High-temperature, short-time (HTST) pasteurization at 72 °C for 15 s almost completely inactivates LPL. Therefore, all milk heat-treated at HTST or more severe conditions, as well as products made from such heated milk, does not contain active LPL. As indicated in the discussion of lipolysis in milk and dairy products below, only lipolysis in raw milk can be due to LPL.

LPL does not exhibit fatty acid specificity but has a strong stereospecificity such that it hydrolyses fatty acids from only the *sn*-1 and *sn*-3 positions of TGs. Thus, the first products formed from TGs are FFAs and *sn*-1,2- and *sn*-2,3-diglycerides, and then these diglycerides are hydrolysed to FFAs and 2-monoglycerides. LPL does not hydrolyse 2-monoglycerides unless they are first isomerised to *sn*-1- or *sn*-3-monoglycerides. LPL can also hydrolyse fatty acids from the *sn*-1-position of phospholipids such as phosphatidylcholine and phosphatidylethanolamine, thus acting as a phospholipase A_1.

When LPL acts in milk and produces FFAs at ~1.5 mmol/100 g fat, some consumers can notice a flavour change, but when the level rises to ~2.0 mmol/100 g fat, most consumers can detect an unpleasant flavour. This is consistent with the report by Wiking *et al.* (2017) that the sensory thresholds for rancid flavour in farm bulk milk was 1.95 mmol/100 g fat. Consumers vary considerably in their ability to detect lipolytic flavour with some being very sensitive to even low lipolysis levels while others cannot detect quite high levels. The foaming capacity of milk also declines with an increase in FFAs; milk with a FFA level of 2.0 mmol/100 g fat will normally have poor foaming capacity. Fortunately, heat treatment (e.g. pasteurization) and homogenization considerably improve the foaming capacity of milk (Deeth and Smith 1983).

Several different methods are used for determining FFA levels (see Section 11.2.6), and these give different values for the same milk. The threshold values given above refer to acid degree values (ADV) determined by the BDI (Bureau of Dairy Industries) method of Thomas *et al.* (1955), but similar numerical levels apply when FFA levels are obtained by solvent extraction followed by alkali titration (Deeth *et al.* 1975). By contrast, values obtained by some laboratories using the copper soap method of Shipe *et al.* (1980) are much lower than those obtained by these methods, with initial values for milk as it leaves the cow of <0.15 mmol/kg (Fromm and Boor 2004; Santos *et al.* 2003a, b; Vanbergue *et al.* 2017) compared with values obtained by extraction-titration methods of 0.5–1.0 mmol/L or mmol/100 g fat. However, the FFA levels obtained by some other laboratories with the copper soap method are comparable to those from extraction-titration methods (e.g. Barbano *et al.* 1983). Because of the variability in consumers' sensory perceptions and of the different methods used for measuring FFA levels and performing sensory evaluation, threshold FFA levels need to be interpreted with care.

11.2.2.2 Spontaneous Lipolysis

The milk of some cows undergoes what is known as spontaneous lipolysis. This milk requires only cooling to <10 °C soon after milking to spontaneously lipolyse during refrigerated storage. Most of the lipolysis occurs in the first few hours of storage, but usually the milk is left for ~12 h to attain close to its maximum FFA level in order to assess its susceptibility to spontaneous lipolysis. The FFA of highly susceptible milk can reach >5 mmol/L; such milks taste very rancid and leave a feathery feeling in the mouth. Curiously, some cows never produce spontaneously lipolytic milk; some produce it sometimes and others always produce it. The incidence varies considerably from herd to herd. This variation has led some researchers to propose a heritability factor since anecdotal evidence indicates the milk of the progeny of certain AI bulls is more prone to spontaneous lipolysis than the progeny of other sires. The involvement of genetics has been demonstrated by Vanbergue *et al.* (2016) who observed significantly higher levels of spontaneous lipolysis in the milk of the KK DGAT1genotype than in the milk of either the KA or AA genotype. DGAT, diacylglycerol acyltransferase, catalyses the transfer of an acyl group from an acyl-CoA to a diglyceride in the final stage of milk triglyceride synthesis. DGAT1 is one of two genes encoding DGAT which have been cloned (Gurr *et al.* 2016). A genetic involvement has also been reported in goats where more spontaneous lipolysis is observed in the milk of goats of the CSN1S1-FF genotype than in milk from goats of the AA genotype; the FF genotype is associated with a low level of α_{S1}-casein in milk, while the AA genotype is associated with a high level of α_{S1}-casein (Chilliard *et al.* 2003, 2013, 2014).

Two other physiological factors increase the occurrence of spontaneous lipolysis: poor feed quality and late lactation. The reasons for this are not entirely clear, but one aspect of late lactation milk could be a weakness in the MFGM, facilitating access by the LPL.

The reasons why some milks spontaneously lipolyse and others do not have been investigated by several researchers. The level of LPL activity in milk has been shown not to be important, as all milk contains an abundance of enzyme to cause extensive lipolysis under the right conditions. The integrity of the MFGM has been shown to be significant in some cases, but the major factor distinguishing between the two types of milk is the balance between activating and inhibiting factors (Deeth and Fitz-Gerald 1975a; Sundheim and Bengtsson-Olivecrona 1987; Cartier and Chilliard 1990). A fact that led to this conclusion is that admixing a normal milk, which does not lipolyse spontaneously, with a "spontaneous" milk soon after milking results in a lower FFA developed during refrigerated storage than if they were combined after storage. This strongly suggests the presence of an inhibiting factor (or factors) in normal milk. The identity of the inhibitor(s) has not been definitively determined, but proteose peptones seem to be probable candidates (Danthine and Blecker 2014; Vanbergue *et al.* 2018). Proteose peptone fraction 3 has been shown to inhibit spontaneous lipolysis (Anderson 1981; Cartier *et al.* 1990). A glycoprotein in the MFGM which is antigenically similar to proteose peptone 3 may also act as an inhibitor (Shimizu *et al.* 1982). Vanbergue *et al.* (2018) have recently suggested that proteose-peptone fraction 5 may be an inhibitor of spontaneous lipolysis.

Milk contains lipoproteins which are immunologically cross-reactive with bovine blood lipoproteins and may be the activators in spontaneous milk (Castberg and Solberg 1974; Anderson 1979). Addition of blood lipoproteins to milk results in relocation of some of the LPL to the MFGM (Bachman and Wilcox 1990). The same occurs in spontaneous milk and it appears to be a prerequisite for lipolysis to occur. This is supported by the fact that the LPL activity associated with the MFGM correlates well with the extent of spontaneous lipolysis (Sundheim and Bengtsson-Olivecrona 1985).

11.2.2.3 Induced Lipolysis

Induced lipolysis usually involves some form of disruption to the MFGM which allows access of LPL to the TG core of the fat globule. Another form of induced lipolysis is through a temperature manipulation process, as discussed below.

Induced lipolysis became significant with the introduction of mechanical milking machines and other mechanical forms of handling milk at the farm and at the factory. In investigations of the milk from farm supplies, it has been found that through either poor design or poor maintenance, milk is subjected to physical stresses which result in damage to the MFGM and development of high levels of FFAs. One design which has been shown to give rise to induced lipolysis is an automated milking system (AMS) (De Marchi *et al.* 2017; Wiking *et al.* 2019). The authors suggested the high frequency of milking associated with this system may also have contributed to the elevated lipolysis level observed. This is supported by a previous report that the high milking frequency associated with AMS was a more important factor than AMS *per se* (Abeni *et al.* 2005).

The root cause of induced lipolysis on-farm is usually mixing air with the warm milk causing foaming of the milk. Formation of the foam exerts high shear forces on fat globules adjacent to the air-liquid interface of the foam and disrupts the MFGM. This can occur if air leaks into the milking machine cluster and causes surging

Table 11.1. Acid degree value (ADV) of milk concentrated by reverse osmosis using three different pressure regulation methods

Concentration factor	ADV of milk in	ADV of milk out
Pressure regulated by high-pressure throttle valve		
1.05	0.88	1.32
1.25	0.72	3.21
1.43	0.92	5.86
Pressure regulated by diaphragm valve		
1.02	1.01	1.06
1.28	1.01	3.76
1.32	1.01	6.02
Pressure regulated by reduction of the diameter of retentate exit pipe		
1.18	0.92	1.18
1.27	0.85	1.18
1.33	0.81	1.05

Data from Barbano et al. (1983).

of milk in the pipe leading to the bulk vat. High riser pipes are often the cause of the surging with foaming. A similar effect can occur where raw milk is pumped, particularly if air can enter the pump through a faulty seal. Pumping itself can also induce lipolysis if it involves severe shear forces such as those involved in pumping long distances and around bends in the piping.

The dairy process which causes the most damage to the MFGM is homogenization. To obviate the risk of lipolysis resulting from homogenization, milk is pasteurized either before or immediately after homogenization. If pasteurization in this manner did not occur, the milk would become rancid in a short time since homogenization creates a fine milk fat emulsion containing a large lipid surface area which is an ideal lipase substrate. A related effect is where (pasteurized) homogenized milk is mixed with raw milk. This also causes rapid lipolysis as the raw milk contains active lipase and the homogenized milk contains the ideal lipase substrate. This mixing should never occur but can occur if operatives are unaware of the consequences of doing it.

Another process which can induce considerable lipolysis is membrane processing. This is exemplified in the work of Barbano *et al.* (1983) in which sub-pasteurized (thermized) milk, heated at 64 °C for 21 s, was concentrated by reverse osmosis. Considerable increases in FFA resulted as the concentration factor was increased when a throttle valve or diaphragm was used for pressure regulation but little or no increase occurred when the pressure was regulated by reduction in the diameter of the exit pipe (Table 11.1). The sub-pasteurization conditions were used because it is a common heat treatment given to raw milk before *Cheddar* cheesemaking in New York state. When the milk was pasteurized at 72 °C for 15 s before being concentrated by reverse osmosis, no lipolysis occurred, due to inactivation of the native milk LPL.

High-pressure homogenization (HPH), a developing non-thermal technology for inactivating bacteria and other applications (Deeth *et al.* 2013), operates at pressures of up to 400 MPa (cf up to 40 MPa in normal homogenization used in the dairy industry). When used on raw milk at around room temperature, it induces considerable lipolysis (Lanciotti *et al.* 2006; Serra *et al.* 2008), but this can be avoided if the temperature reached during the treatment is sufficient to completely inactivate the native LPL. (Due to the effects of turbulence, cavitation, shear stress and impact velocity, HPH raises the temperature of the liquid by 17–24 °C per 100 MPa of pressure.) This was demonstrated by Serra *et al.* (2008) who observed considerable lipolysis when HPH was used at 200 MPa at 30 °C, but when used at 300 MPa at 30 °C, no lipolysis occurred. The outlet temperatures were 73.6 and 79.6 °C, respectively, but with a residence time of <0.7 s, the former being insufficient to completely inactivate the LPL.

Laboratory experimentation has demonstrated the importance of several factors which determine the extent of induced lipolysis. Two of the most important are temperature and intensity of the physical action inducing the lipolysis (Deeth and Fitz-Gerald 1977, 1978). Figure 11.1 illustrates these effects. At low-energy intensity of treatment such as low-speed mixing in partly filled containers or centrifugal pumping, both with air incorporation and bubbling of gas through milk, the FFA developed on storage follows curve (i), which shows maxima at agitation

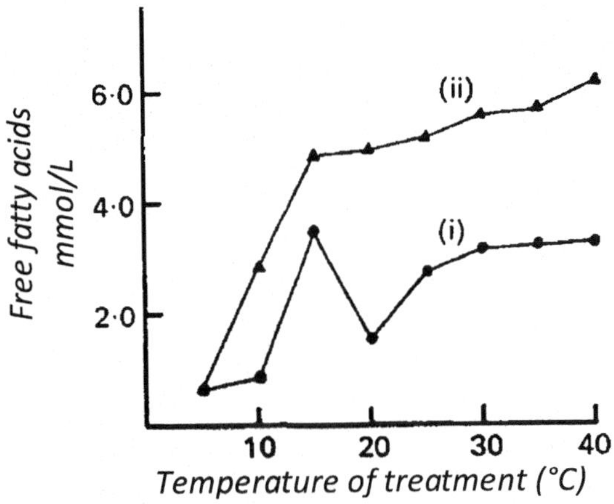

Figure 11.1. Effect of temperature and intensity of mechanical action on induced lipolysis in raw milk. Typical conditions to produce curve of type (i): agitation with Sorvall Omnimixer at 5000 rev/min for 20 s, 100 mL milk, or centrifugal pump, 10 L milk, recirculating for 2 min with incorporation of air; type (ii): Sorvall Omnimixer, 16,000 rev/min for 20 s, 100 mL milk, or single-stage homogenizer at 17000 kPa. (From Deeth and Fitz-Gerald 1978).

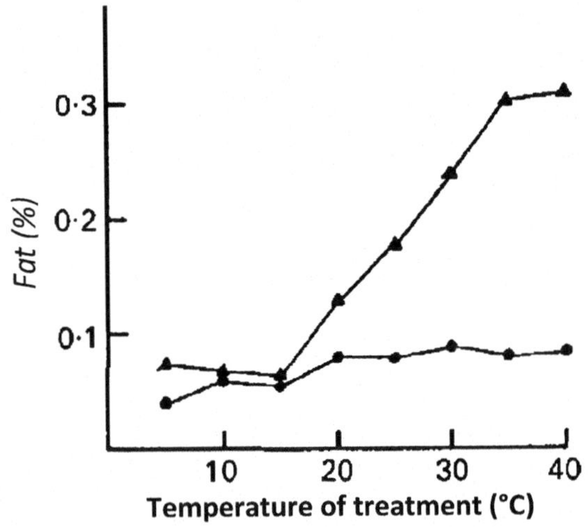

Figure 11.2. Fat in skimmed milk from raw milk mechanically treated as in caption to Figure 11.1.

temperatures of 12–15 °C and ~ 35–40 °C and a distinct minimum at ~20–30 °C. However, when treatments involving high-energy input are used, such as homogenization and high-speed mixing, the resulting FFA with increasing temperature of treatment shows an upward-sloping curve (curve (ii) in Figure 11.1), without maxima or minimum. No minimum occurs when milk with a fat content of less than ~3% fat is treated, and hence the minimum is related to the fat at those temperatures. Interestingly, when milk is foamed over this temperature range, a minimum foamability is observed at a similar temperature to the induced lipolysis minimum in curve (i) in Figure 11.1 (Kamath *et al.* 2008). The minimum is attributable to the milk fat at those temperatures containing a mixture of solid and liquid fats, whereas at lower or higher temperatures, the fat is predominantly solid or liquid, respectively.

Under low-energy treatment conditions, aggregation and coalescence of the fat globules are favoured, while under high-energy treatment conditions, homogenization predominates. As coalescence of fat globules aids separation of the fat by centrifugation, and homogenization reduces creaming and separation, the amount of fat in the skimmed milk from centrifuged mechanically treated milk is an indication of extent of homogenization. This is demonstrated in Figure 11.2 for the milks treated as in Figure 11.1. At all temperatures under low-energy conditions, little homogenization occurred, while in the high-energy-treated milk, homogenization increased almost linearly at treatment temperatures >15 °C (Deeth and Fitz-Gerald 1978).

Wiking *et al.* (2005) reported that pumping milk for 450 s at 31 °C caused more lipolysis than pumping at 4 °C. However, after the milk was incubated at 4° for 60 min, lipolysis produced by pumping at this temperature was increased. They attributed the effect of the incubation to the change in the polymorphic crystal forms of the fat or to greater attachment of LPL to the milk fat globule. This supports the hypothesis that the amount of induced lipolysis in milk is influenced by the state of the fat and the amount of LPL associated with the fat globule.

The amount of induced lipolysis is higher in individual milks susceptible to spontaneous lipolysis than in normal milks (Deeth and Fitz-Gerald 1977). The reason for this is probably the greater association of LPL with the milk fat globule which positions the enzyme favourably for interaction with the fat from disrupted fat globules during mechanical treatments. Interestingly, an important effect of lipolysis induced by mechanically treating raw milk is a redistribution of LPL to the fat globules from the skim phase, a characteristic of spontaneous lipolysis. The amount of lipase transferred to the fat phase can represent over half of the initial lipase present in the milk. The age and the temperature of the milk and the duration of treatment all influence the amount of lipase transferred. The conditions under which maximal transfer of LPL to the cream occurs are fresh milk at 5–10 °C or at ~37 °C and stored milk at 5–10 °C. In practical terms, these temperatures correspond to the temperature of milk when it is most likely to be subjected to induced lipolysis during production and processing, i.e. on the farm during and immediately after milking, and in transit to, and during handling in, the factory. Table 11.2 illustrates the effect of low-energy mechanical agitation at 5, 10 or 15 °C on lipolysis in milk and cream derived from it, as well as the lipase activity transferred to the cream. It shows how mechanical treatment can affect the FFA content of products such as cream and butter; agitation, pumping and other treatments at 5 °C may cause less lipolysis in milk than treatments at 10 or 15 °C but more lipolysis and a higher lipase activity in the cream fraction. Such lipolysis and enzyme activity in the cream are significant when cold milk, which has been subjected to mechanical treatments such as pumping and agitating, is separated and the cream stored for some time before pasteurization. Considerable lipolysis may result in this cream because of the enhanced lipase activity and cause flavour defects in the pasteurized cream and in butter manufactured from it. Furthermore, the transferred enzyme bound to the milk fat globule membrane has enhanced heat stability which may be relevant to the flavour of butter made from pasteurized lipolyzed cream.

As mentioned above, a form of temperature manipulation can induce lipolysis in raw milk which has been cooled. The manipulation involves heating cold raw milk or cream (at ≤5 °C) to 25–35 °C and then cooling to <10 °C. Lipolysis proceeds during cold storage in much the same way as it does in spontaneous milk. Milks vary in their susceptibility to temperature activation, and it has been shown that the MFGM is essential in the process. Milk susceptible to spontaneous lipolysis is more susceptible to lipolysis by temperature manipulation than is normal milk. As with spontaneous lipolysis, lipase becomes attached to the MFGM. However, it differs from spontaneous lipolysis in that the lipase attached to the MFGM is dislodged when the milk is rewarmed; in spontaneous lipolysis, the enzyme remains attached to the MFGM (Claypool 1965; Kon and Saito 1997).

11.2.3 Pasteurized and Extended-Shelf Life (ESL) Milk

Since milk LPL is inactivated by HTST or more severe thermal pasteurization treatments, high levels of FFAs in pasteurized and ESL milk arise from either LPL action in the raw milk or the action by bacterial lipases in the finished product. Ways in which the former can occur have been discussed in the previous sections. Bacterial lipases can cause lipolysis in pasteurized and ESL milk if present in the raw milk at pasteurization and are heat-resistant (most are heat-resistant), or bacteria producing the lipase grow in the heat-treated milk during storage. In the case of pasteurized milk, which has a shelf life

Table 11.2. Free fatty acids (FFAs) in milk and cream and lipase activity in cream following agitation of raw milk in a Waring blender at ~12,000 rpm for 15 s

Agitation temperature	Milk FFA (mmol/100 g fat)	Cream FFA (mmol/100 g fat)	Cream lipase activity (mmol/L/h)
5	2.1	3.9	12.2
10	2.9	3.5	10.9
15	4.6	2.9	4.9
Unagitated control	0.9	0.9	1.9

Data from Deeth and Fitz-Gerald (1977).

of up to ~20 days, several bacterial types grow during storage and some of these are lipolytic. In most cases, lipolysis will not be the sole cause of the spoilage but can contribute to the spoilage (Deeth *et al.* 2002). However, Craven and Macauley (1992) investigated the spoilage patterns of pasteurized milk produced in three factories and concluded that the keeping quality of the milks was largely determined by the standard of hygiene in the factories (as indicated by the initial bacterial count in the pasteurized milk) and the lipolytic activity of the bacterial strains that grow to spoilage levels.

Lipolysis by bacterial lipases is usually due to extracellular lipases produced by psychrotrophic bacteria, particularly *Pseudomonas* species. Little lipase is produced before the bacterial count reaches ~10^5 cfu/mL, and in single cultures of lipolytic bacteria in milk, spoilage occurs when the level reaches 10^6 cfu/mL (Suhren *et al.* 1977) or 10^7 cfu/mL (Overcast and Skean 1959; Shelley *et al.* 1986).

Fromm and Boor (2004) monitored pasteurized milk produced by three processors and found Gram-positive rods, including *Paenibacillus*, *Bacillus* and *Microbacterium* as the predominant organisms. Over 17 days storage of the pasteurized milks, the FFA levels increased from 0.08 to 0.14 mmol/kg to 0.15–0.25 mmol/kg (using the copper soap method; see comments on this method in Section 11.2.2.1).

Santos et al. (2003b) determined that a FFA level (determined by the copper soap method) of 0.25 mmol/kg was the threshold for detection of lipolysis in 2% fat homogenized pasteurized milk. This was based on one-third of sensory panellists detecting an off-flavour at or below this level. Santos *et al.* (2003a) used this threshold value to estimate the effect of different raw milk somatic cell counts on the shelf life of 2% fat homogenized pasteurized milk. They concluded that the somatic cell count had a definite effect on the level of lipolysis in the pasteurized milk but estimated that the pasteurized milk with a low SCC (<100,000 cells/mL) would have a shelf life of >61 days at 6 °C while milk with a high SCC (880,000–1,200,000 cells/mL) would have a shelf life of 35 days at 6 °C. Hence a high somatic cell count appears unlikely to cause significant lipolysis in pasteurized milk during its normal shelf life but may affect the keeping quality of ESL milk which should have a shelf life of >30 days (Deeth 2017). Given that pasteurization inactivates LPL, Santos *et al.* (2003a) suggested that the somatic cell lipase reported by Azzara and Dimick (1985) may be responsible for the lipolysis. Addition of somatic cells to milk has been shown to cause lipolysis (Salih and Anderson 1979), with the level of lipolysis being greater when the cells were disrupted before addition (Jurczak and Sciubisz 1981). However, it is unclear if the somatic cell lipase can withstand pasteurization.

11.2.4 UHT Milk

As UHT processing is a sterilization process, it is highly unlikely that growth of bacteria will cause lipolysis in UHT milk. The only situation where this could conceivably occur is when the ambient storage conditions are high enough to allow the growth of very heat-resistant, lipolytic spore-forming bacteria. This can occur in some tropical regions and in warehouses. However, lipolysis in UHT milk can be caused by heat-resistant bacterial lipases produced by psychrotrophic bacteria, such as *Pseudomonas* species, in raw milk (Blake *et al.* 1995; Celestino *et al.* 1997), although instances of high FFA levels and associated rancid off-flavour in UHT milk are rare. This may be due to the fact that heat-resistant bacterial proteases which cause bitterness and age gelation during storage of UHT milk are more important than lipases in affecting the taste and acceptability of UHT skim milk (Deeth and Lewis 2017).

The bacterial lipases of interest in UHT milk have two important characteristics which differ from those of LPL. They can act on milk fat in intact milk fat globules and are very heat-resistant. Fitz-Gerald and Deeth (1983) reported that lipases in cell-free supernatants of cultures of *Pseudomonas fluorescens*, *Ps. aeruginosa*, *Moraxella* and *Acinetobacter* isolated from raw milk showed comparable rates of lipolysis on tributyrin, emulsified milk fat and cream under optimal conditions of pH 7.5–9.0 and 40–50 °C.

The heat stability of bacterial lipases varies with the species and strain, but some have extremely high heat stability (Cogan 1977). An indication of their significance in UHT products was given by Adams and Brawley (1981). Firstly, they found that a lipase from *Pseudomonas* species MC50, which had been isolated from raw milk, had a $D_{150\,°C}$ of 63 s and a *z*-value of 38.4 °C and hence would be largely resistant to UHT processing. Secondly, the UHT-processed raw milks which had a range of standard plate counts of 7.4×10^3 to 7.8×10^5 cfu/mL increase in ADV by 0.0051–0.0141/day during storage at 25 °C. They predicted that a slight lipolysed flavour would be expected to develop in such milk after storage at room temperature for 32–88 days. This was based on a report by Thomas *et al.* (1955) that lipolyzed flavour can be detected at an ADV of 2 mmol/100 g fat.

Because UHT milk is stored at room temperature for up to 9 months, trace amounts of lipase are sufficient to cause lipolysis during storage, as indicated by the work of Adams and Brawley (1981). The extent of lipolysis depends on both the amount of active residual lipase and the temperature and duration of storage. The amount of active lipase remaining after UHT heat treatment is influenced by the effectiveness of the heat treatment in inactivating the enzyme. Less active enzyme remains after indirect processing than after direct processing (Panfil-Kuncewicz *et al.* 2005). More lipolysis occurs at higher temperatures of storage. Adams and Brawley (1981) reported ~1.9 times more lipolysis during storage at 40 °C than at 25 °C. Similarly, Choi and Jeon (1993) and Singh *et al.* (2004) reported acceleration of lipolysis during storage at higher temperatures.

11.2.5 Dairy Products

11.2.5.1 Cheese

Cheese differs from milk and other dairy products in that FFAs are normal components of the flavour of most varieties (Law 1984). They also act as precursors for several flavour compounds such as methyl ketones, thioesters and esters (McSweeney 2011). In some cheese varieties, such as Pecorino, Provolone, Parmesan and Romano, the fatty acids are so important that lipases, normally lamb, calf or kid pre-gastric esterases/lipases, are added during cheese manufacture to increase the FFA level. These lipases preferentially catalyse the release of short- to medium-chain fatty acids from the milk triglycerides and hence contribute to the strong piquant flavour of these cheeses. However, in most cheeses when excessive lipolysis occurs in the cheese or in the cheese milk before manufacture, rancid flavours may result (Woo *et al.* 1984; Fox and Law 1991).

Lipolysis can occur in cheese due to bacterial lipases from psychrotrophic bacteria which survive the pasteurization process. This occurs during storage of the cheese and can result in cheeses becoming unacceptable creating a substantial economic loss for companies. Law *et al.* (1976) reported that *Cheddar* cheese made from milk with $>10^7$ cfu/mL lipolytic Gram-negative bacteria became rancid after 4 months even though the bacteria had been inactivated by pasteurization.

Because of the high total flavour of cheese, the threshold levels of FFAs for detection of a rancid flavour defect are higher than for milk. The threshold ADV for *Cheddar* cheese is 2.8–3.0 (Deeth and Fitz-Gerald 1975b). *Cheddar* cheese classed as rancid has been shown to have from two to ten times the FFA content of good-quality cheese (Bills and Day 1964; Ohren and Tuckey 1969; Law *et al.* 1976).

In cheese made from pasteurized milk, lipolysis by the native milk LPL does not occur or occurs to a very small extent after manufacture. However, cheese made from raw milk, which contains active LPL, develops a higher level of FFA during ripening than cheese made from pasteurized or thermized milk (Hickey *et al.* 2007). Furthermore, Buffa *et al.* (2001) reported that cheese made from raw goats' milk or raw goats' milk treated with high pressure (500 MPa, 15 min, 20 °C) showed more lipolysis during storage than cheese made from pasteurized goats' milk. These results illustrate the inactivation effect on LPL of thermal pasteurization and the resistance of LPL to high-pressure treatment.

Low-pressure homogenization of the raw cheese milk is sometimes used to enhance the level of FFAs in some cheese varieties, particularly blue vein cheeses. It has also been used to improve the texture of low-fat cheese (Johnson 2011), to increase lipolysis in Feta-type cheese made from ultrafiltered milk (Karami 2017) and in *Cheddar* cheese (Deegan and McSweeney 2006), and to increase early lipolysis and subsequent volatile flavour compound production in Reggianito cheese, a hard Argentinian cheese similar to the Italian cheese Parmigiano Reggiano (Vélez *et al.* 2017).

High-pressure homogenization (HPH) of raw cheese milk has been used to improve cheese yield and characteristics and to accelerate lipolysis and proteolysis, for example, for goats' milk cheese (Guerzoni *et al.* 1999) and the Italian soft cheese, Crescenza (Lanciotti *et al.* 2004). Lanciotti *et al.* (2006) also used HPH in the manufacture of the semi-hard Italian cheese, Caciotta. In that study, they compared cheese made from raw or pasteurized milk with that made from HPH-treated milk. The HPH treatment was at 100 MPa at an inlet temperature of 5–7 °C and an outlet temperature of ~30 °C. The FFA levels at 0, 13 and 27 days for the three cheese types are shown in Table 11.3. This table in high-pressure homogenised milk cheese illustrates the extremely high FFA level in high-pressure homogenised milk cheese at day 0 compared to that of the raw and pasteurized milk cheeses and also the large increase in FFAs in the raw milk cheese during 27 days' storage, about twice that of the pasteurized milk cheese; this difference is largely attributable to the native LPL. Curiously, the FFA level in the HPH cheese did not increase during the storage; Vélez *et al.* (2017) reported a similar result for Reggianito cheese made from low-pressure homogenized milk. The HPH cheeses had a significantly higher overall sensory score than raw and pasteurized milk cheeses.

11.2.5.2 Butter

Hydrolytic rancidity has been considered to be one of the most troublesome defects in butter (Bell and Parsons 1977). It is characterized by off-flavours variously described as rancid, bitter, unclean, wintery, butyric or "lipase". The cause of the off-flavour is FFAs although a good correlation between flavour and FFA levels is sometimes not found (Bell and Parsons 1977). While different threshold ADVs have been published, similar numerical values to those for milk, 1.5–2.0 mmol/100 g fat, appear reasonable. However, the most significant FFAs in butter, butyric (C4) and caprylic (C6), differ from the most significant in milk, capric (C10) and lauric (C12), because of the low flavour threshold of the short-chain water-soluble fatty acids in fat (Patton 1964).

A high level of FFAs in butter can originate pre-manufacture from lipolysis in the raw milk caused by native LPL or in the butter during storage. If it occurs in raw milk pre-manufacture, a high proportion of the most flavoursome short-chain fatty acids, which contribute most to the rancid flavour, is lost in the skimmed milk when the milk is centrifugally separated and in the buttermilk and wash water when the butter is manufactured. This is because these fatty acids, such as butyric (C4) and caprylic (C6), are water-soluble. The FFAs remaining in the butter will be comprised mostly of the fat-soluble, long-chain acids. If, however, lipolysis occurs post-

Table 11.3. Free fatty acids (mg/kg) in Caciotta cheese made from raw and high-pressure homogenized milk

	Storage duration (days) of cheese		
Milk treatment	0	13	27
None (raw milk)	126.6	409.6	972.7
High-pressure homogenized	1154.5	1154.8	1095.2
Pasteurized	217.7	363	628

Data from Lanciotti *et al.* (2006).

manufacture, all the FFAs will remain in the butter. Consequently, determination of the FFA profile chromatographically can provide a good indication of the cause of the lipolysis. McNeill *et al.* (1986) reported that measurement of butyric (C4), caprylic (C6) and caproic (C8) provided the best indication of hydrolytic rancidity in butter.

The different profiles of FFAs in butter arising from pre- and post-manufacture may be the reason for a poor correlation between flavour and measures of FFA. This was demonstrated by Deeth *et al.* (1979) when they compared butters with pre- and post-manufacture lipolysis. Butter produced from lipolysed cream was organoleptically acceptable even when the ADV was abnormally high but butter in which lipolysis occurred post-manufacture was severely downgraded at quite low ADVs.

11.2.6 Analytical Methods

11.2.6.1 Free Fatty Acids

Most analytical methods regarding lipolysis involve analysis of FFAs. The quantification of all FFAs in milk or milk products presents a difficult analytical problem as the acids range from the water-soluble, short-chain acids such as butyric (C_4) and caprylic (C_6) to the water-insoluble, long-chain acids such as palmitic (C_{16}) and stearic (C_{18}). Furthermore, in milk and milk products, they exist with large amounts of fat, largely triglycerides, water-soluble acids such as lactic acid, and phospholipids, some of which are acidic and can interfere with the analyses. Reviews of methods for determining FFAs in milk and milk products have been published by Kuzdzal-Savoie (1980), an IDF/ISO/AOAC Group (Anderson *et al.* 1991), de la Fuente and Juarez (1993), Joshi and Thakar (1994), Collomb and Spahni (1995), Deeth and Fitz-Gerald (2006), Kilcawley (2006), Mannion *et al.* (2016b), Wiking *et al.* (2017) and Amores and Virto (2019).

The methods usually involve an initial step to isolate the fatty acids, with or without the milk fat, followed by a quantification step, with or without derivatization of the acids. However, there are direct methods which do not require any pretreatment step which have some obvious advantages and may be suitable for in-line or at-line FFA monitoring. Commercial milk FT-IR analysis instruments (e.g. MilkoScan™7R [Foss 2018] and Milkoscan FT6000 [Wiking *et al.* 2017]) operating in the mid-IR range are capable of measuring FFAs in milk. According to Wiking *et al.* (2017), FT-IR is now used for analysing FFAs in raw milk because of it is faster and cheaper than traditional methods.

A classic FFA method is the BDI method which involves fat separation followed by titration of the FFAs dissolved in the fat. Evers (2003) published an excellent review of the practical and theoretical aspects of this method. The BDI method measures only fat-soluble acids and hence underestimates the total FFA level (Deeth *et al.* 1975; Van Crombrugge *et al.* 1982). However, for milk, this is of little consequence as the level of the long-chain acids correlates highly with that of the short-chain acids responsible for the flavour. Wiking *et al.* (2017) compared the FFA results obtained by FT-IR (on a Milkoscan FT6000) with those from the BDI method and found a good correlation ($r = 0.90$). In the correlation, they found that an acid degree value of 1.25 mmol/100 g fat by the BDI method corresponded to 1.0 mmol/100 g fat predicted by FT-IR.

A common initial step in FFA analysis is solvent extraction which yields a solution of the fat and the FFAs. Solvents used have been reviewed by Deeth and Fitz-Gerald (2006) and Mannion *et al.* (2016b). The FFAs in the solution can then be titrated with alkali or separated from the fat using various solid supports such as ion exchange resins, neutralized alumina or an aminopropyl resin (Kilcawley 2006). The isolated FFAs can then be analysed chromatographically, by GC or HPLC. According to Mannion *et al.* (2016b), GC analysis is the method of choice for FFA analysis. It is now commonly performed using wall-coated open tubular (WCOT) capillary columns with a coating such as free fatty acid phase (FFAP) which enables all fatty acids to be cleanly separated.

By integrating the areas under the peaks, a quantitative measure of each acid, groups of acids such as the short-chain acids or the total FFAs can be obtained. The fatty acids can be derivatized before analysis. Mannion *et al.* (2019) reported the analysis of butyl esters which they found to be "robust, sensitive, accurate and precise" and overcame some of the limitations of the two most common methods, analysis of the methyl esters and direct on-column injection of the FFAs. The limitations are underestimation of short-chain acids due to the volatility of the methyl ester, and column absorption and deterioration by the free acids. Traditionally, the fatty acids or their esters have been detected on GC by flame ionization detectors. However, mass spectrometer detectors are also used.

Well-resolved chromatograms have been obtained in HPLC analyses of milk FFAs as *p*-bromophenacyl esters (Reed *et al.* 1984; Elliott *et al.* 1989; Garcia *et al.* 1990) and 2-nitrophenylhydrazides (Miwa and Yamamoto 1990). Milk FFAs have also been analysed by HPLC using fluorescent derivatives; Lu *et al.* (2002) used naphythoxyethyl derivatives, while Nishimura *et al.* (2002) used 9-anthrylmethyl derivatives. A search of the recent literature reveals a large number of reports of HPLC analysis of FFA fluorescent derivatives using a wide range of fluorescent reagents which to date have not been used for milk FFAs (e.g. Wang *et al.* 2013; Nishikiori *et al.* 2015; Zeng *et al.* 2017). Another HPLC approach to analysing FFAs along with other lipid classes involves the use of normal-phase HPLC with evaporative light scattering detection (ELSD) (Donot *et al.* 2016). This also has potential for analysis of FFAs in milk and dairy products.

The rapid enzymatic method of Koops *et al.* (1990) involving acyl-CoA synthetase and acyl-CoA oxidase has the potential to become a routine procedure. It has been automated and showed good agreement with the BDI method on farm milk samples. In another enzymatic method, Christmass *et al.* (1998) linked acyl-CoA synthetase, UDP-glucose pyrophosphorylase, phosphoglucomutase and glucose-6-phosphate 1-dehydrogenase and the NADH-luciferase to determine FFA in human milk. The fluorimetric measurement of the resultant NADH overcame the problem of cloudiness due to the added milk which affects colorimetric measurements.

Several publications have compared various methods for determining FFAs in milk and dairy products (Van Crombrugge *et al.* 1982; Chilliard *et al.* 1983; Suhren 1983; Bråthen 1980; Cartier *et al.* 1984; Selselet-Attou *et al.* 1984; Ikens *et al.* 1988; Anderson *et al.* 1991; Joshi and Thakar 1994; Chavarri *et al.* 1997; Mannion *et al.* 2016a; Wiking *et al.* 2017). Wiking *et al.* (2017) compared some novel methods with some standard methods for analysis of FFAs in milk in relation to rancid flavour detection. The methods included proton-transfer reaction-mass spectrometry (PTR-MS), proton nuclear magnetic resonance (NMR) spectrometry, gas chromatography-mass spectrometry (GC-MS) with in-solution derivatization, FTIR, BDI and the acyl-CoA-synthetase-acyl-CoA oxidase (ACS-ACOD) assay. PTR-MS and NMR were used for quantifying individual short-chain FFAs, GC-MS was used for quantifying all individual FFAs and the other three methods were used to estimate the total FFA content of the milks. They found that because all the FFAs were quantitatively related, quantification of individual and total FFA content provided similar information with regard to the likely presence of off-flavour. In general, high correlations are found between the various methods although different methods suffer from different limitations such as overestimation due to inclusion of lactic acid and underestimation of short-chain acids. Most of the methods discussed above have been used with milk, but many can be applied to other products. Some minor variations in extraction procedures such as the inclusion of anhydrous sodium sulphate to eliminate water (Deeth *et al.* 1983) or inclusion of an aqueous acid wash step to remove lactic acid (Chilliard *et al.* 1983) may be necessary.

11.2.6.2 Lipase Activity

Another type of analysis related to lipolysis is of lipase activity. In raw milk, this is seldom measured in practice because the level of LPL is not a useful indicator of the milk's susceptibility to lipolysis. However, a measure of lipase activity can be useful for UHT milk where trace levels of heat-stable bacterial lipases can cause lipolysis during storage. In this case, the low levels of activity make such an analysis very difficult and, currently, no accepted methodology exists. Several methods of determining lipase activity in milk and milk products have been reported, and many of these have been reviewed (Deeth and Touch 2000; Deeth and Fitz-Gerald 2006; Kilcawley 2006; Button 2007).

The lipase assays use a range of substrates including milk fat and other triglycerides, chromogenic and fluorogenic esters and radioactively labelled esters including triglycerides. The methods of analysis of products include alkaline titration of the fatty acids, spectrophotometry, fluorometry, radiometry and gas chromatography. As a general rule, the activity of the lipases against triglyceride substrates matches more closely the activity in milk and milk products than the chromogenic and fluorogenic esters, but the sensitivity of detection is often better with the non-triglyceride substrates. Furthermore, the non-triglyceride esters have a greater tendency to spontaneously hydrolyse during assays, especially during long incubations, which are usually performed at pH 7–8.

Button (2007) investigated lipase assays suitable for detecting very low levels of activity in UHT milk. He concluded that assays using triolein emulsified in gum acacia incubated at 37 °C and pH 7.5 for several hours were more suitable than assays using the chromogenic substrate, β-naphthyl caprylate, or the fluorogenic substrate, 4-methylumbelliferyl oleate. Using the triolein assay, two levels of lipase activity which caused an increase in FFA in UHT milk during storage to a level detectable by most consumers (~2 mmoles FFA/L) were detected in 4 or 24 hours; the corresponding UHT milks reached ~2 mmoles FFA/L in 50 and 100 days, respectively, at 25 °C. The method used involved titration of the fatty acids released with alcoholic KOH. A reduction of incubation time for detection of low levels of activity may be possible with a more sensitive method for analysing FFAs such as gas chromatography.

A novel method for the analysis of residual bacterial lipase activity in UHT milk was recently proposed by Andrewes (2018). It utilizes the lipase-catalysed alcoholysis reaction in which methanol is the acyl acceptor. The amount of fatty acid methyl esters formed is a measure of the lipase activity in the milk.

11.3 Oxidation

11.3.1 Background

Milk fat has a relatively low content of polyunsaturated fatty acids (typically ~3.5% in cows' milk) (MacGibbon and Taylor 2006) and prooxidant metals such as copper and iron. It is therefore somewhat surprising that lipid oxidation is an important issue in the dairy industry. However, certain factors conspire to enhance oxidation in milk and dairy products. These include the presence of photosensitizing components such as riboflavin (~1.75 mg/L) (Dimick 1982), which promotes light-induced oxidation, and the low content of antioxidants. Milk does have some natural protective mechanisms in its structure and composition, but some of these are negated by processing and storage. Furthermore, modifications such as supplementation with ω-3 polyunsaturated acids and conjugated linoleic acid (CLA) can enhance oxidation. On the other hand, the dairy industry has been very conscious of the detrimental effects of oxidation for a long time and has taken steps to reduce these effects. Historically, this includes eliminating copper in materials used for milk handling and developing packaging materials with low oxygen and light permeability.

The major problem associated with oxidation is the production of off-flavours typically called "oxidized" but also referred to by other terms such as cardboardy, metallic, fatty (oily), painty and fishy (Carunchia Whetstine and Drake 2007) as well as rancid and stale. Oxidation of lipids also leads to a reduction in nutritive value, for example, through the concomitant degradation of vitamins (Whited *et al.* 2002) and formation of oxidized cholesterol (Calderon-Santiago *et al.* 2012).

The proteins in milk and dairy products can also undergo oxidation (Dalsgaard *et al.* 2011; Davies 2016; Chen *et al.* 2019). The amino acids phenylalanine, tyrosine, tryptophan and methionine are the major amino acids affected (Chen *et al.* 2019). One result of such oxidation is cross-linking of proteins via di-tyrosine. Protein oxidation is not discussed further here.

Lipid oxidation commonly occurs in dairy products during storage. However, other chemical changes occur at the same time which complicate identification of the compounds due to oxidation alone. Furthermore, there is some evidence that oxidation may occur synergistically with the Maillard reaction, another common reaction in stored dairy products (Stapelfeldt *et al.* 1997; Carunchia Whetstine and Drake 2007).

Lipid oxidation occurs via a free radical chain reaction consisting of initiation, propagation and termination stages. The oxidation target is the unsaturated fatty acids which initially form free radicals. In the propagation step, the fatty acid radical reacts with oxygen to form a peroxide free radical which reacts with another fatty acid to produce another fatty acid radical to continue the reaction. In the termination step, the peroxide radical reacts with oxygen to form a hydroperoxide. Hydroperoxides are considered the primary oxidation products as they decompose to form a variety of secondary products such as aldehydes, ketones and hydrocarbons. Thus, a measure of primary oxidation is the hydroperoxides, often measured as the peroxide value, while secondary oxidation is determined by the level of aldehydes such as hexanal and malondialdehyde (a major component of a mix of aldehydes measured by reaction with thiobarbituric acid (TBA) as TBA-reactive substances or TBARS).

Natural antioxidants in milk, such as α-tocopherol and β-carotene, can block the chain reaction or prevent the initiation; however, they are consumed in this process. It is therefore not surprising that there is a significant correlation between the levels of these compounds and oxidative stability. The antioxidant properties of other milk components including ascorbic acid, thiols (formed during heating of milk) and milk proteins, together with tocopherols and carotenoids, were reviewed by O'Connor and O'Brien (2006).

Lipid oxidation in milk and dairy products has been studied extensively and has been the topic of several reviews. These include those by Greenbank (1948), Weihrauch (1988), O'Connor and O'Brien (1995, 2006), Borle *et al.* (2001), Sieber (2005), Garcia-Martinez and Marquez-Ruiz (2009), Duncan and Webster (2010) and O'Brien and O'Connor (2011). In addition, many reviews of lipid oxidation in general have been published; an extensive list of these was provided by O'Connor and O'Brien (2006). The mechanism of lipid oxidation is covered in detail in these reviews and is not reproduced in detail here.

11.3.2 Raw and Pasteurized Milk

11.3.2.1 Spontaneous Oxidation

Weihrauch (1988) stated that spontaneous oxidation is a major problem of the dairy industry. Milk susceptible to spontaneous oxidation develops an oxidized flavour within 48 h of milking. Some cows always produce milk which is susceptible to spontaneous oxidation, whereas others never produce such milk while others sometimes produce milk which oxidizes spontaneously. The reported proportions of individual milk samples which oxidize spontaneously vary from 23.1 to 38%. The reasons why some milks spontaneously oxidize and some do not have been studied extensively and several hypotheses exist. One relates to the presence of certain enzymes, particularly xanthine oxidase (Aurand *et al.* 1967), although Weihrauch (1988) considered this to be of minor importance. Others relate to the presence of pro-oxidants and deficiency of antioxidants. The most common factor which determines the susceptibility to oxidation is the level of unsaturated lipids, particularly linoleic acid, while the level of copper has been shown to be significant pro-oxidant in some cases (Barrefors *et al.* 1995; Granelli *et al.* 1998; Juhlin *et al.* 2010a, b, 2012). Addition of copper to milk is known to significantly enhance milk fat oxidation (Chawla *et al.* 2003; Panda and Kaur 2007). The most common antioxidant factor associated with the resistance of milk to spontaneous oxidation is the level of antioxidants, particularly α-tocopherol.

An animal's diet can also affect the susceptibility to oxidation of the milk it produces. For example, inclusion of α-tocopherol in the diet reduces the susceptibility of the milk from cows (Chawla *et al.* 2003) and buffaloes (Panda and Kaur 2007) to oxidation. Fauteux *et al.* (2016) reported that the milk from cows fed a diet of alfalfa concentrate had high concentrations of vitamin E (α-tocopherol) and low concentrations of volatile lipid oxidation products. However, in a comparison with the effects of feeding grass silage and corn silage as roughage, Havemose *et al.* (2004) found that the milk from cows fed with grass silage had a higher antioxidative capacity but higher lipid oxidation than milk from cows fed with corn silage. This was attributed to the higher level of linolenic acid in the grass silage milk. Interestingly, the milk from the cows fed with corn silage was more susceptible to protein oxidation, as determined by dityrosine levels, than the milk from cows fed with grass silage.

Recently, there has been considerable interest in milk produced from cows which only graze pasture/grass which has become known as pasture-fed or grass-fed milk (Dairy Farmers of Ontario 2018; Kilcawley *et al.* 2018). This is in contrast to milk from cows fed with a total mixed ration. Grass-fed milk has higher content of ω-3 polyunsaturated fatty acids and conjugated linoleic acid (particularly the *cis*-9, *trans*-11 isomer, rumenic acid) but a lower content of ω-6-polyunsaturated fatty acids than conventional milk (La Terra *et al.* 2010; Elgersma 2015; Dairy Farmers of Ontario 2018). However, grass-fed milk also has a higher concentration of the antioxidants α-tocopherol and β-carotene (La Terra *et al.* 2010; Marino *et al.* 2012, 2014). Urbach (1990) reported that milk from pasture-fed cows is less susceptible to oxidation than milk from cows on dry feed. It appears that the higher level of antioxidants more than compensates for the higher level of unsaturation (Kilcawley *et al.* 2018).

A difference in the significance of certain factors in spontaneous oxidation in the milk from different herds has been observed (Barrefors *et al.* 1995; Granelli *et al.* 1998). This has led to the suggestion of heritability and that the occurrence of spontaneous oxidation may be under genetic control (Juhlin *et al.* 2010a). Juhlin *et al.* (2012) investigated the possible relationship between acyl-coenzyme A:diacylglycerol acyltransferase 1 (DGAT1) K232A polymorphism and the occurrence of spontaneous oxidized flavour in milk. They found a marked effect of the DGAT1 genotype, where the A allele was associated with a higher risk of the milk developing spontaneous oxidized flavour.

The above research findings, and the fact that the susceptibility to spontaneous oxidation correlates well with oxidation-reduction potential of milks, leads to the conclusion that the susceptibility depends on the balance of naturally occurring pro- and antioxidants (Weihrauch 1988) and that this balance may be genetically determined. Interestingly, this indicates that spontaneous oxidation shares certain characteristics with spontaneous lipolysis (discussed in Section 11.2.2.2). Spontaneous lipolysis, which is also exhibited in the milk of some cows and not others, has been attributed to a balance of activating and inhibiting factors; a genetic involvement has also been demonstrated in research on DGAT1 genotypes. A correlation between spontaneous oxidation and spontaneous lipolysis has not been reported.

Spontaneous oxidation in milk is detected by either chemical analysis or sensory evaluation. It is characterized by elevated levels of oxidation products such as hexanal (Barrefors *et al.* 1995). Potts and Peterson (2018) investigated a case of purported spontaneous oxidized flavour in cows' milk in the Midwest dairy region of the United States. GC/MS-olfactometry analysis revealed a range of flavour compounds with flavour attributes of "green", "musty" and "unclean". The flavour compounds and descriptions indicated that the objectionable flavour was microbially derived and was not due to oxidation. The authors cautioned about defining defects without identification of the flavour compounds.

11.3.2.2 Induced Oxidation

Oxidation can be induced in milk and dairy products by light activation, by copper catalysis and by processing. While the importance of copper as a pro-oxidant in milk and dairy products has decreased considerably due to the replacement of copper-containing metals with stainless steel for handling milk, light activation continues to be of concern. Several processes being used in, or developed for, the dairy industry can also influence the oxidation state of milk.

Light-Induced Light-activated flavour in milk is very unpleasant. In a study of the flavour, Walsh *et al.* (2015) found that sensory panellists frequently used the term "disgust" when describing it. However, many other terms have been used to describe this flavour including chemical, burnt, scorched, cabbage and mushroom (Dimick 1982). Walsh *et al.* (2015) suggested that these light-induced flavour defects in fluid milk may be severe enough to be a contributing factor in the reduction in sales of milk. While lipid oxidation makes a major contribution to light-activated flavour, the contribution of protein oxidation should not be overlooked.

The susceptibility of milk and dairy products to light-activated changes has generally been attributed to the presence of the photosensitizer riboflavin, vitamin B2. However, naturally occurring porphyrins and chlorophylls may have a greater effect than riboflavin (Wold *et al.* 2005; Airado-Rodríguez *et al.* 2011). Riboflavin absorbs light in both the visible and ultraviolet (UV) range (200–500 nm (Drossler *et al.* 2002), while the chlorophilic compounds absorb in the visible range, > 575 nm. This leads to the production of active oxygen species which cause oxidation of both lipids and proteins with the resultant production of strong off-flavours. The flavour compounds include aldehydes such as pentanal and hexanal, from oxidation of unsaturated fatty acids, as well as protein-derived compounds such as dimethyl disulphide (DMDS) from methionine.

Light-activated oxidation also causes reduction in some vitamins (A, B2, C, D, E) (Mestdagh *et al.* 2005). For example, Whited *et al.* (2002) studied vitamin A degradation in milk exposed to fluorescent light and concluded that even a brief, moderate light exposure (2 h at 2000 lx) could significantly reduce the vitamin A content and hence the nutritional value of fluid milk products.

Of major concern to the dairy industry has been the storage of retail pasteurized milk packaged in plastic (HDPE) bottles in refrigerated display cases with fluorescent lighting (Bosset *et al.* 1994). In a survey of dairy cases in retail outlets, Chapman *et al.* (2002) found the light intensities to be 750–6460 lx. They chose 2000 lx for an experiment to assess the time of exposure of milk in clear HDPE bottles for the appearance of off-flavours. Flavour defects were detected by trained sensory panellists after 15–30 min exposure and by untrained consumers after 54 min to 2 h, exposure. Given that about half of the milk packages remain in the dairy cabinet for 8 h or more, the flavour of the milk will be impaired for many consumers (Chapman *et al.* 2002). Dimick (1973) reported that a trained sensory panel detected light-activated off-flavour in homogenized milk packaged in glass or plastic bottles after 12 hours' exposure to fluorescent light but no off-flavour was detected in milk packaged in fibreboard containers. Walsh *et al.* (2015) compared the flavour stability at 4 °C of 2%-fat milk packaged in HDPE and HDPE with a foil wrap. The acceptability score of the unprotected milk decreased from 7.2 to 5.85 within 8 h, while the protected milk retained a score of 7.0 over 168 h. Johnson *et al.* (2015) investigated the efficacy of titanium dioxide incorporated into HDPE bottles for protection of milk from light and found that the highest level of TiO_2 (4.3%) afforded similar protection to a foil wrap but a lower percentage of TiO (1.3%) was less effective. The role of packaging in inhibiting light-induced oxidation of milk continues to be of considerable interest. This is demonstrated by a recent report on the degrading effects of light on milk entitled "Packaged Foods Should Be Kept in the Dark: Consumers Shouldn't Be" by a company which has developed a patented method for measuring the light protection capacity of packaging (Noluma International 2018).

Brothersen *et al.* (2016) studied the relative effects of fluorescent and white light LED (light-emitting diode) lighting on the development of light-activated flavour in milk. They used fluorescent light at 2200 lx and LED light at 4000 lx. Despite the lower intensity of the fluorescent light, off-flavours were detected in milk exposed to fluorescent light after 12 h and to LED light after 24 h. The authors suggested the reason for the difference between fluorescent and LED lighting may be the wavelengths emitted >500 nm at which chlorophyll photosensitizers absorb but riboflavin does not.

Process-Induced There is conflicting evidence of whether heating induces oxidation. It could occur through the formation of very reactive alkyl radicals which can combine with molecular oxygen to form peroxy radicals which become involved in the lipid oxidation reaction (O'Brien and O'Connor 2011). Li *et al.* (2019) examined the formation of the oxidized flavour compounds at different heat treatments and found a high correlation ($r = 0.86$) between heat intensity and the concentrations of heptanal, nonanal, 2-heptanone and 2-nonanone in heated milk samples. Heating at 90 °C for 20 min increased the peroxide value and thiobarbituric acid-reactive substances (TBARS) and also reduced the diphenyl picryl hydrazyl (DPPH) free radical scavenging activity of the milk. Garcia-Martinez (2012) reported that sterilization of model liquid milk-based infant formulas did not cause significant oxidation. However, they found a significant decrease in tocopherol concentration which could lead to increased oxidation. Interestingly, they concluded that sodium caseinate in the formulas had a protective effect and reduced tocopherol loss. These results are consistent with those of Calligaris *et al.* (2004) that short-time heat treatments depleted antioxidant levels. However, more severe heat treatments which result in the formation of brown pigments, melanoidins, increase the antioxidative capacity of milk. An antioxidative effect of heat treatment was also reported by Mahmoodani *et al.* (2018) although the reason was not given. Powder samples made from simulated milk which had been heated at 95 °C for 30 s showed lower levels of lipid oxidation products than those without heat treatment.

Pereda *et al.* (2008) investigated the effect of high-pressure homogenization treatment on lipid oxidation in milk and found that homogenization at 300 MPa led to much more secondary oxidation as shown by levels of TBARS and hexanal, than homogenization at 200 MPa. Curiously, the effect on primary oxidation as given by peroxide value showed the opposite effect. Again, the protective effect of casein on the fat globules homogenized at 200 MPa was invoked as a reason for the difference in secondary oxidation.

11.3.3 UHT Milk

The flavour of fat-containing UHT milk changes during storage at room temperature. Immediately after manufacture, UHT milk has a distinct sulphury taste and odour due to the production of volatile sulphur compounds such as hydrogen sulphide and methane thiol during the heating process (Al-Attabi *et al.* 2014). This flavour dissipates over 7–10 days leaving a flavour which is optimal for the product up to 1–2 months. The flavour during that time is due to a range of compounds, of which those produced by the Maillard reaction are prominent. From then, a stale or oxidized flavour begins to appear which intensifies during storage. Other flavours such as bitter and rancid may also develop if residual heat-resistant proteases and lipases, respectively, are present.

The stale or oxidized flavour, which is also characterized as tallowy, coconut-like and cardboardy, is a major cause of criticism of UHT milk flavour in aged UHT milk. It is due to lipid oxidation products, particularly methyl ketones and saturated aldehydes (Mehta 1980). Seven methyl ketones (i.e. 2-pentanone, 2-hexanone, 2-heptanone, 2-octanone, 2-nonanone, 2-decanone, 2-dodecanone) and six aldehydes (i.e. pentanal, hexanal, heptanal, octanal, nonanal, decanal) have been shown to contribute to the stale flavour, all of which increase during ambient storage (Perkins *et al.* 2005b).

While the methyl ketones are produced by decarboxylation of β-keto fatty acids, the saturated aldehydes are formed via oxidation of saturated or unsaturated fatty acids to hydroperoxides. Saturated fatty acid hydroperoxides decompose to aldehydes and short-chain carboxylic acid radicals, which in turn can oxidize, dehydrate and decarboxylate to a produce a further aldehyde molecule (Grosch 1982). Unsaturated fatty acid hydroperoxides decompose to straight-chain aldehydes (Forss 1979). Thus, oleic acid (C18:1) yields nonanal, octanal, decanal and heptanal, while linoleic acid (C18:2) yields predominantly hexanal and, to a lesser extent, pentanal and heptanal (Badings 1970).

Methyl ketones are more abundant in stored UHT milk than aldehydes, but some authors have found the aldehydes to better correlate with stale flavour than the methyl ketones (Jeon *et al.* 1978; Rerkrai *et al.* 1987). In contrast, Moio *et al.* (1994) reported that the methyl ketones 2-heptanone and 2-nonanone were the main contributors to the flavour of UHT milk. Later, Vazquez-Landaverde *et al.* (2005) found 2-heptanone and 2-nonanone to be very important in UHT milk flavour but aldehydes, especially nonanal, decanal, octanal, hexanal, 2-methylbutanal, 3-methylbutanal and 2-methylpropanal, also contribute to it.

The stale or oxidized flavour in UHT milk and, to a lesser extent, in ESL (extended-shelf life) milk can be controlled to a certain extent by the availability of oxygen. During UHT processing, if deaeration is not used, the dissolved oxygen content of indirectly processed milk (i.e. using plate or tubular heat exchangers) remains close to saturation level (6–11 mg/L, Fauteux *et al.* 2016; Deeth and Lewis 2017) but is usually much lower (≤ 1 mg/L) in directly processed milk (i.e. using steam injection or steam infusion). This is because the vacuum treatment step in direct processing removes dissolved oxygen as well as water (Burton 1988). Assuming the packaging material is impermeable to oxygen (certainly true for multilayered paperboard cartons), the change in dissolved oxygen content in UHT milk during storage is determined by the amount of headspace in the package and by the presence of oxidizable material such as sulfhydryl compounds, folic and ascorbic acids and other vitamins, and unsaturated lipids, in milk. Oxygen levels in non-sterile milks decrease with bacterial growth, but sterile milks such as UHT milk should not contain bacteria that can grow under the normal conditions of storage. The volume of headspace in a package depends on the type of package. Perkins *et al.* (2005a) reported that a 1 litreTetra Brik paperboard carton made *in situ* from a roll of paperboard immediately before filling has 7–8 mL, whereas a 1 litre preformed paperboard carton (Combibloc) has 21–40 mL and a 1 litre plastic bottle has 55–63 mL of headspace. If the headspace volume is small, as for Tetra Brik cartons, the concentration of dissolved oxygen in UHT milk decreases during storage (Adhikari and Singhal 1992; Perkins *et al.* 2005b; Al-Attabi *et al.* 2014) due to its consumption in oxidation of oxidizable components. Schroder (1983) showed that oxidized flavour did not develop in UHT milk if the container was "filled to capacity". The dissolved oxygen was initially 6.5 mg/L, but this was reduced to 0 in the first 8 days due to oxidation of sulfhydryl compounds and ascorbic acid. On the other hand, if the headspace is relatively large, the milk will remain saturated with oxygen during storage. One method of reducing the amount of oxygen in the headspace is to flush with nitrogen before closure of the package. This is particularly useful for products containing added oxygen-sensitive ingredients such as ascorbic acid.

To overcome the risk of light-induced flavours, most UHT milk is now packaged in opaque or coloured plastic containers, plastic containers with a light-impervious oversleeve or paperboard cartons. Paperboard containers with an aluminium light barrier are the most effective. However, manufacturers would like to allow consumers to be able to see the product in the container, and for that reason, some milk is packaged in clear plastic or glass bottles. Glass bottles have been found to be suitable for some products such as sterilized flavoured milk; this may be due to the masking of any light-induced flavour by the added flavours and colours, or it may relate to a protective effect of the glass. Brown glass bottles have been shown to be preferable to clear glass bottles for UHT milk (Biewendt and Prokopek 1992).

Rysstad *et al.* (1998) assessed the development of light-induced flavour of UHT milk packaged aseptically in three types of paperboard carton: polyethylene-coated (PE), polyethylene coated with a non-foil paperboard light barrier (X-board) and polyethylene coated with an aluminium foil light barrier (foil). The light transmission percentages through the cartons at 400 nm were 0.2, 0 and 0, respectively, and at 500 nm were 6, 0.2 and 0, respectively. These were all below the IDF recommendations of 2 and 8% at 400 and 500 nm, respectively. However, when the cartons were exposed to strong light at 6 °C, light-induced flavours were detected after 2 and 8 weeks for the PE and X-board cartons, but no change in flavour could be detected in the milk in the foil cartons. These results show clearly the benefit of the aluminium foil light barrier which is a common feature of UHT paperboard-based packaging.

Petrus *et al.* (2009) studied the stability of UHT packaged in HDPE pigmented with titanium dioxide. They concluded that despite the limited light barrier, the milk exhibited good stability with shelf life of 4–11 weeks. Such light protection may be acceptable for ESL milk but not for UHT milk which should have a shelf life of up to 9 months.

11.3.4 Dairy Products

11.3.4.1 Cheese

In terms of susceptibility to oxidation, cheese differs considerably from milk. It has a reduced pH, and, importantly, most is produced with, and contains, microorganisms which absorb oxygen. Consequently, the redox potential (Eh) is reduced from +300–349 mV in milk to −117–200 mV in normal *Cheddar* cheese. The reducing effect of the microorganisms is apparent when cheese is made with GDL (delta-gluconolactone) as an acidulant rather than starter bacteria; it has an Eh of ~300 mV, about the same as milk (Green and Manning 1982; Caldeo and McSweeney 2012).

In addition to the increased reducing environment in cheese, the products of proteolysis during ripening, peptides and amino acids, appear to act as antioxidants on the propagating step in lipid oxidation. Cheese with a high degree of proteolysis has been shown to have a lower level of the oxidation markers, hexanal and heptanal, than cheese with less proteolysis (Dalsgaard *et al.* 2012). Interestingly, protein oxidation as shown by di-tyrosine levels was increased in the cheese with a high proteolysis level. The authors suggested that the formation of di-tyrosine may also contribute to a reduction in lipid oxidation as it is the result of a termination reaction.

Light exposure of cheese initiates oxidation and decreases its sensory acceptability. This occurs particularly with sliced cheese, packaged in transparent packing materials which generally transmit more than 80% of the incident light in the wavelength range 400–800 nm (Juric *et al.* 2003). Such products have a large surface area exposed to light and the surrounding atmosphere and are, thus, more susceptible to oxidatively induced changes. As a result, they suffer discoloration and off-flavour development (Deger and Ashoor 1987). These changes have been reviewed by Mortensen *et al.* (2004).

Autoxidation occurs in cheese during ripening and has been shown to correlate (negatively) with flavour score. It also increases with increasing storage temperature which makes cheese subjected to accelerated ripening more susceptible to oxidation and off-flavour production. Batool *et al.* (2018) demonstrated the effect by ripening *Cheddar* cheese at 18 °C with or without the addition of the antioxidants, vitamin E and selenium (added as selenomethionine). After 6 weeks' storage, cheese with the added antioxidants had much lower levels of oxidation than cheese without antioxidants and had flavour scores similar to cheese stored at 4–6 °C for 9 months. Control cheese which did not contain the antioxidants had unacceptable oxidation levels after 6 weeks at 18 °C.

11.3.4.2 Butter

Oxidation is a major cause of spoilage of butter although its significance has declined since the removal of copper from milk, cream and butter-handling equipment. It is enhanced by light, oxygen, heat and traces of metal (Badings 1970; Kessler 1981). To minimize the risk of oxidation, the air content of butter should be as low as possible. Veberg *et al.* (2007) showed that exposure to nitrogen rather than air greatly reduced oxidation. Therefore, butter packaging materials should be impermeable to light and oxygen and not contain pro-oxidant metals such as copper which act as catalysts for oxidation.

11.3.4.3 Powders

Milk powder is produced using the following steps: milk standardization, heat treatment, concentration (usually by evaporation), homogenization, drying (usually spray-drying) and packaging. Skimmed milk powder (SMP) and whole-milk powder (WMP) are the most common, but several other dairy-based powders are produced. This chapter is specifically concerned with fat-containing powders, particularly WMP.

WMP with a fat content of 25–28% and large surface area is highly susceptible to spoilage during storage, particularly through lipid oxidation. More than 60 aroma-active compounds have been reported in stored WMP with the most important being the result of lipid oxidation: hexanal and other aldehydes and ketones (Clarke *et al.* 2019).

The preheat treatment of the milk varies according to the powder being produced. In WMP manufacture, the heat treatment is 88–95 °C for 15–30 s. The temperature used for WMP is higher than that used for low- and medium-heat SMP. The rationale for the higher temperature is that it allows production of sulphydryl groups which are oxidizable and hence can reduce lipid oxidation of the milk fat. Even so, off-flavours attributable to lipid oxidation can develop after 3–6 months of storage at room temperature (Carunchia Whetstine and Drake 2007).

Stapelfeldt *et al.* (1997) investigated the effect of preheat treatment on the oxidative stability and found high-heat powder (preheat: 88 °C for 20 s) was considerably more stable than low-and medium-heat powders (preheat: 73 °C for 20 s and 88° for 20 s, respectively). This was evident in the level of free radicals, free sulphydryl groups and TBARS. In addition, they reported greater stability at 25 than 45 °C and at water activities of 0.11 and 0.23% than at 0.31%. Romeu-Nadal *et al.* (2007) also reported the greater stability of powders at lower storage temperature. Both non-supplemented powders and powders supplemented with polyunsaturated fatty acids were more stable to oxidation at 25 than at 37 °C. Some authors have found that the rate of oxidation is not significantly reduced by lowering the temperature of storage. This is attributed to the fact that autoxidation is a chemical reaction with a low activation energy (4–5 kcal/mole for the first step and 6–14 kcal/mole for the second step (Hamilton 1983)). The detrimental effect of elevated relative humidity on the oxidative stability of milk powders was also shown for powders containing encapsulated milk fat; they were less stable at 14 and 44% than at 52% (Hardas *et al.* 2002).

Exposure to light during storage of WMP has a dramatic effect on oxidation. This was demonstrated by Ulberth and Roubicek (1995) using a steam distillation-GC method to quantify the aldehydes formed. A comparison of the levels of aldehydes after storage for 80 or 130 days in the dark or in daylight is shown in Table 11.4. Hexanal was over 90 times higher in the light-exposed powder after 130 days, compared with the level in the powder stored in the dark. This highlights the need for packaging which excludes light. Exclusion of oxygen is also important. Lloyd *et al.* (2009) reported that WMP stored under nitrogen at 23 °C did not deteriorate over 12 months whereas air-stored WMP developed painty flavours and had higher peroxide values

Table 11.4. Aldehyde levels (μg/kg) in whole-milk powder during storage in the dark and in daylight

Aldehyde	80 days' storage		130 days' storage	
	In dark at 30 °C	In daylight at 20–22 °C	In dark at 30 °C	In daylight at 20–22 °C
Hexanal	0.07	2.5	0.15	14
Heptanal	0.06	0.75	0.11	5
Nonanal	0.02	0.5	0.05	1.5

Data from Ulberth and Roubicek (1995).

and lipid oxidation volatiles than the nitrogen-flushed WMP.

Oxidation of the fat in milk powders is also accompanied by oxidation of cholesterol and vitamin D. Several authors have studied cholesterol oxidation as some of the products of this reaction have been implicated in the development of some degenerative diseases (Otaegui-Arrazola *et al.* 2010). A significant positive correlation has been found between cholesterol oxidation and overall lipid oxidation in milk powder (McCluskey *et al.* 1997). Sieber (2005) concluded that the levels of cholesterol oxidation products in milk and dairy products are very small unless the products are stored under very harsh conditions. This accords with the finding by Kristensen *et al.* (2001) where cholesterol oxidation in processed cheese was detectable only after long-term storage with exposure to light at high temperature. A significant negative correlation has been found between lipid oxidation and vitamin D concentration in stored simulated milk powders (Mahmoodani *et al.* 2018).

11.3.5 Analytical Methods

Many methods have been used for measurement of lipid oxidation in foods (Blanca *et al.* 2013). Traditional methods include peroxide value and TBARS. Peroxide value is a measure of the primary stage of oxidation, while TBARS is a measure of secondary oxidation products. Mehta *et al.* (2018) reported that peroxide values showed a higher correlation with sensory evaluation of oxidized anhydrous milk fat (ghee) than some other measures which have been used for estimating the result of primary oxidation, namely, weight gain, conjugated dienes, Kreis number and iodine value. In assessing methods to measure secondary oxidation in ghee, Mehta *et al.* (2015) found that carbonyl value showed the highest correlation (r = 0.664) with flavour score, followed by TBARS (r = 0.521) and anisidine value (r = 0.356).

Hexanal, a major secondary oxidation product, and other aldehydes and ketones are now commonly measured as oxidation indicators. Analytical methods for these compounds include steam distillation-gas chromatography (Ulberth and Roubicek 1995) or solid-phase microextraction or dynamic headspace sampling followed by gas chromatography-mass spectrometry (Marsili 1999; Clarke *et al.* 2019). Rohfritsch (2019) recently reported the analysis of oxidation-derived carbonyl compounds in milk powder using ultra-high-pressure liquid chromatography (UPLC) coupled with mass spectrometry after derivatization with 7-(diethylamino)-2-oxochromene-3carbohydrazide (CHH). This method was used successfully to monitor docosahexaenoic acid (DHA)-specific oxidative markers in milk powder enriched with fish oils.

Other instrumental methods include fluorescence spectroscopy and electron spin resonance (ESR) spectroscopy. Front-face fluorescence has been shown to correlate well with sensory evaluation of photooxidation in butter (Veberg *et al.* 2007). The emission spectra of the oxidation products with excitation at 382 nm showed a maximum at 465–470 nm which was absent when oxidation was prevented by exposure of the butter to nitrogen rather than air.

ESR measures free radicals and has also been shown to correlate highly with sensory descriptors for oxidation (Hedegaard *et al.* 2006). According to Stapelfeldt *et al.* (1997), ESR is useful for detection of early events in lipid oxidation in milk powder. Thomsen *et al.* (2005) identified two different free radicals in WMP during storage: the one with a narrow ESR spectrum, which was depleted or removed with oxygen, was attributed to the oxidation process, and the other with a much wider ESR spectrum was linked to late-stage Maillard reaction products.

Another approach to determining the onset of oxidation in milk is to measure the decline in antioxidative capacity. According to Smet *et al.* (2008), oxidation occurs when the balance of anti- and pro-oxidants is disturbed. They demonstrated that, during storage, the antioxidative capacity decreases, and a measure of this capacity, by either the ferric reducing antioxidant power (FRAP) method or the DPPH method, was a useful indicator of susceptibility to oxidation. A similar approach was taken by Wold *et al.* (2005) and Veberg *et al.* (2007) who measured the breakdown of photosensitizers by fluorescence spectroscopy as an indirect measure of the initiation of lipid photooxidation.

References

Abeni, F., Degano, L., Calza, F., Giangiacomo, R. & Pirlo, G. (2005). Milk quality and automatic milking: Fat globule size, natural creaming, and lipolysis. *Journal of Dairy Science, 88*, 3519–3529.

Adams, D. M. & Brawley, T. G. (1981). Heat resistant bacterial lipases and ultra-high temperature sterilization of dairy products. *Journal of Dairy Science, 64*, 1951–1957.

Adhikari, A. K. & Singhal, O. P. (1992). Effect of dissolved oxygen content on the flavour profile of UHT milk during storage. *Australian Journal of Dairy Technology, 47*, 1–6.

Airado-Rodriguez, D., Intawiwat, N., Skaret, J. & Wold, J. P. (2011). Effect of naturally occurring tetrapyrroles on photooxidation in cow's milk. *Journal of Agricultural and Food Chemistry, 59*, 3905–3914.

Al-Attabi, Z., D'Arcy, B. R. & Deeth, H. C. (2014). Volatile sulfur compounds in pasteurised and UHT milk during storage. *Dairy Science & Technology, 94*, 241–253.

Amores, G. & Virto, M. (2019). Total and free fatty acids analysis in milk and dairy fat. *Separations, 6*, 14.

Anderson, M. (1979). Enzyme immunoassay for measuring lipoprotein lipase activator in milk. *Journal of Dairy Science, 62*, 1380–1383.

Anderson, M. (1981). Inhibition of lipolysis in bovine milk by proteose peptone. *Journal of Dairy Research, 48*, 247–252.

Anderson, M., Heeschen, W., Jellema, A., Kuzdzal-Savoie, S., Needs, E.C., Suhren, G. & van Reusel, A. (1991). *Determination of free fatty acids in milk and milk products*. Bulletin 265. Brussels: International Dairy Federation.

Andrewes, P. (2018). Indirect detection of lipase in UHT milk by measuring methyl ester formation. *International Dairy Journal, 79*, 1–4.

Aurand, L. W., Chu, T. M., Singleton, J. A. & Shen, R. (1967). Xanthine oxidase activity and development of spontaneously oxidized flavor in milk. *Journal of Dairy Science, 50*, 465–471.

Azzara, C. D. & Dimick, P. S. (1985). Lipolytic enzyme activity of macrophages in bovine mammary-gland secretions. *Journal of Dairy Science, 68*, 1804–1812.

Bachman, K. C. & Wilcox, C. J. (1990). Effect of blood and high-density lipoprotein preparations upon lipase distribution and spontaneous lipolysis in bovine milk. *Journal of Dairy Science, 73*, 3393–3401.

Badings, H. T. (1970). Cold-storage defects in butter and their relation to the autoxidation of unsaturated fatty acids. *Netherlands Milk and Dairy Journal, 24*(3/4), 147–257.

Barbano, D. M., Bynum, D. & Senyk, G. F. (1983). Influence of reverse osmosis on milk lipolysis. *Journal of Dairy Science, 66*, 2447–2451.

Barrefors, P., Granelli, K., Appelqvist, L. A. & Bjoerck, L. (1995). Chemical characterization of raw milk samples with and without oxidative off-flavor. *Journal of Dairy Science, 78*, 2691–2699.

Batool, M., Nadeem, M., Imran, M., Gulzar, N., Shahid, M. Q., Shahbaz, M., Ajmal, M. & Khan, I. T. (2018). Impact of vitamin E and selenium on antioxidant capacity and lipid oxidation of cheddar cheese in accelerated ripening. *Lipids in Health and Disease, 17*, 79.

Bell, L. I. & Parsons, J. G. (1977). Factors affecting lipase flavor in butter. *Journal of Dairy Science, 60*, 117–122.

Biewendt, H. G. & Prokopek, D. (1992). Light-induced changes and the light protective effects of brown glass bottles for directly heated UHT milk. *Molkerei Zeitung Welt der Milch, 46*(23/24), 676–682.

Bills, D. D. & Day, E. A. (1964). Determination of the major free fatty acids in Cheddar cheese. *Journal of Dairy Science, 47*, 733–738.

Blake, M. R., Weimer, B. C., McMahon, D. J. & Savello, P. A. (1995). Sensory and microbial quality of milk processed for extended shelf-life by direct steam injection. *Journal of Food Protection, 58*, 1007–1013.

Blanca, B., Iciar, A. & Diana, A. (2013). A review of analytical methods measuring lipid oxidation status in foods: A challenging task. *European Food Research and Technology, 236*, 1–15.

Borle, F., Sieber, R. & Bosset, J. O. (2001). Photo-oxidation and photoprotection of foods, with particular reference to dairy products – An update of a review article (1993-2000). *Sciences des Aliments, 21*, 571–590.

Bosset, J. O., Gallmann, P. U. & Sieber, R. (1994). Influence of light transmittance of packaging materials on the shelf-life of milk and dairy products – A review. In M. Mathlouthi (Ed.), *Food packaging and preservation* (pp. 222–268). Glasgow: Blackie Academic and Professional.

Bråthen, G. (1980). Lipolysis in milk. Automated determination of the acidity value with the AutoAnalyzer and assessment of results. *Meieriposten, 69*, 345–352.

Brothersen, C., McMahon, D. J., Legako, J. & Martini, S. (2016). Comparison of milk oxidation by exposure to LED and fluorescent light. *Journal of Dairy Science, 99*, 2537–2544.

Buffa, M., Guamis, B., Pavia, M. & Trujillo, A. J. (2001). Lipolysis in cheese made from raw, pasteurized or high-pressure-treated goats' milk. *International Dairy Journal, 11*, 175–179.

Burton, H. (1988). *Ultra-high temperature processing of milk and milk products*. London: Elsevier/Applied Science.

Button, P. D. (2007). *UHT Milk spoilage owing to lipase and protease from Psychrotrophic pseudomonas species: Characteristics of enzymatic breakdown, improved assays for enzyme detection and sources of the bacteria*. PhD thesis, University of Melbourne, Australia.

Caldeo, V. & McSweeney, P. L. H. (2012). Changes in oxidation-reduction potential during the simulated manufacture of different cheese varieties. *International Dairy Journal, 25*, 16–20.

Calderon-Santiago, M., Peralbo-Molina, A., Priego-Capote, F. & de Castro, M. D. L. (2012). Cholesterol oxidation products in milk: Processing formation and determination. *European Journal of Lipid Science and Technology, 114*, 687–694.

Calligaris, S., Manzocco, L., Anese, M. & Nicoli, M. C. (2004). Effect of heat-treatment on the antioxidant and pro-oxidant activity of milk. *International Dairy Journal, 14*, 421–427.

Cartier, P. & Chilliard, Y. (1990). Spontaneous lipolysis in bovine milk: Combined effects of nine characteristics in native milk. *Journal of Dairy Science, 73*, 1178–1186.

Cartier, P., Chilliard, Y. & Chazal, M.-P. (1984). Dosage de l'activité lipasique et des acides gras libres du lait par titration automatic colorimétrique. *Lait, 64*, 340–355.

Cartier, P., Chilliard, Y. & Paquet, D. (1990). Inhibiting and activating effects of skim milks and proteose-peptone fractions on spontaneous lipolysis and purified lipoprotein lipase activity in bovine-milk. *Journal of Dairy Science, 73*, 1173–1177.

Carunchia Whetstine, M. E. & Drake, M. A. (2007). The flavor and flavor stability of skim and whole milk powders. In K. R. Cadwallader, M. A. Drake & R. McGorrin (Eds.), *Flavor of dairy products* (pp. 217–251). Washington DC: ACS Publishing.

Castberg, H. B. & Solberg, P. (1974). The lipoprotein lipase and the lipolysis in bovine milk. *Meieriposten, 63*, 961–975.

Celestino, E. L., Iyer, M. & Roginski, H. (1997). Reconstituted UHT-treated milk: Effects of raw milk, powder quality and storage conditions of UHT milk on its physico-chemical attributes and flavour. *International Dairy Journal, 7*, 129–140.

Chapman, K. W., Whited, L. J. & Boor, K. J. (2002). Sensory threshold of light-oxidized flavor defects in milk. *Journal of Food Science, 67*, 2770–2773.

Chavarri, F., Virto, M., Martin, C., Najera, A. I., Santisteban, A., Barron, L. J. R. & DeRenobales, M. (1997). Determination of free fatty acids in cheese: Comparison of two analytical methods. *Journal of Dairy Research, 64*, 445–452.

Chawla, R., Kaur, H. & Tyagi, K. (2003). Influence of vitamin E and beta-carotene supplementation on spontaneous oxidised flavour of milk in dairy cows. *The Indian Journal of Animal Sciences, 73*, 807–811.

Chen, Z. F., Leinisch, F., Greco, I., Zhang, W., Shu, N., Chuang, C. Y., Lund, M. N. & Davies, M. J. (2019). Characterisation and quantification of protein oxidative modifications and amino acid racemisation in powdered infant milk formula. *Free Radical Research, 53*, 68–81.

Chilliard, Y., Bauchart, D., Cartier, P. & Chazal, M.-P. (1983). *Étalonnage, comparaison et automatisation de differentes méthodes de dosage des acides gras libres du lait de vache*. INRA-ITEB (France), No. 84031.

Chilliard, Y., Selseletattou, G., Bas, P. & Morandfehr, P. (1984). Characteristics of lipolytic system in goat milk. *Journal of Dairy Science, 67*, 2216–2223.

Chilliard, Y., Ferlay, A., Rouel, J. & Lamberett, G. (2003). A review of nutritional and physiological factors affecting goat milk lipid synthesis and lipolysis. *Journal of Dairy Science, 86*, 1751–1770.

Chilliard, Y., Rouel, J. & Guillouet, P. (2013). Goat alpha-S_1 casein genotype interacts with the effect of extruded linseed feeding on milk fat yield, fatty acid composition and post-milking lipolysis. *Animal Feed Science and Technology, 185*, 140–149.

Chilliard, Y., Toral, P. G., Shingfield, K. J., Rouel, J., Leroux, C. & Bernard, L. (2014). Effects of diet and physiological factors on milk fat synthesis, milk fat composition and lipolysis in the goat: A short review. *Small Ruminant Research, 122*, 31–37.

Choi, I. W. & Jeon, I. J. (1993). Patterns of fatty acids released from milk fat by residual lipase during storage of ultra-high temperature processed milk. *Journal of Dairy Science, 76*, 78–83.

Christmass, M. A., Mitoulas, L. R., Hartmann, P. E. & Arthur, P. G. (1998). A semiautomated enzymatic method for determination of non-esterified fatty acid concentration in milk and plasma. *Lipids, 33*, 1043–1049.

Clarke, H. J., Mannion, D. T., O'Sullivan, M. G., Kerry, J. P. & Kilcawley, K. N. (2019). Development of a headspace solid-phase microextraction gas chromatography mass spectrometry method for the quantification of volatiles associated with lipid oxidation in whole milk powder using response surface methodology. *Food Chemistry, 292*, 75–80.

Claypool, L. L. (1965). *Studies on milk lipase activation*, PhD thesis, University of Minnesota.

Cogan, T. M. (1977). A review of heat resistant lipases and proteinases and the quality of dairy products. *Irish Journal of Food Science and Technology, 1*, 95–105.

Collomb, M. & Spahni, M. (1995). Review of methods used for determination of free fatty acids in milk and milk products. *Lebensm Wissensch Technol, 28*, 355–379.

Craven, H. M. & Macauley, B. J. (1992). Microorganisms in pasteurised milk after refrigerated storage. 3. Effect of milk processor. *Australian Journal of Dairy Technology, 47*, 50–55.

de la Fuente, M. A. & Juarez, M. (1993). Review: Determination of free fatty acids in milk products. *Rev Esp Cie Tecnol Aliment, 33*, 247–269.

Dairy Farmers of Ontario. (2018). *Interim grass-fed milk standard protocol*. https://www.milk.org/Corporate/PDF/GrassFedProtocol.pdf.

Dalsgaard, T. K., Nielsen, J. H., Brown, B. E., Stadler, N. & Davies, M. J. (2011). Dityrosine, 3,4-dihydroxyphenylalanine (DOPA), and radical formation from tyrosine residues on milk proteins with globular and flexible structures as a result of riboflavin-mediated photo-oxidation. *Journal of Agricultural and Food Chemistry, 59*, 7939–7947.

Dalsgaard, T. K., Bakman, M., Hammershoj, M., Sorensen, J., Nebel, C., Albrechtsen, R., Vognsen, L. & Nielsen, J. H. (2012). Light-induced protein and lipid oxidation in low-fat cheeses: Effect on degree of enzymatic hydrolysis. *International Journal of Dairy Technology, 65*, 57–63.

Danthine, S. & Blecker, C. (2014). Interactions of lipases with milk fat globule membrane monolayers using a Langmuir film balance. *International Dairy Journal, 35*, 81–87.

Davies, M. J. (2016). Protein oxidation and peroxidation. *Biochemical Journal, 473*, 805–825.

De Marchi, M., Penasa, M. & Cassandro, M. (2017). Comparison between automatic and conventional milking systems for milk coagulation properties and fatty acid composition in commercial dairy herds. *Italian Journal of Animal Science, 16*, 363–370.

Deegan, K. C. & McSweeney, P. L. H. (2006). Enhancement of lipolysis in cheese during ripening by low pressure homogenisation of cheese milk. *Australian Journal of Dairy Technology, 61*, 212–212.

Deeth, H. C. (2006). Lipoprotein lipase and lipolysis in milk. *International Dairy Journal, 16*, 555–562.

Deeth, H. C. (2011). Milk lipids | lipolysis and hydrolytic rancidity. In J. W. Fuquay, P. F. Fox & P. L. H. McSweeney (Eds.), *Encyclopedia of dairy sciences* (Vol. 3, 2nd ed., pp. 721–726). San Diego: Academic Press.

Deeth, H. C. (2015). Milk lipids: Lipolysis and hydrolytic rancidity. In *Reference module in food sciences* (pp. 1–6). Elsevier. https://doi.org/10.1016/B978-0-08-100596-5.00925-2.

Deeth, H. C. (2017). Optimum thermal processing for extended shelf-life (ESL) milk. *Food, 6*, 102.

Deeth, H. C. & Fitz-Gerald, C. H. (1975a). *Factors governing the susceptibility of milk to spontaneous lipolysis*. Doc 86 (pp. 24–34). Brussels: International Dairy Federation.

Deeth, H. C. & Fitz-Gerald, C. H. (1975b). The relevance of milk lipase activation to rancidity in Cheddar cheese. *Australian Journal of Dairy Technology, 30*, 74–76.

Deeth, H. C. & Fitz-Gerald, C. H. (1977). Some factors involved in milk lipase activation by agitation. *Journal of Dairy Research, 44*, 569–583.

Deeth, H. C. & Fitz-Gerald, C. H. (1978). Effects of mechanical agitation of raw milk on the milk fat globule in relation to the level of induced lipolysis. *Journal of Dairy Research, 45*, 373–380.

Deeth, H. C. & Fitz-Gerald, C. H. (1983). Lipolytic enzymes and hydrolytic rancidity in milk and milk products. In P. F. Fox (Ed.), *Developments in dairy chemistry – 2 lipids* (pp. 195–239). London: Applied Science.

Deeth, H. C. & Fitz-Gerald, C. H. (1995). Lipolytic enzymes and hydrolytic rancidity in milk and dairy products. In P. F. Fox (Ed.), *Advanced dairy chemistry – 2 lipids* (pp. 247–308). London: Chapman Hall Publishers.

Deeth, H. C. & Fitz-Gerald, C. H. (2006). Lipolytic enzymes and hydrolytic rancidity. In P. F. Fox & P. L. H. McSweeney (Eds.), *Advanced dairy chemistry – 2 lipids* (pp. 481–556). New York: Springer.

Deeth, H. C. & Lewis, M. J. (2017). *High temperature processing of Milk and Milk products* (pp. 264–266). Oxford: Wiley Blackwell.

Deeth, H. C. & Smith, R. A. D. (1983). Lipolysis and other factors affecting the steam frothing capacity of milk. *Australian Journal of Dairy Technology, 38*, 14–19.

Deeth, H. C. & Touch, V. (2000). Methods for detecting lipase activity in milk and milk products. *Australian Journal of Dairy Technology, 55*, 153–168.

Deeth, H. C., Fitz-Gerald, C. H. & Wood, A. F. (1975). A convenient method for determining the extent of lipolysis. *Australian Journal of Dairy Technology, 30*, 109–111.

Deeth, H. C., Fitz-Gerald, C. H. & Wood, A. F. (1979). Lipolysis and butter quality. *Australian Journal of Dairy Technology, 34*, 146–149.

Deeth, H. C., Fitz-Gerald, C. H. & Snow, A. J. (1983). A gas chromatographic method for the quantitative determination of free fatty acids in milk and milk products. *New Zealand Journal of Dairy Science and Technology, 18*, 13–20.

Deeth, H. C., Khusniati, T., Datta, N. & Wallace, R. B. (2002). Spoilage patterns of skim and whole milks. *Journal of Dairy Research, 68*, 227–241.

Deeth, H. C., Datta, N. & Versteeg, C. (2013). Non-thermal technologies. In G. Smithers & M. A. Augustin (Eds.), *Advances in dairy ingredients* (pp. 161–216). Oxford: Wiley-Blackwell.

Deger, D. & Ashoor, S. H. (1987). Light-induced changes in taste, appearance, odor, and riboflavin content of cheese. *Journal of Dairy Science, 70*, 1371–1376.

Dimick, P. S. (1973). Effect of fluorescent light on flavor and selected nutrients of homogenized milk held in conventional containers. *Journal of Milk and Food Technology, 36*, 383–387.

Dimick, P. S. (1982). Photochemical effects on flavor and nutrients of fluid milk. *Canadian Institute of Food Science and Technology Journal, 15*, 247–256.

Donot, F., Strub, C., Fontana, A., Jouy, N., Delbes, C., Gunata, Z. & Schorr-Galindo, S. (2016). Rapid analysis and quantification of major neutral lipid species and free fatty acids by HPLC-ELSD from microalgae. *European Journal of Lipid Science and Technology, 118*, 1550–1556.

Driessen, F. M. (1976). Comparative-study of lipase in bovine colostrum and in bovine milk. *Netherlands Milk and Dairy Journal, 30*(3–4), 186–196.

Drossler, P., Holzer, W., Penzkofer, A. & Hegemann, P. (2002). pH dependence of the absorption and emission behavior of riboflavin and aqueous solution. *Chemical Physics, 282*, 429–439.

Duncan, S. E. & Webster, J. B. (2010). Oxidation and protection of milk and dairy products. In E. A. Decker, R. J. Elias & D. J. McClements (Eds.), *Oxidation in foods and beverages and antioxidant applications, Vol 2: Management in different industry sectors* (pp. 121–155). London: Elsevier Applied Science.

Elgersma, A. (2015). Grazing increases the unsaturated fatty acid concentration of milk from grass-fed cows: A review of the contributing factors, challenges and future perspectives. *European Journal of Lipid Science and Technology, 117*, 1345–1369.

Elliott, J. M., Dehaan, B. & Parkin, K. L. (1989). An improved liquid-chromatographic method for the quantitative-determination of free fatty-acids in milk-products. *Journal of Dairy Science, 72*, 2478–2482.

Evers, J. M. (2003). Determination of free fatty acids in milk using the BDI method - some practical and theoretical aspects. *International Dairy Journal, 13*, 111–121.

Fauteux, M. C., Gervais, R., Rico, D. E., Lebeuf, Y. & Chouinard, P. Y. (2016). Production, composition, and oxidative stability of milk highly enriched in polyunsaturated fatty acids from dairy cows fed alfalfa protein concentrate or supplemental vitamin E. *Journal of Dairy Science, 99*, 1111–1126.

Fitz-Gerald, C. H. & Deeth, H. C. (1983). Factors influencing lipolysis by cultures of some psychrotrophic microorganisms. *Australian Journal of Dairy Technology, 38*, 97–103.

Forss, D. A. (1979). Review of the progress of dairy science: Mechanisms of formation of aroma compounds in milk and milk products. *Journal of Dairy Research, 46*, 691–706.

Foss. (2018). *MilkoScan 7 RM brochure*. https://www.fossanalytics.com.

Fox, P. F. & Law, B. A. (1991). Enzymology of cheese ripening. *Food Biotechnology, 5*, 239–262.

Frega, N., Mozzon, M. & Lercker, G. (1999). Effects of free fatty acids on oxidative stability of vegetable oil. *Journal of the American Oil Chemists' Society, 76*, 325–329.

Fromm, H. I. & Boor, K. J. (2004). Characterization of pasteurized fluid milk shelf-life attributes. *Journal of Food Science, 69*, M207–M214.

Garcia, H. S., Reyes, H. R., Malcata, F. X., Hill, C. G. & Amundson, C. H. (1990). Determination of the major free fatty acids in milkfat using a three-component mobile phase for HPLC analysis. *Milchwissenschaft, 45*, 757–759.

Garcia-Martinez, M. C. & Marquez-Ruiz, G. (2009). Lipid oxidation in functional dairy products. *Current Nutrition & Food Science, 5*, 209–216.

Garcia-Martinez, C., Holgado, F., Velasco, J. & Marquez-Ruiz, G. (2012). Effect of classic sterilization on lipid oxidation in model liquid milk-based infant and follow-on formulas. *European Journal of Lipid Science and Technology, 114*, 1373–1380.

Granelli, K., Barrefors, P., Bjorck, L. & Appelqvist, L. A. (1998). Further studies on lipid composition of bovine milk in relation to spontaneous oxidised flavour. *Journal of the Science of Food and Agriculture, 77*, 161–171.

Green, M. L. & Manning, D. J. (1982). Development of texture and flavour in cheese and other fermented products. *Journal of Dairy Research, 49*, 737–748.

Greenbank, G. R. (1948). The oxidized flavor in milk and dairy products – A review. *Journal of Dairy Science, 31*, 913–933.

Grosch, W. (1982). Lipid degradation products and flavour. In I. D. Morton & A. J. MacLeod (Eds.), *Food flavours. Part A. introduction* (pp. 325–398). New York: Elsevier.

Guerzoni, M. E., Vannini, L., Chaves-López, C., Lanciotti, R., Suzzi, G. & Gianotti, A. (1999). Effect of high-pressure homogenization on microbial and chemico-physical characteristics of goat cheeses. *Journal of Dairy Science, 82*, 851–862.

Gurr, M., Harwood, J. L., Frayn, K. N., Murphy, D. J. & Michell, R. H. (2016). *Lipids: Biochemistry, biotechnology and health*. Chichester: Wiley Blackwell.

Hamilton, R. T. (1983). The chemistry of rancidity in foods. In J. C. Allen & R. J. Hamilton (Eds.), *Rancidity in foods* (pp. 1–20). London: Applied Science Publishers.

Hardas, N., Danviriyakul, S., Foley, J. L., Nawar, W. W. & Chinachoti, P. (2002). Effect of relative humidity on the oxidative and physical stability of encapsulated milk fat. *Journal of the American Oil Chemists' Society, 79*, 151–158.

Havemose, M. S., Weisbjerg, M. R., Bredie, W. L. P. & Nielsen, J. H. (2004). Influence of feeding different types of roughage on the oxidative stability of milk. *International Dairy Journal, 14*, 563–570.

Hedegaard, R. V., Kristensen, D., Nielsen, J. H., Frost, M. B., Ostdal, H., Hermansen, J. E., Kroger-Ohlen, M. & Skibsted, L. H. (2006). Comparison of descriptive sensory analysis and chemical analysis for oxidative changes in milk. *Journal of Dairy Science, 89*, 495–504.

Hickey, D. K., Kilcawley, K. N., Beresford, T. P. & Wilkinson, M. G. (2007). Lipolysis in cheddar cheese made from raw, thermized, and pasteurized milks. *Journal of Dairy Science, 90*, 47–56.

Ikens, W. G., Kwak, H. S., Zink, G. S. & Jeon, I. J. (1988). Comparison of methods for quantification of free fatty acids in cheese. *Journal of Food Science, 53*, 1915–1916.

Jeon, I. J., Thomas, E. L. & Reineccius, G. A. (1978). Production of volatile flavour compounds in ultrahigh-temperature processed milk during aseptic storage. *Journal of Agricultural and Food Chemistry, 26*, 1183–1188.

Johnson, M. E. (2011). Preparation of cheese milk. In J. W. Fuquay, P. F. Fox & P. L. H. McSweeney (Eds.), *Encyclopedia of dairy sciences* (Vol. 1, 2nd ed., pp. 544–551). San Diego: Academic Press.

Johnson, D. S., Duncan, S. E., Bianchi, L. M., Chang, H. H., Eigel, W. N. & O'Keefe, S. F. (2015). Packaging modifications for protecting flavor of extended-shelf-life milk from light. *Journal of Dairy Science, 98*, 2205–2214.

Joshi, N. S. & Thakar, P. N. (1994). Methods to evaluate deterioration of milk fat – A critical appraisal. *Journal of Food Science and Technology Mysore, 31*, 181–196.

Juhlin, J., Fikse, F., Lunden, A., Pickova, J. & Agenas, S. (2010a). Relative impact of alpha-tocopherol, copper and fatty acid composition on the occurrence of oxidized milk flavour. *Journal of Dairy Research, 77*, 302–309.

Juhlin, J., Fikse, W. F., Orde-Ostrom, I. L., Barrefors, P. & Lunden, A. (2010b). Factors relating to incidence of spontaneous oxidized flavor and copper in cow's milk. *Acta Agriculturae Scandinavica Section A – Animal Science, 60*, 94–103.

Juhlin, J., Fikse, W. F., Pickova, J. & Lunden, A. (2012). Association of DGAT1 genotype, fatty acid composition, and concentration of copper in milk with spontaneous oxidized flavor. *Journal of Dairy Science, 95*, 4610–4617.

Jurczak, M. E. & Sciubisz, A. (1981). Studies on the lipolytic changes in milk from cows with mastitis. *Milchwissenschaft, 36*, 217–219.

Juric, M., Bertelson, G., Mortenson, G. & Petersen, M. A. (2003). Light-induced colour and aroma changes in sliced, modified atmosphere packaged semi-hard cheeses. *International Dairy Journal, 13*, 239–249.

Kamath, S., Huppertz, T., Houlihan, A. V. & Deeth, H. C. (2008). The influence of temperature on the foaming of milk. *International Dairy Journal, 18*, 994–1002.

Karami, M. (2017). Enhancing the lipolysis of feta-type cheese made from ultrafiltered cow's milk. *LWT Food Science and Technology, 80*, 386–393.

Kessler, H. G. (1981). *Food engineering and dairy technology*. Freising: Verlag A. Kessler.

Kilcawley, K. N. (2006). High performance liquid chromatographic and gas chromatographic methods for lipid analyses. In P. F. Fox & P. L. H. McSweeney (Eds.), *Advanced dairy chemistry – 2 lipids* (pp. 675–696). New York: Springer.

Kilcawley, K., Faulkner, H., Clarke, H., O'Sullivan, M. & Kerry, J. (2018). Factors influencing the flavour of bovine milk and cheese from grass based versus non-grass based milk production systems. *Food, 7*, 37.

Kon, H. & Saito, Z. (1997). Factors causing temperature activation of lipolysis in cow's milk. *Milchwissenschaft, 52*, 435–440.

Koops, J., Klomp, H. & van Hemert, H. (1990). Rapid enzymatic assay of free fatty acids (lipolysis) in farm tank milk by a segmented continuous flow method. Comparison of the results with those obtained by the BDI procedure. *Netherlands Milk and Dairy Journal, 44*, 3–19.

Kristensen, D., Hansen, E., Arndal, A., Trinderup, R. A. & Skibsted, L. H. (2001). Influence of light and temperature on the colour and oxidative stability of processed cheese. *International Dairy Journal, 11*, 837–843.

Kuzdzal-Savoie, S. (1980). *Determination of free fatty acids in milk and milk products*. Bulletin. 118 (pp. 53–66). Brussels: International Dairy Federation.

La Terra, S., Marino, V. M., Manenti, M., Licitra, G. & Carpino, S. (2010). Increasing pasture intakes enhances polyunsaturated fatty acids and lipophilic antioxidants in plasma and milk of dairy cows fed total mix ration. *Dairy Science & Technology, 90*, 687–698.

Lanciotti, R., Lopez, C., Patrignani, F., Paparella, A., Guerzoni, M. E., Serio, A. & Suzzi, G. (2004). Effects of milk treatment with HPH on microbial population as well as on the lipolytic and proteolytic profiles of Crescenza cheese. *International Journal of Dairy Technology, 57*, 19–25.

Lanciotti, R., Vannini, L., Patrignani, F., Lucci, L., Vallicelli, M., Ndagijimana, M. & Guerzoni, M. E. (2006). Effect of high-pressure homogenisation of milk on cheese yield and microbiology, lipolysis and proteolysis during ripening of Caciotta cheese. *Journal of Dairy Research, 73*, 216–226.

Law, B. A. (1984). Flavour development in cheeses. In F. L. Davies & B. A. Law (Eds.), *Advances in the microbiology and biochemistry of cheese and fermented milk* (pp. 187–208). London: Elsevier Applied Science.

Law, B. A., Sharpe, M. E. & Chapman, H. R. (1976). The effect of lipolytic Gram-negative psychrotrophs in stored milk on the development of rancidity in Cheddar cheese. *Journal of Dairy Research, 43*, 459–468.

Li, Y. H., Wang, W. J., Zhang, F., Shao, Z. P. & Guo, L. (2019). Formation of the oxidized flavor compounds at different heat treatment and changes in the oxidation stability of milk. *Food Science & Nutrition, 7*, 238–246.

Lloyd, M. A., Hess, S. J. & Drake, M. A. (2009). Effect of nitrogen flushing and storage temperature on flavor and shelf-life of whole milk powder. *Journal of Dairy Science, 92*, 2409–2422.

Lu, C. Y., Wu, H. L., Chen, S. H., Kou, H. S. & Wu, S. M. (2002). Simple and sensitive analysis of long-chain free fatty acids in milk by fluorogenic derivatization and high-performance liquid chromatography.

Journal of Agricultural and Food Chemistry, 50, 71–73.
MacGibbon, A. K. H. & Taylor, M. W. (2006). Composition and structure of bovine milk lipids. In P. F. Fox & P. L. H. McSweeney (Eds.), *Advanced dairy chemistry – 2 lipids* (pp. 1–42). New York: Springer.
Mahmoodani, F., Perera, C. O., Abernethy, G., Fedrizzi, B. & Chen, H. (2018). Lipid oxidation and vitamin D3 degradation in simulated whole milk powder as influenced by processing and storage. *Food Chemistry, 261*, 149–156.
Mannion, D. T., Furey, A. & Kilcawley, K. N. (2016a). Comparison and validation of 2 analytical methods for the determination of free fatty acids in dairy products by gas chromatography with flame ionization detection. *Journal of Dairy Science, 99*, 5047–5063.
Mannion, D. T., Furey, A. & Kilcawley, K. N. (2016b). Free fatty acids quantification in dairy products. *International Journal of Dairy Technology, 69*, 1–12.
Mannion, D. T., Furey, A. & Kilcawley, K. N. (2019). Development and validation of a novel free fatty acid butyl ester gas chromatography method for the determination of free fatty acids in dairy products. *Journal of Agricultural and Food Chemistry, 67*, 499–506.
Marino, V. M., Schadt, I., La Terra, S., Manenti, M., Caccamo, M., Licitra, G. & Carpino, S. (2012). Influence of season and pasture feeding on the content of alpha-tocopherol and beta-carotene in milk from Holstein, Brown Swiss and Modicana cows in Sicily. *Dairy Science & Technology, 92*, 501–513.
Marino, V. M., Schadt, I., Carpino, S., Caccamo, M., La Terra, S., Guardiano, C. & Licitra, G. (2014). Effect of Sicilian pasture feeding management on content of alpha-tocopherol and beta-carotene in cow milk. *Journal of Dairy Science, 97*, 543–551.
Marsili, R. T. (1999). Comparison of solid-phase microextraction and dynamic headspace methods for the gas chromatographic mass spectrometric analysis of light-induced lipid oxidation products in milk. *Journal of Chromatographic Science, 37*, 17–23.
McCluskey, S., Connolly, J. F., Devery, R., Obrien, B., Kelly, J., Harrington, D. & Stanton, C. (1997). Lipid and cholesterol oxidation in whole milk powder during processing and storage. *Journal of Food Science, 62*, 331–337.
McNeill, G. P., O'Donoghue, A. & Connolly, J. F. (1986). Quantification and identification of flavour components leading to lipolytic rancidity in stored butter. *Irish Journal of Food Science Technology, 10*, 1–10.
McSweeney, P. L. H. (2011). Biochemistry of cheese ripening. In J. W. Fuquay, P. F. Fox & P. L. H. McSweeney (Eds.), *Encyclopedia of dairy sciences* (Vol. 1, 2nd ed., pp. 667–674). San Diego: Academic Press.
Mehta, R. S. (1980). Milk processed at ultra-high temperatures a review. *Journal of Food Protection, 43*(3), 212–225.
Mehta, B. M., Aparnathi, K. D. & Darji, V. B. (2015). Comparison of different methods of monitoring the secondary stage of oxidation of ghee. *International Journal of Dairy Technology, 68*, 589–594.
Mehta, B. M., Kumar Jain, A., Darji, V. B. & Aparnathi, K. D. (2018). Evaluation of different methods to monitor primary stage of oxidation of heat clarified milk fat (ghee). *Journal of Food Processing & Preservation, 42*, e13688.
Mestdagh, F., De Meulenaer, B., De Clippeleer, J., Devlieghere, F. & Huyghebaert, A. (2005). Protective influence of several packaging materials on light oxidation of milk. *Journal of Dairy Science, 88*, 499–510.
Miwa, H. & Yamamoto, M. (1990). Liquid chromatographic determination of free and total fatty acids in milk and milk products as their 2-nitrophenylhydrazides. *Journal of Chromatography, 523*, 235–246.
Moio, L., Etievant, P., Langlois, D., Dekimpe, J. & Addeo, F. (1994). Detection of powerful odorants in heated milk by use of extract dilution sniffing analysis. *Journal of Dairy Research, 61*, 385–394.
Mortensen, G., Bertelsen, G., Mortensen, B. & Stapelfeldt, J. (2004). Light-induced changes in packaged cheeses - a review. *International Dairy Journal, 14*, 85–102.
Nishikiori, M., Iizuka, H., Ichiba, H., Sadamoto, K. & Fukushima, T. (2015). Determination of free fatty acids in human serum by HPLC with fluorescence detection. *Journal of Chromatographic Science, 53*, 537–541.
Nishimura, K., Suzuki, T. & Itabashi, Y. (2002). Determination of free fatty acids in milk and dairy products by reversed-phase HPLC with fluorescence detection. *Journal of Food Hygienic Society of Japan, 43*, 230–233.
Noluma International. (2018). *Packaged foods should be kept in the dark: Consumers shouldn't be*. https://mma.prnewswire.com/media/798797/Noluma_White_Paper_FINAL_12_13.pdf?p=pdf.
O'Brien, N. M. & O'Connor, T. P. (2011). Lipid oxidation. In J. W. Fuquay, P. F. Fox & P. L. H. McSweeney (Eds.), *Encyclopedia of dairy sciences* (Vol. 1, 2nd ed., pp. 716–720). San Diego: Academic Press.
O'Connor, T. P. & O'Brien, N. M. (1995). Lipid oxidation. In P. F. Fox (Ed.), *Advanced dairy chemistry – 2 lipids* (2nd ed., pp. 309–348). London: Chapman and Hall.
O'Connor, T. P. & O'Brien, N. M. (2006). Lipid oxidation. In P. F. Fox & P. L. H. McSweeney (Eds.), *Advanced dairy chemistry – 2 lipids* (3rd ed., pp. 557–600). New York: Springer.
Ohren, J. A. & Tuckey, S. L. (1969). Relation of flavor development in Cheddar cheese to chemical changes in the fat of the cheese. *Journal of Dairy Science, 52*, 598–607.
Olivecrona, T., Egelrud, T., Castberg, H. B. & Solberg, P. (1975). *Is there more than one lipase in bovine milk*. Bulletin 86 (pp. 61–72). International Dairy Federation, Brussels.
Olivecrona, T., Vilaro, S. & Olivecrona, G. (2003). Lipases in milk. In P. F. Fox & P. L. H. McSweeney (Eds.), *Advanced dairy chemistry – 1, proteins* (3rd

ed., pp. 473–494). New York: Kluwer Academic/ Plenum Publishers.
Otaegui-Arrazola, A., Menéndez-Carreño, M., Ansorena, D. & Astiasarán, I. (2010). Oxysterols: A world to explore. *Food and Chemical Toxicology, 48*, 3289–3303.
Overcast, W. W. & Skean, J. D. (1959). Growth of certain lipolytic microorganisms at 4°C and their influence on free fat acidity and flavor of pasteurized milk. *Journal of Dairy Science, 42*, 1479–1485.
Panda, N. & Kaur, H. (2007). Influence of vitamin E supplementation on spontaneous oxidized flavour of milk in dairy buffaloes. *International Journal of Dairy Technology, 60*, 198–204.
Panfil-Kuncewicz, H., Kuncewicz, A. & Juskiewicz, M. (2005). The influence of the sterilisation method on the changes in UHT milk fat fraction. *Milchwissenschaft, 60*(1), 33–36.
Patton, S. (1964). Flavor thresholds of volatile fatty acids. *Journal of Food Science, 29*, 679–680.
Pereda, J., Ferragut, V., Quevedo, J. M., Guamis, B. & Trujillo, A. J. (2008). Effects of ultra-high-pressure homogenization treatment on the lipolysis and lipid oxidation of milk during refrigerated storage. *Journal of Agricultural and Food Chemistry, 56*, 7125–7130.
Perkins, M. L., D'Arcy, B. R., Lisle, A. & Deeth, H. C. (2005a). Solid phase microextraction of stale flavour volatiles from the headspace of UHT milk. *Journal of the Science of Food and Agriculture, 85*, 2421–2428.
Perkins, M. L., Elliott, A. J., D'Arcy, B. R. & Deeth, H. C. (2005b). Stale flavour volatiles in Australian commercial UHT milk during storage. *Australian Journal of Dairy Technology, 60*, 231–237.
Petrus, R. R., Walter, E. H. M., Faria, J. A. F. & Abreu, L. F. (2009). Sensory stability of ultra-high temperature milk in polyethylene bottle. *Journal of Food Science, 74*(1), S53–S57.
Potts, D. M. & Peterson, D. G. (2018). Identification of objectionable flavors in purported spontaneous oxidized flavor bovine milk. *Journal of Dairy Science, 101*, 10877–10885.
Reed, A. W., Deeth, H. C. & Clegg, D. E. (1984). Liquid chromatographic method for quantitative determination of free fatty acids in butter. *Journal of the Association of Official Analytical Chemists, 67*, 718–721.
Rerkrai, S., Jeon, I. J. & Bassette, R. (1987). Effect of various direct ultra-high temperature heat treatments on flavour of commercially prepared milks. *Journal of Dairy Science, 70*, 2046–2054.
Rohfritsch, Z., Schafer, O. & Giuffrida, F. (2019). Analysis of oxidative carbonyl compounds by UPLC-high-resolution mass spectrometry in milk powder. *Journal of Agricultural and Food Chemistry, 67*, 3511–3520.
Romeu-Nadal, M., Chavez-Servin, J. L., Castellote, A. I., Rivero, M. & Lopez-Sabater, M. C. (2007). Oxidation stability of the lipid fraction in milk powder formulas. *Food Chemistry, 100*, 756–763.
Rysstad, G., Ebbesen, A. & Eggestad, J. (1998). Sensory and chemical quality of UHT-milk stored in paperboard cartons with different oxygen and light barriers. *Food Additives and Contaminants, 15*, 112–122.
Salih, A. M. A. & Anderson, M. (1979). Observations on the influence of high cell count on lipolysis in bovine milk. *Journal of Dairy Research, 46*, 453–462.
Santos, M. V., Ma, Y. & Barbano, D. M. (2003a). Effect of somatic cell count on proteolysis and lipolysis in pasteurized fluid milk during shelf-life storage. *Journal of Dairy Science, 86*, 2491–2503.
Santos, M. V., Ma, Y., Caplan, Z. & Barbano, D. M. (2003b). Sensory threshold of off-flavors caused by proteolysis and lipolysis in milk. *Journal of Dairy Science, 86*, 1601–1607.
Schroder, M. J. A. (1983). Light and copper catalyzed oxidized flavors in stored milk. *Journal of the Society of Dairy Technology, 36*, 8–12.
Schwartz, D. P. & Parks, O. W. (1974). The lipids in milk: Deterioration part 1 lipolysis and rancidity. In B. H. Webb, A. H. Johnson & J. A. Alford (Eds.), *Fundamentals of dairy chemistry* (2nd ed., pp. 220–272). Westport: AVI Publishing.
Selselet-Attou, G., Chilliard, Y., Bas, P. & Morand-Fehr, P. (1984). Comparaison de deux méthodes de dosage des acides gras libres totaux du lait de chèvre. *Lait, 64*, 72–84.
Serra, M., Trujillo, A. J., Pereda, J., Guamis, B. & Ferragut, V. (2008). Quantification of lipolysis and lipid oxidation during cold storage of yogurts produced from milk treated by ultra-high pressure homogenization. *Journal of Food Engineering, 89*, 99–104.
Shelley, A. W., Deeth, H. C. & MacRae, I. C. (1986). Growth of lipolytic pschrotrophic pseudomonads in raw and ultra-heat-treated milk. *Journal of Applied Bacteriology, 61*, 395–400.
Shimizu, M., Miyaji, H. & Yamauchi, K. (1982). Inhibition of lipolysis by milk fat globule membrane materials in model milk fat emulsion. *Agricultural and Biological Chemistry, 46*, 795–799.
Shipe, W. F., Senyk, G. F. & Fountain, K. B. (1980). Modified copper soap solvent extraction method for measuring free fatty acids in milk. *Journal of Dairy Science, 63*, 193–198.
Sieber, R. (2005). Oxidised cholesterol in milk and dairy products. *International Dairy Journal, 15*, 191–206.
Singh, R. R. B., Patil, G. R. & Balachandran, R. (2004). Kinetics of lipid hydrolysis during storage of UHT milk. *Journal of Food Science and Technology Mysore, 41*(2), 139–142.
Smet, K., Raes, K., De Block, J., Herman, L., Dewettinck, K. & Coudijzer, K. (2008). A change in antioxidative capacity as a measure of onset to oxidation in pasteurized milk. *International Dairy Journal, 18*, 520–530.
Stapelfeldt, H., Nielsen, B. R. & Skibsted, L. H. (1997). Effect of heat treatment, water activity and storage temperature on the oxidative stability of whole milk powder. *International Dairy Journal, 7*, 331–339.
Suhren, G. (1983). Beziehungen zwischen den Ergibnissen der FFA-Bestimmung nach der AutoAnalyzer- und BLM-Methode. *Dtsch Milchwirtsch, 34*, 214.

Suhren, G., Heeschen, W. & Tolle, A. (1977). Automated measurement of free fatty acids in continuous flow analysis. *Milchwissenschaft, 32*, 641–643.

Sundheim, G. & Bengtsson-Olivecrona, G. (1985). Lipolysis in milk induced by cooling or by heparin: Comparisons of amount of lipoprotein lipase in the cream fraction and degree of lipolysis. *Journal of Dairy Science, 68*, 589–593.

Sundheim, G. & Bengtsson-Olivecrona, G. (1987). Methods to assess the propensity of milk fat globules toward lipolysis and the ability of skim milk to inhibit lipolysis. *Journal of Dairy Science, 70*, 2040–2045.

Thomas, E. L., Nielsen, A. J. & Olson, J. C. (1955). Hydrolytic rancidity in milk. A simplified method for estimating the extent of its development. *American Milk Review, 17*(1), 50–52. 85.

Thomsen, M. K., Lauridsen, L., Skibsted, L. H. & Risbo, J. (2005). Two types of radicals in whole milk powder. Effect of lactose crystallization, lipid oxidation, and browning reactions. *Journal of Agricultural and Food Chemistry, 53*, 1805–1811.

Ulberth, F. & Roubicek, D. (1995). Monitoring of oxidative deterioration of milk powder by headspace gas-chromatography. *International Dairy Journal, 5*, 523–531.

Urbach, G. (1990). Effect of feed on flavor in dairy foods. *Journal of Dairy Science, 73*, 3639–3650.

Van Crombrugge, J., Bossuyt, R. & Van Renterghem, R. (1982). Évaluation de quelques méthodes de détermination de la lipolyse dans le lait cru. *Rev Agric, 35*, 3313–3325.

Vanbergue, E., Peyraud, J. L., Guinard-Flament, J., Charton, C., Barbey, S., Lefebvre, R., Gallard, Y. & Hurtaud, C. (2016). Effects of DGAT1 K232A polymorphism and milking frequency on milk composition and spontaneous lipolysis in dairy cows. *Journal of Dairy Science, 99*, 5739–5749.

Vanbergue, E., Delaby, L., Peyraud, J. L., Colette, S., Gallard, Y. & Hurtaud, C. (2017). Effects of breed, feeding system, and lactation stage on milk fat characteristics and spontaneous lipolysis in dairy cows. *Journal of Dairy Science, 100*, 4623–4636.

Vanbergue, E., Peyraud, J. L., Ferlay, A., Miranda, G., Martin, P. & Hurtaud, C. (2018). Effects of feeding level, type of forage and milking time on milk lipolytic system in dairy cows. *Livestock Science, 217*, 116–126.

Vazquez-Landaverde, P. A., Velazquez, G., Torres, J. A. & Qian, M. C. (2005). Quantitative determination of thermally derived off-flavour compounds in milk using solid-phase microextraction and gas chromatography. *Journal of Dairy Science, 88*, 3764–3772.

Veberg, A., Olsen, E., Nilsen, A. N. & Wold, J. P. (2007). Front-face fluorescence measurement of photosensitizers and lipid oxidation products during the photooxidation of butter. *Journal of Dairy Science, 90*, 2189–2199.

Vélez, M. A., Hynes, E. R., Meinardi, C. A., Wolf, V. I. & Perotti, M. C. (2017). Cheese milk low homogenization enhanced early lipolysis and volatiles compounds production in hard cooked cheeses. *Food Research International, 96*, 215–225.

Walsh, A. M., Duncan, S. E., Potts, H. & Gallagher, D. L. (2015). Comparing quality and emotional responses as related to acceptability of light-induced oxidation flavor in milk. *Food Research International, 76*, 293–300.

Walstra, P., Wouters, J. T. M. & Geurts, T. J. (2006). *Dairy science and technology* (2nd ed.). Boca Raton: CRC Press.

Wang, A. H., Li, G. L., You, J. M. & Ji, Z. Y. (2013). A new fluorescent derivatization reagent and its application to free fatty acid analysis in pomegranate samples using HPLC with fluorescence detection. *Journal of Separation Science, 36*, 3853–3859.

Weihrauch, J. L. (1988). Lipids in milk: Deterioration. In N. P. Wong, R. Jenness, M. Keeney & E. H. Marth (Eds.), *Fundamentals of dairy chemistry* (pp. 216–278). Boston: Springer.

Whited, L. J., Hammond, B. H., Chapman, K. W. & Boor, K. J. (2002). Vitamin A degradation and light-oxidized flavor defects in milk. *Journal of Dairy Science, 85*, 351–354.

Wiking, L., Bertram, H. C., Björck, L. & Nielsen, J. H. (2005). Evaluation of cooling. *Journal of Dairy Research, 72*, 476–481.

Wiking, L., Lokke, M. M., Kidmose, U., Sundekilde, U. K., Dalsgaard, T. K., Larsen, T. & Feilberg, A. (2017). Comparison between novel and standard methods for analysis of free fatty acids in milk – Including relation to rancid flavour. *International Dairy Journal, 75*, 22–29.

Wiking, L., Bjerring, M., Løkke, M., Løvendahl, P. & Kristensen, T. (2019). Herd factors influencing free fatty acids concentration in bulk tank milk. *Journal of Dairy Research*, 1–7. https://doi.org/10.1017/S0022029919000190.

Wold, J. P., Veberg, A., Nilsen, A., Iani, V., Juzenas, P. & Moan, J. (2005). The role of naturally occurring chlorophyll and porphyrins in light-induced oxidation of dairy products. A study based on fluorescence spectroscopy and sensory analysis. *International Dairy Journal, 15*, 343–353.

Woo, A. H., Kollodge, S. & Lindsay, R. C. (1984). Quantification of major free fatty acids in several cheese varieties. *Journal of Dairy Science, 67*, 874–878.

Zeng, Z., Ji, Z. Y., Hu, N., Bai, B., Wang, H. L. & Suo, Y. R. (2017). A sensitive pre-column derivatization method for the analysis of free fatty acids by RP-HPLC with fluorescence detector and its application to Caragana species. *Journal of Chromatography B Analytical Technologies in the Biomedical Life Sciences, 1064*, 151–159.

Physical Characterization of Milk Fat and Milk Fat-Based Products

12

O. J. McCarthy and M. Wong

12.1 Introduction

The sensory properties, especially texture and appearance, of milk fat-based products, such as butter, cream, cheese, ice cream and milk chocolate, are largely dependent on product physical properties, especially properties governed by fat *phase change behaviour*, used here to mean melting and crystallizing behaviour, crystal polymorphism and microstructure (Birker and Padley 1987; O'Brien 2003). The same may be said of the functional properties of milk fat, milk fat fractions and milk fat-based products when these are used as food ingredients.

The term *physical properties* is used here to mean both physical properties (e.g. density) and characteristics that can be measured by physical means (e.g. polymorph type). The measurement of physical properties is essential for the proper understanding and control of processing operations and of final product sensory properties, functionality and shelf stability (Birker and Padley 1987; Rajah 2002; O'Brien 2003).

Techniques for measuring phase change behaviour are described first. These are grouped, somewhat loosely, under the heading *Thermal Properties*, together with techniques for measuring milk fat critical temperatures.

Second, three highly specialized techniques – *Light Spectroscopy, Nuclear Magnetic Resonance (NMR)* and *Ultrasound* – are described in turn. For convenience, light spectroscopy and NMR are introduced together immediately under the heading *Light Spectroscopy*; a full description of NMR is presented separately because in the context of this chapter, NMR relaxometry is perhaps of more importance than NMR spectroscopy, although both are described.

Third, the measurement of *rheological properties* and *density* – mechanical properties – are discussed under the headings *Rheological Techniques* and *Density*.

Then, methods of measuring refractive index, colour, dielectric properties and electrical conductivity are described under the heading *Electromagnetic Properties.* Optical methods for investigating phase change are covered under *Thermal Properties,* rather than here.

The last main section, *Functional Properties*, comprises descriptions of techniques for measuring "ingredient properties" and "end-use properties" of milk fat and milk fat-based products.

The measurement techniques described in this chapter are those of greatest importance and utility in the physical characterization of milk fat and milk fat-based foods for research and food industry purposes.

O. J. McCarthy (✉) · M. Wong
Massey University, Auckland, New Zealand
e-mail: owenmccarthy@extra.co.nz

P. L. H. McSweeney et al. (eds.), *Advanced Dairy Chemistry, Volume 2*,
https://doi.org/10.1007/978-3-030-48686-0_12

12.2 Thermal Properties: Phase Change Behaviour

12.2.1 Melting and Solidification Points: Introduction

Milk fat is a complex and variable mixture of triacylglycerols (TAGs). Each of these, and each polymorph of a given TAG, has its own melting point. The full range of TAG melting points is approximately – 40 °C to 72 °C. However, the melting point range of milk fat itself is about – 40 °C to 37 °C, because higher melting point TAGs dissolve in molten lower melting point ones (Rossell 2003).

Measurements of milk fat melting and solidification points are attempts to quantify the upper end of the milk fat melting point range. The value obtained depends on the method used; a melting point or a solidification point is defined by the technique and conditions of measurement, and the reported value is specific for these (Kaylegian and Lindsay 1994; Firestone 1998).

12.2.2 Melting Points

12.2.2.1 Clear Point

The clear point, also called the *clear melting point*, *complete melting point*, *complete fusion point* or *capillary melting point*, is the temperature at which a fat sample becomes visibly completely clear, indicating the disappearance of all traces of solid fat (Rossell 1986; Stauffer 1996). Its measurement is specified by the American Oil Chemists' Society (AOCS) official methodology 1-25 (Firestone 1998). Samples of the tempered fat contained in at least three vertical glass capillary tubes sealed at their lower ends are attached to a vertical mercury-in-glass thermometer such that their lower ends are level with the lower end of the thermometer's bulb. This assembly is immersed in a water bath and heated, and the temperature at which the sample in each capillary becomes clear is observed. The clear point is taken as the mean of these temperatures.

This measurement is claimed to be more reproducible than the slip point (see below) and is the most widely accepted technique for determining melting point (Rossell 1986; Kaylegian and Lindsay 1994). It is the method usually used for milk fat.

Complete melting points can be determined objectively by dilatometry, differential scanning calorimetry and nuclear magnetic resonance (Mertens 1973; Deman *et al.* 1983). Measured values are not necessarily equivalent either to that obtained using the AOCS Cc 1-25 standard method or to each other. Dilatometry and differential scanning calorimetry are described below.

12.2.2.2 Slip Point

The slip point, sometimes called the *slip melting point*, *softening point*, *rising point* or *open tube melting point* (Rossell 1986; Firestone 1998; O'Brien 2003), is measured using a technique essentially identical to the capillary tube method for clear point (AOCS Cc 1-25), except that the tempered fat is contained in capillary tubes open at both ends, and the temperature at which the fat column begins to rise (slip upwards) in each tube is observed. The slip point is taken as the mean of these temperatures. Details of the basic method are given in AOCS Official Method Cc 3-25 (Firestone 1998). A more sophisticated version of the test, based on International Standard ISO 6321, is described in AOCS Official Method Cc 3b-92 (Firestone 1998). This version has two alternative procedures, one applicable to fats solid at room temperature that do not exhibit polymorphism and one applicable to all fats solid at room temperature. The second procedure is probably the best way of determining the slip point of milk fat and milk fat fractions.

The slip point is lower than the complete melting point and corresponds to a solid fat content of 4–5% (Rossell 1986; Stauffer 1996). The slip points of fats can be used as indicators of relative differences in melt-in-the-mouth characteristics (Mertens 1973).

Slip points are commonly measured for vegetable fats, while clear points are usually measured for milk fat. Care should therefore be taken when comparing reported melting points for these two kinds of fat (Rossell 2003).

12.2.2.3 Dropping Point

The dropping point, commonly called the *Mettler dropping point*, is defined as the temperature at which a fat becomes sufficiently fluid to flow under the conditions of the test (Firestone 1998). It is measured using a proprietary apparatus made by Mettler-Toledo, Inc. In contrast to the clear point and slip point techniques described above, the measurement is objective and automatic.

The molten fat sample is allowed to solidify in a cold sample cup, or a finished product such as butter is placed in the cup, with the cup standing on a clean smooth surface. The assembly is then allowed to stand in a freezer (at ≤ −5 °C) for at least 15 min. Then the cup, which has a small hole in its base, is placed in the furnace of the Mettler apparatus. The temperature of the furnace is automatically ramped upwards. When the fat becomes fluid enough to flow, a drop of partly molten fat falls from the hole in the cup, tripping a photoelectric circuit by temporarily blocking a light beam that keeps the circuit closed. The furnace temperature at which this occurs is automatically displayed as the dropping point (Kaylegian and Lindsay 1994). Details of the technique are specified in AOCS Official Method Cc 18-80. A subjective version of this test is described in British Standard BS 684: Section 1.4: 1976 (British Standards Institution 1976a).

As the dropping point method can be applied to fat fractions and fat-based ingredients, as well as to finished products made from them, it provides a means of studying the influence of manufacturing conditions (Mertens 1973).

The dropping point corresponds to a solid fat content of about 5% (Papalois *et al.* 1996) and corresponds closely to the softening point determined by Barnicoat's method (1944) (Timms 1978; Deman *et al.* 1983; Papalois *et al.* 1996). Barnicoat's method was developed for application to milk fat fractions as an alternative to the tedious Wiley melting point technique. In Barnicoat's method, the softening point is taken as the temperature at which a steel ball bearing, initially on the surface of a column of the fat, sinks halfway through the column. In the Wiley method, historically favoured in America, but now an AOCS "surplus" official method (Firestone 1998), the melting point is taken as the temperature at which a disc of initially solid fat located at the interface between a layer of water and an (upper) layer of 95% ethanol becomes spherical (Mertens 1972). The Barnicoat method is still used in the chocolate industry (Minifie 1999).

Nakae *et al.* (1974) used a temperature gradient apparatus to determine the melting point of butterfat samples as the temperature at the boundary between the solid and liquid parts of the sample. Values obtained were correlated with melting points determined by two conventional methods (unidentified in the English summary of this Japanese language paper).

References to the literature on the measurement of the melting points of milk fat fractions are tabulated by Kaylegian and Lindsay (1994).

12.2.3 Solidification Points

12.2.3.1 Cloud Point

The cloud point is defined as the temperature at which a molten fat sample becomes cloudy, owing to the appearance of fat crystals, while being cooled. Versions of the test, which is a simple subjective one, are described in AOCS Official Method Cc 6-25 (Firestone 1998) and in British Standard BS 684: Section 1.5: 1987 (British Standards Institution 1987). As supercooling is necessary for inducing fat crystallization, the cloud point is always lower than the clear point (Kaylegian and Lindsay 1994).

12.2.3.2 Congeal Point

The congeal point, also called the *setting point* (O'Brien 2003), is a measure of the solidification point of a molten fat under the conditions of the test. The molten fat, contained in a beaker, is cooled until the cloud point is observed and then cooled further until a certain subjectively assessed degree of turbidity (caused by the presence of fat crystals) is reached. The beaker is then kept at an ambient temperature of 20 °C and sample temperature recorded over time. The temperature initially rises owing to the release of latent heat of crystallization and then drops. The maximum temperature reached is recorded as the congeal point.

12.2.4 Dilatometry

Dilatometry is a technique for measuring the solid fat index (SFI), which is an estimate of the mass fraction of fat that is solid at the temperature of measurement.

The *dilatation* of a partly solidified fat sample at a given temperature is the increase in specific volume (volume per unit mass) that occurs on complete melting of the solid fat. The *melting dilatation* of the fat is the increase in specific volume that occurs on complete melting of a fully solidified sample. Dilatation is the result of the expansion that occurs as a result of the phase change solid to liquid; the specific volume of molten fat is about 10% greater than that of solid fat.

The quantity

$$\left(\frac{\text{dilatation}}{\text{melting dilatation}}\right)\times 100\%$$

is an estimate of the solid fat content by mass percentage (SFC). It is called the *solid fat index* (SFI). The relationship between SFI and SFC is a function of fat type, temperature and SFI level (Stauffer 1996).

Dilatation is measured with a specialized pycnometer called a dilatometer. The conventional dilatometer is a glass apparatus consisting of a vertical bulb, equipped with a stopper, connected by a U-tube to a vertical capillary tube graduated in microlitres. A volume of aqueous dye solution (or mercury for very precise measurements) is placed in the bulb. A suitable volume of the molten fat sample is then poured into the bulb to float on the dye solution, pushing a portion of the dye solution into the capillary. The fat sample weight is determined by weighing the dilatometer before and after loading it with the sample and inserting the stopper.

Volume measurements needed for calculating the SFI are made by observing and recording the position of the dye solution meniscus in the graduated capillary tube at one temperature (typically 60 °C) or two temperatures at which the sample is fully molten and at desired measurement temperatures (which will be lower than the sample's clear point). The sample is brought to the first measurement temperature from 0 °C after a standardized tempering procedure. This and subsequent measurement temperatures must be approached from below to avoid supercooling effects. Temperatures are controlled by immersing the whole dilatometer in constant temperature water baths or an ice-water bath.

Dilatation at a melting temperature θ is calculated as:

$$D_\theta = \left[\frac{\left(A_{60} - A_\theta - W_\theta\right)}{m}\right] - V_\theta \qquad (12.1)$$

where:

D_θ = dilatation at θ°(μL g^{-1}).

A_{60} = meniscus reading at 60° (fat fully molten) (μL).

A_θ = meniscus reading at θ° (fat partially solidified) (μL).

W_θ = correction for thermal expansion of the sealing liquid (dye solution) between 60° and θ° (μL).

V_θ = correction for thermal expansion of molten fat.

m = the difference in volume between 1 g of molten fat at 60° and 1 g of supercooled molten fat at θ° (μL g^{-1}). (A table of standard values is used.)

The interpretation of dilatometric volume measurements is complicated by the fact that the molten fat has a coefficient of thermal expansion (about 0.00084 mL g^{-1} K^{-1}) more than twice that of solid fat (about 0.00038 mL g^{-1} K^{-1}). This difference means that the melting dilatation is temperature dependent (Stauffer 1996). In principle, this could be allowed for by measuring the specific volumes of both the fully molten fat and the fully solid fat at the measurement temperature. However, while the specific volume of the molten fat is easily measured using the dilatometer, that of the solid fat is not. This difficulty is circumvented by assuming, by convention, that the temperature dependence of specific volume is the same for both the molten fat and the solid fat (thus making the melting dilatation temperature independent) and that the melting dilation itself has a constant value, independent of fat type, of 0.1 mL g^{-1} (= 0.1 L kg^{-1}). Thus (Hannewijk *et al.* 1964):

$$\mathrm{SFI} = \frac{\text{dilatation}\left(\mathrm{mL\,g^{-1}}\right)}{0.1} \times 100\% = \frac{\text{dilatation}\left(\mathrm{mL\,g^{-1}}\right)}{0.001} \quad (12.2)$$

Dilatometer specifications, and measurement procedures and calculations, are published in the form of standard methods by the AOCS (ACOS Official Method Cd 10-57; Firestone 1998), the International Union of Pure and Applied Chemistry (IUPAC standard method 2.141; Paquot and Hautfenne 1987) and the British Standards Institution/International Organization for Standardization (ISO) (BS 684: Section 1.12:1990/ISO 8293:1990; British Standards Institution 1990). Standard methods incorporate implicit measurement of the temperature dependence of the specific volume of the molten fat and corrections for the thermal expansion of the dye solution and of the dilatometer itself.

The theory and principles of dilatometry are particularly well described by Hannewijk *et al.* (1964).

Dilatometry is accurate and cheap, but time-consuming and laborious (Wan 1991; Kaylegian and Lindsay 1994; Shukla 1995). The technique was accepted worldwide up to about 1970, but a switch to the use of nuclear magnetic resonance (NMR) began in the early 1970s (Shukla 1995). The NMR method is described in 5.8.

12.2.5 Differential Scanning Calorimetry

Differential scanning calorimetry (DSC) is a technique for measuring the thermal behaviour of a substance. In DSC, the difference between the heat flows (J s^{-1}, or W) to or from a sample and a reference material is measured as a function of temperature or time while the sample and the reference material are together subjected to a controlled temperature-time programme (Wright 1984). DSC is thus a form of differential thermal analysis (DTA).

The reference material is chosen to be one that exhibits no thermally induced transitions within the temperature range of interest. Thus, transitions that occur in the sample during the applied temperature programme appear as peaks or troughs, depending on whether they are endothermic or exothermic, on the plot of differential heat flow versus temperature (the thermogram) that is the output of the instrument (Figure 12.1). Commonly used reference materials are listed by Hatakeyama and Quinn (1994).

Plotting the thermogram as the first or the second derivative of the measured heat flow versus temperature data, rather than plotting it as shown in Figure 12.1, can be very useful (Gabbott and Mann 2016). For example, plotting the second derivative can resolve the shoulders of original peaks (or troughs) into otherwise unseen additional peaks (or troughs) attributable to polymorphic forms.

In the context of milk fat, DSC is most widely used to measure temperatures and heats of transitions (phase changes). It is also used to measure specific heat, solid fat content, crystallization kinetics constants and fat purity and in the study of fat crystal polymorphism.

Most commonly used DSC instruments fall into two categories (Laye 2002): power compensation (PC) instruments, for which the term DSC was coined when this type became available in 1963 (Wendlandt 1974), and heat flux (HF) instruments (Figure 12.2). The latter are essentially quantitative DTA instruments; classical DTA is a qualitative or at best semi-quantitative technique (Wright 1984).

In any DSC instrument, the sample and reference material are placed in small individual pans or crucibles, which may be open or hermetically sealed, of 10–20 μL capacity (Harwalkar and Ma 1990). Sample size is therefore small, and accurate weighing is essential. Sometimes the reference is merely an empty pan.

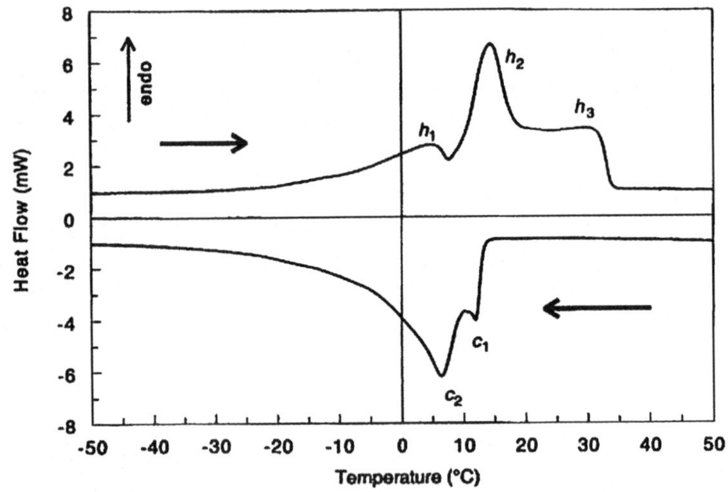

Figure 12.1. Typical DSC thermogram for milk fat. c_1 and c_2 are exothermic crystallization troughs obtained during cooling. h_1, h_2 and h_3 are endothermic peaks obtained during heating. (Reproduced with permission from ten Grotenhuis *W* 1999).

In PC instruments (Figure 12.2a), the sample and reference pans are placed in individual holders (sometimes called furnaces), each equipped with an electric heater and a temperature sensor. During a temperature scan, the sample and reference are maintained at exactly the same temperature at any instant even when a thermal event (transition) is occurring in the sample (Daniels 1973). This null temperature balance is achieved by automatic adjustment of the power supplied by the heater in the sample holder (Daniels 1973). Adjustment occurs when a temperature difference between the sample and reference holders greater than a set threshold value (typically <0.01 K) is detected (Hatakeyama and Quinn 1994). The instrument output signal is the difference between the power supplied to the sample holder and that supplied to the reference holder. The former is higher during an endothermic transition and lower during an exothermic transition.

In HF instruments (Figure 12.2b), the sample and reference pans are heated or cooled in a single temperature controlled furnace. In the most common design, the two pans rest on a horizontal metal plate, the heat flux plate, directly connected to the heating block of the furnace. The heating block is of large thermal mass and often made of silver or other non-corroding metal of high thermal conductivity. The temperature of the block is programmed as required, the good heat flow path provided by the heat flux plate ensuring that the sample and the reference are subjected to the same programme.

When a thermal event occurs in the sample, a temperature difference tends to develop between the sample and the reference. This results in a heat flux between the two via the heat flux plate. This heat flux ensures that the temperature difference between the sample and the reference is always very small. This is important for ensuring that both sample and reference are subjected to essentially the same temperature programme (Bhadeshia 2002).

However, it is this small temperature difference between sample and reference that is measured as a function of temperature during the applied temperature scan. It is converted by the instrument's software, using instrument calibration data, to the difference between the power absorbed or released from the sample and that absorbed or released from the reference. The final instrument output is a thermogram of this differential power plotted against temperature, as for PC instruments.

At least one DSC instrument is available which incorporates features of both PC and HF types (Laye 2002).

When DSC scans are carried out below room temperature, or when the temperature programme is one of decreasing rather than increasing temperature, controlled cooling rather than heating of the sample and reference is required. This is usually achieved by means of an efficient

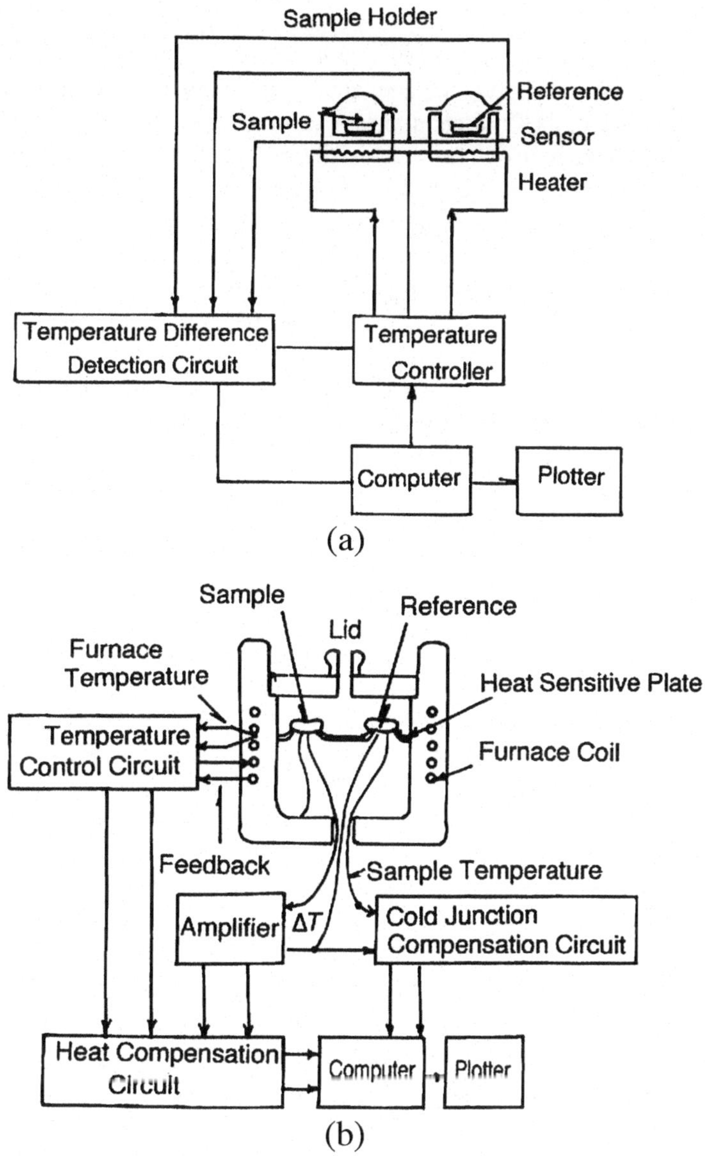

Figure 12.2. Schematic diagrams of (**a**) a power compensation DSC and (**b**) a heat flux DSC. In (**b**), heat-sensitive plate = heat flux plate. (Reproduced with permission from Hatakeyama and Quinn 1994).

refrigeration system (built into the instrument) whose cooling effect is moderated by electric power input via the heater (HF) or heaters (PC) to give the cooling rate required (Daniels 1973; Wunderlich 1990; Laye 2002).

Bashir *et al.* (1998), in a study of phase transitions in a monotropic liquid-crystal polyester, found that PC and HF instruments performed differently in cooling mode. Better resolution of thermal events, i.e. better separation of peaks on the thermogram, was obtained with PC instruments. They pointed out that the performance of any DSC instrument may not be the same during heating as during cooling.

Conventionally, the temperature programme used in DSC, whether heating or cooling, is a linear change in temperature with time. This is most commonly used to investigate thermally driven phase changes. Isothermal DSC, in which the temperature is kept constant at a value at which a transition of interest is known to occur, is especially useful for studying the crystallization of a fat (Simon and Süverkrüp 1995; Blaurock 2000). In this technique the sample (after holding for a time in a fully molten state to erase thermal memory) and the reference are cooled rapidly to sample crystallization temperature, and, after temperature equilibration, the differential power input is measured as a function of time under isothermal conditions. The heat of crystallization released by polymorphs in the fat sample results in exothermic signals, which appear as peaks on the thermogram. The nucleation times (the times at which peaks start forming), times of maximum crystallization rate (the times of peak maxima) and heat of crystallization (proportional to peak areas) can all be determined from the thermogram.

A DSC heating thermogram for the same fat sample can conveniently be measured as soon as isothermal crystallization has gone to completion.

Also, after determining the temporal distribution of crystallization peaks in a complete isothermal experiment, the melting thermogram of a particular polymorph can be determined. First the sample is completely re-melted and held to erase its thermal memory. Then, the isothermal crystallization process is repeated, but only up to the time by which the polymorph of interest has completely solidified (exothermic peak fully formed). A heating scan is then performed immediately (Kawamura 1980, 1981).

The effects of different rates of cooling to isothermal crystallization temperature, different crystallization temperatures and different rates of heating during melting on polymorph formation and transformation are easily studied by DSC. However, it is not possible to unequivocally identify polymorphs by DSC; this must be done by X-ray diffractometry (Rossell 2003) (see below).

Although isothermal DSC is conceptually a good technique for studying fat crystallization, some workers (e.g. Wright *et al.* 2001a) have found it inferior to other techniques such as pNMR, turbidimetry, light-scattering and polarized light microscopy (see below) for studying the crystallization of milk fat. The reason for this was said to be an inherent lack of DSC instrument sensitivity at the high cooling rates needed to maintain isothermal conditions during crystallization.

Breitschuh and Windhab (1996) describe a novel use of DSC for directly investigating or monitoring crystallization processes such as those used in the fractionation of milk fat. In their technique, called "DSC direct", a small sample of fat is taken from the process, immediately placed in the DSC instrument's sample pan and, within seconds, subjected to a heating scan. In this way the melting characteristics of the fat at that particular stage of the process, as determined by processing conditions up to that stage, can be determined directly.

PC instruments are preferable to HF instruments for isothermal studies and in studies in which the temperature scanning rate is high, because the very small sample and reference holders in the PC type have much smaller thermal inertias than the relatively large heating block in the furnace of the HF type (Hatakeyama and Quinn 1994).

More complex temperature programmes are sometimes useful. These might combine periods of variable heating and cooling rates with isothermal periods. For example, stepwise heating can be used to detect the onset of melting under quasi-isothermal conditions (Laye 2002). Modulated temperature DSC (MTDSC), in which the linear temperature scan is perturbed by a sinusoidal, square or saw-tooth wave, or other modulation of temperature, has a number of potential advantages over the conventional linear scan. These include increased sensitivity and resolution and the ability to deconvolute multiple simultaneous thermal events (Laye 2002). Modulated-temperature DSC is described in detail by Kett (2016).

High-sensitivity DSC (HS-DSC) instruments are used for measuring small heats of transition. They were originally developed for measuring heats of denaturation of biopolymers in dilute solution (Hatakeyama and Quinn 1994). The sensitivity of a heat flux instrument, for example, can be improved by increasing sample size, using multiple serially connected thermocouples to measure both sample and reference temperatures, and increasing the heat sink capability of the heat flux plate (Hatakeyama and Quinn 1994).

Gabbott and Mann (2016) briefly describe a number of variants of basic DSC additional to those mentioned above. These include micro-DSC, fast-scanning DSC, UV-DSC, DSC-Raman and DSC-NIR. In UV-DSC, calorimeter design allows the sample, after loading, to be irradiated with UV light to initiate a reaction that is then studied by DSC. In DSC-Raman and DSC-NIR, spectroscopy is used to measure changes in the sample as DSC progresses. These techniques are made possible by the use of fibre-optic cables. The coupling of DSC and X-ray diffraction is discussed in Section 12.2.8.

As already stated, the usual output of a DSC instrument is a thermogram. This is a plot of differential heat flow rate (differential power) versus temperature for a temperature scan, or a plot of differential heat flow rate versus time for an isothermal scan. The thermogram is most logically plotted (by the instrument's software) such that a peak represents an exothermic event in the sample, while a trough represents an endothermic event. However, the thermogram from a PC instrument is sometimes plotted in the opposite sense; this is not entirely illogical since in this type an exothermic event results in the quantity *(power supplied to sample – power supplied to reference)* being negative during an exothermic event and positive during an endothermic event (Figure 12.1). The direction of exothermic and endothermic events should always be indicated on the ordinate of the thermogram.

The instrument software can be used to derive properties of the sample from the thermogram data. These may include specific heat capacity, temperature of transition, heat of transition, solid fat content and reaction kinetics constants.

12.2.5.1 Specific Heat Capacity

When a linear temperature scan has been used, the ordinate of the thermogram can be converted from differential power to sample specific heat capacity if the mass of the sample is known (Wright 1984). The variation of specific heat capacity with temperature is then easily determined. When the temperature of determination falls within a range in which phase transition is occurring, the measured specific heat capacity is an apparent one, as sensible and latent heat changes are then confounded. Usually, specific heat capacity is measured by DSC relative to that of a standard material such as sapphire for which very accurate specific heat capacity data are available (Hatakeyama and Quinn 1994). This avoids inaccuracy owing to inherent errors in the range calibration and programmed heating rate of the DSC instrument (Wright 1984).

12.2.5.2 Temperature and Heat of Transition

DSC is a particularly useful technique for measuring temperatures and heats of transition, two properties of considerable importance with respect to the thermal behaviour of fats.

The properties are determined for a given transition by suitable analysis of the corresponding peak (or trough) on the thermogram. The peak temperature is easily identified but is a less suitable measure of the transition temperature than the onset temperature, the temperature at which transition begins. The onset temperature can be difficult to determine as the thermogram curve often departs only very gradually from the baseline, the path the curve would follow if no transition took place. A number of mathematical strategies are available for dealing with this problem (Wright 1984).

The heat of transition is found by measuring the area under the peak, to which it is directly proportional:

$$\text{Heat of transition} = \text{K} \times \text{peak area} \left(\text{J kg}^{-1}\right) \quad (12.3)$$

where K is an instrument calibration constant established empirically using standard materials with accurately known heats of fusion, and taking into account sample mass. Standards commonly used are listed by Hatakeyama and Quinn (1994).

Provided energy is absorbed or released by the sample as heat only, the system is under a constant pressure, and the only work done is expansion or contraction caused by heating or cooling alone, the heat of transition is equal to the enthalpy of transition (Peyronel 2018).

Determination of the peak area is complicated if the specific heat capacity of the sample changes over the temperature range of the transition, as the form of the baseline is then not known. Approaches to dealing with this problem are described by Daniels (1973) and Wright (1984).

12.2.5.3 Solid Fat Content

A thermogram obtained by completely melting an initially completely solid fat sample using a linear temperature ramp can be used to determine the relationship between solid fat content and temperature. The baseline is assumed to be a straight line between the upper and lower limits of the melting temperature range (Norris *et al.* 1971). The solid fat content at a given temperature within the melting range is calculated as the ratio, to the total area, of the area under the curve between that temperature and the upper limit. Selection of the lower melting limit is difficult because of the small slope of the initial part of the melting curve, but error in this selection does not greatly affect accuracy (Norris *et al.* 1971). This technique is an alternative to dilatometry and pNMR. However, though rapid, its precision and reproducibility are less than those of dilatometry, high sample weighing accuracy is required, and in fact too much information is produced. Thus, the technique appears to be unsuitable as a routine method of measuring solid fat content (O'Brien 2003).

12.2.5.4 Crystallization Kinetics

A number of methods are available for deriving reaction kinetic constants from DSC thermograms (Wright 1984). For example, the thermogram obtained during an isothermal DSC experiment at a temperature at which crystallization of a fat occurs can be analysed in a way similar to that described just above for the determination of solid fat content, but in this case, the evolution of trough area (representing the formation of solid fat crystals) is related to time rather than temperature (Chong 2001).

Blaurock and Carothers (1990) and Blaurock and Wan (1990) describe a simple way, valid for butteroil, of analysing isothermal DSC data to characterize the kinetics of early crystallization in a supercooled oil. This yields a single crystallization-temperature-dependent combined nucleation/crystal growth constant, which they called NG. The temperature dependence of NG could be modelled with the Arrhenius equation.

Wright (1984) has pointed out that most methods for extracting kinetics data from DSC thermograms involve assumptions that should be validated for the system being studied.

Martini (2014) provides an extensive account of the use of DSC to study both the crystallization kinetics and the melting behaviour of lipids and discusses the advantages and disadvantages of using DSC to study lipid crystallization.

12.2.5.5 Adulteration

DSC can be used to identify the fat components of a blend of fats by comparing the thermogram of the blend with thermograms of the individual fats, analysing the DSC data using appropriate chemometrics. This approach has been used to detect the adulteration of fresh cheeses with vegetable fat (Herman-Lara *et al.* 2017), butter with palm oil (Tomaszewska-Gras 2016) and butter with lard (Nurrulhidayah *et al.* 2015).

12.2.5.6 Polymorphism

DSC can be used to demonstrate the presence of polymorphism (Mulder and Walstra 1974). When polymorphism is present, the curves on thermograms for samples of the same fat preconditioned thermally in different ways will have different shapes.

A combination of isothermal and heating DSC scans can be used to study polymorphic behaviour in some detail, as described above.

The polymorphic behaviour of milk fat complicates the interpretation of thermograms obtained for milk fat, milk fat fractions and high-fat dairy products for purposes other than the study of polymorphism itself. The heating or cooling rate used during DSC, as well as the preconditioning regime, can affect the fat's behaviour. Wright (1984) has pointed out that careful choice and control of experimental conditions are necessary for reproducibility and interpretability. The results obtained with DSC while heating are largely independent of effects such as supersaturation, supercooling and diffusion, whereas these phenomena occur during cooling (Norris *et al.* 1971; Wright 1984). In spite of this, perhaps the simplest approach is to melt the sample completely prior to analysis to erase all thermal memory and then carry out a cooling DSC scan. Cooling thermograms have been found to be more reproducible and easier to interpret because during cooling, only exothermic crystallization occurs; heating scans are more complex because polymorphic transformations can occur in addition to endothermic melting (ten Grotenhuis *et al.* 1999). However, it can be argued that DSC analyses should routinely include both heating and cooling scans (Wright 1984).

The theory and principles of DSC; the design, calibration and operation of DSC instruments; and the analysis of experimental data are described by Gaisford *et al.* (2016), Laye (2002), Hatakeyama and Quinn (1994), Wright (1984) and Daniels (1973).

The application of DSC to the study and analysis of edible fats and oils is covered extensively by Marangoni (2018) and Chiavaro (2014).

12.2.6 X-Ray Diffraction

X-ray diffraction (XRD) is the principal means of determining the structures of crystals. It is a technique in which a collimated X-ray beam is directed at a single crystal of the material under investigation or, as is more usual in the study of fat crystals, a sample comprising a large number of randomly orientated crystals. The latter variant, which is the most commonly used, is called powder diffraction (Cullity 1956). Both variants yield essentially the same basic information about structure after data analysis. The following descriptions relate to powder diffraction specifically.

By far the greater proportion of incident X-radiation is transmitted or absorbed by a crystalline sample. However, a small fraction is scattered in all directions by every motif in the material, without change in wavelength. The *motif* is the repeating unit of pattern in a crystal; it is the TAG molecule in the case of a fat crystal. Motifs can be considered to be located at or near the intersections of an imaginary 3D grid. This grid is called the *crystal lattice*, and the intersections are called *lattice points* (Hammond 1997).

The scattered X-rays mutually interfere either destructively or constructively. Almost total destructive interference occurs in all but certain very specific directions in which constructive interference is possible (Sears *et al.* 1982). Constructive interference in a given direction results in a diffraction beam that is intense relative to the sum of all other rays scattered in the same direction.

The planes of motifs in a crystal thus act essentially as diffraction gratings (Bueche 1986). Diffraction of electromagnetic radiation by crystals can occur only when the wavelength of the radiation is of the same order as the regular repeat distance between motifs. This is the basis of the utility of using X-rays to determine crystal structure.

A diffraction beam is produced by constructive interference when the path length difference between reflections from the motifs in any two parallel planes of motifs in a crystal lattice is equal to a whole number of wavelengths. The angle between the incident X-ray beam and the lattice planes (the angle of incidence) is equal to the angle between the diffraction beam and the lattice planes. These phenomena are quantified by the Bragg law (Cullity 1956):

$$n\lambda = 2d\sin\theta \tag{12.4}$$

where:

λ = X-ray wavelength.
d = lattice spacing (the distance between two adjacent parallel planes of motifs).
θ = angle of incidence = angle between the diffraction beam and the lattice planes.
2θ = angle of diffraction.
n = an integer representing a whole number of wavelengths.

If X-rays of known wavelength, λ, are used, θ can be measured and d then calculated. This is known as *structure analysis* (Cullity 1956).

Conventional X-ray sources produce beams that contain rays of different wavelengths.

Equation (12.4) is satisfied by any wavelength for which a set of lattice planes exists in the crystals at an angle θ to the incident beam. As a large number of wavelengths can meet this criterion, the diffraction pattern is potentially quite complex (Bueche 1986).

A crystal structure is solved in three steps (Cullity 1956). Firstly, the size and shape of the unit cell (a crystal lattice consists of identical unit cells) are found from the angular distribution of the diffraction beams. Secondly, the number of molecules per unit cell is computed from the size and shape of the unit cell, the chemical composition of the sample and the sample's measured density. Lastly, the positions of the molecules within the unit cell are deduced from the relative intensities of the diffraction beams. Data analysis, which is complex, is described by Woolfson and Fan (1995) and Clegg (2001).

An X-ray powder diffractometer (Figure 12.3) comprises an X-ray source, the crystalline sample in a sample holder and a detector for detecting and measuring the intensities of the diffraction beams (Cullity 1956; Pecharsky and Zavalij 2003). The incident and diffraction beams are defined and collimated by suitable means. The X-ray source, the sample holder containing a flat powder layer and the detector are at three points of a triangle. The sample holder can be made to rotate about its axis, and the detector can be made to move around the circumference of a circle (the diffractometer circle) whose centre coincides with the middle of the sample face, but whose plane is at right angles to this face. The X-ray beam from the source (which is at a fixed point on the same circumference) strikes the face of the powder sample, and the diffraction beam (if the angle is such that one exists) is reflected into the detector. The rotation of the sample holder and the travel of the detector are mechanically coupled. This allows the angle of diffraction, 2θ (Figure 12.3), to be varied (according to the numerical scale over which the detector passes) while ensuring that the diffraction beam remains in line with the detector. In some instruments the sample holder is non-rotating, and the detector and source travel circumferentially in opposite directions to achieve the

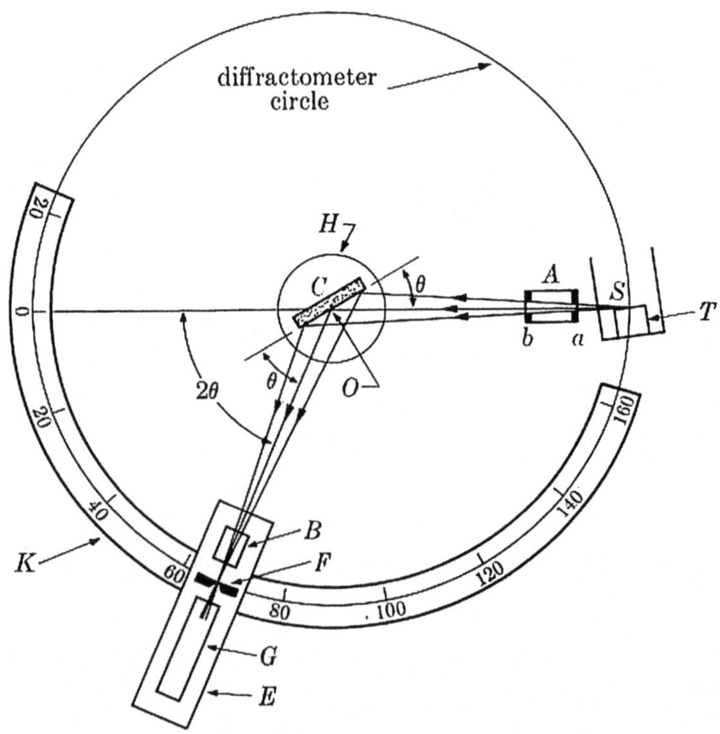

Figure 12.3. Schematic of an X-ray diffractometer. A and B: special slits that define and collimate the incident and diffracted beams. C: powder specimen. E and H: mechanically coupled supports of the detector and powder specimen. F: slit at focal point of diffraction beams. G: detector. K: graduated angle scale. O: diffractometer axis. S: line focal spot on target, T, of the X-ray tube. (Reproduced with permission from Cullity 1956).

same result (Cullity 1956; Pecharsky and Zavalij 2003).

For the study of edible fats, X-ray diffraction is carried out using angles of incidence of 5–20° (wide-angle X-ray diffraction or WAXD), 1.5–4° (small-angle X-ray diffraction or SAXD) and < 1° (ultra-small X-ray diffraction or USAXD) (Peyronel and Pink 2018).

As the angle decreases, the characteristic size of the scattering objects detected in the sample increases; for edible fats, WAXD gives information about primitive unit cell dimensions and polymorphism (2–10 Å), SAXD gives information on lamella spacing and crystal domain size (30–300 Å) and USAXD gives information on supramolecular arrangements of crystal nanoplates (400 nm to 6 μm) (Peyronel 2018; Peyronel and Pink 2018).

In a diffraction experiment, the angle of incidence is varied (scanned) either step by step or continuously, and the intensities of diffraction beams are detected and recorded as a function of angle. Some diffractometers are designed to scan in both SAXD and WAXD modes simultaneously.

The three most commonly used detectors are the gas proportional counter, the scintillation detector and solid-state detectors (Pecharsky and Zavalij 2003). Position-sensitive detectors (also called area detectors), based either on a gas-filled ionization chamber or an image intensifier coupled to a video camera, detect and record diffraction beam intensity in two dimensions simultaneously, a feature that greatly enhances data collection speed (Drenth 1999).

X-rays are generated using either conventional X-ray tubes (most commonly) or synchrotrons. In a synchrotron, an extremely large and expensive facility (Drenth 1999), electrons (or positrons) circulate at extremely high velocity in a large storage ring (ten to several hundred metres in diameter). The particles emit X-rays owing to their continuous centripetal acceleration towards the centre of the ring as their circular direction of travel is maintained by applied magnetic fields. The X-ray beam, of an adjustable and controllable wavelength, can be directed down a straight beam line (tangential to the ring) from the ring to the diffractometer. Synchrotron X-radiation is highly collimated, has extremely high brilliance (a quality related to photon flux) and has a high degree of monochromaticity, among other useful characteristics (Pecharsky and Zavalij 2003; Clegg 2001). A synchrotron X-ray diffractometer is shown schematically in Figure 12.4.

There is no essential difference between conventional X-rays and synchrotron X-rays with respect to their use in the investigation of crystal structure. However, the brilliance of synchrotron X-rays (which is currently some ten orders of magnitude greater than that of conventional X-rays), combined with the use of position sensitive detectors, allows very rapid angle scans to be carried out, which in turn allows the collection of time-resolved diffraction data. This is extremely useful in studying the dynamics of fat crystallization processes at both constant and time-varying temperature (Sato *et al.* 1999; Sato 2001).

The results of a diffraction experiment in which a single angle scan is carried out are presented as a 2D plot of diffraction beam intensity versus angle of diffraction (usually expressed as 2θ) (Figure 12.5). The results of a time-resolved diffraction experiment are presented as a 3D plot in which the x-axis is angle of diffraction, the y-axis is time (and temperature when temperature is varied with time during the experiment) and the z-axis is intensity (Figure 12.6). The plot is essentially a non-continuous response surface formed by the intensity versus angle traces of single scans carried out at successive short intervals of time. Examples of such plots for milk fat are presented by, for example, Sato *et al.* (1999) and Sato (2001).

In either case, the plot can be regarded as a "fingerprint" of the crystal structure and can be used to identify crystals by comparison with stored plots for crystal structures that have been solved. This is very useful in the study of milk fat polymorphism.

The principles of X-ray diffractometry are described by Cullity (1956) and Pecharsky and Zavalij (2003). Peyronel and Pink (2018) describe the principles of USAXD in particular and its use in investigating large crystalline structures in edible fats. Mattice and Marangoni (2018) demonstrate the usefulness of XRD in gaining an

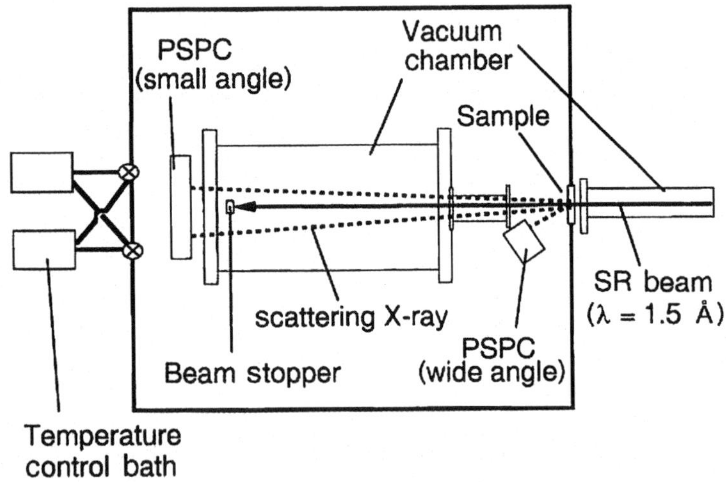

Figure 12.4. Schematic of a synchrotron X-ray diffractometer. PSPC position-sensitive proportional counter; SR beam synchrotron radiation beam. (Reproduced from Ueno *et al.* 2000).

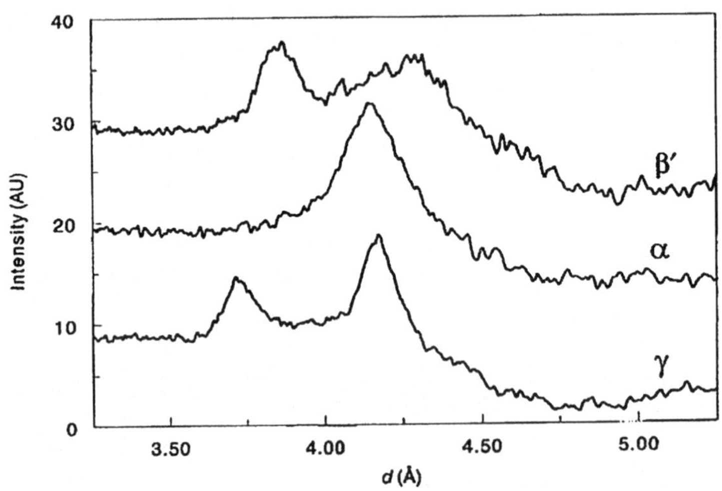

Figure 12.5. Characteristic 2D X-ray powder diffraction patterns of the three polymorphic forms found in milk fat. The pattern of the γ-modification was obtained at – 50°C after cooling at a rate of 20°C per min. The α-diffraction pattern was obtained at – 5°C after cooling at a rate of 5°C per min. The β′–diffraction pattern shown is for milk fat kept at 20°C for 30 min. AU = arbitrary units; d = angle of diffraction. (Reproduced from ten Grotenhuis *et al.* 1999).

understanding of the interaction between fats and other components of food matrices such as cheese. Such understanding is possible only by analysing the intact matrix rather than the bulk fat.

12.2.7 Combined DSC and XRD

DSC and XRD are the methods of choice for studying fat polymorphism and crystalline phase transitions. However, neither is sufficient on its own; they are complementary techniques (Birker and Padley 1987; ten Grotenhuis *et al.* 1999). XRD can be used to measure the amounts of different polymorphs in a mixture, as the intensities of the diffraction beams due to a constituent of a mixture depend on the proportion of that constituent in the mixture (Cullity 1956). Changes in the intensities and angular positions of diffraction beams are used as a measure of polymorphic changes. However, XRD is not sensitive to transitions within the same polymorphic form; changes

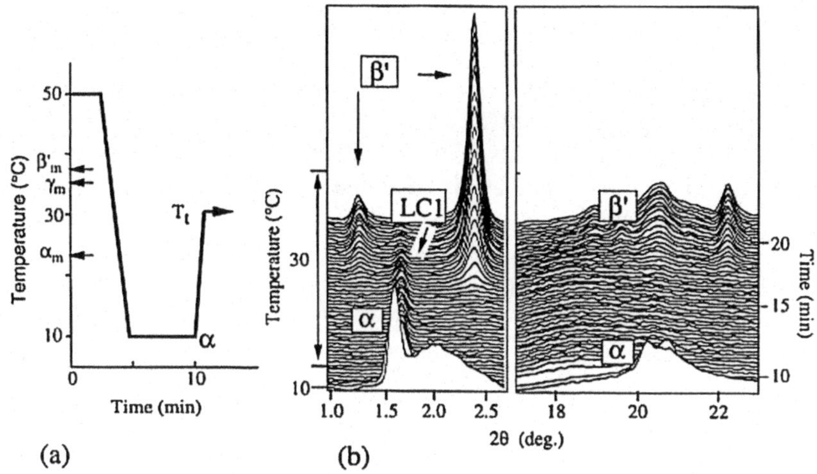

Figure 12.6. (**a**) Temperature variation profile during, and (**b**) 3D plots of the results obtained from, a time-resolved synchrotron X-ray diffraction experiment on SOS (1,3-distearoyl-2-oleoyl-*sn*-glycerol) without annealing. α and β/ polymorphic forms; LC1 liquid crystalline structure 1. (Reproduced with permission from Sato *et al.* 1999).

in molecular composition and crystal size, and thermal shrinkage or expansion effects, tend to be confounded (ten Grotenhuis *et al.* 1999).

DSC, by contrast, is sensitive both to phase transitions within polymorphs and to polymorphic transformations, and the equipment needed is less elaborate than that required for XRD (ten Grotenhuis *et al.* 1999). However, DSC cannot identify crystal forms unequivocally (Birker and Padley 1987), and when several transitions occur simultaneously, it can be difficult to quantify their separate effects. Further, in the case of fats, interpretation of data is complicated by the dependence of fat polymorph heat of transition on TAG composition (ten Grotenhuis *et al.* 1999).

The misinterpretation or under-utilization of data obtained using either XRD or DSC alone can be overcome by using both techniques in a given investigation (ten Grotenhuis *et al.* 1999).

12.2.8 Coupled DSC and XRD

Even when DSC and XRD are both used, quite separate instruments are involved. This leads to difficulties in reconciling results owing to differences in sample thermal history/conditioning, sample dimensions and sample temperature control and uniformity. These difficulties can be entirely overcome by coupling XRD and DSC together in the same instrument and carrying out both types of measurement simultaneously on the same sample.

A custom-made instrument in which high-sensitivity DSC and time-resolved synchrotron SAXD and WAXD can be performed simultaneously on the same sample has been built by a French research group (Keller *et al.* 1998). This instrument is claimed to be an advance over earlier attempts to couple DSC and XRD. The single sample is held in a thin glass capillary tube, and all experimental data are collected by a single computer. Local inhomogeneity in sample temperature caused by absorption of X-ray energy has been shown not to perturb appreciably the measurement of true sample temperature. Details of the design and attributes of the instrument are given by Keller *et al.* (1998).

The instrument has been used extensively to study phase change behaviour in anhydrous milk fat and cream (Lopez *et al.* 2000, 2001a, b, c, d, e, 2002, 2005a). This work showed, *inter alia*, that the brilliance of synchrotron X-radiation and the coupling of both SAXD and WAXD with DSC are essential for interpreting the complex thermal and structural behaviour of fat in dispersed systems such as cream. This has been demonstrated by Lopez *et al.* (2005b) and Karray *et al.* (2004, 2005) who studied the crystallization and melting properties of dromedary milk fat globules and anhydrous dromedary fat, and Amara-Dali *et al.* (2005, 2007, 2008) who carried out equivalent studies of goat's milk fat and fat globules. Further, Lopez *et al.* (2007) investigated the crystallization properties and polymorphism of TAGs in bovine milk fat globules, again using coupled DRC and synchrotron XRD. A description of this technique, with examples of how it is applied, is provided by Calligaris *et al.* (2014).

Coupling of DSC and XRD has considerable potential for improving polymorphism control in the processing of fat-based foods (Lopez *et al.* 2007; Allais *et al.* 2003; Ollivon *et al.* 2001).

12.2.9 Cooling Curves

The cooling curve of a fat is a plot of temperature versus time obtained by measuring the temperature of a sample of the molten fat while the sample is cooled under standard conditions. The cooling curve is a measure of speed of crystallization, or rate of polymorphic change, and is useful in assessing or interpreting the behaviour of a fat or fat-based product in food manufacture. It is a dynamic measure of phase change behaviour and complements solid fat index as measured by dilatometry or solid fat content as measured using pNMR; these are static or equilibrium measures of phase change behaviour (Rossell 1986). The cooling curve is particularly useful in chocolate manufacture (Padley *et al.* 1972; Rossell 1986).

In the International Union of Pure and Applied Chemistry (IUPAC) standard method 2.132 (Paquot and Hautfenne 1987), the cooling curve of the molten sample is obtained by placing the vacuum jacketed flask containing the sample in an ice-water bath and observing sample temperature every minute. The cooling curve is compared with the cooling curve for a sample of soybean oil (which does not crystallize under the conditions of the test) obtained under identical conditions. The important features of the curve are the "prime stay temperature" (the temperature at which the sample and soybean oil curves start to diverge), the temperature minimum (at which the rate of release of latent heat of crystallization equals the rate of heat loss by cooling) and the subsequent temperature maximum (caused by the rate of release of latent heat exceeding for a time the rate of heat loss by cooling). The times at which these temperatures are reached, and the general shape of the curve, are also useful pieces of information.

The British Standard method (BS 684: Section 1.13: 1976; British Standards Institution 1976d) is similar. However, the sample is deliberately stirred during the measurement in such a way that small fat crystals formed in the upper parts of the apparatus are carried down into the sample, where they act as crystallization nuclei. This causes crystallization to occur in the most stable polymorphic form, making the test more suitable than the IUPAC one for assessing the performance of fat in chocolate manufacture (Rossell 1986).

12.2.10 Optical Methods

12.2.10.1 Light and Polarized Light Microscopy

The rate of growth of fat crystals can be measured by photographing growing crystals at regular intervals using a light microscope equipped with a camera and a length-graduated ocular, followed by suitable analysis of crystal size measurements (Kawamura 1979).

van Putte and Bakker (1987) described a system for measuring nucleation rate during fat crystallization in which the number of crystals per unit volume was counted subjectively as a function of time using a light microscope equipped with a Fuchs-Rosenthal counting chamber (commonly used for counting blood cells).

Plane polarized-light microscopy is particularly useful for observing and imaging fat crystals. (The wave motion of plane polarized light occurs in one plane only - that which is perpendicular to the direction of travel.) A polarized light transmission microscope comprises a light source, an optical polarizer (a Nichols prism), a stage on which can be placed a small fat sample held between a temperature controlled glass slide and a coverslip, an optical analyser (another Nichols prism) and a means of image capture (Slayter and Slayter 1992). The polarizer and the analyser are in the "crossed Nichols" orientation, i.e. at 90 ° to one another.

Completely molten fat is optically isotropic, i.e. its refractive index for plane polarized light is the same in all directions. When the sample on the microscope's stage is molten, the crossed Nichols prisms prevent any polarized light leaving the analyser, and no image can be detected. Fat crystals, on the other hand, are optically anisotropic, typically birefringent (having two principal refractive indices) (Slayter and Slayter 1992; Narine and Marangoni 1999). A birefringent material present between polarizer and analyser splits the plane polarized light into two rays, called the ordinary ray and the extraordinary ray. Constructive interference of these rays in the analyser results in light passing through the analyser (James and Tanke 1991; Slayter and Slayter 1992). Thus, as soon as crystallization begins, an image is detectable.

Herrera *et al.* (1999a) measured the change of fat crystal size distribution with time during crystallization using a polarized light microscope equipped with a still camera, by means of which photographs were taken at regular intervals. The photographs were scanned and the size distribution at each time, based on longest crystal dimension, determined using suitable software.

The coupling of polarized light transmission microscopy with image analysis has proved a useful way of measuring the rate of crystallization. For example, Wright *et al.* (2001a) described a system in which a video camera was used to capture the image of the crystallizing fat sample produced by the microscope, at intervals of 15 s. The images were processed using suitable software. After "thresholding", subtracting an initial image from every subsequent image, the numbers of black and white pixels in each image were determined. The number of black pixels, representing crystal mass, was determined as a function of time to give the crystallization rate. Thresholding is critical, as it must be done in such a way that all of the crystal mass is seen as white and all of the background as black, in each image (Narine and Marangoni 1999). Video microscopy is described in detail by Sluder and Wolf (1998).

Peyronel (2018) describes practical aspects of using several types of microscope to study the physical properties of fats.

Other methods of imaging fat crystals and fat crystal networks (not all of them optical) include confocal laser scanning fluorescence microscopy, multiple photon microscopy, atomic force microscopy and electron microscopy (Narine and Marangoni 1999).

12.2.10.2 Laser Light Diffraction Spectrophotometry

Laser light diffraction spectrophotometry and visible light absorption spectrophotometry (Section 12.2.10.3) are described here rather than in Section 12.4 *Light Spectroscopy* because the optical instruments involved are used for purposes other than the acquisition of spectra.

Laser light diffraction spectrophotometry can be used to determine the crystal size distribution in an isothermally crystallizing oil (van Putte and Bakker 1987). Laser light is caused to shine through and on to a sample of the crystallizing mixture by inserting the mixture into the paths of laser beams, and the angles and intensities of light diffracted by the crystals are measured by means of a lens system and light-sensitive detectors. Angle of diffraction is related to particle size, and intensity is related to the proportion present of any particular size. Data analysis is complex.

Fat crystal growth rate can be found by measuring the particle size distribution as a function of time, and calculating growth rate as the rate of increase of maximum crystal size (van Putte and Bakker 1987).

Visible light diffraction is the same phenomenon as X-ray diffraction.

12.2.10.3 Visible Light Absorption Spectrophotometry

The development of turbidity with time during isothermal crystallization of an oil, which is the result of crystal formation, can be used to study nucleation and crystal growth (e.g. Herrera *et al.* 1999b; Dibildox-Alvarado and Toro-Vasquez 1997; Wright *et al.* 2000, 2001a). The sample is temperature-conditioned to erase thermal memory and cooled to crystallization temperature in a temperature-controlled cell or cuvette in a spectrophotometer. The absorbance or transmittance (optical density) of the sample with respect to visible light is then measured as a function of time. The induction or nucleation time is taken as the time from the start of isothermal holding to the time at which the absorbance begins to increase or the transmittance to decrease.

Herrera (1994) described a sophisticated version of this technique in which polarized laser light was used in much the same way as in polarized light transmission microscopy. The laser light beam passed from the source (a helium-neon laser) to a photosensor (a cadmium sulphide photodiode) through, in sequence, an optical polarizer, the crystallizing fat sample in a temperature-controlled glass cell, and an optical analyser in the crossed Nichols position relative to the polarizer. As soon as crystallization began, the optically anisotropic fat crystals that then appeared resulted in light passing through the analyser to be detected by the photosensor. The induction time was taken as the interval between the instant at which the sample reached the isothermal crystallization temperature and the time at which light was first detected.

Similar systems, in which the experimental setups were conveniently in the form of polarized light transmission microscopes, have been described by Koyano *et al.* (1989), Ng (1989) and Herrera *et al.* (1998).

Absorption spectrophotometry is a satisfactory way of measuring induction time, but is quite unsuitable for the kinetic characterization of subsequent crystallization, except in the early stages; minimum transmittance does not correspond to the end of crystallization, transmittance is proportional to the extent of crystallization only when no multiple scattering of the incident light occurs, and an observed decrease in transmittance can be due to light refraction (Marangoni 1998).

12.2.10.4 Visible Light Reflectometry

Wright *et al.* (2001a) described a system for continuously monitoring crystallization in which a beam of visible light was directed on to the crystallizing sample perpendicularly from above. A lens system and a matrix of light detectors were located perpendicularly above the sample. The solid-liquid (crystal-oil) boundaries in the sample scattered the incident light beam, and the intensity of the scattered light was measured by the detectors. The output signal, proportional to crystal mass, was recorded as a function of time.

12.3 Thermal Properties: Critical Temperatures

When a fat or oil is heated, thermal instability may cause decomposition and, depending on the temperature reached, subsequent combustion of volatile gaseous decomposition products (Mehlenbacher 1960). The thermal stability of fats and oils is thus essentially a chemical characteristic. However, stability is characterized by measuring certain critical temperatures, the *smoke*, *flash* and *fire* points, at which certain heat-induced changes become apparent. It is appropriate, therefore, to include methods of measuring these critical points here.

Measurement of smoke, flash and fire points is carried out subjectively by observing the surface of an oil sample while the sample is being heated. (The critical temperatures are higher than the upper limit of the melting point range of a fat.) The smoke point is the temperature at which the sample begins to give off a continuous stream of bluish smoke, observable by means of a light beam directed across the surface of the sample. The flash point is the lowest temperature at which a flash of flame appears at any point on the sample surface when a flame is applied near the surface. The fire point is the lowest temperature at which combustion continues for at least 5 s when a flame is applied (Rossell 1986).

Standard methods for the determination of the critical temperatures are published by the AOCS (AOCS Official Methods Cc 9a-48 (smoke, flash and fire points by an open cup method), and Cc 9b-55 and Cc 9c-95 (flash point by open cup methods); Firestone 1998) and by the British Standards Institution (BS 684: Section 1.8: 1976 (smoke point); British Standards Institution

1976b; and BS 684-1.17:1998/ISO 15267:1998 (flashpoint by a closed cup method); British Standards Institution 1998b).

12.4 Light Spectroscopy

12.4.1 Introduction to Light Spectroscopy and Nuclear Magnetic Resonance Spectroscopy

Spectroscopy methods can be divided into two main classes: *photon spectroscopy* and *particle spectroscopy* (Belton 1994). The first is concerned with interactions between matter and electromagnetic radiation. The second is concerned with molecular, atomic or subatomic particles interacting with or generated from matter. Photon spectroscopy is described here.

Two kinds are considered. The first is *light spectroscopy*: ultraviolet (UV, including luminescence, wavelength = 10 to 4 × 10^2 nm), visible (VIS, 4 × 10^2 to 7 × 10^2 nm) and infrared (IR, 7 × 10^2 to 1 × 10^5 nm) (Hurst 1994).

The second is *pulsed nuclear magnetic resonance* (pNMR, 0.4–5 m for ^{1}H nuclei for spectroscopy, 3.8–20 m for ^{1}H imaging) (https://depts.washington.edu/eooptic/linkfiles/The_Basics_of_NMR.doc. Accessed 26 January 2019).

NMR relaxometry is arguably of greater importance in food analysis than NMR spectroscopy. Both are described in Section 12.5.

Spectroscopy is an extremely useful technique for studying and determining the chemical and physical natures of materials, especially their structure and composition (Harris and Bashford 1987). Its capability rests on the fact that all chemicals and biochemicals absorb, absorb and emit or scatter/reflect energy from at least one region of the electromagnetic spectrum. Absorption occurs when there is a photon energy match between a molecular transition in the sample and the electromagnetic radiation. Absorption and then emission of photons occur in the case of luminescence (fluorescence or phosphorescence) and pNMR. Only inelastic scattering of light is of importance in spectroscopy, as it involves energy interaction between the light and the sample. In the case of elastic scattering no such interaction takes place; the light is merely reflected with no loss or gain of energy.

Because different substances interact with electromagnetic radiation in different ways and to different extents, a unique and specific relationship exists between a substance and its spectrum – the plot of electromagnetic intensity versus frequency, wavelength, wave number or energy. This can be used qualitatively for identification purposes or quantitatively for determination of concentration and other properties (Nixdorf 2018).

One way of classifying spectroscopies is according to the transition that occurs in the sample. Transitions are of four kinds (Belton 1994): three molecular (electronic, vibrational and rotational) and one nuclear (in pNMR).

Electronic transitions in molecules in the sample may be excited by absorption of VIS and UV irradiation. Absorption of UV light can result in fluorescence or phosphorescence. These two types of luminescence differ in the transitions involved, but both result in the emission of VIS or near-UV light.

Molecular vibrational transitions can result in absorption, reflection and scattering of IR, especially mid-IR, radiation. Mid-IR spectroscopy is highly sensitive and gives information about both the chemical and physical states of the molecules represented in spectra. Both absorption and emission IR spectra may be obtained. In Raman spectroscopy, the inelastic Raman scattering of near-infrared (NIR) radiation is used to study vibrational transitions.

Rotational motions in molecules result in the absorption of energy in the far-IR, a region that appears not to be used in spectroscopy as applied to foods.

Nuclear transitions of significance in pNMR are described in Section 12.5. The rest of this section is devoted to light spectroscopy.

Belton (1994) provides a guide to choosing a spectroscopic method in terms of the type of information wanted and sample preparation requirements (if any).

12.4.2 Light Spectrometers

A light spectrometer is an instrument that measures the extent of modification of a beam of electromagnetic radiation resulting from interaction between the beam and the sample. All instruments used in light spectroscopy carry out the same three essential functions (Belton 1994): dispersion of the source radiation into discrete wavelengths/frequencies (or continuous bandwidths in the case of spectrometers that use Fourier transformation), excitation of the sample and detection. In some spectrometers, excitation precedes dispersion. Excitation always precedes detection.

The following sources of information were drawn upon in writing the instrument descriptions below: Franca and Nollet (2018), Aykas and Rodriguez-Saona (2018), Harris and Bashford (1987), Hollas (2004), Lakowicz (2006), Osborne *et al.* (1993), Ozaki *et al.* (2007), Sharma and Schulman (1999) and Wilson (1994).

12.4.3 UV-VIS Spectrometers

The features of light spectrometers mentioned above are exemplified in the UV-VIS absorption spectrometer shown diagrammatically in Figure 12.7, in which dispersion precedes excitation, and in which a mirror is rotated to switch between a UV source (a deuterium lamp) and a VIS source (a tungsten lamp).

Dispersion is achieved using a rotating diffraction grating. This can be set to give a single wavelength or a very narrow bandwidth or can be rotated during measurement to scan through a required wavelength range. The wavelength range is further controlled by the spectrometer's slits. A narrower slit gives a narrower wavelength range, and thus a better spectral resolution, but reduces light beam intensity. The slit should be as wide as possible subject to meeting minimum resolution requirements.

The dispersed beam is sent alternately along two different paths by means of a rotating segmented mirror (sometimes called a *chopper*). The beam is thus directed alternately (every few milliseconds) through the sample (e.g. an analyte in a solvent) and through a reference (e.g. the pure solvent). In each case the beam eventually strikes the single detector. The measured reference signal can be subtracted from the sample signal to give a final spectrum that is representative only of the analyte. This arrangement, known as *alternating in time*, means that the measured difference in absorption between the sample and reference is unaffected by fluctuations in source power.

In a slightly different arrangement, known as *simultaneous in time* or *split beam*, the beam coming from the dispersion step is divided continuously by a beam splitter into two beams of equal intensity. After passing through the sample and reference, respectively, each beam strikes a separate detector.

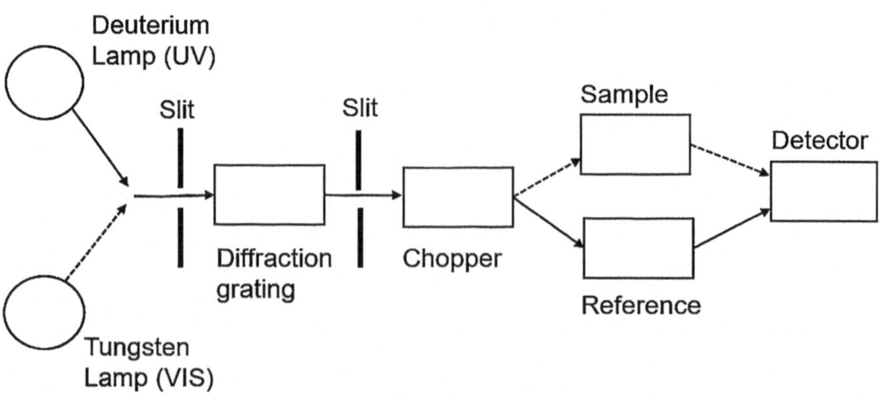

Figure 12.7. Block diagram of a typical double-beam recording visible and near-ultraviolet spectrometer.

In single beam spectrometers, there is no beam chopping or beam splitting; there is just one continuous optical path. This makes the instrument simpler and therefore cheaper. However, the sample and the reference must be measured sequentially, which makes the results subject to error owing to variation in source power or in other factors such as temperature.

Yet another design is the multichannel instrument called the *array spectrophotometer*. In this, polychromatic light from the source passes through the sample (there is no reference) and is then detected by a diode array in which each diode measures a narrow bandwidth of the spectrum.

Aliakbarian *et al.* (2016) showed that UV-VIS spectroscopy could be used to monitor the storage stability of yoghurt.

Hamboyan and Pink (1990) used UV-VIS spectroscopy in developing a way of predicting the likelihood of feathering occurring when cream is added to coffee. The predictive rule developed makes use of the spectroscopically-measured concentration of chlorogenic acid in the coffee.

Bernuy *et al.* (2008) showed that UV-VIS spectroscopy was as good as Raman spectroscopy (see Section 12.4.8) for determining the concentration of conjugated linoleic acid in milk fat.

Parmar and Sharma (2016) reviewed the use of derivatives (up to at least the fourth derivative) of US-VIS spectra in analysing experimental spectral data. They claim that this has invigorated the use of UV-VIS spectroscopy, as it greatly improves spectral resolution. The mathematical software required is now built into UV-VIS spectrometers. The improved resolution allows the identification and quantification of substances in mixtures, a task difficult or impossible using the original spectrum (Nelson 2018).

12.4.4 IR Spectrometers

The emphasis here is on near-infrared (NIR) spectroscopy, as it appears to be far more widely used than mid-infrared (MIR) spectroscopy. However, FT-MIR (Fourier transform MIR) spectroscopy can have significant advantages in terms of spectral assignment, spectral resolution and ease of quantification. Even for complex mixtures, for example, cheese, it is usually possible to assign distinct, well-resolved peaks to specific components without the necessity of establishing correlations, a task that is needed in the case of NIR spectroscopy (Wilson and Goodfellow 1994).

Most modern benchtop NIR spectrometers operate in both *transmission* and *diffuse reflection* modes; some, in addition, operate in *transflection* mode (see below).

A mid-IR and near-IR (MIR and NIR) transmission spectrometer is shown in Figure 12.8. This instrument is similar to the UV-VIS instrument shown in Figure 12.7. The main difference is that excitation precedes dispersion. The detector is a thermocouple. The array spectrophotometer, mentioned above, is an alternative arrangement.

The transmission mode is used for analysing homogeneous samples of relatively high purity such as solutions and fine powders (the latter spread in fine layers on glass slides). The transmittance mode can be used for opaque foods provided samples are appropriately prepared. For example, Karoui *et al.* (2006) prepared Emmental cheese samples for MIR spectroscopy by dispersing 5 g of cheese in 100 ml water, applying 60 μL of the dispersion to a polyethylene card and drying overnight. The wavelength range used for analysis purposely excluded that for MIR absorption by the polyethylene.

Diffuse reflection (Figure 12.9) and transflection (Figure 12.10) modes can be used to analyse solids, both homogeneous and heterogeneous and with any particle size distribution. Transflection combines transmission and reflection: incident NIR light is transmitted by the sample and reflected back by a highly reflective material such as aluminium or gold.

An NIR spectrometer (indeed, any spectrometer) must be calibrated for every different type of material that it will be used to analyse. This is done for a given material by assembling a representative set of samples, analysing them using wet chemistry (or sensory measurement, or rheological measure-

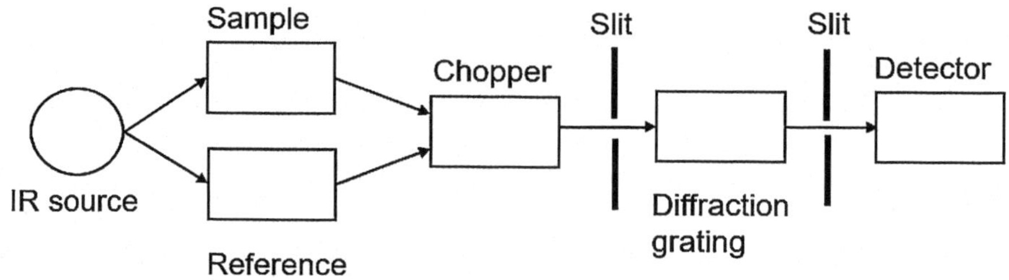

Figure 12.8. Block diagram of a typical double-beam recording mid-infrared and near infrared spectrometer.

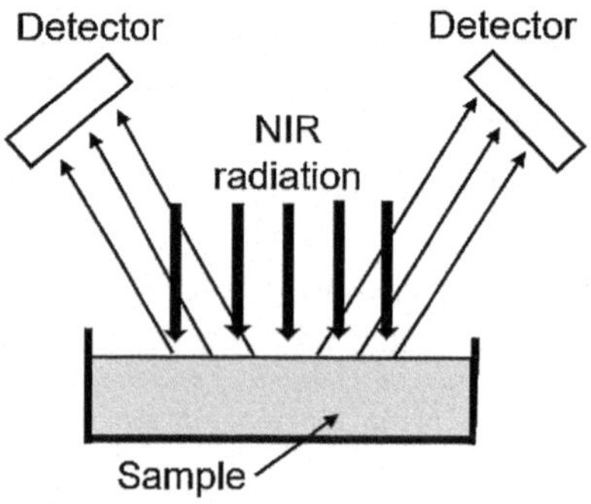

Figure 12.9. Schematic diagram showing the analysis of a solid food sample using NIR in diffuse reflection mode.

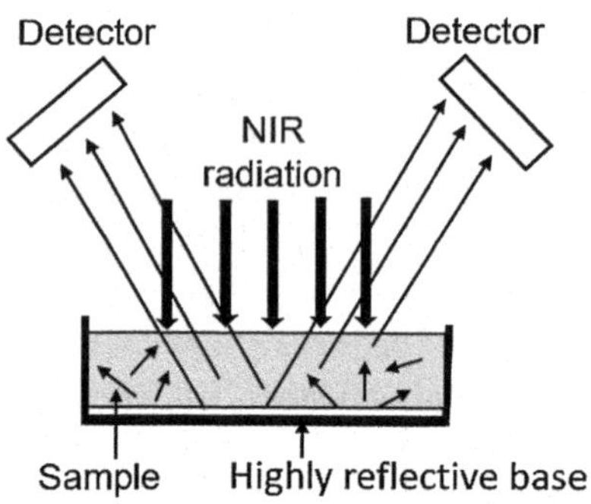

Figure 12.10. Schematic diagram showing the analysis of a food sample using NIR in transflection mode.

ment, etc.), recording their spectra and then processing all of the data obtained, using appropriate chemometrics, to generate calibration equations. The equations can then be used to make predictions of properties of new samples of the material concerned on the basis of spectroscopic measurements. Spectroscopy used in this way is a secondary method of analysis, the primary method being wet chemistry, or sensory measurement, etc.

Chemometrics, which comprises sophisticated mathematics, statistics and computer programming, is essential in processing the very large amount of data contained in spectra. Further, the selectivity of an analytical method such as spectroscopy can be provided as much by the chemometrics used as by the spectroscopic experimental modalities.

12.4.5 NIR Spectroscopy

NIR spectroscopy gives detailed compositional data and is very fast (less than 30 s per sample), non-destructive, non-invasive and highly versatile. It can be used off-line, at-line, on-line and in-line (Holroyd *et al.* 2013). This versatility is made possible by developments in the design of spectrometers, which include the use of fibre-optic cables and probes in place of the classical optical bench. Little or no sample preparation is required, and no reagents are needed. NIR radiation has low reflectivity and low absorptivity, making it highly penetrating and enabling it to interact closely with the sample. NIR in diffuse reflection and transflection modes is especially useful for analysing strongly absorbing and highly scattering samples such as foods.

A semi-quantitative survey of relevant research publications listed in *Food Science and Technology Abstracts* (published by IFIS, Reading, UK) in the last 12 years shows that NIR has been the most widely used light spectroscopic technique in research studies of milk fat and milk fat-based products. It shows also that the most studied product is cheese and that research on this product has been largely in the following areas (confirming the assessments of Holroyd

(2013) and Cattaneo and Holroyd (2013a, b)): *composition and quality, ageing/ripening, physicochemical properties, authentication/proof of origin, adulteration/contamination* and *in-production spectroscopy.*

MIR studies, although much fewer in number than NIR ones, fall roughly into the same categories. Historically, MIR, although very useful for structural elucidation and identification of organic compounds, was not used much for food studies because of various experimental difficulties. These difficulties have been largely mitigated by the introduction of Fourier transform IR, and advantage can now be taken of the excellent spectral assignment, resolution and ease of quantification offered by MIR.

12.4.6 Fourier Transform NIR (FTIR)

In an NIR/MIR spectrometer of the kind shown in Figure 12.8, the source IR radiation is dispersed either before or after passing through the sample. The final signal is recorded – a narrow wavelength range at a time – by the detector, in the form of the usual frequency-domain spectrum (intensity versus frequency). This is relatively slow. Recording the time-domain spectrum instead, and then Fourier transforming it to the usual frequency domain, has important advantages. The following description is based on Abbas and Baeten (2018), Cozzolino (2018), Hollas (2004) and Hsieh (2018).

A typical FTIR spectrometer is shown in Figure 12.11. It is equipped with a Michelson interferometer (the commonest type used) in place of a dispersing element (diffraction grating or prism). The interferometer comprises a beam splitter and two mirrors placed orthogonally to one another. One of the mirrors can be made to travel backwards and forwards along a straight path towards and away from the beam splitter. The beam splitter transmits 50% of incident radiation and reflects the other 50%.

Consider a monochromatic light source. As the moving mirror moves back and forth, the detector sees light and dark bands corresponding to partial and total constructive and destructive interference in the interferometer's output beam. This beam or wave is called a *cosine interferogram.* In printed form it is a plot of intensity versus mirror position.

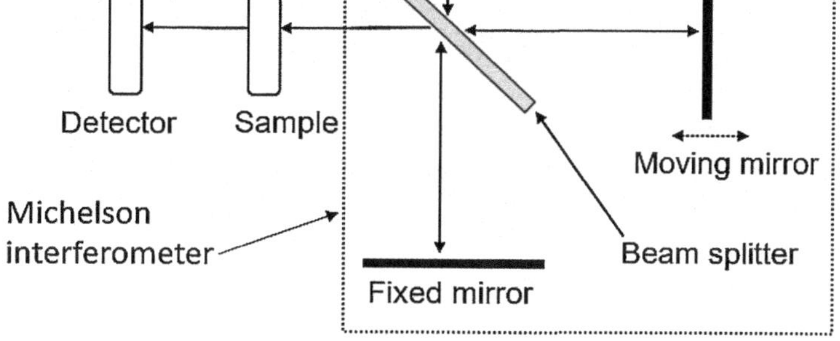

Figure 12.11. Fourier transform near-infrared (FTIR) spectrometer incorporating a Michelson interferometer.

The wave goes from full intensity when both mirrors are equidistant from the beam splitter to zero intensity when the moving mirror has travelled half a wavelength. Full intensity recurs when the moving mirror has travelled a full wavelength and subsequent multiples of a full wavelength. Zero intensity recurs at each odd multiple of half a wavelength of mirror travel.

An IR source is polychromatic, emitting light over a broad range of frequencies. Each individual frequency yields a unique cosine signal; the resulting cosine interferogram is the sum of all the cosine waves generated from the individual infrared frequencies.

Fourier transformation, applied to the recorded data, resolves the frequency and intensity of each cosine wave in the interferogram; it converts the measured intensity-versus-mirror-displacement signal (the interferogram) into a plot of intensity versus frequency (a spectrum).

The interferogram is really a distance domain spectrum, but because the moving mirror moves at a constant speed, the interferogram can be considered also to be a time-domain spectrum. In light spectroscopy a time-domain spectrum cannot be measured directly simply by shining the undispersed source light directly at the sample (source → sample → detector) because no detectors are available capable of responding to the radiation intensity variations at the high frequencies in the IR-VIS-UV range. The interferometer gets around this problem by modulating the high-frequency signal (from the source) to a lower-frequency one without distorting the original signal; frequencies in the modulated signal – the interferogram – are directly proportional to those in the original. (As an interferometer is not a dispersing element of the spectrometer, a diode array detector cannot be used.)

FTIR has several advantages over dispersion spectroscopy. During measurement, all of the frequencies are being recorded all of the time. This means that every point in the interferogram contains information for every wavelength in the input light; if 8000 data points along the interferogram are recorded, every wavelength in the input light is sampled 8000 times. The result is a high signal-to-noise ratio. This is known as the multiplex or Fellgett advantage. The ratio is improved by repeating the back and forth travel of the moving mirror a number of times and averaging the data collected.

FTIR spectroscopy is extremely fast; a full spectrum can be obtained in seconds, as against minutes in the case of a conventional IR spectrometer.

As narrow dispersion slits are not needed (relatively wide apertures may be needed for optical reasons), there is a much larger throughput of light in an FTIR spectrometer. This improves resolution (the Jacquinot advantage).

FTIR spectrometers are highly accurate (Connes advantage); frequency precision is better than 0.01 wavenumbers. Interferometers are equipped with an internal frequency standard, commonly a HeNe (helium-neon) laser (not shown in Figure 12.11). All frequencies in the final output spectrum are calculated relative to the precisely known frequency of the (monochromatic) laser light.

Lastly, the resolution in the measured spectrum is the same for all frequencies. Resolution is improved by increasing the path length travelled by the moving mirror of the interferometer; a longer travel increases the likelihood that the interferometer will successfully discriminate between wave forms that are very similar.

The Fourier transform approach can in principle be used with almost any kind of spectrometer.

12.4.7 Attenuated Total Reflection-FTIR (ATR-FTIR)

ATR-FTIR is one of the commonest and most powerful FTIR techniques; it is non-destructive and has a high sensitivity. The principle of the method is shown in Figure 12.12.

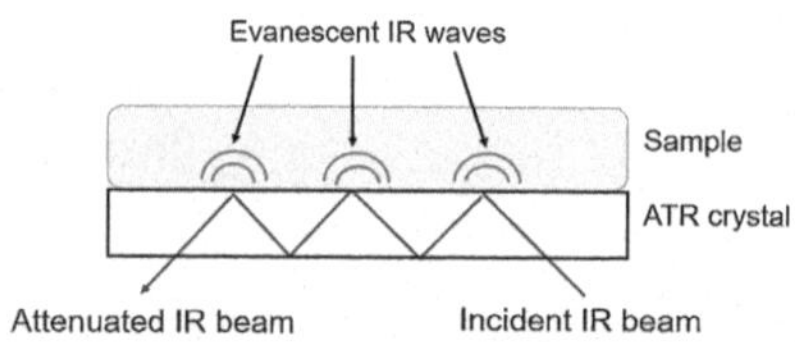

Figure 12.12. Schematic diagram showing the principle of attenuated total reflection (ATR) spectroscopy.

The sample is applied to one surface (sometimes both surfaces) of a crystal (made, e.g. of zinc selenide). The IR beam from the spectrometer's interferometer is directed into one end of the crystal at an angle to the crystal-sample boundary. Provided the crystal has a density (and therefore a refractive index) higher than that of the sample, and provided the angle of incidence of the beam is greater than the critical angle, the beam is totally internally reflected. However, reflection is not quite total: a small amount of IR light energy penetrates into the sample surface by a small distance (0.1–5 μm) in the form of an evanescent (quickly fading) wave. This decays into a tiny amount of heat in the sample. Thus, the IR beam becomes attenuated. The degree of attenuation and its wavelength dependency are recorded in the detected spectrum.

Little or no sample preparation is required for ATR-FTIR; solids, liquids, pastes and powders can all be analysed, usually in their natural state. The main requirement is that the sample be placed in intimate contact with the crystal, to ensure evanescent wave penetration.

An ATR-FTIR accessory can be mounted in the sample compartment of most FTIR spectrometers.

Monaci *et al.* (2007) developed a simple and rapid ATR-FTIR (mid-IR) method of detecting and semi-quantifying the mycotoxin cyclopiazonic acid in mould-ripened ewes' milk cheese. They did this by correlating spectral data with cyclopiazonic acid concentration as determined chromatographically. They pointed out that since no sample preparation is required prior to ATR spectroscopy, analysis time is dramatically shortened compared with that for any chromatographic procedure. This would allow rapid screening of large numbers of cheese samples.

Multiple analytical methods are required for determining cheese composition and flavour quality. These are accurate, but are laborious, expensive and time-consuming. Subramanian *et al.* 2009 established a rapid, simple ATR-FTIR technique by correlating spectral data with *Cheddar* cheese moisture, pH and salt and fat contents. All of these cheese properties, and quality ratings based on spectral data for organic acids, amino acids and short chain fatty acids, could be determined in less than 20 min.

12.4.8 Raman Spectroscopy

Raman spectroscopy is based on Raman scattering of monochromatic light in the NIR-VIS spectral range. When such light is incident on a sample, absorption, reflection and scattering occur to varying degrees. The frequency of scattered light may be equal to that of the incident light (Rayleigh scattering), or it may be of lower frequency (Stokes Raman scattering) or higher frequency (anti-Stokes Raman scattering). Such frequency shifts, which a Raman spectrometer is designed to measure, provide a way of observing the vibrational states of molecular structures in the sample.

In a backscattering Raman spectrometer (Figure 12.13), the sample is illuminated by a laser source with a frequency in the NIR-VIS-near-UV range. The dichroic beam splitter reflects the laser light downwards onto the sample while transmitting the returning scattered Raman frequency-shifted light. The laser light blocking filter ensures that no Rayleigh-reflected laser light reaches the detector; the Raman signal is relatively weak, and care must be taken not to mask it.

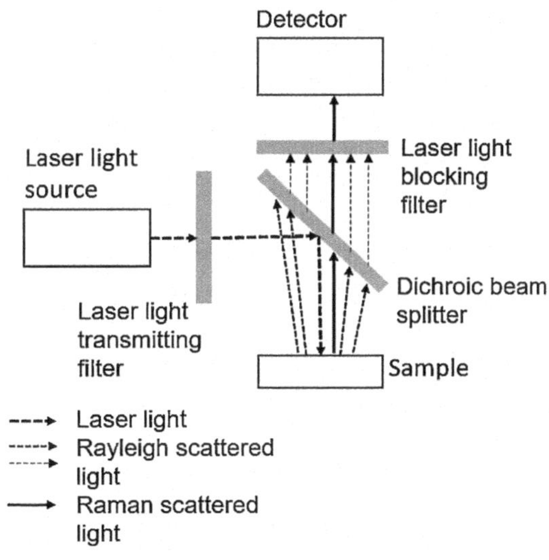

Figure 12.13. Schematic diagram of a Raman spectrometer.

The spectrometer illustrated in Figure 12.13 comprises a wavelength-dispersing device (diffraction grating or prism) or an interferometer (for Fourier transform Raman spectroscopy) and a detector. The laser beam that strikes the sample, and the detector, must be on the same axis. Misalignment results in a poor Raman signal. The laser beam must be normal to the sample surface to avoid misalignment caused by variations in sample height.

Raman spectra constitute unique spectral fingerprints of molecular species, allowing discrimination between closely similar substances. Measurements are simple.

Analysis can be qualitative or quantitative. Only very small samples are needed, and little or no sample preparation is required. Raman spectroscopy is insensitive to water, allowing spectra to be obtained from aqueous solutions and high-water-content materials such as some foods.

Backscattering Raman spectroscopy as described above is just one way of applying Raman spectroscopy. There are a number of variants. Two of these, spatially offset Raman spectroscopy (SORS) and surface-enhanced Raman spectroscopy (SERS), have been found to be useful in certain types of food analysis. They are described by Dhakal *et al.* (2018).

Raman spectroscopy can be used successfully to measure, accurately and rapidly, the extent of adulteration of cream with sunflower, coconut and palm oils (Nedeljkovic *et al.* 2017) and butter with margarine (Nedeljkovic *et al.* 2016; Uysal *et al.* 2013), once appropriate correlations are developed between Raman spectral characteristics and percentage adulteration.

Oliveira *et al.* (2016) developed a correlation that enabled the detection by Fourier transform Raman spectroscopy of starch (an adulterant) in spreadable cheese. The model had a limit of detection of 0.34% (w/w) starch and a limit of quantification of 1.14% (w/w).

Beattie *et al.* (2005) showed that Raman spectroscopy could be used to predict the fatty acid composition of butterfat, and properties such as iodine value and solid fat content at low temperatures, with sufficient accuracy to make the technique good enough for routine quality control purposes. It was suggested that this, together with the fact that no sample preparation is required and measurement time is only 60 s, indicated that high-throughput on-line Raman analysis of butter samples should be possible.

12.4.9 Fluorescence Spectroscopy

Fluorescence is the emission of light in the NIR-VIS-UV region of the electromagnetic spectrum by fluorophores – certain aromatic or highly unsaturated organic compounds. Emission is the result of the relaxation of electrons previously excited by absorption of light from this same region of the spectrum.

The frequency of the spectral peak of the emitted light is generally lower than that of the excitation beam, but quite independent of it. The difference is called the Stokes shift. The intensity of the emitted light is, however, dependent on the excitation peak frequency.

Fluorescence spectra are measured using a spectrometer called a spectrofluorometer. A typical example of this instrument is shown in Figure 12.14. It incorporates two monochromators (diffraction gratings) for setting or scanning wavelength, one in the excitation beam and the other in the emitted beam. This allows two types of spectra to be recorded. An emission spectrum is a plot of the intensity of the emitted light versus emission wavelength at a single constant excitation wavelength. An excitation spectrum is a plot showing the dependence of emission intensity, at a single emission wavelength, when the excitation wavelength is scanned. The emission spectrum is typically a mirror image of the excitation (absorption) spectrum. (Most commercial spectrofluorometers record spectra using the wavelength scale, as opposed to wave number or frequency.)

Not shown in Figure 12.14 is a beam splitter located just after the exit slit of the excitation diffraction grating. This allows some of the excitation beam to be directed to a reference detector. The signals from the reference and sample detectors are processed electronically to give relative fluorescence data.

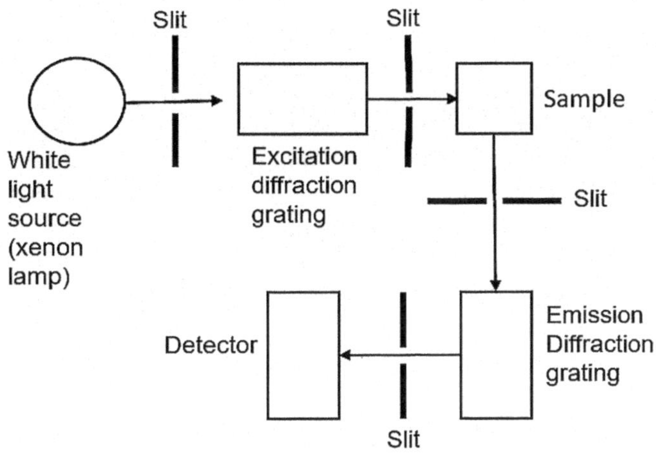

Figure 12.14. Block diagram of a fluorescence spectrometer (spectrofluorometer).

As shown in Figure 12.14, the emitted light leaving the sample cuvette at 90 ° to the excitation beam is collected. The purpose of this is to minimize interference of the emission beam by Rayleigh and Raman scattered light. However, for turbid samples, it may be necessary to use the so-called front face arrangement, in which the excitation beam strikes the face of the sample cuvette at an angle of incidence of about 30 °. In this case the detector collects emitted fluorescence light leaving the cuvette at an angle to the normal of about the same value.

In complex samples containing more than one fluorophore, emission spectral bands may overlap, making complete resolution of spectra impossible. Synchronous fluorescent spectroscopy (SFS) is a technique for mitigating this problem. In SFS, both the excitation and emission monochromators are simultaneously scanned while a constant wavelength interval ($\Delta\lambda = \lambda_{emission} - \lambda_{excitation}$) is maintained between them. This technique decreases spectral overlap by narrowing spectral bands and simplifying spectra by amplifying intense bands. The wavelength interval chosen can be that between the emission peak wavelength and the excitation peak wavelength, which gives a high sensitivity, or a particular interval that yields a compromise between sensitivity and selectivity.

Fluorescence measurements are broadly classified into two types: steady state and time resolved. The spectra defined above are measured under steady-state conditions. In experiments, steady state is reached almost instantaneously owing to the nanosecond timescale of fluorescence.

In a time-resolved measurement, fluorescence decay over time is measured subsequent to excitation by a pulse of light. Although the very high rate of decay necessitates complex and expensive instrumentation, time resolution reveals molecular information that is lost in steady state measurement; steady state measurement provides only an average of the phenomena that occur during the intensity decay of the light coming from the sample. In fact, much of the information that fluorescence can provide is available only via time-resolved measurements.

Fluorescence spectroscopy is orders of magnitude more sensitive than most other methods of detecting organic compounds. There are several reasons for this. First, fluorophores are easily detected even when admixed with substances that do not fluoresce, as there can be no interference from the latter. Second, an analyte that does not fluoresce can be made to do so by chemically attaching to it a moiety that does.

Third, emitted light is in effect being observed against a black background.

Fourth, two wavelengths, the absorption peak intensity wavelength and the emission peak intensity wavelength, are available for identification purposes.

Fluorescence spectroscopy is simple and fast, and the equipment is robust. Relatively low-cost spectrofluorometers with filters instead of monochromators, and with fibre-optic cabling and probes, are available for dedicated at-line or on-line measurements at fixed excitation and emission wavelengths.

Granger *et al.* (2006) used front face fluorescence spectroscopy to measure the emission spectra of tryptophan residues in 16 ice cream formulations. Analysis of the data clearly showed the existence, on a molecular scale, of different protein structures, and different interactions between proteins and fats, depending on formulation.

Larsen *et al.* (2009) used the same technique to show that sensorially measured deterioration of sour cream packaged in cups of varying light barrier properties was well correlated with the degradation of photosensitizers such as riboflavin, hematoporphyrin, protoporphyrin IX and others.

The sensitivity of fluorescence is useful in detecting adulteration of food and in discriminating between samples of the same product that have different provenances.

For example, synchronous fluorescence spectroscopy can detect the adulteration of hard cheese with plant oils down to oil levels of 3.0 and 4.4% for $\Delta\lambda$ intervals of 60 and 80 nm, respectively (Dankowska *et al.* 2015). A similar capability was found in the case of plant oil adulteration of butter (Dankowska *et al.* 2014).

Karoui *et al.* (2005) found that front face fluorescence spectroscopy was able to discriminate between samples of Gruyere PDO (protected designation of origin) cheeses from different locations in Switzerland with 100% accuracy.

Botosoa and Karoui (2013) subjected 20 French Emmental cheeses differing in brand and manufacturing process to physicochemical analyses, mid-IR spectroscopy and front face fluorescence spectroscopy. They found that provided the data from the three analytical approaches were analysed as one set, rather than separately, it was possible to achieve a very high degree of discrimination between individual cheese types.

Useful correlations can be developed between fluorescence spectral data and important properties of cheese, allowing fast and accurate prediction of these properties. Two examples follow.

Karoui *et al.* (2008) demonstrated that front face fluorescence spectroscopy could be used to measure the composition of semi-hard and hard cheeses. They established excellent correlations between tryptophan spectra and vitamin A spectra on the one hand and fat, dry matter, fat in dry matter and total nitrogen on the other. Similar correlations for predicting water-soluble nitrogen and melting point were obtained using the vitamin A spectral data.

Ozbekova and Kulmyrzaev (2017) established correlations between the fluorescence spectra (obtained in front face mode) and cheese rheology, composition and melting temperature. Yield stress and flow stress could be predicted, respectively, from the vitamin A emission spectral data and excitation spectral data. Melting temperature, and moisture, protein and fat contents, could be predicted from the vitamin A emission spectral data. The results show that fluorescence spectroscopy has potential for the accurate, non-destructive and rapid prediction of cheese rheological characteristics.

Applications of fluorescence spectroscopy in dairy processing were reviewed by Shaikh and O'Donnell (2017) under the headings *effects of heat treatment and storage on quality*, *assessment of authenticity and safety* and *monitoring milk coagulation and syneresis*. Fluorescence spectroscopy, in the present context, appears to be the next most widely used spectroscopic technique after NIR.

Sádecká *et al.* (2018) give a detailed account of the basic principles of fluorescence spectroscopy and their application in food analysis.

Tao and Ngadi (2017) review current research progress in the application of NIR spectroscopy, Raman spectroscopy and NMR to fat and fatty acids analysis of dairy foods.

12.5 Nuclear Magnetic Resonance (NMR)

The following description of the principles of nuclear magnetic resonance (NMR) is based mainly on Bovey *et al.* (1987), Goldenberg (2016), Hashemi *et al.* (2018), Jacobsen (2007),

Kirtil and Oztop (2016), Kirtil *et al.* (2017), Lambert and Mazzola (2004) and Nelson (2003). Other sources of information are cited appropriately.

NMR is a spectroscopic technique that makes use of the magnetic properties of the atomic nucleus to determine the structures of organic molecules and biomolecules. It is the most powerful tool available for doing this; X-ray crystallography is the only comparable technique.

12.5.1 Precession and Resonance

In the atoms of many isotopes, the nucleus behaves as if it was spinning about an internal axis; it has a quantum mechanical property called spin, which is characterized by spin angular momentum (Spyros and Dais 2013). Among such isotopes are ^{1}H (the proton) and ^{13}C, whose resonant frequencies are commonly measured in NMR experiments.

As the nucleus is positively charged, its spin creates a very small magnetic field; the nucleus can be thought of as a tiny bar magnet. When such a nucleus is placed in a strong magnetic field, such as the field in an NMR spectrometer (Figure 12.15), the field exerts a torque on the nucleus which causes its spin axis to precess about the field direction – the +z direction shown in Figure 12.15. The precession rate (in Hz) is the resonant frequency of the nucleus. This frequency lies in the radio frequency (RF) range of the electromagnetic spectrum. Its magnitude depends on the strength of the nuclear "magnet" (the magnetogyric ratio) and the strength of the external magnetic field:

$$\upsilon_0 = \frac{\gamma B_0}{2\pi}(\mathrm{Hz}) \qquad (12.5)$$

where:

υ_0 = precession rate = the Larmor frequency = resonant frequency (Hz).
γ = magnetogyric ratio of the type of nucleus concerned (rad s^{-1} T^{-1}).
B_0 = external magnetic field strength (T).

Equation 12.5 shows that the resonant frequency is proportional to the magnetogyric ratio, γ, and to the external field strength, B_0. This is the basis of most NMR phenomena.

Measurement of resonant frequency is accomplished by first placing the sample (a solution or, less commonly, a powder) in the strong magnetic field of an NMR spectrometer (Figure 12.15). Then, the resonant frequency of a given ensemble of like nuclei is found by transmitting RF signals into the sample and

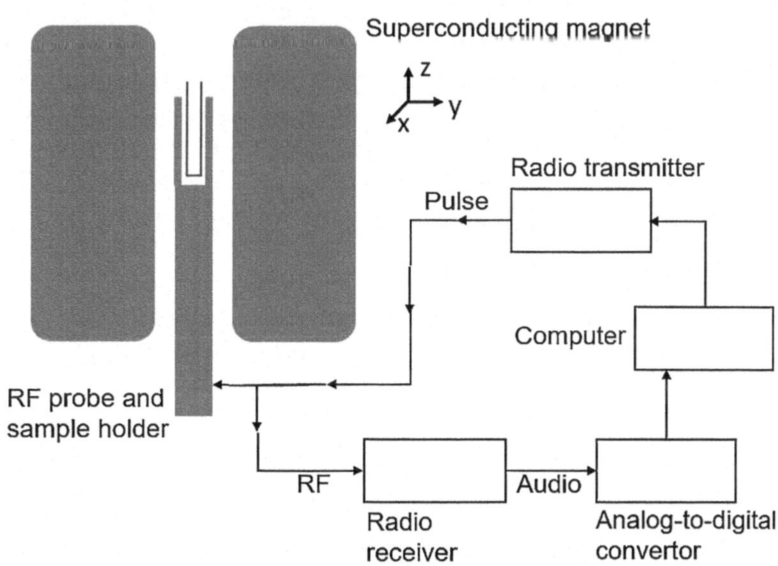

Figure 12.15. Block diagram of an NMR spectrometer.

recording the radio frequencies at which resonance occurs between the RF input and precessing nuclei. Further details are given below. (When the frequency of an excitation signal – the RF input in this case – approaches the natural frequency of a system capable of oscillation, an ensemble of nuclei, the amplitude of the oscillation increases to a maximum. The system is then said to be in resonance with the exciting force.)

12.5.2 The Chemical Shift

Measurement of resonant frequency is useful because the resonant frequency of a given type of nucleus depends on its immediate chemical environment and therefore on its location in the unknown structure being investigated. For example, ^{13}C has a fundamental resonant frequency of 50 MHz at a certain external magnetic field strength in the absence of any outside influences. But a ^{13}C nucleus forming part of an unknown molecular structure will have a resonant frequency that differs from 50 MHz. This is so because bonding electrons in the molecule create small magnetic fields that locally perturb the external field. As a result, the nucleus experiences not B_0 but an effective field strength, B_{eff}. B_{eff} replaces B_0 in Eq. 12.5, and thus the resonant frequency will differ from the fundamental value of 50 MHz. This difference is called the *chemical shift*.

The chemical shift, because it depends on molecular structure, is the main basis of the utility of NMR in determining structures of organic molecules. Consider, for example, the cycloheptanone molecule ($C_7H_{12}O$). The seven carbon atoms in this molecule exhibit between them four different resonant frequencies, indicating four different chemical shifts. From the nuclear resonance point of view, there are four distinct populations of ^{13}C in cycloheptanone, each with its own resonant frequency and chemical shift. Symmetry in the structure of the cycloheptanone molecule results in there being three pairs of ^{13}C nuclei in each of which the two nuclei have the same resonant frequency. The seventh carbon nucleus in the molecule has its own resonant frequency.

Chemical shifts are extremely small relative to the fundamental resonant frequency of the nucleus concerned. For example, the resonant frequencies of carbons 4 and 5 in the cycloheptanone molecule are both 50.00116 MHz, compared with the fundamental frequency of 50.00000 MHz. For this reason, the x-axes of spectra are often labelled in ppm of the fundamental value. Thus carbons 4 and 5 each have a chemical shift of $(50.00116–50.00000)/50 = 23.2$ ppm. The y-axis of an NMR spectrum is a measure of the amplitude of the RF signal induced in the receiving coil of the NMR spectrometer (see Section 12.5.5) by the resonating nuclei.

In addition to chemical shifts, the NMR spectrum yields information on resonance intensities, measured as the areas under resonance peaks in the spectrum. Intensities are found by appropriate integration, which is normally carried out only for ^{1}H spectra. In a ^{1}H spectrum, the total area under the spectral peaks is proportional to the total number of protons in the sample. At a given resonance frequency, the ratio of an individual peak area to the total area is the fraction of the total number of protons that resonate at that frequency.

If the molecular structure of cycloheptanone was not already known, interpretation of the ^{13}C spectrum in the light of the known properties of double bonds, electronegative atoms and other nuclei, together with data obtained using complementary techniques such as light spectroscopy and mass spectrometry, would enable inferences to be made about what the structure might be.

In a basic pNMR (pulsed NMR) experiment, the sample is placed in a strong, uniform magnetic field and subjected to a short pulse of radio frequency energy (Figure 12.15). The purpose of this is explained, and the pNMR spectrometer described, in the following.

12.5.3 Quantum Mechanics

Quantum mechanics, as well as classical physics, must be called upon fully to describe NMR phenomena although, as Jacobsen (2007) has pointed out, NMR is really about statistical mechanics, the behaviour of large groups of nuclei, rather than quantum mechanics, the behaviour of individual nuclei.

The most useful nuclei (they include ^{1}H and ^{13}C) are spin ½ nuclei: they have the spin quantum numbers –½ and + ½. A given nucleus can be either spin –½ (higher energy; oppositional to an external magnetic field) or spin +½ (lower energy; aligned with an external field).

In the absence of an external magnetic field and at thermal equilibrium, these two states have the same energy, and in a large population of nuclei (as is observed in an NMR experiment), there are equal numbers of spin –½ and spin +½ nuclei.

But in the presence of an external magnetic field (in the present context, the field in an NMR spectrometer), the spin +½ nuclei possess a lower energy than, but slightly outnumber, the spin –½ nuclei, as determined by the Boltzmann distribution. The energy gap between the two types of nuclei, ΔE, is given by:

$$E = h\nu_0 = \frac{h\gamma B_0}{2\pi}(\mathrm{J}) \qquad (12.6)$$

where:

h = Planck constant = 6.626 196 $\rho \times 10^{-34}$ J s.
ν_0 = resonant frequency (Hz).

The resonant frequency in Eq. 12.6 has the same value as the Larmor precession rate (Eq. 12.5).

The excess of +½ nuclei that exists in a given population of nuclei (in a sample at equilibrium in the NMR spectrometer's magnetic field) is tiny, about 1 in every 10^6 nuclei, but it is highly significant. It is the only thing that can be observed using NMR, and it is the fundamental phenomenon that makes NMR spectroscopy possible.

The number of excess +½ nuclei in the sample is proportional to ΔE (eq. 12.6), which is in turn proportional to γB_0 and inversely proportional to absolute temperature, T. It can be shown further that the amplitude of the signal received from the sample immediately after the pulse in a pNMR experiment – the amplitude represents the sensitivity of the experiment – is proportional to $N\gamma^3 B_0^2 / T$, where N is the total number of nuclei. This justifies the greater cost of building larger (bigger sample size) and more powerful (higher B_0) NMR spectrometers for some applications. For other applications, for example, the measurement of solid fat content, low resolution (low B_0) instruments may be perfectly adequate.

As explained just above, the sensitivity of an NMR experiment is proportional to the third power of the magnetogyric ratio (γ) of the nucleus being used. The use of a nucleus with a relatively small γ will result in very poor sensitivity. Given this consideration, ^{1}H, ^{13}C and ^{15}N are among the most useful nuclei.

12.5.4 The Vector Model

The so-called vector model is one way of visualizing the importance of the excess +½ nuclei. According to this model, the spin axis of every nucleus can be thought of as a magnetic vector with length equal to the nucleus's magnetic field (γ). All of the vectors precess about the external magnetic field direction at the common resonant frequency. The +½ nuclei are aligned with the field. The –½ nuclei oppose the field.

The vector sum is called the *net magnetization vector*. It is a macroscopic property of a very large ensemble of the target nuclei (e.g. 10^{20} nuclei) and can be thought of as a large magnet. Its magnitude (length) is determined by the excess of +½ over –½ vectors.

All of the vectors precess completely out of phase with each other. This has two consequences.

The first is that the remaining +½ vectors and the –½ vectors, which are present in equal number, cancel each other; only the excess +½ vectors matter.

The second is that the net magnetization vector, which is thus essentially the sum of the excess +½ vectors, lies stationary along the +z axis, the (vertically upwards) direction of the external magnetic field (Figure 12.15). It does not precess, because there is no magnetization in the x-y (transverse) plane. No detectable NMR signal is generated.

However, a pulse of RF energy changes the situation. Every pNMR experiment starts with a pulse being transmitted to the sample at either 90° or 180° to the direction of the spectrometer's magnetic field, that is, in the x-y plane or in the –z direction, respectively (Figure 12.15). This has the effect of bringing all of the excess +½ vectors into phase with one another. The vector sum – the net magnetization vector – then tips away from the z axis. This results in a decrease in longitudinal (z axis) magnetization and the appearance of transverse (x-y) magnetization. This causes the net magnetization vector to start precessing about the +z axis. This results in the absorption by the vector of RF energy supplied in the pulse (provided the radio frequency is exactly equal to the Larmor frequency (Eq. 12.5)) and, as soon as the pulse ends, the induction (by the moving vector) of a voltage (the NMR signal) in the spectrometer's receiving coil.

When the radio frequency is equal to the Larmor frequency, a + ½ nucleus is able to absorb one photon (quantum) of energy equal to ΔE (Eq. 12.6), which is the energy difference between +½ and –½ nuclei, to become a –½ nucleus. There is resonance between the RF input and the nuclei.

But at the same time, –½ nuclei "flip" to become +½ nuclei by absorbing one photon and then emitting two photons. Absorption occurs at a rate proportional to the number of +½ nuclei in the sample, and emission occurs at a rate proportional to the number of –½ nuclei. As initially there is a greater number of +½ nuclei, there will be net absorption of energy from the RF pulse.

If net absorption continued, there would quickly be equal numbers of +½ and –½ nuclei. The absorption rate would then become equal to the emission rate. There would no longer be net absorption of RF energy, and no longer an induced signal in the receiving coil of the spectrometer. This condition is called *saturation*. If saturation always occurred, an NMR "experiment" would consist of a short period of absorption and then nothing (Jacobsen 2007).

Saturation would be unavoidable if the emission of photons was the only way a –½ nucleus could flip to become a + ½ nucleus. But in fact, a –½ nucleus can flip by losing thermal energy instead of emitting photons. This process is called relaxation.

When the pulse ends, relaxation begins (see Section 12.5.6); the energy absorbed by the nuclei starts to be released, and the nuclei start to return to a state of thermal (Boltzmann) equilibrium. Thus, the signal induced in the spectrometer's receiving coil – a sinusoidal voltage large enough to be measurable – begins to fall. The fate of this signal, known as the free induction decay (FID), is described below.

12.5.5 NMR Spectrometer

A simple diagram of an NMR spectrometer is shown in Figure 12.15. The spectrometer comprises a superconducting magnet, a sample holder, a probe that can alternately transmit and receive radio signals, a radio transmitter, a radio receiver, an analog-digital convertor and a computer. The sample holder is continuously rotated about its axis at about 20 revolutions per second to average out any magnetic field inhomogeneity.

Early spectrometers were operated in the so-called continuous wave mode. This could be applied in two ways. In the first, a weak RF signal was continuously transmitted to the sample, its frequency being slowly changed to scan the range in which the resonant frequencies to be measured were thought to lie. Resonant frequencies were determined as frequencies at which peaks occurred in the signal detected by a receiving coil. In the second, the RF input signal was kept constant while the strength, B_0, of the spectrometer's magnetic field was slowly changed. An NMR signal was generated whenever B_0 reached a value that satisfied Eq. 12.5 (in which, in this case, υ_0 was held constant).

All modern NMR spectrometers are pulsed Fourier transform (p-FT-NMR) instruments. In such an instrument, an experiment is initiated by transmitting to the sample via the probe coil a very brief pulse (say, 10 μs) of high-power RF energy (50–300 W). The frequency of the pulse is set to a value at the centre of the range of resonant frequencies expected. Because this range is very narrow and the pulse powerful, all of the target nuclei (e.g. ^{1}H) are excited simultaneously.

12.5.6 Relaxation

When the pulse is switched off, the probe coil acts as a receiver of the macroscopic signal induced in it by the precessing net magnetization vector. Two kinds of exponential relaxation then start.

The first is a phenomenon called transverse or spin-spin relaxation (time constant = T_2). This is the free induction decay, FID (Figure 12.16). Once the pulse has ceased, –½ nuclei start to flip back to the lower energy level by losing thermal energy, and phase unison is lost. The macroscopic signal in the receiving coil dies away exponentially to nothing (Figure 12.16). The signal is thought of as being an "echo" of the pulse. It is a single curve containing all of the resonant frequencies of the nuclei in the sample, combined, as Jacobsen (2007) put it, "in one cacophonous reply [to the RF pulse]".

The FID is a function of time (time domain) but is easily converted, using a Fourier transform, into a frequency-domain spectrum (Figure 12.17).

The second phenomenon is called longitudinal or spin-lattice relaxation (time constant = T_1), which is the exponential growth in the +z component of the net magnetization vector. This growth results from the gradual re-establishment of a Boltzmann distribution between the +½ and –½ nuclei. It is referred to as relaxation because the difference between the increasing +z magnetization and the final equilibrium value decays exponentially. T_2 is always smaller than T_1.

Relaxation, which usually takes a few seconds, is significant for several reasons. First, it determines the time delay required between successive pulse excitations of the same sample. By rule of thumb, that time delay is set at $5T_1$. The real advantage of pNMR is the ability rapidly to repeat the pulse-FT process many times on the sample, summing the data. This results in a high final signal-to-noise ratio, greatly increasing sensitivity compared to the old continuous wave method.

Second, relaxation is the basis of the nuclear Overhauser effect (NOE), which can be used to find the distances between nuclei – one of the most important pieces of information that can be found using NMR.

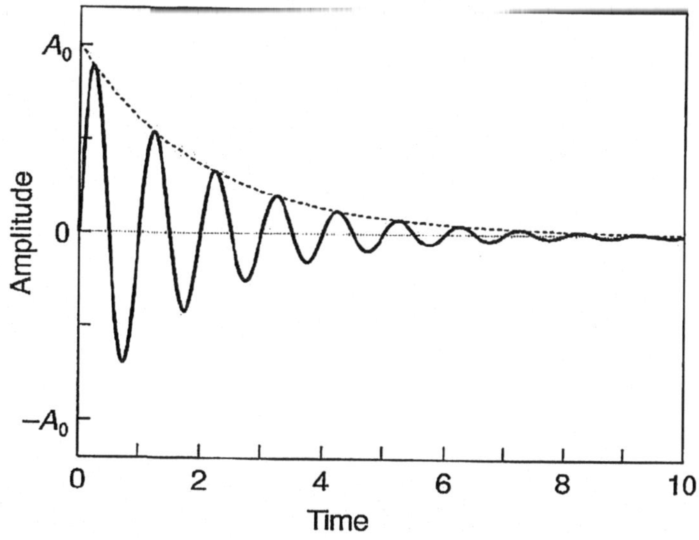

Figure 12.16. Idealized time-domain FID (falling induction decay) curve (dashed line). The line indicates the exponential decay in amplitude with time. (Reproduced with permission from Goldenberg 2016).

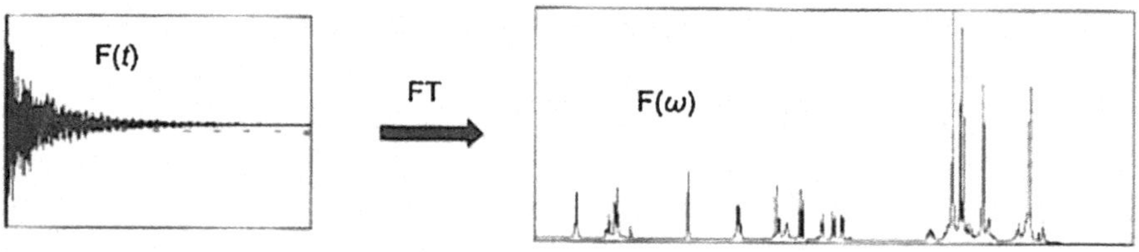

Figure 12.17. Conversion of a (time domain) FID (falling induction decay, F(t)) to a frequency-domain spectrum (F(ω)) by means of a Fourier transform (FT). (Reproduced with permission from Spyros and Dais 2013).

Third, measurement of T_1 and T_2 can provide detailed information about molecular dynamics from the point of view of each nucleus type in the molecule under investigation.

The vector model, although simplistic, is a helpful way of visualizing NMR phenomena. It explains many simple NMR experiments and provides a basis for coming to grips with the range of sophisticated, but highly mathematical, NMR techniques that have been developed. These techniques, which are used in a variety of combinations, include, for example, pulse sequences and delays, selective pulses, long pulses, shaped pulses, composite pulses and pulsed field gradients. 2D, 3D and even multidimensional spectra can be produced.

Generally, the magnetic field strength of an NMR spectrometer has to have a high degree of homogeneity. The purpose of pulsing a gradient into the field during certain procedures is to destroy phase coherences that lead to artifactual signals; these signals obscure the spectrum being measured.

The technology for applying pulsed field gradients comes from the development of magnetic resonance imaging (MRI). (As the impetus for the development of MRI was a medical one, the word *nuclear* has been dropped in order to mitigate against (human) patient antipathy.)

12.5.7 Magnetic Resonance Imaging

A typical MRI spectrometer has three separate sets of gradient coils, orientated so that they can produce gradients in the magnetic field along the x, y and z axes (Figure 12.15). Two or more of the coils can be used together to produce a gradient in any desired direction.

In medical MRI the protons of water in the body (the sample) are the only entity exhibiting chemical shifts. When a gradient is pulsed into the field along a given axis during the acquisition of the FID, the field strength B_0 depends only on position along this axis. In turn, because $\upsilon_0 = \frac{\gamma B_0}{2\pi}$ (Eq. 12.5), the resonant frequency of the protons win the sample also depends on position: the chemical shift scale becomes a distance scale.

During an MRI scan, a pulsed field gradient is used to create slices of the sample – relatively thick non-contiguous slices in so-called 2D imaging (the commonest used), very thin and contiguous slices in 3D imaging. Further pulsed gradients are used to divide the slices into rows and columns of individual voxels (volume elements). The spatial location of each voxel is uniquely described by the resonant frequency at that location (frequency encoding) and phase information (phase encoding), both achieved by applying field gradients. The various gradients are applied in a specific sequence and are synchronized with other events such as the application of pulses and the acquisition of RF signals.

The RF signal data from each voxel are temporarily stored in the so-called k space, from where they are delivered to the corresponding pixels, via Fourier transformation, to give the final image.

In medicine an MRI image is a physical map of the H_2O concentrations in the living tissue being observed. Contrast in the (black and white) image, which allows differentiation between kinds of tissue, is partly due to the fact that the FID signal is directly proportional to proton density. However, proton density variation alone gives relatively poor contrast. Other approaches have been developed. Among these is the analysis of longitudinal and transverse relaxation during the FID. Different tissues exhibit different values of T_1 and T_2. Further, the image brightness is inversely related to T_1 but directly related to T_2. Tissue contrast can be enhanced by suitable processing of the FID (T_2) and transverse relaxation (T_1) data.

Mulas *et al.* (2013) developed a new MRI technique to find a way of discriminating between Sardinian sheep's milk cheese made with heat-treated milk from cheese made with raw milk. Raw milk is a requirement for PDO. The technique involved the use of ^{1}H MRI to generate cross-sectional images of cheese samples. These were used to allow manual selection of areas away from edges (where inhomogeneities might be expected in the external magnetic field) and to allow observation of the internal cheese morphology. Using the MRI spectrometer, measurements of the T_2 relaxation time were made in the selected areas. Pulse sequences were used which supressed signals from fat protons to allow measurement of signals from water protons, and vice versa. For all cheese samples, there was a distribution of T_2 values (signal intensity versus T_2) with peaks at ~9 and ~ 35 ms. This indicated that in the cheese, there were a number of water proton ensembles, including the two principal ones, each with its own T_2. It was found that the area under the distribution curve for heat-treated milk cheese was significantly greater that the corresponding area for raw milk cheese, providing a simple and rapid discriminatory test. (The T_2 distributions for fat protons in the two types of cheese were essentially identical.)

Mulas *et al.* (2016) used a similar approach in developing an MRI method for monitoring the ripening of Grana Padano cheese, for which sensory quality and market value increase with ripening time. Once again, MRI was used to produce cross-sectional images of cheese samples. These were used to select areas in which to measure T_2 of the cheese as a single entity. The intensity versus T_2 distribution showed three peaks, at ~7, ~37 and ~ 90 ms. The area of the first peak, expressed as a fraction of the total area under the distribution curve, proved to be linearly inversely correlated with ripening time.

There have been a significant number of other cheese studies in which MRI has proved to be a valuable technique. Topics include the structure and texture of soft cheeses (Mariette 2003; Mariette and Gollewet 2001; Mariette *et al.* 1999), eye growth in semi-hard cheeses (Grenier *et al.* 2016; Musse *et al.* 2014) and monitoring changes in Feta cheese during brining (Altan *et al.* 2011b). In the three soft cheese studies and in the Feta study, MRI was used in conjunction with relaxometry (see Section 12.5.9).

MRI has been used in several investigations of the spatial and temporal changes that occur during oil migration in chocolate confectionery systems (Cikrikci and Oztop 2017a, b; Rumsey and McCarthy 2012; Maleky *et al.* 2012; Altan *et al.* 2011a). While all of this work was done on dark chocolate, it is expected that the findings would apply at least in part to milk chocolate. Other techniques for studying this problem are differential scanning calorimetry and X-ray diffraction (Maleky 2018).

Wichchukit *et al.* (2005) reported a novel way of measuring the non-Newtonian flow properties of milk chocolate melt. Shear stresses were calculated from pressure drop measurements in tube flow, while shear rates were calculated from MRI velocity profile imaging data measured in the same tube. The melt's flow curve (shear stress versus shear rate) was modelled using the well-known Casson equation. The Casson equation coefficients were incorporated into a successful model for predicting chocolate thickness in the enrobing process. (Exactly the same findings were reported by Karnjanolarn and McCarthy (2006), who used a concentric cylinder rotational viscometer to measure milk chocolate rheological properties; this demonstrates the fact that fundamental rheological behaviour as measured is independent of rheometer geometry and mode of operation, provided stress and strain are evaluated at the same location, as it can be in the two instruments concerned here. (see Section 12.7.4.)

It is noted here that X-ray computed tomography (CT), which utilizes a part of the electromagnetic spectrum far from the radio frequency part, is an alternative to MRI for 3D imaging. In tomography, X-rays are directed at an object from multiple orientations, and the attenuation of the rays along a series of linear paths is measured. The final image is essentially a 3D map of the volumetric distribution of attenuation values, constructed from image "slices". Pinzer *et al.* (2012) and Enyu *et al.* (2018) used X-ray tomography to observe the evolution and stability, respectively, of the microstructure of ice cream. Laverse *et al.* (2011) used the technique to study the microstructure of cream cheese-type products.

12.5.8 Solid Fat Content Determination

Low-resolution (low B_0, often called *low field*) proton pNMR has supplanted dilatometry as the standard technique for measuring solid fat content (*SFC*). Low-resolution spectrometers are not generally used to record spectra but can provide useful information in the form of the time-domain signal – the *FID* – recorded after an RF pulse is delivered to the sample. *SFC* is found by recording the *FID* signal at two times, usually 10 and 70 μs, after the pulse (Colquhoun and Goodfellow 1994). The signal at 10 μs, S_{S+L}, is proportional to the total number of protons in the sample, both solid and liquid fractions. As the decay of the signal from the protons in the solid fraction is much faster than the decay of the signal from the liquid fraction protons, the signal amplitude at 70 μs is proportional to the number of protons in the liquid fraction only (S_L). (At 70 μs, the liquid fraction protons have not yet begun to relax.) Thus, S_S, proportional to the number of protons in the solid fraction, is equal to $(S_{S+L} - S_L)$, and $SFC = S_S/S_{S+L} = (S_{S+L} - S_L)/S_{S+L}$. In practice, (S_{S+L}) has to be corrected by suitable calibration to allow for the fact that the signal from the solid-phase protons decays to a significant extent during the first 10 μs, which is approximately the dead time of the spectrometer.

This procedure for measuring *SFC*, which is considerably more accurate, rapid and convenient than dilatometry, is called the *direct method* (Gribnau 1992). There are two somewhat more complex variations of it called the *indirect method* and the *solid-echo method*, both of which are designed to overcome the dead time problem. They are described by Gribnau (1992) and Peyronel (2018).

The measurement of SFC by proton NMR is widely used in the food industry, and in studies of crystallization, phase behaviour and polymorphism in fat and fat blends, often in alliance with other techniques such as differential scanning calorimetry, NIR and X-ray diffraction. Typical studies are those of Danthine *et al.* (2014) (physicochemical and structural properties of fat blends containing anhydrous milk fat), Janssen and MacGibbon (2007), Ollivon *et al.* (2005), Wiking *et al.* (2009) (crystallization of anhydrous milk fat) and Kaufmann *et al.* (2012) (melting behaviour, texture and microstructure of anhydrous milk fat).

12.5.9 Relaxometry or Time-Domain NMR (TD-NMR)

SFC measurement methods are examples of the use of time-domain NMR data (FID, T_1, T_2) as opposed to the frequency-domain spectral data that could be obtained from it by Fourier transformation. The analysis of time-domain data is called TD-NMR or *relaxometry*.

The three methods for determining SFC described by Gribnau (1992), which use only a very small proportion of the FID data, are suitable only for pure fats and fat blends. For complex mixtures of fats and other materials, for example, foods such as ice cream and cheese, it is necessary to use more complex TD-NMR procedures in which all of the data in the FID are used to yield values of T_1 and T_2.

The relaxation time T_1 is measured using the pulse sequence *180° pulse-delay time-90° pulse-acquisition of FID*, repeated with different delay times, followed by suitable analysis of the data obtained.

The relaxation time T_2 is determined using the spin-echo pulse sequence: *90° pulse-delay time-180° pulse-delay time-acquisition of FID*. This sequence is one of the fundamental building blocks of pulse sequences used in many types of NMR experiment. One version of the spin-echo sequence, called the CPMG (Carr-Purcell-Meiboom-Gill) method, in which the sequence is extended by inserting a number of repeats of the *delay time-180° pulse-delay time-acquisition of FID* part (Jacobsen 2007), has been used successfully to investigate the structures of complex foods. It is designed to overcome loss of signal caused by diffusion of molecules from one part of the sample to another during the experiment. (Recently, faster pulse sequences based on steady-state free precession (SSFP) have been used successfully; these have a much higher signal-to-noise ratio than does the CMPG sequence.)

Using a complex TD-NMR approach involving the CPMG method, and a spectrometer with a unilateral (one-sided) magnet, Veliyulin *et al.* (2008) established a rapid method of measuring the fat content of packaged dairy products (whole cream milk, coffee cream and whipped cream) by establishing a predictive linear relationship between the reciprocal of fat content and T_2. The method has industrial potential.

A unilateral magnet is formed by attaching two separate magnets to an iron yoke. The RF coil for transmitting and receiving signals is placed between the magnets. This configuration allows measurements on large intact samples such as packages of food, which are placed just above the tops of the magnets (Kirtil *et al.* 2017). As the magnetic field homogeneity decreases rapidly with distance from the magnets, measurements represent only a thin layer at the base of the sample (Guthausen and Kamlowski 2009). (Standard NMR spectrometers can take only quite small samples.)

Similarly (but using a conventional spectrometer and ^{1}H TD-NMR) Nascimento *et al.* (2017) used the CPMG method to determine T_2 and then correlated T_2 with the fat content of milk powders as measured by a standard extraction method. An accurate linear relationship was established. A similar correlation was also found between the NIR spectra of the powders and their fat contents. ^{1}H TD-NMR was considered superior to NIR spectroscopy in terms of time efficiency.

Measurement of the fat and water contents of foods has been for many years one of the most widely used applications of TD-NMR (Kirtil *et al.* 2017). Guthausen and Kamlowski (2009) describe two TD-NMR procedures for fat content determination of dairy products and other foods, one for low water content and one for high.

Castell-Palou *et al.* (2013) described a one-step ^{1}H TD-NMR method for simultaneously quantifying the fat and water contents of cheese. The method involved the use of CPMG pulse sequences (among others) in the measurement of T_1 and T_2. Separate models for water and fat were established by correlating NMR data with the results of chemical analyses, using proprietary software. The correlations were accurate over ranges of cheese origins, cheesemakers, milk treatments and ripening stages.

In this work, a cheese sample appears to have been considered as a single component rather than a mixture. In reality, a complex food contains potentially several ensembles of protons, each in its own chemical environment and each with its own relaxation properties (signal intensity, T_1, T_2). The ^{1}H TD-NMR signal from such a food contains relaxation data for all of the proton ensembles in the food. The signal can be quantitatively analysed to yield values for individual ensembles. By carefully examining these values, and taking other relevant data and information (especially product knowledge) into account, it is possible to attribute specific relaxation data to individual components in multi-component multi-phase foods.

This has been well demonstrated by Mariette and Lucas (2005), Lucas *et al.* (2005a, b) in the case of ice cream mix and frozen ice cream. They showed that despite the complexity of the ^{1}H TD-NMR signals from these materials, it was possible to extract a number of relevant parameters useful in studying the effects of product formulation and of process stages on the ice fraction, the crystallized fat fraction and the liquid aqueous fraction.

Recently, Kirtil and Oztop (2016) reviewed dairy applications of TD-NMR (and MRI), while Kirtil *et al.* (2017) reviewed advances in TD-NMR (and MRI) sensors and their applications in food science.

12.5.10 NMR Spectroscopy

TD-NMR gives too low a resolution for analyses requiring the accurate determination of chemical shifts, and concentrations of organic compounds, in the sample. For this, frequency-domain high-resolution (high-B_0, high-field) NMR spectroscopy is required. This has been described earlier. Some examples of its application are presented in the following paragraphs.

12.5.10.1 Milk Fat and Butter

Picariello *et al.* (2013) developed a 1D ^{13}C NMR spectroscopic method for detecting the adulteration of butter with synthetic TAGs. In milk fat TAGs, butyric acid is esterified exclusively at the sn-3 position of the glycerol backbone. Thus, the presence of butyric acid at the sn-2 position, which is easily determined by NMR, is diagnostic of adulteration with synthetic or interesterified TAGs. The method can detect adulteration levels as low as 1%, and quantify them down to 2.5%; 10% is about the lowest level at which fraudulent practice might be economically attractive.

Sacchi *et al.* (2018) showed that it was possible, using the same approach, to detect and quantify milk fat in complex fat blends by using ^{13}C NMR spectroscopy to detect and quantitatively measure the carbonyl (C1) or alpha-carbonyl methylene (C2) resonances of butyrate. NMR measurements of milk fat in experimental blends of anhydrous milk fat, lard and margarine were exactly correlated with data obtained by high-resolution gas chromatography (GC), the standard technique. NMR has advantages over GC in terms of speed, robustness and minimum sample handling. Further, the method can of course indicate when milk fat is *not* present in a fat blend. It can therefore be considered a tool for consumer protection when the consumption of milk-derived material is not allowed for ethical or allergy reasons.

12.5.10.2 Cheese

Schievano *et al.* (2008) found that by measuring the ^{1}H and ^{13}C NMR spectra of an apolar extract of Asiago d'Allevo cheese, and using chemometrics to analyse the data, it was possible to differentiate between artisan cheese made on alpine farms and cheese produced in lowland or mountain factories. No differences between the factory cheeses were detectable. Differentiation was based mainly on a higher content of unsaturated fatty acids in the alpine farm cheeses and a higher concentration of saturated fatty acids in the industrially made cheeses. Conjugated linoleic acid and 1-pentene were further discriminating components. The cheeses from all three sources were classed as PDO.

The work has produced detailed spectral "fingerprints" which could be used to improve the cheese or to authenticate, or otherwise, cheese claiming to be Asiago d'Allevo.

Schievano *et al.* (2009) developed a fast method for measuring the allergen histamine in cheese. An acid extract is made and, without any other treatment being needed, is subjected to ^{1}H NMR spectroscopy. The analysis was proved to be reliable in terms of recovery, linearity, accuracy, repeatability and limits of detection. It was found to be applicable to different types of soft and hard cheeses.

Cagliani *et al.* (2018) reviewed the principles of both time-domain NMR and NMR spectroscopy and their application to foods in general.

12.6 Ultrasound

12.6.1 Principles

An ultrasound wave is a pressure wave, generated by a mechanical vibration, with a frequency between 16 kHz (the highest detectable by the human ear) and 10 MHz. This frequency range can be divided into three parts: power ultrasound (16–100 kHz), high-frequency ultrasound (100 kHz – 1 MHz) and diagnostic ultrasound (1–10 MHz)

Table 12.1. Quantities of importance in the ultrasonic analysis of foods

Quantity	Units	Definition
Amplitude	dBp	Pressure expressed in decibels
Attenuation coefficient	dBp	
Specific acoustic impedance	kg m^{-2} s^{-1}	Resistance to travelling sound wave. Closely analogous to electrical impedance
Intensity	W m^{-2}	Power passing through unit cross-sectional area of sound beam
Power	W	Power passing through total cross-sectional area of sound beam
Reflection and transmission coefficients	Dimensionless	Functions of impedances at boundaries between different materials
Time of flight	s	Elapsed time between excitation of the transmitting transducer and signal reception by the receiving transducer
Sound velocity	m s^{-1}	

Based on Mohammadi *et al.* (2017).

(Patist and Bates 2011). The last, diagnostic or low-power ultrasound, is the focus here.

When a low-power ultrasound pulse is propagated through a sample of a material (e.g. a food), acoustic variables such as ultrasound velocity, attenuation and impedance (Table 12.1) are moderated by properties of, and changes in, the sample. Thus, ultrasound provides a way of probing and measuring sample characteristics.

Indeed, in the case of foods, a given acoustic variable can be a function of a large number of physical, chemical and biochemical properties of the sample. The main advantage of this universal quality is that an ultrasound-based measuring system can be used to characterize a wide variety of foods in a range of contexts. But this universality has a drawback: various changes taking place simultaneously in the sample during an ultrasound experiment can interfere with and influence the acoustic variable being measured. As a consequence, interpretation of acoustic data can be a non-trivial task (Povey 1998; Elvira *et al.* 2017).

However, ultrasound has many advantages, as listed by Kiełczyński (2017): versatility, low cost, ease of use, rapidity, high accuracy, high repeatability, non-destructive and non-invasive. Ultrasound can be computerized and automated. It is suitable for on-line measurements.

Low-power ultrasound measurements are of use in three main areas: measurement of properties and transformations (e.g. crystallization), imaging (mainly in medicine) and process control (Gallego-Juárez 2017).

As intimated above, the basis of using low-power ultrasound to analyse foods (and other materials) is the relationships between foods' acoustic properties and their composition, structure and physical state (McClements 1995). Such relationships can be determined in two ways. The first is to formulate theoretical equations and then establish equation parameters by fitting the equations to experimental data. The second is to generate experimental data and then process it to yield purely empirical correlations and calibration curves. Both approaches yield relationships useful in predicting food properties from ultrasound measurements.

Ultrasound velocity is very sensitive to molecular structure and molecular interactions on the angstrom scale. Its measurement is useful in determining composition, structure and physical state; for studying phase transitions such as crystallization in bulk fats, emulsions and solid lipid nanoparticles; and for the detection of foreign bodies and defects in food (Awad *et al.* 2012; Kiełczyński 2017).

Measurement of ultrasound attenuation and ultrasound impedance can give information about microstructure, phase composition, droplet size distribution in emulsions and rheological characteristics (Awad *et al.* 2012).

Attenuation is the exponential decrease in ultrasonic wave amplitude with distance travelled by the wave. It is characterized by the attenuation coefficient (McClements 1995):

$$A = A_0 e^{-\alpha x} \quad (12.7)$$

where:

A = amplitude at distance x (dB).
A_0 = initial wave amplitude (dB).
α = attenuation coefficient (Np m^{-1}).
x = distance travelled by ultrasound wave.

In ideal materials, attenuation is a consequence solely of the spreading of the wave. In real materials such as foods, attenuation is caused also by conversion of mechanical energy to heat by viscosity effects and by reflection, refraction and scattering by sample structures. The signal data received in ultrasound experiments, especially the amplitude data, have to be carefully analysed to determine the respective contributions from these phenomena to the observed attenuation.

Attenuation is difficult to measure accurately but when analysed carefully can give valuable information about changes on the colloidal particle size scale: nanometres to microns (Kiełczyński 2017). Signal processing can be done in either the time domain or the frequency domain (Elvira *et al.* 2017).

Acoustic impedance, which is closely analogous to electrical impedance, has to do with what happens when an ultrasound wave in one material (1) is incident at the boundary with a different material (2). The wave is partly reflected and partly transmitted. The ratio of the amplitude of the reflected wave (A_r) to that of the incident wave (A_i) is the reflection coefficient (R). For a plane wave incident at a plane boundary (McClements 1995):

$$R = \frac{A_r}{A_i} = \frac{(Z_1 - Z_2)}{(Z_1 + Z_2)} \quad (12.8)$$

where z = specific acoustic impedance (kg m^{-2} s^{-1}).

The greater the difference in impedance between the two materials, the greater is the fraction of incident energy reflected. A transmission coefficient can be defined as (1 – R), since the fraction of incident energy that is transmitted and the fraction reflected must sum to one.

The measurement of impedance, like the measurement of attenuation, is based on the analysis of amplitude data. The analysis is similar to that used for attenuation (Elvira *et al.* 2017). Impedance determination can be useful in the detection of foreign bodies (Awad *et al.* 2012).

12.6.2 Experimental Approaches

Classically, an ultrasound wave is applied to a sample in the form of a single-frequency short pulse (0.2 ms, say). Typical experimental arrangements are shown in Figures 12.18, 12.19 and 12.20. The nature of the commonest wave forms, longitudinal and transverse (shear) waves, is illustrated in Figure 12.21.

In the through-transmission (pitch and catch) configuration, a transmitting transducer (most commonly a piezoelectric one) converts an oscillating electrical signal into an oscillating mechanical one. This is propagated through the sample as an ultrasonic wave. The wave is detected by a receiving transducer which converts the mechanical wave back into an oscillating electrical one. This can then be digitized and data processed to give the acoustic measurement required. The analysis of ultrasound data is generally complicated and requires complex mathematics (Coupland 2001; Povey 1998; Povey and Mason 1998).

In the pulse-echo arrangement, a single transducer both generates the ultrasonic wave and detects the echo. This arrangement is the only one that allows the sample (e.g. a whole cheese) to be tested from one side (using either longitudinal or transverse waves) to detect discontinuities, defects and foreign bodies.

The through-transmission and pulse-echo arrangements can be used to measure the acoustic variables time of flight (TOF), ultrasound velocity, attenuation and impedance.

Ultrasound velocity is usually found by measuring the time of flight (TOF) over a known distance: the distance between the two transducers

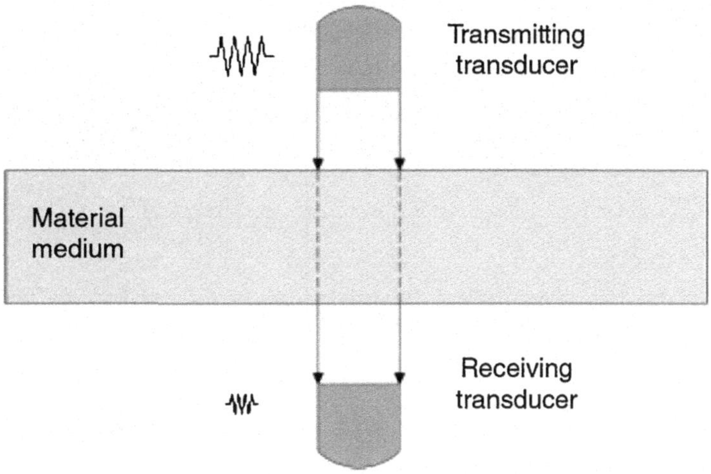

Figure 12.18. Through-transmission application of longitudinal ultrasound waves. (Reproduced with permission from Gallego-Juárez 2017).

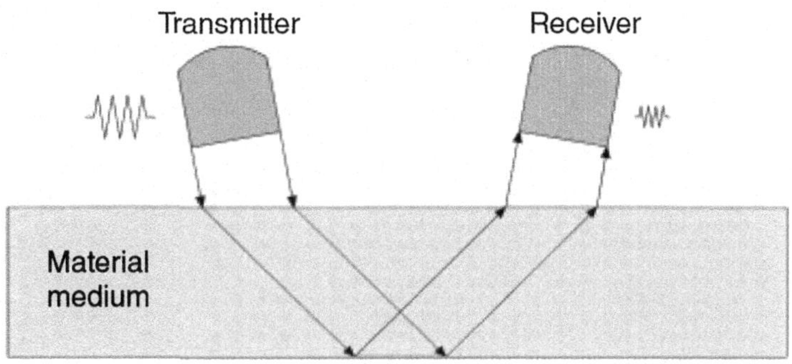

Figure 12.19. Through-transmission application of transverse (shear) ultrasound waves. (Reproduced with permission from Gallego-Juárez 2017).

in the through-transmission arrangement (Figure 12.18) when they are in contact with the sample and twice the sample thickness (T in Figure 12.20) in the pulse-echo arrangement. The distance can be determined accurately by carrying out a TOF measurement using, for example, distilled water at a set temperature, the ultrasound velocity in which is accurately known. The ultrasound velocity in the sample, c_{sample}, is then given by:

$$c_{\text{sample}} = \frac{c_{\text{water}} \times \text{TOF}_{\text{water}}}{\text{TOF}_{\text{sample}}} \tag{12.9}$$

This method is not suitable if the sample is highly dispersive. In that case, sound velocity must be found by analysis of the phase of the wave in the frequency domain.

Sound velocity, attenuation and impedance can also be measured by means of resonant ultrasound spectroscopy (RUS) (Elvira *et al.* 2017). When the TOF of an ultrasonic wave travelling through the sample is small compared with the duration of the excitation signal, there is interference between waves travelling back and forth by reflection. This allows frequency-dependent measurements to be made. Excitation takes the form of a continuous sinusoidal wave with a gradually increasing frequency, a train-burst or a wideband pulse. The first method gives the best resolution. RUS has been used to characterize food texture (Park *et al.* 1994; Liljedahl and Abbot 1994).

In a number of ultrasound studies of edible fats (Häupler *et al.* 2014; Maleky *et al.* 2007; Martini *et al.* 2005a, b, c; Reiner *et al.* 2018; Winkelmeyer *et al.* 2016), a so-called chirp pulse was used in conjunction with a proprietary system that generates the chirp and collects and processes the data produced. A chirp is a burst of ultrasound lasting for perhaps 200 ms and comprising a range of frequencies and amplitudes

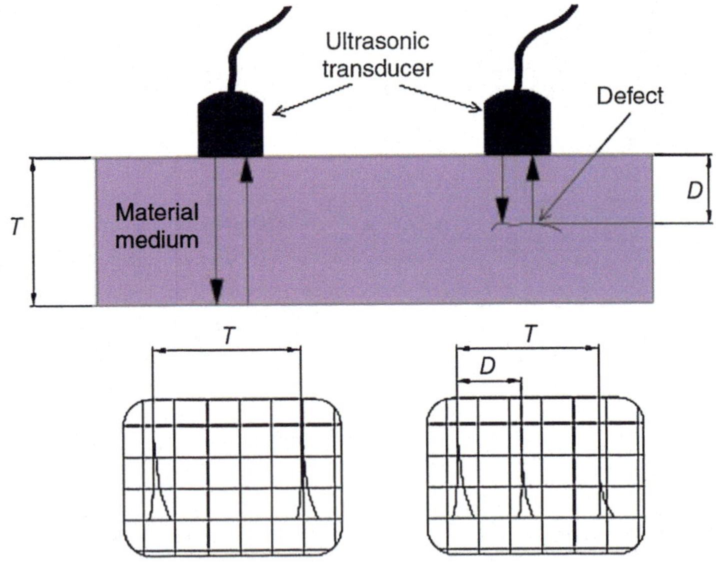

Figure 12.20. Pulse-echo technique for applying ultrasound waves, showing how foreign bodies or structural defects are detected. (Reproduced with permission from Gallego-Juárez 2017).

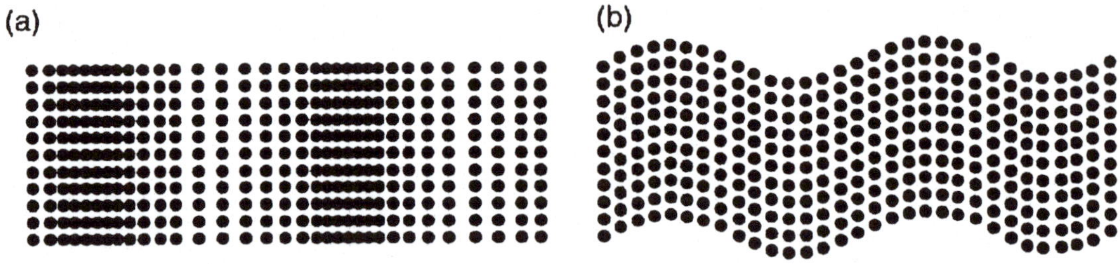

Figure 12.21. Particle displacement by (**a**) a longitudinal wave and (**b**) a transverse (shear) wave. (Reproduced with permission from Gallego-Juárez 2017).

(Figure 12.22). The frequency increases steadily during the duration of the pulse (forming a bandwidth), while the amplitude increases to a maximum and then decreases again. Once the chirp (modified or not by the sample) has reached the receiving transducer, the complicated signal it constitutes is processed to give what is called the Synthetic Impulse™ (SI). The SI contains all of the information needed for analysing the ultrasound characteristics of the sample (Martini *et al.* 2005a).

The SI is produced by convolving the received signal with a replica of the original chirp signal. This allows higher amounts of energy to be put into a pulse, improving the signal-to-noise ratio by overcoming reflection and attenuation limitations set by the sample. It also allows the frequency dependence of the properties of interest to be determined (Winkelmeyer *et al.* 2016).

The SI is characterized by three parameters: the IR (integrated response), the TOF (time of flight) and the FWHM (full-width half-maximum) (Martini *et al.* 2005a; Martini 2007). These are illustrated in Figure 12.23. The IR is a measure of attenuation, the TOF is an indirect measure of ultrasound velocity, and the FWHM is an estimate of the time resolution of the received ultrasound signal. The smaller the FWHM, the better the resolution. Resolution is improved by using wider bandwidths (longer chirps) and higher frequencies.

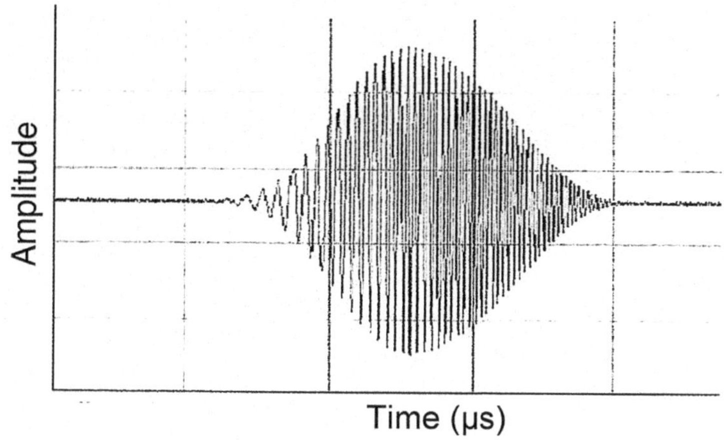

Figure 12.22. Chirp pulse generated by an ultrasound spectrometer using the Synthetic Impulse™ system, showing the typical variation in frequency and amplitude. (Reproduced with permission from Martini 2007).

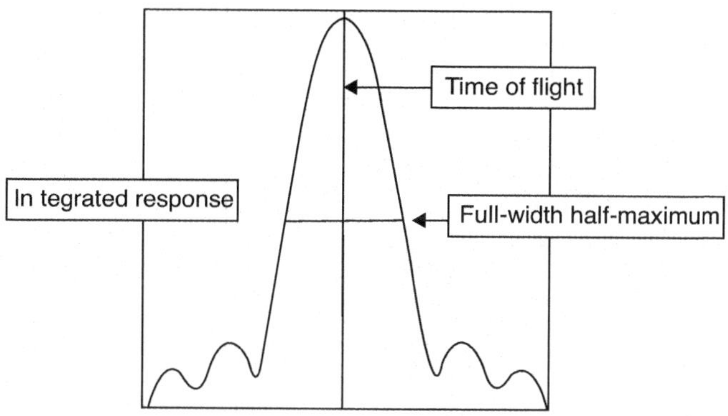

Figure 12.23. Ultrasonic spectrum, generated by an ultrasound spectrometer using the Synthetic Impulse™ system, from which the *integrated response*, *time of flight* and *full-width half-maximum* parameters can be determined. (Reproduced with permission from Martini 2007).

12.6.3 The Use of Ultrasound in the Analysis of High-Fat Dairy Foods

Martini *et al.* (2005a), using the SI system, showed that ultrasonic velocity (derived from the TOF) was highly correlated linearly with solid fat content (SFC, as measured in an unspecified fat using pNMR) up to an SFC of about 20%:

$$c = 2.601 \quad \text{SFC} + 1433.0\left(\text{ms}^{-1}\right) \qquad (12.10)$$

The good correlation between these two factors ($r^2 > 0.9$) was confirmed by Martini *et al.* (2005c). Excessive attenuation precluded measurements at SFC values above 20%.

Martini *et al.* (2005c) found that attenuation of the ultrasound wave depended on crystal size. Larger crystals caused greater attenuation at low and high SFC (in the range 0–20%). Crystal size had no effect at intermediate values (~ 10%).

Singh *et al.* (2002, 2004), using the pulse-echo technique, showed that the ultrasound velocity in anhydrous milk fat (AMF) was related to the SFC content, during both crystallization and melting. The relationship could be modelled using the Miles equation:

$$SFC = 100\left(\frac{\frac{1}{c^2}-\frac{1}{c_L^2}}{\frac{1}{c_S^2}-\frac{1}{c_L^2}}\right)\% \qquad (12.11)$$

where:

c = ultrasound velocity in partly solidified AMF ($m\ s^{-1}$).

c_L = ultrasound velocity in completely molten AMF ($m\ s^{-1}$).

c_S = ultrasound velocity in completely solidified AMF ($m\ s^{-1}$).

However, there were large deviations between prediction and experiment at SFCs greater than 20%, as Martini *et al.* (2005a) had found.

Singh *et al.* (2002, 2004) showed that ultrasound velocity in partly solidified anhydrous milk fat and in cocoa butter was dependent not only on SFC but also on the polymorphic form/microstructure of the solid fat. Therefore, in practical applications, great care would need to be taken to ensure that ultrasound velocity was indicating true values of SFC. Singh *et al.* (2004) suggested that it would be worthwhile to try to extend the Miles equation so that it contained a microstructure parameter. Of course, the sensitivity of ultrasound velocity to solid fat microstructure provides a way of probing that very structure.

Equations 12.10 and 12.11 are good examples of empirical correlations between an acoustic variable (sound velocity) and a property of the material being analysed (SFC). In contrast, Kiełczyński (2017), in characterizing a "model" edible oil (82% diacylglycerols, 18% triacylglycerols), started with a set of theoretical equations for predicting 16 fundamental thermodynamic and physicochemical properties of the oil as functions of ultrasound velocity, density and coefficient of viscosity. He then, by experiment, established empirical equations describing the temperature and pressure dependence of these three variables. (Viscosity was measured using an ultrasound technique.) A value for one of the 16 fundamental properties of the oil could then be predicted for a chosen pressure-temperature combination by substituting into the appropriate theoretical equation calculated values of ultrasound velocity and density, or of ultrasound velocity, density and viscosity, as required by that equation. Results were presented in the form of 3D response surfaces.

Kiełczyński's approach was possible because an edible oil approximates an ideal liquid. For example, it is single-phase and obeys Newton's law of viscosity. The approach could not be applied to partly solidified fats, for instance, or to most other types of food.

The viability of ultrasound for solving problems in cheese manufacture has been investigated by a number of researchers. Derra *et al.* (2018) showed that measurement of the acoustic impedance of cheese milk was an accurate way of determining the coagulation time, defined as the point in time during cheesemaking at which the coagulation phase starts. Impedance increases steadily during the enzymic phase, reaches a maximum at the coagulation time and then decreases steadily as coagulation proceeds. This measurement provides a convenient way of studying the effects of variables such as rennet concentration and temperature. The authors also showed that the second derivative with respect to time of ultrasonic velocity exhibited a clear minimum at the coagulation time, providing an alternative way of determining this time. There was excellent agreement between the two approaches.

Telis-Romero *et al.* (2011) developed an empirical model relating ultrasound velocity (measured using through-transmission) to fat content, water content and temperature for each of two sets of fresh cheeses. The first set comprised 20 commercial fresh cheeses, and the second comprised several mixtures containing different proportions of two of the cheeses. The mixtures were blended to destroy structure. It was found that separate models were needed for the two sets because ultrasound velocity was found to be heavily dependent on cheese structure as well as on composition. The models were formulated in such a way that the influence of the cheese solids not fat was accounted for. The authors concluded that ultrasound velocity measurement had potential as a useful process control tool in cheesemaking.

Nassar *et al.* (2010) examined the possibility of following cheese ripening using measurement of ultrasound velocity and attenuation, variables that would be expected to change as the viscoelastic properties of the cheese change. They found that in unripened cheese (French Comté) through-transmission measurement of ultrasound velocity gave results that correlated very well with rheological measurements of elastic modulus. However, in the case of ripened cheese, the pores and fissures that had developed in the cheese body caused so much reflection of the ultrasound wave that useful data could not be obtained. Because of this, the authors developed a so-called tap-test. In this, a solenoid-driven metal rod was set up to repetitively tap one plane face of the cheese wheel to cause pulses of sound (not ultrasound) to travel across the wheel. A receiving transducer placed at the opposite plane face collected the sound signal. It was found that, regardless of sample, there was always a dominant resonance with a frequency between 20 and 50 Hz. This frequency decreased steadily as the cheese ripened, acting as a measure of ripeness. The tap-test was found to work well in spite of the presence of pores and fissures in ripe cheeses.

Conde *et al.* (2008) developed a tap-test that was successful in detecting fissures (considered defects) in the bodies of Manchego cheese wheels. The transmission of an acoustic wave (not an ultrasound one) was achieved either by letting the head of a gravity-operated hinged mallet fall on to one face of the wheel or by tapping the cheese face manually with a similar mallet. Energy analysis of the sound spectra produced, which were in the frequency range 0–500 Hz, could differentiate between cheese wheels with and without unacceptable fissuring.

Eskelinen *et al.* (2007) used the pulse-echo technique to probe the structural quality of Swiss cheese. Eyes are part of the desired structure of this cheese, whereas fissures and cracks are considered defects. Ultrasonic measurements were made while the transmitter-receiver was moved systematically over the top surface of the cheese sample. These measurements were processed to construct a 3D picture of the interior of the sample. This showed eyes and cracks as dark shadows. The ultrasound images produced by this technique were shown to be accurate by comparing them with 3D images constructed artificially by photographing vertical slices of the same cheese sample and processing the digital data so obtained to give a 3D image that could be directly compared with the ultrasound image. The structural elements (eyes and cracks) in the ultrasonic image showed spatial similarity with the image constructed from the photographs of slices. While attenuation owing to ultrasound reflection by eyes and cracks was still a difficulty, commercially relevant structural elements were identifiable at depths that would allow single-surface probing of entire cheese blocks.

Leemans and Destain (2009) investigated the possibility of using pulse-echo ultrasound (see Figure 12.20) to detect a small foreign body (a short plastic cylinder 3 mm in diameter in this work) in semi-soft cheese. They used both a 0.2 μs pulse and a "homemade" chirp pulse of 200 μs. By analysing TOF patterns of the original pulse and up to four subsequent echoes of it, with or without the foreign body placed in the cheese, it was possible to establish a simple set of rules which an observed echo pattern had to obey for the cheese sample under scrutiny to be accepted as free from contamination.

Aparicio *et al.* (2008) demonstrated that through-transmission measurement of ultrasound velocity in a food being frozen, and suitable correlations, makes possible the determination of the food's initial freezing point, bulk mean temperature during freezing and ice content during freezing. They considered their approach to have considerable potential for on-line monitoring during the manufacture of, for example, ice cream.

Young *et al.* (2008) showed that a Doppler effect-based technique called *ultrasound velocity profiling with pressure difference* (UVP-PD) could be used for accurate in-line measurement of the flow velocity profiles and rheological properties of opaque oil-fat and oil-fat-crystallizer blends, information useful for process control purposes. The authors suggest a way of extending the technique to enable the measurement of SFC in-line and in real time. The ability to do this could potentially lead to a much greater understanding of the physics of fat blends and their ultimate sensory properties.

In summary, it seems that the most promising areas of application of low-power ultrasound are on-line and in-line process control, and quality control, especially in cheesemaking.

Useful reviews on the application of ultrasound to food have been published by Awad *et al.* (2012), Bahram-Parvar (2015), Chen *et al.* (2013), Martini (2007), Mohammadi *et al.* (2017), Otles and Ozyurt (2017) and Villamiel *et al.* (2017).

12.7 Rheological Techniques

12.7.1 Introduction

Rheology is usually defined as *the study of the deformation and flow of matter* (Barnes *et al.* 1989). Rheology is used extensively in investigating and characterizing the nature and microstructure of milk fat and fat-based dairy products. Rheological measurements are useful also in objectively measuring properties related to texture, in measuring functionality, in providing data for process modelling and control and in quality control.

In the following, rheological behaviour is defined, a practical distinction is made between *solids* and *liquids*, and the requirements for the rheological characterization of materials are stated. Then, common *rheometers* (rheological instruments with well-defined geometries) and their use in the measurement of fundamental rheological properties are described. Lastly, applications of rheometers and applications of empirical and imitative rheological instruments and techniques to specific milk fat-based products (milk fat and milk fat fractions, butter, cheese, cream, ice cream, and chocolate) are described. A very wide range of instruments and experimental methods has been used in the rheological study of these products.

Fundamental, *empirical* and *imitative* rheological techniques and measurements, and the distinctions between them, are defined below.

12.7.2 Rheological Behaviour and Material Classification

All real materials fall rheologically between two extremes: the perfectly elastic Hookean solid, for which stress is directly proportional to strain, and the Newtonian liquid, for which (shear) stress is directly proportional to (shear) strain rate. Strain can be defined as deformation relative to a reference length, area or volume (Barnes *et al.* 1989). It is dimensionless. Strain rate (reciprocal time) is the rate of change of strain with time. A stress (force per unit area) must be applied to a material to cause a strain or is set up in a material by the imposition of a strain.

Real materials are neither truly Hookean nor truly Newtonian, though some exhibit Hookean or Newtonian behaviour under certain conditions (Barnes *et al.* 1989). Real materials may exhibit non-linearity, which is a lack of direct proportionality between stress and strain or between stress and strain rate. Real materials may exhibit either predominantly elastic behaviour or predominantly viscous behaviour, or a measurable combination of the two, depending on the stress or strain and the duration of its application (Barnes *et al.* 1989). Such materials are termed viscoelastic. Barnes *et al.* have pointed out that it is better to classify *rheological behaviour* than to classify *materials*; a given material can then be included in more than one rheological class depending on experimental conditions.

However, for practical purposes, it is useful to distinguish between solids and liquids according to the following definitions (Barnes *et al.* 1989):

- A *solid* is a material that will not continuously change in shape when subjected to a given stress.
- A *liquid* is a material that will continuously change in shape (i.e. will flow) when subjected to a given stress, no matter how small that stress is.

A given material may be a solid under some conditions and a liquid under others; for example, a plastic material is a solid at stresses lower than its yield stress but a liquid at higher stresses.

The term *semi-solid* is a convenient though imprecise description of materials that, when unsupported by a container, change shape (deform) under the influence of gravity under ambient conditions. Many fat-based foods are semi-solid.

Viscolelastic materials can be divided into viscoelastic solids and viscoelastic, or simply elastic, liquids. All viscoelastic liquids are non-Newtonian, but not all non-Newtonian liquids are viscoelastic. Non-Newtonian liquids show non-linear rheological behaviour, and this may be time dependent (Barnes *et al.* 1989).

12.7.3 Rheological Characterization of Materials

The fundamental rheological characterization of a material requires the experimental determination of a constitutive equation (a rheological equation of state) that mathematically relates stress and strain or stress and strain rate. The constants in the constitutive equation are the rheological properties of the material.

The necessary experimental data must be obtained using rheometers, instruments in which measurements can be made that yield values of stress, and either strain or strain rate, that are related only by the rheological behaviour of the sample, i.e. are independent of sample shape and size, and instrument geometry. Suitable treatment of these data leads to the constitutive equation that describes the material's rheological behaviour in the ranges of the variables investigated.

Where the complexity of the instrument geometry and/or the sample geometry prohibits analysis of stress-strain conditions during testing, the test and results obtained are *empirical* only; the data cannot be converted into values of fundamental properties. However, such data can still be very useful. Sectilometry, the measurement of the force required to cut with a taut wire, used in characterizing butter, is an example of an empirical approach to measuring rheological behaviour, in this case in terms of "spreadability"; while not a fundamental method, it is simple, fast and inexpensive and gives useful (if rheologically nonfundamental) information.

An *imitative* test is a type of empirical test in which the test geometry and test conditions are designed to mimic those that a product experiences in end use. Such tests can provide very useful means of assessing the rheological functionality of fat-based products. A good example of an imitative test is one for *Mozzarella* cheese stretchability in which the two halves of a divided pizza base covered with the cheese can be moved apart after cooking and the extent of stretching to fracture measured (Guinee *et al.* 1999). Imitative tests that simulate jaw action are useful for measuring product attributes related to texture; texture itself can be defined as the human physiological-psychological perception of a number of rheological and other properties of a food and their interactions, and thus cannot itself be measured by wholly objective means (McCarthy 1987).

Rheometers can be divided into two broad types: *viscometers*, used for measuring the rheological properties of liquids, and what will here be called *solids rheometers*, used for measuring the rheological properties of solids. Viscometers and solids rheometers are not mutually exclusive in application; some viscometer geometries can be used for testing solids, while some solids rheometer geometries can be used for testing (viscous) liquids.

Rheometrical data may be fundamental or empirical in nature depending on the conditions of measurement relative to the sample's rheological behaviour; conditions must be carefully controlled when fundamental data are required.

12.7.4 Viscometers and the Measurement of Fundamental Viscous Properties

A viscometer is an instrument in which shear stress and shear rate (the strain rate in shear) can be measured at the same location in the instrument independently of the properties of the material being tested.

Viscometers can be divided into rotational instruments and axial flow instruments. Rotational instruments include concentric cylinder (cup and bob), cone-and-plate and parallel disc viscometers, while axial flow instruments include capillary, slit and extrusion rheometers.

In rotational instruments, one member (e.g. the cup in a concentric cylinder viscometer) rotates, while the other (e.g. the bob) remains stationary. The sample is held, and sheared, in the gap between the two. In a controlled shear rate measurement, the rotational speed is constant, and the torque on one member caused by the viscous resistance to flow exerted by the sample is measured. In a controlled stress measurement, a constant torque is applied to one member and its speed of rotation measured. Controlled stress instruments are particularly useful for measuring yield stress, the minimum stress causing flow of a plastic material.

In general, shear stress at one location (e.g. the bob surface in a concentric cylinder viscometer) is calculated from the dimensions of the sample gap and the measured or applied torque. Shear rate is calculated at the same location from sample gap dimensions and rotational speed. By making experimental measurements over a range of speeds or torques, the flow curve (shear stress versus shear rate) of the sample can be established. Suitable mathematical treatment of the flow curve data yields the sample's constitutive equation and rheological properties.

In a wide-gap concentric cylinder instrument, the shear rate at a given location depends on the rheological behaviour of the sample. This complicates, but does not prohibit, evaluation of shear rate. In a narrow-gap concentric cylinder instrument (bob radius/cup radius > 0.97), the shear rate may be considered constant at the average value in the gap; shear rate then depends only on radii and rotational speed, making its evaluation easy (Barnes *et al.* 1989).

The software of most commercial concentric cylinder viscometers calculates only an average shear rate, even when the gap is wide according to the above criterion.

Cone-and-plate instruments have the advantage that the shear rate can be considered constant at all points in the gap at a given rotational speed provided the cone angle is less than 4°. They have the disadvantage of being unsuitable for testing materials containing particles, because the gap between cone-and-plate approaches zero at the cone tip.

The parallel disc instrument does not have this limitation, but shear rate is a function of radius, complicating data analysis.

A variation of the concentric cylinder viscometer is the rotating cylinder in an "infinite sample". In this controlled (low) shear rate instrument, the sample is contained in a vessel of such a large diameter relative to the cylinder's diameter that the vessel wall exerts no influence on the shear caused by the cylinder's rotation. The torque required to rotate the cylinder at various controlled speeds is measured (Skelland 1967).

The vane viscometer is yet another form of the concentric cylinder instrument, in which the bob is replaced by a rotor having four blades or vanes each attached by one edge to a vertical shaft, at 90 ° intervals around the shaft. This geometry, which can be used either with a cup or in the "infinite sample" mode, is particularly useful for measuring yield stress, and can also be used for measuring the rheological properties of non-Newtonian liquids. Its advantages are described by Gunasekaran and Ak (2002).

In axial flow viscometers, the sample is made to flow through a duct of regular cross-section. Capillary (circular cross-section) and slit (rectangular cross-section) viscometers are controlled stress instruments: a known pressure difference (which causes shear stress in the sample) is applied over the duct length and the resulting volumetric flow rate measured. In the extrusion viscometer, a controlled shear rate instrument, the sample is extruded through a capillary tube by the action of a constant speed piston, acting on the sample in a cylindrical reservoir to which the capillary is attached. The pressure difference between the ends of the capillary is measured.

In both types of axial flow instrument, the shear stress at the duct wall is calculated from the duct dimensions and the pressure drop along the duct length, while shear rate at the same location is calculated from duct dimensions and volumetric flow rate. The wall shear rate depends on the rheological properties of the sample, complicating but not prohibiting its evaluation.

In viscometry, it is usually necessary to correct for end effects (e.g. entrance and exit effects in axial flow instruments) and for slip between sample and viscometer surfaces.

Most current commercial viscometers worthy of the description *rheometer* are rotational ones, and many are sophisticated and versatile instruments. Axial flow instruments are often user designed and built.

Viscometers of relatively complex geometry, for example, the Ostwald glass U-tube viscometer, can be used for measuring the viscosity of Newtonian liquids, which is shear rate and time independent, after calibration with a Newtonian liquid of known viscosity. Such instruments cannot be used for rheologically characterizing non-Newtonian liquids, and therefore cannot be classed as rheometers, as geometrical complexity prevents evaluation of shear stress and shear rate at a given location independently of sample rheological behaviour.

The principles of rotational and axial flow viscometers, and methods of mathematically manipulating measured experimental data, are decribed by Van Wazer *et al.* (1963), Skelland (1967), Barnes *et al.* (1989) and Steffe (1992).

12.7.5 Solids Rheometers and the Measurement of Fundamental Elastic Properties

Solids rheometers are instruments in which solid samples of regular shape are subjected to well-defined deformations and the forces required to do this measured.

The most common solids rheometers are of the "universal testing machine" (UTM) type (Gunasekaran and Ak 2002). Such an instrument comprises a horizontal stationary base plate and a crosshead, above the base plate, that can move up or down vertically at variable constant speed (Figure 12.24). The crosshead incorporates a load cell for measuring and recording force. The crosshead's speed, position and direction of movement relative to the base plate are accurately controlled and recorded.

The sample, usually in the form of a cylinder, can be subjected to uniaxial compression (the simplest and most common test), uniaxial tension, shear, bending or torsion. In compression, the sample rests on the base plate and is compressed by a horizontal flat plate attached to the crosshead when the crosshead is made to move downwards. For a tension test, the sample, usually in the form of a dumbbell of circular or thin rectangular cross-section, is attached to both the base plate and the crosshead by suitable means, and the crosshead is made to move upwards. For shear and bending tests, suitable assemblies for holding and deforming the sample (which is commonly in the form of a cylinder or a right parallelepiped) are attached to the base plate and crosshead. In the case of torsion testing, the linear motion of the crosshead can be converted to a torque-induced rotation of one end of the sample (which is usually capstan shaped) by means of a line and pulley system; the sample is twisted about its longitudinal axis.

Paradoxically, diametric compression, in which a disc-shaped sample is compressed diametrically between the crosshead plate and the base plate, can be used to measure tensile strength (Gunasekaran and Ak 2002). It is a simpler option than tensile testing because of the difficulty in the

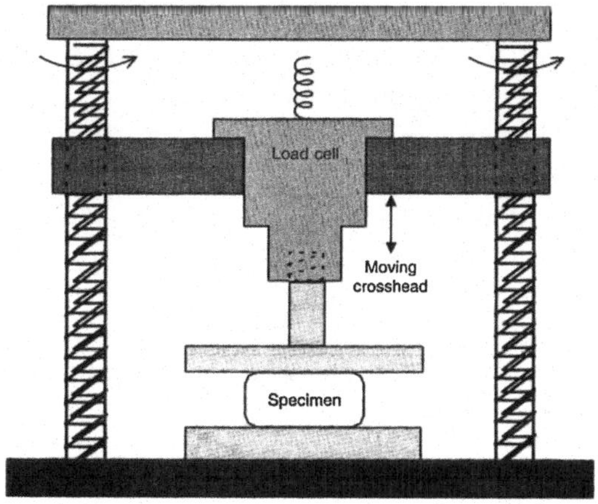

Figure 12.24. Schematic diagram of a universal testing machine (UTM)-type solids rheometer. (Reproduced with permission from Gunasekaran and Ak 2002).

latter of satisfactorily gripping food material samples, which are relatively weak.

A double compression test, in which the same sample is subjected to two consecutive compressions and "unloadings", forms the basis of the Texture Profile Analysis (TPA) technique, which yields mainly empirical rheological data closely related to texture. The test is fully described by Bourne (2002).

Semi-solid foods, such as soft butter and some cheeses, cannot be formed into samples capable of supporting their own weight. For such foods, "compression" testing takes the form of cone or die penetrometry, in which a cone, die, needle or sphere is made to penetrate the sample (held in a suitable container) either under constant load or at constant speed and the penetration depth measured as a function of time. Standard methods for penetrometry of fats are published by the AOCS (AOCS Official Method Cc 16-60; Firestone 1998) and the British Standards Institution (BS 684: Section 1.11: 1976; British Standards Institution 1976c).

Modern solids rheometers are fully computerized. The test parameters (crosshead speed, direction of travel, rate of collection of force-distance data pairs, etc.) can be set up on the instrument's computer. The computer then initiates and controls the mechanical action of the instrument and records the measured data.

The force-distance (force-deformation) data can be converted into stress-strain data using the initial sample dimensions and any relevant instrument factors. The stress-strain data can then be mathematically analysed to yield values of rheological properties such as elastic modulus, tangent modulus, secant modulus, Poisson's ratio, limit of linearity between stress and strain and resilience (the area beneath the stress-strain curve up the limit of proportionality) (Gunasekaran and Ak 2002). These properties can be regarded as quasi-fundamental for real materials provided strains are within the limit of proportionality, and strain rates during testing are relatively high, so that the sample's elastic character predominates (McCarthy 1987). Procedures where these conditions exist may be called *small deformation tests* or *fundamental tests*; they measure well-defined rheological properties, inherent properties of the sample independent of instrument geometry, sample geometry, sample size and stress-strain conditions (McCarthy 1987).

Large deformation tests, in which a solid sample is strained to well beyond its linear limit, and often to fracture, are designed to obtain a quantitative measure of a product's functionality in end use. Many large deformation tests are empirical or imitative, and do not yield fundamental rheological or fracture data. However, such tests can, with some materials, be set up and performed in such a way that fundamental information is obtained (McCarthy 1987).

Fracture properties arguably are not rheological properties, but mechanical properties, given the standard definition of rheology (see above). However, failure of structure in foods, especially high-fat foods, usually begins at strains below the actual fracture point, the point at which the sample begins to break apart into two or more pieces; it is therefore convenient, if not philosophically correct, to include the measurement of fracture properties, e.g. failure stress, in a general discussion of rheological methods.

Details of solids rheometer design, operation and data analysis can be found in Whorlow (1992), Collyer and Clegg (1998) and Gunasekaran and Ak (2002).

12.7.6 Measurement of Linear Viscoelastic Properties

Linear viscoelasticity is the simplest type of viscoelastic behaviour, in which viscoelastic properties are independent of the magnitude of applied stress or strain (Barnes *et al.* 1989; Gunasekaran and Ak 2002). Linear viscoelasticity is usually exhibited by food materials at very small strains (Rao 1992) that cause negligible damage to the food's structure; the phenomenon must therefore be investigated experimentally using small deformation test methods.

Linear viscoelastic properties can be measured in two ways: by static methods or by dynamic methods (Barnes *et al.* 1989).

In static tests the sample is subjected to a step change in stress with the resulting strain being measured as a function of time (creep tests) or a step change in strain with the resulting stress being measured as a function of time (stress relaxation tests). Static tests on solid-like materials, in compression, tension, shear or torsion, can be carried out using solids rheometers of the types described above, while for liquid-like materials, static tests can be carried out in suitably designed concentric cylinder, cone-and-plate and parallel disc viscometers. The parallel disc geometry can be used also for carrying out static tests on solid samples. The sophisticated mechanics, electronics and operating software of many modern viscometers and solids rheometers make creep and stress relaxation experiments easy to carry out.

The most common dynamic method is oscillatory testing, in which the sample is subjected to a sinusoidal oscillatory strain and the resulting oscillatory stress measured. The more sophisticated rotational viscometers have the additional capability of dynamically testing liquid-like materials using small angle oscillatory shear. The use of such an instrument for measuring the linear viscoelastic properties of fats under constant oscillatory or static stress conditions is described by Peyronel (2018). A parallel disc viscometer can be set up for testing solid-like materials, e.g. butter, in oscillatory shear. Some UTM-type solids rheometers, in which the moving crosshead can be made to reciprocate sinusoidally, can be used to test solid-like materials in oscillatory deformation in compression, tension or shear.

A number of highly sophisticated commercial rheometers, rather different in design from the traditional UTM, are now available in which dynamic and static tests on solids in compression, tension, shear and bending can all be carried out in the same instrument.

The four variables in dynamic oscillatory tests are strain amplitude (or stress amplitude in the case of controlled stress dynamic rheometers), frequency, temperature and time (Gunasekaran and Ak 2002). Dynamic oscillatory tests can thus take the form of a strain (or stress) amplitude sweep (frequency and temperature held constant), a frequency sweep (strain or stress amplitude and temperature held constant), a temperature sweep (strain or stress amplitude and frequency held constant) or a time sweep (strain or stress amplitude, temperature and frequency held constant). A strain or stress amplitude sweep is normally carried out first to determine the limit of linear viscoelastic behaviour. In processing data from both static and dynamic tests, it is always necessary to check that measurements were made in the linear region. This is done by calculating viscoelastic properties from the experimental data and determining whether or not they are independent of the magnitude of applied stresses and strains.

The derivation of fundamental linear viscoelastic properties from experimental data obtained in static and dynamic tests and the relationships between these properties are described by Barnes *et al.* (1989), Gunasekaran and Ak (2002) and Rao (1992). In the linear viscoelastic region, the moduli and viscosity coefficients from creep, stress relaxation and dynamic tests are interconvertible mathematically and independent of the imposed stress or strain (Harnett 1989).

12.7.7 Measurement of Non-linear Viscoelastic Properties

Measurement of linear viscolelastic properties is a useful way of gaining information about a food's microstructure and how this influences the food's rheological character (Gunasekaran and Ak 2002; Narine and Marangoni 1999). However, many processing operations, and mastication, involve large, rapid deformations during which viscoelastic behaviour is non-linear.

Although attempts to measure and interpret non-linear behaviour are potentially useful, there are few reports in the literature on the measurement of the non-linear viscoelastic properties of foods. This has been due to a lack of both suitable instrumentation and suitably developed theory; non-linear behaviour, the predominant form of which is the exhibition of normal stresses, and a dependence of viscosity on shear rate, is much more complex than linear behaviour (Gunasekaran and Ak 2002).

Conventional rheometer geometries such as concentric cylinders, cone-and-plate and parallel discs are unsuitable, even when the rheometer is designed to allow measurement of normal forces. Many of the disadvantages of such geometries are overcome in the sliding-plate viscometer (Gunasekaran and Ak 2002). In this instrument (Figure 12.25), the sample (the exact shape and size of which need not be known) is held between a stationary flat plate and a second, parallel, flat plate that is oscillated in its own plane at controlled amplitude and frequency. The sample is subjected to large, uniform deformation in true shear, at high shear rates. Shear stress is measured by means of a force transducer flush mounted in the stationary plate (Gunasekaran and Ak 2002). Gunasekaran and Ak (2002) describe ways of analysing the data obtained with this instrument, specifically in relation to determining the non-linear viscoelastic properties of cheese. Muliawan and Hatzikiriakos (2007) describe the instrument's use in their study of the rheology of *mozzarella* cheese.

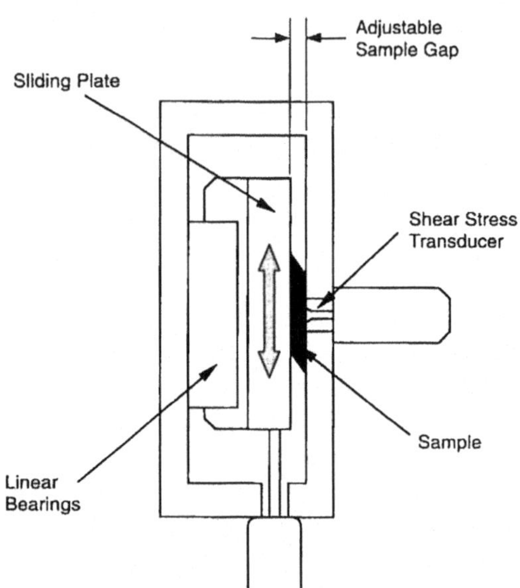

Figure 12.25. Schematic of true shear sliding-plate rheometer. (Reproduced with permission from Gunasekaran and Ak (2002).

The historical lack of reports on the measurement of non-linear viscoelastic properties of foods, mentioned above, is being addressed. For example, Macias-Rodriguez (2018) describes the use of large amplitude oscillatory shear tests for measuring such properties of edible fat systems and suggests that this technique might well become a novel analytical tool in the structure-texture engineering of fat-containing soft materials.

12.7.8 Measurement of Extensional Viscosity

The extensional viscosity of semi-solid fat-based products such as butter, ice cream and some cheeses can be measured by lubricated squeezing flow rheometry (Campanella and Peleg 2002; Gunasekaran and Ak 2002).

In the most suitable test configuration for foods, the sample initially completely fills the gap between two coaxial horizontal discs, one attached to the base and the other to the crosshead, of a UTM type solids rheometer. Lubrication between the sample and disc surfaces is achieved by Teflon coating the surfaces and/or deliberately lubricating the surfaces with a suitable very-low-viscosity liquid.

The sample is subjected to compression by moving the crosshead downwards at constant speed. The sample is extruded from between the two discs, undergoing elongational or biaxial flow; the sample is stretched radially and azimuthally as it flows outwards between the approaching discs. Lubrication ensures that shear flow cannot occur. Elongational viscosity is calculated directly from the measured force-distance data, disc radius and crosshead speed; no rheological model is required (Campanella and Peleg 2002).

The Sentmanat extensional rheometer, and its use in testing *Mozzarella* cheese, is described by Muliawan and Hatzikiriakos (2007). This instrument has two counter-rotating vertical drums, with intermeshing drive gears, between which the sample is clamped and stretched. The torque on one drum, and thus the force needed to stretch the sample, is continuously measured.

Lubricated squeezing flow rheometry (and unlubricated squeezing flow rheometry, in which friction between the sample and discs results in radial shear flow) can be used to measure Newtonian viscosity and the flow properties of non-Newtonian liquids (Campanella and Peleg 2002).

Lubricated squeezing flow rheometry has two significant advantages over more conventional viscometry: slip between sample and instrument surfaces is an advantage rather than a nuisance, and the sample can be mounted in the instrument with minimal damage to its structure. Damage can be avoided altogether by using the so-called imperfect squeezing flow technique. In this, the sample is allowed to form (develop its structure) in a shallow circular container or is collected in the container from, for example, a filling machine. The sample is then compressed in its container by a vertically moving disc, which is concentric with the container but of smaller diameter. Artefacts are minimized by making disc diameter, and the gap between disc and container wall, as large as is practicable and compressing the sample to a very small final height (Campanella and Peleg 2002).

The theory of squeezing flow rheometry assumes that the sample is non-elastic. Tests on viscoelastic samples should therefore be carried out at low strain rates, to minimize elastic response, and results should be reported as *apparent* elongational viscosity.

Good descriptions of the principal ways of carrying out fundamental rheological measurements of all kinds can be found in Collyer and Clegg (1998) and Whorlow (1992).

Gonzalez-Gutierrez and Scanlon (2018) point out that because of the microstructural and rheological complexity of edible fat systems, relatively simple mathematical models of rheological and mechanical (failure) properties tend to be somewhat inadequate. They suggest that finite element analysis, which can cope with non-linear problems intractable to analytical solution, might be a good way of tackling this problem.

12.7.9 Application of Rheological Techniques to Milk Fat and Milk Fat-Based Dairy Products

12.7.9.1 Milk Fat and Butter

Rheological studies of milk fat and butter are mainly concerned with objectively measuring spreadability and texture-related properties. Large deformation tests are particularly useful in this regard, the commonest being disc and cone penetration tests and the sectility test (Wright *et al.* 2001b). In the sectility test, the maximum force required to drive a taut steel wire through the fat or butter sample at constant speed is measured using a UTM-type solids rheometer.

Hardness and apparent yield stress can be calculated from penetration test data, while sectility test data can be converted to a yield stress and a pseudo-Bingham plastic viscosity (Dixon and Williams 1977).

The AOCS has published a standard method (AOCS Cc 16-60) for constant load cone penetrometry of fats and fat emulsions such as butter (Firestone 1998).

Large deformation and failure tests can also be carried out in compression, shear and extrusion (through an orifice) using UTM-type instruments and by means of various imitative techniques (Kawanari *et al.* 1981; Harnett 1989; Rohm and Weidinger 1991; Wright *et al.* 2001b). Uniaxial compression, which is simple and rapid, is a common test. The resulting force-strain curve can be analysed to give a value of the ratio of peak force to work done to peak force, which provides a good way of discriminating between different butter samples. There is some relationship between peak force and sectility force (Frank van de Ven, personal communication, December 2004).

A study (International Dairy Federation 1981) in which disc penetrometry, cone penetrometry, extrusion rheometry and sectilometry were compared as ways of measuring butter firmness (an empirical property) concluded that constant speed cone penetrometry was the best method in terms of ease, speed and cost, though reproducibility was not as good as for sectilometry.

However, the latest standard for measuring butter firmness, ISO 16305/IDF 187:2005 (International Standards Organization 2005), is based on sectility measurement. Strangely, the standard specifies a wire cutting speed of 1.0 mm s^{-1}, almost 17 times slower than the speed found by Rohm (1992) to be necessary for optimal agreement between sectilometric and sensory evaluations of butter firmness.

The texture and spreadability of butter ultimately depend strongly on butter microstructure (Wright *et al.* 2001b). This structure can be probed using static and dynamic measurements of linear viscoelastic properties in compression and shear using suitable geometries such as cone-and-plate (shear), parallel disc (shear), parallel disc (compression) and parallel plate (in the form of the parallel rectangular plate viscoelastometer, designed for creep testing in shear) (Harnett 1989; Rohm 1993a, b; Narine and Marangoni 1999; Wright *et al.* 2001b; Ronholt *et al.* 2012, 2014; Vithanage *et al.* 2009).

Rohm (1993a) obtained agreement between values of the elongational viscosity of butter measured using creep testing and those obtained using lubricated squeezing flow rheometry, as expected from theory.

References to the literature on the measurement of the textural properties of milk fat fractions, mostly by empirical methods, were tabulated by Kaylegian and Lindsay (1994).

The rheological properties of milk fat and butter, and methods of measuring them, were reviewed by Wright *et al.* (2001b).

12.7.9.2 Cheese

Cheeses range from very soft to semisolid to very hard and display a very wide spectrum of rheological behaviour. Consequently, virtually the whole gamut of rheological test methods already described, and many empirical tests, are applied to cheeses.

Cheese structure and fracture behaviour, and texture-related properties, can be investigated using large deformation and failure testing in compression, shear, tension, torsion and bending. Tension testing is relatively rare because of the difficulty in gripping the sample satisfactorily. Torsion testing is also rare because sample preparation is time-consuming, and the test has real advantage only for samples that fail at large strains; cheeses suitable for milling into the required capstan shape are relatively brittle, with small failure strains.

Viscometry (using coaxial cylinder, cone-and-plate and vane rotational viscometers and controlled stress capillary viscometers) can be used for soft cheeses. Lubricated squeezing flow rheometry is particularly useful for measuring cheese meltability.

Static and dynamic linear viscoelastic measurements are used to gain insights into the relationships between cheese structure and rheological behaviour (Everett and Auty 2008). Non-linear viscoelastic measurements have been used to a relatively small degree to measure the response of cheese to large deformations.

Many empirical, mainly imitative, tests have been designed to measure texture-related rheological behaviour, fracture properties, melting properties and stretchability.

Compression testing in a UTM-type solids rheometer, especially in the form of the TPA test, is the commonest mode of rheologically testing cheeses of which self-supporting cylindrical samples can be prepared; sample preparation is straightforward, the test itself is simple, and valuable information about texture-related attributes is obtained (International Dairy Federation 1991; Philip Watkinson, personal communication, December, 2004). TPA testing in penetration mode is used for semisolid cheeses (Philip Watkinson, personal communication, December, 2004).

Measurement of the rheological properties of cheese is covered in detail by the International Dairy Federation (1991), Gunasekaran and Ak (2002) and O'Callaghan and Guinee (2004).

12.7.9.3 Cream

The rheology of cream may be studied using any of the viscometer geometries described above, under controlled shear rate, controlled shear stress or oscillatory shear conditions. Glagovskaia (2000) carried out an extensive study of cream rheology using coaxial cylinder, cone-and-plate and vane viscometry under controlled stress conditions, dynamic oscillatory testing in shear using the cone-and-plate geometry, dynamic testing in squeezing flow between parallel discs and empirical measurements with a Ubbelohde glass capillary viscometer. The dynamic squeezing flow tests were carried out using a CSIRO Micro-Fourier Rheometer. In this instrument, the sample is subjected to controlled small amplitude oscillation, normal to the planes of the discs, with a motion that resembles band-limited random noise. The complete signal spectrum is analysed simultaneously using Fourier transformation to yield values of the viscoelastic properties complex modulus and complex viscosity (CSIRO 2005).

12.7.9.4 Ice Cream

The rheological properties of ice cream are greatly influenced by its microstructure, which can be studied using dynamic oscillatory testing. Temperature sweep oscillatory testing (sometimes called oscillatory thermo-rheometry) is particularly useful, as the sensory properties of ice cream depend to a large extent on the changes in microstructure and rheological properties that occur during melting (Wildmoser *et al.* 2004).

Ice cream hardness can be determined using penetrometry (Marshall *et al.* 2003).

The fundamental rheological properties of unfrozen ice cream mix may be studied in the same ways as those of cream. Simple highly empirical methods of measuring mix viscosity are commonly used in ice cream manufacture for control purposes (Bhandari 2001).

12.7.9.5 Milk Chocolate

Simple empirical viscometers of the orifice and falling ball types, and the controlled shear rate McMichael coaxial cylinder viscometer, have been used traditionally in the chocolate industry. Sophisticated rheometers are now being used increasingly because the economic pressure to reduce the cocoa butter content of chocolate has generated a need for a greater understanding of chocolate rheology (Minifie 1999).

12.8 Density

The density of fat is most commonly measured at a specified temperature at which the fat is completely molten.

Density *per se* is measured by weighing the sample in a volume-calibrated pycnometer or density bottle using, for example, the IUPAC standard method 2.101 (Paquot and Hautfenne 1987).

Density is also measured in terms of the specific gravity and the litre weight in air. The latter is a property of importance in international trade in fats and oils (Firestone 1998).

In AOCS Official Method Cc 10b-25 (Firestone 1998) for determining the specific gravity of solid fats at 25 °C, the specific gravity of an ethanol-water mixture in which a small solid sample of the fat neither sinks nor floats is measured using a Westphal balance (which utilizes the Archimedes buoyancy principle). The specific gravity of the fat sample is equal to this measured value.

The litre weight in air of molten fat is measured using a pycnometer. Standard methods are provided by the AOCS (AOCS Official Method Cc 10c-95; Firestone 1998) and by the Standards Institution (BS 684–1.1: 2000 (British Standards Institution 2000). At temperatures at which the density of water is 1000 kg m^{-3} (0–10 °C), the litre weight in air is identical to specific gravity. The litre weight in air becomes increasingly greater than specific gravity as temperature increases above 10 °C. The difference between the two becomes 4.4% at 100 °C.

12.9 Electromagnetic Properties

12.9.1 Refractive Index

The refractive index of an oil or melted fat is defined for practical purposes as the ratio of the speed of light in air to the speed of light in the sample (Rossell 1986). The difference between the two results in light entering the sample from air, or indeed from any medium of differing refractive index, being refracted.

The commonest instrument for measuring refractive index is the Abbé refractometer, which utilizes the phenomenon of total internal reflection. In this instrument, described in detail by Rheims *et al.* (1997), the sample is sandwiched in a thin layer between an illuminating prism and a measuring prism, both made of glass with a high refractive index (higher than that of the typical sample). The surface of the illuminating prism nearest the light source is ground, causing diffuse light to strike the interface between that prism and the sample at many angles of incidence. At incident angles lower than arcsin(n/N), where n is the refractive index of the sample and N the refractive index of the glass, light incident on the interface between the illuminating prism and the sample is transmitted by the sample, measuring system and focusing eyepiece. At incident angles higher than arcsin (n/N), the critical angle, the incident light is totally internally reflected by the illuminating prism. Thus, at the critical angle, the image seen through the measuring prism exhibits a sharp separation between a bright and a dark area. The observation angle is varied (by rotating the prism pair relative to the eyepiece) until the bright-dark boundary is at the point of intersection of a reticle, usually in the form of an X, seen through the eyepiece of the instrument. Sample refractive index is then read from a fixed scale (observable through a second eyepiece) graduated in refractive index (rather than critical angle).

Measurement temperature must be closely controlled, and the oil sample must be optically clear and free of water.

Standard methods for measuring the refractive index of oils and fats are published by IUPAC (standard method 2.102; (Paquot and Hautfenne 1987), the American Oil Chemists' Society (AOCS Official Method Cc 7-25; Firestone 1998) and the International Standards Organization (International Standard 1739–1975 (E); International Standards Organization 1975). The last specifically applies to the measurement of the refractive index of the fat from butter and was developed jointly with the International Dairy Federation and the American Oil Chemists' Society.

12.9.2 Colour

The Lovibond Tintometer method described in AOCS Official Method Cc 13e-92 (Firestone 1998) is the accepted international standard for measuring the colour of animal and vegetable fats and oils and is widely used in most countries other than the United States (O'Brien 2003). It is based on the British Standard/ISO method BS 684-1.14:1998 ISO 15305:1998 (British Standards Institution 1998a). The method is a subjective one in which the colour of light from a standard source as seen through a specified depth of oil is matched with light from the same source transmitted through a selected set of standard coloured glass slides, the two light beams being viewed side by side in a telescopic eyepiece (Rossell 1986; McGinley 1991). The standard slides are red, yellow, blue and neutral. The slides of any one colour are additive: for example, two 5B slides will give the same blue colour as one 10B slide.

A colour match is obtained using red and yellow slides only if possible. Blue slides are used only if necessary. Results are expressed in Lovibond units (e.g. 5R 50Y), and the path length of light in the oil sample must be reported (McGinley 1991).

The sample must be completely liquid, optically clear, dry and bright. Thus the colour of butter is actually determined on the extracted milk fat (Keen and Udy 1980). The AOCS method 13e-92 requires that if a sample is not liquid at room temperature, it must be heated to a temperature 10 °C above its clear melting point. The operator must not be colour-blind.

Lovibond Tintometer colour can also be measured using objective automated instruments. In one version the intensities of three light beams (red, yellow and white) transmitted by the oil are measured by photoelectric cells and the results displayed as red and yellow colour readings. The white light beam acts as a reference beam and allows compensation for variation in the intensity of the light source (Rossell 1986). AOCS Official Method Cc 13j-97 (Firestone 1998) specifies how an automated tintometer should be used.

However, this standard is valid only for refined oils.

In the United States, the Wesson tintometer method predominates. This subjective method, which is described in AOCS Official Method Cc 13b-45 (Firestone 1998), is experimentally much the same as the Lovibond method, but uses only red and yellow filters, and specifies the yellow number to be used with any selected red number; the method is thus designed to measure the redness of the sample and is more limited than the Lovibond method.

Spectrophotometric objective colour measurement, as described in AOCS Official Method Cc 13-50 (Firestone 1998), gives results that are in general highly correlated with Lovibond colour, although wide discrepancies occur with some oils (O'Brien 2003). Method Cc 13-50 is stated to be applicable to cottonseed, soybean and peanut oils and requires revision to accommodate the use of double beam spectrophotometers. Its applicability to milk fat is not known.

IUPAC also provides a standard spectrophotometric method (standard method 2.103; Paquot and Hautfenne 1987). The results are reported as transmittance values at specified wavelengths or as a plot of transmittance against wavelength.

The spectrophotometric approach has been used to measure the colour of milk fat fractions and butter (Keen and Udy 1980; Norris *et al.* 1971; Keen 1984).

Commercial colorimeters that measure colour by analysing light reflected by the sample surface are readily available, e.g. Konica Minolta (2005).

12.9.3 Dielectric Properties

Dielectric property measurement can be used to measure the moisture and/or salt contents of butter and cheese (Parkash and Armstrong 1969; Parkash 1970; Bosisio and Huy 1976; O'Connor and Synnott 1982; Fagan *et al.* 2004).

Dielectric properties (permittivity, ε'; loss factor, ε''; and loss tangent, $\varepsilon''/\varepsilon'$) can be measured by a variety of techniques that can be grouped into lumped circuit methods, for frequencies lower than 10^8 Hz, and distributed circuit methods for higher frequencies (Hass 1996). The lower-frequency range includes mains electricity and industrial radio frequencies, while the higher includes industrial microwave frequencies.

Lumped circuits are typically of the 4-arm bridge type, commonly the Schering bridge (Figure 12.26), and are evolutions of the 4-arm alternating current bridge counterpart of the well-known direct current Wheatstone bridge (used for measuring an unknown resistance). In the simplest bridge circuit, one arm incorporates the unknown impedance, and the bridge is brought into a null balance (no current flowing) by varying the impedance of one of the other three arms, all of known impedance. For measuring dielectric properties, a test parallel plate capacitor forms part of one arm of the bridge, and its impedance is measured when the sample and when air completely fills the space between the plates. Sample permittivity and loss factor can be calculated from the impedance data obtained (Scaife *et al.* 1971).

Resonating lumped circuits are alternatives to the use of bridge circuits. These involve the injection (by various means) of a voltage or current into an inductor-tuning-capacitor resonant circuit

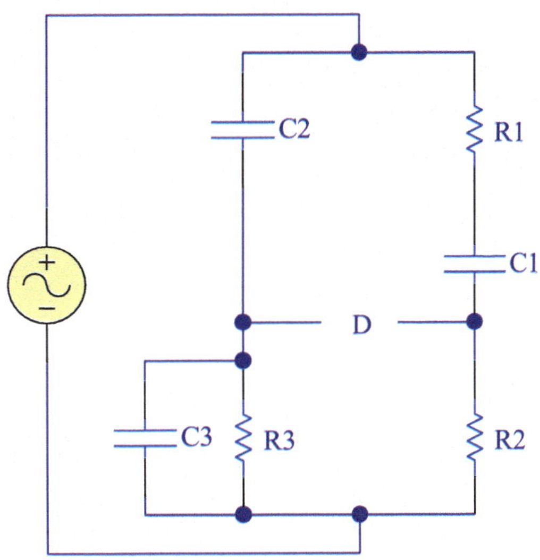

Figure 12.26. Schering bridge circuit. The capacitance being measured (the test capacitance) is represented by C1 and R1 in series. R3 is a fixed resistance. Balance is obtained by adjustment of C3 and either C2 or R2. D is the detector.

and the measurement of the resulting voltage across the tuning capacitor at resonance. The impedance of a test capacitor (filled with the sample and then with air), connected across the resonant circuit's tuning capacitor, is measured by adjusting the latter to restore the resonance that existed in the absence of the test capacitor. The unknown impedance is calculated from the change in the capacitance of the circuit's tuning capacitor and the change in voltage across it (Scaife *et al.* 1971). O'Connor and Synnott (1982) used a resonance method to measure the seasonal variation in the dielectric properties of butter.

Further lumped circuit options include circuits that allow accurate measurement of the current in, and voltage across, the unknown impedance (thus allowing calculation of that impedance) and the auto balancing bridge (Agilent Technologies 2003).

Distributed circuit methods use coaxial lines, waveguides and resonant cavities at microwave frequencies. The circuits are designed for measuring an attenuation factor and a phase factor, from which sample dielectric properties can be calculated. The sample may form the dielectric medium between the two conductors of a coaxial line (Scaife *et al.* 1971) or an open coaxial line is brought into contact with the sample surface (Roussy and Pearce 1995). Fagan *et al.* (2004) used an open coaxial line method to demonstrate that the moisture and salt contents of process cheese could be predicted by measuring dielectric properties over a range of frequencies.

In the waveguide method, the sample is contained by the waveguide in the form of a block closely fitting the waveguide's bore.

Resonant cavities are designed to increase the apparent interaction between microwaves and the sample in order to induce measurable attenuation and are thus particularly useful for measuring the dielectric properties of low loss materials such as fats (Roussy and Pearce 1995).

Measurements in distributed circuit methods are commonly made with network analysers (Roussy and Pearce 1995).

Commercially available impedance measuring instruments are based on circuits described above (Agilent Technologies 2003).

12.9.4 Electrical Conductivity

Electrical conductivity is easily measured by measuring the impedance of the sample in the low-frequency range (see above), but at a frequency above the range in which errors arise owing to electrode polarization effects. For example, measurement of the electrical conductivity of cream, which can be carried out by measuring the impedance between a pair of stainless steel or platinum electrodes immersed in the cream sample, should be done at frequencies of 10^5 Hz or higher (Lawton and Pethog 1993).

12.10 Functional Properties

The general principles of the measurement of rheological properties of milk fat and fat-based products are described above. The measurement of other product-specific functional properties, either *ingredient properties* or *end-use properties*, is described in the following.

12.10.1 Milk Fat and Butter

The functional performance of milk fat, milk fat fractions and butter when used as ingredients in complex food products can be evaluated only by means of empirical trials in which products are made and assessed on the laboratory scale using carefully standardized ingredients, equipment, instruments, procedures and reporting of results (O'Brien 2003). It is commercially important that suppliers of milk fat-based food ingredients be able to demonstrate the efficacy of their products. They can do so by supplying their food industry customers with detailed quantitative advice on how these ingredients could be used to best advantage in customers' products. Functional tests provide the means of accomplishing this (O'Brien 2003).

O'Brien (2003) presents detailed methodologies for assessing fats with respect to their functional properties in cakes, pastry, icing and flour confectionery and biscuit crèmes.

12.10.2 Ice Cream

Air is an important volumetric ingredient of ice cream. The air content is expressed as *overrun*, which is defined as the percentage increase in the volume of the ice cream mix achieved by whipping air into the mix prior to freezing. Ice cream is sold by volume, and its overrun is thus an important property from the regulatory as well as the product quality point of view. Overrun is determined by means of appropriate volume measurements (Marshall *et al.* 2003).

Ice cream *meltdown* at a specified temperature is determined by measuring, as a function of time, the mass dripping from a sample of standard dimensions resting on a screen (Marshall *et al.* 2003).

12.10.3 Chocolate

Particle size is an important property with respect to the sensory quality of chocolate and in chocolate manufacture. It can be measured using laser light diffraction spectrophotometry (see above) and by a variety of other means such as micrometry, microscopy, wet sieve fractionation, sedimentation and Coulter counting (Minifie 1999).

12.10.4 Whole Milk Powders

The bulk density of powders can be measured very simply by weighing a sample in a volume-graduated container, perhaps after using a standardized tapping procedure to uniformly consolidate the samples.

Powder particle density is measured using an air comparison pycnometer. This comprises two piston-cylinder assemblies of equal size, the pistons of which are connected to a differential manometer. A known mass of particles is placed in one cylinder. Both pistons are then moved inwards simultaneously (until the second cylinder hits a stop) in such a way that although the pressure in each cylinder increases, the differential pressure remains equal to zero. The particle volume, and hence particle density, is found from the difference in extents of travel of the pistons and piston diameter.

Powder flowability is measured empirically as the time for a standardized mass of powder to flow out of a horizontal rotating drum through slits in the drum wall. Drum dimensions and rotational speed are standardized.

Powder particle size distribution is normally measured using laser light diffraction spectrophotometry (see above). The mechanical stability of agglomerated particles can be determined by using this technique to measure the fines created by subjecting the sample to a defined mechanical treatment.

Powder characteristics of a physicochemical nature, such as *solubility*, *wettability* and *dispersability*, and measures of the "instant" properties of powders (*sludge*, *slowly dispersible particles*, *hot water sediment* and *coffee test*), are determined by a variety of empirical physical tests, some of which incorporate subjective elements. These and the other tests for whole milk powders identified above are described in detail by Westergaard (1994) and Pisecky (1997)).

Acknowledgements We thank the staff of Massey University Library for their assistance in tracing and acquiring elusive sources of information.

References

Abbas, O. & Baeten, V. (2018). Near-infrared spectroscopy. In A. S. Franca & L. M. L. Nollet (Eds.), *Spectroscopic methods in food analysis* pp. 70–101. Boca Raton: CRC Press.

Agilent Technologies. (2003, December 3). Palo Alto, California, Agilent Technologies Impedance Measurement Handbook. www.optics.arizona.edu/Palmer/OPTI380B/PDFdocs/5950-3000.pdf

Aliakbarian, B., Bagnasco, L., Perego, P., Leardi, R. & Casale, M. (2016). UV-VIS spectroscopy for monitoring yoghurt stability during storage time. *Analytical Methods, 8*(30), 5962–5969.

Allais, C., Keller, G., Lesieur, P., Ollivon, M. & Artzner, F. (2003). X-ray diffraction/calorimetry coupling. A new tool for polymorphism control. *Journal of Thermal Analysis and Calorimetry, 74*, 723–728.

Altan, A., Lavenson, D. M., McCarthy, M. J. & McCarthy, K. L. (2011a). Oil migration in chocolate and almond product confectionery systems. *Journal of Food Science, 76*(6), E489–E494.

Altan, A., Oztop, M. H., McCarthy, K. L. & McCarthy, M. J. (2011b). Monitoring changes in feta cheese during brining by magnetic resonance imaging and NMR relaxometry. *Journal of Food Engineering, 107*(2), 200–207.

Amara-Dali, W. B., Karray, N., Lesieur, P. & Ollivon, M. (2005). Anhydrous goat's milk fat: Thermal and structural behavior. 1. Crystalline forms obtained by slow cooling. *Journal of Agricultural and Food Chemistry, 53*(26), 10018–10025.

Amara-Dali, W. B., Lesieur, P., Artzner, F., Karray, N., Attia, H. & Ollivon, M. (2007). Anhydrous goat's milk fat: Thermal and structural behaviors studied by coupled differential scanning calorimetry and X-ray diffraction. 2. Influence of cooling rate. *Journal of Agricultural and Food Chemistry, 55*(12), 4741–4751.

Amara-Dali, W. B., Lopez, C., Lesieur, P. & Ollivon, M. (2008). Crystallization properties and polymorphism of triacylglycerols in goat's milk fat globules. *Journal of Agricultural and Food Chemistry, 56*(12), 4511–4522.

Aparicio, C., Otero, L., Guignon, B., Molina-García, A. D. & Sanz, P. D. (2008). Ice content and temperature determination from ultrasonic measurements in partially frozen foods. *Journal of Food Engineering, 88*, 272–279.

Awad, T. S., Moharram, H. A., Shaltout, O. E., Asker, D. & Youssef, M. M. (2012). Applications of ultrasound in analysis, processing and quality control of food. *Food Research International, 48*, 410–427.

Aykas, D. P. & Rodriguez-Saona, L. E. (2018). Instrumentation. In A. S. Franca & L. M. L. Nollet (Eds.), *Spectroscopic methods in food analysis* (pp. 237–271). Boca Raton: CRC Press.

Bahram-Parvar, M. (2015). A review of modern instrumental techniques for measurements of ice cream characteristics. *Food Chemistry, 188*, 625–631.

Barnes, H. A., Hutton, J. F. & Walters, K. (1989). *An introduction to rheology*. Amsterdam: Elsevier.

Barnicoat, C. R. (1944). The softening-point of fats. *Analyst (London), 69*, 176–178.

Bashir, Z., Khan, N. & Price, D. M. (1998). Transition-temperature differences on cooling for a monotropic liquid-crystal polyester observed in DSCs of different design. *Thermochimica Acta, 319*, 47–53.

Beattie, J. R., Bell, S. E. J., Borgaard, C., Fearon, A. M. & Moss, B. W. (2005). Multivariate prediction of clarified butter composition using Raman spectroscopy. *Lipids, 39*(9), 897–906.

Belton, P. S. (1994). Introduction – An overview of spectroscopic methods. In R. H. Wilson (Ed.), *Spectroscopic techniques for food analysis* (pp. 1–11). New York: VCH Publishers.

Bernuy, B., Meurens, M., Mignolet, E. & Larondelle, Y. (2008). Performance comparison of UV and FT-Raman spectroscopy in the determination of conjugated linoleic acids in cow milk fat. *Journal of Agricultural and Food Chemistry, 56*(4), 1159–1163.

Bhadeshia, H. K. D. H. (2002). University of Cambridge (January 12, 2005). www.msm.cam.ac.uk/phase-trans/2002/Thermal2.pdf

Bhandari, V. (2001). *Ice cream manufacture and technology*. New Delhi: Tata McGraw-Hill.

Birker, P. J. M. W. L. & Padley, F. B. (1987). Physical properties of fats and oils. In R. J. Hamilton & A. Bhati (Eds.), *Recent advances in chemistry and technology of fats and oils*. London: Elsevier Applied Science.

Blaurock, A. E. (2000). Fundamental understanding of the crystallization of oils and fats. In N. Widlak (Ed.), *Physical properties of fats, oils and emulsifiers* (pp. 1–32). Champaign: AOCS Press.

Blaurock, A. E. & Carothers, A. P. R. (1990). Kinetics of crystallization of simple triglycerides. *INFORM, 1*(4), 351.

Blaurock, A. E. & Wan, P. (1990). Crystallization kinetics of complex oils. *INFORM, 1*(4), 351.

Bosisio, R. G. & Huy, H. H. (1976). On a method of correcting for basis weight variations in dielectric moisture tests on paper, butter fats and grease samples. *Journal of Microwave Power, 12*(4), 309–317.

Botosoa, E. P. & Karoui, R. (2013). Characterization of emmental cheeses with different brand products by combining infrared and fluorescence spectroscopies. *Food and Bioprocess Technology, 6*(9), 2365–2375.

Bourne, M. C. (2002). *Food texture and viscosity* (2nd ed.). San Diego: Academic.

Bovey, F. A., Jerlinski, L. & Mirau, P. A. (1987). *Nuclear magnetic resonance spectroscopy* (2nd ed.). San Diego: Academic.

Breitschuh, B. & Windhab, E. J. (1996). Direct measurement of thermal fat crystal properties for milk-fat fractionation. *Journal of the American Oil Chemists' Society, 73*(11), 1603–1610.

British Standards Institution. (1976a). *British standard methods of analysis of fats and fatty oils, part 1. Physical methods, section 1.4 determination of flow and drop points*. London: British Standards Institution.

British Standards Institution. (1976b). *British standard methods of analysis of fats and fatty oils part 1. Physical methods, section 1.8 determination of smoke point*. London: British Standards Institution.

British Standards Institution. (1976c). *British standard methods of analysis of fats and fatty oils part 1. Physical methods, section 1.11 determination of penetration value*. London: British Standards Institution.

British Standards Institution. (1976d). *British standard methods of analysis of fats and fatty oils, part 1. Physical methods, section 1.13 determination of cooling curve*. London: British Standards Institution.

British Standards Institution. (1987). *British standard methods of analysis of fats and fatty oils, part 1. Physical methods, section 1.5 determination of cloud point*. London: British Standards Institution.

British Standards Institution. (1990). *Methods of analysis of fats and fatty oils, part 1. Physical methods, section 1.12 determination of dilatation*. London: British Standards Institution.

British Standards Institution. (1998a). *Methods of analysis of fats and fatty oils – Part 1: Physical methods – Section 1.14: Determination of Lovibond colour*. London: British Standards Institution.

British Standards Institution. (1998b). *Methods of analysis of fats and fatty oils – Part 1: Physical methods – Section 1.17: Flashpoint limit test using Pensky-Martens closed-cup flash tester*. London: British Standards Institution.

British Standards Institution. (2000). *Methods of analysis of fats and fatty oils – Part 1: Physical methods – Section 1.1: Determination of litre weight in air*. London: British Standards Institution.

Bueche, F. J. (1986). *Introduction to physics for science and engineers*. International Student Edition. New York: McGraw-Hill.

Cagliani, L. R., Scano, P. & Consonni, R. (2018). NMR spectroscopy. In A. S. Franca & L. M. L. Nollet (Eds.), *Spectroscopic methods in food analysis* (pp. 143–188). Boca Raton: CRC Press.

Calligaris, S., Barba, L., Gianmichele, A. & Nicoli, M. C. (2014). Application of DSC-XRD coupled techniques for the evaluation of phase transition in oils and fats and related polymorphic forms. In E. Chiavaro (Ed.), *Differential scanning calorimetry. Applications in fat and oil technology* (pp. 141–161). Boca Raton: CRC Press.

Campanella, O. H. & Peleg, M. (2002). Squeezing flow viscometry for nonelastic semisolid foods – Theory and application. *Critical Reviews in Food Science and Nutrition, 42*(3), 241–264.

Castell-Palou, A., Rosello, C., Femenia, A. & Simal, S. (2013). Simultaneous quantification of fat and water content in cheese by Td-NMR. *Food and Bioprocess Technology, 6*(10), 2685–2694.

Cattaneo, T. M. P. & Holroyd, S. E. (2013a). New applications of near infrared spectroscopy on dairy products. *Journal of Near Infrared Spectroscopy, 21*, 307–310.

Cattaneo, T. M. P. & Holroyd, S. E. (2013b). The use of near infrared spectroscopy for determination of adulteration and contamination in milk and milk powder: Updating knowledge. *Journal of Near Infrared Spectroscopy, 21*, 341–349.

Chen, Q., Zhang, C. Z. & Ouyang, Q. (2013). Recent advances in emerging imaging techniques for non-destructive detection of food quality and safety. *Trends in Analytical Chemistry, 52*, 26–274.

Chiavaro, E. (Ed.). (2014). *Differential scanning calorimetry. Applications in fat and oil technology*. Boca Raton: CRC Press.

Chong, C. L. (2001). Crystallization of palm oil products. In N. Widlak, R. Hartel & S. Narine (Eds.), *Crystallization and solidification properties of lipids* (pp. 110–119). Champaign: AOCS Press.

Cikrikci, S. & Oztop, M. H. (2017a). Mathematical modelling and validation of oil migration on a model chocolate system using magnetic resonance imaging (MRI). *Acta Horticulturae, 1152*, 273–280.

Cikrikci, S. & Oztop, M. H. (2017b). Mathematical modelling and use of magnetic resonance imaging (MRI) for oil migration in chocolate and confectionary systems. *Food Engineering Reviews, 9*(1), 50–70.

Clegg, W. (2001). *Crystal structure analysis. Principles and practice*. Oxford: Oxford University Press.

Collyer, A. A. & Clegg, D. W. (1998). *Rheological measurement*. London: Chapman & Hall.

Colquhoun, I. J. & Goodfellow, B. J. (1994). Nuclear magnetic resonance spectroscopy. In R. H. Wilson (Ed.), *Spectroscopic techniques for food analysis* (pp. 87–145). New York: VCH.

Conde, T., Mulet, A., Clemente, G. & Benedito, J. (2008). Detection of internal cracks in Manchego cheese using the acoustic impulse-response technique and ultrasounds. *Journal of Dairy Science, 91*(3), 918–926.

Coupland, J. N. (2001). Ultrasonic characterization of lipid crystallization. In N. Widlak, R. Hartel & S. Narine (Eds.), *Crystallization and solidification properties of lipids* (pp. 132–145). Champaign: AOCS Press.

Cozzolino, D. (2018). Fourier transform spectroscopy. In A. S. Franca & L. M. L. Nollet (Eds.), *Spectroscopic methods in food analysis* (pp. 103–109). Boca Raton: CRC Press.

CSIRO. (2005, January 21). Lindfield, New South Wales, Australia. http://www.tip.csiro.au/IMP/SmartMeasure/MFR1.htm

Cullity, B. D. (1956). *Elements of X-ray diffraction*. Reading: Addison-Wesley Publishing Company.

Daniels, T. (1973). *Thermal analysis*. New York: Wiley.

Dankowska, A., Malecka, M. & Kowalewski, W. (2014). Application of synchronous fluorescence spectroscopy with multivariate data analysis for determination of butter adulteration. *International Journal of Food Science and Technology, 49*(12), 2628–2634.

Dankowska, A., Malecka, M. & Kowalewski, W. (2015). Detection of plant oil addition to cheese by synchronous fluorescence spectroscopy. *Dairy Science and Technology, 95*(4), 413–424.

Danthine, S., Levebure, E., Trinh, H. N. & Blecker, C. (2014). Effect of palm oil enzymatic interesterification on physicochemical and structural properties of mixed fat blends. *Journal of the American Oil Chemists' Society, 91*(9), 1477–1487.

Deman, J. M., Deman, L. & Blackman, B. (1983). Melting-point determination of fat products. *Journal of the American Oil Chemists' Society, 60*(1), 91–94.

Derra, M., Bakkali, F., Amghar, A. & Sahsah, H. (2018). Estimation of coagulation time in cheese manufacture using an electronic pulse-echo technique. *Journal of Food Engineering, 216*, 65–71.

Dhakal, S., Qin, J., Kim, M. S. & Chao, K. (2018). Raman spectroscopy. In A. S. Franca & L. M. L. Nollet (Eds.), *Spectroscopic methods in food analysis* (pp. 111–142). Boca Raton: CRC Press.

Dibildox-Alvarado, E. & Toro-Vasquez, J. F. (1997). Isothermal crystallization of tripalmitin in sesame oil. *Journal of the American Oil Chemists' Society, 74*(2), 69–76.

Dixon, B. D. & Williams, T. (1977). Measurement of butter firmness by sectility testing. *Australian Journal of Dairy Technology, 32*(4), 177–179.

Drenth, J. (1999). *Principles of protein X-ray crystallography* (2nd ed.). New York: Springer.

Elvira, L., Resa, P., Castro, P., Kant Shukla, S., Sierra, C., Aparicio, C., Durán, C. & Montero de Espinosa, F. (2017). Bioprocess monitoring using low-intensity ultrasound: Measuring transformations in liquid compositions. In M. Villamiel, J. V. García-Pérez, A. Montilla, J. A. Cárcel & J. Benedito (Eds.), *Ultrasound in food processing* (pp. 146–174). Chichester: Wiley Blackwell.

Enyu, G., Kazantsev, D., van Dalen, G., Schuetz, P., Rockett, P., StJohn, D. & Lee, D. (2018). Revealing the microstructural stability of a three-phase soft solid (ice cream) by 4D synchrotron X-ray tomography. *Journal of Food Engineering, 237*, 204–214.

Eskelinen, J. J., Alavuotunki, A. P., Hæggström, E. & Alatossava, T. (2007). Preliminary study of ultrasonic structural quality control of Swiss-type cheese. *Journal of Dairy Science, 90*, 4071–4077.

Everett, D. W. & Auty, M. A. E. (2008). Cheese structure and current methods of analysis. *International Dairy Journal, 18*(7), 759–773.

Fagan, C. C., Everard, C. & O'Donnell, C. P. (2004). Prediction of moisture and salt content of cheese by dielectric sensing. In *Proceedings of ICEF9 international congress on engineering and food*, Montpellier, France, 7–11 March, 2004.

Firestone, D. (Ed.). (1998). *Official methods and recommended practices of the AOCS* (5th ed.). Champaign: American Oil Chemists' Society.

Franca, A. S. & Nollet, L. M. L. (Eds.). (2018). *Spectroscopic methods in food analysis*. Boca Raton: CRC Press.

Gabbott, P. & Mann, T. (2016). Differential scanning calorimetry. In S. Gaisford, V. Kett & P. Haines (Eds.), *Principles of thermal analysis and calorimetry* (2nd ed., pp. 67–103). Cambridge: The Royal Society of Chemistry.

Gaisford, S., Kett, V. & Haines, P. (Eds.). (2016). *Principles of thermal analysis and calorimetry* (2nd ed.). Cambridge: The Royal Society of Chemistry.

Gallego-Juárez, J. A. (2017). Basic principles of ultrasound. In M. Villamiel, J. V. García-Pérez, A. Montilla, J. A. Cárcel & J. Benedito (Eds.), *Ultrasound in food processing* (pp. 3–26). Chichester: Wiley Blackwell.

Glagovskaia, O. (2000). *Rheological study of milks and creams. effects of fat content, fat globule size distribution and temperature*. Master of Applied Science Thesis, RMIT University, Melbourne, Australia.

Goldenberg, D. P. (2016). *Principles of NMR spectroscopy*. An Illustrated Guide. University Science Books, Mill Valley.

Gonzalez-Gutierrez, J. & Scanlon, M. G. (2018). Rheology and mechanical properties of fats. In A. G. Marangoni (Ed.), *Structure-function analysis of edible fats* (2nd ed., pp. 119–167). London: Academic/AOCS.

Granger, C., da Costa, J. P., Toutain, J., Barey, P. & Cansell, M. (2006). Mapping of ice cream formulation using front-face fluorescence spectroscopy. *International Dairy Journal, 16*(5), 489–496.

Grenier, D., Laridon, Y., le Ray, D., Challois, S. & Lucas, T. (2016). *Journal of Food Engineering, 171*, 119–128.

Gribnau, M. C. M. (1992). Determination of solid/liquid ratios of fats and oils by low-resolution pulsed NMR. *Trends in Food Science and Technology, 3*(8/9), 186–190.

Guinee, T. P., O'Callaghan, D. J. & O'Donnell, H. J. (1999). Stretching the limits of cheese testing. *European Dairy Magazine, 11*(4), 28–30.

Gunasekaran, S. & Ak, M. M. (2002). *Cheese rheology and texture*. Boca Raton: CRC Press.

Guthausen, G. & Kamlowski, A. (2009). Developments in time domain and single sided NMR. In M. Guðjónsdóttir, P. Belton & G. Webb (Eds.), *Magnetic resonance in food science. Challenges in a changing world* (pp. 46–55). Cambridge: RSC Publishing.

Hamboyan, L. & Pink, D. (1990). Ultraviolet spectroscopic studies and the prediction of the feathering of cream in filter coffees. *Journal of Dairy Research, 57*(2), 227–232.

Hammond, C. (1997). *The basics of crystallography and diffraction*. Oxford: Oxford University Press.

Hannewijk, J., Haighton, A. J. & Hendrikse, P. W. (1964). Dilatometry of fats. In H. A. Boekenoogen (Ed.), *Analysis and characterization of oils, fats and fat products*. London: Interscience Publishers.

Harnett, M. (1989). *The rheology of butter*. PhD thesis, Massey University, Palmerston North, New Zealand.

Harris, D. A. & Bashford, C. L. (Eds.). (1987). *Spectrophotometry and Spectrofluorimetry – A practical approach*. Oxford: IRL Press.

Harwalkar, V. R. & Ma, C.-Y. (1990). *Thermal analysis of foods*. New York: Elsevier Applied Science.

Hashemi, R. H., Lisanti, C. J. & Bradley, W. G. (2018). *MRI the basics*. Philadelphia: Wolters Kluwer.

Hass, D. D. (1996). *Dielectric sensing of ceramic particle suspensions*. Master of science thesis, University of Virginia, United States of America.

Hatakeyama, T. & Quinn, F. X. (1994). *Thermal analysis fundamentals and applications to polymer science*. Chichester: Wiley.

Häupler, M., Peyronel, F., Neeson, I., Weiss, J. & Marangoni, A. (2014). In situ ultrasonic characterization of cocoa butter using a chirp. *Food and Bioprocess Technology, 7*, 3186–3196.

Herman-Lara, E., Tejeda-Paz, M., Martinez-Sanchez, C. E., Rodriguez-Miranda, J., Ramirez-Rivera, E. J., Hernandez-Santos, B. & Juarez-Barrientos, J. M. (2017). Differential scanning calorimetry coupled with chemometric tools for determining adulteration with vegetable fat in fresh cheeses. *Food Science and Technology, 85*(Part A), 269–274.

Herrera, M. L. (1994). Crystallization behavior of hydrogenated sunflower seed oil: Kinetics and polymorphism. *Journal of the American Oil Chemists' Society, 71*(11), 1255–1260.

Herrera, M. L., Falabella, C., Melgarejo, M. & Anon, M. C. (1998). Isothermal crystallization of hydrogenated sunflower oil. I. Nucleation. *Journal of the American Oil Chemists' Society, 75*(10), 1273–1280.

Herrera, M. L., Falabella, C., Melgarejo, M. & Añón, M. C. (1999a). Isothermal crystallization of hydrogenated sunflower oil: II. Growth and solid fat content. *Journal of the American Oil Chemists' Society, 76*(1), 1–6.

Herrera, M. L., de Gatti, M. L. & Hartel, R. W. (1999b). A kinetic analysis of crystallization of a milk fat model system. *Food Research International, 32*, 289–298.

Hollas, J. M. (2004). *Modern spectroscopy* (4th ed.). Chichester: Wiley.

Holroyd, S. E. (2013). The use of near infrared spectroscopy on milk and milk products. *Journal of Near Infrared Spectroscopy, 21*, 311–322.

Holroyd, S. E., Prescott, B. & McLean, A. (2013). The use of in-line and on-line near infrared spectroscopy for milk powder measurement. *Journal of Near Infrared Spectroscopy, 21*, 441–443.

Hsieh, H.-N. (2018). https://studentportalen.uu.se/uusp-filearea-tool/download.action?nodeId=111007&toolAttachmentId=44350. Accessed 29 Nov 2018.

Hurst, W. J. (1994). Ultraviolet/visible light methods. In R. H. Wilson (Ed.), *Spectroscopic techniques for food analysis* (pp. 221–240). New York: VCH Publishers.

International Dairy Federation. (1981). *Evaluation of the firmness of butter* (International Dairy Federation Bulletin 135). Brussels: International Dairy Federation.

International Dairy Federation. (1991). *Rheological and fracture properties of cheese* (International dairy federation bulletin, 268). Brussels: International Dairy Federation.

International Standards Organization. (1975). *ISO 1739: 1975 butter – Determination of the refractive index of the fat (Reference method)*. Geneva: International Standards Organization.

International Standards Organization. (2005). *ISO 16305:2005 butter – Determination of firmness*. Geneva: International Standards Organization.

Jacobsen, N. E. (2007). *NMR spectroscopy explained. simplified theory, applications and examples for organic chemistry and structural biology*. Hoboken: Wiley-Interscience.

James, J. & Tanke, H. J. (1991). *Biomedical light microscopy*. Dordrecht: Kluwer.

Janssen, P. W. M. & MacGibbon, A. K. H. (2007). Non-isothermal crystallization of bovine milk fat. *Journal of the American Oil Chemists' Society, 84*(9), 871–875.

Karnjanolarn, R. & McCarthy, K. L. (2006). Rheology of different formulations of milk chocolate and the effect on coating thickness. *Journal of Texture Studies, 37*(6), 668–680.

Karoui, R., Bosset, J. O., Mazerolles, G., Kulmyrzaev, A. & Dufour, E. (2005). Monitoring the geographic origin of both experimental French Jura hard cheeses and Swiss Gruyere and L'Etivaz PDO cheeses using mid-infrared and fluorescence spectroscopies: A preliminary investigation. *International Dairy Journal, 15*(3), 275–286.

Karoui, R., Mouazen, A. M., Dufour, E., Pillonel, L., Picque, D., Bosset, J. O. & de Baerdemaeker, J. (2006). Mid-infrared spectrometry: A tool for the determination of chemical parameters in Emmental cheeses produced during the winter. *Lait, 86*(1), 83–97.

Karoui, R., de Baerdemaeker, J. & Dufour, E. (2008). Utilisation of front face fluorescence spectroscopy as a tool for the prediction of some chemical parameters and the melting point of semi-hard and hard cheeses: A preliminary study. *European Food Research and Technology, 226*(5), 1119–1126.

Karray, N., Lopez, C., Lesieur, P. & Ollivon, M. (2004). Dromedary milk fat: Thermal and structural properties. 1. Crystalline forms obtained by slow cooling. *Lait, 84*(4), 399–416.

Karray, N., Lopez, C., Lesieur, P. & Ollivon, M. (2005). Dromedary milk fat: Thermal and structural properties. 2. Influence of cooling rate. *Lait, 85*(6), 433–451.

Kaufmann, N., Andersen, U. & Wiking, L. (2012). The effect of cooling rate and rapeseed oil addition on the melting behaviour, texture and microstructure of anhydrous milk fat. *International Dairy Journal, 25*(2), 73–79.

Kawamura, K. (1979). The DSC thermal analysis of crystallization behaviour in palm oil. *Journal of the American Oil Chemists' Society, 56*(8), 753–758.

Kawamura, K. (1980). The DSC thermal analysis of crystallization behavior in palm oil. II. *Journal of the American Oil Chemists' Society, 57*(1), 48–52.

Kawamura, K. (1981). The DSC thermal analysis of crystallization behavior in high erucic acid rapeseed oil. *Journal of the American Oil Chemists' Society, 58*(8), 826–829.

Kawanari, M., Hamann, D. D., Swartzel, K. R. & Hansen, A. P. (1981). Rheological and texture studies of butter. *Journal of Texture Studies, 12*, 483–505.

Kaylegian, K. E. & Lindsay, R. C. (1994). *Handbook of milkfat fractionation technology and application*. Champaign: AOCS Press.

Keen, A. R. (1984). Seasonal variation in the colour of milkfat from selected herds and two dairy plants. *New Zealand Journal of Dairy Science and Technology, 19*, 263–265.

Keen, A. R. & Udy, C. W. (1980). Seasonal variation in the colour of milkfat from different herds. *New Zealand Journal of Dairy Science and Technology, 15*(1), 91–93.

Keller, G., Lavigne, F., Forte, L., Andrieux, K., Dahim, M., Loisel, C., Ollivon, M., Bourgaux, C. & Lesieur, P. (1998). DSC and X-ray diffraction coupling. *Journal of Thermal Analysis, 51*, 783–791.

Kett, V. (2016). Modulated temperature differential scanning calorimetry. In S. Gaisford, V. Kett & P. Haines (Eds.), *Principles of thermal analysis and calorimetry* (2nd ed., pp. 104–122). Cambridge: The Royal Society of Chemistry.

Kiełczyński, P. (2017). The application of ultrasonics for oil characterization. In M. Villamiel, J. V. Garcia-Pérez, A. Montilla, J. A. Cárcel & J. Benedito (Eds.), *Ultrasound in food processing* (pp. 115–145). Chichester: Wiley Blackwell.

Kirtil, E. & Oztop, M. H. (2016). 1H nuclear magnetic resonance relaxometry and magnetic resonance imaging and applications in food science and processing. *Food Engineering Reviews, 8*, 1–22.

Kirtil, E., Cikrikci, S., McCarthy, M. J. & Oztop, M. H. (2017). Recent advances in time domain NMR & MRI sensors and their food applications. *Current Opinion in Food Science, 17*, 9–15.

Konica Minolta, Japan, (25 February, 2005). http://www.konicaminoltaeurope.com/products/industrial/en/colourm.shtml

Koyano, T., Hachiya, I., Arishima, T., Sato, K. & Sagi, N. (1989). Polymorphism of POP and SOS. II. Kinetics of melt crystallization. *Journal of the American Oil Chemists' Society, 66*(5), 675–679.

Lakowicz, J. R. (2006). *Principles of fluorescence spectroscopy* (3rd ed.). New York: Springer.

Lambert, J. B. & Mazzola, E. P. (2004). *Nuclear magnetic resonance spectroscopy*. Upper Saddle River: Pearson Prentice Hall.

Larsen, H., Geiner Tellefsen, S. B. & Veberg Dahl, A. V. (2009). Quality of sour cream packaged in cups with different barrier properties measured by fluorescence spectroscopy and sensory analysis. *Journal of Food Science, 74*(8), S345–S350.

Laverse, J., Mastromatteo, M., Frisullo, P. & del Nobile, M. A. (2011). X-ray microtomography to study the microstructure of cream cheese-type products. *Journal of Dairy Science, 94*(1), 43–50.

Lawton, B. A. & Pethog, R. (1993). Determining the fat content of milk and cream using AC conductivity measurements. *Measurement Science and Technology, 4*, 38–41.

Laye, P. G. (2002). Differential thermal analysis and differential scanning calorimetry. In P. J. Haines (Ed.), *Principles of thermal analysis and calorimetry*. Cambridge: Royal Society of Chemistry.

Leemans, V. & Destain, M.-F. (2009). Ultrasonic internal defect detection in cheese. *Journal of Food Engineering, 90*, 333–340.

Liljedahl, L. A. & Abbot, J. A. (1994). Changes in sonic resonance of delicious and Golden delicious apples undergoing accelerated ripening. *Transactions of the ASAE, 37*(3), 907–912.

Lopez, C., Lesieur, P., Keller, G. & Ollivon, M. (2000). Thermal and structural behaviour of milk fat. Unstable species of cream. *Journal of Colloid and Interface Science, 229*(1), 62–71.

Lopez, C., Lesieur, P., Keller, G. & Ollivon, M. (2001a). Crystallization in emulsion: Application to thermal and structural behaviour of milk fat. In N. Widlak, R. Hartel & S. Narine (Eds.), *Crystallization and solidification properties of lipids* (pp. 190–199). Champaign: AOCS Press.

Lopez, C., Lavigne, F., Lesieur, P., Bourgaux, C. & Ollivon, M. (2001b). Thermal and structural behaviour of milk fat. 1. Unstable species of anhydrous milk fat. *Journal of Dairy Science, 84*(4), 756–766.

Lopez, C., Lavigne, F., Lesieur, P., Keller, G. & Ollivon, M. (2001c). Thermal and structural behaviour of anhydrous milk fat. 2. Crystalline forms obtained by slow cooling. *Journal of Dairy Science, 84*(11), 2402–2412.

Lopez, C., Lesieur, P., Bourgaux, C., Keller, G. & Ollivon, M. (2001d). Thermal and structural behaviour of milk fat. 2. Crystalline forms obtained by slow cooling of cream. *Journal of Colloid and Interface Science, 240*(1), 150–161.

Lopez, C., Bourgaux, C., Lesieur, P., Bernadou, S., Keller, G. & Ollivon, M. (2001e). Thermal and structural behaviour of milk fat. 3. Influence of cooling rate and droplet size on cream crystallization. *Journal of Colloid and Interface Science, 254*(1), 64–78.

Lopez, C., Bourgaux, C., Lesieur, P. & Ollivon, M. (2002). Crystalline structures formed in cream and anhydrous milk fat at 4 deg C. *Lait, 82*(3), 317–335.

Lopez, C., Lesieur, P., Bourgaux, C. & Ollivon, M. (2005a). Thermal and structural behaviour of milk fat. 3. Influence of cooling rate. *Journal of Dairy Science, 88*(2), 511–526.

Lopez, C., Karray, N., Lesieur, P. & Ollivon, M. (2005b). Crystallization and melting properties of dromedary milk fat globules studied by X-ray diffraction and differential scanning calorimetry. Comparison with anhydrous dromedary milk fat. *European Journal of Lipid Science and Technology, 107*(9), 673–683.

Lopez, C., Bourgaux, C., Lesieur, P. & Ollivon, M. (2007). Coupling of time-resolved synchrotron X-ray diffraction and DSC to elucidate the crystallization properties and polymorphism of triglycerides in milk fat globules. *Lait, 87*(4–5), 459–480.

Lucas, T., Wagener, M., Barey, P. & Mariette, F. (2005a). NMR assessment of mix and ice cream. Effect of formulation on liquid water and ice. *International Dairy Journal, 15*, 1064–1073.

Lucas, T., Le Ray, D., Barey, P. & Mariette, F. (2005b). NMR assessment of ice cream: Effect of formulation on liquid and solid fat. *International Dairy Journal, 15*, 1225–1233.

Macias-Rodriguez, B. A. (2018). Nonlinear rheology of fats using large amplitude oscillatory shear tests. In A. G. Marangoni (Ed.), *Structure-function analysis of edible fats* (2nd ed., pp. 169–195). London: Academic/American Oil Chemists' Society.

Maleky, F. (2018). Oil migration though fats – Quantification and its relationship to structure. In A. G. Marangoni (Ed.), *Structure-function analysis of edible fats* (2nd ed., pp. 241–266). London: Academic/AOCS Press.

Maleky, F., Compos, R. & Marangoni, A. (2007). Structural and mechanical properties of fats quantified by ultrasonics. *Journal of the American Oil Chemists' Society, 84*, 331–338.

Maleky, F., McCarthy, K. L., McCarthy, M. J. & Marangoni, A. G. (2012). Effect of cocoa butter structure on oil migration. *Journal of Food Science, 77*(3), E74–E79.

Marangoni, A. G. (1998). On the use and misuse of the Avrami equation in characterization of the kinetics of fat crystallization. *Journal of the American Oil Chemists' Society, 75*(10), 1465–1467.

Marangoni, A. G. (Ed.). (2018). *Structure-function analysis of edible fats* (2nd ed.). London: Academic/AOCS.

Mariette, F. (2003). NMR relaxometry and MRI for food quality control: Application to dairy products and processes. In P. S. Belton, A. M. Gill, G. A. Webb & D. Rutledge (Eds.), *Magnetic resonance in food science. Latest developments* (pp. 209–222). Cambridge: The Royal Society of Chemistry.

Mariette, F. & Gollewet, G. (2001). Relationships between sensory texture of soft cheese and MRI measurements. In G. A. Webb, P. S. Belton, A. M. Gill & I. Degadillo (Eds.), *Magnetic resonance in food science. A view to the future* (pp. 67–74). Cambridge: The Royal Society of Chemistry.

Mariette, F. & Lucas, T. (2005). NMR signal analysis to attribute the components to the solid/liquid phases present in mixes and ice creams. *Journal of Agricultural and Food Chemistry, 53*, 1317–1327.

Mariette, F., Collwet, G., Marchal, P. & Francomi, J. N. (1999). Internal structure characterization of soft cheeses by MRI. In P. S. Belton, B. P. Hills & G. A. Webb (Eds.), *Advances in magnetic resonance in food science* (pp. 24–34). Cambridge: The Royal Society of Chemistry.

Marshall, R. T., Goff, H. D. & Hartel, R. W. (2003). *Ice cream* (6th ed.). New York: Kluwer Academic.

Martini, S. (2007). Ultrasonic spectroscopy in lipid food systems. *Food Technology, 61*(2), 40–43.

Martini, S. (2014). Application of DSC, pulsed NMR, and other analytical techniques to study the crystallization kinetics of lipid models, oils, fats, and their blends in the field of food technology. In E. Chiavaro (Ed.), *Differential scanning calorimetry. Applications in fat and oil technology* (pp. 163–195). Boca Raton: CRC Press.

Martini, S., Bertoli, C., Herrera, M. L., Neeson, I. & Marangoni, A. (2005a). In situ monitoring of solid fat content by means of pulsed nuclear magnetic resonance spectrometry and ultrasonics. *Journal of the American Oil Chemists' Society, 82*(5), 305–312.

Martini, S., Herrera, M. L. & Marangoni, A. (2005b). New technologies to determine solid fat content online. *Journal of the American Oil Chemists' Society, 82*(5), 313–317.

Martini, S., Bertoli, C., Herrera, M. L., Neeson, I. & Marangoni, A. (2005c). Attenuation of ultrasonic waves: Influence of microstructure and solid fat content. *Journal of the American Oil Chemists' Society, 82*(5), 319–328.

Mattice, K. D. & Marangoni, A. G. (2018). Fat crystallization and structure in bakery, meat, and cheese systems. In A. G. Marangoni (Ed.), *Structure-function analysis of edible fats* (2nd ed., pp. 287–311). London: Academic/AOCS.

McCarthy, O. J. (1987). Large deformation testing of solid foods. *Food Technology in New Zealand, 22*(7), 40–41, 43.

McClements, D. J. (1995). Advances in the application of ultrasound in food analysis and processing. *Trends in Food Science and Technology, 6*, 293–299.

McGinley, L. (1991). Analysis and quality control for processing and processed fats. In J. B. Rossell & J. L. R. Pritchard (Eds.), *Analysis of oilseeds, fats and fatty foods*. London: Elsevier Applied Science.

Mehlenbacher, V. C. (1960). *The analysis of fats and oils*. Champaign: The Garrard Press.

Mertens, W. G. (1972). Automatic melting point determination of fats. *Journal of the American Oil Chemists' Society, 49*(6), 366–370.

Mertens, W. G. (1973). Fat melting point determinations: A review. *Journal of the American Oil Chemists' Society, 50*(4), 115–119.

Minifie, B. W. (1999). *Chocolate, cocoa, and confectionery: Science and technology* (3rd ed.). Gaithersburg: Aspen Publishers.

Mohammadi, V., Ghasemi-Varnamkhasti, M. & González, L. A. (2017). Analytical measurements of ultrasound propagation in dairy products: A review. *Trends in Food Science and Technology, 61*, 38–48.

Monaci, L., Vatinno, R. & de Benedetto, G. E. (2007). Fast detection of cyclopiazonic acid in cheese using Fourier transform mid-infrared ATR spectroscopy. *European Food Research and Technology, 225*(3–4), 585–588.

Mulas, G., Roggio, T., Ussau, S. & Anedda, R. (2013). A new magnetic resonance imaging approach for discriminating Sardinian sheep milk cheese made from heat-treated or raw milk. *Journal of Dairy Science, 96*(12), 7393–7403.

Mulas, G., Anedda, R., Longo, D. L., Roggio, T. & Uzzau, S. (2016). An MRI method for monitoring the ripening of Grano Padano cheese. *International Dairy Journal, 52*, 19–25.

Mulder, H. & Walstra, P. (1974). *The milk fat globule emulsion science as applied to milk products and comparable foods*. Wageningen: Commonwealth Agricultural Bureaux, Farnham Royal, and Centre for Agricultural Publishing and Documentation.

Muliawan, E. B. & Hatzikiriakos, S. G. (2007). Rheology of mozzarella cheese. *International Dairy Journal, 17*(9), 1063–1072.

Musse, M., Challois, S., Huc, D., Quellec, S. & Mariette, F. (2014). MRI method for investigation of eye growth in semi-hard cheese. *Journal of Food Engineering, 121*, 151–158.

Nakae, T., Kataoka, K. & Miyamoto, T. (1974). A method for measuring the melting point of butterfat and margarine fat by means of temperature-gradient apparatus. *Japanese Journal of Zootechnical Science [Nihon Chikusan Gakkai-ho], 45*(6), 347–351.

Narine, S. S. & Marangoni, A. G. (1999). Relating structure of fat crystal networks to mechanical properties: A review. *Food Research International, 32*, 227–248.

Nascimento, P. A. M., Barsanelli, P. L., Rebellato, A. P., Pallone, J. A. L., Colnago, L. A. & Pereira, F. M. V. (2017). Time-domain nuclear magnetic resonance (TD-NMR) and chemometrics for determination of fat content in commercial products of milk powder. *Journal of AOAC International, 100*(2), 330–334.

Nassar, G., Lefbvre, F., Skaf, A., Nongaillard, B. & Noel, Y. (2010). Ultrasonic and acoustic investigation of cheese matrix at the beginning and the end of ripening period. *Journal of Food Engineering, 96*(1), 1–13.
Nedeljkovic, A., Rosch, P., Popp, J., Miocinovic, J. & Pudja, P. (2016). Raman spectroscopy as a rapid tool for quantitative analysis of butter adulterated with margarine. *Food Analytical Methods, 9*(5), 1315–1320.
Nedeljkovic, A., Tomasevic, I., Miocinovic, J. & Pudja, P. (2017). Feasibility of discrimination of dairy creams and cream-like analogues using Raman spectroscopy and chemometric analysis. *Food Chemistry, 232*, 487–492.
Nelson, J. H. (2003). *Nuclear magnetic resonance spectroscopy*. Upper Saddle River: Prentice Hall.
Nelson, D. L. (2018). Introduction to spectroscopy. In A. S. Franca & L. M. L. Nollet (Eds.), *Spectroscopic methods in food analysis* (pp. 3–33). Boca Raton: CRC Press.
Ng, W. L. (1989). Nucleation behaviour of tripalmitin from a triolein solution. *Journal of the American Oil Chemists' Society, 66*(8), 1103–1106.
Nixdorf, S. L. (2018). UV-vis spectroscopy. In A. S. Franca & L. M. L. Nollet (Eds.), *Spectroscopic methods in food analysis* (pp. 35–67). Boca Raton: CRC Press.
Norris, R., Gray, I. K., McDowell, A. K. R. & Dolby, R. M. (1971). The chemical composition and physical properties of fractions of milk fat obtained by a commercial fractionation process. *Journal of Dairy Research, 38*, 179–191.
Nurrulhidayah, A. F., Rohman, A., Amin, I., Shuhaimi, M. & Khatib, A. (2015). Detection of butter adulteration with lard using differential scanning calorimetry. *International Food Research Journal, 22*(2), 832–839.
O'Brien, R. D. (2003). *Fats and oils formulating and processing applications*. Boca Raton: CRC Press.
O'Callaghan, D. J. & Guinee, T. P. (2004). Rheology and texture of cheese. In P. J. Fox, P. L. H. McSweeney, T. M. Cogan & T. P. Guinee (Eds.), *Cheese chemistry, physics and microbiology volume 1 general aspects*. Amsterdam: Elsevier/Academic.
O'Connor, J. F. & Synnott, E. C. (1982). Seasonal variation in dielectric properties of butter at 15 MHz and 4°C. *Irish Journal of Food Science and Technology, 6*, 49–59.
Oliveira, K., de Callegaro, L. S., Stephani, R., Almeida, M. R. & de Oliveira, L. F. C. (2016). Analysis of spreadable cheese by Raman spectroscopy and chemometric tools. *Food Chemistry, 194*, 441–446.
Ollivon, M., Loisel, C., Lopez, C., Lesieur, P., Artzner, F. & Keller, G. (2001). Simultaneous examination of structural and thermal behaviours of fats by coupled X-ray diffraction and differential scanning calorimetry techniques: Application to cocoa butter polymorphism. In N. Widlak, R. Hartel & S. Narine (Eds.), *Crystallization and solidification properties of lipids* (p. 34.41). Champaign: AOCS Press.
Ollivon, M., Relkin, P., Michon, C., Kalnin, D. & Mariette, F. (2005). Crystallization of anhydrous milk fat: Influence of polymorphism and emulsifiers. *Sciences des Aliments, 25*(5–6), 397–411.
Osborne, B. G., Fearn, T. & Hindle, P. H. (1993). *Practical NIR spectroscopy with applications in food and beverage analysis*. Harlow: Longman Scientific and Technical.
Otles, S. & Ozyurt, V. H. (2017). Ultrasound spectroscopy in food analysis. In A. S. Franca & L. M. L. Nollet (Eds.), *Spectroscopic methods in food analysis* (pp. 225–236). Boca Raton: CRC Press.
Ozaki, Y., McClure, W. F. & Christy, A. A. (2007). *Near-infrared spectroscopy in food science and technology*. Hoboken: Wiley.
Ozbekova, Z. & Kulmyrzaev, A. (2017). Fluorescence spectroscopy as a non-destructive method to predict rheological characteristics of Tilsit cheese. *Journal of Food Engineering, 210*, 42–49.
Padley, F. B., Paulussen, C. N., Soeters, C. J. & Tresser, D. (1972). The improvement of chocolate using the mono-unsaturated triglycerides SOS and POS. *International Chocolate Review, 27*(9), 226–228-229.
Papalois, M., Leach, F. W., Dungey, S., Yep, Y. L. & Versteeg, C. (1996). Australian milkfat survey – Physical properties. *Australian Journal of Dairy Technology, 51*(2), 114–117.
Paquot, C. & Hautfenne, A. (1987). *Standard methods for the analysis of oils, fats and derivatives* (7th ed.). Oxford: Blackwell Scientific Publications.
Park, B., Whittaker, A. D., Miller, R. K., Hale, D. S., et al. (1994). Ultrasonic spectral analysis for beef sensory attributes. *Journal of Food Science, 59*(4), 697–701.
Parkash, S. (1970). Measurement of the dielectric constant of butter. *Dairy Industries, 35*(10), 688–689.
Parkash, S. & Armstrong, J. G. (1969). Moisture in butter in relation to dielectric constant measurements. *Journal of Dairy Science, 52*(8), 1224–1228.
Parmar, A. & Sharma, S. (2016). Derivative UV-vis absorption spectra as an invigorated spectrophotometric method for spectral resolution and quantitative analysis: Theoretical aspects and analytical applications: A review. *Trends in Analytical Chemistry, 77*, 44–63.
Patist, A. & Bates, D. (2011). Industrial applications of high power ultrasonics. In H. Feng, G. V. Barbosa-Canovas & J. Weiss (Eds.), *Ultrasound Technologies for Food and Bioprocessing* (pp. 599–616). New York: Springer.
Pecharsky, V. K. & Zavalij, P. Y. (2003). *Fundamentals of powder diffraction and structural characterization of materials*. Boston: Kluwer.
Peyronel, F. (2018). Methods used in the study of the physical properties of fats. In A. G. Marangoni (Ed.), *Structure-function analysis of edible fats* (2nd ed., pp. 313–385). London: Academic/AOCS.
Peyronel, F. & Pink, D. A. (2018). Ultra-small angle X-ray scattering: A technique to study soft materials. In A. G. Marangoni (Ed.), *Structure-function analysis of edible fats* (2nd ed., pp. 267–284). London: Academic/AOCS.

Picariello, G., Sacchi, R., Fiero, O., Melck, D., Romano, R., Paduano, A., Motta, A. & Addeo, F. (2013). High resolution 13C NMR detection of short- and medium-chain synthetic triacylglycerols used in butterfat adulteration. *European Journal of Lipid Science and Technology, 115*(8), 858–864.

Pinzer, B. R., Medebach, A., Limbach, H. J., Dubois, C., Stampanoni, M. & Schneebeli, M. (2012). 3D-characterization of three-phase systems using X-ray tomography: Tracking the microstructural evolution in ice cream. *Soft Matter, 8*(17), 4584–4594.

Pisecky, J. (1997). *Handbook of Milk powder manufacture*. Copenhagen: NIRO A/S.

Povey, M. J. W. (1998). Rapid determination of food material properties. In M. J. W. Povey & T. J. Mason (Eds.), *Ultrasound in food processing* (pp. 30–65). London: Blackie Academic and Professional.

Povey, M. J. W. & Mason, T. J. (Eds.). (1998). *Ultrasound in food processing*. London: Blackie Academic and Professional.

Rajah, K. K. (2002). *Fats in food technology*. Sheffield: Sheffield Academic Press.

Rao, M. A. (1992). Classification, description and measurement of viscoelastic properties of solid foods. In M. A. Rao & J. F. Steffe (Eds.), *Viscoelastic properties of foods*. London: Elsevier Applied Science.

Reiner, J., Peyronel, F., Weiss, J. & Marangoni, A. (2018). Monitoring the polymorphic transformation of a palm kernel-based emulsion using ultrasound. *Food and Bioprocess Technology, 11*, 797–808.

Rheims, J., Köser, J. & Wriedt, T. (1997). Refractive-index measurements in the near-IR using an Abbe refractometer. *Measurement Science and Technology, 8*, 601–605.

Rohm, H. (1992). Influence of sectility speed on the relationship between sensory and instrumental evaluation of edible fat firmness. *Z Lebensmittel Untersuch Forsch, 194*, 240–243.

Rohm, H. (1993a). Rheological behaviour of butter at large deformations. *Journal of Texture Studies, 24*, 139–155

Rohm, H. (1993b). Rheological behaviour of butter at small deformations. *Journal of Texture Studies, 24*, 157–172.

Rohm, H. & Weidinger, K.-H. (1991). Correlations between empirical methods for texture assessment of butter. *Milchwissenschaft, 46*(8), 503–506.

Ronholt, S., Kirkensgaard, J. J. K., Pedersen, T. B., Mortensen, K. & Knudsen, J. C. (2012). Polymorphism, microstructure and rheology of butter. Effects of cream heat treatment. *Food Chemistry, 135*(3), 1730–1739.

Ronholt, S., Madsen, A. S., Kirkensgaard, J. J. K., Mortensen, K. & Knudsen, J. C. (2014). Effect of churning temperature on water content, rheology, microstructure and stability of butter during four weeks of storage. *Food Structure, 2*(1–2), 14–26.

Rossell, J. B. (1986). Classical analysis of oils and fats. In R. J. Hamilton & J. B. Rossell (Eds.), *Analysis of oils and fats*. London: Elsevier Applied Science.

Rossell, B. (Ed.). (2003). *Oils and fat volume 3 dairy fats*. Leatherhead: Leatherhead Publishing.

Roussy, G. & Pearce, J. A. (1995). *Foundations and industrial applications of microwave and radio frequency fields*. Chichester: Physical and Chemical Processes Wiley.

Rumsey, T. R. & McCarthy, K. L. (2012). Modelling oil migration in two-layer chocolate-almond confectionery products. *Journal of Food Engineering, 111*(1), 149–155.

Sacchi, R., Fiero, O., Melck, D., Romano, R., Paduano, A., Motta, A. & Addeo, F. (2018). Assessment of milk fat content in fat blends by 13C NMR spectroscopy analysis of butyrate. *Food Control, 91*, 231–236.

Sádecká, J., Uríčková, V. & Jakubíková, M. (2018). Fluorescence spectroscopy. In A. S. Franca & L. M. L. Nollet (Eds.), *Spectroscopic methods in food analysis* (pp. 189–223). Boca Raton: CRC Press.

Sato, K. (2001). Molecular aspects in fat polymorphism. In N. Widlak, R. Hartel & S. Narine (Eds.), *Crystallization and solidification properties of lipids* (pp. 1–16). Champaign: AOCS Press.

Sato, K., Ueno, S. & Yano, J. (1999). Molecular interactions and kinetic properties of fats. *Progress in Lipid Research, 38*, 91–116.

Scaife, B. K. P., Scaife, W. G. S., Bennett, R. G. & Calderwood, J. H. (1971). *Complex permittivity theory and measurement*. London: The English Universities Press.

Schievano, E., Pasini, G., Cozzi, G. & Mammi, S. (2008). Identification of the production chain of Asiago d'Allevo cheese by nuclear magnetic resonance spectroscopy and principle component analysis. *Journal of Agricultural and Food Chemistry, 56*, 7208–7214.

Schievano, E., Guardini, K. & Mammi, S. (2009). Fast determination of histamine in cheese by nuclear magnetic resonance. *Journal of Agricultural and Food Chemistry, 57*(7), 2647–2652.

Sears, F. W., Zemansky, M. W. & Young, H. D. (1982). *University physics* (6th ed.). Reading: Addison-Wesley Publishing Company.

Shaikh, S. & O'Donnell, C. (2017). Applications of fluorescence spectroscopy in dairy processing: A review. *Current Opinion in Food Science, 17*, 16–24.

Sharma, A. & Schulman, S. G. (1999). *Introduction to fluorescence spectroscopy*. New York: Wiley.

Shukla, V. K. S. (1995). Solid fat content: Pulsed NMR versus dilatometry. *Lipid Technology, 7*(6), 135–137.

Simon, K. & Süverkrüp, R. (1995). Comparison of recent techniques to characterize the crystallization behaviour of fatty suppository bases. *Thermochimica Acta, 248*, 271–287.

Singh, A. P., McClements, D. J. & Marangoni, A. (2002). Comparison of ultrasonic and pulsed NMR techniques for determination of solid fat content. *Journal of the American Oil Chemists' Society, 79*, 431–437.

Singh, A. P., McClements, D. J. & Marangoni, A. (2004). Solid fat content determination by ultrasonic velocimetry. *Food Research International, 37*, 545–555.

Skelland, A. H. P. (1967). *Non-Newtonian flow and heat transfer*. New York: Wiley.
Slayter, E. M. & Slayter, H. S. (1992). *Light and electron microscopy*. Cambridge: Cambridge University Press.
Sluder, G. & Wolf, D. E. (1998). *Methods in cell biology, volume 56, video microscopy*. San Diego: Academic.
Spyros, A. & Dais, P. (2013). *NMR spectroscopy in food analysis*. Cambridge: RSC Publishing.
Stauffer, C. E. (1996). *Fats and oils*. St. Paul: Eagan Press.
Steffe, J. F. (1992). *Rheological methods in food process engineering*. East Lansing: Freeman Press.
Subramanian, A., Harper, W. J. & Rodriguez-Saona, L. E. (2009). Rapid prediction of composition and flavour quality of Cheddar cheese using ATR-FTIR spectroscopy. *Journal of Food Science, 74*(3), 292–297.
Tao, F. & Ngadi, M. (2017). Applications of spectroscopic techniques for fat and fatty acids analysis of dairy foods. *Current Opinion in Food Science, 17*, 100–112.
Telis-Romero, J., Vaquiro, H. A., Bon, J. & Benedito, J. (2011). Ultarsonic assessment of fresh cheese composition. *Journal of Food Engineering, 103*, 137–146.
ten Grotenhuis, E., van Aken, G. A., van Malssen, K. F. & Schenk, H. (1999). Polymorphism of milk fat studied by differential scanning calorimetry and real-time X-ray powder diffraction. *Journal of the American Oil Chemists' Society, 76*(9), 1031–1039.
Timms, R. E. (1978). Automatic determination of the softening point of milk fat, milk fat fractions and butter. *Australian Journal of Dairy Technology, 33*, 143–147.
Tomaszewska-Gras, J. (2016). Rapid quantitative determination of butter adulteration with palm oil. *Food Control, 60*, 629–635.
Ueno, S., Yano, J., Seto, H., Amemiya, Y. & Sato, K. (2000). Synchrotron radiation X-ray diffraction study of polymorphic crystallization in triacylglycerols. In N. Widlak (Ed.), *Physical properties of fats, oils and emulsifiers* (pp. 64–78). Champaign: AOCS Press.
Uysal, R. S., Boyaci, I. H., Genis, H. E. & Tamer, U. (2013). Determination of butter adulteration with margarine using Raman spectroscopy. *Food Chemistry, 141*(4), 4397–4403.
van Putte, K. P. A. M. & Bakker, B. H. (1987). Crystallization kinetics of palm oil. *Journal of the American Oil Chemists' Society, 64*(8), 1138–1143.
Van Wazer, J. R., Lyons, J. W., Kim, K. Y. & Colwell, R. E. (1963). *Viscosity and flow measurement*. New York: Interscience.
Veliyulin, E., Mastikhin, I. V., Marble, A. E. & Balcom, B. J. (2008). Rapid determination of the fat content in packaged dairy products by unilateral NMR. *Journal of the Science of Food and Agriculture, 88*(14), 2563–2567.
Villamiel, M., García-Pérez, J. V., Montilla, A., Cárcel, J. A. & Benedito, J. (Eds.). (2017). *Ultrasound in food processing*. Chichester: Wiley Blackwell.
Vithanage, C. R., Grimson, M. J. & Smith, B. G. (2009). The effect of temperature on the rheology of butter, a spreadable blend and spreads. *Journal of Texture Studies, 40*(3), 346–369.
Wan, P. J. (1991). *Introduction to fats and oils technology*. Champaign: American Oil Chemists' Society.
Wendlandt, W. M. (1974). *Thermal methods of analysis* (2nd ed.). New York: Wiley.
Westergaard, V. (1994). *Milk powder technology evaporation and spray drying*. Copenhagen: NIRO A/S.
Whorlow, R. W. (1992). *Rheological Techniques*. New York: Ellis Horwood.
Wichchukit, S., McCarthy, M. J. & McCarthy, K. L. (2005). Flow behaviour of milk chocolate melt and the application to coating flow. *Journal of Food Science, 70*(3), 165–171.
Wiking, L., Fredrick, E. & Dewettinck, K. (2009). Phospholipids delay crystallization of milk fat. *Milchwissenschaft, 64*(4), 376–379.
Wildmoser, H., Scheiwiller, J. & Windhab, E. J. (2004). Impact of disperse microstructure on rheology and quality aspects of ice cream. *Lebensmittel-Wissenschaft und -Technologie, 37*, 881–891.
Wilson, R. H. (Ed.). (1994). *Spectroscopic techniques for food analysis*. New York: VCH Publishers.
Wilson, R. H. & Goodfellow, B. J. (1994). Mid-infrared spectroscopy. In R. H. Wilson (Ed.), *Spectroscopic techniques for food analysis* (pp. 59–85). New York: VCH Publishers.
Winkelmeyer, C. B., Peyronel, F., Weiss, J. & Marangoni, A. (2016). Monitoring tempered dark chocolate using ultrasonic spectrometry. *Food and Bioprocess Technology, 9*, 1692–1705.
Woolfson, M. & Fan, H.-F. (1995). *Physical and non-physical methods of solving crystal structures*. Cambridge: Cambridge University Press.
Wright, D. J. (1984). Thermoanalytical methods in food research. In H. W.-S. Chan (Ed.), *Biophysical methods in food research*. Oxford: Blackwell Scientific Publications.
Wright, A. J., Hartel, R. W., Narine, S. S. & Marangoni, A. G. (2000). The effect of minor components on milk fat crystallization. *Journal of the American Oil Chemists' Society, 77*(5), 463–475.
Wright, A. J., Narine, S. S. & Marangoni, A. G. (2001a). Comparison of experimental techniques used in lipid crystallization studies. In N. Widlak, R. Hartel & S. Narine (Eds.), *Crystallization and solidification properties of lipids* (pp. 120–131). Champaign: AOCS Press.
Wright, A. J., Scanlon, M. G., Hartel, R. W. & Marangoni, A. G. (2001b). Rheological properties of milkfat and butter. *Journal of Food Science, 66*(8), 1056–1071.
Wunderlich, B. (1990). *Thermal analysis*. Boston: Academic.
Young, N. W. G., Wassell, P., Wiklund, J. & Stading, M. (2008). Monitoring structurants of fat blends with ultrasound based in-line rheometry (ultrasonic velocity profiling with pressure difference). *International Journal of Food Science and Technology, 43*, 2083–2089.

Analytical Methods

13

A. Logan, L. Walter, and P. Watkins

13.1 Introduction

The fat composition of mammalian milk varies with factors such as breed (Kelsey *et al.* 2003), stage of lactation (Konuspayeva *et al.* 2010; Mestawet *et al.* 2012), and season (Heck *et al.* 2009). Bovine milk fat (MF) content normally falls within a range of 2–6% (w/w) and exists as discrete lipid droplets, known as MF globules (MFGs), dispersed within the protein serum phase. Raw milk presents a broad distribution in MFG size, with an average diameter of 2.5–5.7 μm in a typical herd (Logan *et al.* 2014). Each MFG is stabilised by a trilayer membrane composed of polar lipids (PLs) such as glycerophospholipids and sphingolipids, glycolipids, glycoproteins, enzymes, and sterols, with PL accounting for up to around 1.0% (w/w) of the total MF composition (Rombaut *et al.* 2006). The remaining MF comes from the non-polar (or neutral) lipid core, consisting mostly of triacylglycerols (TAGs) along with minor components such as diacylglycerols (DAGs), monoacylglycerols (MAGs), free fatty acids (FFAs), and lipid-soluble antioxidants that include tocopherols and carotenoids. While the concentration of individual fatty acids (FAs) can also vary with diet (Wijesundera *et al.* 2003), season, and stage of lactation (Heck *et al.* 2009; Stoop *et al.* 2009; Walker *et al.* 2013), the relative abundance of major FAs, including C14:0, C16:0, C18:0, and C18:1, remain consistent. Other minor FAs found in bovine MF include a series of short-chain saturated FAs, with an acyl chain that is less than 12 carbons in length, and conjugated linoleic acid (CLA, C18:2 *cis*-9-*trans*-11) which is believed to have a positive impact on the immune system and impart anti-carcinogenic and anti-atherogenic benefits to the consumer (Belury 2002; MacDonald 2000). Within the TAG fraction, Mottram and Evershed (2001) have identified 120 different species which differ in the configuration of FAs along the glycerol backbone of the TAG structure and range from a carbon number (CN) of 28 (e.g. a TAG made up of C4:0, C10:0, and C18:2) to CN54 (e.g. a TAG consisting of three C18:0 FAs). The most abundant TAG structural species were found to be C4:0-C16:0-C18:1 (CN38) and C6:0-C14:0-C16:0 (CN36) (Mottram and Evershed 2001).

Traditional methods for the analysis of complex lipids including MF from milk and other dairy products have relied upon the separation of components for further characterisation, either by gas or liquid chromatography. More recently,

A. Logan (✉) · P. Watkins
CSIRO Agriculture and Food,
Werribee, VIC, Australia
e-mail: amy.logan@csiro.au

L. Walter
CSIRO Agriculture and Food,
Werribee, VIC, Australia

Faculty of Veterinary and Agricultural Sciences, The University of Melbourne, Parkville, VIC, Australia

P. L. H. McSweeney et al. (eds.), *Advanced Dairy Chemistry, Volume 2*,
https://doi.org/10.1007/978-3-030-48686-0_13

the availability of soft ionisation mass spectrometry (MS) techniques has allowed significant progress in the development of direct analysis without the need for separation or derivatisation. Moreover, advances in column specificity, analytical instrumentation, and methods allow for optimum separation and resolution. In this chapter, we will explore a number of different chromatographic and spectroscopic methods for the analysis of milk lipids.

13.2 Milk Fat Extraction and Proximate Analysis

Milk fat can be separated from the protein serum phase through mechanical separation, which mimics traditional processes for the preparation of cream and butter. This involves the concentration of cream through centrifugation or using a plate separator, followed by churning of the cream into butter and buttermilk fractions. Finally, the butter is separated into MF and butter serum using heat (60–70 °C) and allowing the MF to crystallise (Lopez *et al.* 2017; Walter *et al.* 2020; Wijesundera *et al.* 2003). The buttermilk and serum can be pooled together to form the polar lipid (PL) fraction and represent the MFG membrane (MFGM) lipid components. The remaining MF is the neutral lipid fraction representing the MFG lipid core. Another approach is to extract the MF fraction using solvents. Lipids are generally soluble in non-polar organic solvents and insoluble in water, depending upon the interaction strength between the solvent and the hydrophobic or hydrophilic region of the molecule. Non-polar TAGs are soluble in hydrocarbon solvents (e.g. heptane or hexane) and more polar solvents (e.g. chloroform or ethers). Polar lipids, on the other hand, do not dissolve readily in either type of solvent. Straarup *et al.* (2006) utilised the Folch method to extract the total MF fraction from bovine milk. This involves the homogenisation of sample in a chloroform-methanol mixture followed by washing with water or a salt solution. The lipid components are collected in the solvent phase (Folch *et al.* 1957) and evaporated. A modified version of this method has been reported by Mesilati-Stahy *et al.* (2011) using a methanol-chloroform-water (1:2:0.6 v/v) solvent mixture. The Bligh-Dyer method is another traditional approach using chloroform and methanol (Bligh and Dyer 1959); however, this method seems to be less commonly applied to dairy samples, possibly because the extraction yield is less efficient for samples with a fat content of 2% or more compared to the Folch method (Iverson *et al.* 2001). While these methods are well established and yield sufficient lipid recovery rates for major lipid species, they are time-consuming and usually require two steps of lipid extractions followed by solvent removal and reconstitution in a different solvent prior to analysis. In recent years, progress has been made in the development of new extraction methods that allow for high-throughput but otherwise comparable performance (Liu *et al.* 2016; Lopez *et al.* 2008; Shen *et al.* 2011).

Other methods specific for the quantification of fat include Mojonnier ether extraction and Babcock (Barbano *et al.* 1988) and Gerber methods (Kleyn *et al.* 2001). Supercritical fluid extraction has also been applied for the quantitative extraction of fats from foods including dairy, reviewed previously by Rozzi and Singh (2002).

13.2.1 Spectroscopic Methods

The fat content of milk and, in some cases, dairy products can be measured indirectly using spectroscopic methods. These methods are non-destructive and also provide the advantage of speed while being relatively inexpensive, without the need for solvents, aside from the initial capital expenditure to purchase equipment. Moreover, they can be positioned in-line to provide real-time analysis of compositional factors within a farm or factory setting. The angle of the light incident is changed to vary the optical path length, allowing for a range in wavelength to be scanned in a single analysis (Siesler *et al.* 2014) and for the rapid screening of multiple components. Disadvantages include the need for calibration using standards of known composition for different sample matrices.

13.2.1.1 Infrared Spectroscopy

Infrared spectroscopy is based on the absorption of radiation in the infrared region of the electromagnetic spectrum. Analytical tools for applications in MF and dairy products span the mid-infrared (MIR) and near-infrared (NIR) regions:

Mid-infrared This involves analysis within the wavelength range of 2500 nm–25 μm (corresponding to a wavenumber of 4000–400 cm^{-1}) and can measure both proximate and more detailed MF analyses due to the specificity provided to regions within the MIR range, for example, a band located at 966 cm^{-1} corresponding to the C=C and noted as a marker for *trans* FA, at 700 cm^{-1} corresponding to the FA carbonyl (C=O) group, and another between 2800 and 3000 cm^{-1} corresponding to the C-H stretch vibrations of lipid bonds (Bassbasi *et al.* 2014; Karoui *et al.* 2006; Sherazi *et al.* 2009) which allowed for conformation and TAG structure to be identified (Dufour *et al.* 2000). Dufour *et al.* (2000) used this method to examine changes in cheese TAG composition as a function of ripening time. This approach was also used by Soyeurt *et al.* (2011) to predict the FA profile for a range of bovine milk sources. Karoui *et al.* (2006) compared NIR and MIR analysis for determining the composition of Emmental cheese samples and found NIR was most appropriate for the determination of fat and total nitrogen. On the other hand, MIR more accurately predicted NaCl and non-protein nitrogen content in a selection of cheese samples that varied in composition from 0.19 to 1.36% and 0.202 to 0.982%, respectively. Fourier transform infrared (FTIR) spectroscopy is a form of MIR, commonly used in industry to measure a number of compositional factors in a single run, for example, the quantification of fat, protein, and lactose in milk using a Delta Instruments' 'LactoScope' (Logan *et al.* 2017). This and similar instruments, for example, the FOSS 'Milkoscan' (Schennink *et al.* 2007) and the Bentley Instruments' milk analyser (Moate *et al.* 2013), are commonly used in industry for real-time and routine analysis of compositional factors for the purpose of milk pricing and quality. This technique has also been applied to measure the acid value of edible oils through a band located at 3535 cm^{-1}, corresponding to the O–H stretch of carboxyl groups within FFAs. It has also been used for the compositional analysis of dairy products, including the work of Bassbasi *et al.* (2014) who modelled FTIR data to predict the geographical origin of butter.

Near-Infrared This involves the analysis of materials within the wavelength range of 780–2500 nm (corresponding to a wavenumber of 12,820–4000 cm^{-1}), reviewed previously by Osborne (2006) for the purpose of food analysis. The NIR spectrum is formed from a combination of tones and overtones of absorption lines from the C–H, N–H, and O–H chemical bonds within the molecular structure, collected as an absorption spectrum through transmission or diffuse-reflectance measurements. As a result, regions of overlap do occur; for example, bands located in the 4600–4255 cm^{-1} region correspond the CH_2 groups of both fat and protein (Karoui *et al.* 2006), so that differentiation of specific structural changes cannot be deduced.

13.2.1.2 Ultraviolet-Visible Spectrophotometry

Ultraviolet-visible spectrophotometry is a traditional technique employed for the analysis of lipids, based on the absorption of energy in the UV-visible region of the electromagnetic spectrum, between 100 and 780 nm. Absorption occurs within the chromophore region of the analyte at a particular wavelength, where absorbed energies elevate electrons into an excited state. In most cases, the flow cell path length is fixed and a wavelength selected to maximise absorptivity, which can be directly related to the analyte concentration by the Beer-Lambert law (Swinehart 1962). A photodiode array detector is an extension beyond a simple UV-visible detector where, as the sample passes through the flow cell, it is scanned by multiple wavelengths and it is possible to select the best wavelength for subsequent analysis. Additionally, it is also a useful approach to identify additional compounds if multiple peaks are observed. Forcato *et al.* (2005) compared the detection of MF using UV-vis detection to the infrared 'Milkoscan' technique, and Xiong *et al.* (2016) have suggested that UV-vis detection is a suitable technique for the determination of MF at very low concentrations, scanning between 200 and 300 nm and using a 1:1.4:1 mixture of water-ethanol-heptane as solvent.

13.2.1.3 Fluorescence Spectrometry

This is another commonly used detection method based on the absorption of energy within fluorophore regions of a sample. Similar to a chromophore, fluorophores absorb energy at a particular wavelength and electrons become excited; a subsequent release of energy occurs at a different (emission) wavelengths resulting in fluorescence. Fluorescence spectrometry is a useful detection technique for measuring lipid-soluble minor components in milk, including chlorophyll (Bhattacharjee *et al.* 2018), β-carotene, and vitamin A (Ullah *et al.* 2017). Moreover, Ozbekova and Kulmyrzaev (2017) used the vitamin A emission spectra to predict rheological properties, including the melting behaviour of a hard cheese.

13.2.2 Physical Characterisation of Milk Fat

The solid fat content and thermal behaviour of MF can be determined using techniques such as nuclear magnetic resonance (NMR) (Mariette 2009; Versteeg *et al.* 2016), differential scanning calorimetry (DSC), ultrasound velocity (Meyer *et al.* 2006; Ozyurt and Otles 2017), and X-ray scattering (Lopez *et al.* 2007; Truong *et al.* 2015; Zychowski *et al.* 2016), which are discussed in Chapter 12.

13.3 Fatty Acid Compositional Analysis

Bovine MF presents a broad FA distribution of saturated and unsaturated species that ranges from C4:0 through to long-chain FAs that are 18 and 20 carbons in length (Table 13.1). In some studies, FAs with a longer carbon length of C22–24 have been measured to very low levels (Delmonte *et al.* 2012; Zou *et al.* 2013) and in milk from cows fed with special diets, such as feeding regimes designed to increase levels of the beneficial FA C22:6 (docosahexaenoic acid, DHA) (Moate *et al.* 2013). More than 400 FA

Table 13.1. Fatty acid composition (as methyl esters) of milk fat (Zychowski *et al.* 2018).

Fatty acid	Average % (w/w)
C4:0	2.09 ± 0.02
C6:0	1.52 ± 0.02
C8:0	1.10 ± 0.01
C9:0	0.05 ± 0.00
C10:0	2.87 ± 0.03
C10:1	0.27 ± 0.00
C12:0	3.74 ± 0.01
C13:0	0.09 ± 0.00
C12:1	0.15 ± 0.01
C13:0	0.10 ± 0.01
C14:0	12.11 ± 0.06
C15:0	2.45 ± 0.01
C14:1	0.93 ± 0.00
C16:0	29.06 ± 0.05
C17:0	0.80 ± 0.03
C16:1 (7c)	0.25 ± 0.00
C16:1 (9c)	1.75 ± 0.01
C17:0	0.64 ± 0.01
C17:1 (9c)	0.09 ± 0.00
C17:1 (11c)	0.22 ± 0.01
C18:0	11.36 ± 0.04
C18:1 (9 t)	0.31 ± 0.00
C18:1 (10-11 t)	3.77 ± 0.01
C18:1 (12-14 t + 6-8c)	0.29 ± 0.01
C18:1 (9c + 15 t + 6-10c)	18.07 ± 0.06
18:1 (11c)	0.44 ± 0.02
C18:1 (12c)	0.14 ± 0.00
18:1 (13c)	0.07 ± 0.00
18:1 (16 t + 14c)	0.63 ± 0.01
C18:1 (15C)	0.28 ± 0.00
18:2 (9 t,12 t)	0.41 ± 0.01
C18:2 ω6 (11 t, 15c)	0.49 ± 0.01
C18:2 ω6 (total c,c)	1.12 ± 0.00
C20:0	0.18 ± 0.00
C18:3 ω3 (c,c,c)	0.64 ± 0.00
CLA	1.37 ± 0.01

c cis conformation, *t trans* conformation, *ω* omega, *CLA* conjugated linoleic acid.

species have been detected in MF; however, only 13about 12 to 15 of these are present at a concentration of 1% or more of the total profile (Bourlieu *et al.* 2009; Lindmark Månsson 2008), and the number of FAs detected during analysis will depend on the sensitivity, separation, and resolution of the methods used. Differences in relative abundance have been noted for the total FA profile of MF extracted from small and large MFGs (Mesilati-Stahy *et al.* 2011; Walter *et al.* 2020), while other studies have observed differences in the distribution of FAs between the PL and neutral fractions (Ménard *et al.* 2010).

Minor FAs of analytical interest include the odd chain C15:0 and C17:0 and their branched isomers. These FAs cannot be synthesised in the human body and are specific to ruminant fat sources. As such, levels in human adipose tissue and serum are often used as biological markers for the level of MF consumption (German and Dillard 2006). A higher content of C15:0 and C17:0 in human blood serum has been associated with a lower risk of a first myocardial infarction (Warensjö *et al.* 2010) and reduced levels of type 2 diabetes (Kröger *et al.* 2010). A study has shown C17:0 to be associated with increased insulin sensitivity in overweight subjects with metabolic syndrome and may reduce risks associated with type 2 diabetes (Nestel *et al.* 2014). Moreover, other studies have shown increased levels of C16:1 *trans*-7 to be inversely associated with insulin resistance and reduced rates of type 2 diabetes (Mozaffarian *et al.* 2010).

Fatty acid analysis can also provide useful information around a cow's desaturation activity, where a saturated FA is converted to an unsaturated form through the inclusion of a double bond along the acyl chain in the *cis*-9 position. This is achieved through a process of enzymatic desaturation catalysed by stearoyl-CoA desaturase or Δ-9-desaturase, found in the mammary gland and believed to play an important role in maintaining membrane fluidity through FA regulation (Chilliard *et al.* 2000). The level of desaturation activity can be indirectly measured using a desaturation index, calculated as the ratio between the relative abundance of the reaction product (unsaturated FA) and its substrate (saturated FA), demonstrated by Couvreur *et al.* (2007), or the ratio of the product and the sum of the product and substrate, demonstrated by Walter *et al.* (2020). Both authors measured higher desaturase indices in cows producing smaller MFGs compared to those producing larger MFG. This correlated with a higher concentration of unsaturated FAs in the small MFG milk, despite a similar diet fed to all cows involved in the study. In other studies, effects of season (Heck *et al.* 2009), breed, and other physiological factors such as days-in-milk and parity (Kelsey *et al.* 2003) have been shown to influence the desaturation index.

13.3.1 Fatty Acid Derivatisation

During sample preparation, FAs may be cleaved from the lipid molecule and chemically altered into a non-reactive derivatised form that will decrease their boiling point (Eder 1995), thereby increasing analyte volatility and enhancing the resolution and sensitivity of the GC analysis. The most commonly used approach is to form FA methyl esters (FAMEs) or ethyl derivatives, though some studies choose to form heavier ester derivatives, for example, butyl esters using butanol as the reactant (Richards *et al.* 2004). Wolff *et al.* (1995) converted FAs to isopropyl esters specific for the analysis of short-chain FAs. Liu *et al.* (2018b) notes that the analysis of short-chain FA may be underestimated in some studies due to the high volatility and potential evaporation of methyl esters of short-chain FA. Using a longer-chain alkyl derivative may reduce this loss, as shown by Sasaki *et al.* (2015) who found the recovery of short-chain FAs can be improved when using either propyl or butyl esterification. There are two main chemical routes used to form FA ester derivatives, either acidic or alkaline catalysis:

Acid-Catalysed Esterification/Trans-esterification The free FAs (FFAs) already present within the lipid phase are esterified, while the cleaved FAs undergo a process of trans-esterification. This reaction is performed under a large excess of methanol solvent in the presence of an acidic catalyst and heat. Methanolic hydrochloride and boron trifluoride in methanol are widely accepted acid catalysts, though boron trifluoride has been associated with artefact formation and may also be unsuitable for the derivatisation of FA with differing functional groups, such as conjugated FAs (Shantha and Napolitano 1992).

Alkaline-Catalysed Transesterification The cleaved FAs are transesterified rapidly in the presence of methanol solvent and a basic catalyst. Under these conditions, FFAs originally present within the lipid phase are not esterified. This may influence results, particularly for systems with high levels of FFA. More common reagents include sodium or potassium methoxide or hydroxide. This approach is often used to prepare methyl esters from MF (Ménard *et al.* 2010; Wijesundera *et al.* 2003; Zychowski *et al.* 2018); however, it will exclude analysis of sphingolipid FAs as they are attached through an amide bond and will not undergo derivatisation (Christie 1982). Other approaches for FA derivatisation include diazomethane methylation (Salvatore *et al.* 2018); however, diazomethane is known to be toxic, corrosive, and a carcinogen (Brondz 2002) and so is not as commonly employed.

13.3.2 Gas Chromatography

Analytical methods for the analysis of MF FAs are well established, most commonly using gas chromatography (GC) flame ionisation detection (FID) due to the high sensitivity and resolution offered. Gas chromatography works on the principle of partitioning, where a compound is distributed between a carrier gas mobile phase and a stationary phase. Depending on the composition and polarity of the stationary phase, the boiling point of the FA derivatives will impact the degree of partitioning and to what degree each component will be retained in the stationary phase. This in turn is influenced by the GC temperature program. For example, Gómez-Cortés *et al.* (2017) examined the influence of start temperature (150–180 °C), accompanied by a 10 °C min^{-1} ramp to 210 °C, on the resolution of branched chain FA derivatives in butter samples using a highly polar SLB-IL111 capillary column. Nowadays, analysis is predominantly performed using fused silica capillary columns, though limited studies have continued to utilise packed columns for FA determination (Chowdhury *et al.* 2007; Singh and Agrawal 2018). The carrier gas is usually hydrogen or helium, and the stationary phase is commonly a non-volatile liquid coating on the capillary column wall. It is desirable that the liquid phase be a thin film in order to increase the surface area accessible to the carrier gas and compounds injected into the column. A detector measures the amount of each FA eluting from the column as a function of time (the retention time) and represents the output as a series of peaks, with the peak height, width, and area used to define the relative abundance of analyte(s) for that given retention time. A sample of mixed FAs will separate as the individual FAs move through the column at differing rates and elute at different retention times, though it is worth noting that co-elution can occur despite the best of efforts to optimise the GC protocol, dependent largely on the column selection and the GC temperature program.

13.3.2.1 Sample Injection

The derivatised FAs are introduced to the GC column in a liquid form by injection into a heated inlet port (usually >250 °C) and immediately volatilised. Most modern GCs provide automated injection systems. This allows for fast and reproducible loading of sample into the inlet, thereby decreasing the potential for bleed and resulting in sharper peaks upon elution. A split injection protocol is often applied, using a split line to divert a small portion of the sample into the capillary column while the remaining sample is vented into the atmosphere. This process is used to facilitate reproducible sample injection decreasing the concentration of analyte introduced to the column, preventing peak overload and increasing peak resolution. Typical split ratios can vary from around 50:1 to 100:1. Splitless injection, on the other hand, can be used when there is a need to inject a larger sample aliquot into the column, for example, during the analysis of trace or minor components or when analytes elute near the tail of the solvent peak. Under these conditions, a lower injector port temperature (<230 °C) and lower boiling point solvents (25 °C below that of the first analyte of interest) are used (Yang *et al.* 1978). This allows the volatilised sample to condense upon entry into the column and to travel along the column as a narrow band within the solvent, leading to sharper peaks and better separation. A third delivery technique referred to as on-column injection is often used for samples that decompose when exposed to temperatures above their boiling point; this technique involves the direct injection of liquid samples directly into the column, bypassing the inlet port.

13.3.2.2 Gas Chromatography Columns

Capillary columns are most commonly used for the analysis of a FA profile using GC analysis. There are a number of different columns suitable for the quantification of MF FAs, ranging in size from around 25 up to 100 m in length. The 100 m SP-2560 column is a high-polar biscyanopropyl capillary column designed specifically for the resolution of *cis/trans* FA isomers. Figure 13.1 presents a partial chromatogram of the separation for the 18:1 and 18:2 *cis/trans* isomer region in MF. The limitation of this column choice is the lengthy analysis time for each sample. If *cis/trans* separation is not a key focus, shorter columns such as the 25–60 m BPX70 columns loaded with 70% cyanopropyl stationary phase can provide adequate resolution for most peaks with a shorter run time. In addition to column length, the column diameter and film thickness are also important aspects to consider as they dictate the partitioning rate of the sample analytes between the gaseous mobile phase and the stationary phase; high efficiencies can be achieved with a narrow column and thin film. Kramer *et al.* (2002) compared the efficacy of the 100 m CP Sil 88 cyanopropyl column with the 60 m SUPELCOWAX 10 polyethylene glycol column. Advantages of the shorter column were enhanced resolution of the shorter-chain FA region and monounsaturated FA from the saturated FA of a similar acyl chain length and the separation of C18:3 from the C20:1 peak which can often co-elute. However, the level of resolution for the *trans* isomer region was noted to be inadequate for the qualification of major milk fat FAs.

Table 13.2 presents a summary of the column type and split ratio used recently in the literature, indicating the number of FA peaks identified where information has been provided.

13.3.2.3 Analyte Detection

There are a number of different detectors that may be coupled with GC analysis for the determination of non-organic and organic analytes. The thermal conductivity detector is based on the transfer of heat (Horrocks *et al.* 1961) and was used extensively as a non-specific detector for the analysis of organic compounds such as FAMEs. However, this has been superseded by the use of the flame ionisation detector (FID), based on the detection of ions formed during the combustion of organic compounds in a hydrogen flame. Analytes elute from the end of the GC column into the FID where they are mixed with hydrogen (the fuel) and oxygen (the oxidant) and passed into the flame. The sample burns to create carbon atoms, creating a positive bias voltage and electrical current that is measured by the detector. The FID remains a popular detection system due to its nearly universal and linear response for organic compounds of differing concentration and low detection limit (Ojanperä and Rasanen 2008).

Mass spectrometry (MS) can also be coupled with GC analysis and is particularly suitable for structural characterisation. In GC-MS, the analyte exits the column to enter a chamber containing an ion source and becomes charged as a result of passing through an electron beam. This is known as electron ionisation (EI) and causes a high degree of analyte fragmentation and considered a hard ionisation technique. The parent and fragment ions are separated according to their mass-to-charge ratio, and the resulting mass spectrum, similar to a fingerprint, is compared to a mass spectral library for identification.

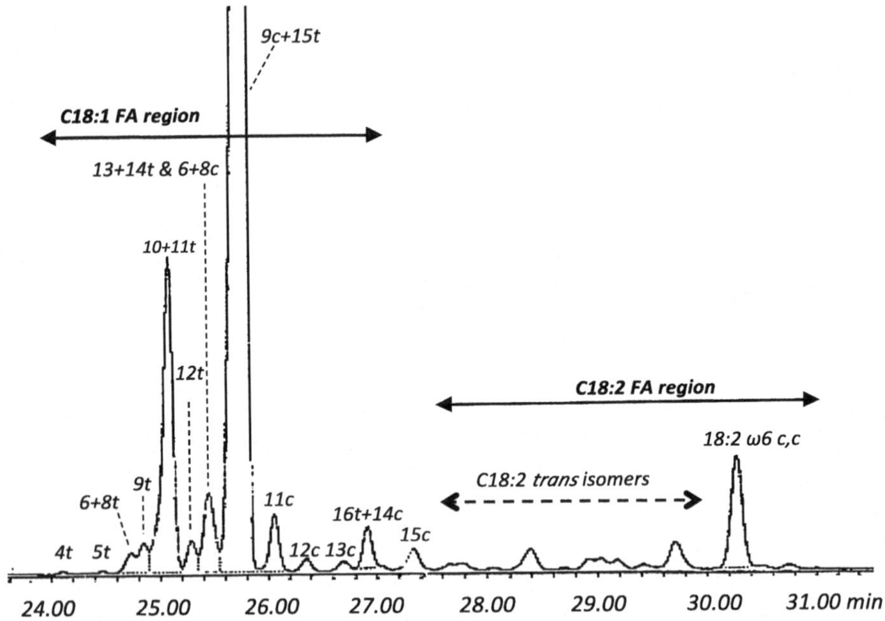

Figure 13.1. Part chromatogram of milk fat fatty acid methyl esters using a Perkin Elmer Model Autosystem XL Gas Chromatograph equipped with an SP-2560 (Supelco, USA) capillary column (100 m × 0.22 mm internal diameter × 0.25 µm film thickness). Temperature profile = 180 °C for 50 min followed by an increase to 200 °C at a rate of 20 °C min^{-1} where the temperature was held for 10 min. Helium was used as the carrier gas at a rate of 1.0 mL min^{-1}. The injection volume was 1 µL and analysis operated with a split ratio of 50:1.

13.3.3 Other Approaches

13.3.3.1 High-Performance Liquid Chromatography

High-performance liquid chromatography (HPLC, see Section 13.4.1) can be applied for the analysis of selected FAs. For example, partial profiles of dominant FAs have been measured in milk by Lu *et al.* (2002) using a fluorogenic FA derivative and a fluorescence detector and by La Nasa *et al.* (2018) using a hydrazide FA derivative and a ESI quadrupole (Q) time-of-flight (TOF) detector. It is possible to measure a number of different components in a single HPLC run, where the order of separation is according to the polarity; saturated FAs and unsaturated FAs will elute first, followed by TAG, FFAs, DAG, and MAG (Brondz 2002). However, this approach does not allow for the detailed analysis of isomeric and minor FAs.

13.3.3.2 Silver Ion Chromatography

Even using a 100 m capillary column, the *cis/trans* isomer regions of MF cannot be completely resolved. Separation may be improved by combining GC with silver ion thin layer chromatography (TLC) or liquid chromatography (LC) pre-fractionation steps, reviewed by Ratnayake (2004) and Delmonte *et al.* (2009). The TLC method uses a silica plate coated with silver nitrate and has been applied to the separation of C18:2 and C18:3 *trans* isomers (Kramer *et al.* 2002) and 18:1 *trans* isomers of MF FAMEs (Cruz-Hernandez *et al.* 2007) into bands of sub-lipid classes based on the degree of unsaturation, followed by the acyl chain length within each subclass. The authors used 2′7′-dichlorofluorescein to stain and visualise the bands under UV light, which were scraped off and extracted into hexane for GC analysis. Others have used the silver ion TLC approach to enrich the concentration of CLA isomers (Christie *et al.* 2001). Separation of the four positional CLA isomers can be achieved using silver ion LC analysis, discussed recently by Holčapek and Lísa (2017), where unsaturated FAs form weak reversible complexes with the silver ions bound to the stationary phase

Table 13.2. Examples of conditions for gas chromatography analysis of milk fat fatty acids

Reference	Column[a]	Split ratio	No. of quantified peaks
Benbrook *et al.* (2018)	SP-2560 (100 × 0.25 × 0.20)	NS	37
Cruz-Hernandez *et al.* (2007)	CP-Sil 88 (100 × 0.25 × 0.20) + silver ion separation	Splitless	57
Delmonte *et al.* (2012)	SLB-IL1111 (2 × 100 × 0.25 × 0.20) + silver ion HPLC separation	100:1/50:1	[b]
Heck *et al.* (2009)	Varian fame select (100 × 0.25 × NS)	NS	34
Kelsey *et al.* (2003)	SP-2560 (100 × 0.25 × 0.20)	100:1	24
Kramer *et al.* (2002)	SUPELCOWAX (60 × 0.25 × 0.20) and CP Sil 88 (100 × 0.225 × 0.20)	Splitless	[b]
Lopez *et al.* (2008, 2014)	BPX70 2x(50 × 0.32 × 0.25)	NS	38/42
Mannion *et al.* (2016a)	CP FFAP CB (30 × 0.25 × 0.32)	20:1	[b]
Mesilati-Stahy *et al.* (2011)	DB-23 (60 × 0.25 × NS)	NS	[b]
Moate *et al.* (2013)	HP INNOWAX (60 × 0.25 × 0.50)	NS	40
Ran-Ressler *et al.* (2011)	BPX70 (25 × 0.22 × 0.25)	Splitless	[b]
Soyeurt *et al.* (2011)	CP Sil-88 (100 × 0.25 × 0.20)	On-column injection	18
Straarup *et al.* (2006)	SP-2380 (60 × 0.25 × 20)	11:1	[b]
Stoop *et al.* (2009)	CP-7420 (100 × 0.25 × NS)	NS	32
Schennink et al. (2007)	Varian fame select (100 × 0.25 × NS)	NS	>30[b]
Walter *et al.* (2020)	BPX70 (30 × 0.25 × 0.25)	50:1	35
Yang *et al.* (2017)	CP Sil-88 (100 × 0.25 × 0.20)	100:1	24
Zou *et al.* (2013)	SLB-IL100 (60 × 0.25 × 0.20)	NS	31
Zychowski *et al.* (2018)	SP-2560 (100 × 0.25 × 0.20)	50:1	36

NS not stated.
[a]Capillary column nomenclature: (length (m) x inner diameter (mm) x film thickness (μm)).
[b]Full profile not presented in paper.

(Turner *et al.* 2011). In this case, the LC and GC results are interpreted in combination to derive a complete FA profile.

13.3.3.3 Supercritical Fluid Chromatography

Less common than GC analysis, supercritical fluid chromatography (SCF) has been reviewed by Senorans and Ibanez (2002) for the determination of FAs and more recently by Laboureur *et al.* (2015) from a lipidomics perspective. This approach involves the use of carbon dioxide as the mobile phase, selected due to a low critical pressure (7.377 MPa) and temperature (304.13 K/30.98 °C) point (Span and Wagner 1996). Beyond this point, the latent heat of vaporisation is 0, and the distinction between the liquid and gas form can no longer be made (Laboureur *et al.* 2015), and the system is a supercritical fluid. The density of a supercritical fluid is similar to a liquid and can act as an efficient solvent when compressed under pressure (Nováková *et al.* 2014). Moreover, polarity can be modified by the addition of a co-solvent into the mobile phase. However, the polarity of a supercritical fluid is heavily influence by density, so that changes in pressure or temperature will affect the analyte retention time. Qu *et al.* (2015) successfully separated the primary FAs of a series of vegetable oils using SCF couple with ESI-MS detection. As separation can occur at lower temperatures than GC, this approach may provide an advantage over GC analysis for heat-sensitive analytes such as polyunsaturated FAs and highly volatile FFAs. In addition to MS detection, SCF is commonly coupled with ultraviolet-visible (UV-vis, discussed in Section 13.7.2.2) and FID detection.

13.3.3.4 Nuclear Magnetic Resonance

Maher and Rochfort (2014) have reviewed the application of NMR to characterise the fat and water components of dairy-based products such as cheese, the solid fat content of MF, and a series of metabolomics products from milk and milk products. Some groups have examined the potential for using ^{1}H NMR for partial FA analysis. Siciliano *et al.* (2013) identified methylated FAs (C18:3, C18:2, C18:1, C18:0) from a pork fat extract and used this information to assess the saturated to unsaturated FA ratio, validated using GC-MS. This approach has also been used to measure the total concentration of CLA in a beef fat extract, though this approach is unable to distinguish between the individual isomers (Prema *et al.* 2015).

13.3.4 Free Fatty Acid Analysis

The accumulation of FFAs can be used to measure the degree of hydrolytic degradation of MF FAs, and the presence of FFAs significantly contributes to the aroma profile of dairy products. Analytical methods to extract and measure FFAs, with an emphasis on GC, have been recently reviewed by Mannion *et al.* (2016b). These authors note the varying solubility of FFA of shorter and longer acyl chain length in MF, so that extraction is more complex compared to that of vegetable oil. For MF, the solvent should ensure that FFA partitioning into both the aqueous and lipid phase is extracted. Moreover, a method that does not require solvent evaporation is preferable, so as not to lose more volatile shorter-chain FFA from the sample matrix (Kilcawley and Mannion 2017). Mannion *et al.* (2018) described a solid-phase extraction approach to separate FFA from MF extracted using heptane and diethyl ether from milk and dairy products, including cheese, at varying degrees of lipolytic degradation. The FFA fraction was converted into butyl ester derivatives for GC FID analysis. This work was built upon an earlier study by Mannion *et al.* (2016a) that involved the derivatisation of FAs using tetramethylammonium hydroxide. Limitations of this method were the co-elution of C4:0 and C6:0 with the solvent front, particularly when in low concentrations. The authors also noted degradation of polyunsaturated FA using this method, believed to be due to the harsher conditions of the tetramethylammonium hydroxide catalyst. Other approaches include ultra-high-performance SCF-MS (Ashraf-Khorassani *et al.* 2015; Qu *et al.* 2015) which can avoid exposure to higher temperatures and is therefore more suitable for heat-sensitive FFAs and HPLC-tandem MS (Li *et al.* 2011).

13.4 Analysis of Lipid Class and Species

Triacylglycerols are the most abundant class of lipids found in plants and animals and make up more than 98% (w/w) of the total MF content. Each TAG is composed of three FAs joined at the carbonyl end to a glycerol backbone. A different chemical environment is experienced by each FA attached to the glycerol structure and numbered (the stereospecific number, *sn*) to differentiate between the *sn*-1, *sn*-2, and *sn*-3 positions (Figure 13.2). The term regiospecific is also used to define the molecular structure. In this case, the chirality of the lipid species is disregarded and the *sn*-1 and -3 positions are considered equal.

The arrangement of FAs can influence the melting profile and solid fat content at a given temperature. For example, a model TAG containing C16:0 in the *sn*-1 and *sn*-3 position and C18:2 in the *sn*-2 position exhibits a melting range of 34–36 °C. In comparison, a TAG of the same FA make-up but with C18:2 in a terminal position decreases the melting range to 25–28 °C

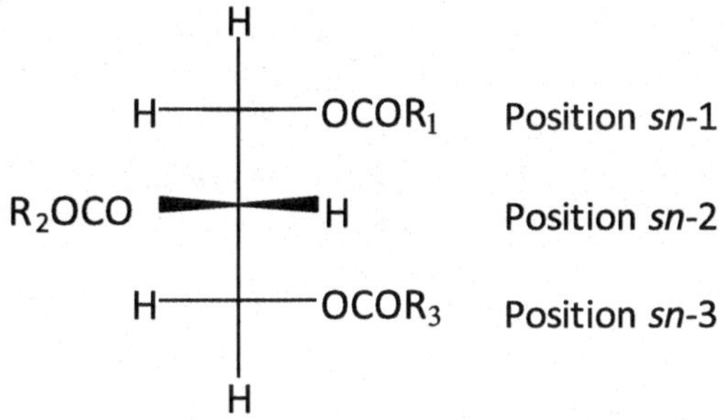

Figure 13.2. The stereospecific triacylglycerol structure.

(Fraser *et al.* 2007). The FA stereospecific position can also affect the nutritional value upon digestion, where human pancreatic lipases selectively liberate FAs attached to the *sn*-1 and *sn*-3 TAG position, leaving a 2-MAG structure to be conserved and absorbed into the body. It is therefore preferential for beneficial FA to be located in the *sn*-2 position. An exception to this rule is human milk substitutes where it is necessary to concentrate the saturated FA C16:0 at the *sn*-2 position to mimic human milk (Straarup *et al.* 2006). Infant formula prepared using palm olein to replicate the saturated fat component of human milk led to a decrease in calcium absorption and a decrease in bone mineral content and bone mass (Koo *et al.* 2006). This is due to the stereospecific structure of palm oil, where the greater majority of C16:0 (~49% of total FA profile) is bound to the terminal positions (Richards *et al.* 2004). The reductions occur when dietary calcium becomes bound into insoluble calcium soaps with cleaved C16:0 and were excreted from the body.

The MAG and DAG lipid classes are structured with one and two FAs attached to the glycerol backbone, respectively. The position of the MAG and DAG FAs can be unstable which undergo acyl migration along the glycerol molecule to neighbouring positions catalysed by heat, acid, and alkali (Mattson and Volpenhein 1962). Monoacylglycerols have been shown to reach an equilibrium of 12% and 88%, respectively, between the 2-MAG and 1-MAG *sn*-position, compared to the comparatively stable equilibrium measured for DAG between the 1(3),2-DAG (42%) and 1,3-DAG (58%) lipid species (Mattson and Volpenhein 1962). Common techniques used for the analysis of lipid class and species in MF are HPLC-MS and GC-MS, reviewed earlier by Andrikopoulos (2002).

13.4.1 High Performance Liquid Chromatography

High performance liquid chromatography is a form of liquid chromatography (LC) where a liquid sample is introduced under pressure into a packed column within a liquid solvent mobile phase. An LC or HPLC column typically consists of spherical particles of silica covalently bonded with either a polar or non-polar stationary phase. The analytes will adsorb to the stationary phase, and the analytical separation and peak resolution governed by the strength of the mobile phase to dissociate the analyte from the stationary phase through competitive adsorption, dictated by the difference in polarity between the mobile and stationary phase. Other factors that will influence peak resolution include the flow rate of the mobile phase through the column, the size of the silica particles and the column length, which typically ranges from 4 to 25 cm in length. Often, a gradient in mobile-phase polarity is employed to increase the dissociation strength, allowing for analytes that adsorb more strongly to the stationary phase to be eluted from the column. This approach can differentiate lipid species within a particular class on the basis of CN and may be able to differentiate between lipid species. The effect of solvent choice for the mobile phase on resolution and detection of compounds with varying polarity has been discussed by Andrikopoulos (2002).

13.4.1.1 Sample Preparation

Silver ion TLC can be used to separate the MF lipids on the basis of class, demonstrated in the work of Robinson and MacGibbon (1998), to separate TAG components from MF. Solid-phase extraction has also been used to separate the neutral and polar lipid fractions from processed milk products (Gallier *et al.* 2010) and to separate MAG and DAG fractions from the TAG components of milk and buttermilk serum (Fagan *et al.* 2004). Derivatisation is not necessary for HPLC analysis; however, some studies convert lipid species to an ammonia adduct for analysis in ESI or APCI 'positive ion' mode. This has been achieved by using ammonium as the nebuliser gas (Linderborg *et al.* 2014) or the addition of ammonium acetate into the mobile phase immediately prior to the ionisation chamber (Berger *et al.* 1999).

13.4.1.2 Sample Injection

The liquid chromatograph is fitted with an injector that consists of a sample loop of a defined volume, typically ranging from 10 μL up to 1 mL. The samples are loaded into a closed loop and injected into the mobile phase by opening a valve. A guard column is often installed immediately before the primary analytical column, consisting of the same stationary phase, and used to capture impurities before they enter the column.

13.4.1.3 Liquid Chromatography Columns

Normal-phase (NP) LC consists of a combination of a polar stationary phase with a less polar mobile phase. Reverse-phase (RP) LC, on the other hand, consists of a non-polar or less polar stationary phase with a more polar mobile phase and is the most commonly used column for the quantification of lipid species in MF. For example, Nagai *et al.* (2015) used LC-ESI-MS/MS equipped with an RP C28 column to separate TAG on the basis of CN and to examine the stereospecific lipid species of MF. This was achieved by converting the precursor ions to ammonium adducts and detecting the precursor and product ion in 'selected reaction monitoring' mode. However, NP columns have proven effective for the preparative separation of different lipid classes, for example, the separation of TAG from FFAs, DAG, and MAG (Andrikopoulos 2002).

13.4.1.4 Analyte Detection

Traditional techniques involved the detection and measurement of lipid class using RP HPLC, where the lipid mixture is separated into fractions and characterised further using GC (Barron *et al.* 1990) or by using a light scattering detector (Holčapek *et al.* 2003; Ruiz-Sala *et al.* 1996). More recently, MS detection has become more common since the technique is most suited to the characterisation of an analytes structure without the need for authentic reference standards (Holčapek *et al.* 2003).

Evaporative Light Scattering Detection For evaporative light scattering detection (ELSD), the eluent exits from the HPLC and is mixed with an inert carrier gas that is passed into a nebuliser to create an aerosol. The aerosol continues through a heated tube, where the mobile phase evaporates resulting in a dried and particulate analyte. The particles are pushed through the tube into the detection zone where the particle interrupts a light beam and the degree of light scattering is measured. As such, ELSD is not limited by solvent flow rate, allowing for faster run rates while maintaining good analytical sensitivity (Rodríguez-Alcalá and Fontecha 2010). The output is usually non-linear and suitable calibrations are needed for quantitative analysis. This form of detection has been used for the analysis of lipid class (Rodríguez-Alcalá and Fontecha 2010) however is more frequently utilised for the quantification of PLs, discussed further in Section 13.4.1.

Mass Spectrometry Nowadays, the most common approach for lipid analysis involves the coupling of LC and HPLC with the growing list of MS detection techniques. Atmospheric pressure chemical ionisation (APCI), matrix-assisted laser desorption and ionisation (MALDI), and electrospray ionisation (ESI) are considered to be soft ionisation approaches that are useful for the determination of lipid class coupled with MS. Gaining in popularity, APCI-MS has been noted to allow for the separation of complex TAG mixtures and the identification of partially or non-resolved peaks and, along with MS/MS, can be used to determine the molecular structure of individual lipid species (Jakab *et al.* 2003; Liu *et al.* 2018b; Mottram and Evershed 2001). As the name suggests, APCI involves the ionisation of analyte molecules under atmospheric pressure conditions. Gastaldi *et al.* (2011) used an RP HPLC system equipped with a C18 column and coupled to an APCI-MS/MS detector to quantify the TAG species present within bovine and other mammalian milk sources, without prior separation. A total of 53 different TAG species were identified in bovine MF, ranging from CN24 (C4:0-C4:0-C16:0) to CN54 (C18:1-C18:0-C18:0) with C4:0-C16:0-C16:0 and C4:0-C16:0-C18:0 the most abundant. A similar approach has been used to quantify levels of C16:0-C4:0-C16:0 and C16:0-C16:0-C4:0 as a measure for MF content in foods (Yoshinaga *et al.* 2013). In contrast, the MALDI approach involves the ionisation of analyte molecules within a laser energy absorbing matrix, leading to the vaporisation of the analyte molecules into cation adducts (Shinn *et al.* 2015). Time-of-flight (TOF) is a form of MS where the mass-to-charge ratio is measured by the time taken for the ionised molecules to travel to the detector through a flight tube, with lower-molecular-weight ions arriving at the detector earlier than the heavier ones. Picariello *et al.* (2007) used MALDI-TOF-MS in combination to characterise the TAG species of a range of different animal fat sources. The matrix solution consisted of crystalline 2,5-dihydroxybenzoic acid in methanol containing 0.1% trifluoroacetic acid. However, MALDI may not be suitable for the analysis of higher CN TAG, due to a decrease in the signal intensity at higher CN (Asbury *et al.* 1999). Picariello *et al.* (2007) determined that hydrogenation prior to analysis may increase the signal intensity, allowing lesser abundant TAG of higher CN number to be characterised. A study by Zhou *et al.* (2014) employed ultra-performance convergence chromatography, a variant of HPLC, with a mobile phase of a 1:1 CO_2/acetonitrile-ethanol, coupled with Q-TOF-MS, to qualify 49 different TAG and 7 different DAG in MF. Analysis with MALDI has also proven useful for the authentication of bovine milk against other mammalian milk sources used for cheese products by identifying lipid species markers (Damário *et al.* 2015). This approach used a treated and porous glass plate to effectively trap lipid components from the cheese samples, in the absence of a matrix, which were subsequently embedded into a MALDI plate for analysis. Electrospray ionisation is probably the most commonly used approach for MS analysis. However, as in the case for MALDI, matrix effects can lead to ion suppression which presents one of the major drawbacks of this technique for the analysis of multiple lipid species.

MS analysers can be fitted with an ion trap (IT) that uses an electric or magnetic field to effective 'trap' ions for analysis with higher sensitivity or storage. Variants include an electrostatic trap, or 'Orbitrap', which consists of an inner spindle and outer barrel electrode that can trap ions in an orbital motion (Clarke, 2017). However, this approach is most commonly applied to the analysis of PLs, which will be discussed in Section 13.4.1. Li *et al.* (2017) recently used a Q-extractive-Orbitrap-MS approach to analyse TAG and DAG lipid species and the polar fraction of a series of mammalian milks, utilising the positive ion and negative ion mode functionality.

Table 13.3 presents a summary of columns, conditions, and detectors used recently in the literature to quantify the different lipid species in MF and dairy products.

13.4.2 Gas Chromatography

While less common, GC methods have been applied to the analysis of MF lipid species which separate according to their CN, reviewed by Andrikopoulos (2002). An advantage of GC over HPLC is better separation of lower-molecular-weight TAG and higher-molecular-weight DAG, which may otherwise overlap. This has proven useful in the characterisation of MAG and DAG as TMS esters using a DB-17ht fused silica capillary column (Fagan *et al.* 2004). This approach has also been used to identify adulterants in MF. For example, Kim *et al.* (2016) identified C52 and C54 as markers for adulteration with other animal fats using an HT-5 aluminium clad capillary, and Tolentino *et al.* (2015) used a HP5 column to characterise TAG in bovine and cap-

Table 13.3. Examples of conditions for liquid chromatography analysis of milk fat lipid species

Reference	Column[a]	Detector	Mobile-phase eluents
Argov-Argaman *et al.* (2014)	Zorbax RX-SIL (4.6 × 250 × 5)	ELSD	NP, solvent – hex:IPA:H_2O:Conc. H_2SO_4 A-97:3:0:0.025, B-75:24:0.9:0.1 Flow rate = 1.5 mL min^{-1} Isocratic 4%B (2 min) linear to 100%B (50 min)
Beccaria *et al.* (2014)	Ascentis express fused-core C18 (150 × 4.6 × 2.7)	IT-TOF-MS	RP, A – ACN, B IPA Flow rate was 1 mL min−1 100% A (0 min) to 30% A (50 min), hold for 5 min
Kallio *et al.* (2017)	BEH C_{18} (100 × 2.1 × 1.7)	ESI MS, MS/MS	RP, A – MeOH/2 mM LiCl, B – IPA/2 mM LiCl Flow rate = 0.2 mL min^{-1}. 0% B (2 min) to 2% B (6 min), isocratic (6 min), linear to 3% B (12–20 min), isocratic for 16 min and linear to 40% B (36–70 min)
La Nasa *et al.* (2018)	Poroshell 120 EC-C18 (3.0 × 50 × 2.7) + Zorbax eclipse guard column	LC-ESI-Q-TOF-MS	RP, A – MeOH/H_2O (85:15) B – IPA Flow rate = 0.3 mL min^{-1}. Isocratic 90% A (5 min), linear to 90%B (30 min), isocratic 90% B (5 min)
Linderborg *et al.* (2014)	Kinetex C18 (100 × 2.1 × 1.7)	APCI–MS/MS, + NH_3 adducts	RP, A – Acet B – ACN Flow rate = 0.4 mL min^{-1} 100% B to 100% A (31 min)
Liu *et al.* (2015b)	Synergi (250 × 4.6 × 4)	LTQ-Orbitrap-MS & TripleQ-MS	RP, A – 5 mM NH_4Form B – ACN, 0.1% FA, IPA (1:1) Flow rate = 0.6 mL min^{-1} Gradient of 75 to 98% B over 25 min
Lu *et al.* (2016a)	Zorbax RX-SIL (4.6 × 250 × 4)	ELSD	NP, hex:IPA:H_2O:Conc. H_2SO_4 A-97:3:0:0.025, B-75:24:0.9:0.1 Flow rate = 1.5 mL min^{-1} Isocratic 4% B (2 min) linear to 100% B (50 min)
Perona and Ruiz-Gutierrez (2003)	Novapack (250 × 4.6 × 4)	ELSD	RP, A – Acet B – ACN Flow rate = 1.0 mL min^{-1} 20% A (0 min) to 45% A (12 min) to 80% A (60 min)
Řezanka *et al.* (2018)	RP Hichrom HIRPB −250A 2x(250 × 2.1 x 5)	APCI-MS	RP, A – ACN B – IPA Flow rate = 1.0 mL min^{-1} Isocratic 99% A (5 min) linear to 30% A (250 min) then isocratic (10 min)
Rodríguez-Alcalá *et al.* (2015)	Zorvax Rx-SIL (250 × 4.5 × 5)	ELSD	NP, complex system of four solvents; see reference for further details
Sun *et al.* (2018)	Lichrospher C18 (250 × 4.6 × 5)	ELSD, APCI-MS	Non-aqueous RP, A – ACN B - IPA Flow rate = 0.8 mL min^{-1} Isocratic 4 0% B (40 min), gradient to 45% B (5 min) and then isocratic (70 min)

ELSD evaporative light scattering detector, *IT* ion trap, *MS* mass spectrometry, *ESI* electrospray ionisation, *Q* quadrupole, *TOF* time-of-flight, *APCI* atmospheric pressure chemical ionisation, *LTQ* linear trap quadrupole, *RP* reverse phase, *NP* normal phase, *Hex* Hexane, *IPA* isopropyl alcohol, *ACN* acetonitrile, *MeOH* methanol, *Acet* acetone, *NH_4Form* ammonium formate, *FA* formic acid.

[a]Column type nomenclature: (length (mm) × inner diameter (mm) × stationary-phase particle size (μm)).

rine milk from different geographical locations within Mexico. Much higher temperatures are required to elute the heavier-molecular-weight components of TAG, DAG, and MAG in MF (~325–370 °C) compared with FAMEs, which will influence column selection dependent upon the maximum temperature limit. For this type of analysis, FID is the most commonly used detection method, considering that MS does not have sufficient range to detect the lipid species after fragmentation.

13.4.2.1 Regio- or Stereospecific Analysis of the Lipid Structure

Selective derivatisation and analysis of FAs cleaved from along the glycerol backbone have allowed for the regio- or stereospecific analysis of TAG by GC, providing useful information regarding the lipid structure. This has been achieved through enzymatic or chemical pathways:

Enzymatic Lipase enzymes are characterised by their ability to catalyse rapidly the hydrolysis of ester bonds during lipolysis. In the presence of excess alcohol, the reaction can be reversed to form esters for regiospecific GC analysis. Based on their positional specificity, lipases can be divided into two groups: non-specific enzymes that hydrolyse all FAs irrespective of position and those that interact with FAs only from the *sn*-1 and *sn*-3 positions. This is due to steric hindrance of the *sn*-2 position, preventing entry into the active site of the enzyme (Willis and Marangoni 2002). Most common lipases consist of an amphiphilic peptide loop that covers the active site when inactive. The enzyme may be activated in the presence of hydrophobic molecules, leading to a change in conformation, thereby uncovering the active site (Bourlieu *et al.* 2009). Long-chain polyunsaturated FAs (>18 carbons) such as C20:5 and C22:6 are noted to be resistant to pancreatic lipase hydrolysis due to the positioning of the double bond closest to the carboxyl group and steric hindrance of the highly unsaturated and kinked structure (Bottino *et al.* 1967). As such, enzymatic regiospecific analysis may not be a viable option for the determination of marine oils but remains a suitable option for the analysis of MF. While earlier studies were exclusively performed utilising pancreatic lipases (Parodi 1982), the extracellular lipase of many microorganisms has been found to match pancreatic lipases for positional specificity, reliability, and relatively low FA selectivity (Aravindan *et al.* 2007). However, the use of pancreatic lipases continues to be a popular approach (Qi *et al.* 2018).

Chemical Analysis that can distinguish between the *sn*-1 and *sn*-3 position of the TAG molecule is possible using a chemical deacetylation approach. Grignard reagent reacts with the FA ester linkage, resulting in the formation of a tertiary alcohol and either the 1,2(2,3)-DAG or 1,3-DAG in a ratio of around 2:1. Ethyl magnesium bromide is a commonly used Grignard reagent producing a tertiary alcohol that is easily separated from the DAG components during GC analysis. Straarup *et al.* (2006) used allyl magnesium bromide to analyse the regiospecific distribution of FAs in human milk and infant formula samples. The *sn*-2 MAG were subsequently separated from the Grignard mixture using TLC on a plate impregnated with boric acid, which were then methylated and analysed using GC. A similar approach was used by Zou *et al.* (2013) to compare the regiospecific FA distribution of human and bovine milk to a number of other mammalian species and Timm-Heinrich *et al.* (2004) for structured lipids.

13.4.3 Other Approaches

Silver ion HPLC has been discussed earlier in terms of FA analysis. This approach has also been applied to the analysis of lipid species and is particularly useful for the separation of regiospecific pairs of the same CN and FA composition (Andrikopoulos 2002).

Supercritical fluid chromatography can similarly be applied to the analysis of lipid species. Tu *et al.* (2017) utilised SFC coupled with Q-TOF-MS to examine the relative concentration of 50 different TAG species detected in infant formula prepared using a mixture of bovine MF and vegetable oil. Both TAG and DAG were measured, providing exact mass information for quasimolecular ions, as ammonia and sodium adducts, and fragment ions in the same run. This mode also allowed for the determination of regiospecific positioning using the fragment ion abundance, where fragment ions of FAs from the *sn*-2 position result in a lower relative abundance compared to the *sn*-1 and *sn*-3 positions. Supercritical fluid has also been coupled with APCI-MS to identify both saturated and unsaturated TAG lipid species in MF from CN26–54 (Laakso and Manninen 1997).

The TAG structure can also be explored by measuring differences in the chemical shift of the acyl chain of TAG FAs using high-resolution ^{13}C nuclear magnetic resonance (NMR). This approach has been used to distinguish between groupings of FA at the *sn*-1,3 and *sn*-2 positions in seed oils (Vlahov *et al.* 2002); however, the depth of information acquired is limited, since the differences between FAs of the same double bond group cannot be distinguished (Gunstone 1994).

13.5 Analysis of the Milk Fat Globule Membrane Material

While some PL compounds, such as DAG and MAG, are found in the lipid core in minor amounts, the larger portion of milk PLs are located within the trilayer membrane of the MFGM. These include phospholipids (glycerophospholipids), including phosphatidylethanolamine (PE), phosphatidylcholine (PC), phosphatidylserine (PS), and phosphatidylinositol (PI); sphingolipids, including sphingomyelin and ceramides; and other minor PL components including glycosphingolipids (Liu *et al.* 2015a). The total PL concentration in MF can account for up to around 1.0% (w/w) of the total and, correlating with the available surface area of the MFG, is found in higher amounts in milk containing smaller MFGs compared to larger MFGs (Logan *et al.* 2014). In addition to PLs, the MFGM consists of lipid-soluble sterols such as cholesterol, which are believed to influence the membrane fluidity (Ipsen *et al.* 1990), enzymes, and proteins.

13.5.1 Polar Lipids

Despite the low abundance of PLs compared to neutral lipid species in milk, PLs are crucial for the technological properties and nutritional value of milk. Equivalent to acylglycerols, phospholipids have a glycerol backbone connected to two FAs and a polar headgroup. Sphingolipids are made of a long-chain sphingoid base, mostly sphingosine, which contains an amine group. Ceramides are the direct precursors or hydrolysis products of sphingomyelin. Sphingomyelin, the major sphingolipid in milk, is formed when a phosphorylcholine headgroup is attached to a hydroxyl group of ceramide. Other minor sphingolipids are gangliosides and cerebrosides, which, despite their low abundance in milk, could have important health benefits (Ortega-Anaya and Jiménez-Flores 2019). The existence of the polar headgroups in phospholipids and sphingolipids results in their amphiphilic structure, responsible for their surface tension-reducing ability and thus emulsifying properties. In terms of their FAs profile, PLs contain almost exclusively long-chain FAs, in contrast to the large proportion of short- and medium-chain FAs found in bovine TAG. The degree of unsaturation depends on the PL class, with SM containing predominantly saturated FAs and PC containing more saturated FAs than PE, a highly unsaturated PL species (Gallier *et al.* 2010). Within the trilayer of the MFGM, PLs are asymmetrically distributed, with PC and SM found mainly in the outer leaflet of the bilayer and PE, PS, and PI predominantly in the inner leaflet of the bilayer and the monolayer surrounding the lipid droplet core (Deeth 1997). Glycolipids such as cerebrosides and gangliosides are also found in the inner leaflets (Deeth 1997). Evidence is mounting about various health benefits related to the PL fraction of milk. These benefits are particularly important for infant nutrition and include a role of SM in neurological development and protection from bacterial infection. Furthermore, similar roles in brain function and immune response have also been suggested for minor lipid species such as cerebrosides and gangliosides, recently reviewed by Ortega-Anaya and Jiménez-Flores (2019).

In milk and milk product, PLs have been studied extensively on the lipid class level. However, the identification of PLs on a species level is only starting to come into focus (Liu *et al.*, 2018b). Accordingly, distinct functions for individual PL species and their origin are rarely discussed, and many minor lipid species are yet to be discovered.

13.5.1.1 Sample Preparation

Since the majority of PLs present in whole milk are found in the MFGM, whole milk can directly be used for PL analysis. The use of whole milk samples is simple and efficient, and the milk can immediately be frozen after collection. Metabolic studies should generally aim to stop metabolic processes by reducing the temperature to a level were metabolites are stable and the resulting lipidomic profiles most closely represent the metabolic state of the animal or product at the point of sample collection.

However, a growing body of work is emerging about the presence of other membranous particles found in milk, which, although present at low concentrations, can contribute to the PL profile. These nanoparticles, called exosomes, are smaller in size than MFG (40–100 nm) and are surrounded by a bilayer lipid membrane (Zempleni *et al.* 2016). Furthermore, it has been suggested that mammary epithelial cells can shed membranous material into the milk serum (Deeth 1997). The use of raw cream instead of raw milk could minimise the impact of exosomes on MFGM analysis but has the disadvantage of excluding the very small fraction of MFGs that remain in the milk serum and can lead to a loss of MFGM material during centrifugation. Particularly, PL species found in the outer layer of the MFGM are predicted to be lost during centrifugation into the milk serum, as demonstrated by Gallier *et al.* (2010) where PC and SM concentrations were lower in the cream samples compared to in raw milk.

Most commonly, separation of MFGM membrane material from the neutral lipid core is achieved through manual churning. This can lead to highly concentrated MFGM extracts and improve the detection of low abundance lipid species. However, some MFGM material can be lost and remain in the serum phase as extraction proceeds. Minimal sample processing is also preferable. For example, Gallier *et al.* (2010) compared the phospholipid composition on a class level between freeze-dried raw milk, freeze-dried processed raw cream, and buttermilk powder samples and found the highest proportion of lysophospholipids in the buttermilk, likely due to disruption caused by pumping or churning (Contarini and Povolo 2013). The MFGM extracts contain high concentrations of neutral lipids mainly due to incomplete separation from the neutral lipid core. For example, Fong *et al.* (2007) reported that 56% of the lipids measured in the MFGM were neutral lipids compared to 41% PL. The presence of contaminating neutral lipids in MFGM extracts needs to be taken into consideration, especially when the abundance of the PL and protein fractions of the MFGM is investigated.

13.5.1.2 Polar Lipid Fractionation

Suitable methods for solvent extraction that are also applicable to PLs have been discussed earlier (Section 13.1.1). This solvent extraction step can be followed by a step to separate and concentrate PLs from the neutral lipid species, most frequently by solid-phase extraction (SPE) (Ali *et al.* 2017; Avalli and Contarini 2005; Donato *et al.* 2011; Gallier *et al.* 2010). Another less common approach is TLC (Fong *et al.* 2007), where bands are scrapped off and reconstituted into another organic solvent for further analysis. Both methods add complexity to the sample preparation and can result in reduced recovery rates of some lipid species (Trenerry *et al.* 2013). As such, sample fractionation prior to analysis is increasingly omitted.

13.5.1.3 Liquid Chromatography

Evaporative Light Scattering Detection High-performance liquid chromatography combined with an ELSD is still the most accurate and commonly used method to quantify PLs on a class level (Argov-Argaman *et al.* 2016; Contarini *et al.* 2017; Donato *et al.* 2011; Lopez *et al.* 2008;

Russo *et al.* 2013; Zou *et al.* 2013). Its main advantage is that, in contrast to MS detectors, the response of the ELSD is not influenced by the CN or number of double bonds, and quantification of PL class can therefore be achieved through the use of standards. Columns used for separation prior to detection are NP columns (Contarini *et al.* 2017; Zou *et al.* 2013) and increasingly hydrophilic interaction liquid chromatography (HILIC) columns (Donato *et al.* 2011; Russo *et al.* 2013). These methods were applied to milk samples from a variety of mammalian milk species and human milk (Russo *et al.* 2013; Zou *et al.* 2013) and several milk products (Rombaut *et al.* 2007). However, HPLC-ELSD is not suitable for lipid species identification. This major drawback, in combination with progress made in the development of protocols for quantification using MS-based methods, has resulted in the decline of this technique in recent years. Nonetheless, it is sometimes applied in parallel with HPLC-MS, where the lipid fractions are split into both an ELSD and MS detector (Contarini *et al.* 2017; Donato *et al.* 2011; Russo *et al.* 2013).

Mass Spectrometry Mass spectrometry detection can achieve the accurate identification of lipids on a species level. Hence, it has emerged as the primary method used both for the identification of PL species and for relative, and in some cases absolute, quantification, as recently reviewed by Liu *et al.* (2018b). Mass spectrometry can be used without prior separation (shotgun MS) or in combination with LC using various types of columns (NP, RP, or HILIC). Normal-phase columns are more suitable for PL class separation; however, RP columns are often used when studying neutral and PLs at the same time. Hydrophilic interaction liquid chromatography columns offer improved ESI efficiencies compared to both NP and RP columns and have become the preferred method for PL species identification and quantification (Liu *et al.* 2018b). Ionisation sources for MS suitable for PL analysis include ESI, APCI, and MALDI. The use of MALDI in combination with TOF-MS for the determination of PLs in raw milk and milk powder has also been reported (Walczak *et al.* 2016). However, although MALDI-TOF-MS is a sensitive technique, it is not suitable for accurate quantification and instead mostly used for fast screening of lipids (Köfeler *et al.* 2012). For example, Walczak *et al.* (2016) identified 34 different PL species in raw milk and 21 different PL species in milk powder, which is less than the numbers reported by others (Table 13.4). Typical mass spectrometers used for PL determination in milk and dairy products include Q-TOF-MS, IT-MS, triple quadrupole (QQQ)-MS, and linear trap quadrupole (LTQ)-Orbitrap-MS. Using these techniques, several authors have achieved the identification of PL species in milk (Table 13.4). However, a large number of PL lipid species remain unidentified. Furthermore, the number of species detected in each lipid class can vary depending on the applied analytical method. This is a reflection of the complexity of PL fraction, which cannot be captured as a whole by a single analytical platform.

Table 13.4 presents a summary of different detectors used recently in the literature to examine PL species of MF from a number of different mammalian sources, including the number of PL species qualified.

13.5.1.4 Other Approaches

^{31}P NMR can be used to differentiate between phosphorous-containing PLs, such as the phospholipids (Diehl 2015). While selective, it is noted to be highly sensitive and presents the advantage that reference standards are not required (Burri *et al.* 2016). Garcia *et al.* (2012) used this approach to quantify 12 different phosphorus-containing PL classes, including sphingomyelin, in bovine, human, and other mammalian milk sources.

13.5.2 Membranous Proteins

Milk fat globule membrane proteins are a diverse group of proteins which make up about 1–2% of total proteins in milk (Lopez 2011). However,

Table 13.4. Number of identified lipid species within the major and minor polar lipid classes in milk and milk products from different mammalian species

	Sample type	Detector	PE	PC	SM	PI	PS	LPC	LPE	Cer	LacCer	GluCer
Ali *et al.* (2017)	Bovine (buffalo and cow), cameline, and caprine milk	UPLC-ESI-Q-TOF-MS	15	22	9	8	8	11	–	–	–	–
Castro-Gómez *et al.* (2017) [a]	Buttermilk	UPLC-Q-TOF-MS	15	16	6	6	6	4	3	–	–	–
Contarini *et al.* (2017)	Equine (donkey) milk	HPLC-ESI-QQQ-MS	–	–	–	–	–	–	–	–	6	6
Donato *et al.* (2011)	Bovine and equine (donkey) milk	HILIC-ESI-IT-TOF-MS	3	8	2	4	3	2	–	–	–	–
Fong *et al.* (2007)	Bovine milk	LC-ESI-QQQ-MS	14	17	18	11	8	–	–	–	9	–
Li *et al.* (2017) [b]	Bovine, caprine, and soy milk	UPLC-Q-Orbitrap-MS	14	21	9	20	14	4	4	5	–	–
Liu *et al.* (2015a)	Bovine milk	HILIC-ESI-LTQ-Orbitrap-MS	7	13	17	6	8	5	–	–	10	4
Liu *et al.* (2017)	Bovine milk	HILIC-ESI-LTQ-Orbitrap-MS	6	11	11	6	8	5	–	–	8	3
Pinto *et al.* (2014) [c]	Buttermilk	MALDI-TOF-MS	8	18	13	3	10	7	3	–	–	–
Russo *et al.* (2013)	Bovine (raw and pasteurised), caprine, equine (donkey), and human milk	HILIC-ESI-IT-TOF-MS	8	14	1	11	3	3	–	–	–	–
Trenerry *et al.* (2013)	Bovine milk	UPLC-ESI-LTQ-Orbitrap-MS	13	18	13	7	–	–	–	–	–	–
Walczak *et al.* (2016)	Bovine milk	HPLC-MALDI-TOF-MS	7	17	5	6	4	7	–	–	–	–

UHPLC ultra-high-performance liquid chromatography, *Q* quadrupole, *MS* mass spectrometer, *HILIC* hydrophilic interaction liquid chromatography, *ESI* electrospray ionisation, *IT* ion trap, *TOF* time-of-flight, *MALDI* matrix-assisted laser desorption ionisation, *PE* phosphatidylethanolamine, *PC* phosphatidylcholine, *SM* sphingomyelin, *PI* phosphatidylinositol, *PS* phosphatidylserine, *LPC* lysophosphatidylcholine, *LPE* lysophosphatidylethanolamine, *Cer* ceramide, *LacCer* lactosylceramide, *GluCer* glucosylceramide.

[a]Ceramide species were only tentatively identified in the study.

[b]Seven phosphatidic acid and nine phosphatidylglycerol species also detected.

[c]Eight species of monosialoganglioside (GM3) also detected.

this minor fraction has important physiological functions. As for the PL fraction, the total MFGM protein content depends on the size of the MFGs, with smaller MFGs delivering more membrane material per unit of fat, and quantification is difficult due to contamination of the protein fraction with highly abundant serum proteins. For example, Fong *et al.* (2007) found that 5% of a MFGM protein fractions obtained through churning was contaminated by caseins from the serum phase.

The major MFGM proteins have been studied extensively, due to their suggested roles in MF globule secretion. The major MFGM proteins comprise xanthine oxidoreductase, butyrophilin, perilipin 2 (adipophilin), lactadherin (PAS 6/7), FA binding protein, and mucin 1 and mucin 15. These proteins, due to their high abundance within the MFGM fraction, can be detected through traditional gel electrophoresis (Le *et al.* 2009; Reinhardt and Lippolis 2008) and can be quantified using densitometry (Holzmüller *et al.* 2016; Le *et al.* 2009). However, the advent of proteomics has led to the identification of hundreds of low abundance proteins in MFGM extracts through more sensitive LC-MS/MS-based techniques (Affolter *et al.* 2010; Lu *et al.* 2011; Lu *et al.* 2016b). A less frequently applied technique to probe relative or absolute abundance of MFGM proteins is using an enzyme-linked immunosorbent assay (Mitrova *et al.* 2014). Within this chapter, we will focus on the specific considerations relevant to the analysis of membrane proteins, namely, their extraction and solubilisation. Following appropriate extraction, the analysis of membrane proteins is identical to that of serum proteins discussed in Chapter "Quantitiation of proteins in milk and milk products" of 'Volume 1A: Proteins' (Dupont *et al.* 2013).

13.5.2.1 Protein Extraction

The preparative considerations discussed earlier for the PL fraction of the MFGM (Section 13.5.1.1) are equally relevant for the study of the protein fraction. One of the main challenges is the lipophilicity of membrane proteins, which complicates solubilisation prior to analysis. On the other hand, proteins with a more hydrophilic nature can be lost into the serum phase during centrifugation or during washing steps. These are important considerations, especially when comparing studies that have used different techniques. To date, a standard protocol for the isolation of membrane material for quantitative or semi-quantitative analysis is lacking. This is not surprising since MFGM content in milk is currently not measured in the dairy industry. However, increasing evidence about the potential health benefits of MFGM components could lead to the development of a standardised procedure in the future.

Some studies have compared the performance of several washing procedures to study the recovery of MFGM material and the extent of contamination by serum milk proteins, during the washing step. For example, Le *et al.* (2009) compared several combinations of washing solutions (buffered salt solutions compared to deionised water), volumes of washing solution (6.75–9 L), temperatures (38–46 °C), and number of washing steps (two or three). Based on gel electrophoretic separation followed by densitometry analysis, the authors concluded that deionised water does not result in increased losses of MFGM material or increase the amount of contaminating serum proteins compared to buffered saline solutions. Moreover, the study showed that the potential benefit of a third washing step to further reduce contamination is accompanied by an increase in MFG coalescence. They also showed that using larger washing volumes is more beneficial than increasing the number of steps from two to three for reducing contamination. However, most studies still apply three washing steps (Table 13.5). The study also revealed that the loss of MFGM material at 46 °C and 42 °C was higher compared to 38 °C, despite increased MFG coalescence at the lower temperature (Le *et al.* 2009). Similarly, Holzmüller *et al.* (2016) also found a shift in MFG size with each additional washing step due to coalescence, particularly in washing solutions with high ionic strength such as simulated milk ultrafiltrate (SMUF). They also found substantial losses of MFGM membrane proteins with each

Table 13.5. Commonly used strategies for the extraction of MFGM proteins, sample preparation, and analytical methods

Reference	Detergent for protein extraction	Sample preparation	Washing	MFGM extract	Analytical method
Affolter *et al.* (2010)	0.1% RapiGest in 50 mM ABC	In-solution digestion with trypsin	NA	Whey protein concentrate, buttermilk concentrate	SCX and/or RP-LC-LTQ-Orbitrap-MS SID-LC-QQQ-MS with MRM
Bianchi *et al.* (2009)	Four different lysis buffer systems. See reference for further details	2D-PAGE and in-gel digestion with trypsin	Thrice in phosphate-buffered saline	Washed cream	MALDI-TOF-MS and IT-MS
Cebo *et al.* (2010)	63 mM Tris-HCl (2% SDS)	SDS-PAGE and in-gel digestion with trypsin	Thrice in 0.9% NaCl	Washed cream	MALDI-TOF-MS
Fong *et al.* (2007)	7 M urea, 2 M thiourea, 4% CHAPS, 0.5% DTT, and 0.5% ampholyte	SDS-PAGE and in gel digestion with trypsin	Thrice in deionised water	Freeze-dried buttermilk	RP-LC-Q-TOF-MS
Fong and Norris (2009)	6.7 M urea, 0.2 M ABC	In-solution digestion with trypsin	NA	Beta-serum[a]	RP-LC-HSRM-MS
Lu *et al.* (2011)	100 mM Tris/HCl, 4% SDS, 0.1 M DTT	FASP and dimethyl labelling	Thrice in deionised water	Washed cream	nanoLC-LTQ-Orbitrap-MS
Lu *et al.* (2016a)	0.4% SDS	FASP and dimethyl labelling	Thrice in PBS	Washed cream	nanoLC-LTQ-Orbitrap-MS

ABC ammonium bicarbonate, *DTE* dithioerythritol, *DTT* dithiothreitol, *PBS* phosphate-buffered saline, *CHAPS* 3-[(3-cholamidopropyl)dimethyl-amino]-1-propanesulfonate, *SDS* sodium dodecyl sulphate, *PAGE* polyacrylamide gel electrophoresis, *FASP* filter-aided sample preparation, *SCX* strong cation exchange, *RP* reverse phase, *LC* liquid chromatography, *LTQ* linear trap quadrupole, *SID* stable isotope dilution, *Q* quadrupole, *QQQ* triple quadruple, *MRM* multiple reaction monitoring, *MS* mass spectrometry, *MALDI* matrix-assisted laser desorption ionisation, *TOF* time-of-flight, *IT* ion trap, *HSRM* high-resolution selected reaction monitoring.
[a]A proprietary product rich in MFGM protein-enriched product (Catchpole *et al.* 2008).

additional washing step, believed to be due to a loss of entire MFGM fragments rather than individual components. This was supported by evidence that suggests it was not only peripheral MFGM proteins, such as lactadherin, lost during the washing procedure but also transmembrane proteins such as BTN. These studies demonstrate the challenges involved with the analysis of MFGM proteins. This can be especially challenging when comparing the MFGM protein abundance in MFGMs isolated from milk with different MFG size distributions. Large MFGs are more susceptible to coalescence and could therefore lose more membrane material during the washing procedures, although it is worth noting that in the study by Le *et al.* (2009), increased coalescence did not relate to membrane protein loss. Furthermore, a loss of MFGM protein could be due to a progressive loss of very small MFGs with each washing step. This has previously been suggested by Zheng *et al.* (2013), who found that higher centrifugal force applied during washing in SMUF resulted in the highest recovery of MFGM proteins, despite also leading to a higher degree of damage to the MFGM, indicated by an increased ζ-potential.

Another important consideration in relation to the extraction of MFGM material based on churning of the washed cream is that the buttermilk obtained during this process only contains around half of the total MFGM proteins. The other half remains in the butter serum (Le *et al.*

2009), which should be separated from the MF component of the butter.

13.5.2.2 Protein Solubilisation

Membrane protein solubilisation is inherently challenging due to the mixture of proteins and lipids found in biological membranes; however, it is crucial for reproducible quantitative analysis. The choice of detergent used for protein solubilisation will depend on the type of membrane extract and the type of qualitative or quantitative downstream application. This is because some detergents have to be removed prior to MS analysis or are incompatible with analytical approaches, such as the Bradford spectrophotometric assay for total protein.

Commonly used solubilisation buffers for the reconstitution of dried MFGM extracts are sodium dodecyl sulphate (SDS) or urea-containing buffers. However, the use of more complex buffers including reducing agents dithiothreitol (DDT), dithioerythritol (DTE), and other surfactants, such as 3-[(3-cholamidopropyl)dimethylamino]-1-propanesulfonate (CHAPS), myristic amidosulfobetaine (ASB-14), and 1,2-diheptanoyl-*sn*-glycero-3-phosphatidiylcholine(DHPC), has also been reported (Table 13.5). Sonication and incubation at room or elevated temperatures (20–30 °C) are also frequently used to aid solubilisation (Affolter *et al.* 2010; Bianchi *et al.* 2009; Fong *et al.* 2007).

A second decision is whether to apply a delipidation step, which increases the purity of the extract but risks losing some of the more lipophilic proteins or proteins attached to the PL fraction. Possible methods to remove lipids are centrifugation at high speeds (Fong *et al.* 2007) or precipitation of the proteins with, for example, chloroform-methanol (Affolter *et al.* 2010; Bianchi *et al.* 2009). Lack of removal of these compounds can lead to streaking or reduced resolution on electrophoretic gels, as shown by Bianchi *et al.* (2009). A delipidation or protein precipitation step may not be necessary when using MFGM extracts, which contain substantially more lipids compared to washed MFGs.

13.5.3 Cholesterol

Cholesterol is found in the MFGM of milk in its free form (90%) or as cholesteryl esters (10%) in the lipid droplet core. Cholesterol is an integral part of biological membranes involved in the maintenance of membrane fluidity and function. Despite further important roles as a precursor for the production of steroid hormones and bile salts, cholesterol is also related to cardiovascular disease, and the determination of its content in dairy foods is therefore of interest.

Cholesterol is extracted along with other milk lipids using solvent-based techniques and can be separated from the total MF through a process of saponification, where it is retained in the unsaponified fraction. This is usually done by hot saponification, although alternative methods are also available (Ahn *et al.* 2012). Most commonly, free cholesterol is quantified by LC or GC. However, an alternative spectrophotometric method had been suggested by Álvarez-Sala *et al.* (2015), where the unsaponified fraction is subjected to an enzymatic process and spectrophotometrically quantified. Moreover, it is possible to detect cholesteryl esters through MS detection, such as shotgun or LC-MS/MS (Murphy *et al.* 2011), which has been used for lipidomic analysis of cholesterol and cholesteryl esters in human breast milk (Villaseñor *et al.* 2014).

13.6 Analysis of Volatile Compounds

Volatile compounds may have a considerable impact on the sensory properties of milk and dairy products, as well as providing information of the overall quality and oxidative stability. The volatile components of dairy products are primarily generated through oxidative and enzymatic pathways associated with the lipid (lipolysis) and protein (proteolysis) fractions, recently reviewed in relation to surface-ripened cheese (Bertuzzi *et al.* 2018).

Sensory descriptive analysis can be performed using trained sensory panellists to characterise aroma and other sensory components that add to the overall sensorial experience, reviewed by Drake (2007), whereas untrained consumer panels are used to determine acceptance. Although not a complete substitute for sensory analysis, analysis of headspace volatiles is often used to measure the concentration of individual analytes within the vapour phase that contribute towards the overall aromatic profile. The method is rapid and efficient, though it should be noted that not all volatiles detected contribute to the odour. The influence of any given volatile on the overall sensory quality is governed by its odour activity value, which is the concentration relative to its olfactory detection threshold (Grosch 1994; Spitzer *et al.* 2013).

13.6.1 Gas Chromatography

Headspace analysis is typically performed using GC in conjunction with both FID and MS detection, where elution order is generally based on the volatility of the component.

13.6.1.1 Direct Injection

The sample is transferred directly into the GC through an insert packed with glass wool or through a temperature-controlled external unit. The injector port is heated to elevated temperatures, and the column is held at ambient temperatures or cooled until all volatiles are trapped at the head of the column. This approach can similarly be coupled with proton transfer reaction (PTR)-TOF-MS and has been used to characterise the volatile components of cheese (Bergamaschi *et al.* 2016).

13.6.1.2 Headspace Analysis

Headspace sampling techniques can be divided into three broad areas: static headspace, dynamic headspace (including 'purge and trap'), and solid-phase microextraction (SPME) analysis.

Static Headspace This involves the heating of a sample within a sealed vial so that volatile components diffuse into the gas phase and reach an equilibrium. Once equilibrated, an aliquot of the headspace sample is taken (usually 1 mL) and injected directly into the GC column.

Dynamic Headspace This approach sweeps the analytes away from the sample matrix using an inert gas, for example, nitrogen, rather than allowing an equilibrium to be reached with the surrounding headspace. This allows more analyte to be released into the headspace and means that higher concentrations can be extracted from the sample beyond the limit of the original headspace volume. After sweeping the matrix, the volatiles are generally passed through an adsorbent collection trap, allowing the inert gas to pass through while retaining the volatile compounds which are either condensed or frozen. The trap can then be heated to desorb or volatilise the analytes for injection into a GC. In general, the term 'purge and trap' is used for liquid sample analysis, while 'dynamic headspace' is used for solid samples (Wampler 2001).

Solid-Phase Microextraction This technique is considered a specialised form of dynamic headspace extraction, reviewed recently by Merkle *et al.* (2015), and involves the placing of an SPME fibre into a sample container that has been heated to allow volatile components to vaporise into the headspace. As such, the analytes distribute between three phases: the sample, the headspace, and the fibre. It is the kinetics of the distribution across these phases that determines the SPME sampling time, which should be sufficient to reach equilibrium. Increasing the sample temperature can decrease the time required to reach equilibrium by driving more volatile components into the headspace. Once inserted into a hot GC injector, the analyte components desorb from the fibre for analysis.

Solid-phase microextraction has become the method of choice for aroma analysis since it offers solvent-free, rapid automated sampling with low-cost and ease of operation. Moreover, analyte losses can occur with conventional methods during extraction and concentration steps. With SPME, sampling, extraction, concentration, and sample introduction can be performed as a simple process (Stashenko and Martínez 2004). Greater sensitivity is also realised for volatile and semi-volatile aroma compounds using SPME compared to normal headspace measurements (Sides *et al.* 2000).

Analyte adsorption to the fibre can be influenced by a number of factors including the selectivity of the fibre coating, extraction time, temperature and agitation conditions, and desorption conditions. It is our experience that there may be fibre to fibre differences, and we suggest results should only be compared between samples analysed using the same fibre. Commercially available fibre coatings range from non-polar (e.g. polydimethylsiloxane) to polar fibres (e.g. Carbowax/divinylbenzene), plus a range with mixed polarities. Selection of an appropriate fibre will depend on the polarity, size, and volatility of the analytes of interest. For example, a non-polar fibre is particularly useful for sampling non-polar compounds. On the other hand, the polar fibre Carboxen®/polydimethylsiloxane is ideal for the analysis of small volatile molecules since Carboxen®, a type of activated carbon, is particularly suitable for the adsorption of low-molecular-weight volatile compounds. Selecting a tri-mixed fibre, such as divinylbenzene/Carboxen®/polydimethylsiloxane, would increase versatility due to the affinity of the divinylbenzene coating for the adsorption of higher-molecular-weight compounds.

13.6.1.3 Gas Chromatography Analysis Columns

Capillary gas chromatography remains the approach of choice for the analysis of aroma compounds given the higher resolution and better separation afforded with these types of columns, compared with shorter packed columns (Table 13.6). In some instances, though, it can be preferable to use packed columns, particularly if the analytical approach is long established and well regarded. Common columns range in polarity and include the low-polar Rtx-5MS (diphenyl dimethyl polysiloxane) and high-polar DB-WAX (polyethylene glycol) (Table 13.6).

13.6.1.4 Gas Chromatography Olfactometry

The combination of GC with olfactometry (O) can be useful for understanding the impact of individual volatile compounds to the overall aroma associated with a dairy product. Gas chromatography allows for the separation of the volatile components, while O uses human assessors to evaluate the aroma profile for components as they elute (Friedrich and Acree 1998). When combined, the GC-O ‘sniffing port’ is located at the detector end where the sample is split between the O and another detector such as an FID or MS. Ikeda *et al.* (2018) used this approach to compare the aroma profile for a series of ready-to-drink milk products. Frank *et al.* (2004) examined the combination of SPME with GC-MS and GC-O to characterise components that contribute to cheese aroma, suggesting that the advantage of GC-O was the analysis of important aroma components that were either close to or below the MS instrumental limits for detection. Similarly, a simultaneous gustometer-olfactometer (SGO) can be used to study the interaction of different taste and aroma components on the perceived intensity or response of a trained analyst. The SGO delivers taste and aroma components directly to the tongue and nose, where the response of different mixtures and analyte concentrations can be simultaneously measured and quantified in the form of a surface response map. Niimi *et al.* (2014) used a SGO to show that cheese flavour intensity is more greatly influenced by changes to the aroma and taste components combined, demonstrating the complexities surrounding any analytical approach for aroma determination.

13.6.2 Other Approaches

The electronic nose utilises an array of solid-state gas sensors to detect aroma compounds. Its application to the analysis of dairy products has been reviewed (Ampuero and Bosset 2003) and an electronic nose was used to determine flavour compounds in milk products using 18 metal oxide sensors in a study by Wang *et al.* (2010).

Table 13.6. Examples of conditions for aroma volatile analysis of milk fat and dairy products

Reference	Milk product	Analytical method	Column[a]	No. of compounds
Ayed *et al.* (2018)	Flavoured ice cream	SPME/GC-MS (various)	DB-wax (30 × 0.25 × 0.50)	14
Breme and Guggenbühl (2014)	Yoghurt drink	SPME/GC-MS/FID-O (DVB/Car/PDMS)	DB-5MS (60 × 0.32 × 1) and VF-WAXMS (60 × 0.32 × 1)	27
Drake *et al.* (2010)	*Cheddar* cheese	SPME/GC-MS (DVB/Car/PDMS)	Rtx-5MS (30 × 0.25 × 0.25)	
El Hatmi *et al.* (2018)	Cameline milk	SHS/GC-MS	Rtx-5MS (specifications not provided)	19
Frank *et al.* (2004)	Cheese	SPME/GC-MS (Car/PDMS)	BP21-polyethylene glycol (30 × 0.32 × 0.25)	>200
Gómez-Torres *et al.* (2016)	Ovine milk cheese	SPME/GC-MS(DVB/Car/PDMS)	ZB-WAXplus (60 × 0.25 × 0.50)	38
Hailu *et al.* (2018)	Cameline cheese	DHS/GC-MS	DB-wax (30 × 0.25 × 0.50)	40
Jo *et al.* (2018a)	Gouda cheese	SPME/GC-MS (DVB/Car/PDMS)	ZB-5MS (30 × 0.25 × 0.25)	97
Jo *et al.* (2018b)	Pasteurised milk	SPME/GC-O (DVB/Car/PDMS)	ZB-5, ZB-wax (30 × 0.25 × 0.25 each column)	55
Mugampoza *et al.* (2019)	Stilton cheese	SPME/GC-MS (Car/PDMS)	Zebron ZB-5 (30 × 0.25 × 1)	21
Perreault *et al.* (2010)	Flavoured skimmed milk	SHS/GC-FID	Chromosorb W-AW 80–100 mesh with a stationary phase of Carbowax 20 M-10% (3 × 2.2)	6
Ribeiro *et al.* (2016)	Acid gels	SHS/GC-FID	DB-wax (30 × 0.32 × 0.5)	8
Rincon-Delgadillo *et al.* (2012)	Various dairy products	SPME/GC-MS (DVB/Car/PDMS)	Rtx-5MS (30 × 0.25 × 0.5)	40
Risner *et al.* (2019)	Cheese whey distillate	SPME/GC-MS	Rxi-5MS (15 × 0.25 × 0.25)	26
Sant'Ana *et al.* (2019)	Caprine milk and cheese	SPME/GC-MS (DVB/Car/PDMS)	VF-5MS (60 × 0.25 × 0.25)	33
Seuvre *et al.* (2008)	Custard	SHS/GC-FID	Chromosorb W-AW 80–100 mesh with a stationary phase of Carbowax 20 M-10% (3 × 2.2)	3 target compounds
Yee *et al.* (2014)	Cheese	DHS (P&T)/GC-MS	Elite 5 M (60 × 0.25 × 1)	45

SPME solid-phase microextraction, *DHS* dynamic headspace, *P&T* purge and trap, *SHS* static headspace, *GC* gas chromatography, *MS* mass spectrometry, *FID* flame ionisation detection, *O* olfactometry, *DVB* divinylbenzene, *Car* Carboxen®, *PDMS* polydimethylsiloxane.

[a]Column type nomenclature: capillary column (length (m) × inner diameter (mm) × film thickness (μm)); packed column (length (m) × inner diameter (mm)).

The PTR approach is based on the rate kinetics and transfer of $[H^+]$ between an ionised volatile product and H_3O^+ and has been coupled with Q-MS and TOF-MS detection to measure the headspace volatiles of milk and dairy products, such as yoghurt (Liu *et al.* 2018a; Soukoulis *et al.*, 2012). It can be used as a stand-alone technique to follow the formation of volatile components in real time and in combination with SPME-GC to confirm identification (Soukoulis *et al.* 2010).

Other more traditional approaches include steam and vacuum distillation (Robards *et al.* 1988), though these remain less common due to the high sensitivity and reproducibility offered by more modern approaches.

Table 13.6 presents a summary of the columns, conditions, and detectors used recently in the literature to quantify the volatile aroma compounds in MF and dairy products.

References

Affolter, M., Grass, L., Vanrobaeys, F., Casado, B. & Kussmann, M. (2010). Qualitative and quantitative profiling of the bovine milk fat globule membrane proteome. *Journal of Proteomics, 73*(6), 1079–1088.

Ahn, J.-H., Jeong, I.-S., Kwak, B.-M., Leem, D., Yoon, T., Yoon, C., Jeong, J., Park, J.-M. & Kim, J.-M. (2012). Rapid determination of cholesterol in milk containing emulsified foods. *Food Chemistry, 135*(4), 2411–2417.

Ali, A. H., Zou, X., Huang, J., Abed, S. M., Tao, G., Jin, Q. & Wang, X. (2017). Profiling of phospholipids molecular species from different mammalian milk powders by using ultra-performance liquid chromatography-electrospray ionization-quadrupole-time of flight-mass spectrometry. *Journal of Food Composition and Analysis, 62*, 143–154.

Álvarez-Sala, A., Garcia-Llatas, G., Barberá, R. & Lagarda, M. J. (2015). Determination of cholesterol in human milk: An alternative to chromatographic methods. *Nutrición Hospitalaria, 32*(4).

Ampuero, S. & Bosset, J. O. (2003). The electronic nose applied to dairy products: A review. *Sensors and Actuators B Chemical, 94*(1), 1–12.

Andrikopoulos, N. K. (2002). Chromatographic and spectroscopic methods in the analysis of triacylglycerol species and regiospecific isomers of oils and fats. *Critical Reviews in Food Science and Nutrition, 42*(5), 473–505.

Aravindan, R., Anbumathi, P. & Viruthagiri, T. (2007). Lipase applications in food industry. *Indian Journal of Biotechnology, 6*(2), 141–158.

Argov-Argaman, N., Mesilati-Stahy, R., Magen, Y. & Moallem, U. (2014). Elevated concentrate-to-forage ratio in dairy cow rations is associated with a shift in the diameter of milk fat globules and remodeling of their membranes. *Journal of Dairy Science, 97*(10), 6286–6295.

Argov-Argaman, N., Hadaya, O., Glasser, T., Muklada, H., Dvash, L., Mesilati-Stahy, R. & Landau, S. Y. (2016). Milk fat globule size, phospholipid contents and composition of milk from purebred and alpine-crossbred mid-eastern goats under confinement or grazing condition. *International Dairy Journal, 58*, 2–8.

Asbury, G. R., Al-Saad, K., Siems, W. F., Hannan, R. M. & Hill, H. H. (1999). Analysis of triacylglycerols and whole oils by matrix-assisted laser desorption/ionization time of flight mass spectrometry. *Journal of American Society of Mass Spectrometry, 10*(10), 983–991.

Ashraf-Khorassani, M., Isaac, G., Rainville, P., Fountain, K. & Taylor, L. T. (2015). Study of ultrahigh performance supercritical fluid chromatography to measure free fatty acids with out fatty acid ester preparation. *Journal of Chromatography B, 997*, 45–55.

Avalli, A. & Contarini, G. (2005). Determination of phospholipids in dairy products by SPE/HPLC/ELSD. *Journal of Chromatography A, 1071*(1–2), 185–190.

Ayed, C., Martins, S. I. F. S., Williamson, A.-M. & Guichard, E. (2018). Understanding fat, proteins and saliva impact on aroma release from flavoured ice creams. *Food Chemistry, 267*, 132–139.

Barbano, D. M., Clark, J. L. & Dunham, C. E. (1988). Comparison of Babcock and ether extraction methods for determination of fat content of milk: Collaborative study. *Journal of the Association of Official Analytical Chemists, 71*(5), 898–914.

Barron, L. J. R., Hierro, M. T. G. & Santa-María, G. (1990). HPLC and GLC analysis of the triglyceride composition of bovine, ovine and caprine milk fat. *Journal of Dairy Research, 57*(4), 517–526.

Bassbasi, M., De Luca, M., Ioele, G., Oussama, A. & Ragno, G. (2014). Prediction of the geographical origin of butters by partial least square discriminant analysis (PLS-DA) applied to infrared spectroscopy (FTIR) data. *Journal of Food Composition and Analysis, 33*(2), 210–215.

Beccaria, M., Sullini, G., Cacciola, F., Donato, P., Dugo, P. & Mondello, L. (2014). High performance characterization of triacylglycerols in milk and milk-related samples by liquid chromatography and mass spectrometry. *Journal of Chromatography A, 1360*, 172–187.

Belury, M. A. (2002). Dietary conjugated linoleic acid in health: Physiological effects and mechanisms of action. *Annual Review of Nutrition, 22*(1), 505–531.

Benbrook, C. M., Davis, D. R., Heins, B. J., Latif, M. A., Leifert, C., Peterman, L., Butler, G., Faergeman, O., Abel-Caines, S. & Baranski. (2018). Enhancing the fatty acid profile of milk through forage-based rations, with nutrition modeling of diet outcomes. *Food Science & Nutrition, 6*(3), 681–700.

Bergamaschi, M., Cecchinato, A., Biasioli, F., Gasperi, F., Martin, B. & Bittante, G. (2016). From cow to cheese: Genetic parameters of the flavour fingerprint of cheese investigated by direct-injection mass spectrometry (PTR-ToF-MS). *Genetics, Selection, Evolution, 48*(1), 89.

Berger, U., Oehme, M. & Kuhn, F. (1999). Quantitative determination and structure elucidation of type A-and B-trichothecenes by HPLC/ion trap multiple mass spectrometry. *Journal of Agricultural and Food Chemistry, 47*(10), 4240–4245.

Bertuzzi, A. S., McSweeney, P. L. H., Rea, M. C. & Kilcawley, K. N. (2018). Detection of volatile compounds of cheese and their contribution to the flavor profile of surface-ripened cheese. *Comprehensive*

Reviews in Food Science and Food Safety, 17(2), 371–390.

Bhattacharjee, U., Jarashow, D., Casey, T. A., Petrich, J. W. & Rasmussen, M. A. (2018). Using fluorescence spectroscopy to identify milk from grass-fed dairy cows and to monitor its photodegradation. *Journal of Agricultural and Food Chemistry, 66*(9), 2168–2173.

Bianchi, L., Puglia, M., Landi, C., Matteoni, S., Perini, D., Armini, A., Verani, M., Trombetta, C., Soldani, P. & Roncada, P. (2009). Solubilization methods and reference 2-DE map of cow milk fat globules. *Journal of Proteomics, 72*(5), 853–864.

Bligh, E. & Dyer, W. J. (1959). A rapid method of total lipid extraction and purification. *Canadian Journal of Biochemistry and Physiology, 37*(8), 911–917.

Bottino, N. R., Vandenburg, G. A. & Reiser, R. (1967). Resistance of certain long-chain polyunsaturated fatty acids of marine oils to pancreatic lipase hydrolysis. *Lipids, 2*(6), 489–493.

Bourlieu, C., Bouhallab, S. & Lopez, C. (2009). Biocatalyzed modifications of milk lipids: Applications and potentialities. *Trends in Food Science and Technology, 20*(10), 458–469.

Breme, K. & Guggenbühl, B. (2014). Aroma profile of a red-berries yoghurt drink by HS-SPME-GC-MS-O and influence of matrix texture on volatile aroma compound release of flavored dairy products. In V. Ferreira & R. Lopez (Eds.), *Flavour science* (pp. 101–106). San Diego: Academic Press.

Brondz, I. (2002). Development of fatty acid analysis by high-performance liquid chromatography, gas chromatography, and related techniques. *Analytica Chimica Acta, 465*(1–2), 1–37.

Burri, L., Hoem, N., Monakhova, Y. B. & Diehl, B. W. K. (2016). Fingerprinting krill oil by 31P, 1H and 13C NMR spectroscopies. *Journal of the American Oil Chemists' Society, 93*(8), 1037–1049.

Castro-Gómez, P., Montero, O. & Fontecha, J. (2017). In-depth lipidomic analysis of molecular species of triacylglycerides, diacylglycerides, glycerophospholipids, and sphingolipids of buttermilk by GC-MS/FID, HPLC-ELSD, and UPLC-QToF-MS. *International Journal of Molecular Science, 18*(3), 605.

Catchpole, O. J., Tallon, S. J., Grey, J. B., Fletcher, K. & Fletcher, A. J. (2008). Extraction of lipids from a specialist dairy stream. *Journal of Supercritical Fluids, 45*(3), 314–321.

Cebo, C., Caillat, H., Bouvier, F. & Martin, P. (2010). Major proteins of the goat milk fat globule membrane. *Journal of Dairy Science, 93*(3), 868–876.

Chilliard, Y., Ferlay, A., Mansbridge, R. M. & Doreau, M. (2000). Ruminant milk fat plasticity: Nutritional control of saturated, polyunsaturated, trans and conjugated fatty acids. *Annales de Zootechnie, 49*, 181–205.

Chowdhury, K., Banu, L. A., Khan, S. & Latif, A. (2007). Studies on the fatty acid composition of edible oil. *Bangladesh Journal of Scientific and Industrial Research, 42*(3), 311–316.

Christie, W. W. (1982). *Lipid analysis* (2nd ed.). Oxford: Pergamon Press.

Christie, W. W., Sébédio, J. L. & Juanéda, P. (2001). A practical guide to the analysis of conjugated linoleic acid (CLA). *Inform, 12*, 147–152.

Clarke, W. (2017). Mass spectrometry in the clinical laboratory: Determining the need and avoiding pitfalls. In H. Nair & W. Clarke (Eds.), *Mass spectrometry for the clinical laboratory* (pp. 1–15). London: Academic Press.

Contarini, G. & Povolo, M. (2013). Phospholipids in milk fat: Composition, biological and technological significance, and analytical strategies. *International Journal of Molecular Sciences, 14*(2), 2808–2831.

Contarini, G., Pelizzola, V., Scurati, S. & Povolo, M. (2017). Polar lipid of donkey milk fat: Phospholipid, ceramide and cholesterol composition. *Journal of Food Composition and Analysis, 57*, 16–23.

Couvreur, S., Hurtaud, C., Marnet, P. G., Faverdin, P. & Peyraud, J. L. (2007). Composition of milk fat from cows selected for milk fat globule size and offered either fresh pasture or a corn silage-based diet. *Journal of Dairy Science, 90*(1), 392–403.

Cruz-Hernandez, C., Kramer, J. K. G., Kennelly, J. J., Glimm, D. R., Sorensen, B. M., Okine, E. K., Goonewardene, L. A. & Weselake, R. J. (2007). Evaluating the conjugated linoleic acid and trans 18: 1 isomers in milk fat of dairy cows fed increasing amounts of sunflower oil and a constant level of fish oil. *Journal of Dairy Science, 90*(8), 3786–3801.

Damário, N., de Oliveira, D. N., Ferreira, M. S., Delafiori, J. & Catharino, R. R. (2015). Cheese lipid profile using direct imprinting in glass surface mass spectrometry. *Analytical Methods, 7*(6), 2877–2880.

Deeth, H. C. (1997). The role of phospholipids in the stability of milk fat gobules. *Australian Journal of Dairy Technology, 52*, 44–46.

Delmonte, P., Kia, A.-R. F., Hu, Q. & Rader, J. I. (2009). Review of methods for preparation and gas chromatographic separation of trans and cis reference fatty acids. *Journal of AOAC International, 92*(5), 1310–1326.

Delmonte, P., Fardin-Kia, A. R., Kramer, J. K. G., Mossoba, M. M., Sidisky, L., Tyburczy, C. & Rader, J. I. (2012). Evaluation of highly polar ionic liquid gas chromatographic column for the determination of the fatty acids in milk fat. *Journal of Chromatography A, 1233*, 137–146.

Diehl, B. W. K. (2015). NMR of polar lipids. In M. U. Ahmad & X. Vu (Eds.), *Polar lipids* (pp. 391–438). Urbana: AOCS Press.

Donato, P., Cacciola, F., Cichello, F., Russo, M., Dugo, P. & Mondello, L. (2011). Determination of phospholipids in milk samples by means of hydrophilic interaction liquid chromatography coupled to evaporative light scattering and mass spectrometry detection. *Journal of Chromatography A, 1218*(37), 6476–6482.

Drake, M. A. (2007). Invited review: Sensory analysis of dairy foods. *Journal of Dairy Science, 90*(11), 4925–4937.

Drake, M. A., Miracle, R. E. & McMahon, D. J. (2010). Impact of fat reduction on flavor and flavor chemistry of Cheddar cheeses. *Journal of Dairy Science, 93*(11), 5069–5081.

Dufour, E., Mazerolles, G., Devaux, M. F., Duboz, G., Duployer, M. H. & Riou, N. M. (2000). Phase transition of triglycerides during semi-hard cheese ripening. *International Dairy Journal, 10*(1–2), 81–93.

Dupont, D., Croguennec, T., Brodkorb, A. & Kouaouci, R. (2013). Quantitiation of proteins in milk and milk products. In P. L. H. McSweeney & P. F. Fox (Eds.), *Advanced dairy chemistry: Volume 1A: Proteins: Basic aspects* (3rd ed., pp. 87–134). New York: Springer.

Eder, K. (1995). Gas chromatographic analysis of fatty acid methyl esters. *Journal of Chromatography B, 671*(1–2), 113–131.

El Hatmi, H., Jrad, Z., Oussaief, O., Nasri, W., Sbissi, I., Khorchani, T. & Canabady-Rochelle, L. L. S. (2018). Fermentation of dromedary camel (*Camelus dromedarius*) milk by *Enterococcus faecium, Streptococcus macedonicus* as a potential alternative of fermented cow milk. *LWT – Food Science and Technology, 90*, 373–380.

Fagan, P., Wijesundera, C. & Watkins, P. (2004). Determination of mono-and di-acylglycerols in milk lipids. *Journal of Chromatography A, 1054*(1–2), 251–259.

Folch, J., Lees, M. & Sloane Stanley, G. H. (1957). A simple method for the isolation and purification of total lipides from animal tissues. *Journal of Biological Chemistry, 226*(1), 497–509.

Fong, B. Y. & Norris, C. S. (2009). Quantification of milk fat globule membrane proteins using selected reaction monitoring mass spectrometry. *Journal of Agricultural and Food Chemistry, 57*(14), 6021–6028.

Fong, B. Y., Norris, C. S. & MacGibbon, A. K. H. (2007). Protein and lipid composition of bovine milk-fat-globule membrane. *International Dairy Journal, 17*(4), 275–288.

Forcato, D. O., Carmine, M. P., Echeverria, G. E., Pécora, R. P. & Kivatinitz, S. C. (2005). Milk fat content measurement by a simple UV spectrophotometric method: An alternative screening method. *Journal of Dairy Science, 88*(2), 478–481.

Frank, D. C., Owen, C. M. & Patterson, J. (2004). Solid phase microextraction (SPME) combined with gas-chromatography and olfactometry-mass spectrometry for characterization of cheese aroma compounds. *LWT – Food Science and Technology, 37*(2), 139–154.

Fraser, B. H., Perlmutter, P. & Wijesundera, C. (2007). Practical syntheses of triacylglycerol regioisomers containing long-chain polyunsaturated fatty acids. *Journal of the American Oil Chemists' Society, 84*(1), 11.

Friedrich, J. E. & Acree, T. E. (1998). Gas chromatography olfactometry (GC/O) of dairy products. *International Dairy Journal, 8*(3), 235–241.

Gallier, S., Gragson, D., Cabral, C., Jiménez-Flores, R. & Everett, D. W. (2010). Composition and fatty acid distribution of bovine milk phospholipids from processed milk products. *Journal of Agricultural and Food Chemistry, 58*(19), 10503–10511.

Garcia, C., Lutz, N. W., Confort-Gouny, S., Cozzone, P. J., Armand, M. & Bernard, M. (2012). Phospholipid fingerprints of milk from different mammalians determined by 31P NMR: Towards specific interest in human health. *Food Chemistry, 135*(3), 1777–1783.

Gastaldi, D., Medana, C., Giancotti, V., Aigotti, R., Dal Bello, F. & Baiocchi, C. (2011). HPLC-APCI analysis of triacylglycerols in milk fat from different sources. *European Journal of Lipid Science and Technology, 113*(2), 197–207.

German, J. B. & Dillard, C. J. (2006). Composition, structure and absorption of milk lipids: A source of energy, fat-soluble nutrients and bioactive molecules. *Critical Reviews in Food Science and Nutrition, 46*(1), 57–92.

Gómez-Cortés, P., Rodríguez-Pino, V., Juárez, M. & de la Fuente, M. A. (2017). Optimization of milk odd and branched-chain fatty acids analysis by gas chromatography using an extremely polar stationary phase. *Food Chemistry, 231*, 11–18.

Gómez-Torres, N., Ávila, M., Delgado, D. & Garde, S. (2016). Effect of reuterin-producing *Lactobacillus reuteri* coupled with glycerol on the volatile fraction, odour and aroma of semi-hard ewe milk cheese. *International Journal of Food Microbiology, 232*, 103–110.

Grosch, W. (1994). Determination of potent odourants in foods by aroma extract dilution analysis (AEDA) and calculation of odour activity values (OAVs). *Flavour and Fragrance Journal, 9*(4), 147–158.

Gunstone, F. D. (1994). 13C NMR of lipids. In J. H. P. Tymm & M. H. Gordon (Eds.), *Developments in the analysis of lipids* (pp. 391–438). Scotland: The Royal Society of Chemistry.

Hailu, Y., Hansen, E. B., Seifu, E., Eshetu, M., Petersen, M. A., Lametsch, R., Rattray, F. & Ipsen, R. (2018). Rheological and sensory properties and aroma compounds formed during ripening of soft brined cheese made from camel milk. *International Dairy Journal, 81*, 122–130.

Heck, J. M. L., Van Valenberg, H. J. F., Dijkstra, J. & Van Hooijdonk, A. C. M. (2009). Seasonal variation in the Dutch bovine raw milk composition. *Journal of Dairy Science, 92*(10), 4745–4755.

Holčapek, M. & Lísa, M. (2017). Silver-ion liquid chromatography–mass spectrometry. In M. Holčapek & W. C. Byrdwell (Eds.), *Handbook of advanced chromatography/mass spectrometry techniques* (pp. 115–140). Urbana: AOCS Press.

Holčapek, M., Jandera, P., Zderadička, P. & Hruba, L. (2003). Characterization of triacylglycerol and diacylglycerol composition of plant oils using high-performance liquid chromatography–atmospheric pressure chemical ionization mass spectrometry. *Journal of Chromatography A, 1010*(2), 195–215.

Holzmüller, W., Müller, M., Himbert, D. & Kulozik, U. (2016). Impact of cream washing on fat globules and milk fat globule membrane proteins. *International Dairy Journal, 59*, 52–61.

Horrocks, L. A., Cornwell, D. G. & Brown, J. B. (1961). Quantitative gas-liquid chromatography of fatty acid methyl esters with the thermal conductivity detector. *Journal of Lipid Research, 2*, 92–94.

Ikeda, M., Akiyama, M., Hirano, Y., Miyaji, K., Sugawara, Y., Imayoshi, Y., Iwabuchi, H., Onodera, T. & Toko, K. (2018). Effects of manufacturing processing conditions on retronasal-aroma odorants from a milk coffee drink. *Journal of Food Science, 83*(11), 2733–2744.

Ipsen, J. H., Mouritsen, O. G. & Bloom, M. J. (1990). Relationships between lipid membrane area, hydrophobic thickness, and acyl-chain orientational order. The effects of cholesterol. *Biophysical Journal, 57*(3), 405–412.

Iverson, S. J., Lang, S. L. C. & Cooper, M. H. (2001). Comparison of the Bligh and Dyer and Folch methods for total lipid determination in a broad range of marine tissue. *Lipids, 36*(11), 1283–1287.

Jakab, A., Jablonkai, I. & Forgács, E. (2003). Quantification of the ratio of positional isomer dilinoleoyl-oleoyl glycerols in vegetable oils. *Rapid Communications in Mass Spectrometry, 17*(20), 2295–2302.

Jo, Y., Benoist, D. M., Ameerally, A. & Drake, M. A. (2018a). Sensory and chemical properties of Gouda cheese. *Journal of Dairy Science, 101*(3), 1967–1989.

Jo, Y., Benoist, D. M., Barbano, D. M. & Drake, M. A. (2018b). Flavor and flavor chemistry differences among milks processed by high-temperature, short-time pasteurization or ultra-pasteurization. *Journal of Dairy Science, 101*(5), 3812–3828.

Kallio, H., Nylund, M., Boström, P. & Yang, B. (2017). Triacylglycerol regioisomers in human milk resolved with an algorithmic novel electrospray ionization tandem mass spectrometry method. *Food Chemistry, 233*, 351–360.

Karoui, R., Mouazen, A. M., Dufour, E., Pillonel, L., Schaller, E., Picque, D., De Baerdemaeker, J. & Bosset, J.-O. (2006). A comparison and joint use of NIR and MIR spectroscopic methods for the determination of some parameters in European Emmental cheese. *European Journal of Lipid Science and Technology, 223*(1), 44–50.

Kelsey, J. A., Corl, B. A., Collier, R. J. & Bauman, D. E. (2003). The effect of breed, parity, and stage of lactation on conjugated linoleic acid (CLA) in milk fat from dairy cows. *Journal of Dairy Science, 86*(8), 2588–2597.

Kilcawley, K. N. & Mannion, D. T. (2017). *Free fatty acids quantification in dairy products* (pp. 209–220). London: IntechOpen.

Kim, H.-J., Park, J.-M., Lee, J.-H. & Kim, J.-M. (2016). Detection for non-milk fat in dairy product by gas chromatography. *Korean Journal of Food Science and Animal Resources, 36*(2), 206.

Kleyn, D. H., Lynch, J. M., Barbano, D. M., Bloom, M. J. & Mitchell, M. W. (2001). Determination of fat in raw and processed milks by the Gerber method: Collaborative study. *Journal of AOAC International, 84*(5), 1499–1508.

Köfeler, H. C., Fauland, A., Rechberger, G. N. & Trötzmüller, M. (2012). Mass spectrometry based lipidomics: An overview of technological platforms. *Metabolites, 2*(1), 19–38.

Konuspayeva, G., Faye, B., Loiseau, G., Narmuratova, M., Ivashchenko, A., Meldebekova, A. & Davletov, S. (2010). Physiological change in camel milk composition (*Camelus dromedarius*) 1. Effect of lactation stage. *Tropical Animal Health and Production, 42*(3), 495–499.

Koo, W. W., Hockman, E. M. & Dow, M. (2006). Palm olein in the fat blend of infant formulas: Effect on the intestinal absorption of calcium and fat, and bone mineralization. *Journal of the American College of Nutrition, 25*(2), 117–122.

Kramer, J. K. G., Blackadar, C. B. & Zhou, J. (2002). Evaluation of two GC columns (60-m SUPELCOWAX 10 and 100-m CP sil 88) for analysis of milkfat with emphasis on CLA, 18:1, 18:2 and 18:3 isomers, and short-and long-chain FA. *Lipids, 37*(8), 823–835.

Kröger, J., Zietemann, V., Enzenbach, C., Weikert, C., Jansen, E. H. J. M., Döring, F., Joost, H.-G., Boeing, H. & Schulze, M. B. (2010). Erythrocyte membrane phospholipid fatty acids, desaturase activity, and dietary fatty acids in relation to risk of type 2 diabetes in the European prospective investigation into Cancer and nutrition (EPIC)–Potsdam study. *American Journal of Clinical Nutrition, 93*(1), 127–142.

La Nasa, J., Degano, I., Brandolini, L., Modugno, F. & Bonaduce, I. (2018). A novel HPLC-ESI-Q-ToF approach for the determination of fatty acids and acylglycerols in food samples. *Analytica Chimica Acta, 1013*, 98–109.

Laakso, P. & Manninen, P. (1997). Identification of milk fat triacylglycerols by capillary supercritical fluid chromatography-atmospheric pressure chemical Ionization mass spectrometry. *Lipids, 32*(12), 1285–1295.

Laboureur, L., Ollero, M. & Touboul, D. (2015). Lipidomics by supercritical fluid chromatography. *International Journal of Molecular Sciences, 16*(6), 13868–13884.

Le, T. T., Van Camp, J., Rombaut, R., Van Leeckwyck, F. & Dewettinck, K. (2009). Effect of washing conditions on the recovery of milk fat globule membrane proteins during the isolation of milk fat globule membrane from milk. *Journal of Dairy Science, 92*(8), 3592–3603.

Li, G., You, J., Suo, Y., Song, C., Sun, Z., Xia, L., Zhao, X. & Shi, J. (2011). A developed pre-column derivatization method for the determination of free fatty acids in edible oils by reversed-phase HPLC with fluorescence detection and its application to *Lycium barbarum* seed oil. *Food Chemistry, 125*(4), 1365–1372.

Li, Q., Zhao, Y., Zhu, D., Pang, X., Liu, Y., Frew, R. & Chen, G. (2017). Lipidomics profiling of goat milk, soymilk and bovine milk by UPLC-Q-Exactive Orbitrap Mass Spectrometry. *Food Chemistry, 224*, 302–309.

Linderborg, K. M., Kalpio, M., Mäkelä, J., Niinikoski, H., Kallio, H. P. & Lagström, H. (2014). Tandem mass spectrometric analysis of human milk triacylglycerols from normal weight and overweight mothers on different diets. *Food Chemistry, 146*, 583–590.

Lindmark Månsson, H. (2008). Fatty acids in bovine milk fat. *Food & Nutrition Research, 52*(1), 1821.

Liu, Z., Moate, P., Cocks, B. & Rochfort, S. (2015a). Comprehensive polar lipid identification and quantification in milk by liquid chromatography–mass spectrometry. *Journal of Chromatography, 978*, 95–102.

Liu, Z., Moate, P., Ezerniks, V., Cocks, B. G. & Rochfort, S. (2015b). Identification and quantification of triacylglycerols containing n-3 long-chain polyunsaturated fatty acids in bovine milk. *Journal of Dairy Science, 98*(12), 8473–8485.

Liu, Z., Rochfort, S. & Cocks, B. G. (2016). Optimization of a single phase method for lipid extraction from milk. *Journal of Chromatography A, 1458*, 145–149.

Liu, Z., Ezernieks, V., Wang, J., Arachchillage, N. W., Garner, J. B., Wales, W. J., Cocks, B. G. & Rochfort, S. (2017). Heat stress in dairy cattle alters lipid composition of milk. *Scientific Reports, 7*(1), 961.

Liu, N., Koot, A., Hettinga, K., De Jong, J. & van Ruth, S. M. (2018a). Portraying and tracing the impact of different production systems on the volatile organic compound composition of milk by PTR-(Quad) MS and PTR-(ToF) MS. *Food Chemistry, 239*, 201–207.

Liu, Z., Rochfort, S. & Cocks, B. (2018b). Milk lipidomics: What we know and what we don't. *Progress in Lipid Research, 71*, 80–85.

Logan, A., Auldist, M., Greenwood, J. & Day, L. (2014). Natural variation of bovine milkfat globule size within a herd. *Journal of Dairy Science, 97*(7), 4072–4082.

Logan, A., Xu, M., Day, L., Singh, T. K., Moore, S. C., Mazzonetto, M. & Augustin, M. A. (2017). Milk fat globule size affects Cheddar cheese properties. *International Dairy Journal, 70*, 46–54.

Lopez, C. (2011). Milk fat globules enveloped by their biological membrane: Unique colloidal assemblies with a specific composition and structure. *Current Opinion in Colloid & Interface Science, 16*(5), 391–404.

Lopez, C., Bourgaux, C., Lesieur, P. & Ollivon, M. (2007). Coupling of time-resolved synchrotron X-ray diffraction and DSC to elucidate the crystallisation properties and polymorphism of triglycerides in milk fat globules. *Lait, 87*(4–5), 459–480.

Lopez, C., Briard-Bion, V., Menard, O., Rousseau, F., Pradel, P. & Besle, J.-M. (2008). Phospholipid, sphingolipid, and fatty acid compositions of the milk fat globule membrane are modified by diet. *Journal of Agricultural and Food Chemistry, 56*(13), 5226–5236.

Lopez, C., Briard-Bon, V. & Ménard, O. (2014). Polar lipids, sphingomyelin and long-chain unsaturated fatty acids from the milk fat globule membrane are increased in milks produced by cows fed fresh pasture based diet during spring. *Food Research International, 58*, 59–68.

Lopez, C., Blot, M., Briard-Bion, V., Cirié, C. & Graulet, B. (2017). Butter serums and buttermilks as sources of bioactive lipids from the milk fat globule membrane: Differences in their lipid composition and potentialities of cow diet to increase n-3 PUFA. *Food Research International, 100*, 864–872.

Lu, C.-Y., Wu, H.-L., Chen, S.-H., Kou, H.-S. & Wu, S.-M. (2002). Simple and sensitive analysis of long-chain free fatty acids in milk by fluorogenic derivatization and high-performance liquid chromatography. *Journal of Agricultural and Food Chemistry, 50*(1), 71–73.

Lu, J., Boeren, S., De Vries, S. C., Van Valenberg, H. J. F., Vervoort, J. & Hettinga, K. A. (2011). Filter-aided sample preparation with dimethyl labeling to identify and quantify milk fat globule membrane proteins. *Journal of Proteomics, 75*(1), 34–43.

Lu, J., Argov-Argaman, N., Anggrek, J., Boeren, S., van Hooijdonk, T., Vervoort, J. & Hettinga, K. A. (2016a). The protein and lipid composition of the membrane of milk fat globules depends on their size. *Journal of Dairy Science, 99*(6), 4726–4738.

Lu, J., Liu, L., Pang, X., Zhang, S., Jia, Z., Ma, C., Zhao, L. & Lv, J. (2016b). Comparative proteomics of milk fat globule membrane in goat colostrum and mature milk. *Food Chemistry, 209*, 10–16.

MacDonald, H. B. (2000). Conjugated linoleic acid and disease prevention: A review of current knowledge. *Journal of the American College of Nutrition, 19*(sup2), 111S–118S.

Maher, A. & Rochfort, S. (2014). Applications of NMR in dairy research. *Metabolites, 4*(1), 131–141.

Mannion, D. T., Furey, A. & Kilcawley, K. N. (2016a). Comparison and validation of 2 analytical methods for the determination of free fatty acids in dairy products by gas chromatography with flame ionization detection. *Journal of Dairy Science, 99*(7), 5047–5063.

Mannion, D. T., Furey, A. & Kilcawley, K. N. (2016b). Free fatty acids quantification in dairy products. *International Journal of Dairy Technology, 69*(1), 1–12.

Mannion, D. T., Furey, A. & Kilcawley, K. N. (2018). Development and validation of a novel free fatty acid butyl ester gas chromatography method for the determination of free fatty acids in dairy products. *Journal of Agricultural and Food Chemistry, 67*(1), 499–506.

Mariette, F. (2009). Investigations of food colloids by NMR and MRI. *Current Opinion in Colloid & Interface Science, 14*(3), 203–211.

Mattson, F. H. & Volpenhein, R. A. (1962). Synthesis and properties of glycerides. *Journal of Lipid Research, 3*(3), 281–296.

Ménard, O., Ahmad, S., Rousseau, F., Briard-Bion, V., Gaucheron, F. & Lopez, C. (2010). Buffalo vs. cow milk fat globules: Size distribution, zeta-potential, compositions in total fatty acids and in polar lipids from the milk fat globule membrane. *Food Chemistry, 120*(2), 544–551.

Merkle, S., Kleeberg, K. & Fritsche, J. (2015). Recent developments and applications of solid phase microextraction (SPME) in food and environmental analysis – A review. *Chromatography, 2*(3), 293–381.

Mesilati-Stahy, R., Mida, K. & Argov-Argaman, N. (2011). Size-dependent lipid content of bovine milk fat globule and membrane phospholipids. *Journal of Agricultural and Food Chemistry, 59*(13), 7427–7435.
Mestawet, T. A., Girma, A., Ådnøy, T., Devold, T. G., Narvhus, J. A. & Vegarud, G. E. (2012). Milk production, composition and variation at different lactation stages of four goat breeds in Ethiopia. *Small Ruminant Research, 105*(1–3), 176–181.
Meyer, S., Rajendram, V. S. & Povey, M. J. W. (2006). Characterization of reconstituted milk powder by ultrasound spectroscopy. *Journal of Food Quality, 29*(4), 405–418.
Mitrova, K., Karpisek, M., Durilova, M., Dragusin, L. G., Nevoral, J. & Bronsky, J. I. (2014). Development of high-sensitive ELISA method for detection of adipophilin levels in human colostrum and breast Milk. *Journal of Clinical Laboratory Analysis, 28*(4), 255–260.
Moate, P. J., Williams, S. R. O., Hannah, M. C., Eckard, R. J., Auldist, M. J., Ribaux, B. E., Jacobs, J. L. & Wales, W. J. (2013). Effects of feeding algal meal high in docosahexaenoic acid on feed intake, milk production, and methane emissions in dairy cows. *Journal of Dairy Science, 96*(5), 3177–3188.
Mottram, H. R. & Evershed, R. P. (2001). Elucidation of the composition of bovine milk fat triacylglycerols using high-performance liquid chromatography–atmospheric pressure chemical ionisation mass spectrometry. *Journal of Chromatography A, 926*(2), 239–253.
Mozaffarian, D., Cao, H., King, I. B., Lemaitre, R. N., Song, X., Siscovick, D. S. & Hotamisligil, G. S. (2010). *Trans*-palmitoleic acid, metabolic risk factors, and new-onset diabetes in US adults: A cohort study. *Annals of Internal Medicine, 153*(12), 790–799.
Mugampoza, D., Gkatzionis, K., Linforth, R. S. T. & Dodd, C. E. R. (2019). Acid production, growth kinetics and aroma profiles of *Lactobacillus* flora from Stilton cheese. *Food Chemistry, 287*, 222–231.
Murphy, R. C., Leiker, T. J. & Barkley, R. M. (2011). Glycerolipid and cholesterol ester analyses in biological samples by mass spectrometry. *Biochimica et Biophysica Acta-Molecular and Cell Biology of Lipids, 1811*(11), 776–783.
Nagai, T., Watanabe, N., Yoshinaga, K., Mizobe, H., Kojima, K., Kuroda, I., Odanaka, Y., Saito, T., Beppu, F. & Gotoh, N. (2015). Abundances of triacylglycerol positional isomers and enantiomers comprised of a dipalmitoylglycerol backbone and short-or medium-chain fatty acids in bovine milk fat. *Journal of Oleo Science, 64*(9), 943–952.
Nestel, P. J., Straznicky, N., Mellett, N. A., Wong, G., De Souza, D. P., Tull, D. L., Barlow, C. K., Grima, M. T. & Meikle, P. J. (2014). Specific plasma lipid classes and phospholipid fatty acids indicative of dairy food consumption associate with insulin sensitivity. *American Journal of Clinical Nutrition, 99*(1), 46–53.
Niimi, J., Eddy, A. I., Overington, A. R., Silcock, P., Bremer, P. J. & Delahunty, C. M. (2014). Cross-modal interaction between cheese taste and aroma. *International Dairy Journal, 39*(2), 222–228.
Nováková, L., Perrenoud, A. G.-G., Francois, I., West, C., Lesellier, E. & Guillarme, D. (2014). Modern analytical supercritical fluid chromatography using columns packed with sub-2 μm particles: A tutorial. *Analytica Chimica Acta, 824*, 18–35.
Ojanperä, I. & Rasanen, I. (2008). Forensic screening by gas chromatography. In M. J. Bogusz (Ed.), *Handbook of analytical separations* (pp. 403–424). London: Elsevier.
Ortega-Anaya, J. & Jiménez-Flores, R. (2019). Symposium review: The relevance of bovine milk phospholipids in human nutrition – Evidence of the effect on infant gut and brain development. *Journal of Dairy Science, 102*(3), 2738–2748.
Osborne, B. G. (2006). Near-infrared spectroscopy in food analysis. In *Encyclopedia of analytical chemistry*. Wiley Online Library.
Ozbekova, Z. & Kulmyrzaev, A. (2017). Fluorescence spectroscopy as a non destructive method to predict rheological characteristics of Tilsit cheese. *Journal of Food Engineering, 210*, 42–49.
Ozyurt, V. H. & Otles, S. (2017). Ultrasound spectroscopy in food analysis. In A. S. Franca & L. M. L. Nollett (Eds.), *Spectroscopic methods in food analysis* (pp. 225–236). Florida: CRC Press.
Parodi, P. W. (1982). Positional distribution of fatty acids in triglycerides from milk of several species of mammals. *Lipids, 17*(6), 437–442.
Perona, J. S. & Ruiz-Gutierrez, V. (2003). Simultaneous determination of molecular species of monoacylglycerols, diacylglycerols and triacylglycerols in human very-low-density lipoproteins by reversed-phase liquid chromatography. *Journal of Chromatography B, 785*(1), 89–99.
Perreault, V., Britten, M., Turgeon, S. L., Seuvre, A.-M., Cayot, P. & Voilley, A. (2010). Effects of heat treatment and acid-induced gelation on aroma release from flavoured skim milk. *Food Chemistry, 118*(1), 90–95.
Picariello, G., Sacchi, R. & Addeo, F. (2007). One-step characterization of triacylglycerols from animal fat by MALDI-TOF MS. *European Journal of Lipid Science and Technology, 109*(5), 511–524.
Pinto, G., Caira, S., Mamone, G., Ferranti, P., Addeo, F. & Picariello, G. (2014). Fractionation of complex lipid mixtures by hydroxyapatite chromatography for lipidomic purposes. *Journal of Chromatography A, 1360*, 82–92.
Prema, D., Turner, T. D., Jensen, J., Pilfold, J. L., Church, J. S., Donkor, K. K. & Cinel, B. (2015). Rapid determination of total conjugated linoleic acid concentrations in beef by 1H NMR spectroscopy. *Journal of Food Composition and Analysis, 41*, 54–57.
Qi, C., Sun, J., Xia, Y., Yu, R., Wei, W., Xiang, J., Jin, Q., Xiao, H. & Wang, X. (2018). Fatty acid profile and the *sn*-2 position distribution in triacylglycerols of breast milk during different lactation stages. *Journal of Agricultural and Food Chemistry, 66*(12), 3118–3126.

Qu, S., Du, Z. & Zhang, Y. (2015). Direct detection of free fatty acids in edible oils using supercritical fluid chromatography coupled with mass spectrometry. *Food Chemistry, 170*, 463–469.

Ran-Ressler, R. R., Sim, D., O'Donnell-Megaro, A. M., Bauman, D. E., Barbano, D. M. & Brenna, J. T. (2011). Branched chain fatty acid content of United States retail cow's milk and implications for dietary intake. *Lipids, 46*(7), 569–576.

Ratnayake, W. M. (2004). Overview of methods for the determination of trans fatty acids by gas chromatography, silver-ion thin-layer chromatography, silver-ion liquid chromatography, and gas chromatography/mass spectrometry. *Journal of AOAC International, 87*(2), 523–539.

Reinhardt, T. A. & Lippolis, J. D. (2008). Developmental changes in the milk fat globule membrane proteome during the transition from colostrum to milk. *Journal of Dairy Science, 91*(6), 2307–2318.

Řezanka, T., Kolouchová, I., Nedbalová, L. & Sigler, K. (2018). Enantiomeric separation of triacylglycerols containing very long chain fatty acids. *Journal of Chromatography A, 1557*, 9–19.

Ribeiro, J. C. B., Granato, D., Masson, M. L., Andriot, I., Mosca, A. C., Salles, C. & Guichard, E. (2016). Effect of lactobionic acid on the acidification, rheological properties and aroma release of dairy gels. *Food Chemistry, 207*, 101–106.

Richards, A., Wijesundera, C., Palmer, M. & Salisbury, P. (2004). Regiospecific triacylglycerol composition of some cooking oils consumed in Australia. *Food Australia, 58*, 365–372.

Rincon-Delgadillo, M. I., Lopez-Hernandez, A., Wijaya, I. & Rankin, S. A. (2012). Diacetyl levels and volatile profiles of commercial starter distillates and selected dairy foods. *Journal of Dairy Science, 95*(3), 1128–1139.

Risner, D., Tomasino, E., Hughes, P. & Meunier-Goddik, L. (2019). Volatile aroma composition of distillates produced from fermented sweet and acid whey. *Journal of Dairy Science, 102*(1), 202–210.

Robards, K., Kerr, A. F. & Patsalides, E. (1988). Rancidity and its measurement in edible oils and snack foods. A review. *Analyst, 113*(2), 213–224.

Robinson, N. P. & MacGibbon, A. K. H. (1998). Separation of milk fat triacylglycerols by argentation thin-layer chromatography. *Journal of the American Oil Chemists' Society, 75*(7), 783–788.

Rodríguez-Alcalá, L. M. & Fontecha, J. (2010). Major lipid classes separation of buttermilk, and cows, goats and ewes milk by high performance liquid chromatography with an evaporative light scattering detector focused on the phospholipid fraction. *Journal of Chromatography A, 1217*(18), 3063–3066.

Rodríguez-Alcalá, L. M., Castro-Gómez, P., Felipe, X., Noriega, L. & Fontecha, J. (2015). Effect of processing of cow milk by high pressures under conditions up to 900 MPa on the composition of neutral, polar lipids and fatty acids. *LWT – Food Science and Technology, 62*(1), 265–270.

Rombaut, R., Camp, J. V. & Dewettinck, K. (2006). Phospho-and sphingolipid distribution during processing of milk, butter and whey. *International Journal of Food Science and Technology, 41*(4), 435–443.

Rombaut, R., Dewettinck, K. & Van Camp, J. (2007). Phospho-and sphingolipid content of selected dairy products as determined by HPLC coupled to an evaporative light scattering detector (HPLC–ELSD). *Journal of Food Composition and Analysis, 20*(3–4), 308–312.

Rozzi, N. L. & Singh, R. K. (2002). Supercritical fluids and the food industry. *Comprehensive Reviews in Food Science and Food Safety, 1*(1), 33–44.

Ruiz-Sala, P., Hierro, M. T. G., Martínez-Castro, I. & Santa-María, G. (1996). Triglyceride composition of ewe, cow, and goat milk fat. *Journal of the American Oil Chemists' Society, 73*(3), 283–293.

Russo, M., Cichello, F., Ragonese, C., Donato, P., Cacciola, F., Dugo, P. & Mondello, L. (2013). Profiling and quantifying polar lipids in milk by hydrophilic interaction liquid chromatography coupled with evaporative light-scattering and mass spectrometry detection. *Analytical and Bioanalytical Chemistry, 405*(13), 4617–4626.

Salvatore, M., Giambra, S., Naviglio, D., DellaGreca, M., Salvatore, F., Burruano, S. & Andolfi, A. (2018). Fatty acids produced by *Neofusicoccum vitifusiforme* and *N. parvum*, fungi associated with grapevine *Botryosphaeria* dieback. *Agriculture, 8*(12), 189.

Sant'Ana, A. M. S., Bessa, R. J. B., Alves, S. P., Medeiros, A. N., Costa, R. G., de Sousa, Y. R. F., Bezerril, F. F., Batista, A. S. M., Madruga, M. S. & Queiroga, R. C. R. E. (2019). Fatty acid, volatile and sensory profiles of milk and cheese from goats raised on native semi-arid pasture or in confinement. *International Dairy Journal, 91*, 147–154.

Sasaki, R., Umezawa, M., Tsukahara, S., Ishiguro, T., Sato, S. & Watanabe, Y. (2015). Assignment of milk fat fatty acid propyl esters by GC-FID analysis with the aid of Ag-ion solid-phase extraction. *Journal of Oleo Science, 64*(12), 1251–1258.

Schennink, A., Stoop, W. M., Visker, M. W., Heck, J. M. L., Bovenhuis, H., Van Der Poel, J. J., Van Valenberg, H. J. F. & Van Arendonk, J. A. M. (2007). DGAT1 underlies large genetic variation in milk-fat composition of dairy cows. *Animal Genetics, 38*(5), 467–473.

Senorans, F. J. & Ibanez, E. (2002). Analysis of fatty acids in foods by supercritical fluid chromatography. *Analytica Chimica Acta, 465*(1–2), 131–144.

Seuvre, A.-M., Turci, C. & Voilley, A. (2008). Effect of the temperature on the release of aroma compounds and on the rheological behaviour of model dairy custard. *Food Chemistry, 108*(4), 1176–1182.

Shantha, N. C. & Napolitano, G. E. (1992). Gas chromatography of fatty acids. *Journal of Chromatography A, 624*(1–2), 37–51.

Shen, Z., Apriani, C., Weerakkody, R., Sanguansri, L. & Augustin, M. A. (2011). Food matrix effects on in vitro digestion of microencapsulated tuna oil powder. *Journal of Agricultural and Food Chemistry, 59*(15), 8442–8449.

Sherazi, S. T. H., Talpur, M. Y., Mahesar, S. A., Kandhro, A. A. & Arain, S. (2009). Main fatty acid classes in vegetable oils by SB-ATR-Fourier transform infrared (FTIR) spectroscopy. *Talanta, 80*(2), 600–606.

Shinn, S. E., Liyanage, R., Lay, J. O. & Proctor, A. (2015). Using MALDI MS for rapid analysis of food lipids. *Lipid Technology, 27*(11), 255–257.

Siciliano, C., Belsito, E., De Marco, R., Di Gioia, M. L., Leggio, A. & Liguori, A. (2013). Quantitative determination of fatty acid chain composition in pork meat products by high resolution 1H NMR spectroscopy. *Food Chemistry, 136*(2), 546–554.

Sides, A., Robards, K. & Helliwell, S. (2000). Developments in extraction techniques and their application to analysis of volatiles in foods. *Trends in Analytical Chemistry, 19*(5), 322–329.

Siesler, H. W., Ozaki, Y., Kawata, S. & Heise, H. M. (2014). *Near-infrared spectroscopy*. Wiley Online Library.

Singh, P. & Agrawal, T. (2018). Analysis of fatty acids composition of the seed oil of the *Albizia lebbeck* (L.) Benth. *International Journal of Science Research Multidisciplinary Studies, 4*(9), 17–21.

Soukoulis, C., Aprea, E., Biasioli, F., Cappellin, L., Schuhfried, E., Märk, T. D. & Gasperi, F. (2010). Proton transfer reaction time-of-flight mass spectrometry monitoring of the evolution of volatile compounds during lactic acid fermentation of milk. *Rapid Communication in Mass Spectrometry, 24*(14), 2127–2134.

Soukoulis, C., Biasioli, F., Aprea, E., Schuhfried, E., Cappellin, L., Märk, T. D. & Gasperi, F. (2012). PTR-TOF-MS analysis for influence of milk base supplementation on texture and headspace concentration of endogenous volatile compounds in yogurt. *Food and Bioprocess Technology, 5*(6), 2085–2097.

Soyeurt, H., Dehareng, F., Gengler, N., McParland, S., Wall, E. P. B. D., Berry, D. P., Coffey, M. & Dardenne, P. (2011). Mid-infrared prediction of bovine milk fatty acids across multiple breeds, production systems, and countries. *Journal of Dairy Science, 94*(4), 1657–1667.

Span, R. & Wagner, W. (1996). A new equation of state for carbon dioxide covering the fluid region from the triple-point temperature to 1100 K at pressures up to 800 MPa. *Journal of Physical and Chemical Reference Data, 25*(6), 1509–1596.

Spitzer, J., Klos, K. & Buettner, A. (2013). Monitoring aroma changes during human milk storage at + 4 °C by sensory and quantification experiments. *Clinical Nutrition, 32*(6), 1036–1042.

Stashenko, E. E. & Martínez, J. R. (2004). Derivatization and solid-phase microextraction. *TrAC Trends in Analytical Chemistry, 23*(8), 553–561.

Stoop, W. M., Bovenhuis, H., Heck, J. M. L. & Van Arendonk, J. A. M. (2009). Effect of lactation stage and energy status on milk fat composition of Holstein-Friesian cows. *Journal of Dairy Science, 92*(4), 1469–1478.

Straarup, E. M., Lauritzen, L., Faerk, J., Høy, C.-E. & Michaelsen, K. F. (2006). The stereospecific triacylglycerol structures and fatty acid profiles of human milk and infant formulas. *Journal of Pediatric Gastroenterology and Nutrition, 42*(3), 293–299.

Sun, C., Wei, W., Zou, X., Huang, J., Jin, Q. & Wang, X. (2018). Evaluation of triacylglycerol composition in commercial infant formulas on the Chinese market: A comparative study based on fat source and stage. *Food Chemistry, 252*, 154–162.

Swinehart, D. F. (1962). The Beer-Lambert Law. *Journal of Chemical Education, 39*(7), 333.

Timm-Heinrich, M., Nielsen, N. S., Xu, X. & Jacobsen, C. (2004). Oxidative stability of structured lipids containing C18:0, C18:1, C18:2, C18:3 or CLA in *sn2*-position–as bulk lipids and in milk drinks. *Innovative Food Science and Emerging Technologies, 5*(2), 249–261.

Tolentino, R. G., León, S. Y. V., Pérez, M. N., Herrera, M. C., Ayala, A. R., González, J. J. P., Bermúdez, B. S., Salinas, R. O., Francisca, M. V. & Martínez, J. G. R. (2015). Composition of triacylglycerols in fats of cow and goat milk produced in four zones of Mexico. *Food and Nutrition Science, 6*(6), 555.

Trenerry, V. C., Akbaridoust, G., Plozza, T., Rochfort, S., Wales, W. J., Auldist, M. & Ajlouni, S. (2013). Ultra-high-performance liquid chromatography–ion trap mass spectrometry characterisation of milk polar lipids from dairy cows fed different diets. *Food Chemistry, 141*(2), 1451–1460.

Truong, T., Morgan, G. P., Bansal, N., Palmer, M. & Bhandari, B. (2015). Crystal structures and morphologies of fractionated milk fat in nanoemulsions. *Food Chemistry, 171*, 157–167.

Tu, A., Ma, Q., Bai, H. & Du, Z. (2017). A comparative study of triacylglycerol composition in Chinese human milk within different lactation stages and imported infant formula by SFC coupled with Q-TOF-MS. *Food Chemistry, 221*, 555–567.

Turner, T., Rolland, D. C., Aldai, N. & Dugan, M. E. R. (2011). Rapid separation of *cis* 9, *trans* 11-and *trans* 7, *cis* 9-18: 2 (CLA) isomers from ruminant tissue using a 30 m SLB-IL111 ionic column. *Canadian Journal of Animal Science, 91*(4), 711–713.

Ullah, R., Khan, S., Ali, H., Bilal, M. & Saleem, M. (2017). Identification of cow and buffalo milk based on Beta carotene and vitamin-A concentration using fluorescence spectroscopy. *PLoS One, 12*(5), e0178055.

Versteeg, C., Logan, A. & Müller, W. (2016). Australian milk fat – Seasonal and regional variation of melting properties. *Journal of Dairy Science, 99*(4), 2571–2587.

Villaseñor, A., Garcia-Perez, I., Garcia, A., Posma, J. M., Fernández-López, M., Nicholas, A. J., Modi, N., Holmes, E. & Barbas, C. (2014). Breast milk metabolome characterization in a single-phase extraction, multiplatform analytical approach. *Analytical Chemistry, 86*(16), 8245–8252.

Vlahov, G., Chepkwony, P. K. & Ndalut, P. K. (2002). ^{13}C NMR characterization of triacylglycerols of *Moringa* oleifera seed oil: An "oleic-vaccenic acid" oil. *Journal of Agricultural and Food Chemistry, 50*(5), 970–975.

Walczak, J., Pomastowski, P., Bocian, S. & Buszewski, B. (2016). Determination of phospholipids in milk using a new phosphodiester stationary phase by liquid chromatography-matrix assisted desorption ionization mass spectrometry. *Journal of Chromatography A, 1432*, 39–48.

Walker, G. P., Wijesundera, C., Dunshea, F. R. & Doyle, P. T. (2013). Seasonal and stage of lactation effects on milk fat composition in northern Victoria. *Animal Production Science, 53*(6), 560–572.

Walter, L., Shrestha, P., Fry, R., Leury, B. & Logan, A. (2020). Lipid metabolic differences in cows producing small or large milk fat globules: Fatty acid origin and the role of membrane proteins. *Journal of Dairy Science, 103*(2), 1920–1930.

Wampler, T. P. (2001). Analysis of food volatiles using headspace-gas chromatographic techniques. In R. Marsili (Ed.), *Flavor, fragrance, and odor analysis* (2nd ed., pp. 41–70). Rockford: Marsili Consulting Group.

Wang, B., Xu, S. & Sun, D.-W. (2010). Application of the electronic nose to the identification of different milk flavorings. *Food Research International, 43*(1), 255–262.

Warensjö, E., Jansson, J.-H., Cederholm, T., Boman, K., Eliasson, M., Hallmans, G., Johansson, I. & Sjögren, P. (2010). Biomarkers of milk fat and the risk of myocardial infarction in men and women: A prospective, matched case-control study. *American Journal of Clinical Nutrition, 92*(1), 194–202.

Wijesundera, C., Shen, Z., Wales, W. J. & Dalley, D. E. (2003). Effect of cereal grain and fibre supplements on the fatty acid composition of milk fat of grazing dairy cows in early lactation. *Journal of Dairy Research, 70*(3), 257–265.

Willis, W. M. & Marangoni, A. G. (2002). Enymatic interesterification. In C. C. Akoh & D. B. Min (Eds.), *Food lipids: Chemistry, nutrition and biotechnology* (2nd ed., pp. 857–893). New York: Marcel Dekker.

Wolff, R. L., Bayard, C. C. & Fabien, R. J. (1995). Evaluation of sequential methods for the determination of butterfat fatty acid composition with emphasis on *trans*-18:1 acids. Application to the study of seasonal variations in French butters. *Journal of the American Oil Chemists' Society, 72*, 1471–1383.

Xiong, S., Adhikari, B., Chen, X. D. & Che, L. (2016). Determination of ultra-low milk fat content using dual-wavelength ultraviolet spectroscopy. *Journal of Dairy Science, 99*(12), 9652–9658.

Yang, F. J., Brown, A. C., Jr. & Cram, S. P. (1978). Splitless sampling for capillary-column gas chromatography. *Journal of Chromatography A, 158*, 91–109.

Yang, Y., Ferreira, G., Teets, C. L., Corl, B. A., Thomason, W. E. & Griffey, C. A. (2017). Effects of feeding hullless barley on production performance, milk fatty acid composition, and nutrient digestibility of lactating dairy cows. *Journal of Dairy Science, 100*(5), 3576–3583.

Yee, A. L., Maillard, M.-B., Roland, N., Chuat, V., Leclerc, A., Pogačić, T., Valence, F. & Thierry, A. (2014). Great interspecies and intraspecies diversity of dairy propionibacteria in the production of cheese aroma compounds. *International Journal of Food Microbiology, 191*, 60–68.

Yoshinaga, K., Nagai, T., Mizobe, H., Kojima, K. & Gotoh, N. (2013). Simple method for the quantification of milk fat content in foods by LC-APCI-MS/MS using 1, 2-dipalmitoyl-3-butyroyl-glycerol as an indicator. *Journal of Oleo Science, 62*(3), 115–121.

Zempleni, J., Aguilar-Lozano, A., Sadri, M., Sukreet, S., Manca, S., Wu, D., Zhou, F. & Mutai, E. (2016). Biological activities of extracellular vesicles and their cargos from bovine and human milk in humans and implications for infants. *Journal of Nutrition, 147*(1), 3–10.

Zheng, H., Jiménez-Flores, R. & Everett, D. W. (2013). Bovine milk fat globule membrane proteins are affected by centrifugal washing processes. *Journal of Agricultural and Food Chemistry, 61*(35), 8403–8411.

Zhou, Q., Gao, B., Zhang, X., Xu, Y., Shi, H. & Yu, L. L. (2014). Chemical profiling of triacylglycerols and diacylglycerols in cow milk fat by ultra-performance convergence chromatography combined with a quadrupole time-of-flight mass spectrometry. *Food Chemistry, 143*, 199–204.

Zou, X., Huang, J., Jin, Q., Guo, Z., Liu, Y., Cheong, L., Xu, X. & Wang, X. (2013). Lipid composition analysis of milk fats from different mammalian species: Potential for use as human milk fat substitutes. *Journal of Agricultural and Food Chemistry, 61*(29), 7070–7080.

Zychowski, L. M., Logan, A., Augustin, M. A., Kelly, A. L., Zabara, A., O'Mahony, J. A., Conn, C. E. & Auty, M. A. E. (2016). Effect of phytosterols on the crystallization behavior of oil-in-water milk fat emulsions. *Journal of Agricultural and Food Chemistry, 64*(34), 6546–6554.

Zychowski, L. M., Logan, A., Augustin, M. A., Kelly, A. L., O'Mahony, J. A., Conn, C. E. & Auty, M. A. E. (2018). Phytosterol crystallisation within bulk and dispersed triacylglycerol matrices as influenced by oil droplet size and low molecular weight surfactant addition. *Food Chemistry, 264*, 24–33.

Index

P. L. H. McSweeney et al. (eds.), *Advanced Dairy Chemistry, Volume 2*,
https://doi.org/10.1007/978-3-030-48686-0

Printed by Printforce, the Netherlands